CHEMISTRY, EMISSION CONTROL, RADIOACTIVE POLLUTION AND INDOOR AIR QUALITY

Edited by **Nicolás A. Mazzeo**

INTECHOPEN.COM

Chemistry, Emission Control, Radioactive Pollution and Indoor Air Quality
Edited by Nicolás A. Mazzeo

CBS, Edition 2016

Published by InTech

Janeza Trdine 9, 51000 Rijeka, Croatia

Publishing Process Manager Natalia Reinic

Technical Editor Teodora Smiljanic

Cover Designer InTech Design Team

Image Copyright Mikael Damkier, 2010. Used under license from Shutterstock.com

Additional hard copies can be obtained from orders@intechopen.com

Chemistry, Emission Control, Radioactive Pollution and Indoor Air Quality,
Edited by Nicolás A. Mazzeo
p. cm.
ISBN 978-953-307-316-3

Contents

Preface

The atmosphere may be our most precious resource. Accordingly, the balance between its use and protection is a high priority for our civilization. Air pollution has been with man since the first fire was lit, although, different aspects have been important at different times. While many of us would consider air pollution to be an issue that the modern world has resolved to a greater extent, it still appears to have considerable influence on the global environment. In many countries with ambitious economic growth targets the acceptable levels of air pollution have been transgressed. Serious respiratory disease related problems have been identified with both indoor and outdoor pollution throughout the world. In this century there has come to significant developments in science, technology and public policy of air pollution.

The 25 chapters of this book deal with several air pollution issues grouped into the following sections: a) air pollution chemistry; b) air pollutant emission control; c) radioactive pollution and d) indoor air quality.

The first section includes only one chapter prepared by an expert from China. This chapter describes how the introduction of aluminium oxide phase can effectively enhance textural properties and thermal stability, resulting in an improvement in photocatalytic activity over the hierarchically macro/mesoporous Fe/TiO_2 photocatalysts. This chapter shows that the hierarchical macro/mesoporous Fe/TiO_2 photocatalysts are effective visible-light-driven photocatalytic functional materials for air purification.

The second section includes four chapters. Their authors are from Netherlands, USA, China and Spain. Chapter 2 provides an overview of the existing scientific base and insights into ongoing and needed scientific research and development on several aspects (as emission, capture, transport and storage, air quality policy) of carbon dioxide, one of the most important greenhouse gases. Chapter 3 discusses techniques of the conversion of municipal waste plastics to liquid hydrocarbon fuel. Chapter 4 describes the use of non-thermal plasma technology in air pollution control in the abatement of hazardous air pollutants such as volatile organic compounds. Chapter 5 presents studies conducted to assess environmentally friendly biotechnologies, such as biofilters and biotrickling filters, for VOC abatement in air.

The third section has two chapters, which have been prepared by authors from Slovenia and USA. Chapter 6 presents the results of parallel monitoring of radon decay products and general aerosols that was performed in air of the Postojna Cave (size range 10–1100 nm) and in a dwelling (size range 5–350 nm) in a suburban area. Chapter 7 presents comparisons of wind frequencies among four five-year periods for various locations where the possibility of radionuclide releases exist and the comparison among test cases for these periods involving a dose assessment model used to estimate dose following short-term atmospheric releases.

Seventeen chapters constitute the fourth section. Their authors are from United Kingdom, Spain, Italy, USA, Portugal, USA, Germany, Romania, Tunisia, Greece, Finland, USA-France, Australia, USA, Estonia, India, Denmark and Sweden. Chapter 8 gives a historical review on the role of carbon dioxide as an indicator of air quality inside buildings. It serves to strengthen the case for an upgrade of regulations pertaining to air quality, which would require both consistent design standards and a new model for post occupancy evaluation or building performance evaluation. Chapter 9 studies the relation between indoor air conditions with fungi and bacteria growth, using a well known statistical technique and considering parameters as indoor and outdoor temperature and relative humidity, pets' presence and localised humidity problems. Chapter 10 describes how the indoor air quality assessment and control is necessary to evaluate the occupants' discomfort and health effects and to develop guidelines and standards. The chapter focuses on the indoor air quality assessment of VOCs (identification of sources, sampling methods and analysis of data). Chapter 11 evaluates the effectiveness of air sampling in detecting differences in fungal and bacterial bioaerosols in a building with environmental fungal and bacterial contamination. Chapter 12 summarises the main elements of a distributed smart sensing network for indoor air quality assessment. This system may provide an intelligent assessment of air conditions for risk factor reduction of asthma or chronic obstructive pulmonary disease. Chapter 13 describes an exposure model for identifying health risk due to environmental microbial contamination in hospitals, based on the American Industrial Hygiene Association Exposure Assessment Strategy. Chapter 14 deals with different methods that can be used to determine the air change rate between indoor and outdoor using tracer gas measurements. This chapter also includes a discussion on the dependence of air change from the prevailing weather conditions, such as the current wind and temperature conditions. In chapter 15 an olfactory comfort analysis in buildings is performed. This chapter describes the development of a computational model for indoor air quality numerical simulation and a methodology to determine the outside airflow rate and to verify the indoor air quality in enclosed spaces. Chapter 16 presents the results of the assessment of solvent exposure and the evaluation of neuro-psychological effects related to chronic exposure to solvents, obtained from an epidemiological survey carried out in a printing company for flexible packaging where large quantities of organic solvents are used. Chapter 17 focuses on the pollutants common to indoor and outdoor air environments and those who are measured more often in indoor environments and whether likely levels of exposure are hazardous to human health and the environment. Furthermore, indoor moisture is an important factor decreasing indoor air quality and limiting the building service life. In this sense, chapter 18 presents a mathematical method to predict indoor moisture generation rate and to determine indoor moisture generation levels that can be used in predicting building heat and moisture transfer. Chapter 19 presents the 3D particle tracking velocimetry method applied to the monitoring of air displacements and pollutant dispersion in rooms. Chapter 20 describes the use of environment friendly wind driven ventilation to improve the quality and comfort of human existence. This chapter focuses on wind driven ventilation systems that utilize wind as a natural energy to provide improved air quality within buildings. Chapter 21 studies the ability of the desiccant unit to remove IAQ-related microorganisms from the air. The ability of active desiccants to remove particulates, bioaerosols, chemical pollutants, and water vapor from the airstream delivered to a building provides a unique opportunity to view active desiccant technology as a viable control strategy for enhancing and maintaining a favorable IAQ in cooling climates. Chapter 22 studies indoor climate (temperature and humidity conditions) and energy use and the factors that affect them in typical apartment buildings erected from prefabricated concrete elements in Estonian cold climate. Chapter 23 illustrates the air pollution problem in rural areas. This chapter mentions the major air pollutants in rural areas, in indoor and outdoor environments, their natural and anthropogenic sources and health impacts. It clearly declares the necessity to strengthen both the quantity and quality of evidence linking air pollution and various health outcomes, especially for developing countries and for health conditions with weak or no evidence. Chapter 24 illustrates the use of computational fluid dynamics methods to design ventilation systems that reduce the ammonia concentration in pig housing and simultaneously reduce the required ventilation capacity. Chapter 25 describes differences in indoor air quality in horse stables under winter and summer conditions and studies correlations between selected components of stable air and indices of respiratory health in people and in stabled horses spending considerable time in the stable environment. Results contribute to the identification of suitable biomarkers to monitor the indoor horse stable environment and respiratory health in humans and horses.

This book provides a source of material for all those involved in the field, whether as a student, scientific researcher, industrialist, consultant, or government agency with responsibility in this area.

It should be emphasized that all chapters have been prepared by professionals who are experts in their research fields. The content of each chapter expresses the point of view of its authors who are responsible for its development. All chapters have been submitted to reviews in order to improve their presentation following several interactions between the Editor-Publisher-Authors. In this sense, the Editor, the Publisher and hard-working air quality professionals have worked together as a team to prepare a book that may become a reference in the field next years. This will have been achieved, mainly, thanks to the group of experts in their research fields joined as authors of this book.

Nicolás A. Mazzeo
National Scientific and Technological Research Council
National Technological University
Argentina

Part 1

Air Pollution Chemistry

1

Al_2O_3-enhanced Macro/Mesoporous Fe/TiO_2 for Breaking Down Nitric Oxide

Dieqing Zhang and Guisheng Li
Department of Chemistry, Key Laboratory of Resource Chemistry of Ministry of Education, Shanghai Normal University, Shanghai 200234, China

1. Introduction

High air contaminant levels in the indoor environment come from either the ambient air or from indoor sources. (Cao, 2001) Nitrogen oxide is one of the most common gaseous pollutants found in the indoor environment with the concentration in the range of 70-500 parts-per-billion (ppb) levels. This has serious implications on the environment and health of the mankind. (Huang et al., 2009) Conventional techniques to treat nitric oxide in industrial emission mainly include physical adsorption, biofiltration, and thermal catalysis methods. However, these methods usually suffer from some disadvantages, such as the low efficiency for pollutants at the parts per billion level and the difficulty in solving the postdisposal and regeneration problems. (Huang et al., 2008)

As a promising environmental remediation technology, semiconductor-mediated photocatalytic technology has been widely used to purify contaminated air and wastewater. (Fox & Dulay, 1993) Titanium dioxide is the most widely used photocatalyst because of its superior photoreactivity, nontoxicity, long-term stability and low price. Recently, great attention has been paid to macro/mesoporous TiO_2 for its interconnected macroporous and mesoporous structures. Such hierarchical material may enhance properties compared with single-sized pore materials due to increased mass transport through the material and minimized pressure drop over the monolithic material.(Yuan et al. 2006) Meanwhile the macroporous channels could serve as light-transfer paths for the distribution of photon energy onto the large surface of inner photoactive mesoporous frameworks. Therefore, higher light utilization efficiency could be obtained for heterogeneous photocatalytic systems including photooxidation degradation and solar cells. In addition, the hierarchical structure-in-structure arrangement of mesopore and macropore is benefit for the molecule traffic control and for the resistance of the photocatalyst to poisoning by inert deposits.(Rolison 2003)

Though such structure contributes great advantages to TiO_2, such as a readily accessible pore-wall system and better transport of matter compared to the traditional TiO_2 photocatalysts, the anatase TiO_2 semiconductor has a relatively large band gap of 3.2 eV, corresponding to a wavelength of 388 nm.(Yu et al., 2006) The requirement of UV excitation impedes the development of solar-driven photocatalytic systems. As a promising way, doping method can effectively extend the light absorption of TiO_2 to the visible region and reduce the recombination of photoinduced electrons and holes.(Zhu et al., 2007) Among various dopants, the Fe^{3+} dopant is most frequently employed owing to its unique half-filled electronic configuration, which might narrow the energy gap through the formation of new intermediate energy levels and also diminish recombination of photoinduced electrons and holes by capturing photoelectrons. However, calcination of the photocatalysts at high temperature is usually indispensable for removing organic templates, enhancing structural crystallization, and allowing doped ions to enter into the frameworks of TiO_2.(Wang et al., 2009) Such treatment at high temperature will result in great loss of surface area and destroying the pore systems owing to the grain growth, especially for porous materials. Thus, the photoactivity of the calcined samples with low specific area will be greatly reduced for the poor light-harvesting capability.(Yu et al., 2006) Fortunately, using inorganic structure stabilizers (SiO_2, ZrO_2, and Al_2O_3) could allow the anti-sintering properties of porous materials to be promoted greatly enough for application in high temperature environment, such as treating automotive exhaust.(Wang et al., 1999)

In this chapter, we describe a detailed study of the effect of Al_2O_3 as a promoter in enhancing a macro/mesoporous visible-light photocatalyst, Fe/TiO_2, for the oxidation of nitric oxide (NO). The photocatalysts are synthesized through directing the formation of inorganic phases (Al_2O_3-Fe/TiO_2) with multidimensional pore systems through the self-assembly of a single surfactant under hydrothermal conditions. The experimental results showed that doping Fe^{3+} into the framework of TiO_2 can effectively extend the optical absorption spectrum to visible light range. Introducing highly dispersed amorphous Al_2O_3 species into the Fe/TiO_2 system could greatly increased the thermal stability of the Fe/TiO_2 framework with higher surface area and larger pore volume. It is surprising that the Al_2O_3-Fe/TiO_2 sample treated at 700 °C possessed a high specific surface area (ca. 130 m^2/g), about 6 times of that of the Al_2O_3-free sample. The photooxidation of NO in air over the 3D macro/mesoporous Al_2O_3-Fe/TiO_2 photocatalysts was studied. These products were utilized to remove gaseous NO at 400 parts-per-billion level in air under visible-light irradiation. These Al_2O_3-Fe/TiO_2 photocatalysts exhibited very strong ability to oxidize the NO gas in air under visible-light irradiation. Importantly, these 3D macro/mesoporous Al_2O_3-Fe/TiO_2 photocatalysts showed excellent stability and maintained a high level of photocatalytic activity after multiple reaction cycles.

2. Experiment section

2.1 Preparation of 3D macro/mesoporous Al_2O_3-Fe/TiO_2 photocataltsts

Brij 56 [$C_{16}(EO)_{10}$], titanium isopropoxide, aluminum sec-butoxide, and ferric (III) nitrate are purchased from Aldrich. All chemicals were used as received. In a typical synthesis of macro/mesoporous visible light photocatalysts Al_2O_3-Fe/TiO_2, required amount of ferric (III) nitrate was dissolved in a aqueous solution of Brij 56 (15 wt %) with pH = 2 adjusted by sulfuric acid under ultrasonic irradiation in an ultrasonic clean bath (Bransonic ultrasonic cleaner, model 3210E DTH, 47 kHz, 120 W, USA). 18 ml mixture of aluminum sec-butoxide and titanium isopropoxide with a metal-to-metal molar ratio (M_{Al}/M_{Ti} = 20:100) was added drop by drop into the above medium under stirring, followed by further stirring for 0.5 h. The obtained mixture was then transferred to a Teflon-lined autoclave and heated at 80 °C for 36 h under static condition during which the inorganic precursor hydrolyses and polymerizes into a metal oxide network. Finally, the as-prepared white samples were clacined at 400-700 °C for 8 h at 1 °C/min to remove the surfactant species and improve the crystallinity. The as-prepared Al_2O_3-Fe/TiO_2 samples were denoted as Al-Fe/TiO_2-400, Al-

Fe/TiO_2-500, Al-Fe/TiO_2-600 and Al-Fe/TiO_2-700, where 400-700 refers to the calcinations temperature. For comparation, macro/mesoporous photocataltsts, pure TiO_2 and Fe/TiO_2, were also prepared by the same procedure. The molar ratio of Fe/Ti is 0.25 % for all the Fe doped samples.

2.2 Characterization

X-ray diffraction (XRD) measurements were carried out using a Bruker D8 Advance X-ray diffractometer (Cu $K\alpha_1$ irradiation, $\lambda = 1.5406$ Å) at a scanning rate of 0.02 Degree/Second. The Schrerrer equation ($\Phi = K\lambda/\beta\cos\theta$) was used to calculate the crystal size.(Machida, Norimoto et al. 1999) In the above equation, λ (0.154 nm) is the wavelength of the X-ray irradiation, K is a constant of 0.89, β is the peak width at half-maximum height after subtraction of the instrumental line broadening using silicon as a standard, and $2\theta = 25.3\,^{\circ}$ and $2\theta = 27.4\,^{\circ}$ for anatase and rutile. The phase composition was estimated using the following equations: rutile % = $100 \times (0.884A/R + 1)^{-1}$,(Machida, Norimoto et al. 1999) where A is the peak area of anatase (101) and R is the peak area of rutile (110). The intensity of both of the two peaks is the most intense reflection in the diffractograms. The number of 0.884 is the coefficient of scattering. The morphology and the surface roughness of as-prepared samples were examined by a LEO 1450 VP scanning microscope. Standard transmission electron microscopy images were recorded using a CM-120 microscope (Philips, 120 kV). High-resolution transmission electron microscopy (HRTEM) was recorded in JEOL-2010F at 200 kV. A trace amount of sample was suspended in ethanol solution. After sonication for 10 min, carbon-coated copper grids were used to hold the samples followed by drying. Nitrogen adsorption-desorption isotherms were analyzed at 77 K using Micromeritics ASAP 2010 equipment. The reflectance spectra of the samples over a range of 200-700 nm were recorded by a Varian Cary 100 Scan UV-vis system equipped with a Labsphere diffuse reflectance accessory. Labsphere USRS-99-010 was employed as a reflectance standard. FT-IR spectra on pellets of the samples mixed with KBr were recorded on a Nicolet Magna 560 FT-IR spectrometer.

2.3 Photocatalytic activity testing

The photocatalytic experiments for the removal of NO gas in air were performed at ambient temperature in a continuous flow rectangular reactor (10 H cm*30 L cm*15Wcm). A 300Wcommercial tungsten halogen lamp (General Electric) was used as the simulated solar light source. A piece of Pyrex glass was used to cut off the UV light below 400 nm. Four minifans were used to cool theflowsystem. Photocatalyst (0.2 g) was coated onto a dish with a diameter of 12.0 cm. The coated dish was then pretreated at 70 °C to remove water in the suspension. The NO gas was acquired from compressed gas cylinder at a concentration of 48 ppm NO (N_2 balance, BOC gas) with traceable National Institute of Stands and Technology (NIST) standard. The initial concentration of NO was diluted to about 400 ppb by the air stream supplied by a zero air generator (Thermo Environmental Inc. model 111). The desired humidity level of the NO flow was controlled at 70% (2100 ppmv) by passing the zero air streams through a humidification chamber. The gas streams were premixed completely by a gas blender and the flow rate was controlled at 4 L.min-1 by a mass flow controller. After the adsorption-desorption equilibrium among water vapor, gases and photocatalysts was achieved, the lamp was turned on. The concentration of NO was continuously measured by a chemiluminescence NO analyzer (Thermo Environmental Instruments Inc. model 42c), with a sampling rate of 0.7 L/min.

3. Results and discussion

3.1 X-ray diffraction and N_2 sorption

The crystal composition, thermal stability and mesoporous structure of the as-prepared samples were investigated by X-ray diffraction (XRD) and N_2 sorption analyses. Figure 1a shows the wide-angle XRD patterns of the Fe-doped TiO_2 calcined at different temperatures. For the 400 °C sintering sample, a broad peak corresponding to (101) diffraction of anatase-TiO_2 (JCPDS 21-1272) was observed. The broadening of the diffraction peak may have been caused by the small crystalline grain size (6.1 nm). Upon increasing the temperature to 500 °C, the intensity of this peak became stronger and sharper, indicating that larger particles (8.0 nm) were formed. However, when the calcination temperature was increased to 600 °C, the intensity of the anatase-TiO_2 diffraction peak decreased. Meanwhile, weak peaks indexable as diffractions of rutile-TiO_2 (JCPDS 87-920) appeared. About 4.5 % of the anatase-TiO_2 was converted to rutile-TiO_2.

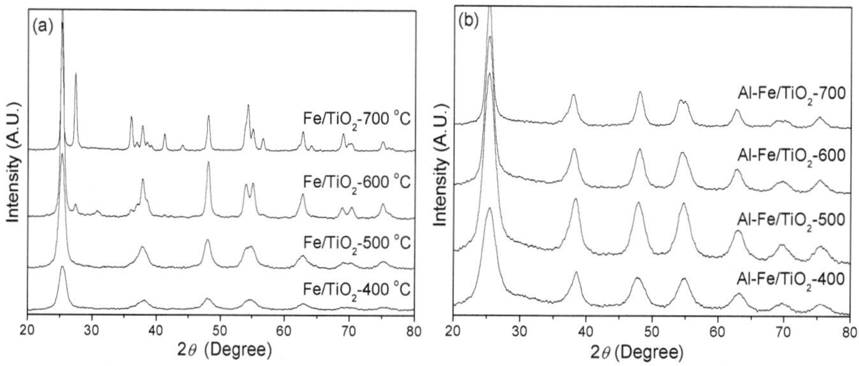

Fig. 1. XRD patterns of the as-prepared Fe/TiO_2 and $Al\text{-}Fe/TiO_2$ samples calcined at different temperatures.

After calcinations at 700 °C, the size of anatase-TiO_2 increased dramatically to 20.0 nm, and nearly 43.3 % of the anatase-TiO_2 was converted to rutile-TiO_2. Nevertheless, introducing Al_2O_3 species into the Fe-doped TiO_2 system greatly inhibited the crystal growth and the phase-transition. As shown in Figure 1b, all of the Al_2O_3 modified Fe/TiO_2 samples exhibited pure anatase phase at different temperatures. More interestingly, the crystal size of the anatase TiO_2 was greatly decreased after modifying Al_2O_3. Even after 700 °C calcination, the grain size can be maintained about 7.2 nm, much smaller than that (20.0 nm) of Fe/TiO_2 sample calcined at 700 °C. These wide-angle XRD results revealed that thermal treatment induced the growth of crystal size and subsequent phase transition could be effectively prohibited by doped Al_2O_3 species acting as structural agents. As known, thermal-induced changes in crystal composition and size also had remarkable effects on the textural properties of TiO_2 framework. N_2 sorption analyses were utilized to confirm the change of textural properties.

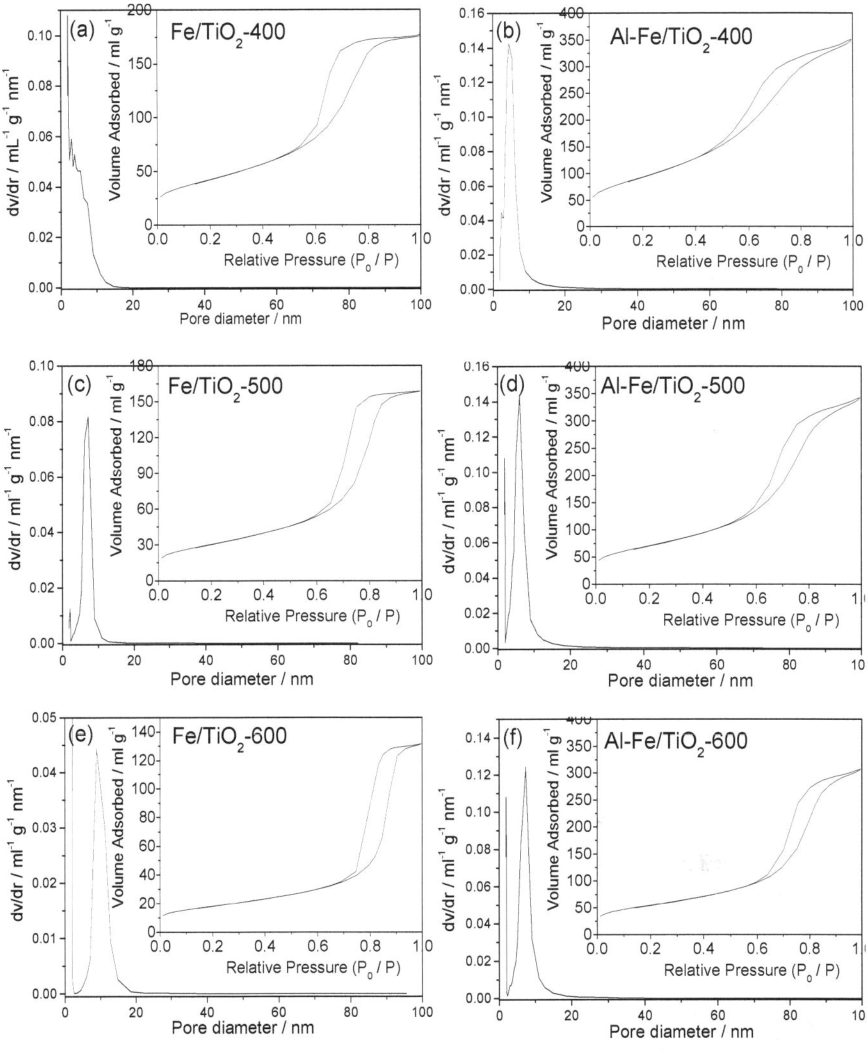

Fig. 2. N_2 adsorption-desorption isotherms (a, c, and e) and pore size distributions (b, d, and f) of the as-prepared Fe/TiO_2 and Al-Fe-TiO_2 samples calcined at different temperatures.

N_2 sorption analyses were utilized to investigate the change of textural properties of the as-prepared products. Figure 2 shows the N_2 sorption isotherms and pore size distributions for the modified titanium dioxide calcined at different temperatures. Upon 400 °C calcinations, both of Fe/TiO_2-400 and Al-Fe/TiO_2-400 samples exhibited stepwise adsorption and desorption (type IV isotherms) as shown in Figure 2a and b, indicative of a typical mesoporous structure within the as-prepared samples. As shown in Table 1, Al-Fe/TiO_2-400 sample possesses a surface area of 340.0 m^2/g, much higher than that (153.6 m^2/g) of Fe/TiO_2-400 owing to the anti-agglutination effect of induced Al_2O_3 species. The modification of Al_2O_3 also contributed a super lager pore volume (0.53 cm^3/g) to Al-Fe/TiO_2-400 sample. It is about two times of that (0.27 cm^3/g) of Fe/TiO_2-400. Such high surface area and large pore volume will make this material an excellent photocatalyst for its strong adsorption capability. With increase the calcinations temperature, the mesoporous structure of Fe/TiO_2 sample was destroyed. When the calcinations temperature was increased to 700 °C, the surface area was decreased to 22.6 m^2/g, and the pore volume was decreased to 0.11 cm^3/g. This is an indication of the collapse of the pore. However, Al-Fe/TiO_2 still owns a high surface area of 131.8 m^2/g and pore volume of 0.44 cm^3/g. To the case of the pore-size distribution, the modification of Al_2O_3 also inhibited the pore size changing owing to its porous structure-stabilizing capability.

Sample	Surface Area/m^2/g^a	Pore Volume/mL/g^b	Pore Size/nm^c	Rutile content/Wt %	Crystal Size/nm^d
Fe/TiO_2-400	153.6	0.27	5.9	0	6.1
Fe/TiO_2-500	109.5	0.24	7.6	0	8
Fe/TiO_2-600	64.9	0.20	11.3	4.5	13.4
Fe/TiO_2-700	22.6	0.11	31.2	43.3	20.0
Al-Fe/TiO_2-400	340.0	0.53	5.3	0	3.2
Al-Fe/TiO_2-500	254.7	0.52	6.8	0	4.0
Al-Fe/TiO_2-600	197.1	0.46	8.1	0	5.3
Al-Fe/TiO_2-700	131.8	0.44	12.4	0	7.2

[a] BET surface area is calculated from the linear part of the BET plot (p/p_0 = 0.1-0.2).[b] The total pore volumes are estimated from the adsorbed amount at a relative pressure of p/p_0 = 0.99.[c] The pore-size distributions (PSD) are derived from the adsorption branches of the isotherms by using the Barrett-Joyner-Halenda (BJH) method.[d] Crystal size was calculated based on XRD results.

Table 1. Textural properties and crystalline structures and of the prepared porous samples.

3.2 Scanning Electron Microscopy (SEM)

The N_2 sorption analyses could provide mesoporous structure information of the as-prepared materials. To the case of the macroscopic properties, scanning electron microscopy (SEM) should be utilized to examine the macrostructure of the modified TiO_2 monolithic particles. Meanwhile, the high-resolution state of SEM images could also give information on the mesoscopic properties. As shown in Figure 3a, Al-Fe/TiO_2-400 is typically in a large monolithic form (> 30 μm), and exhibits macroscopic network structure with relatively homogeneous macropores of 1~2 μm (size) and about 20 μm (length) in dimension as shown Figure 3b. It is more interesting that these ultralong macroscopic channels are arranged parallel to each other. Figure 3b also demonstrates the extension of the parallel-arrayed macropores completely through the material from the side view of the sample. Such open-ended tubelike macrochannels could serve as ideal light-transport routes for introducing more photoenergy into the interior of the framework of TiO_2. Meanwhile, the high-resolution SEM images (Figure 3c) shows that the walls of the macroporous TiO_2 frameworks are composed of small interconnected TiO_2 particles. The mesoporous structure of the as-prepared materials is probably partly due to the intraparticle porosity and partly due to the interparticle porosity of these fine particulates.(Wang, Yu et al. 2005) The macro/mesoporous structure nearly can be maintained even after 600 °C calcinations.

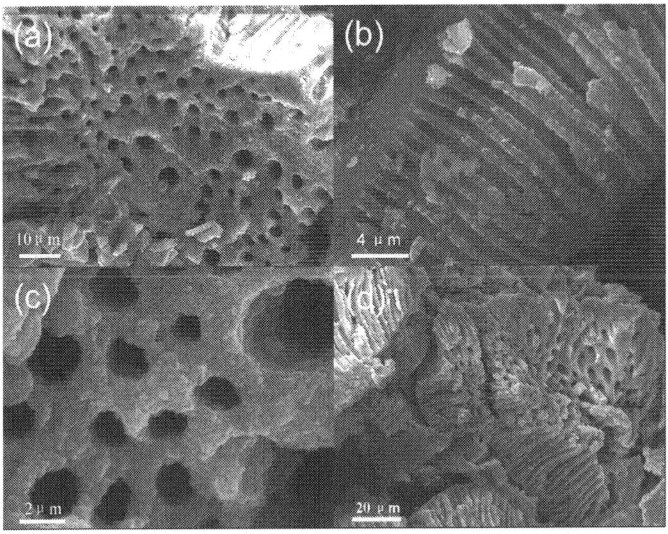

Fig. 3. SEM images of the Al-Fe/TiO_2 samples calcined at 400 °C (a, b, c), and 600 °C (d).

3.3 UV-vis spectra

UV-visible diffuse reflectance spectroscopy (DRS) was utulized to investigate the electronic states of the as prepared samples. Figure 4a shows the UV-visible absorption spectra of TiO_2-400, Fe/TiO_2-400, Al-Fe/TiO_2-400 and Al-Fe/TiO_2-700 samples. For large energy gap of anatase (3.2 eV), TiO_2-400 sample has no significant absorbance for visible-light. Upon doping Fe^{3+} ions, the light absorption edge of Fe/TiO_2-400 sample was extended to visible light region (λ < 650 nm) attributed to the formation of Fe-intermediate energy levels, resulting in a decrease in the energy band. The other three samples exhibit a broad absorption bands from 200 to 600 nm with respect to the pure TiO_2, indicating the effective photo-absorption property for this macro/mesoporous structure oxide composite photocatalyst system. This is because Fe-doping induces the absorbance for visible light owing to, leading to a decrease in the energy band gap.(Nahar, Hasegawa et al. 2006)
Compared to Fe/TiO_2-400 sample, Al-Fe/TiO_2 shows a higher light-absorbance ability located in 200~400 nm. This is because the macro/mesoporous structure, enlarged surface area and multiple scattering enable it to harvest light much more efficiently.(Yu, Wang et al. 2004) This enhanced light-trapping effect is the result of the reflection or transmission of the light scattered by the macroporous tunnels or mesopores implanted in the body of Al, Fe co-doped TiO_2 matrix. It is also noted that the modification of Al_2O_3 did not change the light

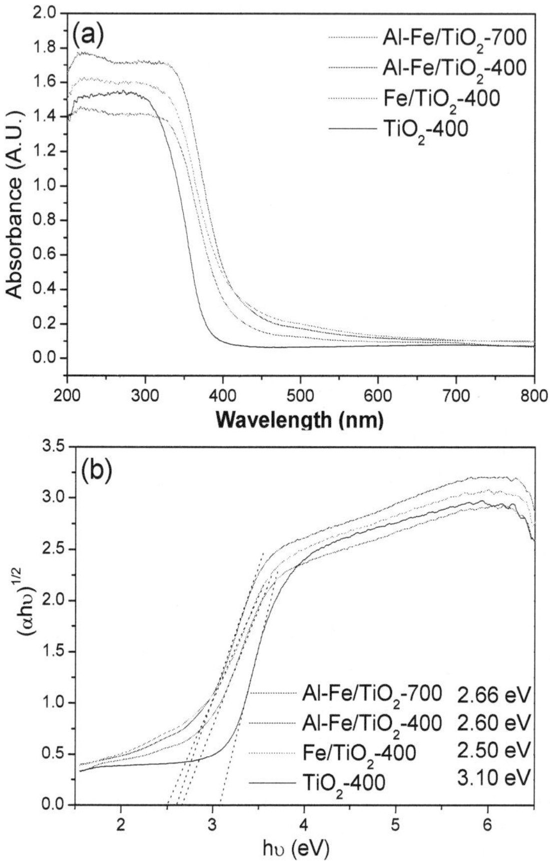

Fig. 4. UV-visible absorption spectra of (a) and determination of indirect interband transition energies (b) for pure TiO_2, Fe/TiO_2-400, Al-Fe/TiO_2-400, and Al-Fe/TiO_2-700 samples.

sensitization region. However, a very weak blue shift to short wavelength was observed for Al-Fe/TiO_2 sample after 700 °C calcining treatment. The band energy gap of the as-prepared samples could be calculated by using $(\alpha h\nu)^n = k(h\nu - E_g)$, where α is the absorption coefficient, k is the parameter that related to the effective masses associated with the valence and conduction bands, n is 1/2 for a direct transition, $h\nu$ is the absorption energy, and E_g is the band gap energy.(Li, Zhang et al. 2009) Plotting $(\alpha h\nu)^{1/2}$ versus $h\nu$ based on the spectral response in Figure 4a gave the extrapolated intercept corresponding to the E_g value (see Figure 4b). The optical band energies of the macro/mesoporous TiO_2-400, Fe/TiO_2-400, Al-Fe/TiO_2-400 and Al-Fe/TiO2-700 samples (3.10 eV, 2.50 eV, 2.60 eV, and 2.66 eV respectively) exhibit obvious red-shifts with respect to that of TiO_2-400 sample (3.10 eV). The results of this study therefore indicate that the enhanced ability to absorb visible-light of this type of macro/mesoporous Al-Fe/TiO_2 makes it a promising photocatalyst for solar-driven applications.

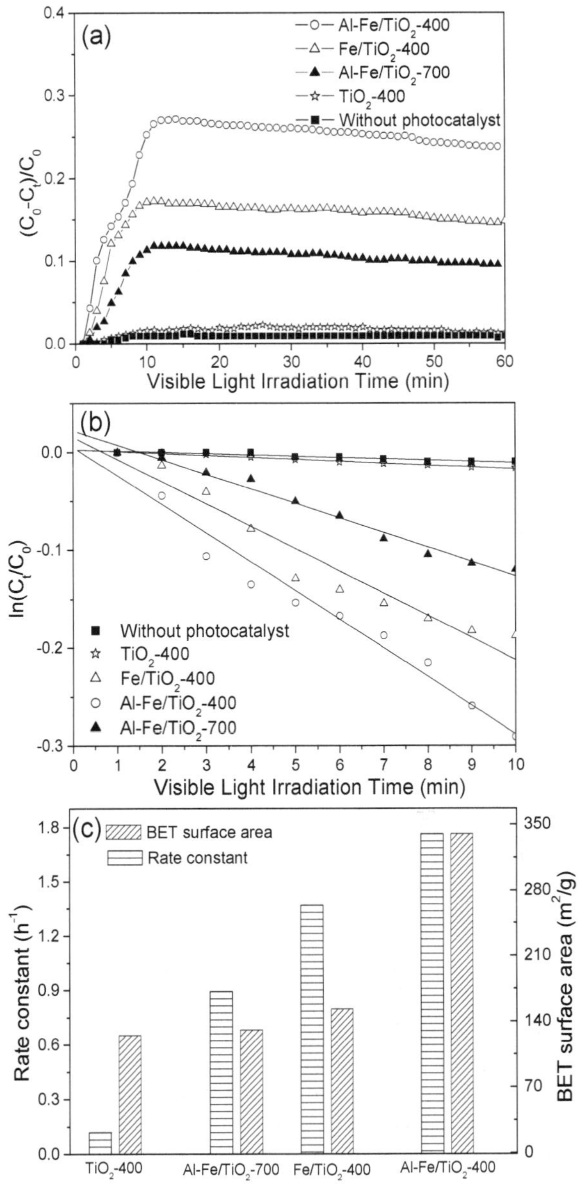

Fig. 5. (a) Plots of the removal of NO concentration vs irradiation time in the presence of the as-prepared products with visible-light irradiation ($\lambda > 400$ nm). (b) Dependence of $\ln(C/C_0)$ on irradiation time. (c) Relationship between rate constant and BET surface area over the as-prepared products.

To evaluate the photocatalytic performance of the as-prepared materials. The photo-oxidation of NO gas under visible light irradiation (λ > 400 nm) in a single pass flow was used as a photoreaction probe. Figure 5a shows the relative of NO removal rate against irradiation time in the presence of photocatalysts under visible-light irradiation. In the absence of the photocatalyst, no obvious removal rate of NO can be observed. The Photocatalytic performance of pure TiO_2 can be nearly neglected. The NO removal rate over Fe/TiO_2-400 sample reaches 17 % after 20 min irradiation, indicating the promotion effect of Fe-doping. Compared to Fe/TiO_2-400, Al-Fe/TiO_2-400 exhibits a much higher removal rate (about 28 % after 20 min irradiation). Such high photocatalytic performance maybe attributed to the high surface area, large pore volume. Besides, 3D connected pore tunnels are also very important because they can allow the NO molecule to transport very conveniently in the body of the catalyst. Further increasing the calcinations temperature to 700 °C quickly decreased the removal rate to about 10 %.
For a clear quantitative comparison, we use the Langmuir-Hinshelwood model (L-H) to describe the initial rates of photocatalytic removal of NO. The photocatalytic oxidation of NO was recognized to follow a first-order-kinetics approximately as a result of low concentration target pollutants, as evidenced by the linear plot of ln (C/C_0) versus photocatalytic reaction time t (Figure 5b). The rate constants of the TiO_2-400, Fe/TiO_2-400, Al-Fe/TiO_2-400 and Al-Fe/TiO_2-700 samples are 0.119 h^{-1}, 1.368 h^{-1}, 1.762 h^{-1} and 0.893 h^{-1} respectively. Figure 5c shows the relationship between reaction rate constants and BET surface areas. Al-Fe/TiO_2-400 sample owns the highest surface area of 340 m^2/g, resulting in an excellent photocatalytic performance in oxidation of NO. Though the surface area of Al-Fe/TiO_2-700 is similar to that of Fe/TiO_2-400, the reaction rate constant of the formed is much lower that of the latter. Except for the effect of surface area, other factors such as the band gap and light adsorption capability also play an important role in controlling the photocatalytic performance of the catalysts. As shown in Figure 4b, the band gap of Al-Fe/TiO_2-700 is 2.66 eV is higher than that of Fe/TiO_2-400. Meanwhile, the light adsorption intensity of Al-Fe/TiO_2-700 is much lower than that of Fe/TiO_2-400.

4. Conclusions

Macro/mesoporous Fe/TiO_2 was fabricated by soft-chemical synthesis in the presence of surfactants, followed by calcination. Such materials have been proved as a good photocatalyst for treating NO at air conditions for its special macro/mesoporous structures. The modification of Al_2O_3 can effectively increase the thermal stability of Fe/TiO_2 with a very high surface area, resulting in an excellent photocatalytic performance during the oxidation of 400 ppb level of NO in air under visible light irradiation. The present work demonstrates that the hierarchical macro/mesoporous Fe/TiO_2 photocatalysts are effective visible-light-driven photocatalytic functional materials for air purification.

5. Acknowledgments

This work was supported by the Program for Professor of Special Appointment (Eastern Scholar) at Shanghai Institutions of Higher Learning, the National Natural Science Foundation of China (21007040, 21047009), Natural Science Funding of Shanghai

(11ZR1426300), the Research Fund for the Doctoral Program of Higher Education (20103127120005), the Project supported by the Shanghai Committee of Science and Technology (10160503200), and by a Scheme administrated by Shanghai Normal University (8K201104).

6. References

Chen, X., Wang, X. & Fu, X. (2009). Hierarchical Macro/Mesoporous TiO_2/SiO_2 and TiO_2/ZrO_2 Nanocomposites for Environmental Photocatalysis. Energy & Environmental Science, Vol.2, No.8, (May 2009), pp.872-877, ISSN: 1754-5692

Li, G., Zhang, D. & Yu, J. (2009). Thermally Stable Ordered Mesoporous CeO_2/TiO_2 Visible-Light Photocatalysts. Physical Chemistry Chemical Physics, Vol.11, No.19, (February 2009), pp.3775-3782, ISSN: 1463-9076

Machida, M., Norimoto, K., Watanabe, T., Hashimoto, K. & Fujishima, A. (1999). The Effect of SiO_2 Addition in Super-Hydrophilic Property of TiO_2 Photocatalyst. Journal of Materials Science, Vol.34, No.11, (June 1999), pp.2569-2574. ISSN: 0022-2461

Nahar, S., Hasegawa, K. & Kagaya, S. (2006). Photocatalytic Degradation of Phenol by Visible Light-Responsive Iron-Doped TiO_2 and Spontaneous Sedimentation of the TiO_2 Particles. Chemosphere, Vol.65, No.11, (December 2006), pp.1976-1982. ISSN: 0045-6535

Rolison, D. (2003). Catalytic Nanoarchitectures - The Importance of Nothing and the Unimportance of Periodicity. Science, Vol.299, No.5613, (May 2003), pp.1698-1701. ISSN: 0036-8075

Wang, X., Yu, J., Chen, Y., Wu, L. & Fu, X. (2006). ZrO_2-Modified Mesoporous Manocrystalline $TiO_{2-x}N_x$ as Efficient Visible Light Photocatalysts. Environmental Science & Technology, Vol,40, No.7, (April 2006), pp.2369-2374, ISSN: 0013-936X

Wang, X., Yu, J., Ho, C., Hou, Y. & Fu, X. (2005). Photocatalytic Activity of a Hierarchically Macro/Mesoporous Titania. Langmuir, Vol.21, No.6, (March 2005), pp.2552-2559, ISSN: 0743-7463

Wu, N., Wang, S. & Rusakova, I. (1999). Inhibition of Crystallite Growth in the Sol-Gel Synthesis of Nanocrystalline Metal Oxides. Science, Vol.285, No.5432, (August 1999), pp.1375-1377, ISSN: 0036-8075

Yu, J., Li, G., Wang, X., Hu, X. Leung, C. & Zhang, Z. (2006). An Ordered Cubic Im3m Mesoporous Cr-TiO_2 Visible Light Photocatalyst. Chemical Communications, Vol.25, (September 2006), pp.2717-2719, ISSN: 1359-7345

Yu, J. C., Wang, X., & Fu, X. (2004). Pore-Wall Chemistry and Photocatalytic Activity of Mesoporous Titania Molecular Sieve Films. Chemistry of Materials, Vol.16, No.8, (April 2004), pp.1523-1530, ISSN: 0897-4756

Yuan, Z. &. Su, L. (2006). Insights into Hierarchically Meso-Macroporous Structured Materials. Journal of Materials Chemistry, Vol.16, No.7, (Febrary2006), pp.663-677, ISSN: 0959-9428

Zhu, J., Ren, J., Huo, Y., Bian, Z. & Li, H. (2007). Nanocrystalline Fe/TiO_2 Visible Photocatalyst with a Mesoporous Structure Prepared via a Nonhydrolytic Sol-Gel Route. Journal of Physical Chemistry C, Vol.111, No.51, (December 2007), pp.18965-18969, ISSN: 1932-7447

Part 2

Air Pollutant Emission Control

2

Carbon Dioxide Capture and Air Quality

Joris Koornneef[1], Toon van Harmelen[2], Arjan van Horssen[2] and Andrea Ramirez[3]
[1]*Ecofys Netherlands, Utrecht,*
[2]*TNO, Utrecht,*
[3]*Copernicus Institute, Utrecht University, Utrecht, Netherlands*

1. Introduction

Carbon dioxide (CO_2) is one of the most important greenhouse gases (GHG). The most dominant source of anthropogenic CO_2 contributing to the rise in atmospheric concentration since the industrial revolution is the combustion of fossil fuels. These emissions are expected to result in global climate change with potentially severe consequences for ecosystems and mankind. In this context, these emissions should be restrained in order to mitigate climate change.

Carbon Capture and Storage (CCS) is a technological concept to reduce the atmospheric emissions of CO_2 that result from various industrial processes, in particular from the use of fossil fuels (mainly coal and natural gas) in power generation and from combustion and process related emissions in industrial sectors. The Intergovernmental Panel on Climate Change (IPCC) regards CCS as "an option in the portfolio of mitigation actions" to combat climate change (IPCC 2005).

However, the deployment of CO_2 capture at power plants and large industrial sources may influence local and transboundary air pollution, i.e. the emission of key atmospheric emissions such as SO_2, NO_X, NH_3, Volatile Organic Compounds (VOC), and Particulate Matter ($PM_{2.5}$ and PM_{10}). Both positive as negative impacts on overall air quality when applying CCS are being suggested in the literature. The scientific base supporting both viewpoints is rapidly advancing.

The potential interaction between CO_2 capture and air quality targets is crucial as countries are currently developing GHG mitigation action plans. External and unwanted trade-offs regarding air quality as well as co-benefits when implementing CCS should be known before rolling out this technology on a large scale.

The goal of this chapter is to provide an overview of the existing scientific base and provide insights into ongoing and needed scientific endeavours aimed at expanding the science base. The chapter outline is as follows. We first discuss the basics of CO_2 capture, transport and storage in section 2. In section 3, we discuss the change in the direct emission profile of key atmospheric pollutants when equipping power plants with CO_2 capture. Section 4 expands on atmospheric emissions in the life cycle of CCS concepts. We provide insights in section 5 into how air quality policy and GHG reduction policy may interact in the Netherlands and the European Union. Section 6 focuses on atmospheric emissions from post-combustion CO_2

capture. We highlight in section 7 the most important findings and provide outlook on (required) research and development.

2. Carbon dioxide capture, transport and storage

2.1 CO_2 capture

The first step of the CCS chain is the capture process. A major element of this process comprises the separation of CO_2 from a gas stream. This can be the separation from produced natural gas, which often contains acid gases such as H_2S and CO_2. It also can be separated during the production of ammonia and during refining processes in the hydrocarbon industry. There is considerable less experience with removing CO_2 from flue gases at atmospheric pressure. This entails flue gases from power plants as well as industrial plants producing, for instance, steel, cement or iron. These large point sources form the largest potential for applying CO_2 capture. There are four approaches to capture CO_2 from large point sources: 1) Post-combustion capture; 2) Pre-combustion capture; 3) Oxyfuel combustion capture; 4) Capture from industrial processes.

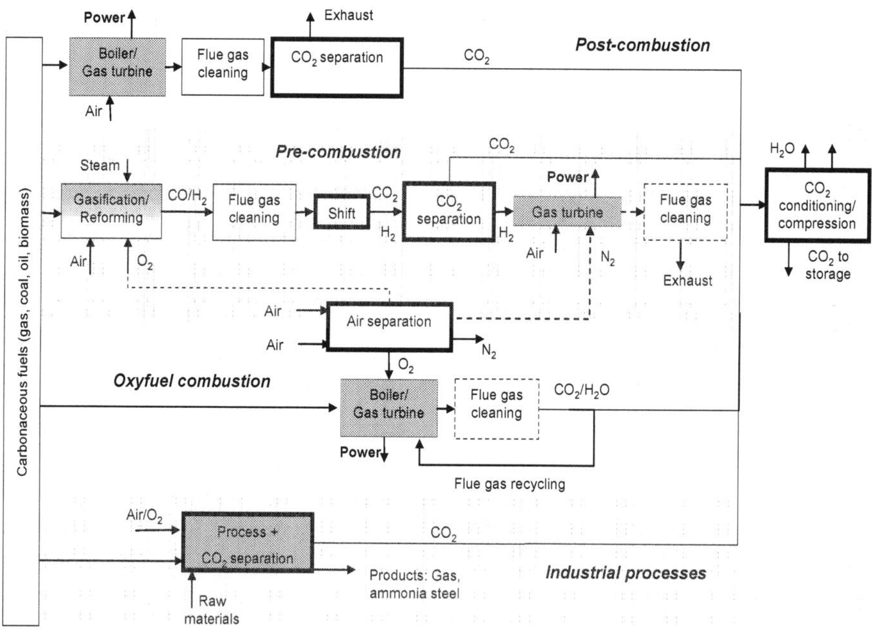

Fig. 1. Simplified overview of the three CO_2 capture systems for power plants: post-, pre- and oxyfuel combustion. Grey components indicate power generation processes. Components with highlighted borders indicate processes causing a drop in generating efficiency. Components with dashed borders indicate optional processes. Note that natural gas reforming using steam is an endothermic process and therefore not a power generation process, hence the altered shading.

2.1.1 Post-combustion capture

CO_2 can be captured from the flue gas of a combustion process. This can be flue gas coming from any (pressurized) combustion in a boiler, gas turbine or industrial process yielding CO_2. Various capture mechanisms, or combinations of them, can be applied, being: phase separation, selective permeability and sorption. The last mechanism, sorption, is the most widely suggested mechanism to be used at large point sources. This mechanism encompasses chemical or physical absorption and also adsorption. In the CO_2 capture processes based on this mechanism a sorption medium, or a sorbent, is used. When these sorbents are in solution they are called solvents. The current research, development and demonstration (RD&D) focus is on using chemical and physical solvents to separate the CO_2 from the gas stream. Retrofitting existing power plants with CO_2 capture will highly likely be done with a chemical absorption based post-combustion capture technology.

The RD&D focus in post-combustion capture is mainly aimed at reducing energy requirement and capital cost trough developing and adapting solvents, optimizing the required process installations and integrating the capture system with the power generation process. The application of the capture process on contaminated flue gases, e.g. flue gases from coal fired power plants, is already commercially applied (Strazisar, Anderson et al. 2003). However, large-scale CO_2 capture as well as dealing with the contaminants in the flue gas remains a challenge.

2.1.2 Pre-combustion capture

Pre-combustion capture comprises a group of technologies that removes CO_2 before the combustion of the fuel. This requires a carbonaceous fuel to be broken down into hydrogen (H_2) and carbon monoxide (CO), i.e. syngas. To make CO_2 capture with high efficiencies possible, the syngas that is formed after steam reforming or partial oxidation/gasification has to be shifted after it is cleaned. The 'shift reaction', or 'water gas shift' (WGS) reaction, yields heat and a gas stream with high CO_2 and H_2 concentrations. The CO_2 can then be removed with chemical and physical solvents, adsorbents and membranes.

For the near-term it is expected that chemical or physical solvents (or a combination) are used for the CO_2 removal. The CO_2 removal step yields relative pure CO_2 and a gas stream with a high hydrogen and low carbon content. The latter can be used for power production in for example a (modified) gas turbine. The gas with reduced carbon content can (after further purification) also be used in the production of synfuels, the refining of hydrocarbons or for the production of chemicals. (IPCC 2005)

For solid and liquid fuels, pre-combustion CO_2 capture can be applied in an IGCC (Integrated Gasification Combined Cycle) power plant. For gas fired power generation with pre-combustion capture other concepts are being studied (Ertesvag, Kvamsdal et al. 2005;Kvamsdal and Mejdell 2005;IEA GHG 2006c;Kvamsdal, Jordal et al. 2007).

The technology to capture CO_2 from the syngas generated in a gasifier can be considered proven technology, is commercially available and used for several decades in other applications than for electricity production. Examples are hydrogen, ammonia and synthetic fuel production (Nexant Inc. 2006). Also, reforming and partial oxidation of (natural) gas are already widely applied, e.g. for the production of hydrogen in the ammonia production process.

The pre-combustion concept has not yet been proven in an IGCC power plant. Proving its reliability and effectiveness in power plant concepts is therefore one of the main RD&D targets. In addition, improving the efficiency of the WGS step and integration of this process with CO_2 capture is also an area of research.

2.1.3 Oxyfuel combustion

Oxyfuel combustion is based on denitrification of the combustion medium. The nitrogen is removed from the air through a cryogenic air separation unit (ASU) or with the use of membranes. Combustion thus takes place with nearly pure oxygen. The final result is a flue gas containing mainly CO_2 and water. The CO_2 is purified by removing water and impurities. The production of oxygen requires a significant amount of energy, which results in a reduction of the efficiency of the power plant. Further, the purification and the compression of the CO_2 stream also require energy.

The combustion with oxygen is currently applied in the glass and metallurgical industry (Buhre, Elliott et al. 2005;IPCC 2005;M. Anheden, Jinying Yan et al. 2005). Oxyfuel combustion for steam and power production using solid fuels has been at present only proven in test and pilot facilities. Oxyfuel combustion can also be applied in natural gas fired concepts. Power cycles for gaseous and solid fuels, however, vary significantly.

Although there are no significant differences compared to air firing of solid fuels, the combustion process and optimal configuration of the burners are considered to be the most important hurdles to overcome. In addition, the design and configuration of the flue gas cleaning section and CO_2 purification section are challenges for the short-term. For the gas fired concepts, system integration and development of critical components hinder direct application on a commercial scale. Examples of critical components are the turbines and combustors for the near- and medium-term options and, additionally, the fuel reactors for the concepts in the longer term.

2.1.4 Capture from industrial processes

This group of technologies is often mentioned as the early opportunity for CCS at relative low cost. The total reduction potential due to CO_2 capture from these point sources is however considered to be rather limited. Examples for industrial processes are: the production of cement, iron and steel, ethylene (oxide), ammonia and hydrogen. In addition, CO_2 can be captured from natural gas sweetening processes and from refineries (IPCC 2005). The capture processes applied are in general the same technologies as already described above.

2.1.5 Increased primary energy use

When applying CO_2 capture, energy is needed to separate the CO_2 and compress the CO_2 to pressures required for transport. This energy consumption results in a reduction of the overall efficiency of for instance a power plant. This reduction is called the efficiency penalty, or energy penalty. Table 1 shows typical energy penalties for power generation concepts with CO_2 capture.

Post-combustion CO_2 capture and capture using oxyfuel combustion of solid fuels show about equal increases in primary energy use. For post-combustion this increase is mainly determined by the heat requirement in the capture process. In oxyfuel combustion the separation of oxygen from the air is the main factor causing a drop in efficiency, i.e. about half of the efficiency penalty when considering a coal fired power plant (Andersson and Johnsson 2006). Both systems require significant compressor power to boost the CO_2 from atmospheric to transport pressures (i.e. > 100 bar). This compressor power is substantially lower in the pre-combustion technology as the CO_2 is removed under pressures higher than atmospheric. The required steam and the removal of chemical energy from the syngas in the process prior to CO_2 removal, the water gas shift reaction, contributes the most to the

Capture process	Conversion technology[a]	Generating efficiency[b] (%)	Energy penalty of CO_2 capture (% pts.)	Capture efficiency (%)
Post-combustion (chemical absorption)	PC	30-40	8-13	85-90
	NGCC	43-55	5-12	85-90
Oxyfuel	PC	33-36	9-12	90-~100
	GC and NGCC	39-62	2-19	50-~100
Pre-combustion	IGCC	32-44	5-9	85-90
	GC	43-53	5-13	85-~100

Table 1. Simplified overview of energy conversion and CO_2 capture efficiencies of power plants equipped with various CO_2 capture technologies, after (Damen, Troost et al. 2006; Hetland and Christensen 2008)[a]PC = Pulverized Coal, NGCC= Natural Gas Combined Cycle, PFBC = Pressurized Fluidized Bed Combustion, GC = Gas Cycle, IGCC = Integrated Gasification Combined Cycle. [b]Efficiencies are reported based on the Lower Heating Value (LHV) and assuming a CO_2 product pressure of 110 bar.

increase in primary energy use. The CO_2 removal itself requires less energy in this concept. Overall, the relative increase in primary energy is the lowest for the pre-combustion capture concepts. For the gaseous fuel fired concepts, the increase in primary energy requirement is relatively lower because of the lower carbon content per unit of primary energy.

2.2 CO_2 transport

The captured CO_2 can be transported as a solid, gas, liquid and supercritical fluid. The desired phase depends on whether the CO_2 is transported by pipeline, ship, train or truck. Of these options, transport by pipeline is considered the most cost-effective one. The transport of CO_2 by pipeline in the gas phase is not favourable for projects that require the transport of significant amounts of CO_2 over considerable distances. The disadvantageous economics (large pipeline diameter) and relative high energy requirement (due to the large pressure drop) are the reasons for this (IPCC 2005;Zhang, Wang et al. 2006). Increasing the density of CO_2 by compression renders the possibility to transport the CO_2 with less infrastructural requirements and lower cost.

There is worldwide experience in transporting CO_2 using the transport media mentioned above in the oil industry for enhanced oil recovery (EOR) by injecting CO_2 into an oil field. CO_2 transport by ship is being conducted on a small scale, but is being researched as a possibility to reach offshore storage capacity or as a temporary substitute for pipelines (IEA GHG 2004;Aspelund, Molnvik et al. 2006). Transport by ship can be economically favourable when large quantities have to be transported over long distances (>1000 km) (IPCC 2005). It requires the compression and liquefaction of the CO_2.

2.3 CO_2 storage in geological formations

The last step in the CCS chain is the injection of CO_2 into geological formations. Alternatives to injection in geological formations are injection into the deep ocean and sequestration through mineral carbonation, but the current research focus is on storage in geological formations. CO_2 storage in these geological formations encompasses the injection of CO_2

into porous rocks that may hold or have held gas and or liquids. In literature, several storage media are proposed, especially: deep saline formations (aquifers); (near) empty oil reservoirs, possibly with enhanced oil recovery (EOR); (near) empty gas reservoirs, possibly with enhanced gas recovery (EGR) and deep unminable coal seams combined with enhanced coal bed methane production (ECBM). (Van Bergen, Pagnier et al. 2003;IPCC 2005) The total CO_2 storage capacity ranges between 2 and 11 Tt. It should be stressed that high uncertainties still persist regarding the estimation of storage capacity due to the use of incomplete data or simplified assumptions on geological settings, rock characteristics, and reservoir performance (Bradshaw, Bachu et al. 2006). Despite the uncertainty of these estimates, the figures suggest that there is enough storage potential to support CO_2 emissions reduction with CCS for considerable time. In practice, matching the temporal and geographical availability of sources and sinks may become a bottleneck.

3. Change in key atmospheric emissions due to CO_2 capture

Key direct atmospheric emissions of specific interest for biomass and coal fired concepts are CO_2, NO_x, NH_3, SO_2, HCl, HF, VOC, PM, Hg, Cd, and other heavy metals. For gas fired concepts CO_2 and NO_x are the most dominant atmospheric emissions. Equipping power plants with CO_2 capture technologies affects both the formation and fate of many of these emissions. We limited our study to three main capture systems for the removal of CO_2 depicted in Fig. 1: post-combustion, pre-combustion and oxyfuel combustion.

The chemical absorption technologies that we reviewed in detail include technologies using alkanolamines, such as monoethanolamine (MEA), Fluor's Econamine FG+ and MHI's KS-1 solvent. Other technologies reviewed are based on absorption using chilled ammonia (NH_3), alkali salts (i.e. potassium carbonate -K_2CO_3) and amino salts. The post-combustion system can be applied to various energy conversion technologies. In this study we focus on its application to Pulverized Coal (PC), Natural Gas Combined Cycle (NGCC) and Pressurized Fluidized Bed Combustion (PFBC) power plants. The energy conversion technology that is envisaged using pre-combustion that is mainly investigated in this study is the Integrated Gasification Combined Cycle (IGCC) power plant. The energy conversion technologies using oxyfuel combustion that have been reviewed in this study more extensively are rather conventional PC and NGCC power plants. Advanced technologies briefly touched here include, for instance, chemical looping combustion.

A summary of emission factors for key atmospheric emissions reported in literature for these technologies is presented in Fig. 2. The main effects of CO_2 capture on atmospheric emissions are summarized below for the key atmospheric emissions.

3.1 Carbon dioxide

CO_2 emissions predominantly depend on the type of fuel, on the efficiency of the energy conversion and of the removal efficiency of CO_2. The removal efficiency for the oxyfuel combustion concept is found to be the highest on average (95-98%), yielding the lowest CO_2 emissions for the gas fired conversion technologies (0-60 g/kWh). Post- and pre-combustion show about equal removal efficiencies of 87-90% and 89-95%, respectively. The typically higher conversion efficiency for gasification or reforming results however in typically lower net CO_2 emissions for the pre-combustion concepts (21-97 g/kWh) compared to the post-combustion concepts (55-143 g/kWh).

3.2 Sulphur dioxide

In the coal fired power plants equipped with *post-combustion* CO_2 capture, SO_2 emissions are reduced significantly compared to a power plant without capture. One reason is that power plants with CO_2 capture should be equipped with improved flue gas desulphurization (FGD) facilities (Tzimas, Mercier et al. 2007). Furthermore, additional removal in the post-combustion capture process is expected. Koornneef et al (2010) summarized reported values in literature and show that the minimum expected additional reduction per $MJ_{primary}$ compared to a power plant without CO_2 capture is approximately 40%; on average it is 85%. For the amine based concept it is required to reduce the concentration of SO_x in the inlet gas of the CO_2 capture facility as these compounds may react with the solvent, which leads to the formation of salts and solvent loss. Knudsen et al. (2006;2008) for instance reported a 40-85% uptake of total sulphur depending on the type of solvent[1] used. Iijima et al. (2007) and Kishimoto et al. (2008) report that a minimum of 98% of the SO_2 is additionally removed[2] before entering the CO_2 capture process. They state that then 'almost all' of the still remaining SO_2 is removed from the flue gas as salts. In literature studies additional SO_2 reductions of 90-99.5% are assumed (Rao and Rubin 2002;IEA GHG 2006a;Tzimas, Mercier et al. 2007;Koornneef, van Keulen et al. 2008).

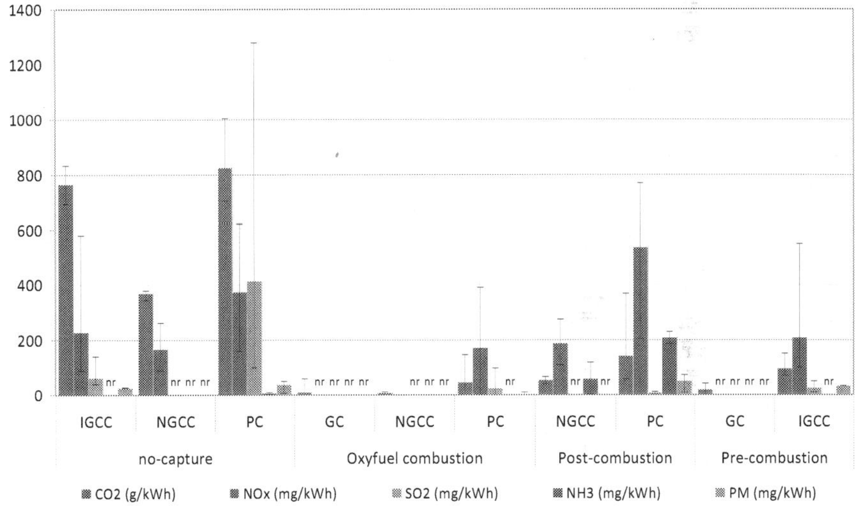

Fig. 2. Atmospheric emissions of substances CO_2, NO_x, SO_2, NH_3 and particulate matter for various conversion technologies with and without CO_2 capture, adapted from (Koornneef, Ramirez et al. 2010). Ranges indicate maximum and minimum values reported. Note that emissions are based on various fuel specifications and on the configuration and performance of the power plant and CO_2 capture process. 'nr' = 'not reported'.

[1] During field tests MEA (monoethanolamine) and the amine based 'Castor 1' and 'Castor 2' solvents were tested. The Castor 2 solvent resulted in the 40 % uptake of sulphur compared to 80% for MEA.

[2] This additional reduction succeeding the conventional FGD is achieved by the reaction of SOx with caustic soda in the flue gas cooler which cools the flue gas before it enters the absorber.

The other post-combustion technology considered here uses chilled ammonia as solvent to remove the CO_2 from the flue gas. Remaining SO_2 in the flue gas can according to Yeh and Bai (1999) react with the ammonia solution to form the recoverable ammonium sulphate. All in all, it is expected that most of the acid gases can be removed from the flue gas when a proper design of the scrubbing process is applied (Yeh and Bai 1999). However, at present no quantitative estimates for additional SO_2 reduction in the CO_2 absorption process based on chilled ammonia are available.

For *oxyfuel combustion* technologies the SO_2 emissions will generally decrease compared to conventional coal fired power plants. The reduction can be the result of several mechanisms: increased ash retention, enhanced efficiency of conventional FGD, co-injection and the possibility for new SO_x removal technologies.

According to Buhre et al. (2005) and Anheden et al. (2005) the amount of SO_x formation per tonne of coal combusted is essentially unchanged when applying oxyfuel combustion. However, the composition and concentration of SO_x, constituting SO_2 and SO_3, does change as the flue gas stream is reduced in both volume and mass. A higher SO_x concentration in the flue gas may pose equipment corrosion problems. A possible positive effect is that it also may enhance the capture efficiency of the electrostatic precipitator (ESP) (Tan, Croiset et al. 2006). Another expected positive side effect is that a higher SO_x concentration may increase the removal efficiency[3] of FGD technologies. Moreover, the reduced flue gas stream allows for smaller equipment. (Marin and Carty 2002;Chatel-Pelage, Marin et al. 2003;Chen, Liu et al. 2007;WRI 2007)

The issues, challenges and design considerations taken into account when designing the flue gas cleaning section for *oxyfuel* combustion are presented in (Yan, Anheden et al. 2006). There, possible configurations for flue gas cleaning are predominantly based on (adapted) conventional flue gas cleaning technologies. The additions compared to a conventional configuration consisting of an SCR, ESP and FGD, are a flue gas cooler (FGC) and CO_2 compression & purification process. The FGC is aimed to reduce the temperature, acidic substances (SO_2 between 93 and 97%, SO_3 between 58 and 78%), water content (>85%) and particulates (>90%) in the flue gas prior to compression. In the following compression & purification step additionally NO_x, SO_x, HCl, water and heavy metals are removed as condensate from the compressors, and with the use of an activated carbon filter and an adsorber (Burchhardt 2009;Thébault, Yan et al. 2009;Yan, Faber et al. 2009). Overall, a deep reduction of SO_2 and NO_x emissions is expected to be possible with oxyfuel combustion, although R&D is required to better understand the behaviour of these substances in the CO_2 compression & purification process.

Co-injection of sulphur compounds into the underground together with the CO_2 is technically possible. Another possibility is the removal of sulphur compounds in condensate streams after compression of the flue gas. Both options would make the FGD section redundant. As suggested by White et al. (2008) the SO_2 may be recovered from the CO_2 stream in the form of sulphuric acid (H_2SO_4) through reaction with NO_2. Experiments indicate SO_2 conversion efficiencies between 64 and ~100% depending on process conditions (White, Torrente-Murciano et al. 2008).

---

[3] Tests in a research facility indicate that SOx removal was improved in the case of oxygen rich combustion, which can partly be explained by longer gas residence time in the FGD (Marin and Carty 2002;Chatel-Pelage, Marin et al. 2003;Chen, Liu et al. 2007;WRI 2007).

In circulating fluidized bed boilers often limestone is injected into the furnace to control SO_x emissions. In the case of oxygen firing the in-furnace desulphurization efficiency with limestone is found to be between 4 and 6 times higher compared to air firing (Buhre, Elliott et al. 2005;ZEP 2006).
The variance shown in Fig. 2 is due to parameters that may vary case by case, e.g. the sulphur content in the coal, uncontrolled SO_x emission (including ash retention), removal efficiency of the FGD section, removal in CO_2 purification section and the degree of co-injection.
IGCC power plants have low SO_2 emissions, either with or without *pre-combustion* CO_2 capture. This is due to the high (typically > 99%) removal efficiencies of sulphur compounds (H_2S and COS) in the acid gas removal section and adjoined facilities. The application of pre-combustion CO_2 capture in an IGCC is assumed to enhance the SO_2 removal. The application of CO_2 capture is likely to result in a decrease of the emission of SO_2 per $MJ_{primary}$, but depending on the efficiency penalty may result in an increase per kWh. Both increase and decrease per kWh have been reported in literature. The reduction per $MJ_{primary}$ is expected to be lower compared to the post-combustion and oxyfuel combustion technologies (see Fig. 2). With pre-combustion it is also possible to yield a stream of CO_2 with H_2S and co-inject this into the underground. This may however complicate the transport and storage process. Also, it may be prohibited by national law and varies per country.

3.3 Nitrogen oxides

If an amine based solvent is used for *post-combustion* capture, the reduction of NO_x emissions per $MJ_{primary}$ is expected to be small, i.e. between 0.8 and 3%[4] (Knudsen, Vilhelmsen et al. 2006;Kishimoto, Hirata et al. 2008). CO_2 capture requires a significant increase in primary energy use resulting in a net increase in NO_x emissions per kWh. For the chilled ammonia technology, the NO_x emissions are not known to be affected by the CO_2 absorption process. It is, therefore, likely that emissions will increase proportionally with the increase in primary energy use.
For *oxyfuel* combustion, in general, net NO_x emissions per $MJ_{primary}$ are likely to decrease compared to conventional coal fired power plants. The two most important factors are that coal fired oxyfuel power plants are likely to show lower levels of NO_x formation in the combustion process and that further high degree of removal of NO_x in the CO_2 treatment train is possible.
NO_x emission reduction and underlying mechanisms are fairly well understood for the oxyfuel combustion technology. NO_x formation during oxyfuel combustion is found to be lower as thermal NO_x formation is suppressed and fuel NO_x is reduced (Croiset and Thambimuthu 2001;Buhre, Elliott et al. 2005;Tan, Croiset et al. 2006;WRI 2007). Overall, the reduction potential for NO_x formation of oxyfuel combustion is according to several experiments in the range of 60-76% (Chatel-Pelage, Marin et al. 2003;Buhre, Elliott et al. 2005;Farzan, Vecci et al. 2005;Andersson 2007;Yamada 2007). However, also no reduction has been found in some experiments (Anheden, Jinying Yan et al. 2005).

---

[4] The main fraction of NOx is formed by NO which is expected to be unaffected by the CO2 capture process. NO2 fraction of NOx, which is typically about 5-10%, may react with the solvent resulting in a reduction of NOx emission per MJprimary. However, also not all of the NO2 is expected to react, i.e. only 25 % (Rao and Rubin 2002;IPCC 2005).

The final emission of NO_x depends also on the flue gas treatment section. The flue gas has a high CO_2 concentration, but also contains NO_x, Ar, N_2, O_2 and SO_2 when it enters the CO_2 treatment train. There are several options for the treatment of the raw CO_2 stream. None of them requires a $DeNO_x$ facility like SCR or SNCR[5] (DOE and NETL 2007). The first option is to co-inject the NO_x together with the CO_2. This requires only compression and drying of the flue gas stream. The second option is to purify the CO_2 with multiple auto-refrigeration flash steps. The gaseous pollutants are, in that case, separated from the CO_2 stream to a high degree and vented into the atmosphere. The remaining fraction is co-injected. A $DeNO_x$ installation may be used to clean the vent stream (IEA GHG 2006b). Another concept is suggested and tested by White et al. (2006;2008) and incorporates compression of the flue gas and removal of NO_x in the form of nitric acid (HNO_3) through a series of reactions[6]. Preliminary results suggest that 48-90% of the NO_x is converted to nitric acid[7] and can consequently be removed from the CO_2 stream.

The oxyfuel combustion variant shows no NO_x emissions from gas fired power plants equipped with CO_2 capture. This estimate is based on one literature source only, i.e. see (Davison 2007). This may result in an underestimation of NOx emissions. As the purity of the oxygen stream is in practice not 100%, some nitrogen may still be present in the combustion air, causing some NO_x formation (IEA GHG 2006c). Whether this is co-injected or separated depends on process configuration.

During normal operation of the IGCC with *pre-combustion* CO_2 capture, NO_x will be mainly formed during the combustion of the hydrogen rich gas with air in the gas turbine. The application of CO_2 capture in an IGCC will decrease the NO_x emissions per $MJ_{primary}$ as relatively less gas is combusted in the gas turbine per unit of primary energy input. This outcome strongly depends on the assumption that the issue of NO_x formation in a gas turbine fired with fuel gas with a high hydrogen content is solved by turbine manufacturers. The flame temperature is namely dependent on the gas composition and heating value. Both of these will change when applying CO_2 capture. If dilution with steam or nitrogen is not applied, the flame temperature during firing of hydrogen rich fuel will increase resulting in an increase in NO_x formation. Consequently, emissions per kWh can also become higher when applying CO_2 capture. The uncertainty is thus higher than the range indicated in Fig. 2. This is however not quantified. (Chiesa, Lozza et al. 2005;IEA GHG 2006c;Davison 2007;DOE/NETL 2007;Tzimas, Mercier et al. 2007)

For gas fired concepts equipped with pre-combustion capture, NO_x emissions are uncertain but expected to be typically higher than for conventional state-of-the-art NGCC cycles (Kvamsdal and Mejdell 2005).

Further NO_x emission reduction can be achieved by adding an SCR process. A possible trade-off for SCR application is the emission of unreacted ammonia, or ammonia slip. This is especially the case when the SCR is applied on exhaust gases with low NO_x concentrations. Ammonia slip from a SCR are however very small (<5 ppmv) and is assumed comparable to

---

[5] S(N)CR = Selective (Non) Catalytic Reduction; a technology to reduce NOx emissions by converting NOx into N2 with the use of reactants, such as fun instance ammonia or urea.

[6] These reactions are (taken from (White, Torrente-Murciano et al. 2008)): 2 NO2 + H2O ↔ HNO2 + HNO3 and 3 HNO2 ↔ HNO3 + 2 NO + H2O.

[7] A potential by-product of this process may be mercuric nitrate (NO3)-2Hg2+ which is formed due to a reaction between the nitric acid and mercury in the flue gas. Although this substance is highly toxic it means that mercury is effectively removed from the flue gas.

normal air combustion in a pulverized coal power plant and a NGCC power plant. An optimum between NO_x reduction and ammonia slip is however to be determined (Rao 2006).

3.4 Ammonia

Ammonia slip from DeNOx facilities is the main source of NH_3 emissions from conventional fossil fuel fired power plants without CCS.

A significant increase of NH_3 emissions may be caused by oxidative degradation of amine based solvents that possibly will be used in *post-combustion* CO_2 capture. In the chilled ammonia technology, the unwanted emission of NH_3 from the CO_2 capture process is a serious challenge (Yeh and Bai 1999). This emission is expected to be significantly reduced by adding a water wash section at the outlet of the CO_2 capture process and by adaptations in the capture process (Yeh and Bai 1999;Corti and Lombardi 2004;Kozak, Petig et al. 2008).

As indicated, the uncertainty regarding the estimation of NH_3 emissions can be considered high as the scientific literature reports a variety of values (Rao and Rubin 2002;IEA GHG 2006a;Knudsen, Jensen et al. 2008). Furthermore, new solvents and additional treatment options are possible to prevent or mitigate the emission of ammonia. The ranges shown in Fig. 2 are thus rather conservative estimates.

For *oxyfuel* combustion, no quantitative estimates for ammonia emissions are known to be reported.

Ammonia formed during gasification is effectively removed in the gas cleaning section in an IGCC with *pre-combustion*. Therefore, emissions are considered negligible.

3.5 Volatile organic compounds

No quantitative estimates for VOC emissions could be derived due to a lack of quantitative information in the pertaining literature.

It is possible that VOC emissions are not significantly influenced by the *post-combustion* CO_2 capture process. In that case the VOC emissions will increase with the increase in primary energy use. However, degradation of amine based solvents may result in the emission of volatile substances, e.g. formaldehyde, acetone, acetaldehyde (Knudsen, Jensen et al. 2008). New solvents are being developed and tested that do not show these degradation products (Hopman. 2008; Knuutila, Svendsen et al. 2009).

No clear information was found on the effect of *oxyfuel* combustion on the formation, reduction and final emission of VOC. However, the oxygen rich conditions during combustion may have an effect on VOC formation. The fate of the formed VOC is uncertain, but it is plausible that a part of the VOCs is either co-injected or vented from the CO_2 purification section (Harmelen, Koornneef et al. 2008).

In IGCC power plants there are two main origins of VOC emissions: the gas turbine section and the fuel treatment section. The formation of VOC in the first is expected to be reduced due to *pre-combustion* CO_2 capture and the associated higher hydrogen content of the fuel gas. Quantitative estimates for the reduction of VOC are however not available. The emissions from the fuel treatment section are expected to remain equal per $MJ_{primary}$. VOC emission reporting for an IGCC operating in the Netherlands does not provide decisive insights into which section is the dominant source of VOC (NUON 2005;NUON 2006). The net effect of both may thus be an increase or decrease per kWh. For gas fired cycles, the replacement of natural gas with hydrogen rich fuel gas is expected to lower the emission of VOC.

3.6 Particulate matter

Often no distinction is made in the consulted literature between various sizes[8] of emitted particulate matter in emission reporting. In this review, therefore also no distinction could be made between size fractions.

The high variance for *post-combustion* capture technologies for solid fuel fired power plants stands out in Fig. 2. The variance represents the varying assumptions in literature. On the one hand, some scholars assume a deep reduction of PM due to the application of post-combustion CO_2 capture; on the other hand, other scientists assume that it will not have an effect on PM emissions. Results from an amine based post-combustion capture demonstration project however indicate a decrease in emission of particulate matter of 64-80%[9] per $MJ_{primary}$ (Kishimoto, Hirata et al. 2008). Also Kozak et al. (2008) suggest a decrease with the use of chilled ammonia technology[10]. An increase in emission per $MJ_{primary}$ is never assumed. Together with the energy penalty due to CO_2 capture, PM emissions may however increase per kWh.

The low particulate matter emissions found for the *oxyfuel* combustion technology are partly due to the enhanced removal efficiency of the ESP[11] that is possible during oxyfuel combustion. Particulates may also be partially co-injected with the CO_2 stream. Another possibility is that particulates are vented from the CO_2 treatment section. Yet another option is that PM is removed with the condensate stream that is formed when SO_2 and NO_x are removed as sulphuric and nitric acid, as mentioned earlier. All together, PM emissions are estimated to be very low.

IGCC power plants are assumed to have lower PM emission factors compared to other conversion technologies and types of power plants. *Pre-combustion* CO_2 capture has virtually no influence on the emission of PM (per $MJ_{primary}$) from an IGCC.

Although no quantitative estimates are available, it may be possible that PM emissions, in specific $PM_{2.5}$ emissions, will be lower due to the enhanced capture of sulphur compounds from the syngas, which is expected to reduce the formation of sulphates, which are characterized as $PM_{2.5}$.

3.7 Other atmospheric emissions of interest

Fig. 1 shows that the *post-combustion* CO_2 capture process is situated after the flue gas cleaning section. Depending on the type of solvent that is used, impurities need to be removed from the flue gas in order to limit operational problems. When MEA is used, its consumption in the capture process is mainly caused by degradation by oxygen and impurities in the flue gas. Important impurities are sulphur oxides (SO_x), nitrogen dioxide (NO_2), hydrogen chloride (HCl), hydrogen fluoride (HF) and particulate matter as they react

[8] Particulate matter can be subdivided into particles with a diameter larger than 10 microns (>PM10) and smaller than 10 microns (PM10). PM10 can then be further subdivided into the size categories 'Coarse' (PM2.5-10) and 'fine' (PM2.5).

[9] 'Dust' (not further specified as PM10 or PM2.5) emissions are reduced by 40-50% in the flue gas cooler prior to the absorption process in which another 40-60% of the particulates is removed from the flue gas.

[10] They do not report a quantitative estimate but suggest that the flue gas cooler will result in a deep reduction of particulate matter entering the absorption process.

[11] The efficiency of the Electrostatic Precipitator is possibly improved as a larger share of SOx is represented by SO3 (Tan, Croiset et al. 2006).

with the MEA or cause foaming of the solvent. This may result in reduction of HF and HCl emissions. Estimates in literature vary but are as high as 90-95%.

Power plants equipped with CO_2 capture should thus be equipped with highly efficient flue gas desulphurization (FGD), $DeNO_x$ installations and electrostatic precipitators (ESP) and/or fabric filters to remove PM. Also, the flue gas typically requires cooling before it is processed in the CO_2 capture installation. In the CO_2 capture process also some of these substances are partially removed. The capture process is thus expected to affect (i.e. lower) the emission of these air pollutants directly and indirectly.

The consumption of solvent in the capture process is an important driver for solvent development as solvent loss deteriorates operational economics and has environmental consequences. The consumption of the solvent varies per type of solvent but is for post-combustion typically in the order of 1-2 kg/tonne CO_2 captured. Recent pilot plant test campaigns report solvent consumption rates for MEA at 0.3 kg/tonne captured (Moser et al. 2011). Typically, the consumption of MEA is higher compared to its alternatives. Moreover, the consumption of solvent used in IGCC with or without pre-combustion concepts can be considered very low, although an increase is expected when CO_2 capture is applied.

The higher oxygen concentration in the flue gas from natural gas combustion possibly results in higher oxidative degradation of solvents. MEA is to instance susceptible for this type of degradation (Supap, Idem et al. 2009). However, as other impurities such as SO_2 and PM are virtually not present in the flue gas, overall degradation and consumption is considerably lower compared to coal fired power plants.

For some post-combustion variants additional atmospheric emissions are expected. More specifically, the emission of solvent or degradation products of the solvent are currently of high interest (see section 6). For MEA based solvents this may be direct MEA emissions. The exact quantity of this 'MEA slip' (estimates range between 1 and 4 ppmv) and possible effects on the environment, including human safety, are not fully known but are intensively researched. In addition, solvent additives (e.g. corrosion inhibitors) may result in trace emissions of heavy metals (Thitakamol, Veawab et al. 2007).

For the chilled ammonia process solvent emissions may be NH_3 (Yeh and Bai 1999;Corti and Lombardi 2004;Kozak, Petig et al. 2008). The alkanolamine based solvents may result in the emission of VOC and NH_3 due to the degradation of the solvent (Strazisar, Anderson et al. 2003;Rao, Rubin et al. 2004;Knudsen, Jensen et al. 2008). Korre et al. (Korre, Nie et al. in press) report that NH_3 emission from using the MHI KS-1 solvent is expected to be higher than from using MEA or potassium carbonate[12]. Contrarily, IEA GHG (IEA GHG 2006a) reports lower values for NH_3 emissions for the MHI KS-1 process compared to Fluor's process based on MEA.

For a K_2CO_3 based sorbent the slip into the atmosphere is considered negligible. Furthermore, this substance is considered to be less toxic to the environment (Oexmann and Kather 2009;Smith, Ghosh et al. 2009). K_2CO_3 may however require the addition of promoters to increase the reaction rate. Some promoters, like arsenic trioxide and piperazine, are known to be toxic (Smith, Ghosh et al. 2009).

Allaie and Jaspers (Allaie and Jaspers 2008) claim that the use of amino salts does not result in ammonia formation, losses due to evaporation and virtually nihil emissions of the solvent.

[12] In this case the Piperazine, an amine, is added to the potassium carbonate sorbent as an activator to increase reaction rate.

Mercury (Hg) and other heavy metals may be partially removed in the CO_2 capture process. Measurements on reclaimer bottoms have indicated that mercury is present in the bottoms (Strazisar, Anderson et al. 2003). A recent study however indicates that a combined process of removing mercury and CO_2 would not lead to significant reductions, i.e. below 15% (Cui, Aroonwilas et al. 2010).

In coal and biomass fired *pre-combustion* concepts using solvents, no solvent emission to the air is expected during normal operation as any slip of the solvent would be combusted in the gas turbine or end up in the CO_2 stream. For this technology, co-sequestration of H_2S is a technical possibility and co-injection is common in for instance Canada (Thomas and Benson 2005).

Carbon monoxide emissions from an IGCC with capture are reduced as the WGS process is introduced, converting CO and H_2O into H_2 and CO_2. A second source of CO emissions is the pre-preparation of the fuel, i.e. storage and grinding. The latter emissions will not be directly affected by the CO_2 capture process, although per kWh those emissions will probably increase as more fuel has to be stored and handled per kWh.

No solvents are used in the *oxyfuel* combustion system. Additional gaseous emissions per primary energy input are thus not expected. In literature it is however suggested that in oxyfuel concepts, due to higher oxygen concentrations, a larger part of elemental mercury (Hg) is converted to ionized Hg species, which will possibly result in higher capture efficiencies of Hg in flue gas cleaning sections ($DeSO_x$ and DeHg). (Marin and Carty 2002;Chatel-Pelage, Marin et al. 2003) (WRI 2007) This may be an additional benefit of CO_2 capture with oxyfuel combustion.

For the advanced oxyfuel concepts, chemical looping combustion[13], the loss of metal oxides to the atmosphere may be a concern. Metal oxides are used as oxygen carriers to physically separate the reduction and oxidising step in this power cycle. These metal oxides may contribute to the environmental impacts of energy supply with this concept as it might bring forward direct environmental impacts, i.e. some metals are considered toxic. Also, these oxygen carriers may bring forward environmental impacts in their life cycle, e.g. during mining, treatment and disposal.

4. Atmospheric emissions across the value chain of CCS

Life Cycle Assessment (LCA) is today one of the most used tools for evaluating the potential environmental impact of products and materials. LCA is a technique for assessing the environmental aspects and potential impacts associated with inputs and outputs of a product system. In the case of CCS, a full LCA includes the production of the fuel carrier (e.g., mining of coal), fuel transport, power production, CO_2 capture, CO_2 transport and CO_2 storage (see Fig. 3). Note that for most studies found in the literature, including those on CCS, emissions from the infrastructure and the extraction of raw materials other than fuel tend to be excluded since they are assumed to be relatively small in comparison to primary burdens or there is lack of data that does not allow for a reliable analysis.

[13] Chemical looping concepts typically include an oxidizing reactor (OX) were a metal oxide (oxygen carrying metals that are considered are: Cu, Co, Ni, Fe and Mn) is formed through the exothermic reaction of a metal with oxygen. This is the oxygen carrier that transports the oxygen to the reduction reactor (RED). In the reduction reactor the fuel reacts (oxidizes) with the oxygen from the metal oxide (MeO).

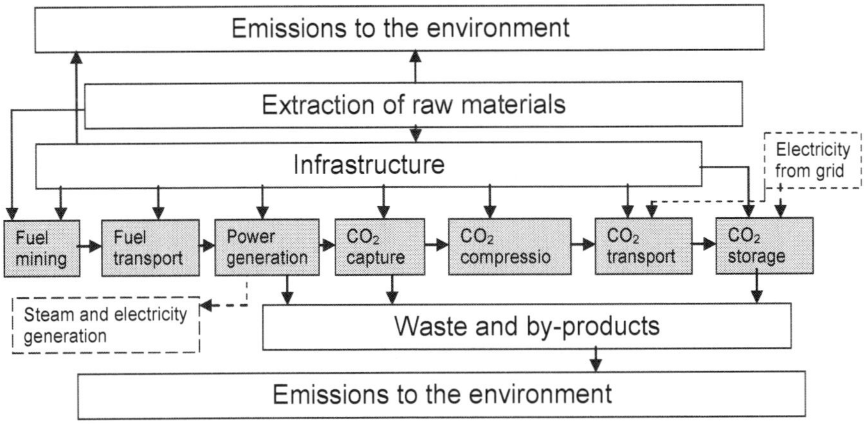

Fig. 3. Schematic overview of step chains of fossil fuel production with CCS and interactions with other systems.

In this section, we consider the results found by 25 studies published between 1995 and 2009. Each reviewed study typically addresses different impact categories. The focus of this section is on the atmospheric emissions of CCS during its life cycle[14] and, therefore, only the following categories will be examined in detail: CO_2 emissions, SOx/NOx and Particulate matter. The literature used to assess the emissions is listed in Table 2.

Capture technology type	Literature sources
PC with CCS	(Rao and Rubin 2002;Spath and Mann 2004;IEA GHG 2006a;Khoo and Tan 2006;Tzimas, Mercier et al. 2007;Viebahn, Nitsch et al. 2007;Weisser 2007;Koornneef, van Keulen et al. 2008;Odeh and Cockerill 2008;RECCS 2008;Bauer and Heck 2009;Markewitz, Schreiber et al. 2009;NEEDS 2009;Nie 2009;Pehnt and Henkel 2009;Schreiber, Zapp et al. 2009;Korre, Nie et al. 2010)
IGGC with CCS	(Waku, Tamura et al. 1995;Akai, Nomura et al. 1997;Lombardi 2003;Tzimas, Mercier et al. 2007;Viebahn, Nitsch et al. 2007;Weisser 2007;Odeh and Cockerill 2008;RECCS 2008;Pehnt and Henkel 2009)
Oxyfuel with CCS	(Viebahn, Nitsch et al. 2007;RECCS 2008;Bauer and Heck 2009;NEEDS 2009;Nie 2009;Pehnt and Henkel 2009)
Natural gas with CCS	(Lombardi 2003;Spath and Mann 2004;Sundkvist, Klang et al. 2004;IEA GHG 2006a;Hertwich, Aaberg et al. 2008;Odeh and Cockerill 2008;RECCS 2008;Bauer and Heck 2009;NEEDS 2009)

Table 2. Literature sources used to assess the emissions of the CCS value chain.

[14] NOx and SOx emissions lead to the formation of acid gases, which can lead to acidification, eutrophication and smog formation nd not in the consequence of a given emission. For instance, SO2 and NOx can cause acidification.

4.1 CO_2 equivalent emissions

The main goal of CCS is to reduce CO_2 emissions and consequently, Global Warming Potential (GWP). For pulverized coal-fired power plants with post-combustion capture technology using MEA a range in GWP over the life cycle of 79-275 gCO_2eq/kWh is reported (range for PC without CCS is in the range 690 to 1100 gCO_2eq/kWh). Where PCs without CCS have a share of power plant operation in life cycle GWP of about 80-95%, installing CO_2 capture decreases this share to about 43-60%. Thus, the deployment of CCS results in a pronounce increase in the share of indirect CO_2eq. emissions in the complete life cycle[15].

In the case of IGCCs with pre-combustion CO_2 capture, GWP values reported are in the range 110 to 181 gCO_2eq/kWh (the range for IGCCs without CCS is 666 to 870 gCO_2eq/kWh). Lignite-fired IGCCs with CCS have almost 20% less absolute emissions compared to hard coal-fired IGCCs with CCS. Installing CCS results in a reduction of about 82 to 87% for lignite-fired IGCCs with CCS relative to IGCCs without CCS, while for hard coal-fired IGCCs the relative differences are in the range of 69 to 81%.

Interestingly, hard coal-fired power plants with CCS technology are reported as having between 20% (IGCC with CCS) and 30% (PC with CCS) more GHG emissions than similar lignite-fired power plants with CCS, while without CCS technology the hard coal-fired power plants have about 10% lower emissions than lignite-fired power plants. This is due to a typically larger share of upstream emissions (e.g. fuel extraction and processing) for hard coal-fired power plants than for lignite-fired power plants. Lignite-fired power plants are often directly located at the mining site ('mine-mouth' operated) which results in lower transport emissions. As these upstream emissions are not reduced by CCS technology, but mainly increase due to the energy penalty, the GHG emissions from mine-mouth based power plants can be reduced further when implementing CCS.

The range in GWP for oxyfuel power plants with CCS is 25-176 gCO2eq/kWh. The relative decrease in GWP ranges from 78% to 97%. Specifically for hard coal-fired power plants, the range in the relative difference is smaller (78% to 87%).

For NGCCs with post-combustion using MEA, the GWP are in the range 75-245 gCO2eq/kWh, which is about 51-80% less than the values reported for NGCCs without CCS. The two studies (Spath and Mann 2004;Odeh and Cockerill 2008) reporting the lower value range (51-58%) also report about 25% indirect emissions for NGCCs without CCS technology compared to about 12-15% reported in other studies. Both studies state that in the case of NGCCs the amount of methane leakage from natural gas extraction and transport has a significant effect on life cycle GHG emissions and is more inopportune in the case with CCS due to the increase in primary energy consumption. It is, however, unclear if the other studies include this methane leakage in the reported values.

4.2 Particulate matter

PM_{10} emissions reported for the life cycle of PC power plants with post-combustion using MEA range between 0.013 and 0.434 gPM_{10}/kWh while $PM_{2.5}$ for the same type of plants are reported between 0.05 to 0.07 gPM2.5/kWh. PC plants without CCS report PM_{10} in the

[15] Two outliers are identified. Markevitz et al., (2009) shows a significantly smaller (23%) share of power plant operation after capture while Pehnt and Henkel (2009) showed a larger share (79%). The origin of the differences cannot be identified from the data reported.

range 0.009 to 0.35 gPM_{10}/kWh and $PM_{2.5}$ in the range 0.009 to 0.35 gPM_{10}/kWh. Contrary to the results found for GWP, no clear difference is reported for hard coal-fired and lignite-fired power plants.

Only two studies (Viebahn, Nitsch et al. 2007;RECCS 2008) report the contribution of the different part of the CCS chains. In these studies, the contribution of the PC plant with CO_2 capture is estimated at 33% and 45%, which is lower than the estimated contribution of a similar PC plant without CCS (60% and 65%, respectively).

The amount of studies reporting PM emissions for other CO_2 capture technologies is limited. One study by Odeh and Cockerill (2008) reports PM emissions for IGCCs with and without CCS technology (0.004 g/ kWh in both cases). The value is lower than those reported for PCs due to the high removal of PM in the syngas (to avoid detrimental effects in the turbine). NEEDS (2009) reports PM values for NGCCs equipped with MEA based post-combustion technology in the order of 0.005-0.006 gPM_{10}/kWh and 0.009-0.010 $gPM_{2.5}/kWh$ (the values for a NGCC without CCS are in the range of 0.003-0.012 gPM_{10}/kWh and 0.007-0.008 $gPM_{2.5}/kWh$). In this case, PM emissions are mainly associated with NOx emissions (which are PM precursors) from the power plant and the winning of natural gas. Values for oxyfuel power plants with CO_2 capture are also reported by NEEDs (2009). For PM_{10} the range reported is 0.012 to 0.025 gPM_{10}/kWh while for $PM_{2.5}$ this is 0.07 to 0.36 $gPM_{2.5}/kWh$.

4.3 NO_x and SO_x

Table 3 shows the ranges found in the literature for the life cycle of power plants with and without CCS. The number of studies reporting specific results, particularly for IGCC and oxyfuel plants with CCS, is quite limited making it difficult to draw robust conclusions.

Results for PC with post-combustion capture using MEA indicate an increase in the amount of NOx during the life cycle. The partial removal of NOx during the capture process is not large enough to offset the increase in emissions caused by the additional fuel needed to compensate the energy penalty. Literature appears less clear on the impact of post-combustion on SOx levels, with some studies showing a decrease in emissions while others indicating an increase on emissions. Koornneef et al., (2008) indicates that a decrease in both NOx and SOx emissions associated with the transport of coal can be expected since stringent regulations are anticipated to reduce sulphur content in marine fuel and to limit NOx emissions during ship transport.

In the case of NGCC plants with post-combustion MEA, SOx emissions are reported to increase over the life cycle. The level however remains well below those of PC plants due to the very low sulphur content of natural gas.

Type of plant	NOx			SOx		
	With CCS (gNOx/kWh)	% change	N	With CCS (gSO_2/kWh)	% change	N
PC	0.58- 1.39	+13% to+49%	7	0.08-0.84	-60% to +20%	6
IGGC	0.10	-16%	1	0.33	+10	1
Oxyfuel	0.27-0.60	---	1	0.11-0.35	---	1
NGCC	0.13-0.30	-50% to +15%	5	0.01-0.16	+14 % to +100%	3

Table 3. Emissions of SOx and NOx reported in the literature and relative change compared to a similar plant without CCS (N= number of studies).

5. Case studies – Netherlands and Europe

In this section we discuss two case studies showing the consequences of implementing CCS in the power and heat sector on non-GHG emissions that affect air quality. We consider two geographical regions: the Netherlands and the European Union.

5.1 Scenario study for the Dutch power and heat sector

The BOLK research programme was initiated by the Dutch Ministry of Housing, Spatial Planning and the Environment (VROM) to acquire more detailed information on the synergy and/or trade-offs of GHG mitigation policies and transboundary air pollution (AP) policies (Harmelen, Koornneef et al. 2008;Horssen, Kuramochi et al. 2009). Part of that programme was aimed at assessing CCS technologies and includes scenario analyses for key atmospheric emissions (NO_x, SO_2, PM_{10} and NH_3) from the power sector. Three scenarios were developed and compared to the 2006 emission level (see Fig. 4):

1. Power sector emissions without CCS : No CCS is applied to the power plant sector in 2020
2. 2020 with CCS -S1: CCS is applied to two new coal fired power plants, 1 with post-combustion and 1 with pre-combustion.
3. 2020 with CCS -S2: CCS is applied to all new coal fired power plants, 3 with post-combustion and 1 with pre-combustion.

The impact of applying CO_2 capture technologies to the power generating sector is assessed in that study for several view years (i.e. up to 2050). For the view year 2020 the scenario is based on the actual and planned power plants.CO_2 emissions from the sector are assumed to increase from 38 in 2006 to 62 Mtonne in 2020, without the introduction of CCS. In 2020, up to 24 Mtonne CO_2 could be avoided when equipping new coal fired power plants with CO_2 capture.

The results of that scenario study are shown in Fig. 4. Emissions of NO_x, SO_2, PM_{10} and NH_3 in the sector are estimated to increase in the reference scenario for the year 2020, due to the increase in Dutch coal fired capacity without CCS. The introduction of CCS (only post- and pre-combustion) is expected to lead to a further increase of NO_x (up to 1.5 ktonne), PM (up to 70 tonne) and NH_3 emissions (up to 0.7 ktonne). SO_2 emissions decrease below the 2006 level. The introduction of CCS leads to a relative large increase (from 0.1 ktonne to up to 0.8 ktonne) in the overall low contribution of the power generation sector to the NH_3 emissions. For the scenario analyses, the emission data in relative old publications associated with the use of ethanolamines were used. Developments are still going on to reduce the solvent degradation and with it the emissions from solvents. With the improvement of the solvent technology, NH_3 emissions will be strongly reduced.

The additional cost of mitigating key atmospheric emissions are roughly assessed and compared to the reference scenario. The results showed that the cost mainly consist out of the increased costs for air pollution control needed to counteract the projected capacity increase of the power plant sector. The additional mitigation costs due to CCS are estimated to be small compared to those costs. The costs of NO_x and PM_{10} dominate the overall mitigation costs of approximately 50 million Euros per year.

The mitigation costs of NO_2, PM_{10} and NH_3 compared to the reference scenario are positive, because of the increase of the emissions in the CCS scenarios. Mitigation cost for SO_2 are negative due to the emission reduction of this substance in CCS equipped power plants.

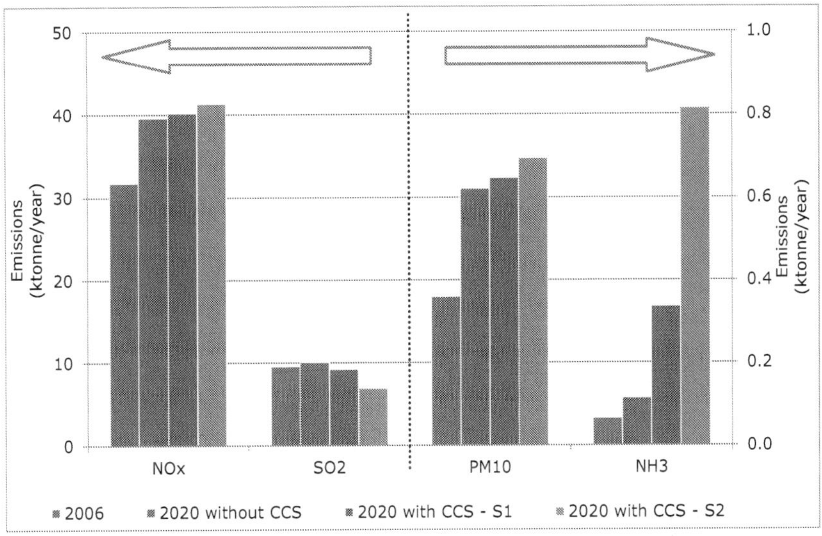

Fig. 4. Emissions of NO_2, SO_2 (ktonne/year) on left axis and of PM_{10}, NH_3 (ktonne/yr) on right axis, in the Dutch power plant sector in 2020.

5.2 Scenario study for the European power and heat sector in 2030

In a study by Koornneef et al (2010) the trade-offs and synergies between climate and air quality policy objectives for the European power and heat (P&H) sector were quantified. The analysis includes assessing the impact of applying CO_2 capture in the European P&H sector on the emission level of key air pollutants in 2030. A model was developed with the assumption that all power plants built between 2020 and 2030 are equipped with CO_2 capture and that all plants built between 2010 and 2020 are retrofitted[16] with CO_2 capture before 2030. Four scenarios were investigated: one without CCS (baseline) and three with CCS. Each one focuses on a different CO_2 capture system (post-combustion, oxyfuel combustion and pre-combustion).

The first scenario without climate measures was drawn entirely from the GAINS[17] model developed by the IIASA (IIASA 2008). The emission levels of NECD substances in 2030 were defined by combining sector activity and emission factors for P&H plants with and without CO_2 capture.

In the three scenarios with CO_2 capture the share of power plants equipped with CO_2 capture technology was determined by estimating for each country the additional sector activity (in primary energy use) in the baseline scenario per combination of conversion technology and fuel for the periods 2010-2020 and 2020-2030.

[16] From 2010 onwards it is more likely that the power plants will be built capture ready. In this study it was therefore assumed that only power plants built between 2010 and 2020 are retrofitted between 2020 and 2030.

[17] Greenhouse gas - Air pollution Interactions and Synergies model. The GAINS model is developed to analyse trade-offs and co-benefits of strategies aimed at the reduction of air pollution and greenhouse gases on the medium-term, i.e. until 2030.

Results show a reduction in GHG emissions compared to the baseline scenario between 7% and 16% for scenarios with CCS penetration in the European power and heat sector. This comes with an increase in total primary energy use in the EU of approximately 1-5%. In the power and heat sector alone this increase is between 2% and 17%.

SO_2 emissions are estimated to be very low for all scenarios that include large-scale implementation of CO_2 capture in 2030, i.e. a reduction varying between 27% and 41%. This holds especially for the scenario with a large share of oxyfuel combustion technology.

Further, it was found that NO_x emissions from the P&H sector could be 15% higher in a scenario with predominantly post-combustion CO_2 capture compared to the baseline scenario without CCS. A reduction in NO_x emissions is expected when oxyfuel combustion (-16%) or IGCC with pre-combustion CO_2 capture (-20%) is mainly applied.

Large-scale implementation of the post-combustion technology in 2030 may also result in significant higher NH_3 emissions compared to scenarios without CCS and with other CO_2 capture options, although uncertainty in this estimate is substantial. If these emissions are not controlled properly, NH_3 emissions from the P&H sector change from an insignificant contribution of 0.5% towards a possible very significant contribution of 13% of the EU total for all sectors together.

Direct particulate matter emissions are likely to be lower in the scenarios with CO_2 capture. The scenario with implementation of the oxyfuel combustion technology shows the largest (i.e. 59%) reduction in PM emissions in the P&H sector followed by the scenario with a significant share allocated to pre-combustion CO_2 capture showing a reduction of 31%. Post-combustion capture may show an increase in PM emissions due to a limited removal and a larger increase in primary energy use. The scenario with post-combustion capture resulted in PM emissions varying between 35% reduction and 26% increase. No robust conclusions could however be drawn on how CO_2 capture influences the emissions of various PM size categories (i.e. $PM_{2.5}$, PM_{10} and $>PM_{10}$) as this is not satisfactorily addressed in pertaining literature.

6. Special highlight topic: atmospheric emissions from post-combustion capture

6.1 Solvent emissions

Due to the low partial CO_2 pressure in flue gases from the power sector, the use of chemical solvents is preferred in post-combustion capture. Chemical solvents seem to be the preferred option for the short-term, since this technology is relatively mature, commercially available at industrial scale (though not yet power plant scale), and post-combustion can be used to retrofit existing power stations (as end-of-pipe treatment). The disadvantages of using amines as chemical solvent are high costs for energy (energy penalty), space (due to large gas volumes) and equipment. Furthermore, amines and degradation products are found to be emitted from the stack, causing potential environmental impacts (Horssen, Kuramochi et al. 2009).

Amines can leave the power plant with the CO_2 captured gas to be stored in the deep underground where it is considered to have limited environmental impact. Emissions of amines to the air can take place when the residuals are taken out of process when recycling the amine at the top of the absorber (especially relevant for safety of workers) and with the cleaned flue gas at the top of the absorber where they are emitted into the atmosphere (relevant for the public and environment).

Finally, amines can degrade, e.g. into ammonia (also treated in the previous sections). The amines can also react with oxidized nitrogen in the atmosphere to form potentially harmful compounds such as nitrosamines, nitramines, aldehydes and amides. The environmental impacts are not easy to assess since there are a large number of degradation products which not only depend on the degradation mechanisms occurring in the capture process, but also on the type of amines used (Knudsen 2008).

6.2 Environmental impacts of amines

In 2007, the Norwegian Institute for Air Research (NILU) conducted a screening study to understand more about atmospheric amine chemistry and to evaluate the environmental effects of amine emissions and degradation products such as nitrosamines, nitramines, aldehydes and amides.

The amines studied are monoethanolamine (MEA), piperazine, aminomethylpropanol (AMP) and methyldiethanolamine (MDEA). Among these amines, piperazine has been through a thorough evaluation and classification in the EU system. There are several experimental studies available on MEA, but most of them were conducted during 1960s and -70s. For AMP and MDEA the toxicological data are rare. High quality inhalation studies are lacking. For piperazine and MEA indications exist of reproductive and developmental toxicity. In addition, one study suggests similar effects of AMP. None of the amines have been reported to be carcinogenic, but this should also be evaluated further with additional studies (Låg, Andreassen et al. 2009).

Låg et al. concluded that amines themselves are most likely causing little risk to human health, but the emissions contribute to the nitrogen load and potentially to eutrophication which could have impacts on sensitive terrestrial ecosystems.

Nitrosamines (N-nitrosamines) are a large and diverse family of synthetic and naturally occurring compounds described by the formula (R1)(R2) N-N=O, where R1 and R2 is an alkyl or aryl group. Nitrosamines are typically liquids, oils or volatile solids. Nitrosamines occur in the diet, through use of tobacco, cosmetics, pharmaceutical products and agricultural chemicals. Nearly all commercially available alkylamines are generally contaminated by small quantities of their corresponding N-nitroso analogues. Industrial installations producing or using amines might be a source of nitrosamine pollution (Tricker, Spiegelhalder et al. 1989).

Exact data on concentration levels in power plants using amine based carbon capture are very sparse and very hard to find in the public literature. However, nitrosamines are considered of particular concern because of their toxic and carcinogenic properties at extremely low levels.

Nitramines are also of concern as they are suspected to be carcinogenic, though considerably less than the nitrosamines. However, the longer lifetime in the atmosphere may lead to higher exposure values. Modelling also indicates that amines lower the surface tension of water droplets, which under appropriate climatic conditions can be a trigger for rain with the potential of causing negative impacts to the local environment. Worst case studies for a generic full scale amine plant with conditions representing the west coast of Norway show that the predicted concentrations of photo oxidation compounds are at the same level as the proposed "safety limits", implying that risks to human health and natural environment cannot be ruled out (Knudsen 2009).

With regard to aldehydes, Låg et al. (2009) concluded that at airborne levels for which the prevalence of sensory irritation is minimal, both in incidence and degree (<1.2 mg/m^3), risks of respiratory tract cancer are considered to be negligibly low. Acetamide may induce skin irritation. The irritating potential of the aldehydes and amides might in this context be the most relevant adverse health effect of these compounds, as the amines probably to be used in CO_2 capture also have such effects. Therefore, all these compounds have to be evaluated together with respect to irritating potential of the air around the gas plants (Marit, Instanes et al. 2009).

These statements highlight the necessity for further testing and analysis of amine effects in order to limit the risks, especially for nitrosamines and nitramines. At least as important is the measurement of amine related substances from CCS equipped power plants in order to assess the exact concentration level of specific amine related species. This is needed to assess the risks for public and workers as well as to understand the chemical formation processes as a basis for the development of countermeasures against amine effects due to CCS.

6.3 CO_2 capture solvents and regulations

The health and environmental properties of a number of CO_2 capture compounds have been evaluated by StatoilHydro in the light of the REACH regulation. REACH is the new chemical legislation in the EU. REACH stands for Registration, Evaluation and Authorisation of CHemicals. Industrial CO_2 capture plants are covered by REACH and the IPPC, the EU directive restricting polluting discharges from industry. An important item is the discharge permission based on the comparison with the Best Available Technologies (BAT). The current BREF (Reference Document on Best Available Techniques) for large combustion Plants (IPPC 2006) does not contain information on solvent related emissions from CO_2 capture.

Svanes (2008) shows in his study that the selected compounds (mainly amines) and degradation products (ammonia) are not on the restricted list. Most of the compounds are classified as harmful to health and/or the environment. Using the compounds will not be severely restricted by REACH. The study did not incorporate the degradation products as nitrosamines. A more comprehensive study is recommended containing exposure studies and mapping of degradation products.

Based on the available literature Låg et al. (2009) suggested exposure guidelines for four amines; particularly for AMP and MDEA there are few high quality studies. The guidelines presented are therefore just indicative. The uncertainty factors were chosen in accordance with EU guidelines. Based on inhalation exposure risk, the general population, over time, should not be exposed to levels in the air higher than:

- MEA: 10 µg/m³
- AMP: 6 µg/m³
- MDEA: 120 µg/m³
- Piperazine: 5 µg/m³

Finally, it has been stated that it is highly relevant to know which precise amine is used in CCS, because each individual amine has different effects and potential risks. Furthermore, use of more than one amine infers that the exposure guidelines should be evaluated again, since amines seem to have similar adverse effects and might therefore also show additive or synergistic effects.

7. Conclusions and way forward

Depending on the applied CO_2 capture technology, trade-offs and synergies can be expected for key atmospheric emissions, being: NO_x, SO_2, NH_3, particulate matter, Hg, HF and HCl. For all three (pre-, post- and oxyfuel combustion) capture systems it was found that SO_2, NO_x and PM emissions are expected to be reduced or remain equal per unit of primary energy input compared to power plants without CO_2 capture. Increase in primary energy input as a result of the energy penalty for CO_2 capture may for some technologies and substances result in a net increase of emissions per kWh output. The largest increase is found for the emission of NO_x and NH_3 when equipping power plants with post-combustion capture. A decrease is expected for SO_2 emissions, which are low for all power plants with CO_2 capture.

Additional research (measurements and modelling) and regulatory efforts (norm setting) are required to cope with 'new' emissions from predominantly post-combustion CO_2 capture technologies. Laboratory and field experiments are necessary to obtain more precision in the estimates of emission levels, as little information exists in open literature. For this, accurate sampling and analysis methods have to be developed and validated for low concentrations.

Also, for post-combustion capture using amines it is necessary to identify and quantify the specific compounds that will be emitted or formed post-emission, where particular focus should be put on nitrosamines and nitramines. It is recommended to focus research on the determination of atmospheric degradation paths, precise degradation yields, and degradation products' lifetime in the atmosphere. Development of models is necessary to quantify the mass fluxes and chemical interactions, and finally to integrate them in a dispersion model to quantify the load and possible environmental consequences.

Furthermore, research should be focused on the assessment of toxicity levels of these substances, as a basis for the development of both acute and chronic human toxicity exposure limits for amines and associated substances, both for workers and the public.

This is needed to further compile data and information to create a relative ranking of amines with respect to potential environmental and health effects and toxicity and to find efficient ways to mitigate formation of nitrosamines and nitramines.

We recommend to set up extensive environmental monitoring programmes at currently planned CO_2 capture (demonstration) plants aimed at creating a better understanding of the formation and fate of solid, liquid and atmospheric pollutants. Emissions that should be monitored are: SO_x, NO_x, HF, HCl, Hg, PAH, dioxins, hydrocarbons, heavy metals, NH_3, MEA and PM. For particulate matter it is especially of interest to discern the removal efficiencies for the various sizes of particulate matter. For heavy metals it is of interest to measure to what extent the transposition occurs from atmospheric emission to waste water effluent and solid waste. Monitoring programmes should help to quantify emissions in further detail and share its knowledge with the wide research community.

Life cycle effects of implementing CO_2 capture options should not be neglected when reviewing the environmental performance of complete CCS chains, from cradle to grave. Recent studies namely indicate that for some substances (e.g. SO_x) direct (atmospheric) emissions may decrease due to CO_2 capture; but that additional life cycle emissions by up- and downstream process may result in a deterioration of the overall environmental performance of the CCS chain compared to a power plant without CCS; except for the global warming potential.

For a number of environmental impact categories no agreement exists on the exact direction (positive, negative) or the level of the life cycle impacts due to CCS deployment. The sometimes large ranges often indicate that specific regional or technical issues influence the overall environmental performance of a chain. This requires further research.

We also see a high value in screening next generation CO_2 capture technologies at an early stage on their environmental performance in order to facilitate the optimization of CCS chains not only on its cost and GHG emission performance, but also taking into account other environmental themes.

8. References

Akai, M., et al. (1997). "Life-cycle analysis of a fossil-fuel power plant with CO2 recovery and a sequestering system." Energy 22(2-3): 249.

Allaie, S. and D. Jaspers (2008). TNO CATO pilot plant at E-On Maasvlakte (confidential report). Delft, the Netherlands, TNO: 29

Andersson, K. (2007). Combustion tests and Modelling of the oxy-fuel process-an overview of research activities at Chalmers University, Sweden. 2nd International Oxy-Combustion Network Meeting Windsor, Connecticut, USA IEA GHG, Alstom Power.

Andersson, K. and F. Johnsson (2006). "Process evaluation of an 865 MWe lignite fired O2/CO2 power plant." Energy Conversion and Management 47(18-19): 3487.

Anheden, M., et al. (2005). "Denitrogenation (or Oxyfuel Concepts)." Oil & Gas Science and Technology Rev. IFP 60(3): 485-495.

Aspelund, A., et al. (2006). "Ship transport of CO2 - Technical solutions and analysis of costs, energy utilization, exergy efficiency and CO2 emissions." Chemical Engineering Research & Design 84(A9): 847-855

Bauer, C. and T. Heck (2009). Carbon capture and storage: Life cycle assessment and external costs of future fossil power generation. . International Conference on Life Cycle Management 2009.

Bradshaw, J., et al. (2006). CO2 storage capacity estimation: issues and development of standards. Greenhouse Gas Control Technologies 8, Trondheim.

Buhre, B. J. P., et al. (2005). "Oxy-fuel combustion technology for coal-fired power generation." Progress in Energy and Combustion Science 31(4): 283.

Burchhardt, U. (2009). Experiences from Commissioning and Test Operation of Vattenfall's Oxyfuel Pilot Plant. 1st International Oxyfuel Combustion Conference, Cottbus, Germany.

Chatel-Pelage, F., et al. (2003). A pilot-scale demonstration of oxy-combustion with flue gas recirculation in a pulverized coal-fired boiler. 28th international conference on coal utilization & fuel systems - March 2003, Clearwater, Florida,.

Chen, J.-C., et al. (2007). "Emission characteristics of coal combustion in different O2/N2, O2/CO2 and O2/RFG atmosphere." Journal of Hazardous Materials 142(1-2): 266.

Chiesa, P., et al. (2005). "Using Hydrogen as Gas Turbine Fuel." Journal of Engineering for Gas Turbines and Power 127: 73-80.

Corti, A. and L. Lombardi (2004). "Reduction of carbon dioxide emissions from a SCGT/CC by ammonia solution absorption - preliminary results." International Journal of Thermodynamics 7(4): 173-181.

Croiset, E. and K. V. Thambimuthu (2001). "NOx and SO2 emissions from O2/CO2 recycle coal combustion." Fuel 80(14): 2117.

Cui, Z., et al. (2010). "Simultaneous Capture of Mercury and CO2 in Amine-Based CO2 Absorption Process." Industrial & Engineering Chemistry Research 49(24): 12576-12586.

Damen, K., et al. (2006). "A comparison of electricity and hydrogen production systems with CO2 capture and storage. Part A: Review and selection of promising conversion and capture technologies." Progress in Energy and Combustion Science 32(2): 215.

Davison, J. (2007). "Performance and costs of power plants with capture and storage of CO_2." Energy 32(7): 1163.

DOE and NETL (2007). Pulverized Coal Oxycombustion Power Plants -Volume 1: Bituminous Coal to Electricity.

DOE/NETL (2007). Cost and Performance Baseline for Fossil Energy Plants- Volume 1: Bituminous Coal and Natural Gas to Electricity, National Energy Technology Laboratory, US department of Energy.

Ertesvag, I. S., et al. (2005). "Exergy analysis of a gas-turbine combined-cycle power plant with precombustion CO2 capture." Energy 30(1): 5.

Farzan, H., et al. (2005). Pilot-Scale Evaluation of Coal Combustion in an Oxygen-Enriched Recycled Flue Gas. The 30th International Conference on Coal Utilization and Fuel Systems, Clearwater, Florida, U.S.A.

Harmelen, T. v., et al. (2008). The impacts of CO2 capture technologies on transboundary air pollution in the Netherlands. Utrecht, the Netherlands, TNO, Group Science, Technology and Society

Hertwich, E., et al. (2008). "Life-cycle assessment of carbon dioxide capture for enhanced oil recovery." Chinese Journal of Chemical Engineering 16(3): 343-353

Hetland, J. and T. Christensen (2008). "Assessment of a fully integrated SARGAS process operating on coal with near zero emissions." Applied Thermal Engineering 28(16): 2030-2038.

Hopman. (2008). Personal Communication. TNO, Delft

Horssen, A. v., et al. (2009). The impacts of CO2 capture technologies in power generation and industry on greenhouse gases emissions and air pollutants in the Netherlands. Dutch Policy Research Programme on Air and Climate (BOLK). Utrecht, Utrecht University, TNO: 160

IEA GHG (2004). Ship transport of CO_2. Cheltenham, International Energy Agency Greenhouse Gas R&D Programme

IEA GHG (2006a). Environmental impact of solvent scrubbing of CO_2 R. v. Gijlswijk, et al, TNO, IEA Greenhouse Gas R&D Programme

IEA GHG (2006b). international Oxy-combustion Research Network for CO2 capture; Report on Inaugural workshop

IEA GHG (2006c). Near zero emission technology for CO2 capture from power plant IEA Greenhouse Gas R&D programme

IIASA. (2008). "Greenhouse Gas and Air Pollution Interactions and Synergies (GAINS) ".

Iijima, M., et al. (2007). "Long-Term Demonstration of CO_2 Recovery from the Flue Gas of a Coal-Fired Power Station " Technical review - Mitsubishi Heavy Industries 44(2).

IPCC (2005). IPCC Special Report on Carbon Dioxide Capture and Storage. Prepared by Working Group III of the Intergovernmental Panel on Climate Change. B. Metzet al. Cambridge University Press, Cambridge, United Kingdom and New York, NY, USA,: 442.

IPPC (2006). Reference Document on Best Available Techniques for Large Combustion Plants, Integrated Pollution Prevention and Control Bureau,.

Khoo, H. H. and R. B. H. Tan (2006). "Life Cycle Investigation of CO_2 Recovery and Sequestration." Environ. Sci. Technol. 40(12): 4016-4024.

Kishimoto, S., et al. (2008). Current Status of MHI's CO2 Recovery Technology and Optimization of CO2 Recovery Plant with a PC Fired Power Plant. 9th International Conference on Greenhouse Gas Control Technologies (GHGT-9), Washington DC, USA.

Knudsen, J. N., et al. (2008). Experience with CO2 capture from coal flue gas in pilot-scale: Testing of different amine solvents. 9th International Conference on Greenhouse Gas Control Technologies (GHGT-9), Washington DC, USA.

Knudsen, J. N., et al. (2006). CASTOR 1 t/h CO_2 absorption pilot plant at the Elsam Kraft A/S Esbjerg power plant - First year operation experience. Greenhouse Gas Control Technologies 8, Trondheim, Norway.

Knudsen, S., et al. (2009). Summary Report: Amine Emissions to Air during Carbon Capture, Phase I: CO2 and Amines Screening Study for Effects to the Environment, NILU report N-108068

Knudsen, S., et al. (2008). Environmental impact of amines from CO2 capture, NILU report O-107124.

Knuutila, H., et al. (2009). "CO2 capture from coal-fired power plants based on sodium carbonate slurry; a systems feasibility and sensitivity study." International Journal of Greenhouse Gas Control 3(2): 143.

Koornneef, J., et al. (2010). "The impact of CO2 capture in the power and heat sector on the emission of SO2, NOx, Particulate Matter, Volatile Organic Compounds and NH3 in the European Union." Atmospheric Environment 44: 1369-1385

Koornneef, J., et al. (2008). "Life cycle assessment of a pulverized coal power plant with post-combustion capture, transport and storage of CO2." International Journal of Greenhouse Gas Control 2(4): 448.

Korre, A., et al. (2010). "Life cycle modelling of fossil fuel power generation with post-combustion CO2 capture." International Journal of Greenhouse Gas Control 4: 289-300.

Korre, A., et al. (in press). "Life cycle modelling of fossil fuel power generation with post-combustion CO2 capture." International Journal of Greenhouse Gas Control In Press, Corrected Proof.

Kozak, F., et al. (2008). Chilled Ammonia Process for CO2 Capture. 9th International Conference on Greenhouse Gas Control Technologies (GHGT-9), Washington DC, USA.

Kvamsdal, H. M., et al. (2007). "A quantitative comparison of gas turbine cycles with CO2 capture." Energy(32): 10-24.

Kvamsdal, H. M. and T. Mejdell (2005). Tjeldberodden power/methanol - CO2 reduction efforts SP2: CO2 capture and transport, Sintef energy research

Låg, M., et al. (2009). Health effects of different amines relevant for CO2 capture, NILU report N-108068

Lombardi, L. (2003). "Life cycle assessment comparison of technical solutions for CO_2 emissions reduction in power generation." Energy Conversion and Management 44(1): 93.

M. Anheden, et al. (2005). "Denitrogenation (or Oxyfuel Concepts)." Oil & Gas Science and Technology Rev. IFP 60(3): 485-495.

Marin, O. and R. Carty (2002). Demonstration study of high sulfur coal combustion in oxygen enriched flue gas Air Liquide, Illinois Clean Coal Institute.

Marit, L., et al. (2009). Health effects of possible degradation products of different amines relevant for the CO2 capture, NILU report N-108068

Markewitz, P., et al. (2009). "Environmental impacts of a Germand CCS strategy." Energy Procedia 1: 3763-3770.

Moser, P., S. Schmidt and K. Stahl. (2011). Investigation of trace elements in the inlet and outlet streams of a MEA-based post-combustion capture process Results from the test programme at the Niederaussem pilot plant. Greenhouse Gas Control Technologies 10, Amsterdam.

NEEDS (2009). NEEDS: New Energy Externalities Developments for Sustainability. Deliverable no. 6.1 External costs from emerging electricity generation technologies, Sixth Framework Programme 2009

Nexant Inc. (2006). Environmental Footprints and Costs of Coal-Based Integrated Gasification Combined Cycle and Pulverized Coal Technologies. Washington, U.S. Environmental Protection Agency.

Nie, Z. (2009). Life cycle modeling of carbon dioxide capture and geological storage in energy production. PhD thesis Imperial College London

NUON (2005). Milieujaarverslag 2004 Nuon Power Buggenum B.V. -Willem-Alexander Centrale

NUON (2006). Milieujaarverslag 2005 Nuon Power Buggenum B.V. -Willem-Alexander Centrale

Odeh, N. A. and T. T. Cockerill (2008). "Life cycle GHG assessment of fossil fuel power plants with carbon capture and storage." Energy Policy 36(1): 367.

Oexmann, J. and A. Kather (2009). "Post-combustion CO2 capture in coal-fired power plants: Comparison of integrated chemical absorption processes with piperazine promoted potassium carbonate and MEA." Energy Procedia 1(1): 799-806.

Pehnt, M. and J. Henkel (2009). "Life cycle assessment of carbon dioxide capture and storage from lignite power plants." International Journal of Greenhouse Gas Control 3(1): 49.

Rao, A. (2006). Implications of CO2 Sequestration for Gas Turbines - section 1.2.2 The Gas Turbine Handbook R. Dennis, U.S. Department of Energy, National Energy Technology Laboratory.

Rao, A. B. and E. S. Rubin (2002). "A technical, economic, and environmental assessment of amine-based CO_2 capture technology for power plant greenhouse gas control." Environmental Science and Technology 36(20): 4467.

Rao, M., et al. (2004). An integrated modeling framework for carbon management technologies. Volume 1-Technical documentation: Amine-based CO_2 capture and storage systems for fossil fuel power plant. Pittsburg, United States, Carnegie Mellon University.

RECCS (2008). RECCS: Ecological, economic and structural comparison of renewable energy technologies with carbon capture and storage - an integrated approach, Federal Ministry for the Environment, Nature Conservation and Nuclear Safety (BMU)

Schreiber, A., et al. (2009). "Environmental assessment of German Electricity generation from coal-fired power plants with amine-based carbon capture." International Journal of Life Cycle Assessment 14: 547-559

Smith, K., et al. (2009). "Recent developments in solvent absorption technologies at the CO2CRC in Australia." Energy Procedia 1(1): 1549-1555.

Spath, P. and M. Mann (2004). Biomass power and conventional fossil systems with and without CO2 sequestration-comparing the energy balance, greenhouse gas emissions and economics. Golden, Colorado, National Renewable Energy Laboratory: 28 p.

Strazisar, B. R., et al. (2003). "Degradation Pathways for Monoethanolamine in a CO_2 Capture Facility." Energy Fuels 17(4): 1034-1039.

Sundkvist, S. G., et al. (2004). AZEP gas turbine combined cylce power plants -thermal optimisation and LCA analysis. 7th International Conference on Greenhouse Gas Control Technologies, Vancouver, Canada.

Supap, T., et al. (2009). "Kinetics of sulfur dioxide- and oxygen-induced degradation of aqueous monoethanolamine solution during CO2 absorption from power plant flue gas streams." International Journal of Greenhouse Gas Control 3(2): 133-142.

Svanes, E. (2008). Evaluation of direct toxic impact of CO2 capture chemicals in view of REACH regulation, Østfoldforskning report

Tan, Y., et al. (2006). "Combustion characteristics of coal in a mixture of oxygen and recycled flue gas." Fuel 85(4): 507.

Thébault, C., et al. (2009). Behaviors of NO_X and SO_X in CO_2 Compression and Purification Processes - Experience at 30 MWth Oxy-Coal Combustion CO_2 capture Pilot Plant. 1st Oxyfuel Combustion Conference. Cottbus, Germany

Thitakamol, B., et al. (2007). "Environmental impacts of absorption-based CO_2 capture unit for post-combustion treatment of flue gas from coal-fired power plant." International Journal of Greenhouse Gas Control 1(3): 318.

Thomas, D. C. and S. M. Benson (2005). Carbon dioxide capture for storage in deep geological formations - Results from the CO_2 capture project. Volume 2: Geologic storage of carbon dioxide with monitoring and verification. Amsterdam, Elsevier.

Tricker, A. R., et al. (1989). "Environmental exposure to preformed nitroso compounds " Canc. Surv. 8: 251-272

Tzimas, E., et al. (2007). "Trade-off in emissions of acid gas pollutants and of carbon dioxide in fossil fuel power plants with carbon capture." Energy Policy 35(8): 3991.

Van Bergen, F., et al. (2003). Feasibility study on CO2 sequestration and enhanced CBM production in Zuid-Limburg, Novem: 76.

Viebahn, P., et al. (2007). "Comparison of carbon capture and storage with renewable energy technologies regarding structural, economic, and ecological aspects in Germany." International Journal of Greenhouse Gas Control 1(1): 121.

Waku, H., et al. (1995). "Life cycle analysis of fossil power plant with CO2 recovery and sequestering system." Energy Conversion and Management 36(6-9): 877.

Weisser, D. (2007). " A guide to life-cycle greenhouse gas (GHG) emissions from electric supply technologies." Energy 32: 1543-1559

White, V., et al. (2006). Purification of oxyfuel-derived CO2 for sequestration for EOR. Greenhouse Gas Control Technologies 8, Trondheim.

White, V., et al. (2008). Purification of Oxyfuel-Derived CO2. 9th International Conference on Greenhouse Gas Control Technologies (GHGT-9), Washington DC, USA.

WRI (2007). Development of a Novel Oxygen Supply Process and its Integration with an Oxy-Fuel Coal-Fired Boiler. Laramie, Wyoming, Western Research Institute.

Yamada, T. (2007). Pilot scale experiments giving direct comparison between air and oxy-firing of coals and implication for large scale plant design; IHI, Japan. . 2nd International Oxy-Combustion Network Meeting Windsor, Connecticut, USA IEA GHG, Alstom Power.

Yan, J., et al. (2006). Conceptual Development of Flue Gas Cleaning for CO2 Capture from Coal-fired Oxyfuel Combustion Power Plant. Greenhouse Gas Control Technologies 8, Trondheim.

Yan, J., et al. (2009). Flue-gas Cleaning Processes for CO_2 Capture from Oxyfuel Combustion - Experience of FGD and FGC at 30 MWth Oxyfuel Combustion Plant. 1st Oxyfuel Combustion Conference. Cottbus, Germany

Yeh, A. C. and H. Bai (1999). "Comparison of ammonia and monoethanolamine solvents to reduce CO2 greenhouse gas emissions." The Science of The Total Environment 228(2-3): 121

ZEP (2006). Power Plant and Carbon Dioxide Capture -The final report from Working Group 1, The European Technology Platform for Zero Emission Fossil Fuel Power Plants (ZEP)

Zhang, Z. X., et al. (2006). "Optimization of pipeline transport for CO2 sequestration." Energy Conversion and Management 47(6): 702.

3

Municipal Waste Plastic Conversion into Different Category of Liquid Hydrocarbon Fuel

Moinuddin Sarker
Natural States Research, Inc.
USA

1. Introduction

Plastics were first invented in 1860, but have only been widely used in the last 30 years. Plastics are light, durable, modifiable and hygienic. Plastics are made up of long chain of molecules called polymers. Polymers are made when naturally occurring substances such as crude oil or petroleum are transformed into other substances with completely different properties. These polymers can then be made into granules, powders and liquids, becoming raw materials for plastic products. Worldwide plastics production increases 80 million tons every year. Global production and consumption of plastics have increased, from less than 5 million tons in the year 1950 to 260 million tons in the year 2007. Of those over one third is being used for packaging, while rest is used for other sectors. Plastic production has increased by more than 500% over the past 30 years. Per capita consumption of plastics will increase by more than 50% during the next decades. In the Western Europe total annual household waste generation is approximately 500 kg per capita and 750 kg per capita in the United States; 12% of this total waste is plastics. The global total waste plastic generation is estimated to be over 210 million tons per year. US alone generate 48 million tons per year (Stat data from EPA). The growth in plastics use is due to their beneficial characteristics; 21st century Economic growth making them even more suitable for a wide variety of applications, such as: food and product packaging, car manufacturing, agricultural use, housing products and etc. Because of good safety and hygiene properties for food packaging, excellent thermal and electrical insulation properties, plastics are more desirable among consumers. Low production cost, lower energy consumption and CO_2 emissions during production of plastics are relatively lower than making alternative materials, such as glass, metals and etc. Yet for all their advantages, plastics have a considerable downside in terms of their environmental impact. Plastic production requires large amounts of resources, primarily fossil fuels and 8% of the world's annual oil production is used in the production of plastics. Potentially harmful chemicals are added as stabilizers or colorants. Many of these have not undergone environmental risk assessment and their impact on human health and environment is currently uncertain. Worldwide municipal sites like shops or malls had the largest proportion of plastic rubbish items. Ocean soup swirling the debris of plastics trash in the Pacific Ocean has now grown to a size that is twice as large as the continental US. In 2006, 11.5 million of tons of plastics were wasted in the landfill. These types of disposal of the waste plastics release toxic gas; which has negative impact on environment.

Most plastics are non-biodegradable and they take long time to break down in landfill, estimated to be more than a century. Plastic waste also has a detrimental impact on wild life; plastic waste in the oceans is estimated to cause the death of more than a million seabirds and more than 100,000 marine mammals every year (UN Environmental Program Estimate). Along with this hundreds of thousands of sea turtles, whales and other marine mammals die every year eating discarded waste plastic bags mistaken for food. Setting up intermediate treatment plants for waste plastic, such as: plastic incineration, recycle, or obtaining the landfill for reclamation is difficult. The types of the waste plastics are LDPE, HDPE, PP, PS, PVC, PETE, PLA and etc. The problems of waste plastics can't be solved by landfilling or incineration, because the safety deposits are expensive and incineration stimulates the growing emission of harmful green house gases, e.g COx, NOx, SOx and etc. By using NSR's new technology we can convert all types of waste plastics into liquid hydrocarbon fuel by setting temperature profile 370 degree C to 420 degree C, we can resolve all waste plastic problems including land, ocean, river and green house effects. Many of researcher and experts have done a lot of research and work on waste plastics; some of the thesis's are on thermal degradation process [1-10], pyrolysis process [11-20] and catalytic conversion process [21-30]. Producing fuels can be alternative of heating oil, gasoline, naphtha, aviation, diesel and fuel oil. We also produce light gaseous (natural gas) hydrocarbon compound (C_1-C_4), such as: methane, ethane, propane and butane. This process is profitable because it requires less production cost per gallon. We can produce individual plastic to fuel, mixed waste plastic to fuel and that produced fuel can make different category fuels by using further fractional distillation process. This NSR technology will not only reduce the production cost of fuel, but it will also reduce 9% of foreign oil dependency, create more electricity and new jobs all over the world. To mitigate the present world market demand, we can substitute this method as a potential source of new renewable energy.

2. Experimental section

2.1 Waste plastics properties

A plastic has physical and chemical properties. Different types of plastics displayed distinguishable characteristics and properties. Many kinds of plastics are appeared like LDPE, HDPE, PP, PS, PVC &PETE etc. Several individual plastics properties are elaborated in shortly, that's given below in Table-1, Table-2, Table-3 and Table-4.

Quantity	Value	Units
Thermal expansion	110 - 130	e-6/K
Thermal conductivity	0.46 - 0.52	W/m.K
Specific heat	1800 - 2700	J/kg.K
Melting temperature	108 - 134	°C
Glass temperature	-110 - -110	°C
Service temperature	-30 - 85	°C
Density	940 - 965	kg/m^3
Resistivity	5e+17 - 1e+21	$Ohm.mm^2/m$
Shrinkage	2 - 4	%
Water absorption	0.01 - 0.01	%

Table 1. HDPE-2 Plastic Properties

Quantity	Value	Units
Thermal expansion	150 - 200	e-6/K
Thermal conductivity	0.3 - 0.335	W/m.K
Specific heat	1800 - 3400	J/kg.K
Melting temperature	125 - 136	°C
Glass temperature	-110 - -110	°C
Service temperature	-30 - 70	°C
Density	910 - 928	kg/m^3
Resistivity	5e+17 - 1e+21	$Ohm.mm^2/m$
Breakdown potential	17.7 - 39.4	kV/mm
Shrinkage	1.5 - 3	%
Water absorption	0.005 - 0.015	%

Table 2. LDPE-4 Plastic Properties

Quantity	Value	Units
Thermal expansion	180 - 180	e-6/K
Thermal conductivity	0.22 - 0.22	W/m.K
Melting temperature	160 - 165	°C
Glass temperature	-10 - -10	°C
Service temperature	-10 - 110	°C
Density	902 - 907	kg/m^3
Resistivity	5e+21 - 1e+22	$Ohm.mm^2/m$
Breakdown potential	55 - 90	kV/mm
Shrinkage	0.8 - 2	%

Table 3. PP-5 Plastic Properties

Quantity	Value	Units
Thermal expansion	60 - 80	e-6/K
Thermal conductivity	0.14 - 0.16	W/m.K
Specific heat	1300 - 1300	J/kg.K
Glass temperature	80 - 98	°C
Service temperature	-10 - 90	°C
Density	1040 - 1050	kg/m^3
Resistivity	1e+22 - 1e+22	$Ohm.mm^2/m$
Breakdown potential	100 - 160	kV/mm
Shrinkage	0.3 - 0.7	%

Table 4. PS-6 Plastic Properties

2.2 Pre analysis of Gas Chromatography & Mass Spectrometer (GC/MS) analysis

Before starting the fuel production experiment, we have analyzed each of the individual raw waste plastics. Types of analyzed raw waste plastics are following, HDPE-2 (High Density Polyethylene), LDPE-4 (Low Density Polyethylene), PP-5 (Polypropylene) and PS-6 (Polystyrene).

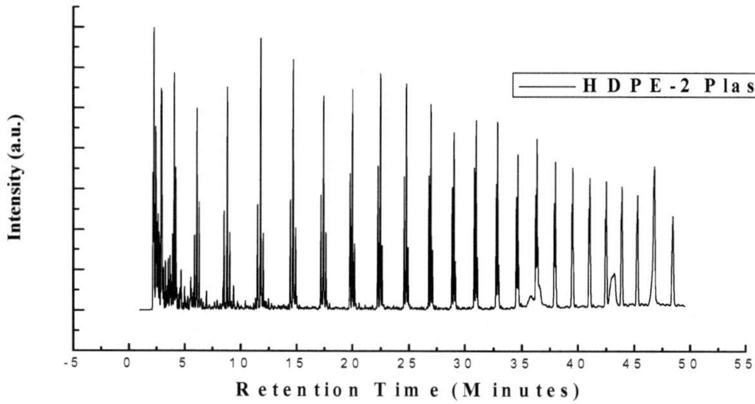

Fig. 1. GC/MS Chromatogram of HDPE-2 Raw Waste Plastic

Retention Time	Compound Name	Formula	Retention Time	Compound Name	Formula
2.14	Propane	C_3H_8	22.62	Tetradecane	$C_{14}H_{30}$
2.23	3-Butyn-1-ol	C_4H_6O	24.57	1,13-Tetradecadiene	$C_{14}H_{26}$
17.61	Dodecane	$C_{12}H_{26}$	40.94	1,19-Eicosadiene	$C_{20}H_{38}$
19.78	1,13-Tetradecadiene	$C_{14}H_{26}$	41.02	1-Docosene	$C_{22}H_{44}$
20.00	1-Tridecene	$C_{13}H_{26}$	42.48	1-Docosene	$C_{22}H_{44}$
20.19	Tridecane	$C_{13}H_{28}$	43.89	1-Tetracosanol	$C_{24}H_{50}O$
22.24	1,13-Tetradecadiene	$C_{14}H_{26}$	45.28	9-Tricosene, (Z)-	$C_{23}H_{46}$
22.45	Cyclotetradecane	$C_{14}H_{28}$	46.76	17-Pentatriacontene	$C_{35}H_{70}$

Table 4. GC/MS Compound List of HDPE-2 Waste Plastic

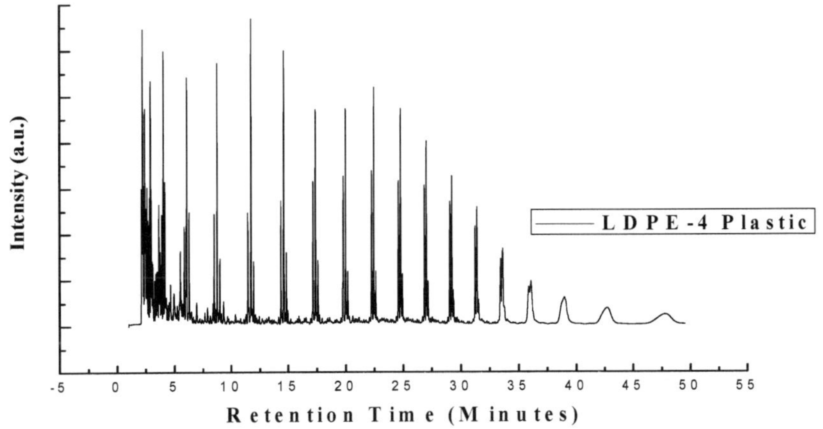

Fig. 2. GC/MS Chromatogram of LDPE-4 Raw Waste Plastic

Retention Time (Minutes)	Compound Name	Formula	Retention Time (Minutes)	Compound Name	Formula
2.11	Propane	C_3H_8	17.13	1,11-Dodecadiene	$C_{12}H_{22}$
2.19	Cyclopropyl carbinol	C_4H_8O	17.37	Cyclododecane	$C_{12}H_{24}$
11.44	1,9-Decadiene	$C_{10}H_{18}$	33.62	1-Nonadecene	$C_{19}H_{38}$
11.73	Cyclodecane	$C_{10}H_{20}$			
11.95	Decane	$C_{10}H_{22}$	35.87	1,19-Eicosadiene	$C_{20}H_{38}$
14.35	1,10-Undecadiene	$C_{11}H_{20}$	36.08	1-Heneicosyl formate	$C_{22}H_{44}O_2$
14.61	1-Undecene	$C_{11}H_{22}$	42.76	1-Docosanol	$C_{22}H_{46}O$
14.84	Undecane	$C_{11}H_{24}$	47.91	9-Tricosene, (Z)-	$C_{23}H_{46}$

Table 5. GC/MS Chromatogram Compound list of LDPE-4 Raw Waste Plastic

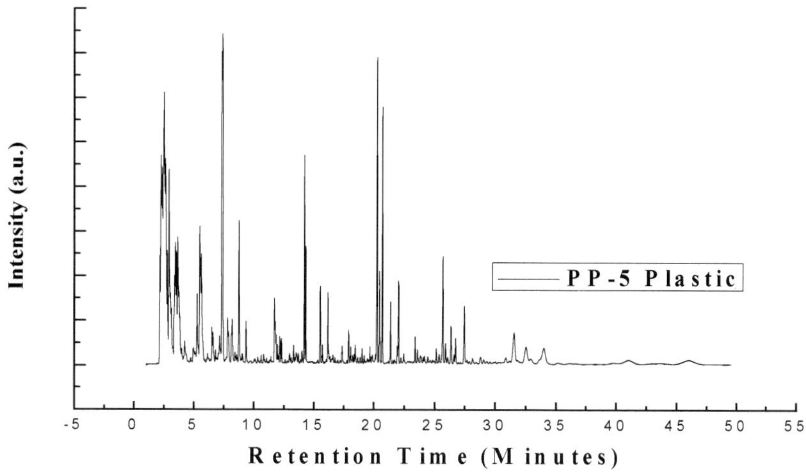

Fig. 3. GC/MS Chromatogram of PP-5 Raw Waste Plastic

Retention Time (Minutes)	Compound Name	Formula	Retention Time (Minutes)	Compound Name	Formula
2.13	Cyclopropane	C_3H_6	12.29	Decane, 4-methyl-	$C_{11}H_{24}$
2.26	1-Butyne	C_4H_6	14.18	2-Dodecene, (E)-	$C_{12}H_{24}$
9.36	1,6-Octadiene, 2,5-dimethyl-, (E)-	$C_{10}H_{18}$	26.35	1-Hexadecanol, 3,7,11,15-tetramethyl-	$C_{20}H_{42}O$
11.71	Nonane, 2-methyl-3-methylene-	$C_{11}H_{22}$	31.52	1-Heneicosyl formate	$C_{22}H_{44}O_2$
11.78	1-Ethyl-2,2,6-trimethylcyclohexane	$C_{11}H_{22}$	32.51	1-Nonadecanol	$C_{19}H_{40}O$
12.17	Nonane, 2,6-dimethyl-	$C_{11}H_{24}$	33.98	1,22-Docosanediol	$C_{22}H_{46}O_2$

Table 6. GC/MS Chromatogram Compound List of PP-5 Raw Waste Plastic

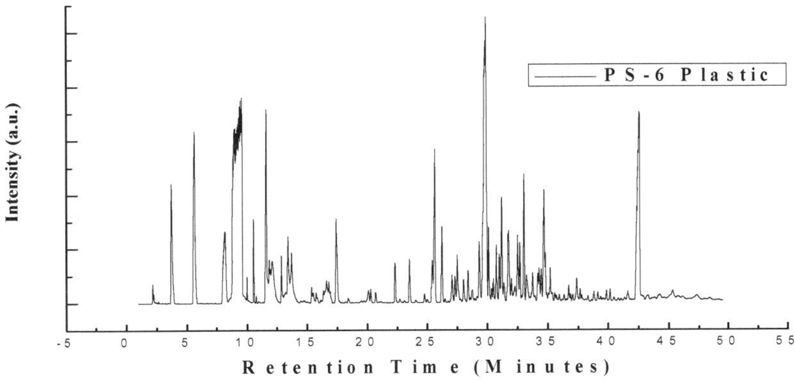

Fig. 4. GC/MS Chromatogram of PS-6 Raw Waste Plastic

Retention Time (Minutes)	Compound Name	Formula	Retention Time (Minutes)	Compound Name	Formula
2.17	Cyclopropane	C_3H_6	24.78	1,1'-Biphenyl, 3-methyl-	$C_{13}H_{12}$
2.24	Methylenecyclopropane	C_4H_6	25.64	1,2-Diphenylethylene	$C_{14}H_{12}$
5.52	Toluene	C_7H_8	27.30	1,2-Diphenylcyclopropane	$C_{15}H_{14}$
20.09	1,4-Methanonaphthalene, 1,4-dihydro-	$C_{11}H_{10}$	37.35	Naphthalene, 1-(phenylmethyl)-	$C_{17}H_{14}$
20.28	Benzocycloheptatriene	$C_{11}H_{10}$	37.63	p-Terphenyl	$C_{18}H_{14}$
20.67	Naphthalene, 1-methyl-	$C_{11}H_{10}$	38.79	Fluoranthene, 2-methyl-	$C_{17}H_{12}$
22.32	Biphenyl	$C_{12}H_{10}$	39.83	Benzene, 1,1'-[1-(ethylthio)propylidene]bis-	$C_{17}H_{20}S$
23.52	Diphenylmethane	$C_{13}H_{12}$	40.13	Benzene, 1,1',1'',1'''-(1,2,3,4-butanetetrayl)tetrakis-	$C_{28}H_{26}$

Table 7. GC/MS Chromatogram of PS-6 Raw Waste Plastic Compound List

Individual raw waste plastics of GCMS pre-analysis in accordance with their numerous retention times many compound are found, some of them are mentioned shortly. In HDPE-2 raw waste plastics on retention time 2.14, compound is Propane (C_3H_8), on retention time 22.45, compound is Cyclotetradecane and finally on retention time 46.76 obtained compound is Pentatriacotene ($C_{35}H_{70}$) [Shown above Fig.1 and Table-4]. In LDPE-4 raw waste plastics on retention time 2.11, compound is Propane (C_3H_8), on retention time 14.84, compound is Undecane ($C_{11}H_{24}$) and finally on retention time 47.91 obtained compound is 9-Tricosene (Z)-($C_{23}H_{46}$) [Shown above Fig.2 and Table-5]. In PP-5 initially on retention time 2.13 compound is Cyclopropane (C_3H_6) and finally on retention time 33.98 obtained compound is 1, 22-Docosanediol ($C_{22}H_{46}O_2$) [Shown above Fig.3 and Table-6]. Accordingly in PS-6 on retention time 2.17 found compound is Cyclopropane and eventually on retention time 40.13 obtained compound is Benzene, 1,1',1'',1'''-(1,2,3,4-butanetetryl)tetrakis[Shown above Fig.4 and Table-7].

2.3 Sample preparation

We take municipal mixed waste plastics or any other source of mixed waste plastics; we initially sort out the foreign particles, clean the waste plastics and clean wash them with detergent. After clean up all waste plastics spread in the open air for air dry. When dried out we shred them by scissors, now shredded plastics are grinded by grinding machine. Grinded samples structure are granular form small particles and that easy to put into the reactor. In our laboratory facility we can utilize 400g to 3kg of grinding sample for any experimental purposes.

3. Process description

3.1 Individual plastic to fuel production process

The process has been conducted in small scales with individual plastics in laboratory, on various waste plastics types; High-density polyethylene (HDPE, code 2), low-density polyethylene (LDPE, code 4), polypropylene (PP, code 5) and polystyrene (PS, code 6). These plastic types were investigated singly. For small-scale laboratory process the weight of input waste plastics ranges from 400 grams to 3kg. These waste plastics are collected, optionally sorted, cleaned of contaminants, and shredded into small pieces prior to the thermal liquefaction process. The process of converting the waste plastic to alternative energy begins with heating the solid plastic with or without the presence of cracking catalyst to form liquid slurry (thermal liquefaction in the range of 370-420 °C), condensing the vapor with standard condensing column to form liquid hydrocarbon fuel termed "NSR fuel". Preliminary tests on the produced NSR fuel have shown that it is a mixture of various hydrocarbons range. The produced fuel density varies based on individual plastic types. In equivalent to obtaining the liquid hydrocarbon fuel we also receive light gaseous hydrocarbon compounds (C_1-C_4) which resembles natural gas. Further fractional distillation based on different temperature is producing different category fuels; such as heating oil, gasoline, Naphtha (chemical), Aviation, Diesel and Fuel Oil. Experiment diagram given below in Fig.5.

3.2 Mixed waste plastic to fuel production process

Mixed waste plastics to fuel production process performed in the laboratory on various waste plastics types; High-density polyethylene (HDPE, code 2), low-density polyethylene

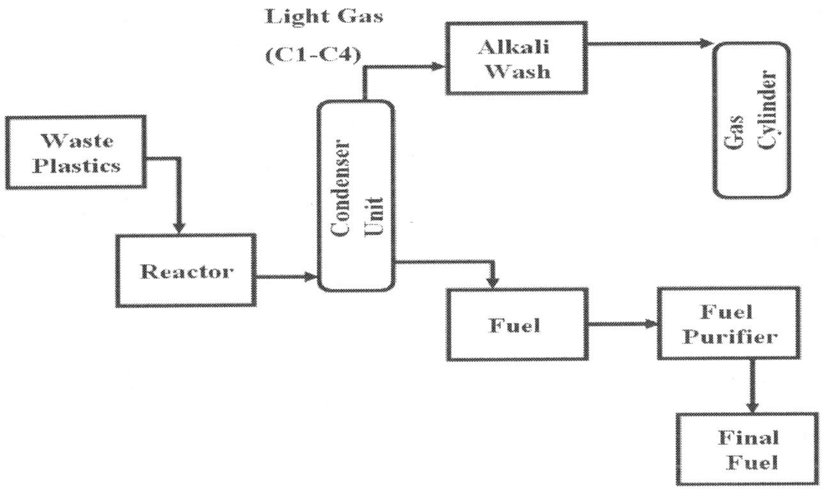

Fig. 5. Individual & Waste Plastic to Fuel Production Process Diagram

(LDPE, code 4), polypropylene (PP, code 5) and polystyrene (PS, code 6). These processes were investigated with mixture of several plastics such as HDPE-2, LDPE-4, and PP-5 &PS-6. These waste plastics are collected, optionally sorted, cleaned of contaminants, and shredded into small pieces prior to the thermal degradation process. The experiment could be randomly mixture of waste plastics or proportional ratio mixture of waste plastics. For small-scale laboratory process the weight of input waste plastics ranges from 300 grams to 3kg. In the laboratory processes our present reactor chamber capacity is 2-3 kg. We put 2 kg of grinding sample into the reactor chamber to expedite the experiment process. At the starting point of experiment reactor temperature set up at 350 °C for quick melting, after melted temperature maintained manually from "reactor temperature profile menu option" by increasing and decreasing depending to the rate of reaction. The optimum temperature (steady & more fuel production state) is 305 °C. From 2kg of waste plastics obtained fuel amount is 2 liter 600 ml (2600 ml), fuel density is 0.76 g/ml. We defined the fuel as heating oil named "NSR fuel". The experiment additionally produced light gases Methane, Ethane, Propane and Butane as well as few amount of carbon ashes as a remaining residue. These light gases would be the alternative source of natural gases. Mixed waste plastic to produced fuel preliminary test indicated that the hydrocarbon compound rage from C_3 to C_{27}.

3.3 Fractional distillation process

Fractional distillation process has been conducted according to the laboratory scale. We measured 700 ml of NSR fuel called heating fuel and took the weight of 1000 ml boiling flask (Glass Reactor). Subsequently fuel poured into the boiling flask, after that we put filled boiling flask in 1000 ml heat mantle as well as connected variac meter with heat mantle. Attached distillation adapter, clump joint, condenser and collection flask with high temperature apiezon grease and insulated by aluminum foil paper. Initially we ran the experiment at 40 °C to collect gasoline grade, after gasoline collection subsequently we raised the temperature to 110 °C for naphtha (Chemical), 180 °C for aviation fuel, 260 ° C for diesel fuel and eventually at 340 °C we found fuel oil. At the end of the experiment remaining residual fuel was less, approximately amount 10-15 ml. Out of 700 ml NSR fuel we collected 125 ml of gasoline; density is 0.72 g/ml, 150 ml of naphtha; density is 0.73, 200 ml of aviation fuel; density is 0.74, 150 ml of diesel fuel; density is 0.80 g/ml and 50-60 ml of fuel oil; density is 0.84.

4. Fuel production yield percentage

After all experiment done on behalf of each experiment we calculated the yield percentages of fuel production, light gases and residue. In addition described the physical properties of each fuel such as fuel density, specific gravity, fuel color and fuel appearance respectively. Similarly, individual fuel production yield percentages & properties are given below in Table 8 (a) & 9 (a) and Mixed Waste Plastics to fuel Yield percentages & properties are also given below in Table 8(b) & 9 (b).

Waste Plastic Name	Fuel Yield %	Light Gas %	Residue %
HDPE-2	89.354	5.345	5.299
LDPE-4	87.972	5.806	6.221
PP-5	91.981	2.073	5.944
PS-6	85.331	4.995	9.674

Table 8. (a): Individual Fuel Production Yield Percentage

Sample Name	Fuel Yield %	Light Gas %	Residue %
HDPE,LDPE,PP&PS	90	5	5

Table 8. (b): Mixed Waste Plastic to Fuel Yield Percentage

Name of Waste Plastic Fuel	Fuel Density gm/ml	Specific Gravity	Fuel Color	Fuel Appearance
LDPE-4	0.771	0.7702	Yellow, light transparent	Little bit wax and ash content
HDPE-2	0.782	0.7812	Yellow, no transparent	Wax, cloudy and little bit ash content
PP-5	0.759	0.7582	Light brown, light transparent	Little bit wax and ash content
PS-6	0.916	0.9150	Light yellow, not transparent	Wax, cloudy and little bit ash content

Table 9. (a): Individual Plastic to Fuel Properties

Name of Fuel	Density g/ml	Specific Gravity	Fuel Color	Fuel
Mixed Plastic to Fuel	0.775	0.7742	Yellow light transparent	Ash contain present

Table 9. (b): Mixed Waste Plastic to Fuel Properties

4.1 Fuel analysis and result discussion

4.2 Gas Chromatography and Mass Spectrometer (GC/MS) analysis

Analysis of Individual waste plastics (HDPE-2, LDPE-4, PP-5, and PS-6) to individual fuel:

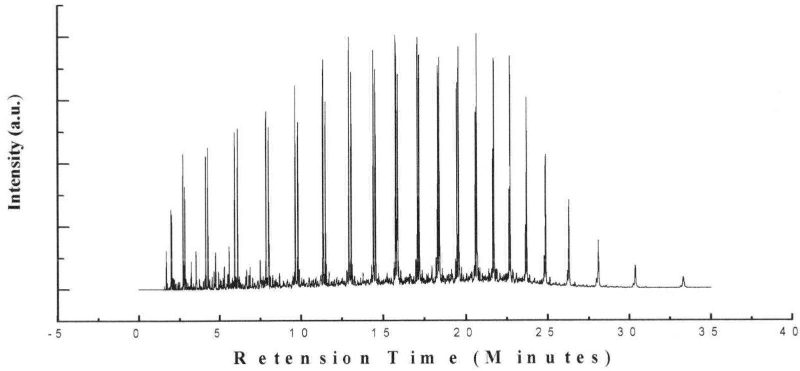

Fig. 6. GC/MS Chromatogram of HDPE-2 Waste Plastic to Fuel

Retention Time (Minutes)	Compound Name	Formula	Retention Time (Minutes)	Compound Name	Formula
1.56	Propane	C_3H_8	12.18	Cyclopentane, hexyl-	$C_{11}H_{22}$
1.66	2-Butene, (E)-	C_4H_8	12.92	1-Dodecene	$C_{12}H_{24}$
1.68	Butane	C_4H_{10}	13.05	Dodecane	$C_{12}H_{26}$
1.96	Cyclopropane, 1,2-dimethyl-, cis-	C_5H_{10}	13.76	Cyclododecane	$C_{12}H_{24}$
9.65	1-Decene	$C_{10}H_{20}$	27.98	1-Docosene	$C_{22}H_{44}$
9.80	Decane	$C_{10}H_{22}$	28.09	Tetracosane	$C_{24}H_{50}$
11.35	1-Undecene	$C_{11}H_{22}$	30.24	1-Docosene	$C_{22}H_{44}$
11.49	Undecane	$C_{11}H_{24}$	30.38	Octacosane	$C_{28}H_{58}$

Table 10. GC/MS Chromatogram Compound List of HDPE-2 Waste Plastic to Fuel

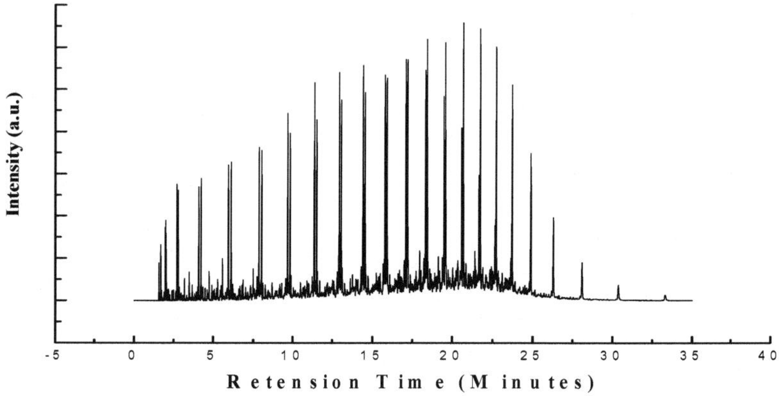

Fig. 7. GC/MS Chromatogram of LDPE-4 Waste Plastic to Fuel

Retention Time (Minutes)	Compound Name	Compound Formula	Retention Time (Minutes)	Compound Name	Compound Formula
1.55	Cyclopropane	C_3H_6	12.92	1-Dodecene	$C_{12}H_{24}$
1.68	Butane	C_4H_{10}	13.06	Dodecane	$C_{12}H_{26}$
1.96	2-Pentene, (E)-	C_5H_{10}	13.76	Cyclododecane	$C_{12}H_{24}$
1.99	Pentane	C_5H_{12}	14.40	1-Tridecene	$C_{13}H_{26}$
10.48	Cyclodecane	$C_{10}H_{20}$	24.88	Heneicosane	$C_{21}H_{44}$
10.89	Cyclohexene, 3-(2-methylpropyl)-	$C_{10}H_{18}$	26.31	Heneicosane	$C_{21}H_{44}$
11.35	1-Undecene	$C_{11}H_{22}$	28.09	Tetracosane	$C_{24}H_{50}$
11.49	Undecane	$C_{11}H_{24}$	33.21	Octacosane	$C_{28}H_{58}$

Table 11. GC/MS Chromatogram Compound List of LDPE-4 Waste Plastic to Fuel

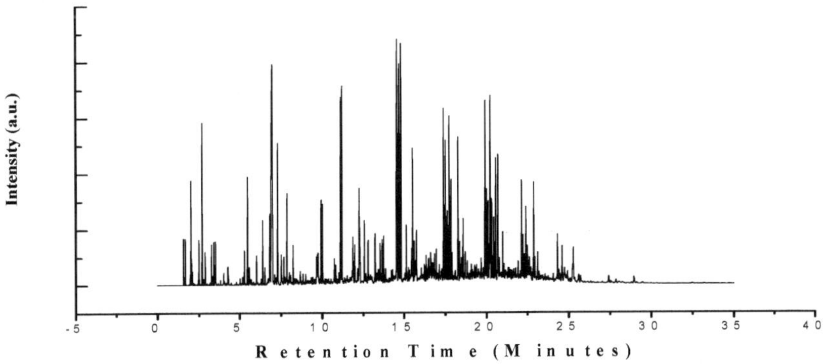

Fig. 8. GC/MS Chromatogram of PP-5 Waste Plastic to Fuel

Retention Time (Minute)	Compound Name	Formula	Retention Time (Minute)	Compound Name	Formula
1.55	Cyclopropane	C_3H_6	11.13	Cyclooctane, 1,4-dimethyl-, cis-	$C_{10}H_{20}$
1.66	1-Propene, 2-methyl-	C_4H_8	11.20	1-Tetradecene	$C_{14}H_{28}$
1.99	Pentane	C_5H_{12}	11.86	1-Dodecanol, 3,7,11-trimethyl-	$C_{15}H_{32}O$
2.48	Pentane, 2-methyl-	C_6H_{14}	12.25	(2,4,6-Trimethylcyclohexyl) methanol	$C_{10}H_{20}O$
9.64	Nonane, 2-methyl-3-methylene-	$C_{11}H_{22}$	23.13	Dodecane, 1-cyclopentyl-4-(3-cyclopentylypropyl)-	$C_{25}H_{48}$
9.74	3-Undecene, (Z)-	$C_{11}H_{22}$	25.72	Cyclotetradecane , 1,7,11-trimethyl-4-(1-methylethyl)-	$C_{20}H_{40}$
9.92	Octane, 3,3-dimethyl-	$C_{10}H_{22}$	28.95	Dodecane, 1-cyclopentyl-4-(3-cyclopentylypropyl)-	$C_{25}H_{48}$
10.73	3-Decene, 2,2-dimethyl-, (E)-	$C_{12}H_{24}$			

Table 12. GC/MS Chromatogram Compound List of PP-5 Waste Plastic to Fuel

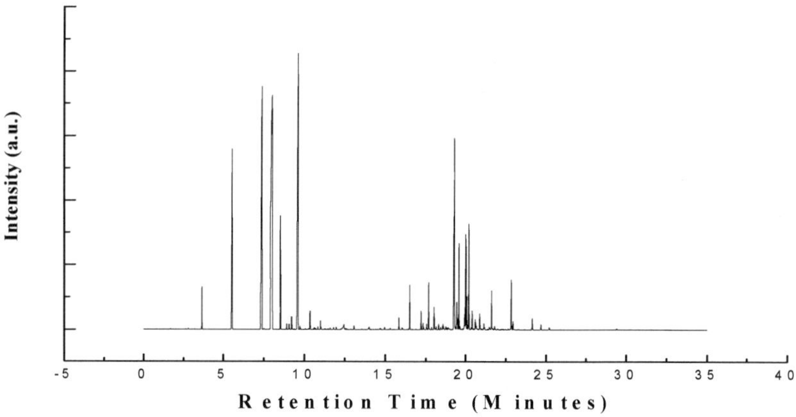

Fig. 9. GC/MS Chromatogram of PS-6 Waste Plastic to Fuel

Retention Time (Minute)	Compound Name	Formula	Retention Time (Minute)	Compound Name	Formula
3.65	1,5-Hexadiyne	C_6H_6	17.68	Benzene, 1,1′-(1,2-ethanediyl)bis-	$C_{14}H_{14}$
5.54	Toluene	C_7H_8	18.03	Benzene, 1,1′-(1-methyl-1,2-ethanediyl)bis-	$C_{15}H_{16}$
7.94	Styrene	C_8H_8	19.30	Benzene, 1,1′-(1,3-propanediyl)bis-	$C_{15}H_{16}$
11.00	Acetophenone	C_8H_8O	21.61	Naphthalene,1-phenyl-	$C_{16}H_{12}$
13.07	Naphthalene	$C_{10}H_8$	21.81	o-Terphenyl	$C_{18}H_{14}$
15.84	Biphenyl	$C_{12}H_{10}$	22.83	2-Phenylnaphthalene	$C_{16}H_{12}$
16.51	Diphenylmethane	$C_{13}H_{12}$	24.14	9-Phenyl-5H-benzocycloheptene	$C_{17}H_{14}$
17.22	Benzene,1,1′-ethylidenebis-	$C_{14}H_{14}$	24.67	p-Terphenyl	$C_{18}H_{14}$

Table 13. GC/MS Chromatogram Compound List of PS-6 Waste Plastic to Fuel

From GCMS analysis of Individual HDPE-2, LDPE-4, PP-5, and PS-6 fuel, in accordance with their numerous retention times many compounds are found, some of them are mentioned shortly. In HDPE-2 fuel at retention time 1.56, compound is Propane (C_3H_8), and finally at retention time 30.38 obtained compound is Octacosane ($C_{28}H_{58}$), [Shown above, Fig.6 & Table-10]. In LDPE-4 fuel at retention time 1.55, compound is Cyclopropane (C_3H_6), and finally at retention time 33.21 obtained compound is Octacosane ($C_{28}H_{58}$) [Shown above,

Fig.7 & Table-11]. In PP-5 initially at retention time 1.55 compound is Cyclopropane (C_3H_6) and finally at retention time 28.95 obtained compound is Dodecane,-1-Cyclopentyl-4-(3-Cyclopentylpropyl) ($C_{22}H_{46}O_2$) [Shown above, Fig.8 & Table-12] . Accordingly in PS-6 at retention time 3.65 found compound is 1, 5-Hexadiyne and eventually at retention time 24.67 obtained compound is p-Terphnyl ($C_{18}H_{14}$) [Shown above, Fig.9 & Table-13].

Analysis of Mixed Waste Plastics to Fuel (Heating Oil):

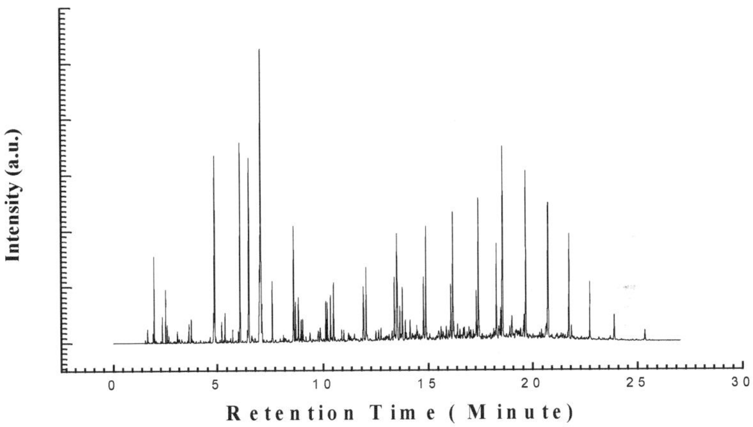

Fig. 10. GC/MS Chromatogram of Mixed Waste Plastic to Fuel (Heating Oil)

Compound Name	Formula	Compound Name	Formula
Cyclopropane	(C3H6)	Dodecane	(C12H26)
2-Butene, (E)-	(C4H8)	Decane, 2,3,5,8-tetramethyl-	(C14H30)
Pentane	(C5H12)	1-Tridecene	(C13H26)
Pentane, 2-methyl-	(C6H14)	Tridecane	(C13H28)
Cyclopropane, 1-heptyl-2-methyl-	(C11H22)	Heneicosane	(C21H44)
Undecane	(C11H24)	Nonadecane	(C19H40)
1-Dodecanol, 3,7,11-trimethyl-	(C15H32 O)	Benzene, hexadecyl-	(C22H38)
1-Dodecene	(C12H24)	Heptacosane	(C27H56)

Table 14. GC/MS Chromatogram Compound List of Mixed Waste Plastic to Fuel (Heating Oil)

From GCMS analysis of NSR fuel (Called Heating Fuel) primarily we found long chain hydrocarbon of compound. In the GCMS data we have noticed that the obtained compounds are Cyclopropane (C_3H_6) to Heptacosane ($C_{27}H_{56}$) including long and short chain of hydrocarbon compound [Shown above, Fig.10 & Table-14].

GCMS Analysis of Mixed Waste Plastics to Fractional Distillation Fuel:

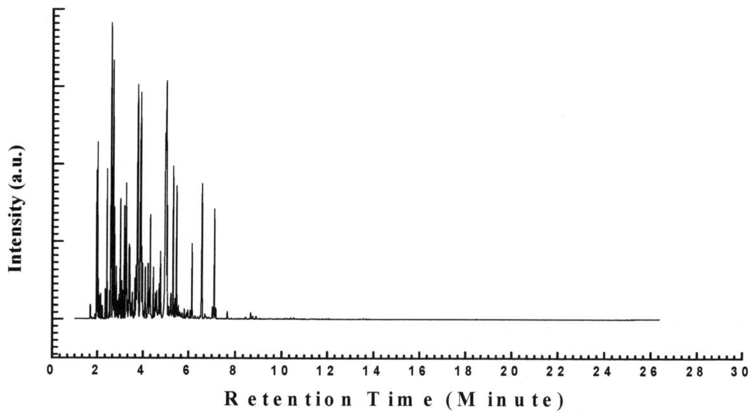

Fig. 11. GC/MS Chromatogram of Mixed Waste Plastic Fuel to 1st Fractional Fuel (Gasoline)

Compound Name	Formula	Compound Name	Formula
1-Propene,2-methyl-	(C4H8)	Heptane	(C7H16)
Butane	(C4H10)	1,4-hexadiene,4-methyl-	(C7H12)
2-Pentene	(C5H10)	1,4-Heptadiene	(C7H12)
2-Pentene,(E)	(C5H10)	Cyclohexane,methyl-	(C7H14)
Cyclohexane	(C6H12)	1-Nonane	(C9H18)
Hexane,3-methyl	(C7H16)	Styrene	(C8H8)
Cyclohexene	(C6H10)	Nonane	(C9H20)
1-Hexene,2-methyl-	(C7H14)	Benzene,(1-methylethyl)-	(C9H12)
1-Heptane	(C7H14)		

Table 15. GC/MS Chromatogram compound list of Mixed Waste Plastic Fuel to 1st Fractional Fuel (Gasoline)

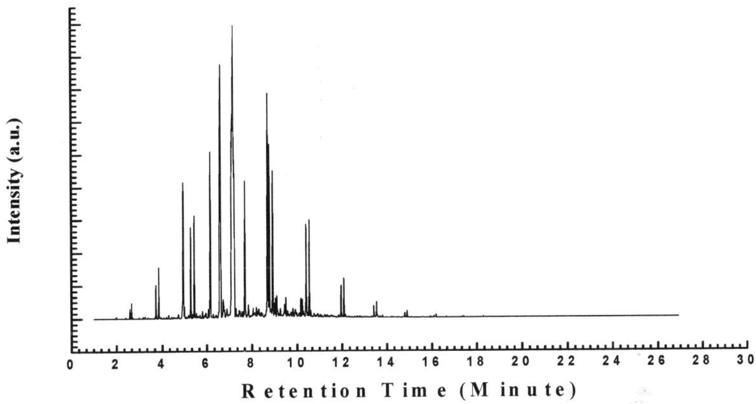

Fig. 12. GC/MS Chromatogram of Mixed Waste Plastic Fuel to 2nd Fractional Fuel (Naphtha, Chemical)

Compound Name	Formula	Compound Name	Formula
1-Hexene	(C6H12)	Cyclopentane-butyl-	(C9H8)
Hexane	(C6H14)	Benzene,propyl	(C9H12)
1-Heptene	(C7H14)	a-methylsyrene	(C9H10)
Heptane	(C7H16)	1-Decene	(C10H20)
2,4-dimethyl-1-heptene	(C9H18)	Cyclopropane,1-heptyl-2-methyl-	(C11H22)
Ethylbenzene	(C8H10)	Undecane	(C11H24)
1-Nonene	(C9H18)	1-Dodecene	(C12H24)
Styrene	(C8H8)	Dodecane	(C12H26)
1,3,5,7-Cyclooctatetraene	(C8H8)	Tridecane	(C13H28)
Nonane	(C9H20)	Tetradecdane	(C14H30)

Table 16. GC/MS Chromatogram Compound List of Mixed Waste Plastic Fuel to 2nd Fractional Fuel (Naphtha, Chemical)

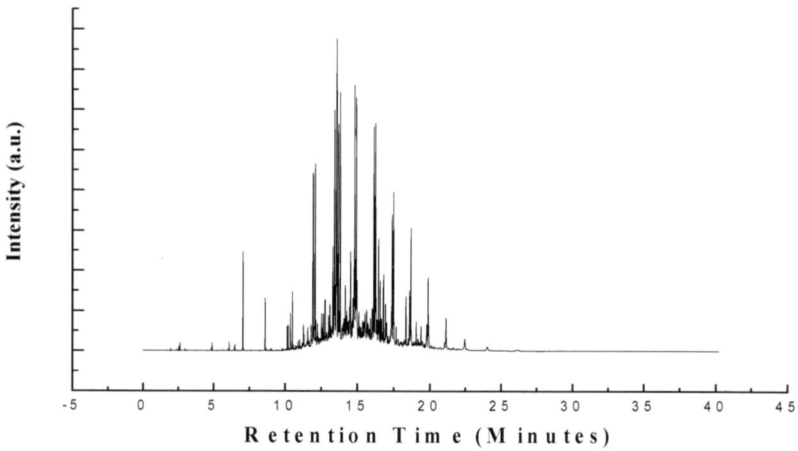

Fig. 13. GC/MS Chromatogram of Mixed Waste Plastic Fuel to 3rd Fractional Fuel (Aviation)

Retention Time (Min.)	Compound Name	Formula	Retention Time (Min.)	Compound Name	Formula
7.04	Styrene	C8H8	14.93	Tetradecane	C14H30
8.60	a-Methylstyrene	C9H10	16.12	Cyclopentadecane	C15H30
10.18	Cyclooctane,1,4-dimethyl-,cis-	C10H20	16.23	Pentadecane	C15H32
10.38	1-Undecene	C11H22	17.37	1-Hexadecene	C16H32
12.07	Dodecane	C12H26	19.80	E-15-Heptadecanal	C17H32O
13.42	1-Tridecene	C13H26	19.89	Octadecane	C18H38
13.56	Tridecane	C13H28	21.13	Nonadecane	C19H40
14.81	Cyclotetradecane	C14H28	22.45	Eicosane	C20H42

Table 17. GC/MS Chromatogram Compound list of Mixed Waste Plastic Fuel to 3rd Fractional Fuel (Aviation)

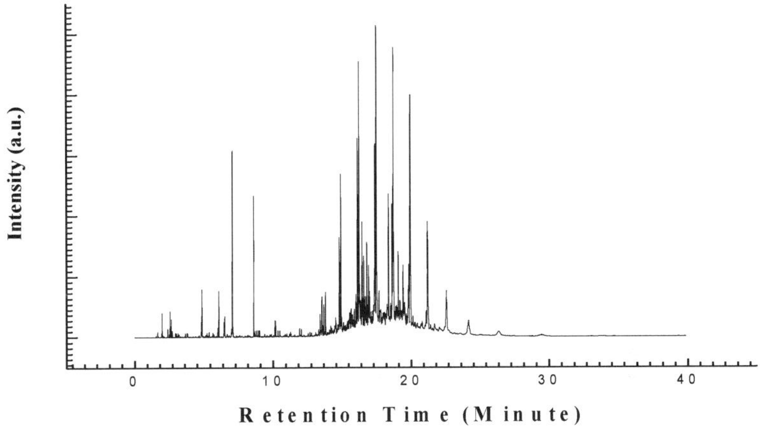

Fig. 14. GC/MS Chromatogram of Mixed Waste Plastic Fuel to 4th Fractional Fuel (Diesel)

Compound Name	Formula	Compound Name	Formula
Pentane	(C5H12)	1-Pentadecene	(C15H30)
1-Pentene, 2-methyl-	(C6H12)	Pentadecane	(C15H32)
Heptane, 4-methyl-	(C8H18)	1-Nonadecanol	(C19H40 O)
Toluene	(C7H8)	1-Hexadecene	(C16H32)
E-14-Hexadecenal	(C16H30 O)	Eicosane	(C20H42)
4-Tetradecene, (E)-	(C14H28)	Heneicosane	(C21H44)
Tetradecane	(C14H30)	Octacosane	(C28H58)

Table 18. GC/MS Chromatogram Compound List of Mixed Waste Plastic Fuel to 4th Fractional Fuel (Diesel)

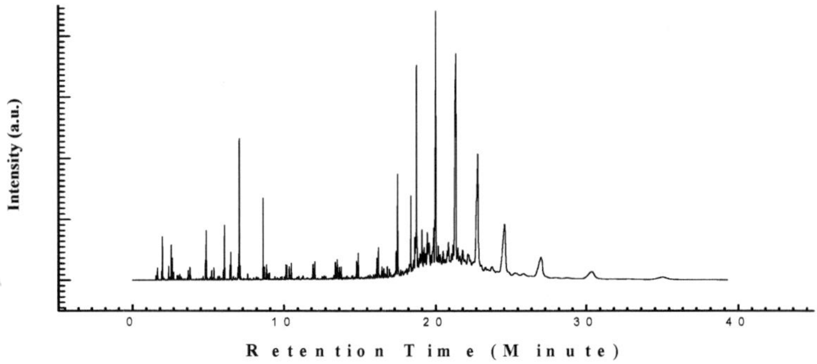

Fig. 15. GC/MS Chromatogram of Mixed Waste Plastic Fuel to 5th Fractional Fuel (Fuel Oil)

Compound Name	Formula	Compound Name	Formula
1) 1-Propene, 2-methyl-	(C4H8)	16) Tridecane	(C13H28)
2) Pentane	(C5H12)	17) Tetradecane	(C14H30)
3)1-Pentene, 2-methyl-	(C6H12)	18) Pentadecane	(C15H32)
4) Hexane	(C6H14)	19) Hexadecane	(C16H34)
5) Heptane	(C7H16)	20) Benzene, 1,1'-(1,3-propanediyl)bis-	(C15H16)
6) à-Methylstyrene	(C9H10)	27) Heneicosane	(C21H44)
7) Decane	(C10H22)	28) Tetracosane	(C24H50)
8) Undecane	(C11H24)	29) Heptacosane	(C27H56)

Table 19. GC/MS Chromatogram Compound list of Mixed Waste Plastic Fuel to 5th Fractional Fuel (Fuel Oil)

GC/MS analysis of fractional distillation fuel, a lot of compound is appeared in each individual fuel. Some of those compounds are mentioned, such as in Gasoline (1ST Fraction) we found Carbon range C_4 to C_9 and compound is 1-Propene-2-Methyl (C_3H_8) to Benzene, (1-methylethyl) - (C_9H_{12}) [Shown above, Fig.11 & Table-15]. In naphtha (2nd Fraction) Carbon range is C_6 to C_{14} and compound is 1- Hexene (C_6H_{12}) to Tetradecane ($C_{14}H_{30}$) [Shown above, Fig.12 & Table-16]. In Aviation fuel (3rd Fraction) Carbon range is C_8 to C_{20} and compound is Styrene (C_8H_8) to Eicosane ($C_{20}H_{42}$) [Shown above, Fig.13 & Table-17]. In Diesel (4th Fraction) Carbon range is C_5 to C_{28} and compound is pentane (C_5H_{12}) to Octacosane ($C_{20}H_{58}$) [Shown above, Fig.14 & Table-18].Eventually in Fuel oil (5th Fraction) Carbon range is C_4 to C_{27}, and compound is 1-Propene-2-methyl (C_4H_8) to Heptacosane ($C_{27}H_{56}$) [Shown above, Fig.15 & Table-19].

4.3 FTIR (Spectrum-100) analysis

Analysis of Individual waste plastics (HDPE-2, LDPE-4, PP-5, and PS-6) to individual fuel:

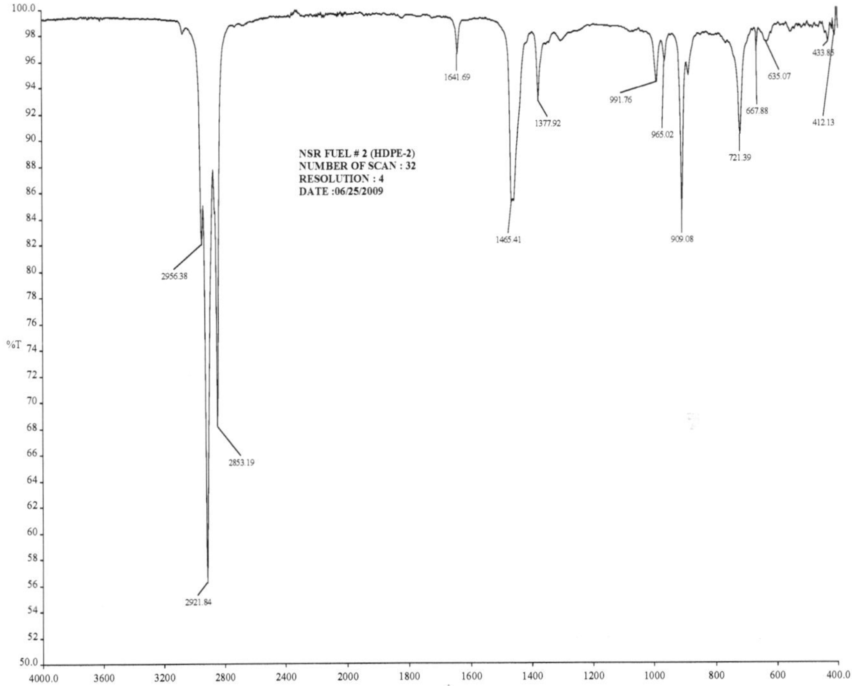

Fig. 16. FTIR Spectra of HDPE-2 Plastic to Fuel

Band Peak Number	Wave Number (cm^{-1})	Compound Group Name
1	2956.38	$C-CH_3$
2	2921.84	$C-CH_3$
3	2853.19	CH_2
4	1641.69	Non-Conjugated
5	1465.41	CH_3
6	1377.92	CH_3
7	991.76	$-CH= CH_2$
8	965.02	-CH=CH-(Trans)
9	909.08	$-CH= CH_2$
10	721.39	-CH=CH-(Cis)
11	667.88	-CH=CH-(Cis)

Table 20. FTIR Spectra of HDPE-2 Plastic to Fuel Functional Group Name

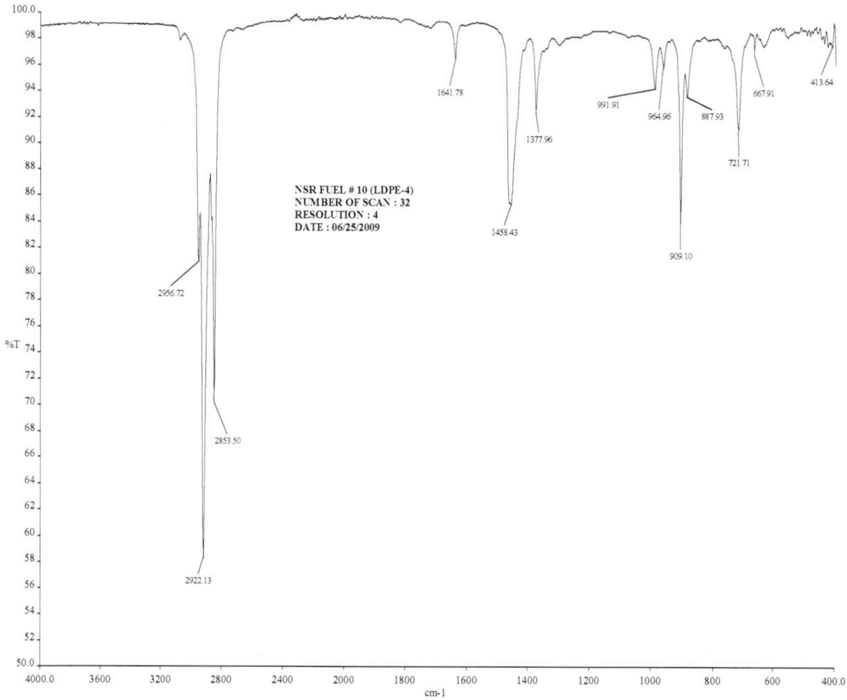

Fig. 17. FTIR Spectra of LDPE-4 Plastic to Fuel

Band Peak Number	Wave Number (cm^{-1})	Functional Group Name
1	2956.72	$C-CH_3$
2	2922.13	$C-CH_3$
3	2853.50	CH_2
4	1641.78	Non-Conjugated
5	1458.43	CH_3
6	1377.96	CH_3
7	964.96	$-CH= CH_2$
8	909.10	-CH=CH-(Trans)
9	887.93	$-CH= CH_2$
10	721.71	-CH=CH-(Cis)
11	667.91	-CH=CH-(Cis)

Table 21. FTIR Spectra of LDPE-4 Plastic to Fuel Functional Group Name

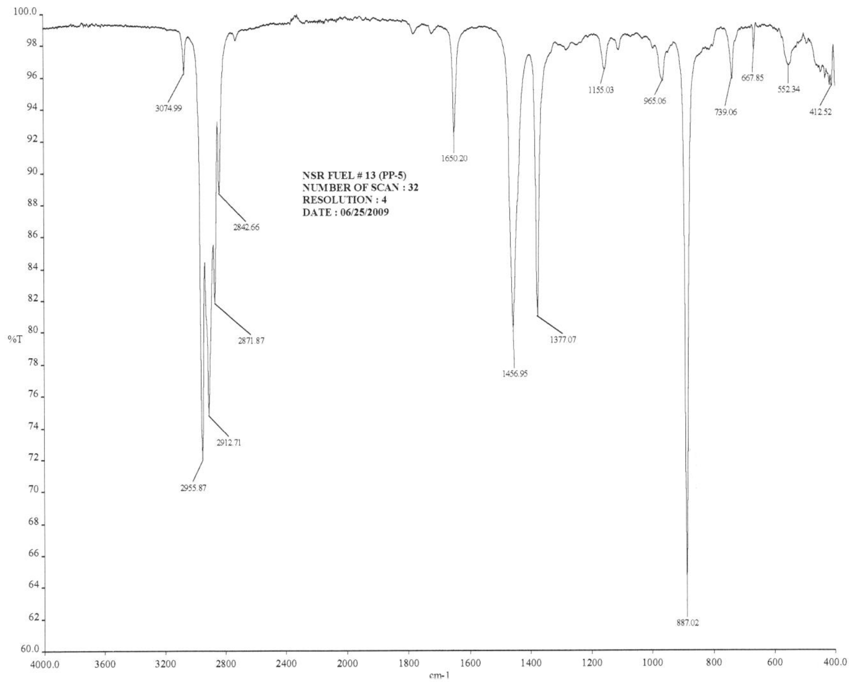

Fig. 18. FTIR Spectra of PP-5 Plastic to Fue.

Band Peak Number	Wave Number (cm^{-1})	Compound Group Name	Band Peak Number	Wave Number (cm^{-1})	Compound Group Name
1	3074.99	H Bonded NH	8	1377.07	CH_3
2	2955.87	C-CH_3	9	1155.03	
3	2912.71	C-CH_3	10	965.06	-CH=CH-(Trans)
4	2871.87	C-CH_3	11	887.02	C=CH_2
5	2842.66	C-CH_3	12	739.06	-CH=CH-(Cis)
6	1650.20	Amides	13	667.85	-CH=CH-(Cis)
7	1465.95	CH_2			

Table 22. FTIR Spectra of PP-5 Plastic to Fuel Functional Group Name

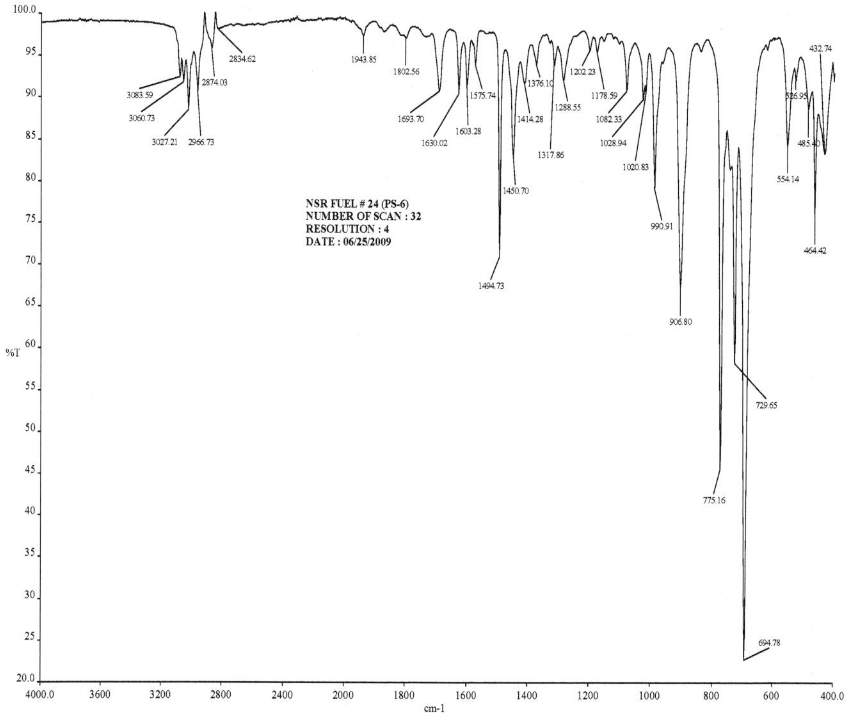

Fig. 19. FTIR Spectra of PS-6 Plastic to Fuel

Band Peak Number	Wave Number (cm⁻¹)	Compound Group Name	Band Peak Number	Wave Number (cm⁻¹)	Compound Group Name
1	3083.59	=C-H	15	1414.28	CH_2
2	3060.73	=C-H	16	1376.10	CH_3
3	3027.21	=C-H	17	1317.86	
4	2966.73	C-CH_3	18	1288.55	
5	2874.03	C-CH_3	19	1202.23	
6	2834.62	C-CH_3	20	1178.59	
7	1943.85		21	1082.33	
8	1802.56	Non-Conjugated	22	1028.94	Acetates
9	1693.70	Conjugated	23	1020.83	Acetates
10	1630.02	Conjugated	24	990.91	-CH=CH_2
11	1603.28	Conjugated	25	906.80	-CH=CH_2
12	1575.74		26	775.16	
13	1494.73		27	729.65	-CH=CH-(Cis)
14	1450.70	CH_3	28	694.78	-CH=CH-(Cis)

Table 23. FTIR Spectra of PS-6 Plastic to Fuel Functional Group Name

In FTIR analysis of HDPE-2 fuel obtained functional groups are C-CH_3, CH_2, Non-Conjugated, CH_3,-CH=CH_2,-CH=CH- (Cis) and -CH=CH-(Trans) [Shown above, Fig.16&Table-20].In LDPE-4 analysis functional groups are C-CH_3, CH_2, Non-Conjugated, CH_3,-CH=CH_2,-CH=CH- (Cis) and -CH=CH-(Trans)[Shown above, Fig.17 &Table-21].In PP-5 analysis functional groups are CH_3,C-CH_2,-CH=CH- (Cis) and,-CH=CH- (Trans). [Shown above, Fig.18 &Table-22] Subsequently in PS-6 analysis obtained functional groups are CH2, CH3, Acetates,-CH=CH2 and -CH=CH-(Cis) etc. [Shown above, Fig.19 & Table-23].

FTIR Analysis of Mixed Waste Plastics to Fuel:

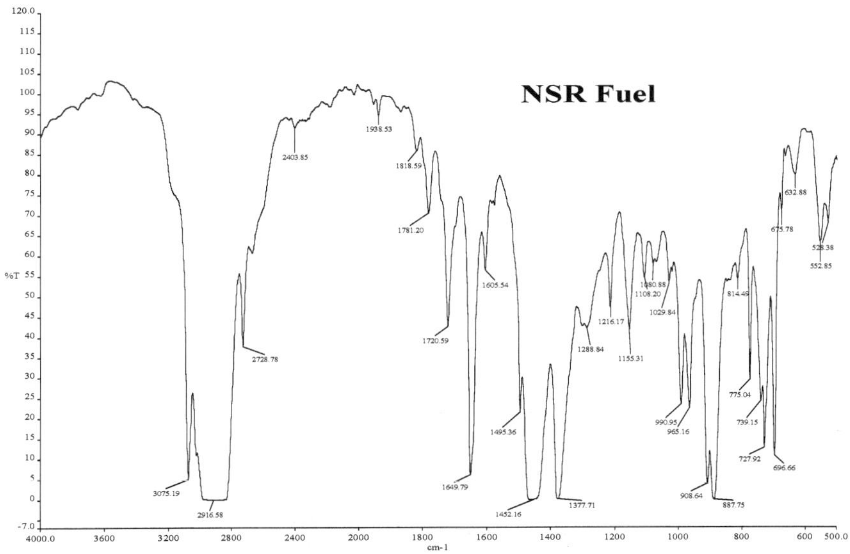

Fig. 20. FTIR Spectra of Mixed Waste Plastic to Fuel

Band Peak Number	Wave Number (cm^{-1})	Functional Group Name	Band Peak Number	Wave Number (cm^{-1})	Functional Group Name
1	3075.19	H Bonded NH	13	1377.71	CH_3
2	2916.58	CH_2	19	1029.84	Acetates
3	2728.78	C-CH_3	20	990.95	Secondary Cyclic Alcohol
5	1938.53	Non-Conjugated	21	965.16	-CH=CH- (trans)
6	1818.59	Non-Conjugated	22	908.64	-CH=CH_2
7	1781.20	Non-Conjugated	23	887.75	C=CH_2
8	1720.59	Non-Conjugated	26	739.15	-CH=CH- (cis)
9	1649.79	Amides	27	727.92	-CH=CH- (cis)
10	1605.54	Non-Conjugated	28	696.66	-CH=CH- (cis)
12	1452.16	CH_2	29	675.78	-CH=CH- (cis)

Table 24. FTIR Spectra of Mixed Waste Plastic to Fuel Functional Group Name

In FTIR analysis of mixed waste plastics to NSR fuel obtained functional groups are: CH_3, Acetates, Secondary Cyclic Alcohol,-$CH=CH_2$, $C=CH_2$,-CH=CH-(Cis) and -CH=CH-(Trans) etc. [Shown above, Fig. 20 &Table-24].

FTIR Analysis of Mixed Waste Plastics to Fractional Distillation Fuel:

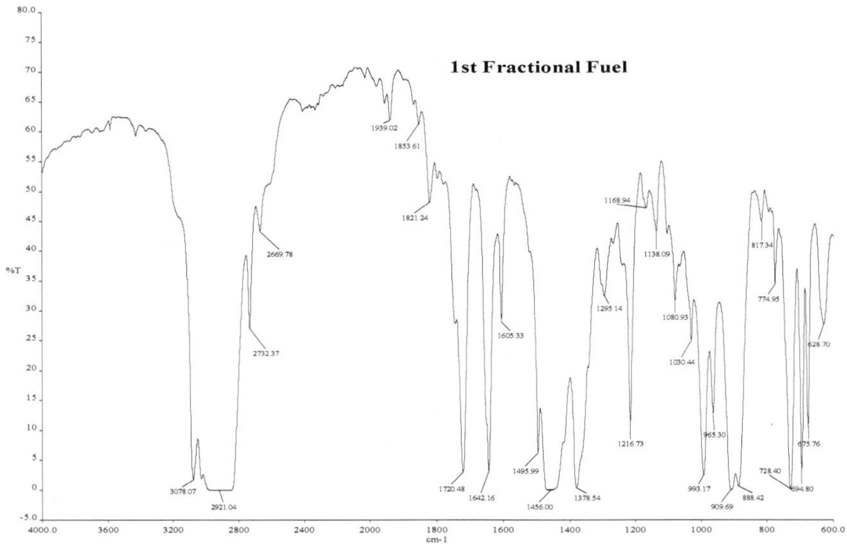

Fig. 21. FTIR Spectra of Mixed Waste Plastic Fuel to 1st Fractional Fuel (Gasoline)

Band Peak Number	Wave Number (cm^{-1})	Functional Group Name	Band Peak Number	Wave Number (cm^{-1})	Functional Group Name
1	3078.07	H Bonded NH	13	1378.54	CH_3
2	2921.04	C-CH_3	19	1030.44	Acetates
3	2732.37	C-CH_3	20	993.17	Secondary Cyclic Alcohol
4	2669.78	C-CH_3	21	965.30	-CH=CH- (trans)
6	1853.61	Non-Conjugated	22	909.69	-$CH=CH_2$
7	1821.24	Non-Conjugated	23	888.42	$C=CH_2$
8	1720.48	Non-Conjugated	26	728.40	-CH=CH- (cis)
9	1642.16	Conjugated	27	694.80	-CH=CH- (cis)
10	1605.33	Conjugated	28	675.76	-CH=CH- (cis)
12	1456.00	CH_3	29	628.70	-CH=CH- (cis)

Table 25. Mixed Waste Plastic Fuel to 1st Fractional Fuel (Gasoline) FTIR Functional Group List

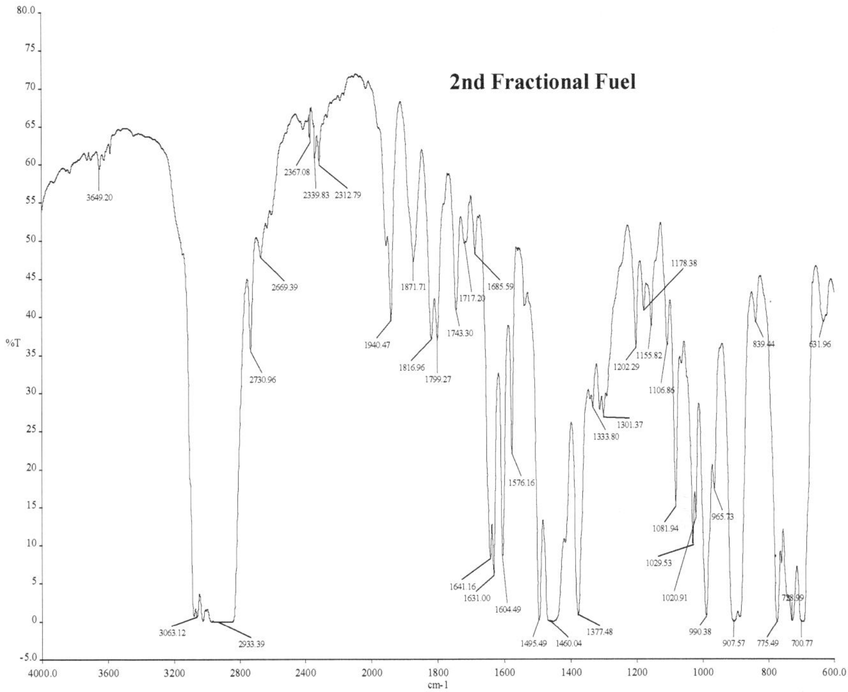

Fig. 22. FTIR Spectra of Mixed Waste Plastic Fuel to 2nd Fractional Fuel (Naphtha, Chemical)

Band Peak Number	Wave Number (cm^{-1})	Functional Group Name	Band Peak Number	Wave Number (cm^{-1})	Functional Group Name
2	3063.12	=C-H	16	1641.16	Non-Conjugated
3	2933.39	C-CH_3	17	1631.00	Non-Conjugated
4	2730.96	C-CH_3	21	1460.04	CH_3
5	2669.39	C-CH_3	22	1377.48	CH_3
9	1940.47	Non-Conjugated	30	1029.53	Acetates
10	1871.71	Non-Conjugated	31	1020.91	Acetates
11	1816.96	Non-Conjugated	32	990.38	-CH=CH_2
12	1799.27	Non-Conjugated	33	965.73	-CH=CH- (trans)
13	1743.30	Conjugated	34	907.57	-CH=CH_2
14	1717.20	Non-Conjugated	37	728.99	-CH=CH- (cis)
15	1685.59	Conjugated	38	700.77	-CH=CH- (cis)

Table 26. Mixed Waste Plastic Fuel to 2nd Fractional Fuel (Naphtha) FTIR Functional Group List

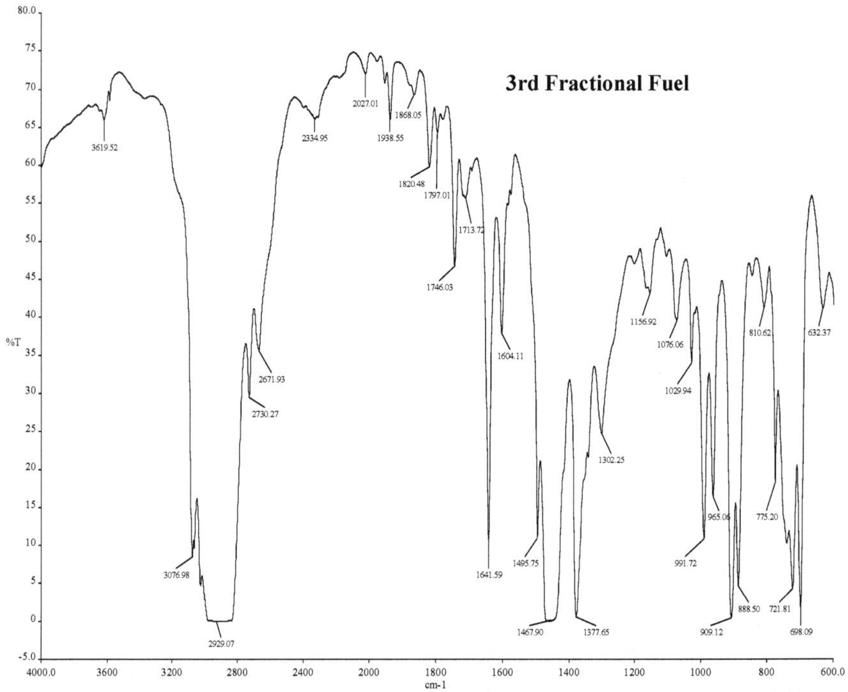

Fig. 23. FTIR Spectra of Mixed Waste Plastic Fuel to 3rd Fractional Fuel (Aviation)

Band Peak Number	Wave Number (cm^{-1})	Functional Group Name	Band Peak Number	Wave Number (cm^{-1})	Functional Group Name
3	2929.07	C-CH_3	17	1467.90	CH_3
4	2730.27	C-CH_3	18	1377.65	CH_3
5	2671.93	C-CH_3	22	1029.94	Acetates
8	1938.55	Non-Conjugated	23	991.72	-CH=CH_2
9	1868.05	Non-Conjugated	24	965.06	-CH=CH- (trans)
10	1820.48	Non-Conjugated	25	909.12	CH=CH_2
11	1797.01	Non-Conjugated	26	888.50	C=CH_2
12	1746.03	Non-Conjugated	29	721.81	-CH=CH- (cis)
13	1713.72	Non-Conjugated	30	698.09	-CH=CH- (cis)
14	1641.59	Non-Conjugated			

Table 27. Mixed Waste Plastic Fuel to 3rd^{t} Fractional Fuel (Aviation) FTIR Functional Group List

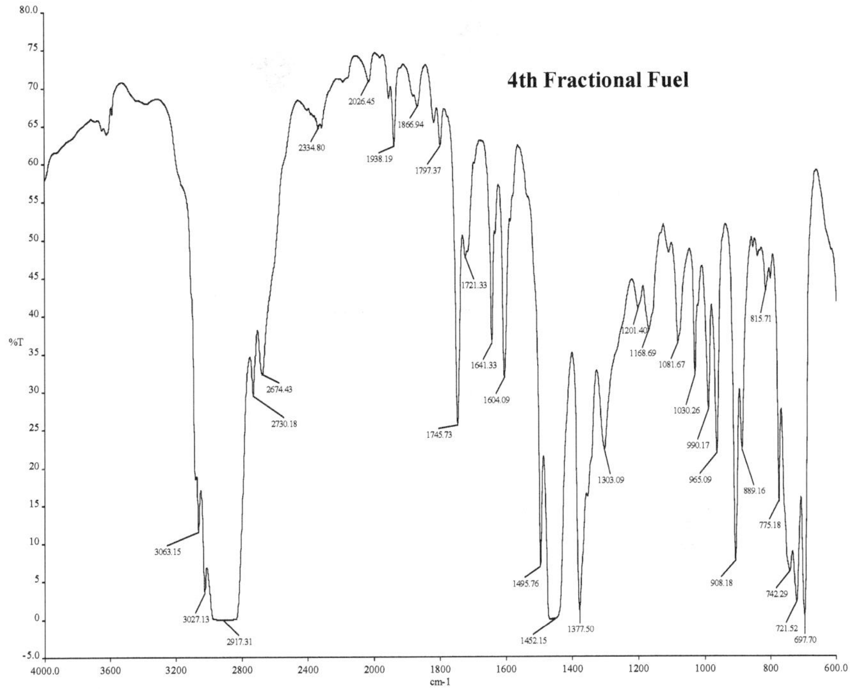

Fig. 24. FTIR Spectra of Mixed Waste Plastic to Fuel (Diesel)

Band Peak Number	Wave Number (cm^{-1})	Functional Group Name	Band Peak Number	Wave Number (cm^{-1})	Functional Group Name
1	3063.15	=C-H	16	1452.15	CH_2
2	3027.13	=C-H	17	1377.50	CH_3
3	2917.31	CH_2	22	1030.26	Acetates
4	2730.18	C-CH_3	23	990.17	-CH=CH_2
5	2674.43	C-CH_3	24	965.09	-CH=CH- (trans)
8	1938.19	Non-Conjugated	25	908.18	-CH=CH_2
9	1866.94	Non-Conjugated	26	889.16	C=CH_2
10	1797.37	Non-Conjugated	29	742.29	-CH=CH- (cis)
11	1745.73	Non-Conjugated	30	721.52	-CH=CH- (cis)
12	1721.33	Non-Conjugated	31	697.70	-CH=CH- (cis)
13	1641.33	Non-Conjugated			

Table 28. Mixed Waste Plastic Fuel to 4th Fractional Fuel (Diesel) FTIR Functional Group List

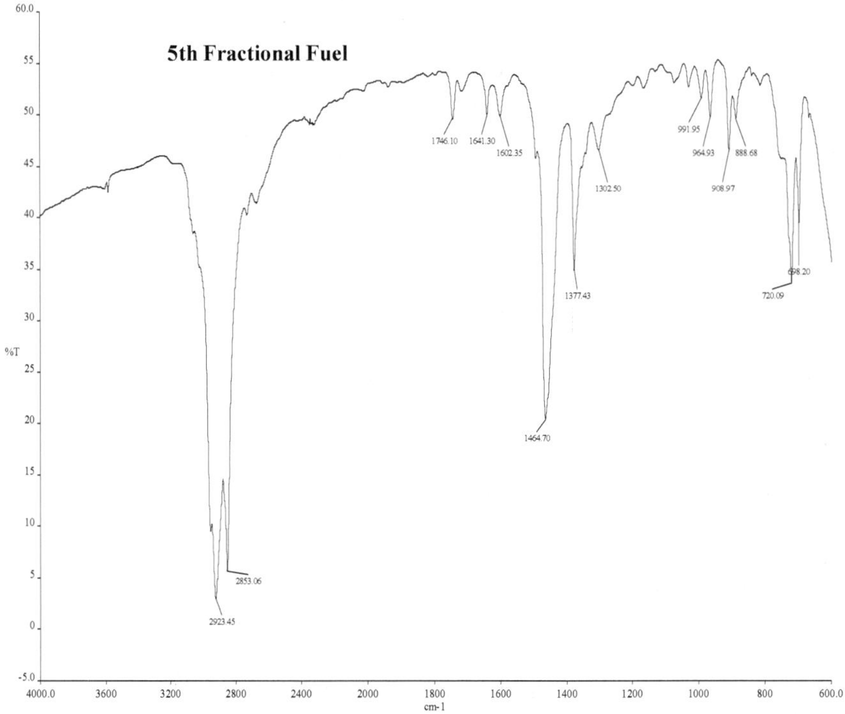

Fig. 25. FTIR Spectra of Mixed Waste Plastic to Fuel (Fuel Oil)

Band Peak Number	Wave Number (cm^{-1})	Functional Group Name	Band Peak Number	Wave Number (cm^{-1})	Functional Group Name
1	2923.45	CH_2	9	991.95	Secondary Cyclic Alcohol
2	2853.06	CH_2	10	964.93	-CH=CH- (trans)
3	1746.10	Non-Conjugated	11	908.97	$-CH=CH_2$
4	1641.30	Non-Conjugated	12	888.68	$C=CH_2$
5	1602.35	Non-Conjugated	13	720.09	-CH=CH- (cis)
6	1464.70	CH_2	14	698.20	-CH=CH- (cis)
7	1377.43	CH_3			

Table 29. Mixed Waste Plastic Fuel to 5th Fractional Fuel (Fuel Oil) FTIR Functional Group List

In FTIR analysis of fractional distillation fuel such as in 1ST Fraction Fuel (Gasoline) obtained functional groups are CH_3, Acetates, Secondary Cyclic Alcohol, $-CH=CH_2$,

C=CH_2,nad -CH=CH- (Cis). [Shown above, Fig.21 & Table-25]. In 2nd Fraction Fuel (Naphtha) analysis functional groups are CH_3, Non-Conjugated, Acetates,-CH=CH_2,-CH=CH- (Cis) and -CH=CH-(Trans). [Shown above, Fig.22&Table-26]. In 3rd Fraction Fuel (Aviation) analysis functional groups are CH_3, Acetates, C-CH_2,-CH=CH- (Cis) and -CH=CH-(Trans) [Shown above, Fig.23&Table-27]. In 4th Fraction Fuel (Diesel) analysis functional groups are CH_2, CH_3, Acetates,-CH=CH_2, C=CH_2 and,-CH=CH- (Cis) [Shown above, Fig.24 &Table-28]. Subsequently in 5th Fraction Fuel (Fuel Oil) analysis obtained functional groups are Secondary Cyclic Alcohol,-CH=CH_2, C=CH_2, -CH=CH (Trans) and -CH=CH-(Cis) etc. [Shown above, Fig.25&Table-29].

5. Electricity production from waste plastic fuel

Both NSR fractional fuels (NSR fractional 1st Fractional Fuel and NSR 4th Fractional Fuel) have been used to produce electricity by the help of conventional internal combustion generator. A flow diagram illustrating the process of energy production and consumption from NSR Fuel (Heating Oil) is shown below in Fig.26.

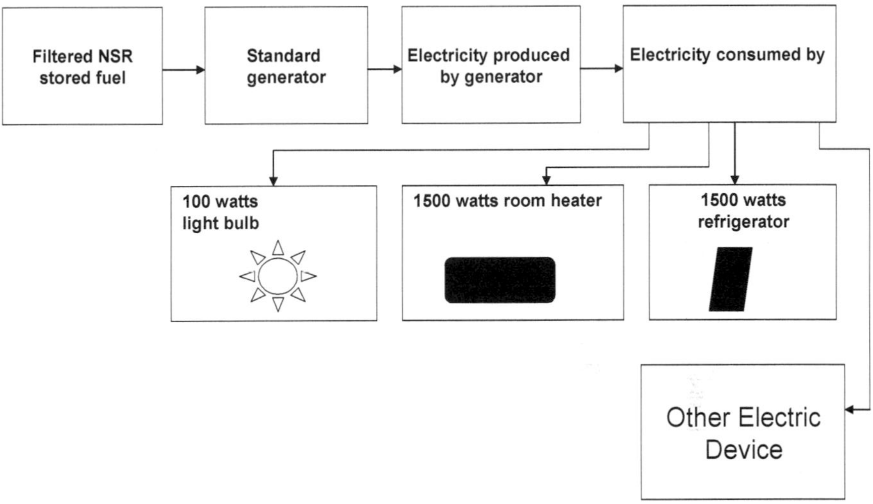

Fig. 26. Flow diagram of electricity generation consumption

NSR fractional 1st collection fuel was used in a gasoline generator with max 4.0 kW and volt output of 120. ~1 litter of fractional fuel was injected in the generator and with ~2900 watt constant demand; the generator ran a total of 42 minutes. A similar test was performed with commercial gasoline (87). ~1 litter of commercial gasoline (87) was injected and with the same ~ 2900 watt, constant demand the generator ran a total of 38 minutes. The difference in time occurs because NSR fraction 1st collection fuel has longer Carbon content than that of the commercial gasoline (87).

NSR fractional 4th collection fuel was used in a diesel generator with a max 4.0 kW and an output of 120 volt. ~1 litter of NSR fractional 2nd collection fuel was injected in the generator and with a constant demand of 3200 watt; the generator ran a total of 42 minutes. The same test was conducted with commercial diesel, and with the same demand the generator ran for 34 minutes.

A diagram [Fig.27] is provided below showing the produced electricity consumption of commercial gasoline (87) and NSR fractional fuel 1st collection.

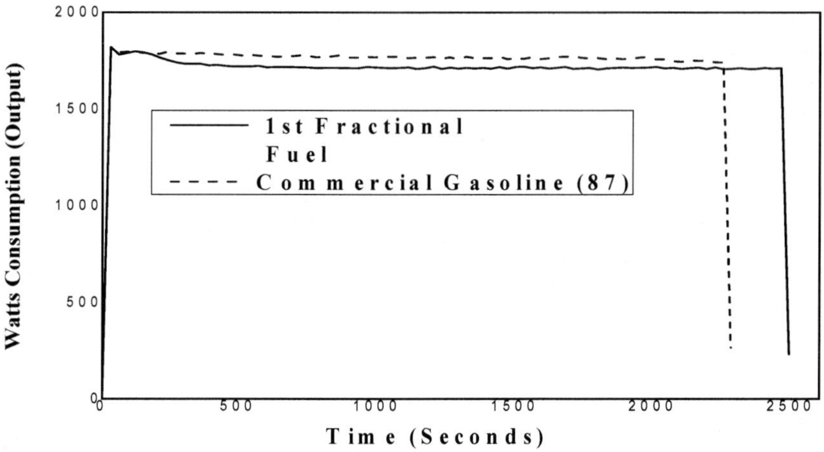

Fig. 27. Electricity Consumption and run time monitored by EML 2020 logger system for 1st Fractional Fuel (Gasoline) and Commercial Gasoline87.

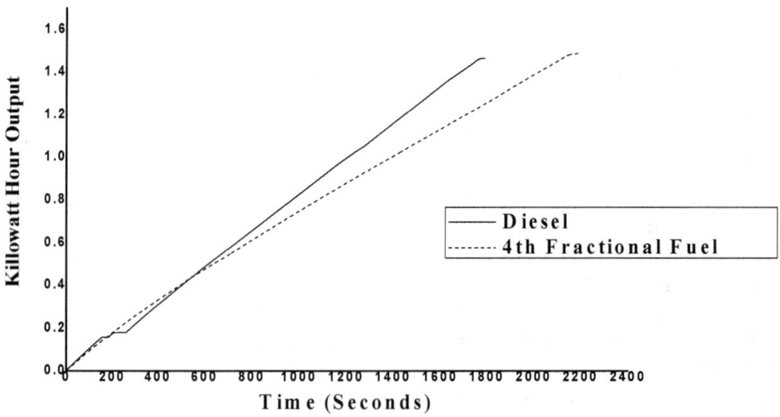

Fig. 28. Electricity Output Comparison Graph of Waste Plastic Fuel to 4th Fractional Fuel and Commercial Diesel Fuel

Fuel Name	Generator	Fuel Amount	Duration	kWh
4th Fractional Fuel (Diesel)	AMCO	1 Liter	37 min	2.028
Commercial Diesel	AMCO	1 Liter	28 min	1.463

Table 30. Comparison Table of 4th Fractional Fuel (Diesel) and Commercial Diesel

Comparison of NSR 4th fraction fuel and commercial diesel was conduced using an AMCO Diesel Generator. Above, **Fig. 28** and **Table 30** demonstrate the comparative results between the two fuels. The results indicate that the NSR-2 fuel provided a longer run time of the generator than the diesel. This is due to the NSR fuel having longer carbon chains than the diesel fuel.

6. Automobile test driving

Both NSR fractional 5th collection fuel and commercial gasoline (87) was used for a comparison automobile test. A 1984 Oldsmobile vehicle (V-8 powered engine) was used for the test-drive and one gallon of fuel was used for both cases after complete drainage of the pre-existing fuel in the fuel tank. The test-drive was done on a rural highway with an average speed of 55 mph.

Based on the preliminary automobile test-drive, the NSR fuel has offered a competitive advantage in mileage over the commercial gasoline-87. NSR fuel showed better mileage performance of 21 miles per gallon (mpg) compared to 18 mpg with commercial gasoline (87).

It is expected that NSR double condensed fuel will show even higher performance with more fuel-efficient car such as V-4 engine and hybrid vehicles. Additional test-driving is going to be conducted in the near future to verify the results.

7. Conclusion

The conversion of municipal waste plastics to liquid hydrocarbon fuel was carried out in thermal degradation process with/without catalyst. Individually we ran our experiment on waste plastics such as: HDPE-2, LDPE-4, PP-5 & PS-6. Each of those experiment procedures are maintained identically, every ten (10) minutes of interval experiment was monitored and found during the condensation time changes of individual waste plastics external behavior different because of their different physical and chemical properties. Similarly, we ran another experiment with 2kg of mixture of waste plastics in stainless steel reactor. Initial temperature is 350 °C for quick melting and optimum temperature is 305 °C. For glass reactor every experiment temperature was maintained by variac meter, when experiment started variac percent was 90% (Tem-405 °C) for quick melting, after melted variac percent decreased to 70% (Tem- 315 °C) due to smoke formation. Average (optimum) used variac percent in this experiment 75% (337.5 °C).Gradually temperature range was maintained by variacmeter with proper monitoring. In fractional distillation process we separated different category of fuel such as gasoline, naphtha, jet fuel, diesel and fuel oil in accordance with their boiling point temperature profile.

8. Acknowledgement

The author acknowledges the support of Dr. Karin Kaufman, the founder and President of Natural State Research, Inc (NSR). The author also acknowledges the valuable contributions NSR laboratory team members during the preparation of this manuscript.

9. Reference

[1] J. Aguado,* D. P. Serrano, G. Vicente, and N. Sa´nchez, Enhanced Production of r-Olefins by Thermal Degradation of High-Density Polyethylene (HDPE) in Decalin Solvent: Effect of the Reaction Time and Temperature, *Ind. Eng. Chem. Res.* 2007, *46,* 3497-3504

[2] Antonio Marcilla,* AÄ ngela N. Garcı´a, and Maria del Remedio Herna´ndez, Thermal Degradation of LDPE-Vacuum Gas Oil Mixtures for Plastic Wastes Valorization, *Energy & Fuels* 2007, *21,* 870-880

[3] Achyut K. Panda [a,b,*], R.K. Singh [a,1], D.K. Mishra [b,2], Thermolysis of waste plastics to liquid fuel A suitable method for plastic waste management and manufacture of value added products—A world prospective, Renewable and Sustainable Energy Reviews 14 (2010) 233–248

[4] N. Miskolczi[a], L. Bartha[a], G. Dea´k[a], B. Jo´ ver[b], Thermal degradation of municipal plastic waste for production of fuel-like hydrocarbons, Polymer Degradation and Stability 86 (2004) 357-366

[5] Miguel Miranda [a,*], Filomena Pinto [a], I. Gulyurtlu [a], I. Cabrita [a], C.A. Nogueira [a], Arlindo Matos [b], Response surface methodology optimization applied to rubber tyre and plastic wastes thermal conversion, Fuel 89 (2010) 2217–2229

[6] M. Stelmachowski *, Thermal conversion of waste polyolefins to the mixture of hydrocarbons in the reactor with molten metal bed, Energy Conversion and Management 51 (2010) 2016–2024

[7] Karishma Gobin, George Manos*, Polymer degradation to fuels over microporous catalysts as a novel tertiary plastic recycling method, Polymer Degradation and Stability 83 (2004) 267–279

[8] Weibing Ding, Jing Liang, and Larry L. Anderson*, Hydrocracking and Hydroisomerization of High-Density Polyethylene and Waste Plastic over Zeolite and Silica-Alumina-Supported Ni and Ni-Mo Sulfides, *Energy & Fuels* 1997, *11,* 1219-1224

[9] Anthony Warren, Mahmoud El-Halwagi *, An economic study for the co-generation of liquid fuel and hydrogen from coal and municipal solid waste, Fuel Processing Technology 49 (1996) 157-166

[10] Wei-Chiang Huang [a,b,c], Mao-Suan Huang [c,d,1], Chiung-Fang Huang [a,b,c], Chien-Chung Chen [c,e,*], Keng-Liang Ou [c,e,f,**], Thermochemical conversion of polymer wastes into hydrocarbon fuels over various fluidizing cracking catalysts, Fuel 89 (2010) 2305–2316

[11] Valerio Cozzani, † Cristiano Nicolella,‡ Mauro Rovatti,‡ and Leonardo Tognotti*,†, Influence of Gas- Phase Reactions on the Product Yields Obtained in the Pyrolysis of Polyethylene, *Ind. Eng. Chem. Res.* 1997, *36,* 342-348

[12] A. Marcilla,* M. I. Beltra´n, and R. Navarro, Evolution with the Temperature of the Compounds Obtained in the Catalytic Pyrolysis of Polyethylene over HUSY, *Ind. Eng. Chem. Res.* 2008, *47*, 6896–6903

[13] Ma del Remedio Herna´ndez, AÄ ngela N. Garcı´a, Amparo Go´mez, Javier Agullo´, and Antonio Marcilla*, Effect of Residence Time on Volatile Products Obtained in the HDPE Pyrolysis in the Presence and Absence of HZSM-5, *Ind. Eng. Chem. Res.* 2006, *45*, 8770-8778

[14] Levent Ballice,* Mithat Yu¨ ksel, and Mehmet Sag˘lam, Classification of Volatile Products from the Temperature-Programmed Pyrolysis of Low- and High-Density Polyethylene, *Energy & Fuels* 1998, *12*, 925-928

[15] Paula A. Costa,*,† Filomena J. Pinto,† Ana. M. Ramos,‡ Ibrahim K. Gulyurtlu,† Isabel A. Cabrita,† and Maria. S. Bernardo‡, Kinetic Evaluation of the Pyrolysis of Polyethylene Waste, *Energy & Fuels* 2007, *21*, 2489-2498

[16] Biswanath Saha† and Aloke K. Ghoshal*, Hybrid Genetic Algorithm and Model-Free Coupled Direct Search Methods for Pyrolysis Kinetics of ZSM-5 Catalyzed Decomposition of Waste Low-Density Polyethylene, *Ind. Eng. Chem. Res.* 2007, *46*, 5485-5492

[17] R. W. J. Westerhout, J. Waanders, J. A. M. Kuipers,* and W. P. M. van Swaaij, Recycling of Polyethene and Polypropene in a Novel Bench-Scale Rotating Cone Reactor by High-Temperature Pyrolysis, *Ind. Eng. Chem. Res.* 1998, *37*, 2293-2300

[18] R. W. J. Westerhout, J. Waanders, J. A. M. Kuipers,* and W. P. M. van Swaaij, Development of a Continuous Rotating Cone Reactor Pilot Plant for the Pyrolysis of Polyethene and Polypropene, *Ind. Eng. Chem. Res.* 1998, *37*, 2316-2322

[19] R. W. J. Westerhout, R. H. P. Balk, R. Meijer, J. A. M. Kuipers,* and W. P. M. van Swaaij, Examination and Evaluation of the Use of Screen Heaters for the Measurement of the High Temperature Pyrolysis Kinetics of Polyethene and Polypropene, *Ind. Eng. Chem. Res.* 1997, *36*, 3360-3368

[20] Lan Tang,* H. Huang, Zengli Zhao, C. Z. Wu, and Y. Chen, Pyrolysis of Polypropylene in a Nitrogen Plasma Reactor, *Ind. Eng. Chem. Res.* 2003, *42*, 1145-1150

[21] George Manos,*,† Arthur Garforth,‡ and John Dwyer§, Catalytic Degradation of High-Density Polyethylene on an Ultrastable-Y Zeolite. Nature of Initial Polymer Reactions, Pattern of Formation of Gas and Liquid Products, and Temperature Effects, *Ind. Eng. Chem. Res.* 2000, *39*, 1203-1208,

[22] George Manos,*,† Arthur Garforth,‡ and John Dwyer§, Catalytic Degradation of High-Density Polyethylene over Different Zeolitic Structures, *Ind. Eng. Chem. Res.* 2000, *39*, 1198-1202

[23] Yoshio Uemichi,* Junko Nakamura, Toshihiro Itoh, and Masatoshi Sugioka, Conversion of Polyethylene into Gasoline-Range Fuels by Two-Stage Catalytic Degradation Using Silica-Alumin and HZSM-5 Zeolite, *Ind. Eng. Chem. Res.* 1999, *38*, 385-390

[24] George Manos,*, † Isman Y. Yusof, ‡ Nikos Papayannakos,§ and Nicolas H. Gangas§, Catalytic Cracking of Polyethylene over Clay Catalysts. Comparison with an Ultrastable Y Zeolite,*Ind. Eng. Chem. Res.* 2001, *40*, 2220-2225

[25] Jose´ M. Arandes,*,† In˜ aki Abajo,† Danilo Lo´ pez-Valerio,§ Inmaculada Ferna´ ndez,† Miren J. Azkoiti,‡ Martı´n Olazar,† and Javier Bilbao†, Transformation of Several Plastic Wastes into Fuels by Catalytic Cracking, *Ind. Eng. Chem. Res.* 1997, *36*, 4523-4529

[26] Selhan Karago¨z,†,§ Jale Yanik,*,‡ Suat Ucüar,† and Chunshan Song§, Catalytic Coprocessing of Low-Density Polyethylene with VGO Using Metal Supported on Activated Carbon, *Energy & Fuels* 2002, *16,* 1301-1308

[27] Toshiyuki Kanno, Masahiro Kimura, Na-oki Ikenaga, and Toshimitsu Suzuki*, Coliquefaction of Coal with Polyethylene Using Fe (CO) 5-S as Catalyst, *Energy & Fuels* 2000, *14,* 612-617

[28] Mohammad Nahid Siddiqui [a],*, Halim Hamid Redhwi [b], Catalytic coprocessing of waste plastics and petroleum residue into liquid fuel oils, J. Anal. Appl. Pyrolysis 86 (2009) 141–147

[29] Ikusei Nakamura *, Kaoru Fujimoto, Development of new disposable catalyst for waste plastics treatment for high quality transportation fuel, Catalysis Today 27 (1996) 175-179

[30] A.G. Buekens *, H. Huang, Catalytic plastics cracking for recovery of gasoline-range hydrocarbons from municipal plastic wastes, Resources, Conservation and Recycling 23 (1998) 163–181

4

Removal of VOCs Using Nonthermal Plasma Technology

Tao Zhu
School of Chemical and Environmental Engineering,
China University of Mining& Technology, Beijing 100083,
China

1. Introduction

Volatile organic compounds (VOCs) are liquids or solids that contain organic carbon (carbon bonded to carbon, hydrogen, nitrogen, or sulfur, but not carbonate carbon as in $CaCO_3$ nor carbide carbon as in CaC_2 or CO_2), which vaporize at significant rates. VOCs are probably the second-most widespread and diverse class of emissions after particulates.

VOCs are a large family of compounds. Some (e.g., benzene) are toxic and carcinogenic, and are regulated individually as hazardous pollutants. The control of VOCs in the atmosphere is a major environmental problem now. Toluene and benzene are two of the typical VOCs. They effluents in some industries, such as paints, paint thinners, fingernail polish, lacquers, adhesives, rubber, and some printing and leather tanning processes, have attracted more and more attention of researchers. The traditional methods of VOCs removal such as absorption, adsorption, and incineration and so on, which are referred to the new environmental condition have many technical and economic disadvantages. In these years, some new technologies, such as biologic process, photo-catalysis process, and plasma technology, were paid more and more attention.

As an emerging technology for environmental protection, there have been extensive researches on using non-thermal plasma (NTP) over the past 20 years. The major advantages of NTP technology include the moderate operation conditions (normal temperature and atmospheric pressure), moderate capital cost, compact system, easy operations and short residence times, Etc., compare to the conventional technologies). In the field of air pollution control, the NTP technology has been tested for the abatement of various types of hazardous air pollutants such as volatile organic compounds (VOCs), SO_2, NOx, CFCs, odors, mercury, etc.

In this chapter, we will introduce a new synergy technology basing on non-thermal plasma for VOCs decomposition.

2. Experimental setup

The reaction system was a tube-wire packed-bed reaction system at ambient temperature and atmospheric pressure. The schematic diagram of the NTP system is shown in Fig.1. Dry air (78.5% N_2, 21.5% O_2) was used as a balance gas for VOCs decomposition. Air supplied from an air compressor was divided into two airflows and each flow rate was controlled with a mass flow meter. One airflow was introduced into a VOCs liquid bottle (3), which contained liquid VOCs. The air with a mass of saturated VOCs vapor was mixed with the other airflow in a blender (4) and the gaseous phase VOCs was diluted to a prescribed concentration. A wire-tube DBD reactor with packed materials as shown in Fig.2 was used.

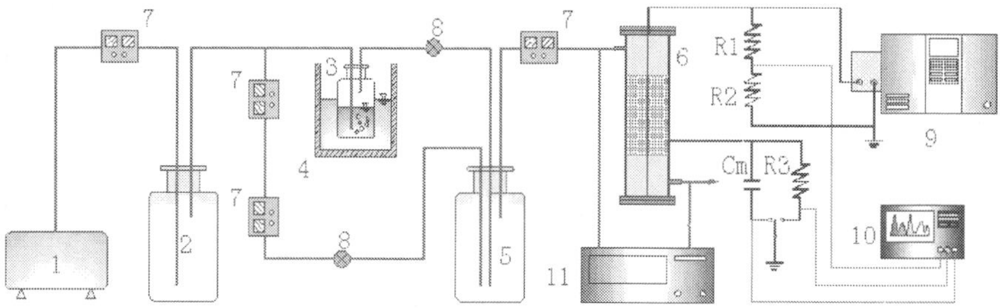

1.air compressor 2.buffer 3.toluene liquid bottle 4.attemperator 5.blender 6.NTP reactor 7.mass flow meter 8.needle valve 9.high voltage 10.oscukkograph 11.gas chromatograph

Fig. 1. Schematic diagram of the NTP system

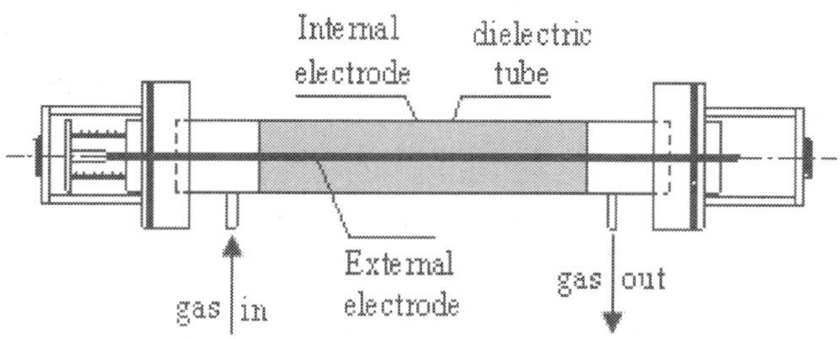

Reactor: organic-glass tube (i.d.50mm, length 150mm)
Internal electrode: tungsten filament (i.d.0.5mm)
External electrode: dense steel mesh

Fig. 2. The NTP reactor

An AC power supply of 150 Hz was employed in the NTP reactor. The AC voltage was applied to the reactor in the radial direction and the voltage extension changed from 0 kV to 50 kV. The voltage and current of the discharge process were detected by an oscillograph (manufactured by American Tektronix Co., TDS2014). Primary power values were measured with the voltage-charge (V-Q)Lissajous method in the plasma reactors. The circuit diagram on power measurement with Lissajous developed and used in this study is shown in figure 3.

To investigate the electric characteristics of dielectric barrier discharge (DBD), the voltage applied to the reactor was sampled by a voltage divider with a ratio of 12500:1. Also, the current was determined from the voltage drop across a shunt resistor (R3 =10kΩ) connected in series with the grounded electrode. In order to obtained the total charge and discharge power simultaneously, a capacitor (Cm =2μF) was inserted between the reactor and the ground. The electrical power provided to the discharge was measured using the Q-V

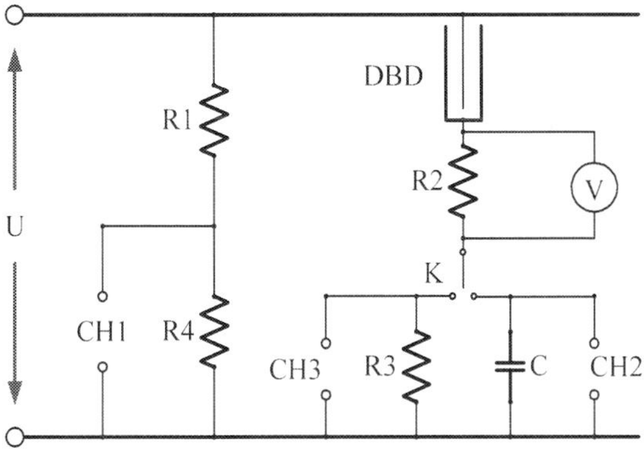

*R*1=250MΩ, *R*2=10KΩ, *R*3=10KΩ, *R*4=20KΩ, *C*=0.33μF,
K—Switch, CH1—Voltage sampling of oscillograph,
CH2—Lissajous sampling of oscillograph,
CH3—Current sampling of oscillograph

Fig. 3. Circuit diagram on power measurement with Lissajous

Lissajous diagram. Typical Lissajous diagram represents to be a parallelogram, and we could calculate power though calculated the area of parallelogram.
VOCs analysis was carried out by gas chromatography (manufactured by Aglient Co., HP6890N) with a flame ionization detector (FID). The byproducts were detected by GC-MS (manufactured by American Thermo Finnegan Co.) using EI mode, 70 eV and full scan. Ozone concentration produced in the NTP reactor was measured by an iodine-titration method. The plasma reactor employed an AC power supply of 50-500Hz scanning from 0 kV to 100 kV was applied to the reactor in the radial direction.
As evaluation criteria, the VOCs removal efficiency, reactor energy density, energy efficiency, the selectivity of CO_2 and the selectivity of CO (eg., benzene) in the gas phase were calculated as follows:
VOCs removal efficiency (η):

$$\eta(\%) = \frac{[VOCs]_{inlet} - [VOCs]_{outlet}}{[VOCs]_{inlet}} \times 100\% \tag{1}$$

Reactor energy density (RED):

$$RED(kJ / L) = \frac{input \cdot power(W)}{gas \cdot flow \cdot rate(L / \min)} \times 60 \times 10^{-3} \tag{2}$$

Energy efficiency (ς):

$$\varsigma(g / kWh) = \frac{[VOCs]_{inlet} \times \eta}{RED} \times 3.6 \times 10^{-3} \tag{3}$$

Selectivity of CO_2 (ζ):

$$\zeta(\%) = \frac{[CO_2]}{6([Benzene]_{inlet} - [Benzene]_{outlet})} \times 100\% \tag{4}$$

Selectivity of CO (ξ):

$$\xi(\%) = \frac{[CO]}{6([Benzene]_{inlet} - [Benzene]_{outlet})} \times 100\% \quad (5)$$

2.1 Adsorption-enhanced Non-thermal Plasma

We previously reported the oxidative decomposition of formaldehyde, benzene, VOCs and odorin air using a plasma reactor packed with various materials. The frequency of the applied high voltage alternating current (AC) had an important influence on energy efficiency and intermediate frequency of the applied AC power was beneficial for VOCs removal, especially 150 Hz.

Urashitnal et al. used NTP packed with active carbon to decompose VOCs and trichoroethylene (TCE) with the discharge energy efficiencies of 26 g/(kW·h) and 13 g/(kW·h), respectively. The TCE removal efficiency was 40% using NTP technology alone and it was up to 90% using NTP packed with active carbon. Ogata et al. found MS-4A molecular griddle played a special role in VOCs removal using NTP technology and reported that the VOCs removal efficiency in the dielectric barrier discharge (DBD) reactor packed with MS-4A molecular griddle and $BaTiO_3$ particles was 1.4~2.1 times higher than that in DBD reactor packed with $BaTiO_3$ particles.

In this experiment, the mechanism of adsorption-enhanced NTP for volatile organic compounds (Toluene) removal was discussed. A sorbent was packed into the space of discharge plasma so that reaction time was prolonged between VOCs molecules and NTP and removal efficiency of VOCs was improved without increasing the size of NTP reactor. The sorbent was helpful for enhancing discharge energy efficiency due to VOCs molecules enrichment on the surface of the sorbent.

2.2 Adsorption kinetics

The pellets of γ-Al_2O_3, 5~7mm in diameters, was used as the sorbent to be packed into the NTP reactor. Figure 4 shows the VOCs concentration profile as a function of time on γ-Al_2O_3 pellets in the NTP reactor before plasma was applied (VOCs: 800 mg/m^3; flow rate: 2 mL/min; dry air). VOCs was adsorbed on the surface of γ-Al_2O_3 and its concentration was gradually increased with time. After 150 min, the VOCs concentration reached the adsorption-desorption equilibrium. So the decomposition tests were initiated after the adsorption-desorption equilibrium.

The collision frequency between adsorption rate (Ra) and adsorbate molecules on the adsorbent surface is proportional to fraction of the vacant active sites on the adsorbent surface (1-θ):

$$R_a = k_{a0} T^{-1/2} (1-\theta) y^0 \quad (6)$$

where $k_{a0} = a/(R/2\pi M)^{1/2}$.

The desorption rate (Rd) is proportional to the fraction of the occupied active sites (θ):

$$R_d = k_{a0} \theta exp(-E_d/RT) \quad (7)$$

And the net-desorption rate of adsorbate molecules should be the difference between the desorption rate and the adsorption rate:

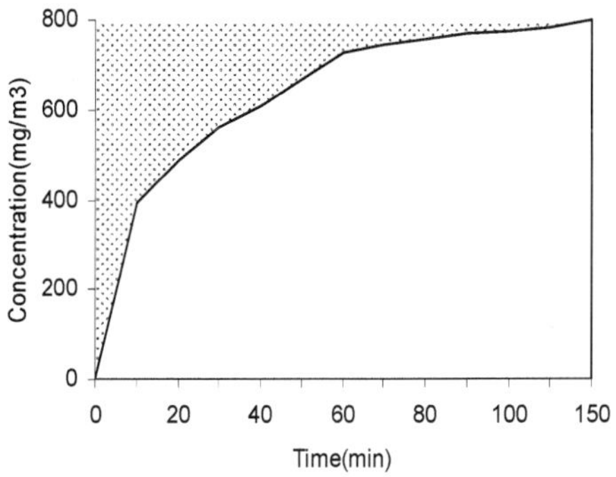

Fig. 4. VOCs concentration on the surface of γ-Al_2O_3 pellets

$$-d\theta/dt=R_d-R_a=k_d\theta-k_a(1-\theta)y^0 \quad (8)$$

where $k_a=k_{a0}T^{-0.5}$ and $k_d=k_{d0}exp(-E_d/RT)$

According to the reaction status on the surface of catalyst in the NTP reactor, adsorbate molecules are in the slow diffusion in the adsorbent channels and the balance between adsorbent surface molecules and gas phase would be established for a long time. In this case, the net adsorption rate can be expressed as:

$$-d\theta/dt=k(y^0-y) \quad (9)$$

During a certain period of time dt, the adsorption process mass balance between the gas and solid phase adsorbent is:

$$F_y M=m_s q_m(-d\theta/dt) \quad (10)$$

The adsorbate molecules in the pores of adsorbent diffuse slowly and the mass transfer rate constant k is smaller. As we known, the desorption temperature is a linear function of the frequency of high voltage and the reaction time:

$$T=T_0+a_H f+\beta_H t \quad (11)$$

By combining Eq.(9), (10) and (11), the coverage rate of adsorbate on the surface of adsorbent as a function of temperature can be obtained:

$$d\theta/dt=(k_d\,\theta/a_H\beta_H)/\{[-k_a(1-\theta)/k]+[k_a(1-\theta)b-1]\} \quad (12)$$

Where $b=-m_s q_m/FM$.

As mentioned earlier, assuming that the mass transfer coefficient is small enough, the contaminants adsorbed on the surface would be desorped and degraded immediately on the surface of catalyst, that is, $k_a(1-\theta)/k >> [k_a(1-\theta)b-1]$. Then eq.(12) becomes:

$$d\theta/dt=-k\,k_d\,\theta/[a_H\beta_H\,k_a(1-\theta)] \quad (13)$$

When the desorption rate is maximum and $-d^2\theta/dT^2=0$, eq.(13) can be obtained:

$$-d\theta/dT|_{T=Ti}=\theta_i(1-\theta_i)\,(1/2T_i+E_d/RT_i^2) \quad (14)$$

In the situation of γ-Al_2O_3 adsorption interaction with the plasma, the VOCs concentration in the surface of γ-Al_2O_3 reached adsorption equilibrium. The removal amount of VOCs equaled to 25% of the inlet total concentration of VOCs by adsorption ofγ-Al_2O_3.

2.3 Effect of packed materials on removal efficiency

Figure 5 shows the relationship between reactor energy density (RED) and VOCs removal efficiency with different packed materials in the NTP reactor. The VOCs removal efficiency increased with RED and was in an order of no padding < common packed materials < γ-Al_2O_3. On one hand, γ-Al_2O_3, as a sorbent, could adsorb short-living free redicals. In gas discharge process, these free redicals would accelerate decomposition reactions on the surface of microhole structure of the sorbent. The surface of the cellularity particles could also become active sites with electrons stricking. On the other hand, due to their higher permittivity of 11, γ-Al_2O_3 pellets gained more electric charges and enhanced the local discharge and the discharge current. As a result, γ-Al_2O_3 pellets were helpful for the VOCs removal reaction.

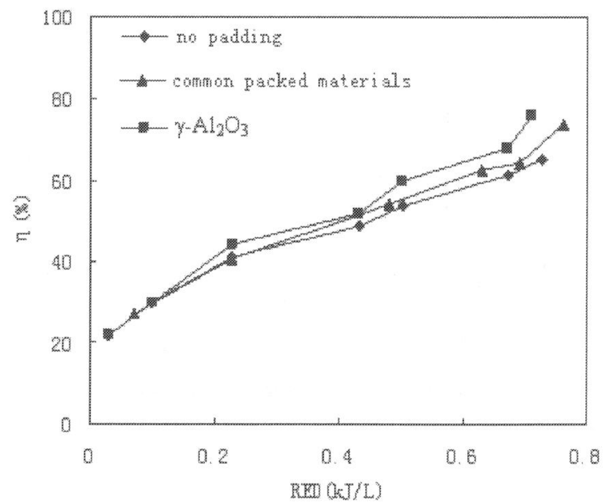

Fig. 5. Relationship between RED and removal efficiency of VOCs with different packed materials

2.4 Effect of packed materials on ozone concentration

Figure 6 shows the relationship between the packed materials and ozone concentration. The effect on the ozone concentration was in an order of γ-Al_2O_3 < no padding < common packed materials at various REDs. The ozone concentration reached the peak at RED of 0.7 kJ/L with the same packed materials in the NTP reactor as shown in figure 6.

High energy electrons produced by gas discharge ionized the VOCs molecules which were adsorbed on the γ-Al_2O_3 surface to produce more positive ions including N^+, N_2^+, O_2^+ and H_3O^+. These positive ions activated VOCs molecules for VOCs decomposition though electric discharge transferring reactions. In the NTP reactor, the VOCs molecules were decomposed by the radicals of $O^\bullet$, $OH^\bullet$ and $N^\bullet$, etc. Evans et al. believed that $O^\bullet$ played a key role for VOCs decomposition in the NTP process. Ozone as the main long-living radical was transported to the surface of γ-Al_2O_3 and could take part in oxidation reaction. The pathways of reaction were stated as follows:

$$e + O_2 \rightarrow 2O + e \tag{15}$$

$$O + O_2 + M \rightarrow O_3 + M \tag{16}$$

$$O + O_3 \rightarrow 2O_2 \tag{17}$$

$$e + O_3 \rightarrow O + O_2 + e \tag{18}$$

The active oxygen species formed during the O_3 decomposition would also be helpful for the VOCs removal.

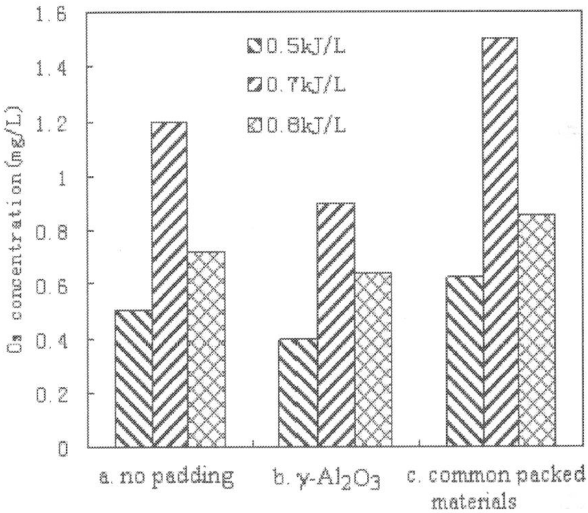

Fig. 6. Relationship between the packed materials and ozone concentration

2.5 Effect of packed materials on energy efficiency

Figure 7 shows the relationship between RED and electric energy efficiency with different packed materials in the NTP reactor. The energy efficiency for VOCs removal decreased with increasing RED and was in an order of no padding < common packed materials < γ-Al_2O_3. Ions, electrons, excited neutral molecules and metastable free radicals in the plasma process attacked the surface of γ-Al_2O_3 pellets and made many kinds of reciprocity possible:

1. The plasma brought sub-electrons launching.
2. The plasma induced chemical reactions on the surface of the γ-Al_2O_3 pellets and produced active atoms, molecules and free radicals, which would take part in chemical reactions for the further VOCs decomposition.
3. Many active oxidation groups in plasma were adsorbed on the surface of the γ-Al_2O_3 pellets. The plasma induced the desorption of VOCs molecules and active oxidation

groups so that further reactions took place. As a result, the whole reaction process was accelerated and more reaction routes were induced.

4. VOCs removal efficiency and electric energy efficiency were improved.

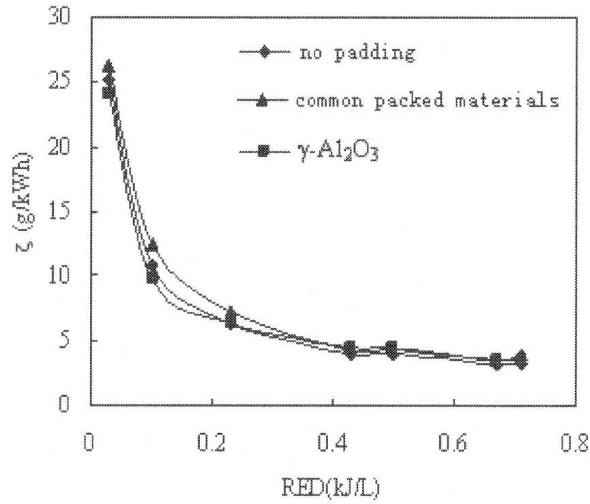

Fig. 7. Relationship between RED and electric energy efficiency with different packed materials

3. Effect of modified ferroelectric on nonthermal plasma process

The major bottleneck of developing NTP with catalysis technology is the reduction of energy consumption. If this requirement is not satisfied, the non-thermal plasma process may lose its potential for commercial applications. In order to resolve this problem, Ayrault et al. used platinum (Pt)-based catalyst supported on an alumina wash-coated honeycomb monolith by means of a high voltage bi-polar pulsed excitation. The energy efficiency was 0.14 mol/kWh at an energy density of 200 J/L for 2-Heptanone decomposition. For a comparison, the energy efficiency decreased to 0.029 mol/kWh using an uncoated monolith even at a higher energy density of 500 J/L.

In this investigation, we developed a new ferroelectric packed bed NTP reactor and prepared a sample of $Ba_{0.8}Sr_{0.2}Zr_{0.1}Ti_{0.9}O_3$ to serve as modified ferroelectrics. The permittivity of $Ba_{0.8}Sr_{0.2}Zr_{0.1}Ti_{0.9}O_3$ was 10^4, 12 times higher than that of the pure phase of $BaTiO_3$, while dielectric loss was 1/6 in room temperature. The experimental results show that this type of modified ferroelectrics packed into the NTP reactor could both reduce the energy consumption and raise energy efficiency significantly. Compared with $BaTiO_3$, $Ba_{0.8}Sr_{0.2}Zr_{0.1}Ti_{0.9}O_3$ had better ferroelectric physical properties to improve NTP process for VOCs control.

3.1 Materials and methods

In the experiment, three kinds of packed materials, including ceramic rings, $BaTiO_3$ rings and $Ba_{0.8}Sr_{0.2}Zr_{0.1}Ti_{0.9}O_3$ rings (hollow cylinder shape, 5 mm i.d., 1 mm wall thick, and 10 mm length), were used to pack into the NTP reactor.

Nano-size $Ba_{0.8}Sr_{0.2}Zr_{0.1}$ $Ti_{0.9}O_3$ powder was prepared using the method of water-thermal composite action at atmospheric pressure. Inorganic salts, including $TiCl_4$, $Ba(OH)_2 \cdot 8H_2O$ and $Sr(OH)_2 \cdot 8H_2O$, were the precursors for $Ba_{0.8}Sr_{0.2}Zr_{0.1}$ $Ti_{0.9}O_3$ formation. Firstly, a proper quantity of $TiCl_4$ was added to 100mL water as the precursor solution and a ammonia used to adjust pH to 7. By strictly controlling the reaction conditions in a ventilation cabinet, the precursor solution hydrolyzed to α-H_2TiO_3. And then, Cl^- was removed by hot water washing and filtrated by decompression and boiled at 100 ℃ for 4 hours. Certain amounts of $Ba(OH)_2 \cdot 8H_2O$ and $Sr(OH)_2 \cdot 8H_2O$ dropped into H_2TiO_3 and ammonia adjusted pH to 6~6.5 and the solution was shielded from air and agitated for hours. During the preparation, if needed, water was added to keep the balance of the liquid quantity. Whereafter, gained solid (nano-$Ba_{0.8}Sr_{0.2}Zr_{0.1}Ti_{0.9}O_3$) was ground to powders and dried at 100 °C in a crucible. The powder was made into rings (5 mm i.d., 1 mm wall thick, and 10 mm length) which were placed in a muffle furnace to calcine at 1200 °C for two hours. The calcined product was cooled to ambient temperature and served as the packed materials in the NTP reactor. At the same time, a $BaTiO_3$ (powders made in Beijing Research Institute of Chemical Engineering & Metallurgy) ring was also made with the same weight as the $Ba_{0.8}Sr_{0.2}Zr_{0.1}Ti_{0.9}O_3$ ring.

The crystal structure and the surface shape of the $Ba_{0.8}Sr_{0.2}Zr_{0.1}Ti_{0.9}O_3$ sample were detected by XRD (manufactured by Germany Bruker Co., D8 ADVANCE) and SEM (manufactured by Japan, JEOL-JSM-6500F) and the BET surface area determined by Micromeritics (manufactured by American Quantachrome Co., NOVA 1000). The relative permittivity of the $Ba_{0.8}Sr_{0.2}Zr_{0.1}Ti_{0.9}O_3$ sample was measured using an LCR automatism test instrument (manufactured by China, 4210).

3.2 Characteristic of modified ferroelectric

The crystal structure of $Ba_{0.8}Sr_{0.2}Zr_{0.1}Ti_{0.9}O_3$ detected by XRD as shown in Figure 8 should be similar to cube crystal structure of calcium-titanium oxide. $Ba_{0.8}Sr_{0.2}Zr_{0.1}Ti_{0.9}O_3$ was a type of ferroelectric like $BaTiO_3$. The average diameter of sample particulates was of 59 nm. Fig.9 shows that crystal shape of $Ba_{0.8}Sr_{0.2}Zr_{0.1}Ti_{0.9}O_3$ was spherical. The BET surface area of the

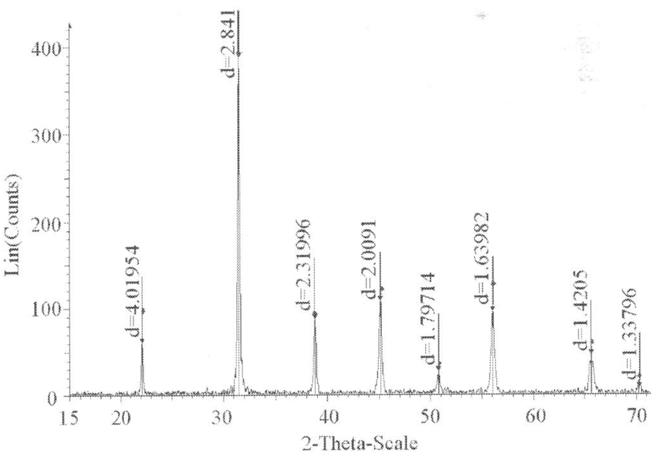

Fig. 8. XRD testing results of $Ba_{0.8}Sr_{0.2}Zr_{0.1}Ti_{0.9}O_3$

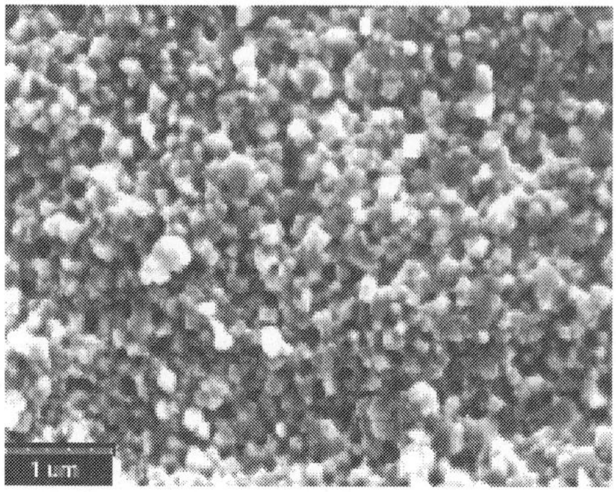

Fig. 9. SEM testing results of $Ba_{0.8}Sr_{0.2}Zr_{0.1}Ti_{0.9}O_3$

$Ba_{0.8}Sr_{0.2}Zr_{0.1}Ti_{0.9}O_3$ powders are 8.8 m^2/g, and Longmuir surface area detected by Micromeritics are 12.3 m^2/g respectively. The relative permittivity of $Ba_{0.8}Sr_{0.2}Zr_{0.1}Ti_{0.9}O_3$ detected by LCR is about 12000.

3.3 Effect of $Ba_{0.8}Sr_{0.2}Zr_{0.1}Ti_{0.9}O_3$ on removal efficiency of toluene

Fig.10 shows the effect of different packed materials in the NTP reactor on the removal efficiency (η) of toluene. The removal efficiency of toluene increases with reactor energy density (RED) and is in the order of without packed materials < with $BaTiO_3$ < with $Ba_{0.8}Sr_{0.2}Zr_{0.1}Ti_{0.9}O_3$, at the same RED.

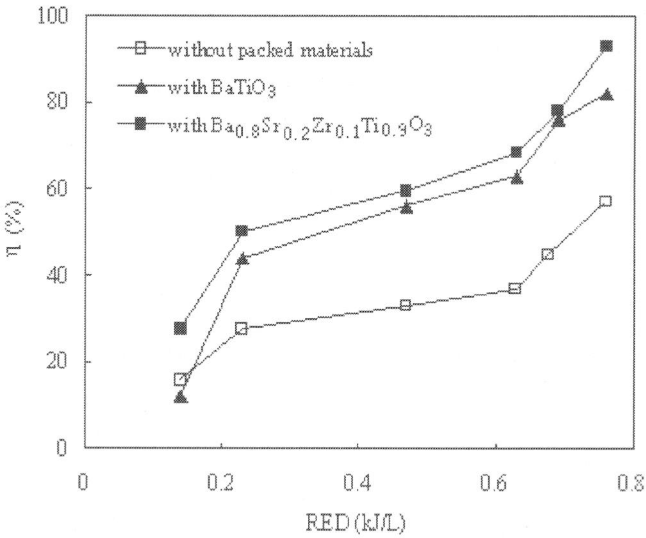

Fig. 10. Effect of packed materials on removal efficiency (toluene concentration: 1000mg/m^3 or so; gas flow rate: 2L/min; AC frequency: 150Hz)

The influence on toluene removal efficiency of some other parameters than could be taking into account such as the permittivity of packed materials, surface specific area, adsorption properties or catalytic effect:

1. It is well known that the presence of solid material in the electrode gap enhances the NTP efficiency likely by favoring the formation of homogeneous plasma rather than a filamentous one. Eliasson et al. reported that the packed materials in NTP reactors played a key role for the proper functioning of DBD and generating more high energy electrons. As the removal efficiency was proportional to the numbers of high energy electrons, because these high energy electrons could destroy the molecular structure of toluene and decompose toluene molecules into CO_2, CO and H_2O by effective collisions taking place between the high energy electrons and the toluene molecules.
2. $Ba_{0.8}Sr_{0.2}Zr_{0.1}Ti_{0.9}O_3$ has higher permittivity than $BaTiO_3$. The electric field strength is positive to the permittivity of packed materials in the NTP reactor.
3. It is well known that the surface specific area is positive to the adsorption properties of the adsorbent. The rings of $BaTiO_3$ and $Ba_{0.8}Sr_{0.2}Zr_{0.1}Ti_{0.9}O_3$ have bigger surface specific area of 59 and 57 or so after the rings are calcined at 1200℃. The adsorption properties can adsorb radicals to accelerate decomposition reaction on the surface of the sorbent and the surface of the cellular rings can also become active sites with electrons striking. So the NTP reactor with packed materials can obtain higher toluene removal efficiency than that without packed materials.

Fig.11 shows the voltage and current waveforms detected by oscillograph with and without packed materials ($Ba_{0.8}Sr_{0.2}Zr_{0.1}Ti_{0.9}O_3$ rings) at electric field strength of 10 kV/cm. Ricketts et al.[16] believed that the pulse peak numbers of gas discharge were directly proportion to the removal efficiency of VOCs. As shown in figure 6, the pulse peak numbers are higher with packed materials than those without in the NTP reactor. So packed materials increase the pulse peak numbers of DBD and help for increasing the removal efficiency.

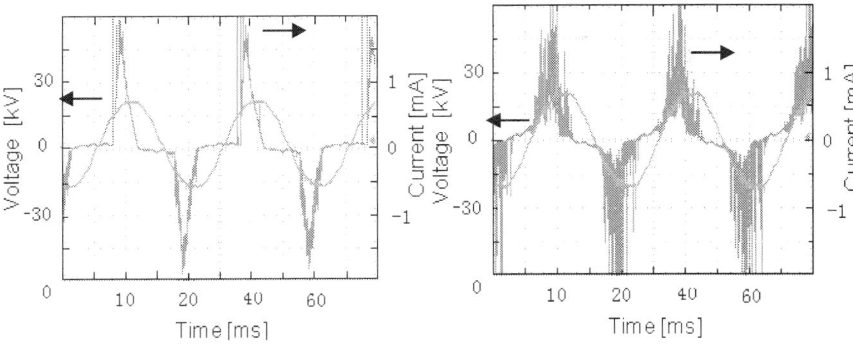

(a) Without packed materials: (b) With packed materials:

Fig. 11. Voltage and current waveforms

Fig.12 shows V-Q Lissajous diagram with or without packed materials ($Ba_{0.8}Sr_{0.2}Zr_{0.1}Ti_{0.9}O_3$ rings) at electric field strength of 10 kV/cm. BC and AD shown in fig.12 represent DBD courses. The packed materials enhance the discharge intensity of BC and AD courses and produce higher the pulse peak numbers. Pulse peak numbers are directly proportion to RED and increase the removal efficiency of toluene.

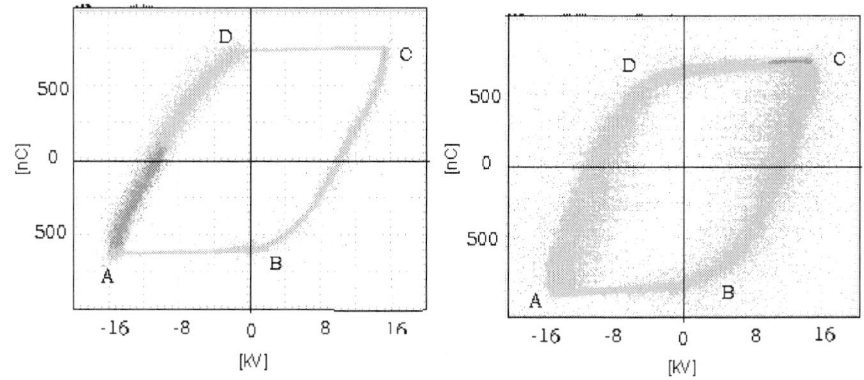

(a) Without packed materials: (b) With packed materials

Fig. 12. V-Q Lissajous diagram

Therefore, the removal efficiency was higher with packed materials than that without packed materials, in agreement with the results shown in fig.10. During the preparation of $Ba_{0.8}Sr_{0.2}Zr_{0.1}Ti_{0.9}O_3$ sample, strontium (Sr) and zirconium (Zr) ions were adulterated into the powder particles and crystal boundary. These metal ions enter crystal lattices of $BaTiO_3$ equably and lower the Curie temperature (Tc). As a result, the permittivity of $Ba_{0.8}Sr_{0.2}Zr_{0.1}Ti_{0.9}O_3$ is 12000, 8 times higher than that of pure phase of $BaTiO_3$ (1500) in room temperature. According to Yamamoto et al., the dielectric constant had a significant influence on the discharge energy of the NTP reactor. The electric field strength is calculated as follows:

$$E_r = \frac{3\varepsilon}{\varepsilon + 2} E_0 \cos\theta \tag{19}$$

Where E_r is the local electric field strength after dielectric polarization, E_0 the local electric field strength before dielectric polarization, and ε the relatively permittivity. As shown in formula (19), E_r is in direct ratio with E_0. E_r equals to 3 times of E_0 (θ=0) with ε close to infinity. So, the electric field strength is positive to the relative permittivity of packed materials in the NTP reactor. RED increases with the electric field strength, improving the removal efficiency of toluene. Therefore, $Ba_{0.8}Sr_{0.2}Zr_{0.1}Ti_{0.9}O_3$ leads to better the removal efficiency of 97% for toluene decomposition.

3.4 Effect of $Ba_{0.8}Sr_{0.2}Zr_{0.1}Ti_{0.9}O_3$ on energy efficiency

Fig.13 shows the change of energy efficiency (ζ) for toluene removal with and without the packed materials. At the identical RED, the energy efficiency is in the order of without packed materials < with $BaTiO_3$ < with $Ba_{0.8}Sr_{0.2}Zr_{0.1}Ti_{0.9}O_3$. The energy efficiency is 15 g/kWh with $Ba_{0.8}Sr_{0.2}Zr_{0.1}Ti_{0.9}O_3$, 11 g/kWh with $BaTiO_3$, and 6 g/kWh without packed materials at RED of 0.23 kJ/L in the NTP reactor. The results show that $Ba_{0.8}Sr_{0.2}Zr_{0.1}Ti_{0.9}O_3$

has a better ferroelectric property to improve energy efficiency and reduce energy consumption in the NTP process for VOCs control, compared with $BaTiO_3$.

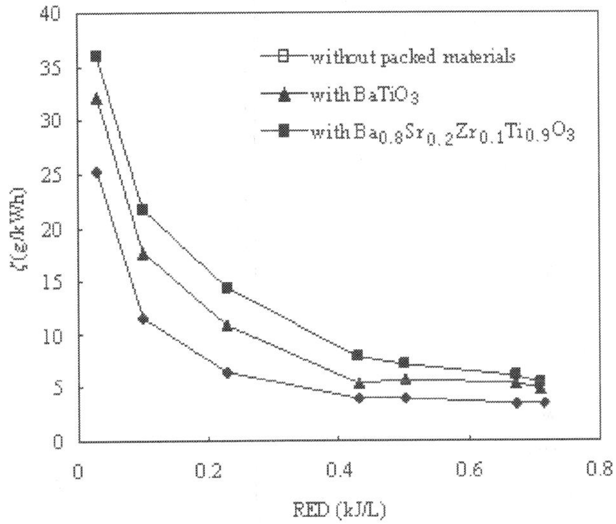

Fig. 13. Effect of packed materials on energy efficiency (toluene concentration: 1000mg/m^3 or so; gas flow rate: 2L/min; AC frequency: 150Hz)

3.5 Effect of $Ba_{0.8}Sr_{0.2}Zr_{0.1}Ti_{0.9}O_3$ on ozone formation

Fig.14 shows the ozone (O_3) concentration with and without the packed materials. O_3 concentration is the highest with $Ba_{0.8}Sr_{0.2}Zr_{0.1}Ti_{0.9}O_3$ and is in the order of without the packed materials < with $BaTiO_3$ < with $Ba_{0.8}Sr_{0.2}Zr_{0.1}Ti_{0.9}O_3$ at the identical RED.

O_3 as the main long-living radical was generated and transported to the packed materials and could take part in oxidation reaction on the packed materials' surface. The pathways of reaction were stated as follows:

$$e + O_2 \rightarrow 2O + e \tag{20}$$

$$O + O_2 + M \rightarrow O_3 + M \tag{21}$$

$$O + O_3 \rightarrow 2O_2 \tag{22}$$

$$e + O_3 \rightarrow O + O_2 + e \tag{23}$$

In figure 9, it also shows that O_3 concentration increases with the REDs at the first stage from 0 to 0.7 kJ/L and reaches the maximum at the RED of 7 kJ/L or so. This pattern of ozone production had also been reported by Yamamoto et al. In this experiment, because $Ba_{0.8}Sr_{0.2}Zr_{0.1}Ti_{0.9}O_3$ has higher relative permittivity than $BaTiO_3$, the electric field strength and RED are enhanced significantly in the NTP process with $Ba_{0.8}Sr_{0.2}Zr_{0.1}Ti_{0.9}O_3$ as the

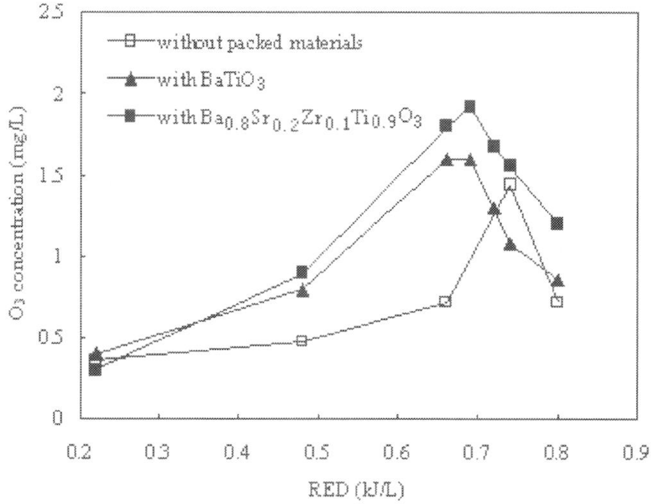

Fig. 14. Effect of packed materials on O_3 concentration (toluene concentration: $1000mg/m^3$ or so; gas flow rate: 2L/min; AC frequency: 150Hz)

packed materials. As a result, O_3 concentration increases according to Equation (21) (RED ≤ 0.7 kJ/L). While RED ≥ 0.7 kJ/L, the superfluous high-energy electrons accelerate the decomposition of O_3 to O_2 according to Equation (22) and Equation (23). The active oxygen species formed during the O_3 decomposition would also be helpful for the toluene removal on the surface of packed materials.

4. Decomposition of benzene in dry air by super-imposed barrier Discharge NonThermal plasma–photocatalytic system

In this section, NTP coupled with nano-titania (TiO_2) photo-catalyst for benzene decomposition to further reducing the energy consumption and harmful byproducts in plasma process.

O_2 and H_2O are adsorbed on the surface of TiO_2 to form adsorption oxygen and adsorption water:

$$O_2(g) \rightarrow O_2(ads) \tag{24}$$

$$H_2O(g) \rightarrow H_2O(ads) \tag{25}$$

The molecules of VOC are also adsorbed on the surface of TiO_2 to form adsorption matter:

$$VOCs \rightarrow (VOCs)_{ads} \tag{26}$$

Discharge plasma as a driving force of photocatalyst furnished a mess of UV light. According to Kim's report, hole-electron pairs are produced by supplying energy larger than the band-gap energy of TiO_2 (3.2 eV for anatase crystal type).

$$TiO_2 + hv \rightarrow TiO_2(e^- + h^+) \quad (27)$$

And then, OH radicals come into being:

$$h^+ + H_2O(ads) \rightarrow {}^{\bullet}OH + H^+ \quad (28)$$

$$h^+ + OH^- \rightarrow {}^{\bullet}OH \quad (29)$$

High-energy particles, such as electrons, excited molecules, and radicals may transfer their energy to TiO_2 by bombardment when TiO_2 is placed in a NTP reactor. Various chemical reactions are induced on the excited TiO_2 surface through the following reactions:

$$e^- + O_2(ads) \rightarrow {}^{\bullet}O_2^-(ads) \quad (30)$$

$$^{\bullet}O_2^-(ads) + H^+ \rightarrow HO_2^{\bullet} \quad (31)$$

$$2HO_2^{\bullet} \rightarrow O_2 + H_2O_2 \quad (32)$$

$$H_2O_2 + {}^{\bullet}O_2^-(ads) \rightarrow {}^{\bullet}OH + OH^- + O_2 \quad (33)$$

At last, the molecules of VOC are decomposed as follow:

$$^{\bullet}OH_{ads}(HO^{\bullet}_{2ads}, O_{ads} \text{ or } h^+, \text{etc.}) + (VOCs)_{ads} \rightarrow (\text{active intermediate products}) \rightarrow\rightarrow\rightarrow CO_2 + H_2O + CO \quad (34)$$

This section illuminates the experimental results of the effect of packed materials on benzene decomposition using NTP generated by dielectric barrier discharge (DBD) coupled with nano-TiO_2 catalyst. The effects of A and B packed materials on benzene removal were compared in the paper. The results show that removal effect was visible by B packed materials in NTP reactor. At the same time, we got higher removal efficiency and a better selectivity of carbon dioxide or carbon monoxide with B packed materials coated with nano-TiO_2. Above all that means this technology of self-support ray polarization would have a great potential for application in the future.

4.1 Materials and methods

There are two kinds of packed materials of A and B (i.d.2, 4, 5, 7 mm, thickness 3 mm, length 10 mm, raschig ceramic ring) in NTP reactor. The structure characteristics of packed materials A and B were detected by XRD as shown in fig.15 (manufactured by Japan, D/MAX-RA). According to fig.15, the results of physics characteristic of two packed materials were indicated in tab.1. Non-crystal content of A packed materials was up to 70% and non-crystal content of B packed materials was 50%. It means B packed materials interstitial rate was higher than A packed materials, and adsorption capability was bigger than A packed materials.

The packed materials coated with nano-TiO_2 catalyst were packed into the reactor. The reactor was made of ceramic tube (i.d.50 mm, reaction length 500 mm), inner axes electrode (i.d.0.8 mm, stainless steel wire), and outer electrode (80 circles stainless steel wire). The characteristic of packed materials includes volume density is 21.7 g/cm^3, hole rate is 12.7 % and bibulous is 5.9 %.

Packed materials	Component (%)			Interstitial rate (%)	Hygroscopic coefficient (%)
	quartz	Al_2O_3	non-crystal		
A	15	15	70	1.8	0.8
B	15	35	50	19.5	9.4

Table 1. Physics characteristic of two packed materials

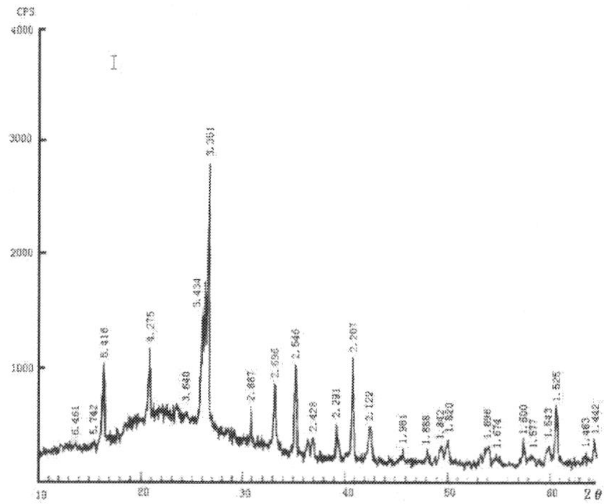

(a) A packed materials

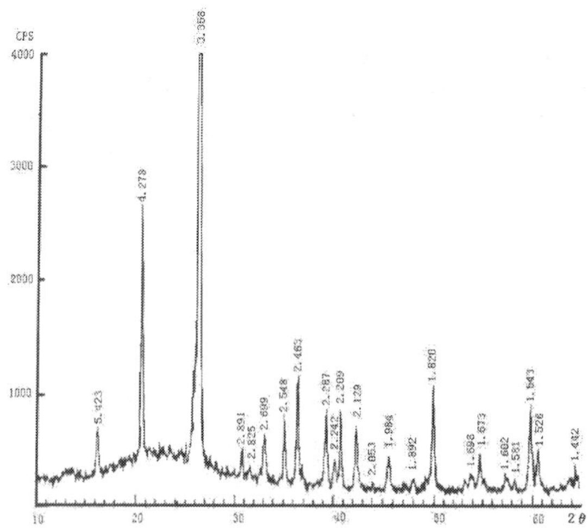

(b) B packed materials

Fig. 15. XRD pattern of packed materials

Nano-TiO_2 films were prepared by the Sol-Gel method in the experiment. Flow chart on preparing nano-TiO_2 thin film by Sol-Gel method referred to Fig.16.

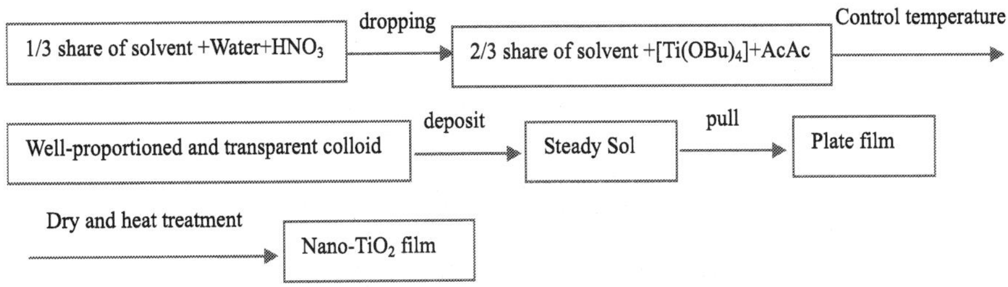

Fig. 16. Flow chart on preparing nano-TiO_2 thin film by Sol-Gel method

The nanometer TiO_2 thin film was inspected and analyzed by Scan Electric Mirror (SEM, Made in Japan, S-2700). The results of SEM micrograph show that average particulate diameters of TiO_2 were less than 100 nm. SEM micrograph of the samples referred to Fig.17.

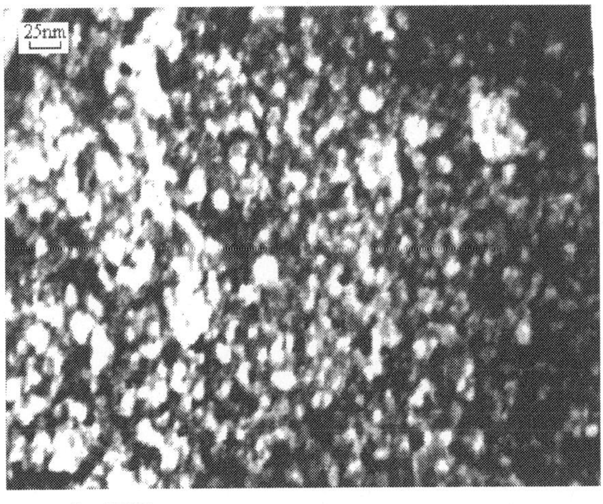

Fig. 17. SEM micrograph of TiO_2

The plasma reactor employed an alternating current (AC) power supply of 60 Hz (designed by ourselves). The AC voltage was applied to the reactor in the radial direction, and the AC voltage extension lied from 0 kV to 30 kV. The benzene concentration was determined on a gas chromatography (manufactured by American Thermo Finnegan Co., TRACE-GC ULTRA) with a flame ionization detector (FID) and a capillary column of DB-1. Separately, another GC (SC—1001) equipped with an FID detector and a methane converter was used to analyze concentration of CO_2 and CO. Reaction gas samples were taken by a syringe from the sampling ports of the reactor. The byproducts were identified by GC-MS with a 30-m-long wide-bore capillary column (DB-1). The experimental condition was in atmospheric pressure (760 mmHg) and temperature (20°C).

4.2 Relationship packed materials between and removal efficiency

Fig.18 & Fig.19 showed the effect of A & B packed materials on removal efficiency of benzene (benzene concentration of 600 mg/m^3, gas flux is 100 L/h, dry air, A & B packed materials coated with nano-TiO_2). With increasing electric field strength, the removal efficiency of benzene increased. During an impulse cycle in NTP reactor, a mass of high-energy electrons were produced in discharge space. When effective collisions between high energy electrons and benzene molecules took place in NTP reactor, electron energy would destruct molecular structure of benzene and benzene molecules could be converted into inorganic little molecules like carbon dioxide (CO_2), carbon monoxide (CO) and water (H_2O) at last. Thus, removal efficiency of benzene was proportional to the number of electrons, while the electrons' number was positive to electric field strength. It had come to light that TiO_2 was helpful for generating higher concentrations of different types of active oxygen species in non-thermal plasma. So the hybrid system would have an effective utilization of active oxygen species in benzene removal.

As shown in fig.18, removal efficiency with B packed materials (i.d.2 mm) was higher than that with A packed materials (i.d.2 mm). Firstly, compared with the component of A & B packed materials, non-crystal type being in existence had influence on dielectric polarization. Secondly, B packed materials possessed higher interstitial rate and hygroscopic coefficient and bigger surface area. These factors were helpful for benzene molecule adsorption so that they prolonged reaction time between benzene molecule and high electrons or free radicals. Thirdly, the surface of B packed materials was rough and could assemble more polarization electric charge to form more local electric field, so that electrons in NTP gained higher energy to improve reaction efficiency.

As shown in fig.19, the size of B packed materials (i.d.2, 4, 5, 7 mm) had effect on removal efficiency of benzene. Removal efficiency increased with the size of raschig ceramic ring decreasing. The removal efficiency was an order of i.d.2 mm > i.d.4 mm > i.d.5 mm > i.d.7 mm. The removal efficiency with B packed materials of i.d.2 mm was 81% with electric field

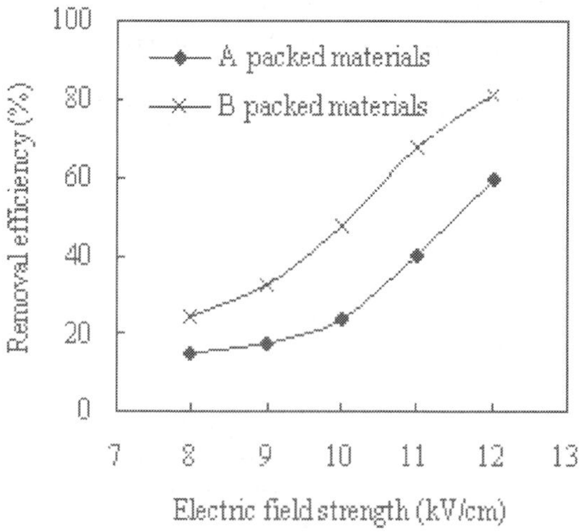

Fig. 18. Effect of different packed materials on removal efficiency

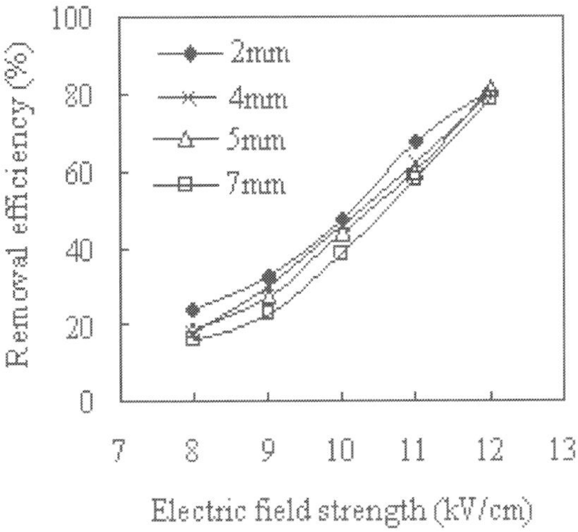

Fig. 19. Effect of different diameters of B packed materials on removal efficiency

strength of 12 kV/cm. With the size of raschig ceramic ring decreasing, packed materials density increased and interspace between packed materials reduced. These factors were helpful for dielectric polarization and enhanced electric field strength. The same result was gained by Ogata *et al.*

4.3 Relationship between packed materials and ozone concentration

Fig.20 and fig.21 showed ozone concentration increased with increasing electric field strength (benzene concentration of 600 mg/m^3, gas flux of 100 L/h, dry air, A & B packed materials coated with nano-TiO_2). When electric field strength increased, more high-energy electrons and radicals were generated in the early discharge phase. They possessed high energy compared to the dissociation energy of O_2 so that a series of reaction took place in NTP reactor. The oxygen dissociation was the most important radical formation reaction.

$$e + O_2 \rightarrow e + O + O \tag{35}$$

Electronically excited atomic oxygen ($O(^1D)$) was a very short-lived radical, whereas ground state atomic oxygen (O) and hydroxyl (OH) had a longer lifetime. $O(^1D)$ reacted with H_2O resulting in formation of OH radicals. O^- and OH^- radicals were consumed by O_3 formation. O_3 as the main long-living radical was transported to packed materials and could take part in oxidation reaction on packed materials' surface. The pathways of reaction were stated as follows:

$$O + O_2 + M \rightarrow O_3 + M \tag{36}$$

$$O + O_3 \rightarrow 2O_2 \tag{37}$$

$$e + O_3 \rightarrow O + O_2 + e\,. \tag{38}$$

As shown in Fig.20, ozone concentration with A packed materials (i.d.2 mm) was higher than that with B packed materials (i.d.2 mm) in NTP reactor. Because the adsorption capability of B packed materials in NTP reactor was higher than that of A packed materials. It meant B packed materials would be helpful for decreasing ozone concentration though prolonging reaction time on the surface of B packed materials followed equation (37) to (38). As shown in fig.21, ozone concentration was an order of B packed materials of i.d.2 mm > i.d.3 mm > i.d.5 mm > no padding. It was obvious that B packed materials was helpful for increasing ozone concentration. The reaction took place just like equation (36).

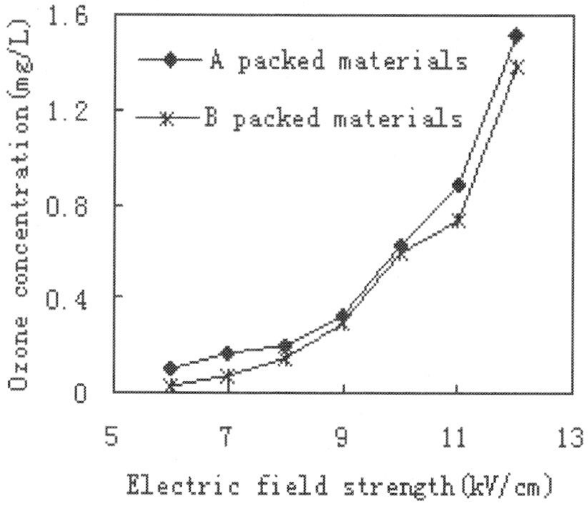

Fig. 20. Effect of different packed materials on ozone concentration

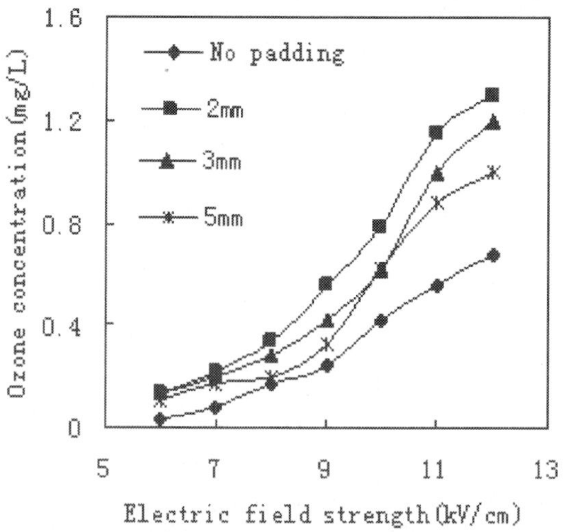

Fig. 21. Effect of different diameters of B packed materials on ozone concentration

4.4 Relationship between power consumption and removal efficiency

Fig.22 showed the biggest electric power consumption was up to 120 W with electric field strength changing from 0 to 12 kV/cm with B packed materials coated with nano-TiO_3. Kuniko Urashima *et al.* got the similar conclusion at the same experimental conditions.

Fig.23 showed that power consumption was 13.5 W if removal efficiency was up to 85% and benzene concentration was 600 mg/m^3. Through calculating, electric energy consumption was 2.25×10^{-4} kWh to treat benzene quality of 1 mg. If benzene concentration was 1500 mg/m^3, power consumption was 20 W for the same removal efficiency. Through calculating, electric energy consumption was 1.33×10^{-4} kWh to treat benzene quality of 1 mg.

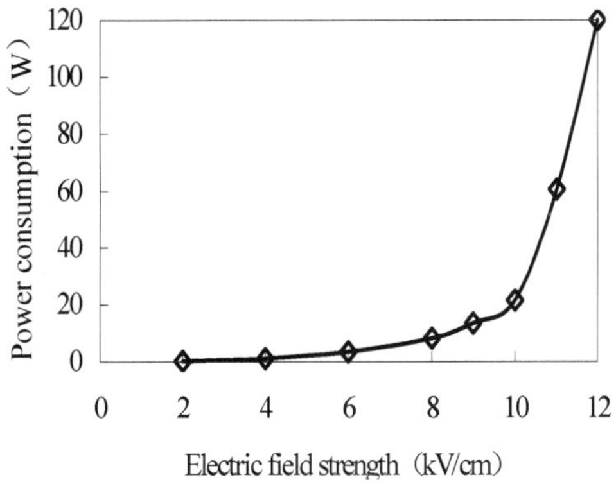

Fig. 22. Relationship between electric field strength and power (B packed materials coated with nano-TiO_2)

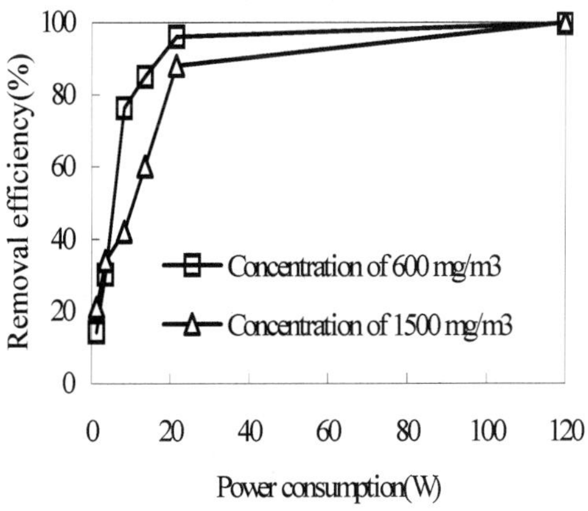

Fig. 23. Relationship between power consumption and removal efficiency (B packed materials coated with nano-TiO_2)

4.5 Photocatalyst and ozone effect

In the next experiment, we chose B packed materials. The packed materials(5 mm i.d., 1 mm wall thick, 10 mm length) were divided into two groups, coated with photocatalyst or without photocatalyst.

Fig.24 shows the relationship between removal efficiency of benzene and electrostatic field strength in the plasma reactor with or without packed materials.

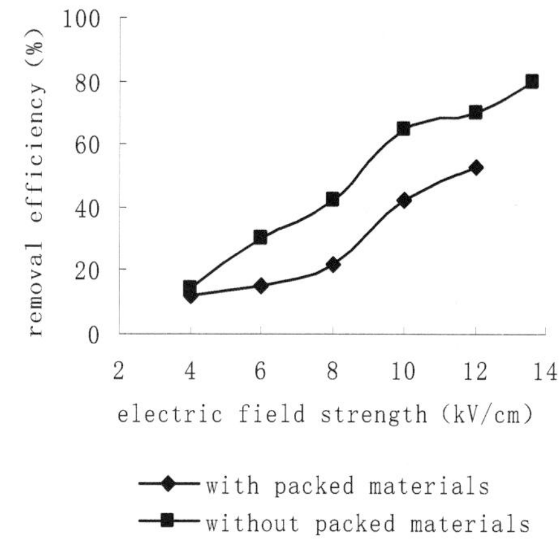

Fig. 24. The effect of removal efficiency with or without packed materials

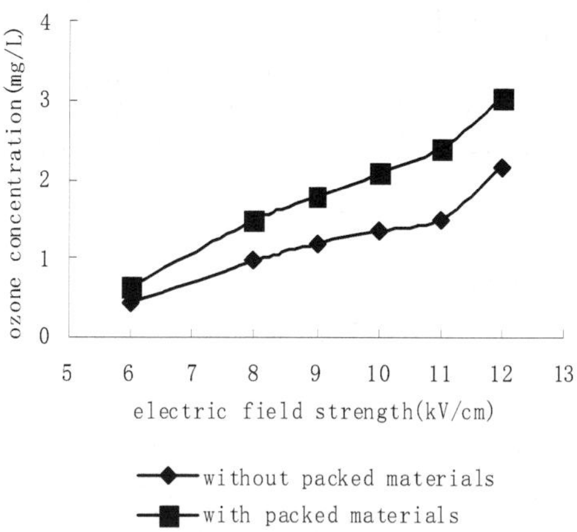

Fig. 25. The effect of ozone concentration with or without packed materials

Fig. 25 shows the relationship between ozone concentration and electrostatic field strength in the plasma reactor with or without packed materials.

With increasing electrostatic field strength, the removal efficiency of benzene increases. When initial concentration of benzene is 1300 mg/m^3, the average electrostatic field strength is 13.6 kV/cm and gas flux is 100 L/h, the removal efficiency of benzene arrives at 80% in the reactor with packed materials as shown in Fig.24.

In the reactor, the space occupied by contamination air is always full of high energy electrons. When effective collisions between high energy electrons and benzene molecules take place in the reactor, electron energy will destruct molecular structure of benzene and benzene molecules will be converted into inorganic little molecules like carbon dioxide (CO_2), carbon monoxide (CO) and water (H_2O). Thus, removal efficiency of benzene are proportional to the electrons. In Fig.3, when packed materials placed in the plasma reactor, with electrostatic field strength increasing, more and more high energy electrons are produced due to mediums polarization of packed materials. So packed materials in the reactor increases the removal efficiency of benzene.

Fig.25 shows when initial concentration of benzene is 1300 mg/m^3, the average electrostatic field strength is 12 kV/cm and gas flux is 100 L/h, ozone concentration is about 3.04 mg/L with packed materials and ozone concentration is about 2.16 mg/L without packed materials. Ozone concentration with packed materials heightens 1 mg/L than that without packed materials in the plasma reactor. It is obvious that packed materials in the reactor is helpful of increasing ozone concentration. The reason is high energy electrons and radicals are generated in the early discharge phase. They possess high energy compared to the dissociation energy of O_2 so that a series of reaction takes place in the plasma. The oxygen dissociation is the most important radical formation reaction.

$$e + O_2 \rightarrow e + O + O \tag{39}$$

Electronically excited atomic oxygen ($O(^1D)$) are very short-lived radicals, whereas ground state atomic oxygen (O) and hydroxyl (OH) have a longer lifetime. $O(^1D)$ reacts with H_2O resulting in formation of OH radicals. O^- and OH^- radicals are removed by formation of O_3.

In Fig.24 and Fig.25, the test results also indicate ozone is helpful for benzene removal, at least, ozone acts as the oxidant precursor. With ozone concentration increasing, the removal efficiency of benzene increases. Because ozone as a kind of oxidative species produced by the initial oxidation just like OH radical, has an effect on further reaction of benzene.

The humidities of contaminated air in the reactor have influence on ozone concentration as shown in Fig.26a. Ozone concentration without vapor is higher 35% than that with relative humidity 67%, and ozone concentration decreases with humidity increasing. Because H_2O molecule have electronegative, it will consume the electrons in the plasma. At the same time, H_2O will react with O (^1D) which is the origin of formation of O_3.

$$H_2O + O(^1D) \rightarrow 2OH \tag{40}$$

So humidity counteracts the formation of ozone.

In Fig.26b, the findings show the removal efficiency reduces with humidity increasing. Probably, the active sites for benzene removal are reduced by water vapor occurring competitive adsorption on the surface of nano-TiO_2.

Thus, humidity affects the benzene removal in three ways: deactivation of high energy electrons, inhibition of ozone formation, and suppression of the catalyst activity of nano-TiO_2 for benzene oxidation with ozone in the plasma reactor.

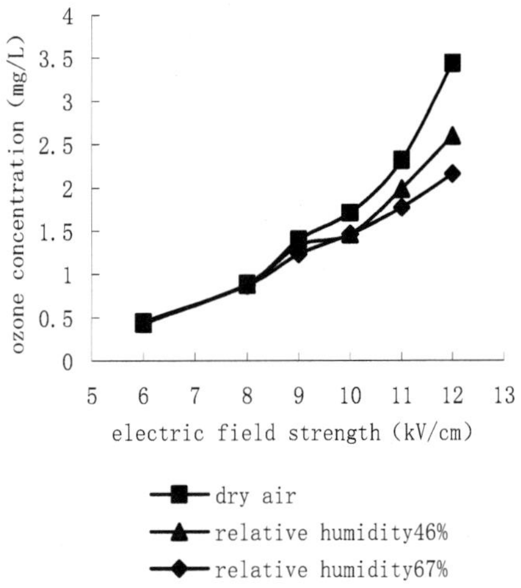

Fig. 26a. The effect of humidity on ozone concentration without photocatalyst

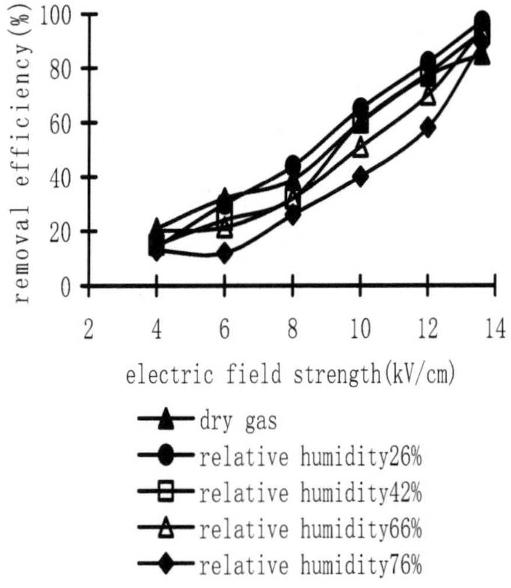

Fig. 26b. The effect of humidity on removal efficiency with photocatalyst

Fig.27 shows the relationship between ozone concentration and gas flux with catalyst or without catalyst in the plasma reactor under three initial concentration of benzene. When gas passes the reactor, and electrostatic field strength is 10kV/cm, ozone concentration increases with gas flux increasing as shown in Fig.27a, 27b and 27c, regardless of with or without photocatalyst.

Fig.27d and 27e show benzene concentration reduces with initial concentration of benzene increasing, regardless of with or without photocatalyst.

The relationship between benzene degradation and electrostatic field strength with or without photocatalyst is shown in Fig.28 where benzene initial concentrations changes from 600 mg/m^3 to 1500 mg/m^3.

When initial concentration is 600 mg/m^3, the average electrostatic field strength is 10 kV/cm, and gas flow rate is 14 mm/s, the removal efficiency of benzene attains 98% in the reactor with photocatalyst, but the removal efficiency of benzene attains 78% in the reactor

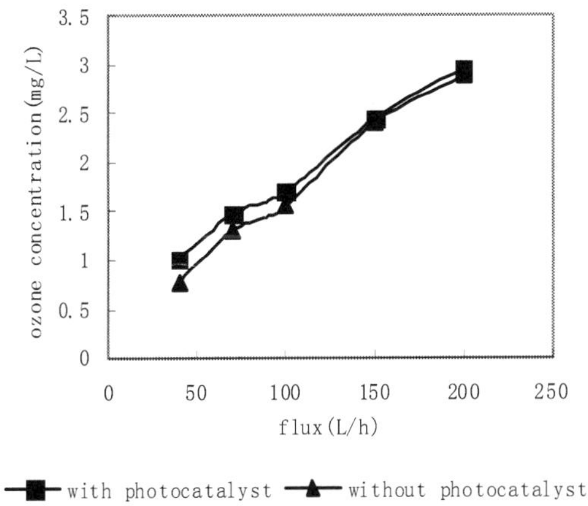

Fig. 27a. Relationship between ozone concentration and flux with or without catalyst when benzene concentration is 0mg/m^3

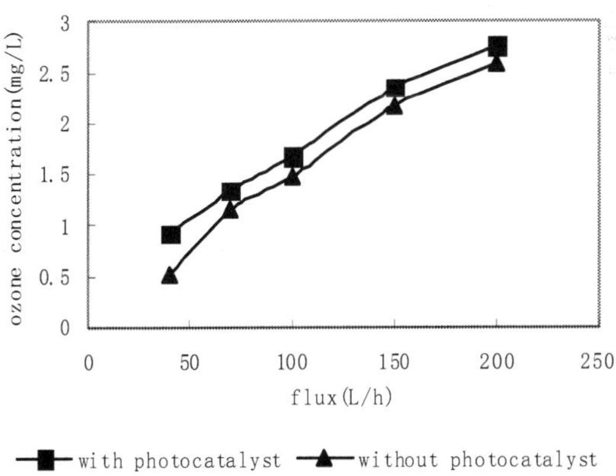

Fig. 27b. Relationship between ozone concentration and flux with or without catalyst when benzene concentration is 700mg/m^3

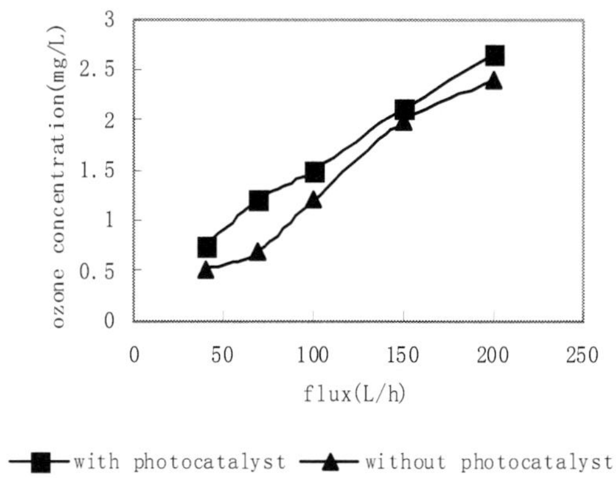

Fig. 27c. Relationship between ozone concentration and flux with or without catalyst when benzene concentration is 2000mg/m³

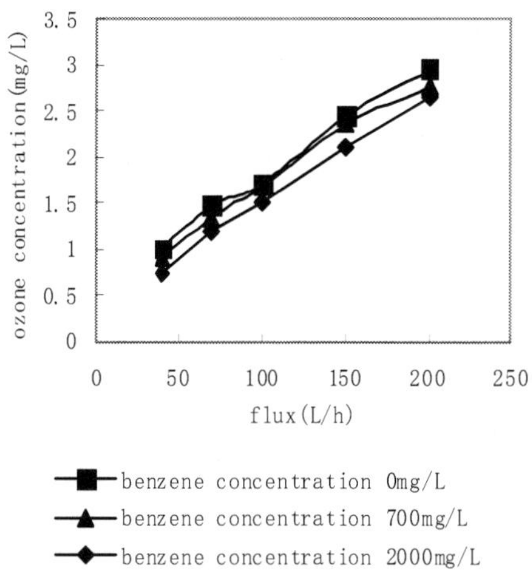

Fig. 27d. Relationship between ozone concentration and flux with catalyst

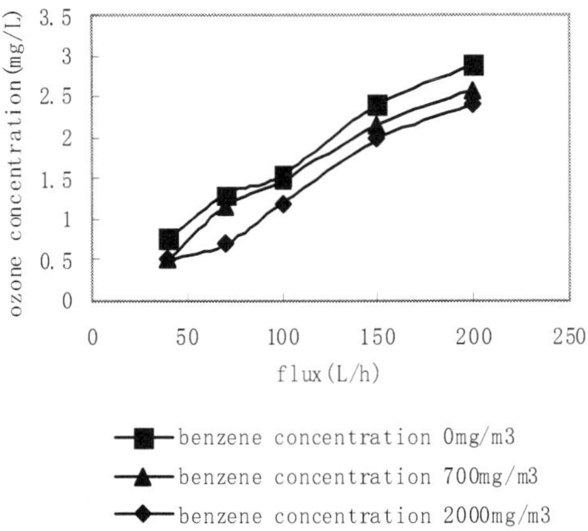

Fig. 27e. Relationship between ozone concentration and flux without catalyst

without photocatalyst as shown in Fig.28a. When initial concentration is 1500 mg/m^3, the average electrostatic field strength is 12 kV/cm and gas flow rate is 14 mm/s, the removal efficiency is higher 19% with photocatalyst than without photocatalyst in the plasma reactor as shown in Fig.28b. The results indicate photocatalyst enhanced the benzene removal efficiency obviously with ozone. When both photocatalyst and ozone coexist, there will be an improved removal efficiency of benzene in the plasma reactor. As you know, TiO_2 is a photocatalyst material of 3.2 eV band gap. If it absorbs bigger energy than band gap, it makes photo-excited electron-hole pairs that could oxidize benzene. At same time, the surface hydroxyl groups are oxidized to form composition of benzene in the photocatalytic reactions. So we have thought that it was advantageous to use photocatalyst in plasma system to control of oxidation step of benzene.

The influence of the catalyst on ozone formation is presented in Fig.27a, 27b and 27c. It shows the catalyst could reduce the ozone formation to a certain extent. This is because ozone as the main long-living radical can capture free electrons which are produced by photocatalysis and produce OH radical. It not only avoids hole-electron pairs compounding but increases photons efficiency. Further more, OH radical is a kind of good oxidant and it can transform organism into mineral.

$$O_3 + e^- \rightarrow O_3^- \bullet \tag{41}$$

$$H^+ + O_3^- \bullet \rightarrow HO_3 \bullet \tag{42}$$

$$HO_3 \bullet \rightarrow O_2 + \bullet OH \tag{43}$$

$$O_3 + hv \rightarrow \bullet O + O_2 \tag{44}$$

$$\bullet O + H_2O \rightarrow 2 \bullet OH \tag{45}$$

From (41)~(45), we can arrive at conclusions. On the one hand, ozone increases photons efficiency of photocatalusis so that is helpful of benzene removal. On the other hand, photocatalyst promotes ozone to separate into OH radical and reduces ozone output.Complete oxidation of benzene to CO_2 is the final goal of the experiment, and the catalytic effect on the product distribution had been investigated. Photo-oxidation activity should be expressed as selectivity to CO_2 because other byproducts except CO_2 were emitted in plasma process.

4.6 The compare of photocatalyzed characteristic by different heat treatment

The packed materials with phot ocatalyst were to calcine at 450°C, 600°C and 700°C in the muffle. Then, they were packed into the plasma reactor. The samples of packed materials were detected by X-Ray and testified that nano-TiO_2 was anatase at 450 °C and nano-TiO_2 was mixture of anatase and rutileat at 600°C and nano-TiO_2 was rutile mostly at 700°C. The experimental results are shown in Fig.28.

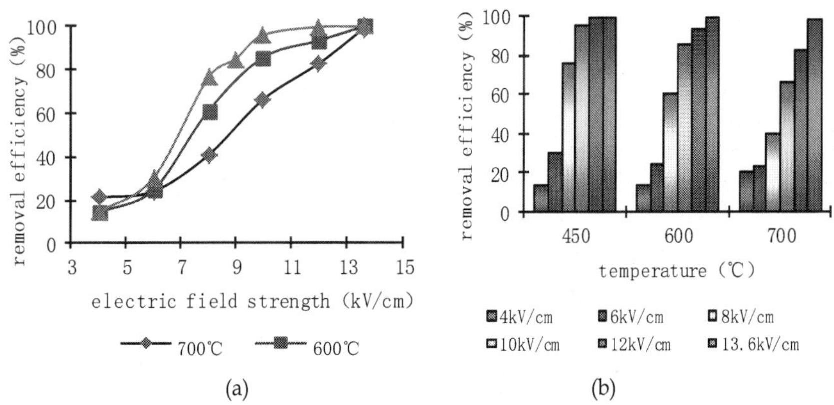

Fig. 28. Compare with catalyst characteristic of different treatment temperature

The decomposition efficiency of benzene is best with anatase photocatalyst in the plasma reactor. Next, the decomposition efficiency is better with the mixture photocatalyst of anatase and rutileat, and the last to rutileat photocatalyst. The above all are shown in Fig.28(a).

Fig.28(b) shows the decomposition efficiency of benzene reduces gradually when the packed materials with photocatalyst by heat treatment from 450°C to 700 °C. On one hand, when sinter temperature is raised, the surface areas of catalyst reduce, and the surface adsorption capacity decrease. On the other hand, nano-TiO_2 catalyst will transform from anatase to rutileat. In fig.28(a), the test shows the reaction activity of anatase catalyst is higher than that of rutileat catalyst. There are four reasons.

a. Because of structural difference, two type of catalyst have different quality densities and different structure of energy gap of electron. The quality density of anatase of 3.894g/cm^3 is less than that of rutileat of 4.250g/cm^3. The energy of energy gap of anatase who is 3.2 eV is higher than that of rutileat who is 3.1 eV. The higher energy of energy gap leads to the higher reaction activity for catalyst.

b. The surfaces of anatase possess symmetrical structure with the molecular structure of benzene, so it can adsorb benzene effectively.
c. The hydroxyl of surface of rutileat is not more than that of anatase. Because the hydroxyl of surface is helpful for benzene removal, anatase is better than rutileat on benzene degradation.
d. The surface area of rutileat catalyst declines sharply because a large number of particals converge under high temperature. The adsorption capacity of rutileat of TiO_2 is bad for O_2, so catalyst activity is low.

So nano-TiO_2 photocatalyst of anatase crystal was employed for next experiment.

4.7 Analysis of reaction products

Though GC-MS, the main products in the plasma reactor were CO_2, H_2O, and a small quantity of CO. Ozone was the only byproduct, and no other byproducts could be detected in the tail gas. In addition, certain brown-yellow products that were observed in the plasma reactor regardless of with or without catalyst appeared. The composition of the brown-yellow products was indistinct, and maybe it was aromatic polymer detected by GC-MS.
The minimum of CO/CO_2 is 0.286 and CO/CO_2 decreases with electrostatic field strength increasing as shown in Fig.29a&b. There are no products except CO_2 and H_2O at 11 kV/cm and 12 kV/cm. CO/CO_2 of byproducts is lower 8.2% with catalyst than that without catalyst. These findings show the plasma reactor packed with materials with catalyst has a better selectivity of CO_2 than that without catalyst.

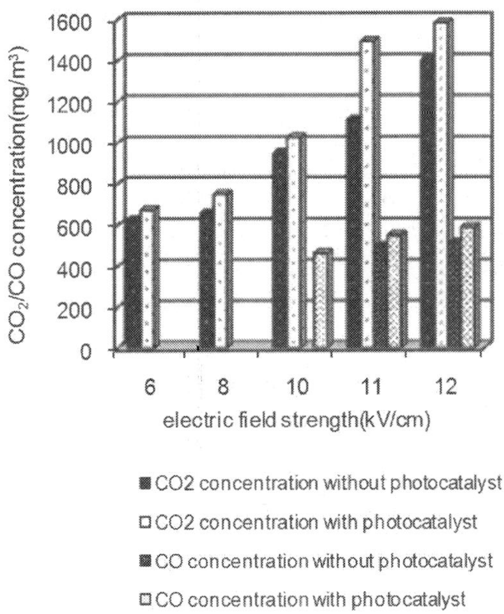

Fig. 29a. Results of byproducts detection when benzene concentration is 750mg/m³ with or without catalyst

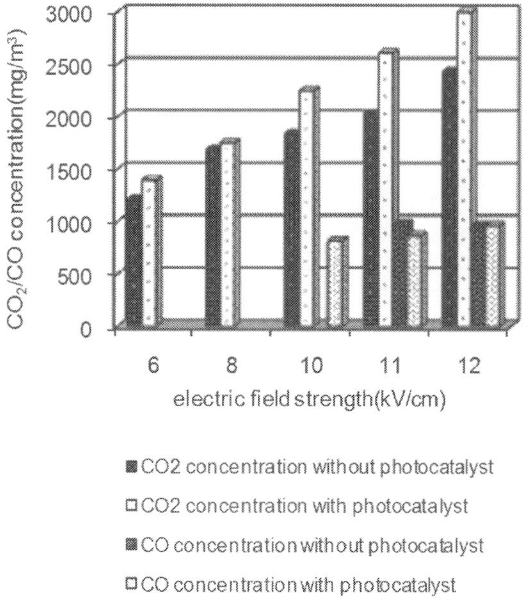

Fig. 29b. Results of byproducts detection when benzene concentration is 1500mg/m^3 with or without catalyst

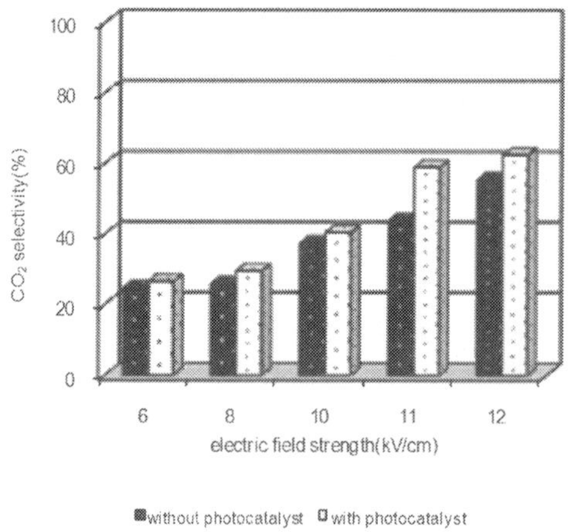

Fig. 29c. CO_2 selectivity when benzene concentration is 750mg/m^3 with or without catalyst

From Fig.29c, it is found the selectivity of CO_2 ranges from 65% to 69% in the plasma reactor without catalyst, while the selectivity of CO_2 ranges from 68% to 73% in the reactor with catalyst. The selectivity of CO_2 is independent of electrostatic field strength. The selectivity of CO_2 is enhanced due to the benzene oxidation near or on the photocatalyst surface. For that, it could be thought that intermediates and secondary products are more oxidized to CO_2 on photocatalyst surface. With benzene concentration increasing, the total output of CO_2 increases.

Fig.30 showed the change of CO_2 and CO selectivity in NTP reactor with nano-TiO_2 catalyst. The CO_2 and CO selectivity were 61% and 30%, while removal efficiency was 94% at electric field strength of 12 kV/cm, benzene concentration of 1500 mg/m^3 and gas flux of 100 L/h. According to calculating, the total carbon was up to 91%, close to removal efficiency of 94%. It is obvious that NTP coupled with nano-TiO_2 catalyst resulted in a higher CO_2 selectivity and a more thorough removal effect in NTP processing, i.e. the final reaction products were almost CO_2, CO and H_2O.

Fig.31 showed the reaction products detected by GC-MS. According to GC-MS patterns, the main reaction products were CO_2, CO and H_2O, including a very little mass of aldehyde, ketone, acylamide and acetic acid, etc.

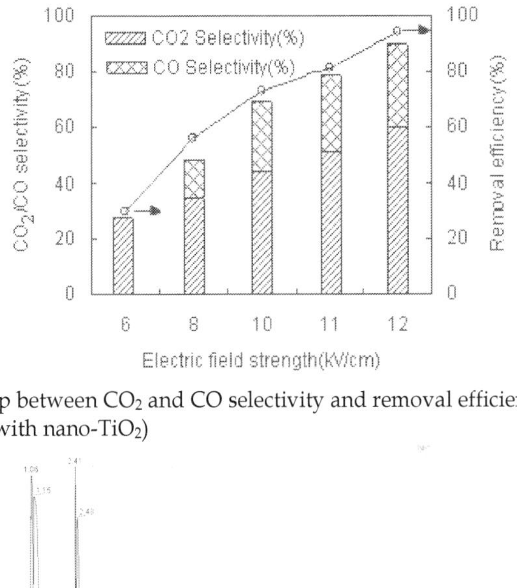

Fig. 30. Relationship between CO_2 and CO selectivity and removal efficiency (B packed materials coupled with nano-TiO_2)

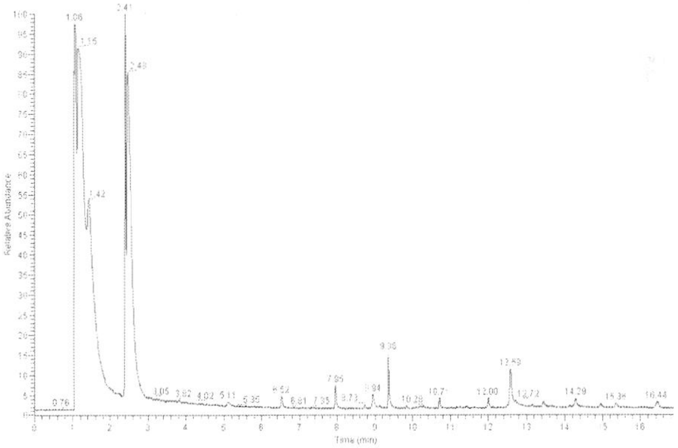

Fig. 31. GC-MS patterns of reaction products (electric field strength of 10 kV/cm, benzene concentration of 1500 mg/m^3, gas flux of 100 L/h, B packed materials of i.d.2 mm coated with nano-TiO_2)

As shown in fig.29-31, the reaction mechanism could be speculated as follows.
Free radicles formation in the surface of nano-TiO_2:

$$\text{plasma} + O_2 \longrightarrow 2O\cdot \tag{46}$$

$$\text{plasma} + H_2O \longrightarrow H\cdot + \cdot OH \tag{47}$$

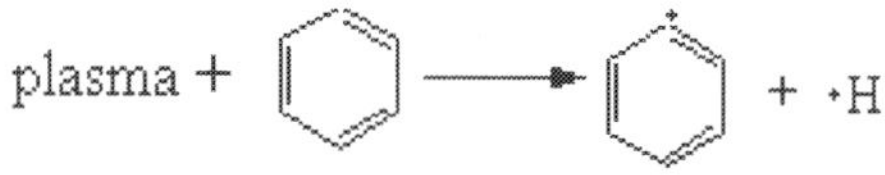

(48)

$$\text{plasma} + N_2 \longrightarrow 2N\cdot \tag{49}$$

$$\text{plasma} + CO_2 \longrightarrow O\cdot + \cdot CO \tag{50}$$

The reaction between free radicles and benzene molecules:

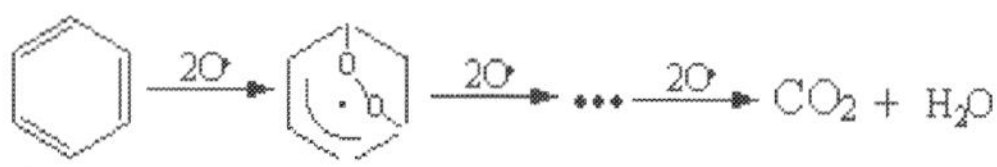

(51)

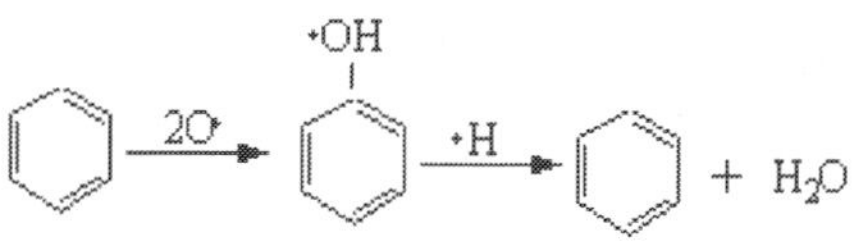

(52)

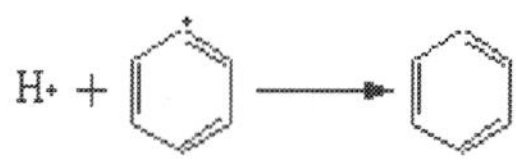

(53)

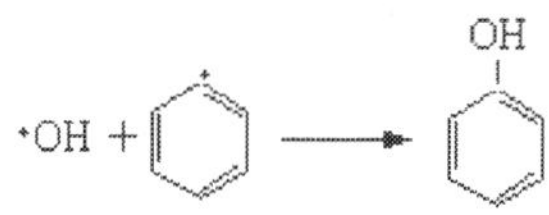

(54)

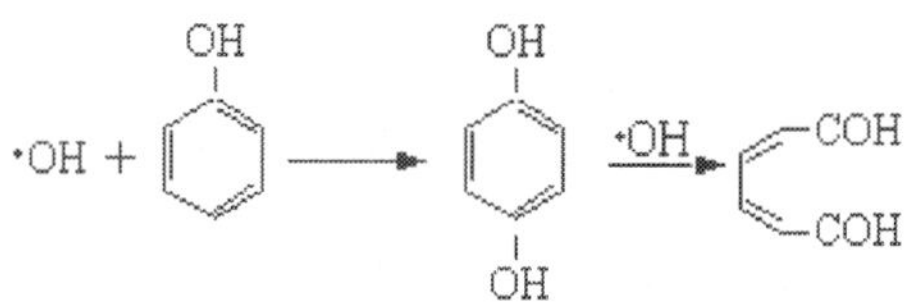

(55)

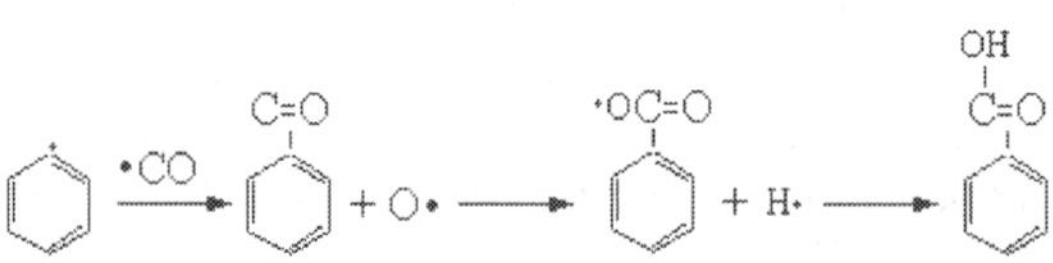

(56)

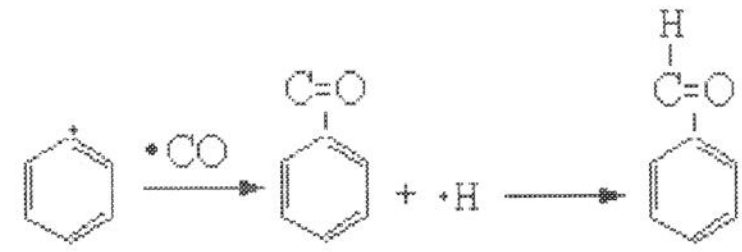
(57)

(58)

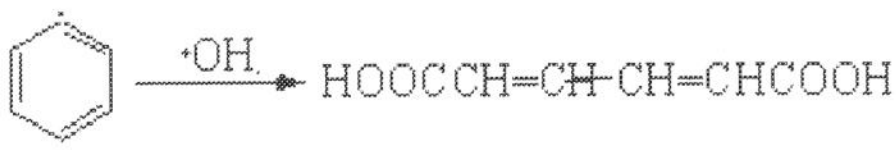
(59)

$$CH_3COOH \rightarrow CH_2{=}C{=}O \begin{cases} \rightarrow CH_3COOH \\ \rightarrow CH_3COOOCCH_3 \\ \rightarrow CH_3COOC_2H_5 \end{cases} \quad (60)$$

(61)

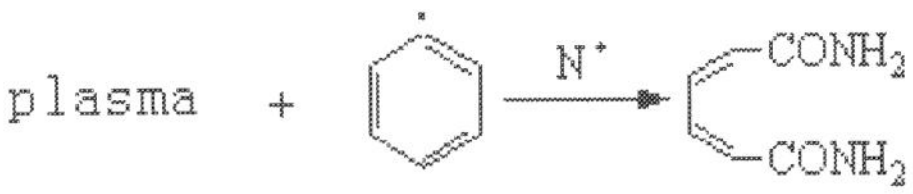
(62)

(63)

At last, the final reaction products should be CO_2, CO and H_2O if there was enough energy in NTP reactor from AC high voltage.

5. Synergistic effect of catalyst for oxidation removal of toluene

In our opinion, the best way for removing VOCs is by a combination of plasma and catalytic treatments. Ayrault et al. used platinum (Pt)-based catalyst supported on an alumina wash-coated honeycomb monolith by means of a high voltage bipolar pulsed excitation. The energy removal efficiency of 2-Heptanone was of 0.14 mol/kWh at an energy density of 200 J/L. They also found that energy efficiency decreases to 0.029 mol/kWh when using an uncoated monolith even at an energy density of 500 J/L. Sekiguchi studied degradation of VOCs with an ozone-decomposition catalyst under conditions of UV irradiation. They found that TiO_2 has a lower VOCs removal ratio without UV irradiation and MnO_2-based catalyst has higher VOCs removal ratio at a higher H_2O humidity. MnO_2 is a catalyst for the decomposition of ozone. The researches on the synergistic action of NTP and catalyst have been carried out for more than 10 years, but there are only a few researches, which have been done adequately, involving byproducts (e.g. O_3) and showing energy efficiency, especially those using MnO_2/γ-Al_2O_3 in AC of 50~500Hz.

In the present work, the synergistic effect of NTP and catalyst for VOCs removal has been studied. It was found that the catalyst could solve the problem of O_3 formation and improve energy efficiency and at the same time increase significantly the removal efficiency of NTP decomposition.

5.1 Materials and methods

VOCs decomposition was studied by using a manganese - alumina catalyst. The manganese catalyst was prepared by intrusion of pellet type alumina, with granule diameter of being 5 - 7 mm and BET (Brunauer Emmett Teller) surface of 228 m^2/g (Detected by Micromeritics, Amercian Quantachrome Co., NOVA 1000).

The specific surface of the catalysts (as determined by the BET method), is given in table 2. BET surface area has not changed too much with catalysts onγ-Al_2O_3.

Catalyst	BET Surface Area(m^2/g)
γ-Al_2O_3	228
TiO_2/γ-Al_2O_3	203
5wt%MnO_2/γ-Al_2O_3	218
10wt%MnO_2/γ-Al_2O_3	202

Table 2. BET surface areas of the catalysts

5.2 Effect of mass percentage of MnO_2 on γ -Al_2O_3

Fig.32 shows the effect of mass percentage of MnO_2 on removal efficiency of VOCs (VOCs concentration: 1000mg/m^3 or so; gas flow rate: 2L/min; AC frequency: 150Hz). The removal efficiency of MnO_2/γ-Al_2O_3 increased with increasing RED and was of about 10 wt% or 15 wt% >5 wt% at the same RED. However, the removal efficiency of 10 wt% was practically equal to that of 15 wt% at the same RED.

Fig.33 shows the effect of the mass percentage of MnO_2 catalyst in NTP reactor on the concentration of VOCs and ozone in the gas exhaust (RED: 0.5 kJ/L). As the mass percentage of MnO_2 catalyst increased, ozone and VOCs concentrations were diminished, especially in the case of 10 wt% of MnO_2 onγ-Al_2O_3. It is clear that manganese oxides accelerated the decomposition of O_3 to O_2 in gas phase. The active oxygen species formed during the O_3 decomposition must be helpful for VOCs removal by MnO_2/γ-Al_2O_3.

5.3 Effect of catalyst on removal efficiency and energy efficiency

VOCs removal efficiency is shown in fig.34 as a function of RED with or without TiO_2 or MnO_2 on γ-Al_2O_3. The removal efficiency increased with increasing RED and was in the order of MnO_2/γ-Al_2O_3 > TiO_2/γ-Al_2O_3 > γ-Al_2O_3 at the same RED. It was obvious that MnO_2 and TiO_2 played a role in VOCs oxidation to a certain extent. γ-Al_2O_3 possessed sorbent characteristic, so it could improve VOCs concentration on the catalyst surface and increase the reaction time. MnO_2 is known as metal oxide catalyst and was found to possess a potential activity in redox reactions. MnO_2 surface has been found to expose metal (Mn^{n+}), oxide (O^{2-}) and defect sites of various oxidation states, degrees of coordination unsaturation, and acid and base properties. Furthermore, the d–d electron exchange

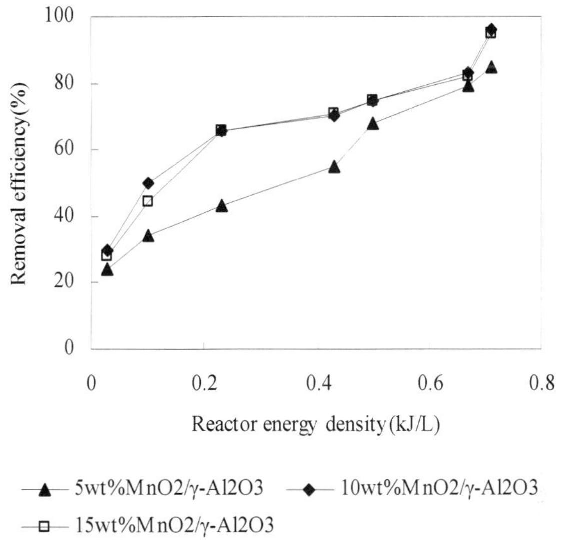

Fig. 32. Effect of mass percentage of MnO_2 on removal efficiency of VOCs

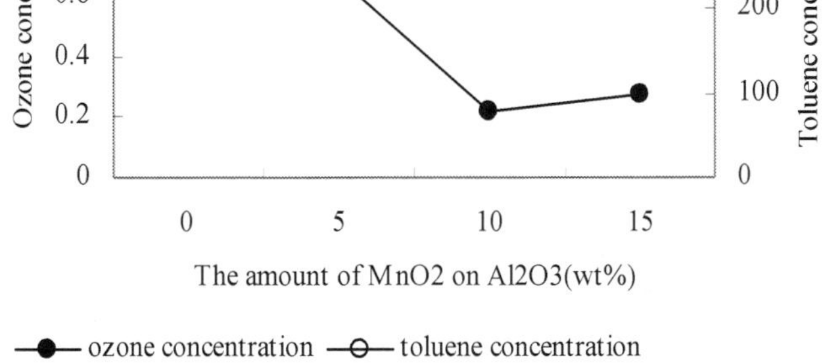

Fig. 33. The changes in the concentration of VOCs and ozone with the amount of MnO_2 on Al_2O_3

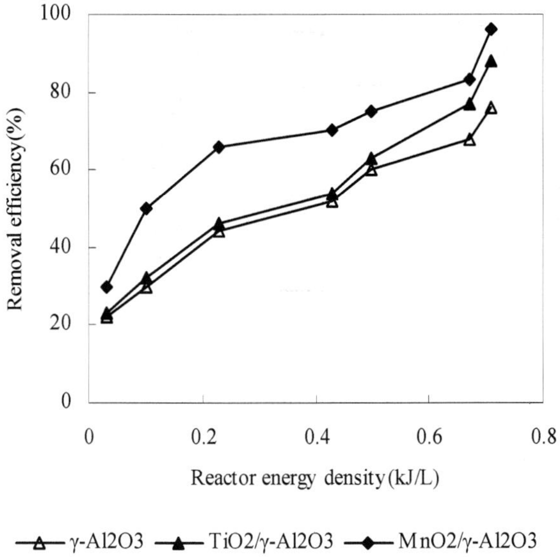

Fig. 34. Effect of RED on removal efficiency with NTP and catalyst combined reactor (10wt% $MnO_2/\gamma\text{-}Al_2O_3$; VOCs concentration: 1000mg/m^3 or so; gas flow rate: 2L/min; AC frequency: 150Hz)

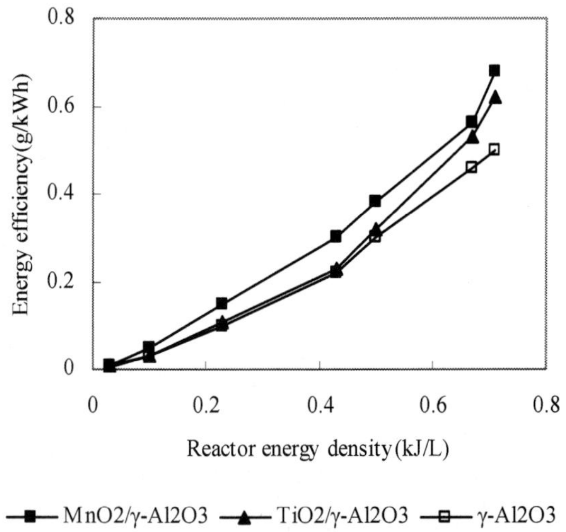

Fig. 35. Relationship between RED and energy efficiency with NTP and catalyst combined reactor (10wt% $MnO_2/\gamma\text{-}Al_2O_3$; VOCs concentration: 1000mg/m^3 or so; gas flow rate: 2L/min; AC frequency: 150Hz)

interactions between intimately coupled manganese ions of different oxidation states [Mn^{n+}-O-$Mn^{(n+1)+}$] furnish the electron-mobile environment necessary for the surface redox activity. These factors would be helpful for VOCs removal.

At the same experimental conditions, the change of energy efficiency is shown in fig.35. The energy efficiency increased with increasing RED and was in the order of MnO_2/γ-Al_2O_3>TiO_2/γ-Al_2O_3>γ-Al_2O_3 at the same RED. The result indicate that NTP coupled with MnO_2/γ-Al_2O_3 catalyst saved more energy to decompose the same amount of VOCs.

6. Synergistic effect of a combination of catalysts with nonthermal plasma

Many researcher found that for VOCs control, ferroelectric could improve energy efficiency significantly, but ozone concentration increased due to ferroelectric presence. Ogata *et al.* investigated the effects of alumina and metal ions in plasma discharge using NTP reactors packed with a mixture of $BaTiO_3$ and porous Al_2O_3 pellets. The results indicated that the oxidative decomposition of benzene was enhanced by concentrating benzene on the Al_2O_3 pellets. The selected catalyst of MnO_2 was well known for high potentials to decompose ozone. Futamura *et al.* tested catalytic effects of TiO_2 and MnO_2 with NTP. The results showed that the ozone generated from gaseous oxygen is decomposed by MnO_2, but not by TiO_2.

A series of experiments were performed for toluene decomposition from a gaseous influent at normal temperature and atmospheric pressure. In this section, the prepared nano-$Ba_{0.8}Sr_{0.2}Zr_{0.1}Ti_{0.9}O_3$ catalyst was used in the plasma reactor. Doped some ions (Sr & Zr) into the powder particles and crystal boundary in the experiment. The metal ions such as strontium, zinc and zirconium entered into crystal lattices of $BaTiO_3$ equably and the Curie temperature (Tc) fell. As a result, the permittivity of nano-$Ba_{0.8}Sr_{0.2}Zr_{0.1}Ti_{0.9}O_3$ was up to 10^4 which were 12 times higher than that of pure $BaTiO_3$, while dielectric loss reduced to 1/6 in normal temperature. This study found that this nano-material could reduce the energy consumption and increase energy efficiency significantly.

The oxidative decomposition of toluene was enhanced by concentrating toluene on the Al_2O_3 pellets. The selected catalyst of MnO_2 was well known for high potential to decompose ozone. In the experiment, the prepared MnO_2/γ-Al_2O_3 was used as catalyst to reduce the byproducts and toluene concentrations --- also justify about 10 wt%. The objective of this study was to use a combination of catalysts (MnO_2/γ-Al_2O_3 coupled with modified ferroelectric of nano-$Ba_{0.8}Sr_{0.2}Zr_{0.1}Ti_{0.9}O_3$) in the NTP process for toluene decomposition in order to enhance toluene decomposition efficiency and increase energy efficiency and reduce byproducts for commercial applications.

6.1 Materials and methods

An alternating current (AC) of 150 Hz was supplied to the NTP reactor in the radial direction, and the voltage extension changed from 0 kV to 50 kV. The experimental parameters of the process of discharge were detected by an oscillograph (model TDS2014, manufactured by American Tektronix Co.). The primary power values were measured with the voltage-charge (V-Q) Lissajous method in the plasma reactor.

Toluene decomposition was studied with a combination of catalysts including MnO_2/γ-Al_2O_3 and nano-$Ba_{0.8}Sr_{0.2}Zr_{0.1}Ti_{0.9}O_3$ catalysts (volume ratio of 1:1). The manganese oxide catalysts (5wt%, 10wt%, 15wt%) were prepared by impregnation of pellet type γ-alumina with the granules diameter of 5~7 mm and BET surface area of 228 m^2/g detected by Micromeritics (model NOVA 1000, manufactured by American Quantachrome Co.).

Nanometer-sized $Ba_{0.8}Sr_{0.2}Zr_{0.1}Ti_{0.9}O_3$ powders were prepared with inorganic salts, such as $TiCl_4$ and $Ba(OH)_2$, as the raw materials by a water-thermal method at normal pressure. Particulate diameters of $Ba_{0.8}Sr_{0.2}Zr_{0.1}Ti_{0.9}O_3$ was 59 nm which was detected by XRD (model D8 ADVANCE, manufactured by Germany Bruker Co.) and BET surface area was 8.8 m^2/g. The relative permittivity of nano-$Ba_{0.8}Sr_{0.2}Zr_{0.1}Ti_{0.9}O_3$ was about 10^4 (detected by LCR automatism test instrument 4210).

The toluene concentration was determined using a gas chromatography (model HP6890N, manufactured by Agilent Co.) with a flame ionization detector (FID) and a capillary column of HP-5 (internal diameter of 0.32 mm, length 30 m). The byproducts such as aldehyde, alcohols, amide, hydroxybenzene and polymerization products, etc, were identified by GC-MS (manufactured by American Thermo Finnegan Co.) and FT-IR (model Vertex 70, manufactured by Germany). Ozone concentration was measured by a chemical titration method of iodine.

6.2 Effect of combined catalysts on toluene removal efficiency

As the $MnO_2/\gamma\text{-}Al_2O_3$ catalyst has the best effect for ozone decomposition but not for toluene decomposition, and nano-$Ba_{0.8}Sr_{0.2}Zr_{0.1}Ti_{0.9}O_3$, a type of developmental material on base of pure $BaTiO_3$ (typical ferroelectric), enhances energy efficiency because of its higher relative permittivity of 10^4, a combination of nano-$Ba_{0.8}Sr_{0.2}Zr_{0.1}Ti_{0.9}O_3$ with $MnO_2/\gamma\text{-}Al_2O_3$ as a combined catalyst was tested in this study.

The effect of various catalysts such as multiple catalyst, nano- $Ba_{0.8}Sr_{0.2}Zr_{0.1}Ti_{0.9}O_3$, $MnO_2/\gamma\text{-}Al_2O_3$ and no padding on removal efficiency is shown in Fig.36. The removal efficiency increased significantly with the catalysts than that without. The removal efficiency increased in the order of: combined catalyst >nano-$Ba_{0.8}Sr_{0.2}Zr_{0.1}Ti_{0.9}O_3$ >$MnO_2/\gamma\text{-}Al_2O_3$ >no padding.

The best removal efficiency of 98.7% was achieved in the NTP process. It indicated that the combination of catalysts exhibited a synergistic effect for toluene decomposition.

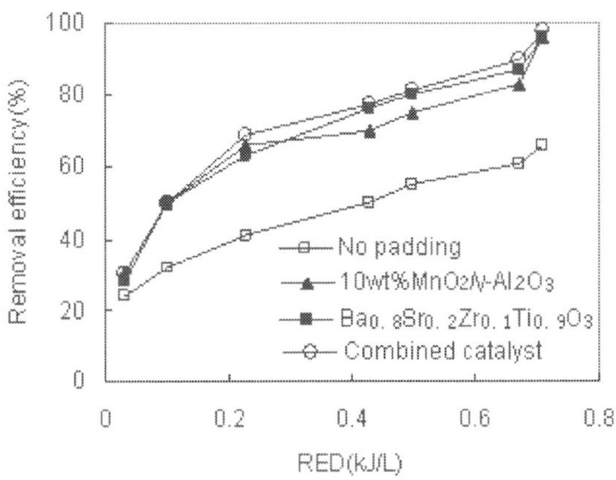

Fig. 36. The change of removal efficiency with various padding (toluene concentration: 800-1000 mg/m^3; gas flow rate: 2 L/min; AC frequency: 150 Hz)

6.3 Effect of combined catalysts on ozone formation

Fig.37 shows the influence of various catalysts on ozone formation with the order of: combined catalyst > $MnO_2/\gamma\text{-}Al_2O_3$ > no padding > nano-$Ba_{0.8}Sr_{0.2}Zr_{0.1}Ti_{0.9}O_3$ at RED of 0.5 kJ/L. This result suggested that $MnO_2/\gamma\text{-}Al_2O_3$ in the combination of catalysts should have a main effect on ozone decomposition.

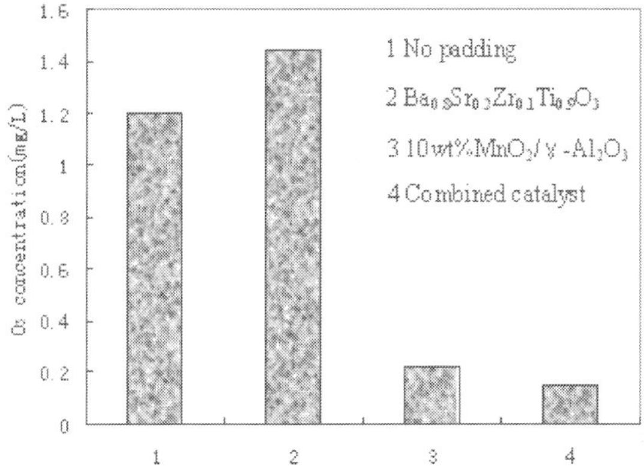

Fig. 37. The change of O_3 concentration with various padding

6.4 Effect of the combination of catalysts on energy efficiency

Fig.38 shows the influence of various catalysts on energy efficiency with the order of: combined catalyst > nano-$Ba_{0.8}Sr_{0.2}Zr_{0.1}Ti_{0.9}O_3$ > $MnO_2/\gamma\text{-}Al_2O_3$ > no padding at the same SED. These results indicated that the nano-$Ba_{0.8}Sr_{0.2}Zr_{0.1}Ti_{0.9}O_3$ in the combination of catalysts should play an important role for improving energy efficiency.

As a result, the combination of catalysts shows the best removal efficiency of toluene, the best decomposition effect of ozone and the best energy efficiency for toluene removal.

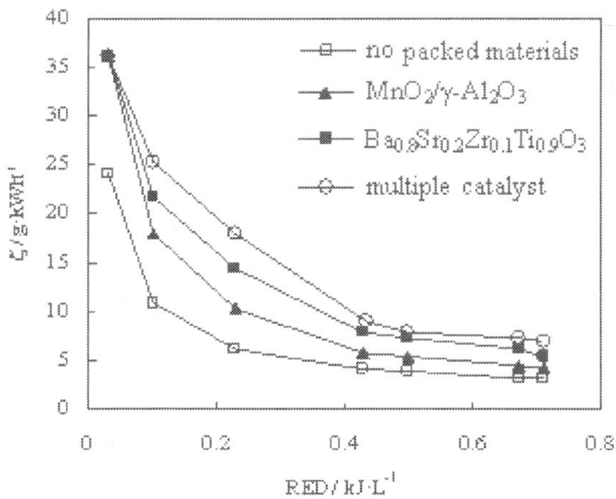

Fig. 38. The energy efficiency with various padding

6.5 Byproducts and decomposition pathways of toluene

Non-thermal plasma has high potential in air cleaning technology, but in some cases unwanted byproducts are formed which could be more harmful than the original VOCs. Fig.39 shows the FT-IR spectrum of the byproducts of toluene decomposition and Fig.40 shows the FT-IR spectrum of the byproducts on the surface of the combination of catalysts.

As shown in Fig.40(a), the -NH- and -NH_2 peak appeared at 3350 cm^{-1} while the peak of 2730 cm^{-1} N=C-N was absent. The peak -NH- with benzene ring appeared at 3450 cm^{-1}, -OH at 3400 cm^{-1}, -CH_3/-CH_2 at 2900 cm^{-1}, benzene derivative (hydroxybenzene, polymerization products, etc) at 1700~1100 cm^{-1}, and CO_2 and CO separately at the rang of 2300~2100 cm^{-1} and 700~500 cm^{-1}. So the byproducts on the surface of the combination of catalysts involved aldehyde, alcohols, amide, and benzene derivative. However, when the combination of catalysts were packed into the NTP reactor, the byproducts on the surface of the packed materials in the NTP reactor reduced greatly as shown in Fig.40(b). Except of amine, CO_2 and CO, no other byproducts were detected on the surface of catalysts. It illuminated that the synergic effect of the NTP with the combination of catalysts could control byproducts effectively.

In Fig.39, the products of toluene decomposition included CO_2, CO and H_2O. At the same time, there are a mass of ozone (strong peak at 1000 cm^{-1}), and several amide and benzene derivatives. Compared spectrum 'a' with 'b' in Fig.39, the benzene derivatives and ozone concentration reduce while the amounts of CO_2 and H_2O increase with the increase of the electric field strength.

A large number of high-energy electrons, ions and free radicals were produced in the NTP reaction process. Firstly, the high-energy electrons could take part in reaction with oxygen in air as follow:

$$\vec{e} + O_2 \rightarrow e + O_2^{\bullet} (A^3 \textstyle\sum_u^+) \rightarrow e + O(^3P) + O(^3P) \tag{64}$$

$$\vec{e} + O_2 \rightarrow e + O_2^{\bullet} (B^3 \textstyle\sum_u^-) \rightarrow e + O(^1D) + O(^3P) \tag{65}$$

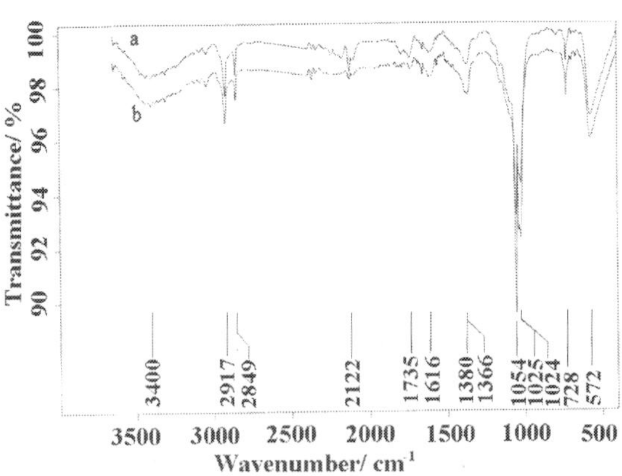

Fig. 39. FT-IR spectrum of the products from toluene decomposition (a. electric field strength of 10 kV/cm; b. electric field strength of 13kV/cm)

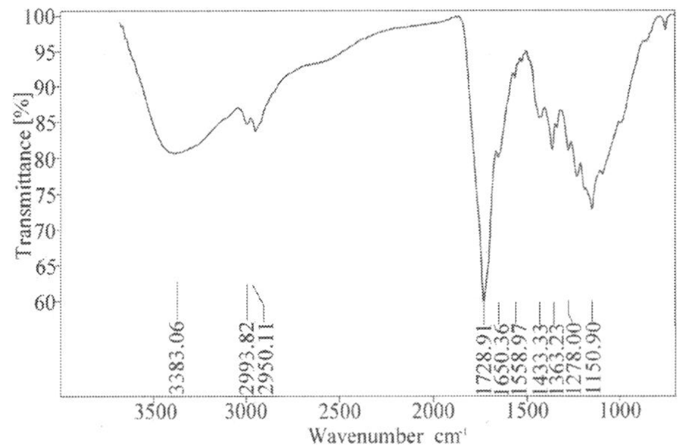

(a) Without catalyst in the NTP reactor

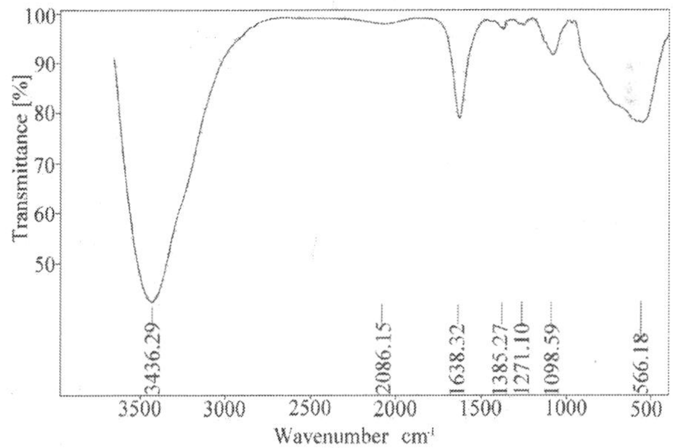

(b) With the combination of catalysts in the NTP reactor

Fig. 40. FT-IR spectrum of the byproducts on the surface of the combination of catalysts

The oxygen free radical groups react with oxygen and other molecules to form ozone:

$$O + O_2 + M \rightarrow O_3^{\bullet} + M \rightarrow O_3 + M \quad (66)$$

At the same time, the high-energy electrons react with H_2O and N_2 in gaseous phase:

$$\vec{e} + H_2O \rightarrow OH^{\bullet} + H^{\bullet} \quad (67)$$

$$\vec{e} + H_2O \rightarrow 2H^{\bullet} + O^{\bullet} \quad (68)$$

$$H_2O + O(^1D) \rightarrow 2OH^{\bullet} \quad (69)$$

$$\vec{e} + N_2 \rightarrow 2N^{\bullet} \quad (70)$$

Toluene bond energy between the carbon of benzene ring and the carbon of the substituent radical is 3.6 eV, which is lower than that of carbon-carbon bond or hydrocarbon bond. As a hydrogen atom in a benzene ring is replaced by a methyl radical to form toluene, the newly formed bond is less stable and the most vulnerable. Of course, the other bonds are also likely to be destroyed by high energy electrons. Formulas 71 to 76 are the possible reaction equations of the process of toluene removal.

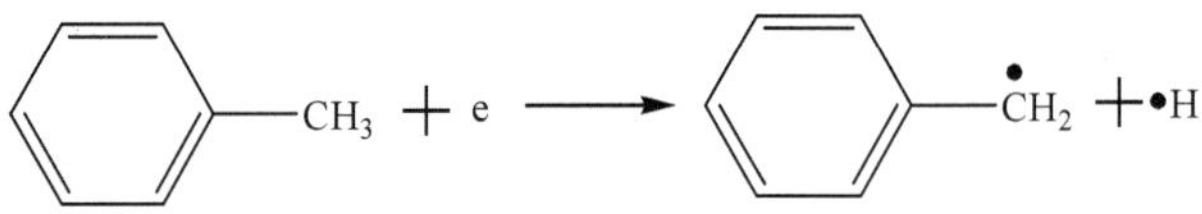

(71)

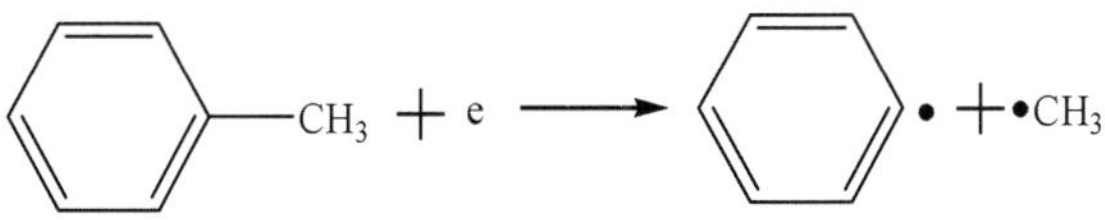

(72)

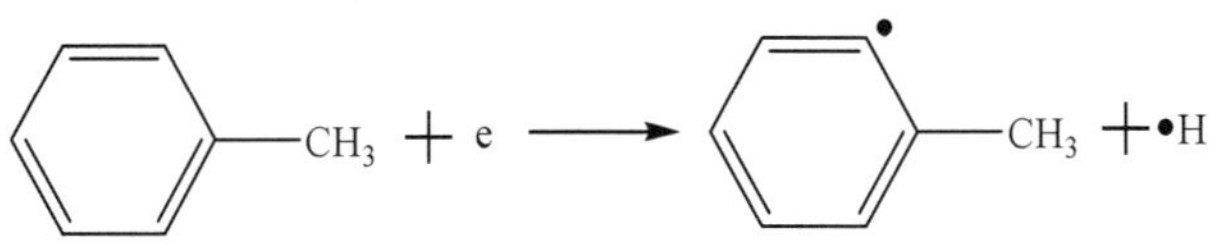

(73)

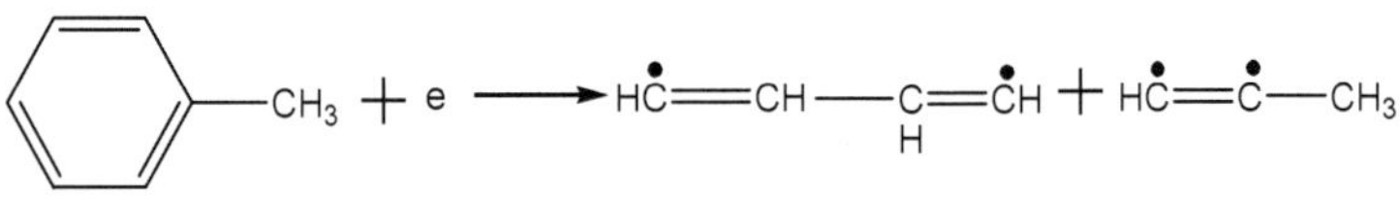

(74)

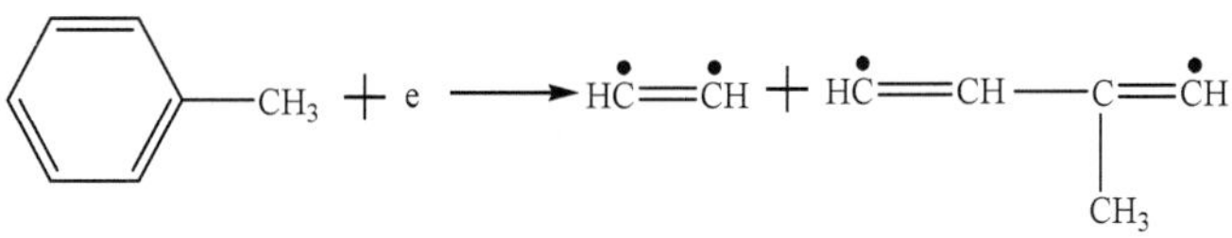

(75)

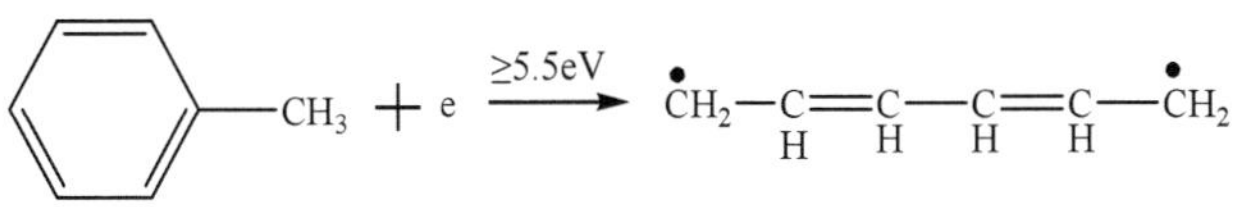

(76)

According to the FT-IR spectrums (Fig.40), the author speculated the reaction pathways for toluene decomposition with the NTP and the combination of catalysts (Fig.41). The oxygen and hydroxyl free radicals of should be the inducement during the process of toluene oxidation. The oxidation process of toluene may involve many reactions and these reactions cooperate and interact with each other for toluene decomposition. Firstly, a series of chain reactions take place between OH radicals and toluene molecules due to the higher oxidation ability of OH radicals than that of oxygen radicals:

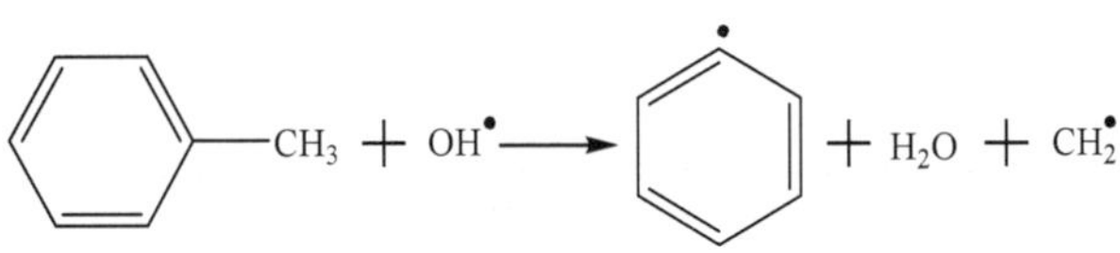

(77)

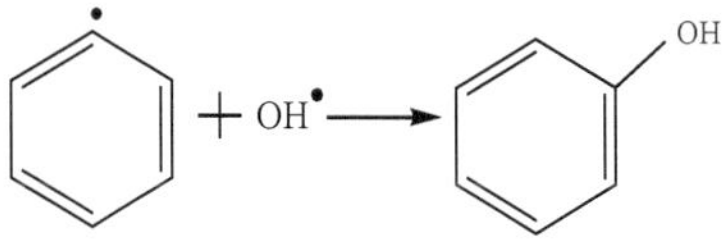

(78)

(79)

Then, the idiographic reactions occur because of oxygen free radicals during the subsequent oxidation reaction as follows:

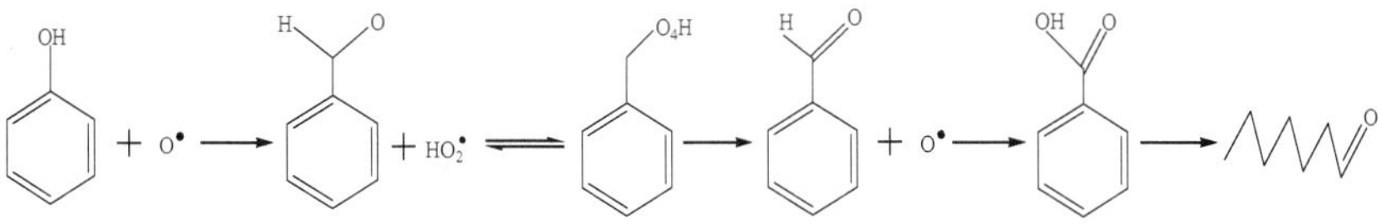

(80)

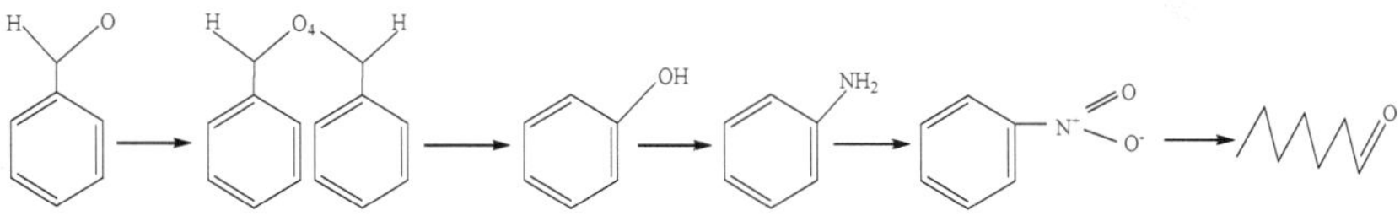

(81)

The binding bonds inside the benzene ring break down after the bonds outside the benzene ring break as follows:

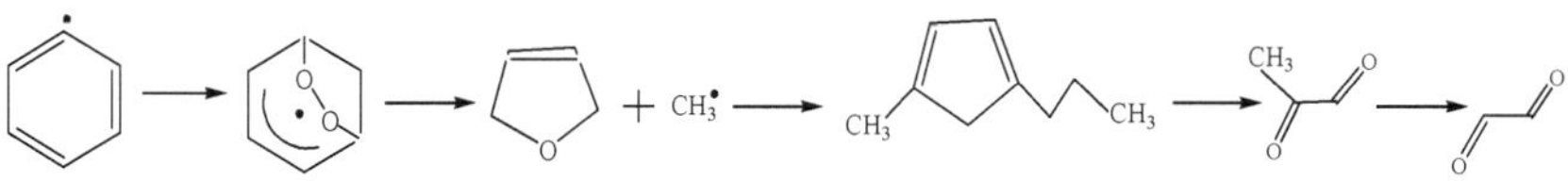

(82)

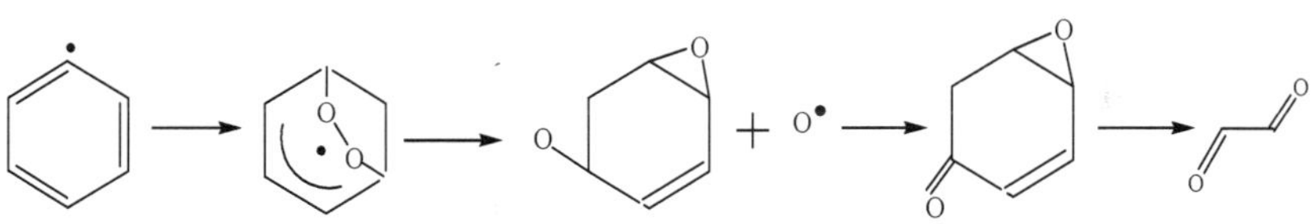

(83)

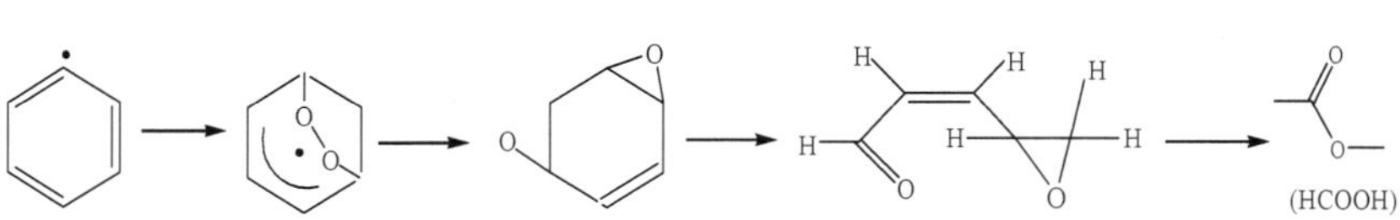

(84)

At last, the byproducts were oxidized to CO_2 and H_2O with increasing RED and the help of catalysis.

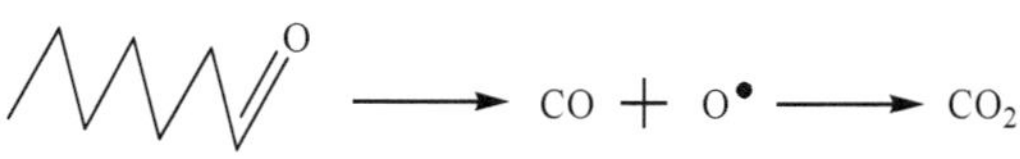

(85)

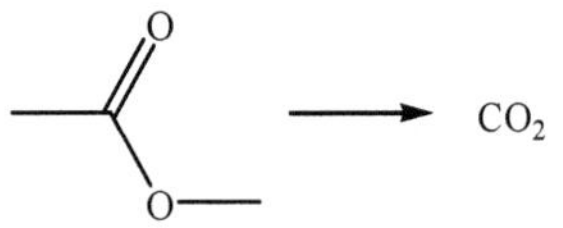

(86)

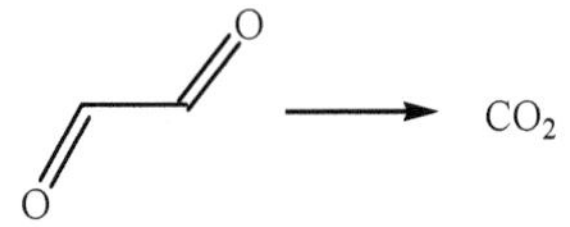

(87)

The byproducts of toluene decomposition were detected using GC-MS at electric field strength of 8 kV/cm and the results show the peak of these products in Fig.42 (a). Products including aldehyde, alcohols, amide, and benzene derivative have been identified.

Fig.42 (b) shows a minor amount of toluene and trace amounts of the products exist at electric field strength of 14 kV/cm. Chang et al. claimed that VOC removal depended on two main mechanisms: direct electrons attack on VOC molecules and indirect reaction between VOC molecules and radicals. These radicals involved oxygen plasma, free radical groups, ozone, etc., which were reactive and could react with toluene molecules to form less hazardous products. If the electric field strength was strong enough or RED was high enough, the toluene molecules would be oxidized to form CO_2, CO and H_2O as the final products.

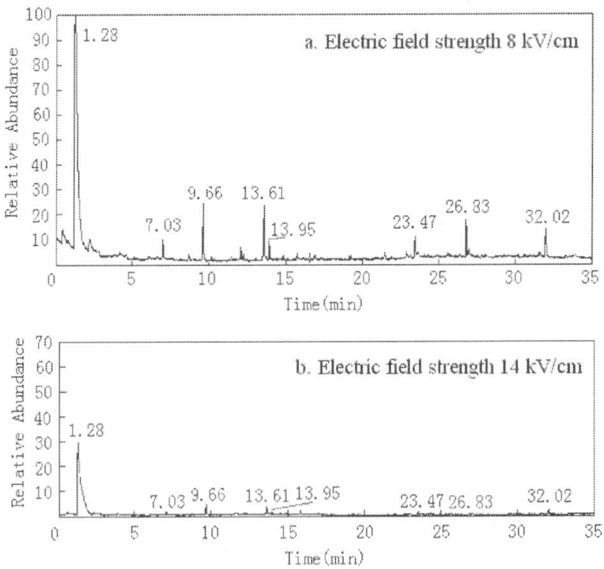

Fig. 42. Mass spectrum of byproducts of toluene decomposition

Atkinson *et al.* (1977) reported that aromatic compounds react with OH radicals by two pathways: hydrogen atom abstraction and OH addition to the aromatic ring. Reaction control pathways I-XII were illustrated in Fig.43. The results showed in a complex oxidation

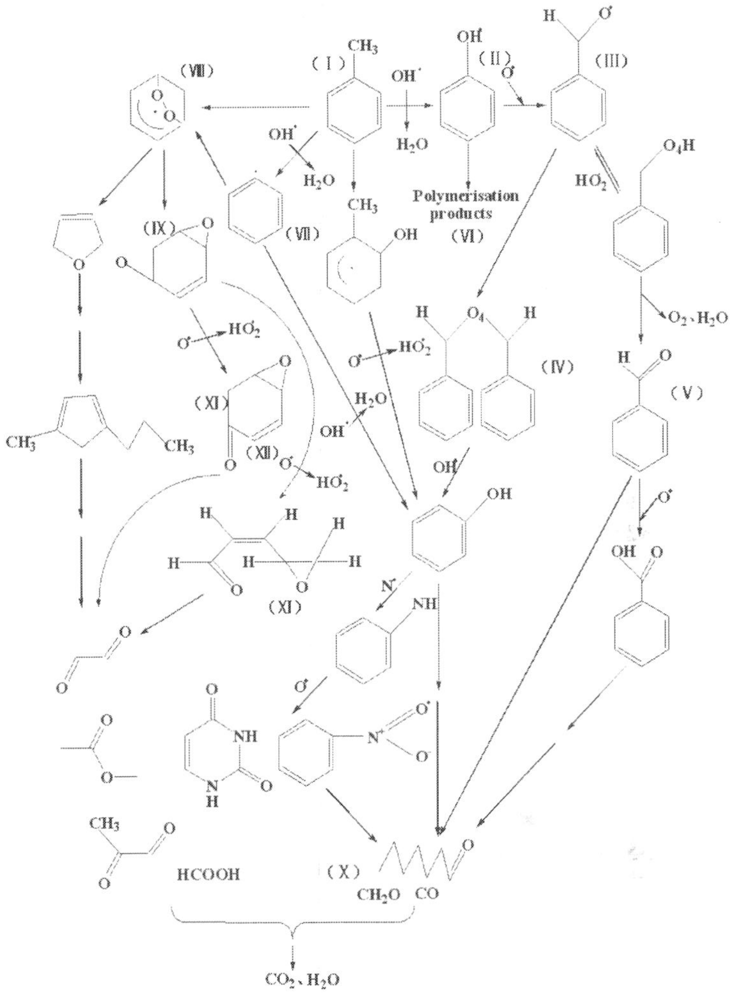

Fig. 43. Abatement pathways of toluene by NTP with the combination of catalysts

mechanism of toluene via several pathways, producing either ring-retaining or ring-opening products. The final products were CO_2 and H_2O.

The synergistic effect of the combination of catalysts with the NTP reactor is presented in Fig.44. The catalyst carrier of γ-Al_2O_3 possesses sorbent characteristic, so it could improve toluene concentration on the catalyst surface and increase the reaction time. MnO_2 is known as a metal oxide catalyst and has been reported to possess a potential activity in redox reactions. MnO_2 surface has been found to expose metal (Mn^{n+}), oxide (O^{2-}) and defect sites of various oxidation states, present degrees of coordination instauration, and exhibit acid and base properties. Furthermore, the d－d electrons exchange interactions between intimately coupled manganese ions of different oxidation states $[Mn^{n+} - O - Mn^{(n+1)+}]$ furnish the electron-mobile environment necessary for the surface redox activity:

$$O_3 + Mn^{n+} \rightarrow O_2^{2-} + Mn^{(n+2)+} + O_2 \quad (88)$$

$$O_3 + O^{2-} + Mn^{(n+2)+} \rightarrow O_2^{2-} + Mn^{(n+2)+} + O_2 \quad (89)$$

$$O_2^{2-} + Mn^{(n+2)+} \rightarrow Mn^{n+} + O_2 \quad (90)$$

These factors would be helpful for toluene decomposition. Radhakrishnan reported that ozone decomposed to O^{2-} and O_2^{2-} in the surface of MnO_2. Naydenov *et al.* believed that O^- existed in the surface of MnO_2 according to the oxidation of benzene in the surface of MnO_2. As a modified ferroelectric, nano-$Ba_{0.8}Sr_{0.2}Zr_{0.1}Ti_{0.9}O_3$ has a higher dielectric constant than $BaTiO_3$ and is polarized at lower electric field strength. More high energy electrons and active radicals are generated to accelerate the reaction between NTP and toluene molecules.

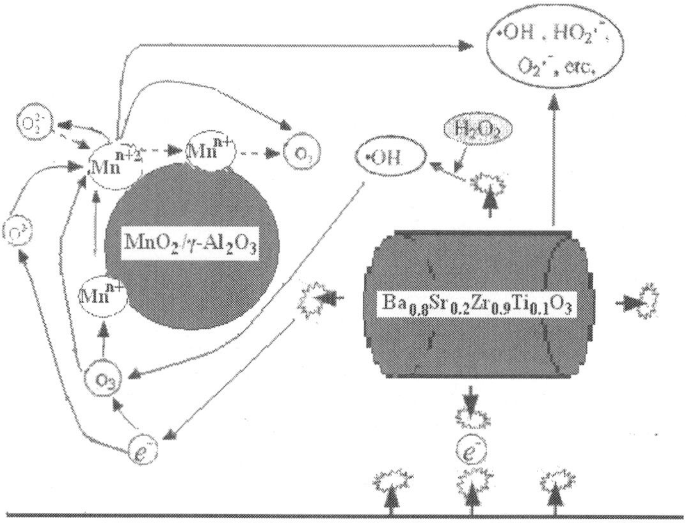

Fig. 44. Catalysis chart of the combination of catalysts in the process of gas discharge

7. Conclusions

In section secend, the adsorption kinetics was studied and the removal amount of VOCs was 25% or so through adsorption of γ-Al_2O_3. In the experiment, with the non-thermal plasma reactor size fixed, the immediate advantage of adsorption of the packed materials into the space of air discharge is the longer reaction time of VOCs with plasma and higher removal efficiency. The functions of γ-Al_2O_3 in plasma reactor were to adsorb free radicals and VOCs molecules and to provide reaction surface for VOCs decomposition and to release reaction products. Plasma decomposed air molecules and provided free radicals for catalysis reactions on the surface of the γ-Al_2O_3 pellets. So adsorbent γ-Al_2O_3 enhanced NTP technology and resulted in higher VOCs removal efficiency and energy efficiency and a better inhibition for O_3 formation in the gas exhaust. For the study in the future, some catalysts should be considered to add into the NTP reactor.

In section third, a series of experiments for the effect of NTP technology were performed to abate toluene from a gaseous influent at room temperature and atmospheric pressure. Three types of NTP reactors were used in the NTP process for toluene removal with and without packed materials. A new modified ferroelectric material of $Ba_{0.8}Sr_{0.2}Zr_{0.1}Ti_{0.9}O_3$ as the packed materials was prepared by us in laboratory. Compared with the two packed materials in terms of removal efficiency of toluene, RED, energy efficiency and ozone concentration, the experimental results were obtained as follows: Packed materials with $Ba_{0.8}Sr_{0.2}Zr_{0.1}Ti_{0.9}O_3$ enhanced removal efficiency of toluene and energy efficiency than those with $BaTiO_3$. $Ba_{0.8}Sr_{0.2}Zr_{0.1}Ti_{0.9}O_3$ had better ferroelectric than $BaTiO_3$. By operating at the RED of 0.76 kJ/L, removal efficiency was up to 97% and the energy efficiency was 6.48 g/kWh when the packed materials of $Ba_{0.8}Sr_{0.2}Zr_{0.1}Ti_{0.9}O_3$ are used. O_3 concentration had a maximum value at the RED of 0.7 kJ/L or so. The hybrid NTP technology should be more effective to improve energy efficiency for VOCs removal than the simple technology of NTP. Based on the above results, we would consider how to control the formation of ozone in the further experiments. Our research would provide a reference to improve energy efficiency for the commercial applications of the NTP technology.

In section fourth, the laboratory-scale plasma reactor was used for benzene removal in an air stream and the following conclusions are obtained. B packed materials was better than A packed materials for benzene removal. Compared different size of packed materials, B packed materials of i.d.2 mm was better than the others' size for benzene decomposition.

In section fifth, nano-TiO_2 packed bed reactor is used to decompose benzene. The experimental results show as follows: With ozone concentration increasing, the removal efficiency of benzene increases. Ozone concentration with packed materials is heigher than that without packed materials in the plasma reactor. Water vapor reduces ozone concentration, and occurring competitive adsorption on the surface of TiO_2 Ozone concentration increases with gas flux increasing, and the removal efficiency of benzene reduces with initial concentration of benzene increasing. When both photocatalyst and ozone coexist, there will be an improved removal efficiency of benzene in the plasma reactor. Effective utilization of active oxygen species is essential in VOCs removal, and TiO_2 can generate higher concentrations of different types of active oxygen species in non-thermal plasma. It is facile and promising to simultaneously hybridize plasma with TiO_2 based on the data presented. The plasma reactor packed with catalyst (B packed materials coated nano-TiO_2) showed a better selectivity of CO_2. Detected by GC-MS, the main products in the plasma reactor are CO_2, H_2O and a small quantity of CO. The plasma reactor packed with catalyst shows a better selectivity of CO_2 than that without catalyst. The selectivity of CO_2 is independent of electrostatic field strength. The selectivity of CO_2 is enhanced due to the benzene oxidation near or on the photocatalyst surface. With benzene concentration increasing, the total output of CO_2 increases. The hybrid system comprising a non-thermal plasma reactor and nanometer TiO_2 catalyst, not only in the gas phase but on the catalyst surface, resulted in the higher energy efficiency and enhanced performance for the oxidative removal of benzene with lower medium reactivities and higher CO_2 selectivity in non-thermal plasma.

In section sixth, the synergistic effect of NTP and catalyst for VOCs removal is tested in the experiment. The results show that removal efficiency increased with increasing RED and was in the order of 10wt% MnO_2/ γ-Al_2O_3≈ 15wt% MnO_2/ γ-Al_2O_3 >5wt% MnO_2/ γ-Al_2O_3 at the same RED. As the mass percentage of MnO_2 catalyst increased, ozone and VOCs concentrations were decreased, especially for 10 wt% MnO_2/ γ-Al_2O_3. The removal efficiency and energy efficiency increased with increasing RED and was in the order of MnO_2/ γ-Al_2O_3>TiO_2/ γ-Al_2O_3> γ-Al_2O_3 at the same RED. So we could draw a conclusion that MnO_2/ γ-Al_2O_3 has a better potential than the other catalysts in the experiment to improve the energy efficiency and reduce O_3 formation.

In section seventh, a series of experiments basing on above all researches, were performed for removal of toluene gaseous influent at room temperature and atmospheric pressure. The self-prepared combined catalyst was used to improve the NTP process and to take the catalytic advantages of both MnO_2/γ-Al_2O_3 and nano-$Ba_{0.8}Sr_{0.2}Zr_{0.1}Ti_{0.9}O_3$. From the view of materials application, the authors adopted NTP coupled with the combination of catalysts technology to decompose VOCs in there. The catalyst materials could be prepared easily and cheap, and at the same time, this combined technology resolved the key bottlenecks effectively, i.e. saving energy consumption and reducing byproducts what we don't want. Therefore, the combination of catalysts technology could advance to the NTP technology and improve applications in the industry in the future.

8. Acknowledgements

This work was supported by the Youth Research Funding of China University of Mining & Technology (Beijing), and China College of innovative pilot projects (No.101141309), and the Fundamental Research Funds for the Central Universities (No.2009QH03), and Environmental Protection Commonweal Industry Research Special Projects (No.201009052-02), and the Doctoral Program of Higher Education of China (20040005009).

9. References

Atkinson R., Pitts J.N. (1977). Jr. "Kinetics and mechanism of the gas phase reaction of hydroxyl radicals with aromatic hydrocarbons over the temperature range 296-473 K". *J. Phys. Chem.* 81, 296–304.

Ayrault C., Barrault J., Blin-Simiand N., Jorand F., Pasquiers S., Rousseau A., Tatibouet J.M. (2003): Oxidation of 2-heptanone in air by a DBD type plasma generated within a honeycomb monolith supported Pt-based catalyst, Cata. Today. 89: 75-83.

Chang M.B., Balbach J.H., Rood M.J., Kushner M.J. (1991). "Removal of SO_2 from gas streams using a dielectric barrier discharge and combined plasma photolysis". *J. Appl. Phys.*, 69, 4409-4418.

Chang M.B., Lee C.C. (1995): Destruction of formaldehyde with dielectric barrier discharge plasmas, Environ. Sci. Tech. 29: 181-186.

Chang M.B., Tseng T.D. (1996). "Gas-phase removal of H_2S and NH_3 with dielectric barrier discharges". *J. Environ. Eng.*, 122, 41-46.

Delagrange S., Pinard L., Tatiboue¨t J.M. (2007): Combination of a non-thermal plasma and a catalyst for toluene removal from air: Manganese based oxide catalysts, Appl. Catal. B: Environ. 68: 92-98.

Ding S.W., Wang J., Qin J.L. (2001). "The structure and performance of complex of nanometer $BaTiO_3$". Sci. China, Ser. B. 31, 525-529.

Eliasson B., Kogelschatz U. (1991): Nonequilibrium volume plasma chemical processing, IEEE Trans. Plasma Sci. 19: 1063-1068.

Einaga H., Lbusuki T., Futamura S., 2001. Performance evaluation of a hybrid system comprising silent discharge plasma and manganese oxide catalysts for benzene decomposition. *IEEE Tran. Ind. Appli.* 37 (5), 858-863.

Futamura S., Zhang A.H., Yamamoto T. (1997). "The dependence of nonthermal plasma behavior of VOCs on their chemical structures". *J. Electrostat.*, 42, 51-62.

Futamura S., Einaga H., Kabashima H. (2004). "Synergistic effect of silent discharge plasma and catalysts on benzene decomposition". *Catal. Today*, 89, 89-95.

Gal A., Ogata A., Futamura S. (2003). "Mechanism of the dissociation of chlorofluorocarbons during nonthermal plasma processing in nitrogen at atmospheric pressure". *J. Phys. Chem. A*, 107, 8859-8866.

Guo Y.F., Ye D.Q., Chen K.F. (2006). "Toluene decomposition using a wire-plate dielectric barrier discharge reactor with manganese oxide catalyst in situ". *J. Molecular Cata. A: Chem.* 245, 93-100.

Guo Y.F., Ye D.Q., Chen K.F., He J.C. (2007). Toluene removal by a DBD-type plasma combined with metal oxides catalysts supported by nickel foam. Catal. Today, 126(3-4), 328~337.

Kim H.H. (2006). "Effect of different datalysts on the decomposition of VOCs using flow-type plasma-driven catalysis". *IEEE Trans. Plasma Sci.*, 34, 984-995.

Kim H.H., Ogata A., Futamura S. (2008). "Oxygen partial pressure-dependent behavior of various catalysts for the total oxidation of VOCs using cycled system of adsorption and oxygen plasma". *Appl. Catal. B: Environ.* 79, 356-367.

Kohno H., Berezin A.A., Chang J.S. (1998). "Destruction of volatile organic compounds used in a semiconductor industry by a capillary tube discharge reactor". *IEEE Trans. Ind. Appl.* 34 953-966.

Krawczyk K., Mlotek M. (2001). Combined plasma-catalytic processing of nitrous oxide. Appl. Catal. B: Environ., 30(3-4), 233-245.

Li C.Y., Li M.Q. (1996). Study on Sol-Gel transition of TiO_2. J. Chinese Ceramic Society, 24(3), 338-341.

Li G.W., Fan Q.J., Liu Q. (1998). The control technique over the pollution caused by VOCs. J. Xi'an Univ. of Arch & Tech. 30(4), 399-402.

Li J., Ma G.D., 2000. The mechanismic analysis and experiment on controlling volatile organic compounds (VOCs) with corona discharge. *J. Xi'an Univ. of Arch & Tech.*, 32 (1), 24-27.

Li R., Tang Q., Yin S., Sato T. (2006). Plasma catalysis for CO_2 decomposition by using different dielectric materials. Fuel Proc. Tech., 87(7), 617–622.

Li W., Han C., Liu W., Zhang M.H., Tao K.Y. (2007). Expanded graphite applied in the catalytic process as a catalyst support. Catal. Today, 125(3-4), 278-281.

Liang Y.H., Zhu T., Ma G.D., 2006. Study of nano-TiO_2 thin film derived by sol-gel process. *J. Xi'an Univ. of Arch. & Tech.*, 38 (6), 799-803.

Ma G.D. (2003). Air Pollution Control Engineering (second edition). Beijing, China Environ. Sci. Press. 105-134.

Magureanu M., Mandache N.B., Hu J.C., Richards R., Florea M., Parvulescu V.I. (2005): Plasma-assisted catalysis for volatile organic compounds abatement, Appli. Cata. B: Environ. 61: 12-20.

Magureanu M., Mandache N.B., Subrahmanyam C., Kiwi-Minsker L. (2007): Improved performance of non-thermal plasma reactor during decomposition of trichloroethylene: Optimization of the reactor geometry and introduction of catalytic electrode, Appl. Catal. B: Environ. 74: 270-283.

Magureanu M., Mandache N. B., Hu J.C., Richards R., Florea M., Parvulescu V.I. (2007). Plasma-assisted catalysis total oxidation of trichloroethylene over gold nano-particles embedded in SBA-15 catalysts. Appl. Catal. B: Environ.,76(3-4), 275-281.

Malik M. A., Minamitani Y., Schoenbach K. H. (2005). "Comparison of catalytic activity of aluminum oxide and silica gel for decomposition of volatile organic compounds (VOCs) in a plasmacatalytic reactor". *IEEE Trans. Plasma Sci.*, 33 (1), 50-56.

Masuda S., 1988. Pulse corona induced plasma chemical process: a horizon of new plasma chemical technologies. *Pure Appl. Chem.*, 60 (5), 727-731.

Masuda S., Hosokawa S., Tu X.L., Tsutsumi M., Ohtani T., Tsukahara T., Matsuda N., 1993. The performance of an integrated air purifier for control of aerosol, microbial, and odor. *IEEE Trans. Ind. Appl.*, 29 (4), 774-780.

Maria D., Hernandez A., Juan M.C., Maira A.J., Soria J., Loddo V., Augugliaro V. (2002). Ozone enhanced activity of aqueous titanium dioxide suspensions for photocatalytic oxidation of free cyanide ions. Appl. Catal. B: Environ., 39(3), 257-267.

Mizuno A., Clements J.S., Davis R.H., 1986. A method for the removal of sulfur dioxide from exhaust gas utilizing pulsed streamer corona for electron energization. *IEEE Trans. Ind. Appl.*, 22 (3), 516-522.

Mizuno A., Yamazaki Y., Obama S., 1993. Effect of voltage waveform on partial discharge in ferroelectric pellet layer for gas cleaning. *IEEE Trans. Ind. Appl.*, 29 (2), 262-267.

Muhamad A.M., Jiang X.Z., 2000. Catalyst assisted destruction of trichloro ethylene and toluene in corona discharges. *J. Environ. Sci.*, 12 (1), 7-11.

Naydenov A., Mehandjiev D.. (1993). "Complete oxidation of benzene on manganese dioxide by ozone". App1.Cata1.A: Gen, 97, 17-22.

Noel D.N. (2000). Air Pollution Control Engineering (second edition). Beijing, McGraw-Hill. 329-382.

Nunez C M, Ramsey G H, Ponder W H, Abbott J H, Hamel L E, Kariher P H. (1993). "Corona destruction: An innovative control technology for VOCs and air toxics". Air & Waste, 43 (2), 242-247.

Ogata A., Yamanonchi K., Mizuno K. (1999). "Decomposition of benzene using alumina-hybrid and catalyst-hybrid plasma reactors". *IEEE Tran. Ind. Appli.* 35, 1289-1295.

Ogata A., Einaga H., Kabashima H. (2003). "Effective combination of nonthermal plasma and catalysts for decomposition of benzene in air". *Appl. Catal. B: Environ.* 46, 87-95.

Ohkubo T., Kanazawa S., Nomoto Y., 1994. NOx removal by a pipe with nozzle-plate electrode corona discharge system. *IEEE Trans. Ind. Appl.*, 30 (4), 856-861.

Park J.Y., Jung J.G., Kim J.S. (2003). "Effect of nonthermal plasma reactor for CF decomposition". *IEEE Trans. Plasma Sci.*, 31, 1349-1354.

Rachel F. (1998): Automated system for power measurement in the silent discharge, IEEE Trans. Ind. Appl. 34: 563-570.

Radhakrishnan R., Oyama S.T., Chen J.G. (2001). "Electron transfer effects in ozone decomposition on supported manganese oxide". J.Phys.Chem.B, 105, 4245-4253.

Ricketts C.L., Wallis A.E., Whitehead J.C. (2004): A mechanism for the destruction of CFC-12 in a nonthermal, atmospheric pressure plasma, J. Phys. Chem. A. 108: 8341-8345.

Subrahmanyam Ch., Renken A., Kiwi-Minsker L. (2007). "Novel catalytic non-thermal plasma reactor for the abatement of VOCs". Chem. Eng. J. 134, 78-83.

Thevenet F., Guaitella O., Puzenat E., Herrmann J.M., Rousseau A., Guillard C. (2007). Oxidation of acetylene by photocatalysis coupled with dielectric barrier discharge. Catal. Today, 122(1-2), 186-194.

Tonkyn R.G., Barlow S.E., Orlando T.M. (1996). "Destruction of carbon tetrachloride in a dielectric barrier/packed-bed corona reactor". *J. Appl. Phys.*, 80, 4877-4886.

Tonkyn R.G., Barlow S.E., Hoard J. (2003). "Reduction of NOx in synthetic diesel exhaust via two-step plasma-catalytic treatment". *Appl. Catal. B: Environ.*, 40, 207-217.

Urashima K., Chang J. (2000). "Removal of volatile organic compounds from air streams and industrial flue gases by non-thermal plasma technology". *IEEE Trans. Dielectr. Electr. Insul.*, 7, 602-614.

Van Durmea J., Dewulfa J., Sysmansa W., Leysb C., Van Langenhove H. (2007). „Efficient toluene abatement in indoor air by a plasma catalytic hybrid system". *Appl. Cata. B: Environ.* 74, 161-166.

Van Durme J., Dewulf J., Leys C., Langenhove H. (2008): Combining non-thermal plasma with heterogeneous catalysis in waste gas treatment: A review, Appl. Cata. B: Environ. 78: 324-333.

Wallis A.E., Whitehead J.C., Zhang K. (2007). The removal of dichloromethane from atmospheric pressure air streams using plasma-assisted catalysis. Appl. Catal. B: Environ., 72(3-4), 282-288.

Yamamoto T., Ramanatiran K, Lawless P. A., Ensor D.S., Nwesome J. R., Plaks N., Ramsey G. H. (1992). "Control of Volatile organic compound by ac energized ferroelectric pellet reactor and a pulsed corona reactor". *IEEE Trans Ind Appl*, 28(3), 528-533.

Yamamoto T., Mizuno K., Tamori I., Ogata A., Nifuku M., Michalska M., Prieto G., (1996). "Catalysis-assisted plasma technology for carbon tetrachloride destruction". *IEEE Trans Ind Appl*, 32(1), 100-106.

Yamamoto T. (1997): Methods and apparatus for controlling toxic compounds using catalysis-assisted non-thermal plasma, Environ. Int. 23: 3-10.

Yoichi I., Junya S., Takashi N., Shigeyoshi M. (2008). Synthesis of visible-light active TiO_2 photocatalyst with Pt-modification: Role of TiO_2 substrate for high photocatalytic activity. Appl. Catal. B: Environ., 79(2), 117-121.

Young S.M., Mirosław D., Jerzy M. (2004). "Effect of reaction temperature on NOx removal and formation of ammonium nitrate in nonthermal plasma process combined with selective catalytic reduction". *IEEE Trans. Plasma Sci.*, 32, 799-807.

Zhu T., Li J., Dou B.J., Liang W.J., Jin Y.Q., 2007. Study on treatment tobacco odor with non-thermal plasma technology. *J. Xi'an Univ. of Arch. & Tech.*, 39 (6), 862-866.

Zhu T., Li J., Liang W.J., Dou B.J., Jin Y.Q. (2007). Gaseous phase toluene decomposition by non-thermal plasma. Environ. Pollution & Protection, 29(12), 920-924.

Zhu T., Li J., Dou B.J., Liang W.J., Jin Y.Q. (2007). Study on treatment tobacco odor with non-thermal plasma technology. J. Xi'an Univ. of Arch & Tech., 39(6), 862-866.

Zhu T., Li J., Liang W.J., Jin Y.Q. (2008). Research Progresses in treatment of waste gas containing volatile organic compounds by combined plasma technology. Environ. Protec. Chem. Ind., 28(2): 121-125.

Zhu T., Li J., Liang W.J., Dou B.J., Jin Y.Q., 2008. Research Progresses in Treatment of Waste Gas Containing Volatile Organic Compounds by Combined Plasma Technology. *Environ. Protec. Chem. Ind.* 28 (2), 121-125.

Zhu T., Li J., Jin Y.Q., Liang Y.H., Ma G.D. (2008). "Decomposition of benzene by non-thermal plasma processing-photocatalyst and ozone effect". *Int. J. Environ. Sci. Tech.* 5, 375-384.

Zhang D.Y., Pan X.L., Li L.H., Song W.J., Wu G.Y., Liu J. (2010) Glow plasma degradation of aqueous direct black 38, Fresenius Environ. Bulletin.19: 1-7.

Zhu T., Li J., Jin Y.Q., Liang Y.H., Ma G.D. (2009). „Gaseous phase benzene decomposition by nonthermal plasma coupled with nano-titania catalyst". *Int. J. Environ. Sci. Tech.* 6, 141-152.

Zhu T., Li J., Liang W.J., Jin Y.Q. (2009). "Synergistic effect of catalyst for oxidation removal of toluene". *J. Hazardous Materials.* 165, 1258-1261.

Zhu T., Xu D.Y., He X.W., Shu X.Q. (2010): Decomposition of benzene in dry air bu super-imposed barrier discharge nonthermal plasma-photocatalytic system, Fresenius Environ. Bulletin. 19: 1275-1282.

Zhu T., Wan Y.D., He X.W., Xu D.Y., Shu X.Q. (2011). "Effect of modified ferroelectric on nonthermal plasma process for toluene decomposition". *Fresenius Environmental Bulletin.* 20(1), 149-155.

Lab-scale Evaluation of Two Biotechnologies to Treat VOC Air Emissions: Comparison with a Biotrickling Pilot Unit Installed in the Plastic Coating Sector

F. Javier Álvarez-Hornos, Feliu Sempere, Marta Izquierdo and Carmen Gabaldón
GI²AM Research Group, Department of Chemical Engineering, University of Valencia, Burjassot, Spain

1. Introduction

Volatile organic compounds (VOCs) are one of the top five atmospheric pollutants, and, according to an EC directive, are defined as *"all organic compounds arising from human activities, other than methane, which are capable of producing photochemical oxidants by reactions with nitrogen oxides in the presence of sunlight"* (Council Directive 2001/81/EC). This definition highlights the fact that VOCs play a vital role in the formation of tropospheric ozone, which causes photochemical smog. Short-term exposure to photochemical smog affects respiratory function and has adverse effects on plants (World Health Organization, 2004). The distinction between biogenic and anthropogenic VOCs in the atmosphere is far from straightforward, because many VOC species are produced by both sources (Popescu & Ionel, 2010). Anthropogenic sources of VOCs include air emissions from wastewater treatments plants, motor vehicles, gasoline storage facilities and transportation, dry cleaning and other industrial sources (D.J. Kim & H. Kim, 2005). In this sense, the main sectors involved in non-methane VOC emissions in the EU-27 are solvent and product use (41%), road and non-road transportation (18%), and commercial, institutional, and household associated emissions (14%) (European Environment Agency, 2010). Regarding the industrial sources, Fig. 1 illustrates the contributions from various industrial sectors to EU-27 non-methane VOC industrial emissions in 2008 (European Pollutant Release and Transfer Register, 2008). The three most important industrial sources are: energy (41%); the chemical industry (22%); and coating and surface treatment activities (18%). In fact, over the past decade, emerging European Union environmental policy has focused on abatement of VOCs from industrial emissions, in an effort to protect environmental and public health. As a result of these initiatives, new European VOC emission limits have been established in the VOC Solvent Emissions Directive (Council Directive 1999/12/EC) for a wide range of industrial sectors. Currently, VOC concentration limits range from 50 to 150 mg C/Nm^3, depending on the application and solvent consumption.

Although process changes and the substitution of solvent-based products for water-based ones have the potential to minimise VOC emissions, stringent VOC emission limits require

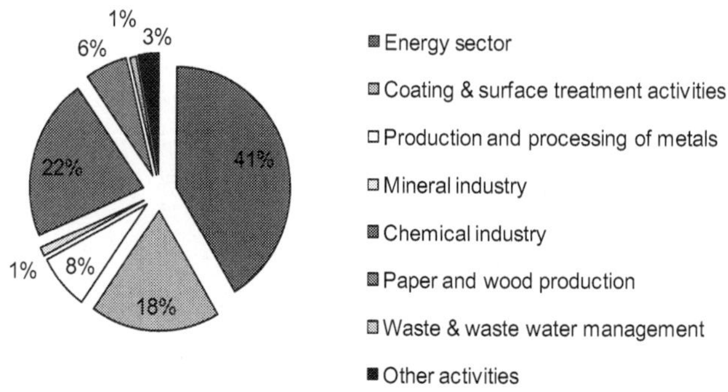

Fig. 1. Distribution of non-methane VOC industrial emissions in the EU-27 in 2008 by industrial sector (adapted from European Pollutant Release and Transfer Register, 2008).

additional treatment technologies, better known as 'end-of-pipe' techniques. For years, these techniques have been primarily based on non-biological methods, such as condensation, adsorption, absorption/scrubbing and thermal destruction. However, over the last decade, vapour-phase biotechnologies, including biofilters, biotrickling filters and bioscrubbers, have proven to be both efficient and environmentally-friendly for the treatment of VOC emissions, and have been classified as best available technologies (BATs) for the reduction of VOC emissions in the chemical sector by the European IPPC Bureau (European Commission, 2003). Thus, biotechnologies are a potential alternative to conventional physicochemical processes for the removal of VOCs from high flow rate emissions streams with relatively low VOC concentrations: conditions which are common in painting, coating and printing processes. For example, replacing a conventional thermal oxidiser with a biotreatment system for the control of VOC emissions from a panel board press reduced greenhouse gas emissions by 60 to 80 percent, and operating costs by 90 percent (Boswell, 2009).

Vapour-phase biotechnologies are based on the capability of microorganisms of utilizing their metabolism to transform the organic pollutants to less toxic compounds. However, because VOC pollutants are in the air, they must first be transferred from the gas phase to an aqueous phase, where biodegradation can occur. Through biodegradation, contaminants are used as energy and carbon sources for microbial growth, and are converted to carbon dioxide and water. The main vapour-phase biotechnologies available for the treatment of VOC emissions include biofilters (BF), biotrickling filters (BTF) and bioscrubbers (BS). Although the basic VOC removal mechanisms are similar in these systems, there are notable differences with respect to the aqueous phase and microorganism growth, which are summarised in Table 1.

Biotechnology system	*Microorganism growth*	*Aqueous phase*
Biofilter	Attached growth	Stationary
Biotrickling filter	Attached growth	Flowing
Bioscrubber	Suspended growth	Flowing

Table 1. Classification of vapour-phase biotechnology systems.

Biofilters work by passing polluted pre-humidified air through a porous packed bed of natural organic material, in which a culture of pollutant-degrading microorganisms is developed (Fig. 2 a). The packing material is a key factor for the successful application of biofilters: it is necessary to choose media with adequate physical and chemical properties, such as high surface area, long-term stability, low pressure drop, low-cost, good moisture retention, pH buffering capacity, appropriate adsorbent capacity and nutrients (Shareefdeen & Singh, 2005). In addition, because moisture content control is also critical, biofilter systems usually incorporate some kind of water addition. Use of an occasional nutrient supply is also advisable. For biotrickling filters (Fig. 2 b), the polluted air is passed through inert packing material while a liquid stream is re-circulated over the bed. In this case the biofilm is developed on the packing surface, with the liquid phase providing nutrients to the biofilm, allowing for greater pH control and yielding a more stable operation in comparison with biofilters. These characteristics, along with a larger air/liquid specific surface area, lead to higher VOC removal rates than those obtained with conventional biofilters (Koutinas et al., 2005), suggesting that smaller biotrickling filters can be installed with lower capital investment for industrial applications.

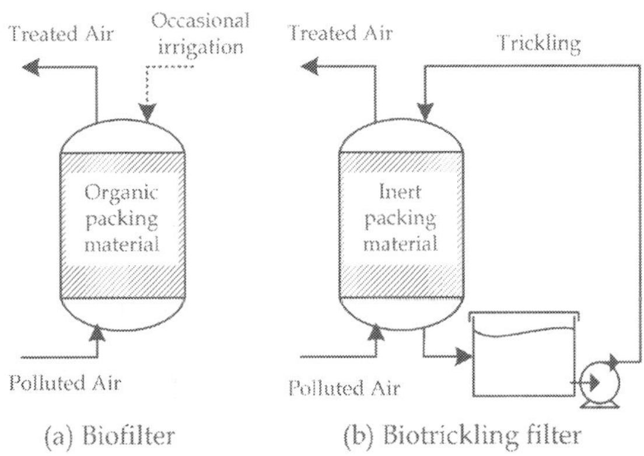

Fig. 2. Schematic of vapour-phase biotechnology systems.

The number of laboratory studies on the control of continuous VOC emissions using biofilters (BF) and biotrickling filters (BTF) has significantly increased over the last two decades. These studies have mainly focused on evaluating process performance as a function of the operational conditions (e.g. gas empty bed residence time, inlet load or nutrient formulation and concentration), and/or packing material characteristics, with the aim of obtaining valuable information from these biological technologies. BFs and BTFs have been applied in the treatment of streams contaminated with a wide variety of pollutants, including oxygenated compounds (Cox & Deshusses, 2001; Steele et al., 2005), aromatics (Hwang et al., 2008; Kennes et al., 1996), or its mixtures (Paca et al., 2006; Sempere et al., 2008).

However, studies dealing with the removal of complex VOC mixtures from painting and coating processes using BFs or BTFs are limited, especially for pilot/full-scale units. Emissions from these processes typically result in streams with high flow rates and relatively low VOC concentrations, which usually contain a complex mixture of hydrophobic (e.g. toluene and xylenes) and hydrophilic (e.g. n-butyl acetate, ethyl acetate and methyl propyl ether) compounds. Concerning laboratory scale experiments, Mathur & Majumder (2008) investigated methyl ethyl ketone (MEK), toluene, n-butyl acetate and o-xylene elimination using a coal-based BTF, and reported a maximum elimination capacity (EC) of 185 g m^{-3} h^{-1} for an inlet load (IL) of 278 g m^{-3} h^{-1}, working at an empty bed residence time (EBRT) of 42.4 s. Similarly, Higuchi et al. (2010) observed a maximum EC of 87.5 g m^{-3} h^{-1} for removal of 2-butanone, butyl acetate, butoxyl, toluene, ethylbenzene and xylene, using a BTF packed with poly-vinyl formal (PVF) material operated at an inlet concentration of 0.43 g m^{-3} and an EBRT of 12 s. These data indicate that, although biofiltration is suitable in terms of overall VOC removal and has shown significant potential, it still requires additional optimisation. It is thus essential to assess BFs and the BTFs in real situations using pilot scale units, with the aim of obtaining valuable and useful information necessary for scale-up. Webster et al. (1999) installed a 0.47 m^3-volume BTF pilot unit to treat off-gases from two spray paint booths, achieving removal efficiency (RE) higher than 70% working at an EBRT between 11 and 39 s for the target pollutants: MEK, methyl isobutyl ketone, o-xylene, m-xylene, p-xylene and n-butyl acetate. Martinez-Soria et al. (2009) evaluated introduction of an activated carbon prefilter in the treatment of VOC emissions from spray paint booths in the wood furniture industry, using a 0.75 m^3-volume BTF pilot unit: the prefilter buffered VOC fluctuations, ensuring that legal limits were met while working at an EBRT of 24 s.

This chapter presents studies conducted to assess environmentally friendly biotechnologies, such as biofilters and biotrickling filters, for VOC abatement in air at two scales. First, a laboratory-scale study was designed to investigate the use of a biofilter (BF) and a biotrickling filter (BTF) under continuous feeding conditions, for VOC removal from air contaminated with three compounds commonly found in air emissions from paint and coating processes: n-butyl acetate, toluene and m-xylene (a 2:1:1 weight mixture was used to simulate exhaust gases). These compounds have been previously identified as representative VOCs in paints (Boswell et al., 2001). Second, a biotrickling filter pilot unit was used to assess treatment of exhaust gases from a robotic spray paint booth at a plastic coating facility (located in Soria, Spain), which is a supplier of car mirrors to the automotive sector. The performance of this pilot-scale BTF was compared with results obtained from the lab-scale systems.

2. Materials and methods

Two experimental phases were carried out during this research: (1) laboratory scale experiments using both biofilter and biotrickling filter, and (2) pilot-scale operation of a biotrickling unit connected to a robotic spray paint booth at a plastic coating facility.

2.1 Laboratory-scale systems

Treatment of air polluted with a 2:1:1 (wt) n-butyl acetate:toluene:m-xylene mixture was studied both in a biofilter and in a biotrickling filter at the laboratory scale.

A schematic of the laboratory-scale biofilter set-up is shown in Fig. 3. The lab-scale BF was made of methacrilate, with a total length of 97 cm and an internal diameter of 13.6 cm. The BF was equipped with 5 equidistant sampling ports to measure VOC concentrations, and four additional ports for temperature measurements and filter bed sampling. Fibrous peat (ProEco Ambiente, Spain) was used as the filter material. Because the peat was acidic, the pH was adjusted to neutral using a diluted sodium hydroxide solution. Compressed, filtered and dried-air was passed through two serial-humidifiers, to assure a relative humidity of ≥ 90%. The EBRT was adjusted using a mass flow controller (Bronkhorst Hi-Tec, The Netherlands). Pollutant was introduced to the air stream using a syringe pump (New Era, infusion/withdraw NE 1000 model, USA) and then, air polluted with the VOC mixture was flowed downward into the bed.

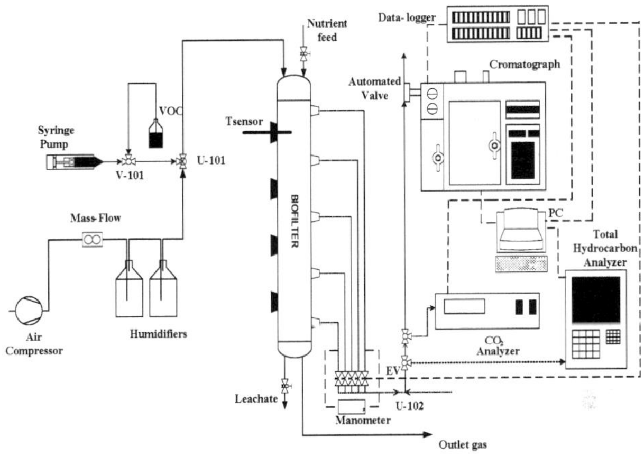

Fig. 3. Schematic of the laboratory-scale biofilter (BF).

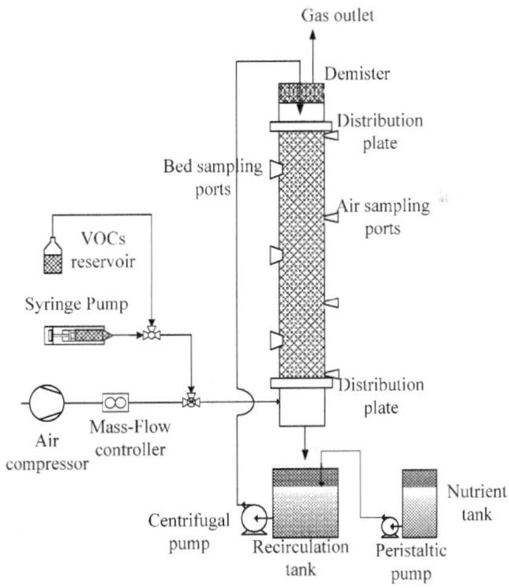

Fig. 4. Schematic of the laboratory-scale biotrickling filter (BTF).

A schematic of the laboratory-scale biotrickling filter is shown in Fig. 4. The lab-scale BTF was composed of 3 cylindrical modules of Plexiglas, with a total bed length of 120 cm and an internal diameter of 14.4 cm, and was equipped with a recirculation tank (effective volume, 10 L). The BTF was randomly filled with 1-inch of nominal diameter Flexiring™

polypropylene rings (Koch-Glistch B.V.B.A., Belgium). Compressed, filtered and dried-air air was polluted with VOCs as above, using a syringe pump (New Era, infusion/withdraw NE 1000 model, USA). Then, VOC contaminated air was introduced through the bottom of the column, and the flow rate was adjusted using mass flow controllers (Bronkhorst Hi-Tec, The Netherlands). The recirculation stream was introduced counter to the air flow, and was partially renewed (50 - 100% of total volume) every week.

The main properties of the peat and Flexiring™ ring packing materials used in the BF and the BTF set-ups, respectively, are shown in Table 2.

Fibrous peat	
Organic content, % wt	95
BET specific surface area, $m^2\ g^{-1}$	13.4
pH	4.8
Bulk density, kg m^{-3}	133
Water-holding capacity, % wt	88
Shape	Fibres
Fibre length, cm	2 - 8
Flexiring™	
Specific surface area, $m^2\ m^{-3}$	207
Void fraction, %	92
Bulk density, kg m^{-3}	71
Shape	Rings
Ring diameter, cm	24.5

Table 2. Physical and chemical properties of the packing materials.

The necessary macro and micronutrients were incorporated using a pH buffered nutrient solution containing N (3 g L^{-1}), P (0.6 g L^{-1}), Ca, Fe, Zn, Co, Mn, Mo, Ni and B at trace doses. The nutrient solution was incorporated in the biofilter by directly pouring 100 mL per day on the top the column and was supplied to the biotrickling filter using a peristaltic pump, whose flow rate was adjusted to keep nitrogen concentrations > 10 mg L^{-1} in the recirculation solution.

The two bioreactors were inoculated with a mixed microbial culture (obtained from activated sludge which was adapted to the compounds to be treated). For acclimation of the microbes, the VOC mixture was continuously fed to an activated sludge from the secondary clarifier of Carraixet Wastewater Treatment Plant (located in Alboraya, Spain), for a period of at least two months in order to obtain an adapted inoculum. The Carraixet treatment plant receives urban sewage from Alboraya town and pollutants from the Alboraya industrial site. Inoculation of the BF was performed by mixing the peat with a 1 L of the inoculum. For the BTF, 1.5 L of the inoculum was added into the recirculation tank.

The BF and the BTF were operated under continuous feeding conditions for a total period of 4 months and 3 months, respectively. For the BF, an EBRT of 60 s (gas flow rate of 0.85 $m^3\ h^{-1}$) and an IL of 19 g C $m^{-3}\ h^{-1}$, corresponding to an inlet total VOC concentration of 300 mg C Nm^{-3}, were applied. For the BTF, 6 experiments, each with duration of 2 weeks, were performed at an EBRT between 60 and 15 s (gas flow rate between 1.2 and 4.7 $m^3\ h^{-1}$) for IL varying from 11 to 72 g C $m^{-3}\ h^{-1}$, corresponding to VOC concentrations between 160 mg C Nm^{-3} and 350 mg C Nm^{-3}.

2.1.1 Analytical techniques

The total VOC concentration was measured by using a total hydrocarbon analyser equipped with an FID detector (Nira Mercury 901 model, Spirax-Sarco, Spain). The composition of the gas streams were monitored using a gas chromatograph (7890 model, Agilent Technologies, USA) equipped with a 1.0 mL automated gas valve injection system, a flame ionisation detector and an HP-5 capillary column (30 m × 0.32 mm × 0.25 µm, Agilent Technologies, USA). The gas carrier was helium, and a flow-rate of 9.4 ml min^{-1} was used. The injector, oven and detector temperatures were 180, 50 and 250 °C, respectively.

CO_2 concentrations in the influent and effluent gas streams were periodically determined using a CARBOCAP® carbon dioxide analyser (GM70 model, Vaisala, Finland). The pressure drop was monitored daily with a digital manometer (KIMO, MP101 model, Spain). Temperature and pH were also measured daily for the biofilter leachate. The moisture content of the biofilter media was determined once a week at two locations (upper and lower), using the dry weight method. Similarly, the conductivity and pH of the BTF recirculation solution were analysed daily. Soluble chemical oxygen demand (COD), suspended solids, and nitrate concentrations in the recirculation solution were periodically measured.

2.2 Pilot-scale unit

The pilot-scale plant was supplied by Pure Air Solutions B.V. (The Netherlands). Pure Air Solutions has developed a biotrickling pilot-scale unit, using its innovative abatement technology (VOCUS™ Biotrickling Filter System). A schematic of this unit is presented in Fig. 5. The biological reactor, a column with a volume of 0.75 m^3, was randomly filled with two inches of nominal diameter Flexiring™ propylene rings (Koch-Glistch B.V.B.A., Belgium), with a 93% void fraction. The bioreactor was operated in counter-current mode. VOC polluted air from the factory was introduced below the column at a flow rate between 34 and 90 m^3 h^{-1}. Recirculated water was poured on the top of the filter media at a flow rate of 1.2 m^3 h^{-1} and the spraying frequency was fixed at 20 minutes per hour. The trickled

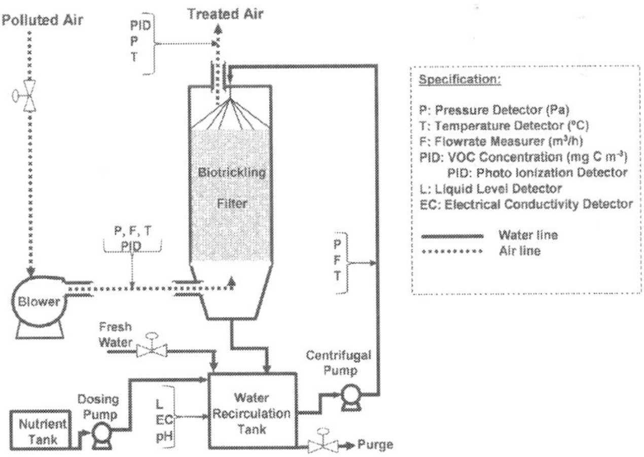

Fig. 5. Schematic of the pilot-scale biotrickling filter unit (VOCUS™).

water was collected in a 0.4 m^3 recirculation tank. The liquid level of this tank was controlled, so that fresh water was added when the level of water decreased as a consequence of evaporation. Recycled water was fully drained and replaced with fresh water once a month. A nutrient dosing system composed of a 100 L nutrient vessel and a dosing pump was set-up in the pilot unit. The nutrient solution (pH = 7, 3 g N L^{-1}, 0.6 g P L^{-1}, trace elements) was added to the recirculation tank at a maximum rate of 0.15 L h^{-1} in order to achieve a $C_{degraded}/N_{supplied}$ mass ratio above 30 - 40. The system was equipped with a programmable logic controller, and a set of sensors and devices enabling control and monitoring of the plant via modem communication.
The pilot-scale biotrickling system was inoculated with activated sludge from the municipal Wastewater Treatment Plant (located in Soria, Spain) without further acclimation, in order to simulate operational protocols at an industrial site. An inoculum volume of 100 L was added to the recirculation tank and continuously flowed through the bed for 72 hours.
At the plastic coating facility, rear view mirrors for cars are coated in three serial robotic spray booths (primer, base coat and clear coat layers). From the three existing sources of VOC emissions, the clear coating spray booth was selected as the focus for this work. The gaseous emissions from this booth were piped to the pilot unit, and contained a mixture of oxygenated and aromatic compounds coming from the applied product formulation (a mixture of a specific clear coat with its thinner) in the coating process. The pilot-scale unit was operated at EBRTs between 30 and 80 seconds (air flows between 34 and 90 m^3 h^{-1}) over a period of 3 months, in order to determine the minimum EBRT value that enables legal regulations to be met (emission limit value (ELV): an average value of 75 mg C Nm^{-3} for all valid readings, with none of the hourly averages exceeding the ELV by more than a factor of 1.5).

2.2.1 Analytical techniques

Inlet and outlet gas temperatures, inlet and outlet total VOC concentrations (measured using two photo ionisation detectors, or PIDs), air flow, the pressure drop between the gas inlet and the outlet of the media bed, tank levels, and the conductivity, pH and temperature of the trickling solution were continuously monitored. A total hydrocarbon analyser (Nira Mercury 901 model, Spirax-Sarco, Spain) was periodically used to check and calibrate the PIDs sensors. Samples of the recirculated water were collected to analyse COD, suspended solids and nitrogen and phosphorus content.

2.3 Parameters for characterisation of biodegradation performance

In general, the performance of the above described biotechnologies is evaluated using the parameters defined below:
Empty bed residence time (s):

$$EBRT = V_f / Q \tag{1}$$

where Q = air flow rate (m^3 s^{-1}) and V_f = filter bed volume (m^3).
Removal efficiency (%):

$$RE = 100 \cdot (1 - C_o / C_i) \tag{2}$$

where C_i and C_o = inlet and outlet pollutant concentration, respectively (g Nm^{-3}).

Inlet load (g m-3 h-1):

$$IL = C_i Q / V_f \tag{3}$$

Elimination capacity (g m-3 h-1):

$$EC = (C_i - C_0) Q / V_f \tag{4}$$

3. Results and discussion

3.1 Laboratory-scale systems

The performance parameters of both biotechnologies are summarised in Table 3 for the different stages of the experimental plan. The performance of each unit during the entire operation period is shown in Fig. 6, illustrating the different stages described in Table 3. The evolution of inlet and outlet total VOC concentrations and the total removal efficiency (RE) for the biofilter and biotrickling filter are plotted in Fig. 6 (a) and 6 (b), respectively. The average temperature in the reactors was 25.3 ± 1.6 and 21.2 ± 2.3°C for BF and BTF, respectively. Biofilter leachate pH was maintained at approximately 5.8 ± 0.2, while the pH and conductivity of the BTF recirculation solution were 8.7 ± 0.3 and 4.3 ± 1.4 mS cm^{-1}, respectively. Pressure drop ranged from 330 to 500 Pa m^{-1} for the BF unit, and from 40 to 280 Pa m^{-1} for the BTF system, indicating the absence of clogging. Soluble COD values for the BTF recirculation solution were periodically analysed, and were stable at approximately 154 ± 56 mg L^{-1}, representing less than 2% of the weekly inlet load fed to the system. Therefore, the organic carbon quantity removed in the purge was considered negligible. For the BF unit, the moisture content of the media (a key parameter for optimum biofilm development), was maintained at appropriate values and a slight stratification was observed: values (wet basis, %wt) varied between 78.4% ± 2.9% in the upper zone and 80.7% ± 1.7% in the lower zone of the bed.

	Days	EBRT, s	IL, g C m^{-3} h^{-1}	EC, g C m^{-3} h^{-1}	RE, %
Biofilter					
Stage BF1	0 - 112	60.0 ± 1.3	18.8 ± 1.5	17.7 ± 1.4	94.1 ± 2.7
Biotrickling filter					
Stage BTF1	0 - 12	59.1 ± 0.1	11.2 ± 0.8	8.5 ± 0.8	74.6 ± 6.4
Stage BTF2	12 - 36	59.1 ± 0.1	21.1 ± 1.9	11.5 ± 1.7	54.8 ± 7.6
Stage BTF3	36 - 50	30.0 ± 0.3	19.6 ± 1.4	13.0 ± 1.4	66.0 ± 2.5
Stage BTF4	50 - 63	30.0 ± 0.3	31.7 ± 2.9	21.2 ± 3.3	66.6 ± 4.4
Stage BTF5	63 - 79	15.1 ± 0.7	39.6 ± 1.7	21.2 ± 2.4	53.5 ± 5.4
Stage BTF6	79 - 92	15.1 ± 0.7	72.4 ± 1.2	32.7 ± 1.3	45.3 ± 2.2

Table 3. Operational and performance parameters for both systems on the different stages.

After inoculation, both systems were operated at an approximate EBRT of 60 s, with moderate ILs (18.8 and 11.2 g C m^{-3} h^{-1} for BF and BTF, respectively). High and stable REs were reached in 2 - 5 days for both bioreactors, indicating appropriate development of the inoculum. For the BF unit, total REs remained at very high values (as high as 97%) throughout the entire experimental period. In the case of the BTF unit, no complete removal was obtained, even for the lower applied IL (stage BTF1), due to the low removal of toluene and m-xylene in comparison with the complete degradation of n-butyl acetate. Kinney & Moe (2004) also observed lower biodegradability of aromatic compounds in the treatment of gas emissions contaminated with 2-pentanone, n-butyl acetate, ethyl 3-ethoxypropianate, toluene and p-xylene, using a biotrickling filter. The decrease in the EBRT caused a progressive drop in the total RE of the BTF system. For example, at an inlet VOC concentration of approximately 170 mg C Nm^{-3}, average RE values decreased from 75 to 54% for EBRTs ranging from 59 to 15 s (stages BTF1 and BTF5, respectively). At this concentration, the BTF was able to meet legal limits (ELV), even for the lowest EBRT (IL of 39.6 g C m^{-3} h^{-1}, stage BTF5). Similarly, the total RE was negatively affected when the inlet VOC concentration was duplicated. In this case, total RE values increased slightly, from 45 to 54%, when IL was dropped from 72.4 to 39.6 g C m^{-3} h^{-1} (stages BTF6 and BTF5, respectively).

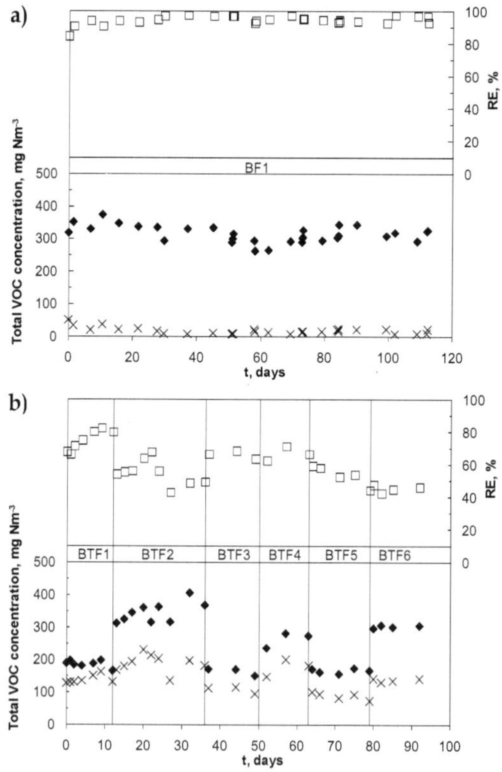

Fig. 6. Evolution of inlet VOC concentrations (♦), outlet VOC concentrations (X) and total RE values (□) in (a) BF and (b) BTF with time.

The variation of EC with IL at different EBRTs for the biofilter and the biotrickling filter is plotted in Fig. 7. ECs were calculated from the top of the bioreactors to each sampling port (first quarter, half, three-quarters and total bed volume for the BF; and first third, two-thirds and total bed volume for the BTF). Maximum ECs for both systems were estimated from data taken from the first section of the bed. EC values were nearly directly proportional to the total IL of the VOC mixture: up to 50 g C m^{-3} h^{-1} for the BF, and 20 g C m^{-3} h^{-1} for the BTF. No significant differences between both biotechnologies were observed for the maximum EC values. A maximum EC value of 45 g C m^{-3} h^{-1} was obtained for both systems. It is worth noting the slight influence of the applied EBRT (which ranged between 15 and 60 s) on EC values for the applied IL in the biotrickling filter.

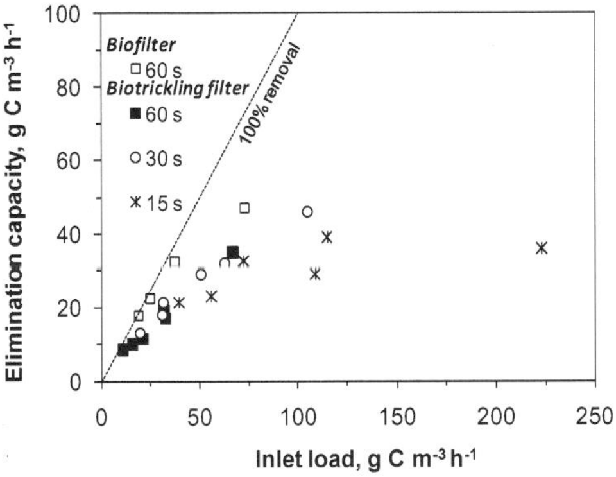

Fig. 7. Variation of EC vs. IL at different EBRTs for the biofilter and biotrickling filter.

The evolution of CO_2 production as a function of elimination capacity (EC) is presented in Fig. 8 for both biotechnologies. A proportional ratio exists between EC and CO_2 production. As can be seen, the biofilter datum follows the same trend as the biotrickling filter results. Linear regression of these data yielded a value of 1.74 g CO_2 g C^{-1}. Assuming a general biomass composition formula of $C_5H_7O_2N$, and negligible organic and inorganic carbon removal during the periodic purges, overall yield coefficient (defined as g C of dry biomass synthesised per g C of substrate consumed) can be determined from the following biodegradation reaction balance:

$$C_xH_yO_z + aO_2 + NH_3 \longrightarrow bC_5H_7O_2N + cCO_2 + dH_2O \quad (5)$$

yielding a value of 0.53.

The degradation of each of the three compounds of the VOC mixture is shown in Fig. 9, where normalised pollutant gas concentration profiles along the bed length of the BF and BTF have been plotted. For the BTF, concentration profiles from the three highest ILs are shown. As can be seen in the Fig. 9, complete removal of n-butyl acetate was always observed in the first two-thirds of the bioreactors for all stages. Greater penetration was observed for aromatic compounds, with individual removal efficiencies of approximately 30% for each aromatic compound in the first quarter of the BF, whereas no degradation of these compounds was obtained in the first third of the BTF. Among aromatic compounds, greater emissions were observed for m-xylene, which is less biodegradable than toluene. Previous studies have reported similar phenomena (Álvarez-Hornos et al., 2007; Paca et al., 2006).

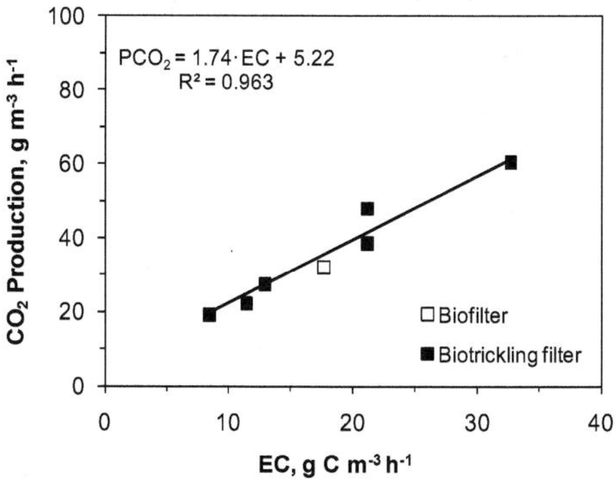

Fig. 8. The relationship between EC and CO_2 production for both biotechnologies.

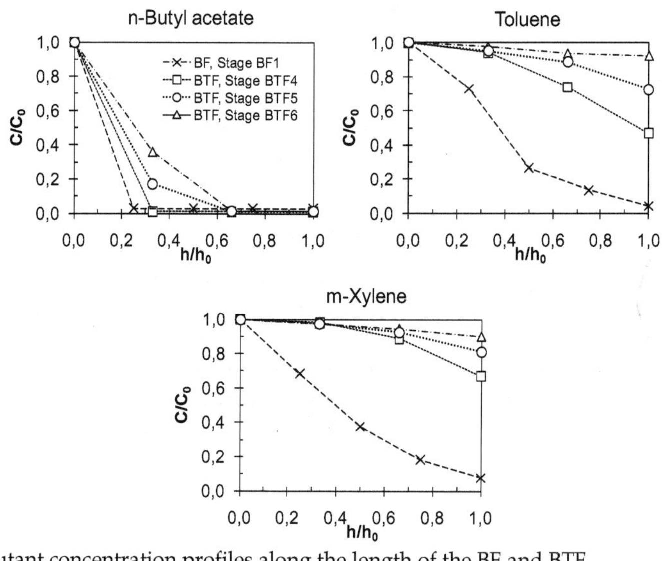

Fig. 9. Pollutant concentration profiles along the length of the BF and BTF.

3.2 Pilot unit

3.2.1 Characterisation of emissions from the spray paint booth

The industrial site operates full-time, 24-hours a day, from 6:00 am on Monday to 6:00 am on Saturday, with shut down periods during weekends. Gas flow rate for the emissions coming from the clear coating spray booth was approximately 35 000 $Nm^3\ h^{-1}$, with daily average emission temperatures ranging between 8 and 22 °C throughout the experimental period. The VOC concentration was relatively stable during working hours, with hourly average values ranging between 100 and 450 mg C Nm^{-3}, and with a daily average concentration of 235 ± 57 mg C Nm^{-3}. A typical VOC concentration emission pattern from the booth system over one working day is shown in Fig. 10. As can be seen, because the inlet hourly average VOC concentration exceeded the ELV by a factor greater than 1.5 (maximum hourly legal limits), treatment of booth emissions may be required.

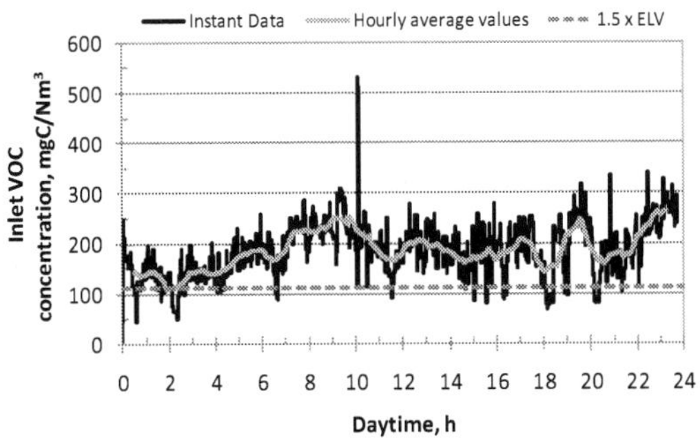

Fig. 10. Typical instant and hourly VOC concentration emission patterns from the booth system. The discontinuous line represents a factor of 1.5 of the ELV.

GC-MS analyses of both emission samples and samples of the solvents used in the coating process detected the following compounds: oxygenated compounds such as butyl acetate, butyl glycol acetate and methyl acetate (comprised > 70%); and solvent naphtha including small proportions of aromatics compounds. The two product formulations used by the facility during the experimental period are summarised in Table 4.

	Product A	*Product B*
Clear coat *(70% wt)*	xylene 10-25% n-butyl acetate 10-25% 2-methoxy-1-methylethyl acetate 10-25% Isobutyl acetate 10-25%	butyl glycol acetate 25-50% naphtha, light aromatic 10-25% naphtha, heavy aromatic 10-25% 1,2,4-trimethyl benzene 2.5-10%
Thinner *(30% wt)*	n-butyl acetate 100%	n-butyl acetate 25-50% butyl glycol acetate 10-25% 1,2,4-trimethyl benzene 2.5-10% naphtha, light aromatic 2.5-10% xylene 2.5-10%

Table 4. Product formulations used for the clear coating process during the experimental period.

3.2.2 Pilot unit performance

After inoculation, the pilot unit was operated at an approximate EBRT of 20 s. During the first month, a progressive increase in removal efficiency was observed until stable performance was achieved. Once the system was stable, the air flow rate through the system was periodically adjusted to values of 33.8, 60, 41.5 and 90 $Nm^3\ h^{-1}$, corresponding to EBRTs of 80, 65, 45 and 30, respectively. In Fig. 11, (a) to (d), the inlet and outlet hourly average VOC concentrations obtained during a typical day of operation are plotted for the four EBRTs applied.

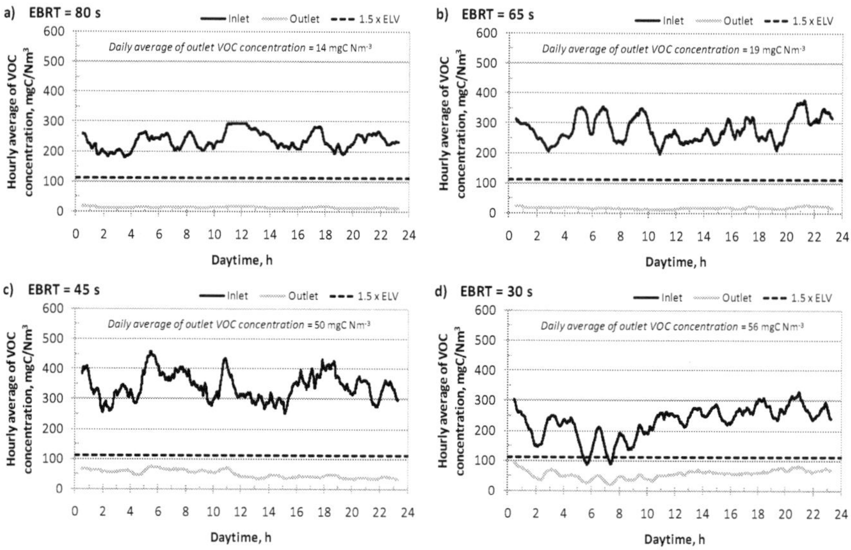

Fig. 11. Daily pilot unit monitoring at four EBRTs: hourly average values working at an EBRT of (a) 80 s, (b) 65 s, (c) 45 s and (d) 30 s. The discontinuous line represents a factor of 1.5 times the ELV.

As can be seen, outlet emissions met legal regulations for all of the tested EBRT, with daily average values of 14, 19, 50 and 56 mg C Nm^{-3}, for 80, 65, 45 and 30 s of EBRT, respectively. Therefore, with respect to legal compliance, bioreactor performance was not significantly affected by the decrease in the EBRT. It is worth noting that outlet emissions did not exceeded the legal limit after the facility started the coating process on Monday mornings after weekend shutdown periods (data not shown here).

Fig. 12 summarises the performance of the pilot unit during the test period. The variation of overall RE versus the four applied EBRTs is presented in Fig. 12 (a). The variation of the daily average outlet concentration versus the daily average inlet concentration as a function of the tested EBRT is shown in Fig. 12 (b). As can be seen in Fig. 12 (a), the overall removal efficiency was not significantly affected (RE values were between 85 and 95%) when the pilot unit was operated with EBRTs ranging between 45 and 80 s. However, RE dropped to approximately 70% when the EBRT decreased to 30 s. Still, the lowest EBRT (30 s) was enough to maintain outlet VOC concentrations below legal limits (75 mg C Nm^{-3}), despite the observed decrease in RE (see Fig. 12 (b)). Furthermore, when the pilot unit was operated at an EBRT > 45 s, the outlet concentration was always in compliance with the ELV over the entire inlet concentration range. Therefore, the minimum EBRT value that would allow legal regulations to be met within an adequate safety margin is between 35 and 40 s.

The pressure drop in the pilot unit was low (< 60 Pa m^{-1}) during the entire test period, indicating that the short-term starvation periods during weekend closures, combined with the relatively low VOC inlet load (~ 25 g C m^{-3} h^{-1}) avoided clogging episodes in the reactor. Soluble COD values measured in the recirculation liquid varied between 200 and 300 mg COD L^{-1}, representing < 1% of the inlet organic carbon fed to the pilot unit. The pH and conductivity average values of the recirculation liquid were 7.2 and 1.0 mS cm^{-1}, respectively.

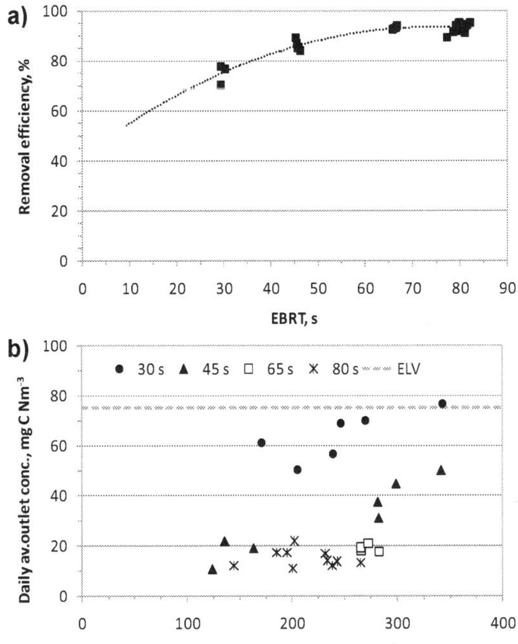

Fig. 12. Summary of pilot unit performance during the entire experimental period: (a) Overall RE vs. EBRT and (b) Daily average outlet concentration vs. daily average inlet concentration.

3.2.3 Comparison with laboratory-scale biotechnologies

A comparison of the pilot-scale unit performance with results obtained from the laboratory-scale biofilter and biotrickling filter (used to treat a continuous VOC emission polluted with a 2:1:1 (wt) n-butyl acetate:toluene:m-xylene mixture) is shown in Fig. 13. The variation in RE versus EBRT for the pilot unit, biofilter and biotrickling filter, for every section of the bed height, is presented in Fig. 13 a. Results from the biotrickling filter correspond to an inlet VOC concentration of approximately 200 mg C Nm^{-3} (stage BTF1, BTF3 and BTF5), similar in magnitude to the industrial VOC emission at the pilot unit. The variation of EC with IL for all data obtained from the three systems is plotted in Fig. 13 (b).

As can be seen, results obtained from the pilot-scale unit match those from the laboratory-scale experiments. The slightly better performance of the pilot unit versus the lab-scale BTF may be due to the following: (1) the oxygenated compound composition tested in the pilot-scale industrial emission (> 70%) was higher than that tested in the laboratory-scale BTF unit (50%); (2) greater variability in the organic composition of the industrial emission could have derived from a more complex microbial ecosystem; and (3) the short-term starvation periods due to weekend closures at moderate operational conditions could help to improve the activity of the biological system (Wright et al., 2005). Importantly, comparison between the laboratory-scale and pilot-scale units indicate that laboratory studies could be a timesaving tool for obtaining valuable data for scale-up.

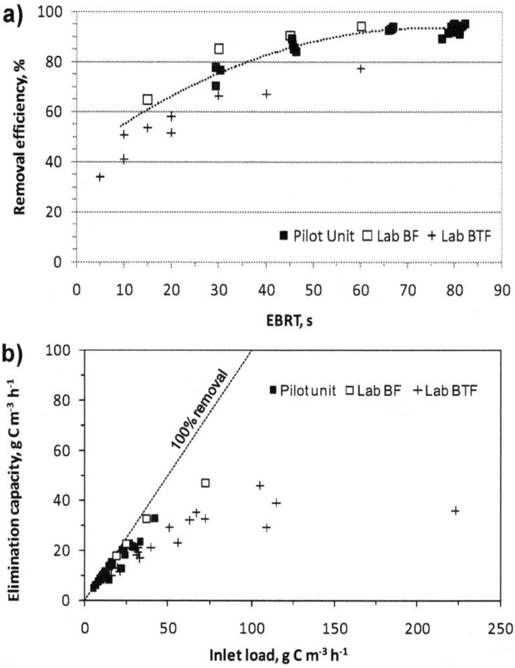

Fig. 13. Comparison of the performance of the pilot unit with results obtained from the laboratory-scale biofilter and biotrickling filter: (a) Overall RE vs. EBRT and (b) EC vs. IL.

4. Conclusions

Biological removal of VOCs from exhaust gases from a robotic spray paint booth at an industrial plastic coating facility has been studied. The present work evaluated the performance of environmentally friendly biotechnologies, such as biofilters and biotrickling filters, on a laboratory-scale; for comparison with a biotrickling filter pilot-scale unit connected to an industrial facility. Treatment of air contaminated with three compounds typically found in air emissions from paint and coating processes (n-butyl acetate, toluene and m-xylene) was conducted using the laboratory-scale systems. Although the biofilter displayed a higher removal efficiency than the biotrickling filter during the test period, a similar maximum elimination capacity of 45 g C m^{-3} h^{-1} was estimated for both laboratory biotechnologies. Results from the laboratory-scale experiments demonstrate the slight influence of the applied EBRT (which ranged between 15 and 60 s) on the EC for the applied IL in the biotrickling filter. However, application of EBRTs lower than 60 s to the BF would cause greater drying of the organic packing material, resulting in a decrease in the removal efficiency. Because of the higher EBRTs required, industrial scale biofilter applications would require greater sizes, and therefore higher investment costs. Moreover, BTF units are compatible with 8 - 10 m high bioreactors, enabling construction of units with smaller footprints than comparable BF units. In addition, it is also worth noting that the pressure drop in the BTF unit (< 60 Pa m^{-1}) was lower than that observed in the BF (around 400 Pa m^{-1}), due to the larger void fraction of the inert packing material: this difference in pressure drop is significant, because lower values require less power consumption from the blower equipment. Although complete removal of n-butyl acetate was achieved in both cases, aromatic compounds penetrated throughout the bed, indicating that mechanisms of gas removal for hydrophilic and hydrophobic compounds may be significantly different. Our evaluation of the operation of the pilot-scale plant for 3 months demonstrates the suitability and robustness of the biological process for controlling VOC emissions from a robotic spray paint booth. Importantly, legal emission limits were always achieved for the four tested empty bed residence times (EBRTs). The minimum EBRT required to meet legal regulations appears to be between 35 and 40 s. From an engineering perspective, comparison of the laboratory-scale and pilot-scale units indicates that laboratory-scale studies could be a timesaving tool for obtaining valuable data to establish safe operation limits which should allow legal requirements to be met.

5. Acknowledgments

Financial support from the Ministerio de Ciencia e Innovación (Spain, research projects CTM2010-15031 and TRA2009_0135) is acknowledged. F. J. Álvarez-Hornos acknowledges Ministerio de Educación, Spain, for providing a post-doctoral contract by means of Programa Nacional de Movilidad de Recursos Humanos del Plan Nacional de I+D+i 2008-2011. The authors would like to give special thanks to Pure Air Solutions and FICOMIRROR, SA (Ficosa International Group) for their consistent collaboration and support.

6. References

Álvarez-Hornos, F.J., Gabaldón, C., Martínez-Soria, V., Marzal, P., Penya-roja, J.M. & Izquierdo, M. (2007). Long-term performance of peat biofilters treating ethyl acetate, toluene, and its mixture in air. *Biotechnology & Bioengineering*, Vol.96, pp. 651-660, ISSN 0006-3592

Boswell, J.T., John, P.C., Stewart, B., Forrest, S., Branchik, R. & Morgan, S. (2001). Biofiltration of VOCs from Paint Manufacturing, *Proceedings of A&WMA's 94th Annual Meeting & Exhibition*, Orlando, USA, June 24-28, 2001

Boswell, J. (2009). Clearing the air: the use of bio-oxidation for industrial air emissions control. *Engineered Wood Journal*, Fall, pp. 26-28

Cox, H.H.J. & Deshusses, M.A. (2001). Biotrickling filters, In: *Bioreactors for waste gas treatment*, C. Kennes & M.C. Veiga (Ed.), 99-131, Kluwer Academic Publisher, ISBN: 978-0-7923-7190-8, Dordrecht, The Netherlands

European Commission (2003). *IPPC Reference document on Best Available Techniques in common waste water and waste gas treatment/management systems in the chemical sector*, Sevilla, Spain

European Environment Agency (2010). *European Union emission inventory report 1990–2008 under the UNECE Convention on Long-range Transboundary Air Pollution (LRTAP)*. EEA Technical report 7/2010, ISBN 978-92-9213-102-9, Copenhagen, Denmark

European Pollutant Release and Transfer Register (2008). Available from http://prtr.ec.europa.eu/

Higuchi, T., Morita, Y. & Minato, R. (2010). Biofiltration of the exhaust gas contaminated by six VOC compounds for testing the feasibility of its application to a painting process, *Proceedings of the 2010 Duke-UAM Conference on Biofiltration for Air Polllution Control*, pp. 273-278, Washington, USA, October 28-29, 2010

Hwang, J.W., Choi, C.Y. & Park, S. (2008). Biodegradation of gaseous styrene by Brevibacillus sp. using a novel agitating biotrickling filter. *Biotechnology Letters*, Vol.30, pp. 1207-1212, ISSN 0141-5492

Kennes, C., Cox, H.H.J., Doddema, H.J. & Harder, W. (1996). Design and performance of biofilters for the removal of alkylbenzene vapors. *Journal of Chemical Technology and Biotechnology*, Vol.66, pp. 300-304, ISSN 1097-4660

Kim, D.J. & Kim, H. (2005). Degradation of toluene vapor in a hydrophobic polyethylene hollow fiber membrane bioreactor with Pseudomonas putida. *Process Biochemistry*, Vol.40, pp. 2015-2020, ISSN 1359-5113

Kinney, K.A. & Moe, W.M. (2004). *Optimization of an innovative biofiltration system as a VOC control technology for aircraft painting facilities*, Final report - SERDP Project CP 1104. Air Force Research Laboratory, Tyndal AFB, EEUU.

Koutinas, M., Peeva, L.G. & Livingston, A.G. (2005). An attempt to compare the performance of bioscrubbers and biotrickling filters for degradation of ethyl acetate in gas streams. *Journal of Chemical Technology and Biotechnology*, Vol.80, pp. 1252-1260, ISSN 1097-4660

Martínez-Soria, V., Gabaldón, C., Penya-roja, J.M., Palau, J., Álvarez-Hornos, F.J., Sempere, F. & Soriano, C. (2009). Performance of a pilot-scale biotrickling filter in controlling the volatile organic compound emissions in a furniture manufacturing facility. *Journal of the Air & Waste Management Association*, Vol.59, pp. 998-1006, ISSN 1047-3289

Mathur, A.K. & Majumder, C.B. (2008). Biofiltration and kinetic aspects of a biotrickling filter for the removal of paint solvent mixture laden air stream. *Journal Hazardous Materials*, Vol.152, pp. 1027-1036, ISSN 0304-3894

Paca, J., Klapkova, E., Halecky, M., Jones, K. & Webster, T.S. (2006). Interactions of hydrophobic and hydrophilic solvent component degradation in an air-phase biotrickling filter reactor. *Environmental Progress*, Vol.25, pp. 365-372, ISSN 1944-7450

Popescu, P. & Ionel, I. (2010). Anthropogenic air pollution sources, In: *Air quality*, K. Ashok, (Ed.), 1-22, Sclyo, ISBN 978-953-307-131-2, Rijeka, Croatia

Sempere, F., Gabaldón, C., Martínez-Soria, V., Marzal, P., Penya-roja, J.M. & Álvarez-Hornos, F.J. (2008). Performance evaluation of a biotrickling filter treating a mixture of oxygenated VOCs during intermittent loading. *Chemosphere*, Vol.73, pp. 1533-1539, ISSN 0045-6535

Shareefdeen, Z. & Singh, A. (2005). *Biotechnology for odor and air pollution control*. Springer, ISBN 978-3-540-74049-0, Heidelberg, Germany

Steele, J.A., Ozis, F., Fuhrman, J.A. & Devinny, J.S. (2005). Structure of microbial communities in ethanol biofilters. *Chemical Engineering Journal*, Vol.113, pp. 135-143, ISSN 1385-8947

Webster, T.S., Togna, A.P., Guarini, W.J. & McKnight, L. (1999). Application of a biological trickling filter reactor to treat volatile organic compound emissions from a spray paint booth operation, *Metal Finishing*, Vol.97, pp. 20-26, ISSN 0026-0576

World Health Organization (2004). *Health aspects of air pollution*, WHO Regional Office for Europe, Copenhagen, Denmark

Wright, W.F., Schroeder, E.D. & Chang, D.P.Y. (2005). Regular transient loading response in a vapor-phase flow-direction-switching biofilter. *Journal Environmental Engineering*, Vol.131, pp. 1649-1658, ISSN 0733-9372

Part 3

Radioactive Pollution

6

Nano Aerosols Including Radon Decay Products in Ambient Air

Janja Vaupotič
Jožef Stefan Institute, Ljubljana
Slovenia

1. Introduction

Ambient air is composed of major constituents, nitrogen and oxygen, and small amounts of argon and carbon dioxide. Water vapour and trace gases are also present. It is, in fact, not a gas but an aerosol with suspended particulate matter. The particle sizes ranges from several nm for molecular clusters to about 100 µm for fog droplets and dust particles. Particles larger than 100 µm cannot stay suspended in air and may therefore not be considered as aerosols (Colbeck, 1998). Classification of aerosols with respect to their particle size differs depending on the purpose of use as well as the author (Wen, 1996; Colbeck, 1998; Turner & Colbeck, 2008; Busek & Adachi, 2008; Kumar et al., 2010b). Terms of ultrafine (<100 nm), fine (<1000 nm) and coarse (>1000 nm) particles are the preferred terms used by toxicologists and regulatory bodies. Aerosol scientists, on the other hand, refer mostly to different modes (Kumar et al., 2010b): nucleation (1–30 nm), Aitken (20–100 nm) and accumulation mode (90–1000 nm). Nevertheless, the borders are not strictly fixed and may differ from author to author. The term nano is not applied univocally either. Although it may refer to any particle of <1 µm size, it is reserved for particles of <300 nm (Kumar et al., 2010b), <100 nm (Busek & Adachi, 2008; Nowack & Bucheli, 2001), <50 nm (Morawska et al., 2008), or even smaller (Anastasio & Martin, 2001; Shi et al., 2001). In any case, monodisperse aerosols are very rare. They are mostly polydisperse with a particle size distribution which is mathematically described by either a differential or an integral distribution function $P^{\mathrm{N}}(d)$ (Wen, 1996). The latter is more often applied and, for number size distribution, expressed as:

$$P^{\mathrm{N}}(d) = -\frac{\mathrm{d}C_{\mathrm{gen}}^{\mathrm{N}}}{\mathrm{d}d} , \tag{1}$$

where $C_{\mathrm{gen}}^{\mathrm{N}}$ is the number concentration of particle with diameter d. Because of the wide range of sizes in polydisperse systems, logarithm of diameter is used instead, and Eq. 1 converts to:

$$P_{\mathrm{L}}^{\mathrm{N}}(d) = -\frac{\mathrm{d}C_{\mathrm{gen}}^{\mathrm{N}}}{\mathrm{d}(\ln d)} . \tag{2}$$

Similarly as for the number concentration, also size distributions of other property can be expressed, such as particle surface, volume, mass, activity (for radioactive aerosols) and others. For the same aerosol they appear to differ considerably from one another, in respect to both the number of peaks and peak positions in the particle size region (Colbeck, 1998; Turner & Colbeck, 2008; Busek & Adachi, 2008). For instance, the plot of the number distribution shows a maximum at smaller particle sizes than the plot of the volume (or mass) distribution, because the smaller particles, though abundant in number, have smaller volumes (smaller cube of diameter). For the same reason, the maximum in the surface distribution appears at larger particle sizes than that in the number distribution.

1.1 General aerosols

The term 'general' is used here to comprise all particles, carrying or not carrying radioactive elements (in our case radon short-lived decay products), although the contribution of the latter is minimal, as will be seen later. Atmospheric particles have natural and anthropogenic sources (Baltensperger & Nyeki, 1998; Nowack & Bucheli, 2001; Dobos et al., 2009; Kumar et al., 2010b). Forests and sea surfaces are permanent generators of aerosol particles (O'Dowd et al., 1997). Particles are also produced continuously in the atmosphere by condensation of semi-volatile organics (O'Dowd et al., 2002), and caused by photochemical reactions and gas → particle conversion (Vakeva et al., 1999; Holmes, 2007; Kumar et al., 2009). Volcanic eruptions (Amman & Burtcher, 1990), forest fires (Makkonen et al., 2010) and dust storms (Schwikowski et al., 1995) are, temporally, very strong sources, but mostly local and of short duration.

Particulates are emitted from a number of various human activities (Baltensperger & Nyeki, 1998; Kumar et al., 2010b). They are released at huge amounts through the chimneys of thermal power plants burning fossil fuel or biomass, incinerators, mineral mining and milling facilities and some others. In urban areas an important or even major particle source is traffic (Shi et al., 2001; Han et al., 2005; Agus et al., 2007; Young & Keeler, 2007; Busek & Adachi, 2008; Morawska et al., 2008; Kumar et al., 2010b). While particles emitted by vehicles using conventional and bio fuel are mostly smaller than 150 nm, debris generated by the tyre-road interaction fall into the 2.5–10 µm range. Nanoparticles are also produced intentionally (Kumar et al., 2010a) to be used as constituents in electronics, medicines, pharmaceuticals, cosmetics, paints and a variety of other consumer products. Nanotechnology is increasing fast and so is the possibility for the nanoparticles to appear in air of workplaces and to be released into the outdoor atmosphere and enter the living environment (Mackay & Henry, 2009).

Apart from size, aerosol particles also differ markedly in their shape, structure and chemical composition, which is beyond the scope of this paper.

When breathing air, aerosol particulates are deposited partly on the walls of the respiratory tract. Mathematical simulations have shown that their deposition depends strongly on the particle size (Hofmann & Koblinger, 1990; Alföldy et al., 2009). Thus, for instance (Oberdörster et al., 2005), about 90 % of the inhaled 1 nm particles are deposited in the nasopharyngeal region and the rest, in the tracheobronchial region with no deposition in the alveolar region. 5 nm particles are almost equally deposited in all three regions. On the other hand, half of the 20 nm particles are deposited in the alveolar region and the remaining half in the other two regions equally. Physical translocation and clearance in the respiratory tract are also size-dependent. Aerosol particles enter the body also by ingestion and absorption by skin. This uptake is more effective for smaller particles than for larger ones, nonetheless minor compared with inhalation. Because the ratio of the numbers of surface to bulk atoms increases exponentially with reducing size, smaller particles are expected to be chemically and biochemically more reactive, and thus potentially more toxic, than larger ones (Oberdörster et al., 2005). It has been recognised that nanoparticles cause oxidation stress, pulmonary inflammation and cardiovascular events, although the mechanisms of these detrimental effects are not entirely understood (Nowack & Bucheli, 2001; Oberdörster et al., 2005; Nel et al., 2006; Kumar et al., 2010b).

1.2 Radioactive aerosols – radon short-lived decay products

Atoms of several radionuclides attach to the atmospheric particles which thus appear as radioactive aerosols (Schery, 2001; Papastefanou, 2008). They are of different origin and may be classified as follows (Papastefanou, 2008): (i) cosmogenic: ^{7}Be, ^{22}Na, ^{32}P and ^{35}S; (ii) radon and thoron short-lived decay products; (iii) fission products: ^{89}Sr, ^{90}Sr, ^{103}Ru, ^{131}I, ^{132}Te, ^{137}Cs and ^{140}Ba; (iv) originated from high-energy accelerators: ^{7}Be, ^{22}Na, ^{24}Na and ^{52}Mn; (v) plutonium isotopes from nuclear weapons testing and reactor accidents; and (vi) mine particulate matter containing radionuclides of the uranium, thorium and actinium radioactive decay chains. We shall focus here on radon short-lived decay products.

There are three primordial radioactive decay chains in the earth crust (Nero, 1988): the thorium chain starting with ^{232}Th, the uranium chain starting with ^{238}U, and the actinium chain starting with ^{235}U. In each chain, radioactive transformation of radium results in formation of radon, the only gas in the entire decay chains. Thus, ^{220}Rn (thoron, half-life $t_{1/2}$ = 55.6 s), ^{222}Rn (radon, 3.82 days) and ^{219}Rn (actinon, 3.9 s) radon isotopes, respectively, are created. Only a fraction of radon atoms succeed in leaving the mineral grain due to their recoil energy, thus entering the void space. From there, radon travels through the medium either by diffusion or, more effectively and to longer distances, carried by gas or water (Etiope & Martinelli, 2002). On its way, it accumulates in underground rooms (mines, karst caves, fissures, basements) and eventually enters the atmosphere, and appears in the air of living and working environments. We will deal here with ^{222}Rn (hereafter called radon, Rn) which, usually, only appears at significant levels in the environmental air because of its relatively long half-life, as compared with the half-life of ^{220}Rn and especially that of ^{219}Rn.

Radon α-transformation is followed by a radioactive chain of its short-lived decay products (RnDP): ^{218}Po (α, 3.05 min) → ^{214}Pb (β, 26.8 min) → ^{214}Bi (β, 19.7 min) → ^{214}Po (α, 164 μs) (Nero, 1988). Initially, they appear mostly as positive ions (Chu & Hopke, 1988; Porstendörfer & Reineking, 1992; Hopke, 1996) which react with molecules of trace gases and vapours (mostly water) in air, are partially oxidized and form small charged clusters. They also get neutralized (Goldstein & Hopke, 1985; Porstendörfer & Reineking, 1992; Hopke, 1996) by (i) recombination with ion pairs produced by α, β and γ emissions and recoil atoms during radioactive transformations of airborne radionuclides, as well as by background γ and cosmic rays, (ii) electron scavenging by OH radicals formed by radiolysis of water molecules, and (iii) charge transfer from molecules of lower ionisation potential. The state of ^{218}Po and $^{218}PoO_2$ at the moment of its α-transformation is decisive because it determines the initial characteristics and behaviour of the subsequent members in the chain. For ^{218}Po clusters in 50 % humid air at an ionisation rate of 3.2 pC kg^{-1} s^{-1} (45 μR h^{-1}), the rates of the above three processes of neutralisation are 0.07×10^{-2} s^{-1}, 1.07×10^{-2} s^{-1} and 0.4×10^{-2} s^{-1}, respectively, and more than half of the clusters are neutral (Dankelmann et al.,

2001). These processes are accompanied and followed by attachment of clusters (Hopke, 1996; Dankelmann et al., 2001; Schery, 2001; Pagelkopf & Porstendörfer, 2003; Papastefanou, 2008), both charged and already neutralised, to the background atmospheric aerosol particles. Because of the recoil energy gained during α-transformation of the attached parent atoms, a considerable proportion of the RnDP atoms desorbs from the aerosol particles (Mercer, 1976; Porstendörfer, 1984), the phenomenon being more pronounced for smaller aerosol particles (Tu et al., 1994).

The total concentration of RnDP in air is reported as the equivalent equilibrium activity concentration (C^{A}_{RnDP}, Bq m^{-3}) expressed as (Nero, 1988):

$$C^{A}_{RnDP} = 0.1065\, C^{A}_{218Po} + 0.515\, C^{A}_{214Pb} + 0.379\, C^{A}_{214Bi}\,, \tag{3}$$

where C^{A} (Bq m^{-3}) stands for the individual activity concentrations of ^{218}Po, ^{214}Pb and ^{214}Bi. Because of its short life time, ^{214}Po activity is equal to the activity of ^{218}Bi and is therefore already included in the last term of Eq. 3. Due to air movement and deposition of RnDP onto surfaces, individual activity concentrations of RnDP are always lower than that of radon (C^{A}_{Rn}) and secular equilibrium between radon and RnDP is never reached in the ambient air, and the degree of equilibrium is defined as $F = C^{A}_{RnDP} / C^{A}_{Rn}$.

On the world average, RnDP contribute about half to the effective dose (radon contribution is minor) a member of the general public receives from all natural radioactivity (United Nations Scientific Committee on the Effects of Atomic Radiation [UNSCEAR], 2000, thereafter UNSCEAR, 2000), and are a major cause of lung cancer, second only to cigarette smoking (Darby et al., 2005). For this reason, their levels in living and working environments are of great social concern and scientific challenge.

In contrast to general aerosols, the health detrimental effects of radioactive aerosols have been known for a long time and their dosimetry well elaborated. For general purposes of radiation protection, the International Commission of Radiological Protection in its Publication 65 recommends for radon dosimetry a dose conversion factor (D_{CFE}) of 4 mSv WLM^{-1} at home and 5 mSv WLM^{-1} at the workplace (International Commission on Radiological Protection [ICRP], 1994, thereafter ICRP, 1994), as the values deduced from epidemiologic studies. Above, 1 WLM (working-level-month) is the exposure gained by 170 hour breathing air in which the potential α-energy concentration of RnDP ($E_{\alpha RnDP}$) is 1.3×10^{8} MeV m^{-3}. $E_{\alpha RnDP}$ (MeV m^{-3}) is expressed through the activity concentrations (C^{A}, Bq m^{-3}) of ^{218}Po, ^{214}Pb and ^{214}Bi, as (Nero, 1988):

$$E_{\alpha RnDP} = 3690 C^{A}_{218Po} + 17830 C^{A}_{214Pb} + 113120 C^{A}_{214Bi}\,. \tag{4}$$

On the other hand, Birchall and James (Birchall & James, 1994; Marsh et al., 2002) elaborated the dosimetric approach of calculating the dose conversion factor (D_{CFD}) which is used mainly for research purposes. They also showed that the parameter mostly affecting D_{CFD} is the fraction (f^{un}) of the unattached RnDP, defined as (Nero, 1988):

$$f^{un} = \frac{C^{Aun}_{RnDP}}{C^{A}_{RnDP}} \tag{5}$$

where C^{Aun}_{RnDP} is the equivalent equilibrium concentration of the unattached RnDP, obtained if activity concentrations of only the unattached ^{218}Po, ^{214}Pb and ^{214}Bi are taken into Eq. 3. Further, they expressed D_{CFD} based on f^{un} with an empirical formula:

$$D_{CFD} = 11.35 + 43\, f_{un}. \quad (6)$$

In addition, Porstendörfer (1996) proposed relationships between D_{CFD} and f_{un} for 0.8 nm RnDP:

$$D_{CFD} = 101\, f_{un} + 6.7(1 - f_{un}) \text{ for nasal breathing} \quad (7)$$

and

$$D_{CFD} = 23\, f_{un} + 6.2(1 - f_{un}) \text{ for mouth breathing.} \quad (8)$$

Because f_{un} is the key factor in radon dosimetry models it is of great interest to know the parameters influencing its values and time variation in various environments.

1.3 Size distribution of radon short-lived decay products

For RnDP, as for other radioactive aerosols (Schery, 2001; Papastefanou, 2008), the activity size distribution $P_L^A(d)$ is most often reported, rather than the number size distribution $P_L^N(d)$ (Eq. 2); it is expressed as:

$$P_L^A(d) = -\frac{dC^A}{d(\ln d)}. \quad (9)$$

The two distributions differ because the attachment coefficient $\beta(d)$ of RnDP to aerosol particles is a function of size; the relationship between them is (Porstendörfer, 1996; Porstendörfer et al., 2000):

$$P_L^A(d) = \frac{C^A}{X}\beta(d) P_L^N(d). \quad (10)$$

Here, X is the attachment rate and expresses the adsorption velocity of RnDP to aerosol particles at their number concentration C^N, and $\beta(d)$ is determined by the diffusion coefficient, velocity and mean free path of RnDP. Under different environmental conditions, values of X were found to range from 7 h^{-1} to 1660 h^{-1} (Reineking et al., 1985; Tu et al., 1991) and those of $\beta(d)$, from 0.2×10^{-3} $cm^3\ h^{-1}$ to 14.1×10^{-3} $cm^3\ h^{-1}$ (Reineking et al., 1985; Porstendörfer, 1984; Tokonami, 2000).

According to a review by Porstendörfer & Reineking, (1992), the activity median diameter (AMD) of the RnDP clusters falls in the range of 0.9 nm to 30 nm, while the activity median aerodynamic diameter (AMAD) of the aerosol particles carrying RnDP attached, falls in the range of 50 nm to 500 nm. In a radon chamber containing carrier aerosol, AMD values of 0.82, 0.79, 1.7 and 0.82 nm were obtained for unattached ^{218}Po, ^{214}Pb, ^{214}Bi and ^{214}Po, respectively (Fukutsu et al., 2004). The border between unattached and attached is not fixed. Thus, for indoor air, RnDP associated to particles smaller than 20 nm, grouped around 5 nm (Tu et al., 1991), and particles in the 0.5–1.5 nm size range may be considered as unattached RnDP (Hopke et al., 1992). Measurements in indoor air also showed that within the unattached region of <10 nm, two (with AMD of 0.8 and 4.2 nm) or even three activity distribution peaks (0.6, 0.85 and 1.25 nm) may appear (Porstendörfer, 1996; Porstendörfer, 2001). In addition, RnDP appeared in the nucleation mode (attached to particles of 14–40 nm), the accumulation mode (210-310 nm), and the coarse mode (3000–5000 nm) (Porstendörfer, 2001). In an intercomparison experiment carried out in a test chamber, the

AMD values of the unattached RnDP were found in the range from 0.53 to 1.76 nm, followed by a gap until about 50 nm when the attached RnDP appeared (Cheng et al., 2000). As reviewed by Porstendörfer & Reineking, (1992), f^{un} differs substantially from place to place and, depending on the environmental conditions, its value range from 0.006 to 0.83. Generally, it is inversely proportional to the number concentration of general aerosol particles (Butterweck et al., 1992; Cheng et al., 1997), the relationship being approximated by (Porstendörfer, 1996):

$$f^{un} = \frac{400}{C_{gen}^{N} / cm^{-3}} . \tag{11}$$

Thus, f^{un} is very low in mines with high aerosol concentrations (Butterweck et al., 1992) and high in karst caves with very clean air (Butterweck et al., 1992; Cheng et al., 1997; Sainz et al., 2007; Thinová & Burian, 2008).

Elevated f^{un} values have been also observed in the Postojna Cave (Butterweck et al., 1992; Vaupotič, 2008b). Because the behaviour of RnDP aerosols, including f^{un} values, is determined by the characteristics of general aerosols, measurements of the number concentration and size distribution of general aerosol particles in the 10–1100 nm size range were recently introduced in the cave (Iskra et al., 2010). This kind of measurements has been also extended to selected indoor and outdoor places. In the present paper, measurements in the Postojna Cave and in a dwelling in a suburban area are described, and results are presented and commented on.

2. Experimental

2.1 Measuring techniques

2.1.1 Radon decay products

Individual activity concentrations (C^A, Bq m^{-3}) of ^{222}Rn, ^{218}Po, ^{214}Pb, ^{214}Bi/^{214}Po have been measured using the EQF3020-2 device (Sarad, Germany). Air is pumped for 6 minutes at a flow rate of 2.4 dm^3 min^{-1} over a metal mesh grid on which aerosols smaller than 5 nm (considered as unattached RnDP) are separated from those above this size (considered as attached RnDP). The two fractions are deposited electrostatically on two separate 150 mm^2 semiconductor detectors. Gross alpha activity is measured during three consecutive intervals within 110 minutes after the end of pumping and, applying the Markov method (Markov, 1962), individual activity concentrations of ^{218}Po, ^{214}Pb and ^{214}Bi in the unattached and attached fractions are obtained. The device also gives radon activity concentration (C_{Rn}^{A}), equivalent equilibrium activity concentration of RnDP (C_{RnDP}^{A}), equilibrium factor between Rn and RnDP (F), fraction of the unattached RnDP (f^{un}), potential α-energy concentration of RnDP ($E_{\alpha RnDP}$), as well as air temperature and humidity.

In order to facilitate comparison of number concentrations of general aerosols and RnDP aerosols, the activity concentrations (C^A) of radionuclides are converted into their number concentrations (C^N), by applying the radioactivity law equation: $C^A = \lambda \times C^N$, with $\lambda = \ln 2 / t_{1/2}$. The obtained number concentrations of ^{218}Po, ^{214}Pb and ^{214}Bi atoms (cm^{-3}) are denoted by C_{218Po}^{Nun}, C_{214Pb}^{Nun} and C_{214Bi}^{Nun}, for the unattached, and C_{218Po}^{Natt}, C_{214Pb}^{Natt} and C_{214Bi}^{Natt}, for the attached RnDP.

2.1.2 General aerosols

Number concentration and size distribution of general aerosol particles have been measured with a Grimm Aerosol SMPS+C instrument, Series 5.400. A long Vienna DMA unit was used for 10–1100 nm and a medium DMA unit, for 5–350 nm size range. The DMA unit separates charged particles into 44 channels, based on their electrical mobility which depends on the particle size and electrical charge: the smaller the particle and the higher its electrical charge the higher is its mobility. Particles enter the CPC unit containing a heater saturator in which alcohol vapour molecules condense onto the entering particles, thus causing them to grow into droplets. The droplets are then detected with a laser beam (DLS detection) and counted. The frequency of measurement is one in seven minutes for the long, and one in four minutes for the medium DMA unit. The instrument gives the total number concentration $C_{\mathrm{gen}}^{\mathrm{N}}(\mathrm{tot})$, geometric mean of particle diameter (GM_{pd}), and size distribution $\mathrm{d}C_{\mathrm{gen}}^{\mathrm{N}}(d)\,/\,\mathrm{d}\ln d$, where d is the electrical mobility equivalent particle diameter.

2.2 Site description

Two experimental sites were selected: the Postojna Cave and a dwelling in the suburban area.

2.2.1 Postojna Cave

The Postojna Cave is the largest show cave in Slovenia. Like the majority of karst caves, it is only naturally ventilated. Its air temperature is practically constant around 9 °C all the year round and relative air humidity is close to 100 %. It is a practically horizontal cave. The air flow differs considerably in summer and winter. In winter, when the cave temperature is higher than outside, cave air is released from the cave into the outdoor atmosphere due to the air draught caused by the 'chimney effect' (Hakl et al., 1997; Kertész et al., 1999), thus allowing fresh and cold outdoor air to enter the cave through low lying openings. This effect is not operative in summer, when the outside temperature is higher than in the cave, and air draught is minimal or reversed. The air speed is low, never exceeding 1.5 m s^{-1} along the main entrance corridor and being much less or zero in other parts. As a result of different natural ventilation regimes, the cave interior, at least along the main air paths, is less wet in winter than in summer.

The cave is open for visitors every day from 10h to 16h in winter and from 9h to 18h in other seasons. The daily number of visitors is around 3500 in summer and from fifty to several hundred in winter, totalling about half a million a year. Visitors ride an electric train for the first 2 km from the entrance to the train stop in the cave, walk 1.8 km on an 8-shaped route and return back to the train in about an hour and a half. The lowest point, our study site, is in the middle of the walking route. The walking path is paved with a special concrete containing silica sand to prevent slippery steps when wet.

The major source of particulate matter in the cave air is the inflow of fresh outside air in winter time. Deposition of dust is observed on the surfaces in the main corridor all the way from the entrance to the train stop, but not further. Another source is the railway. Particles originate from rusting of iron parts, grinding of sand, and rotting and damages of wooden sleepers. The particles are lifted and resuspended by the air draught caused by the train running at a speed of 1.6 m s^{-1} through the narrow corridors and galleries. Also human activity, both visits and maintenance work at the cave infrastructure, is a potential source.

In the cave, radon and RnDP have been monitored systematically for years in order to estimate radiation doses for the personnel and to keep them below an acceptably low level (Vaupotič, et al., 2001). In addition to elevated radon levels, as found in many caves worldwide (Hakl et al., 1997; Cigna, 2005; Field, 2007; Vaupotič, 2010; Kávási et al., 2010), f_{un} values are also high (Vaupotič, 2008b), as also in some other caves (Cheng et al., 1997; Sainz et al., 2007; Thinová & Burian, 2008). They are considerably higher than in Slovenian kindergartens (Vaupotič, 2007), schools (Vaupotič & Kobal, 2006) and wineries (Vaupotič, 2008a). Preliminary measurements showed (Butterweck et al., 1992) that the activity size distribution of RnDP was bimodal, with the unattached fraction in the range 1.3–6.0 nm and the attached, 119–289 nm.
Previous radon measurements (Vaupotič, et al., 2001; Vaupotič, 2008b; Gregorič et al., 2011) pointed out that the difference in air temperature outside and in the cave played a dominant role in governing both diurnal and seasonal variations of the environmental conditions in the cave. Radon levels were lowest in winter when the cave temperature is higher than outdoors, because radon-rich air is released from the cave into the outdoor atmosphere and fresh outdoor air with low radon concentration is driven into the cave. This effect is not operative in summer, when the outdoor air temperature is higher than in the cave, and the resulting radon levels in the cave air are higher. As an example, at the lowest point the following average values in summer and in winter, respectively, were obtained in 1999 (Vaupotič, 2008b): C^{A}_{Rn} : 4540±600 Bq m^{-3} and 2070±1160 Bq m^{-3}, C^{A}_{RnDP} : 1580±280 Bq m^{-3} and 1130±530 Bq m^{-3}, F: 0.35±0.09 and 0.58±0.11, and f_{un}: 0.60±0.16 and 0.14±0.08.
Several 5–10 day measurements were carried out with the EQF devices in summer 2009 and winter 2010 at the lowest point along the guided walking route in the cave. Within these periods, the SMPS+C instrument with the long DMA unit was also used, but for several hours only during morning visits, because the instrument is not designed for practically 100 % air humidity and therefore its operation was minimized.
Bearing in mind that RnDP in the Postojna Cave are attached to aerosols larger than 100 nm (Butterweck et al., 1992), we selected 50 nm as the border between the unattached and attached RnDP. We were thus interested in concentrations of particles smaller than 50 nm ($C^{N}_{gen}(\langle 50)$) and larger than 50 nm, as well as the fraction of the smaller ones, defined as $x_{gen}(\langle 50) = C^{N}_{gen}(\langle 50) / C^{N}_{gen}(tot)$.

2.2.2 Dwelling

A farm in Zalog, a suburb of Ljubljana, was selected for our experiment. It comprised a residential house and several accompanying farm buildings. They were built in 1987 of concrete and brick. The family lives in the ground floor of the residential house, while the first floor is left unoccupied. One person lives temporarily in a small 20 m^2 flat in the basement, its floor lying 1.2 m below the courtyard level. It consists of a kitchen, living room, bathroom and corridor. A door and a window of the kitchen look at the courtyard in front of the house, while the other door connects it to other rooms. Central heating using hot water radiators is based on burning wood. There is no air conditioning. The farm is situated at the end of a small farmers' village, at a distance of about 20 m from the Ljubljanica River and with the nearest neighbour about 50 m away. The village is surrounded by fields. Across the river at a distance of about 500 m, the main railway Ljubljana - Zagreb runs along a hill covered by forest. The Ljubljana waste water purification plant is about 400 m out of the village and several small industrial plants are several kilometres away.

Measurements were performed in the kitchen of the basement flat (Smerajec et al., 2010) in which a preliminary radon survey had shown elevated radon activity concentration, ranging from 600 Bq m^{-3} to 1000 Bq m^{-3}, as compared with the national winter average of 121 Bq m^{-3} for a thousand dwellings (Humar et al., 1992). Both instruments were operated simultaneously from October 2010 to January 2011, covering temporal human activities, e. g., cooking, washing, opening window and door, etc. Here, the medium DMA unit was used. Because it is designed for >5 nm sizes, the size distribution of the unattached RnDP could not be evaluated. Therefore the assumption (Tu et. al., 1991; Shimo & Saito, 2000) was adopted that the attached RnDP in indoor air are associated with particles larger than 20 nm. Thus, the fraction of the general aerosol particles related to the unattached RnDP was expressed by $x_{\text{gen}}(\langle 20) = C^{\text{N}}_{\text{gen}}(\langle 20) / C^{\text{N}}_{\text{gen}}(\text{tot})$.

3. Results and discussion

Results are presented separately for the Postojna Cave in summer and in winter, and for the dwelling.

3.1 Postojna Cave

3.1.1 Summer results

For one of the measurement carried out in summer, Fig. 1 shows the size distribution of general aerosols (a) before visits started and (b) during regular visits. Their geometric mean values (GM_{pd}) of 26 and 31 nm did not differ significantly.

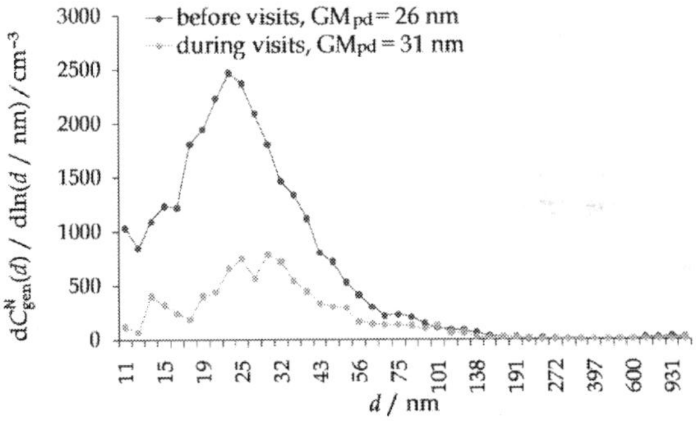

Fig. 1. Number size distribution ($dC^{\text{N}}_{\text{gen}}(d) / d\ln(d / \text{nm})$) of general aerosol particles at the lowest point in the Postojna Cave in summer 2009: a) before visits started and b) during regular visits

The decrease in total number concentration ($C^{\text{N}}_{\text{gen}}(\text{tot})$) from 2700 cm^{-3} to 700 cm^{-3} is mainly due to the smaller particles while concentrations of larger ones remained practically constant during our measurement (Fig. 2). As seen in Fig. 3, the concentration of the >50 nm particles remained practically constant during visits, while that of <50 nm ones, contributing 80-90 % of the total concentration, is steadily decreasing. This is presumably because the smaller particles are preferentially deposited on the cave surfaces (Holub et al., 1988;

Porstendörfer, 1984; Morawska & Jamriska, 1996; Papastefanou, 2008), and caught by cloths (Balcázar et al., 1999) and taken up by the lungs (Hofmann & Koblinger, 1990; Alföldy et al., 2009) of visitors walking, and thus enhancing air turbulence (Vanmarcke et al., 1991), through the narrow corridor at the lowest point of a cross section of less than 20 m^2.

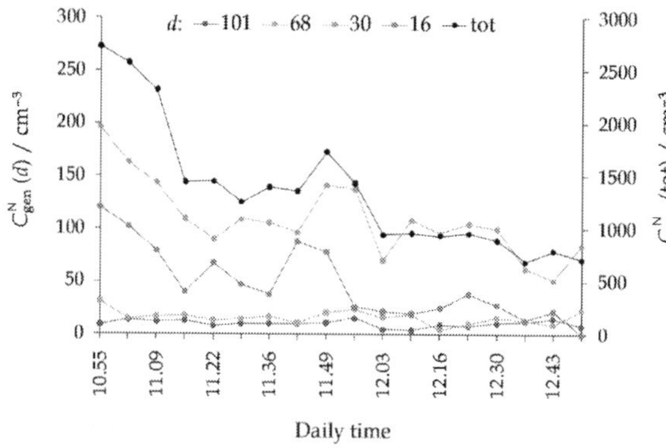

Fig. 2. Time run of the total number concentration of general aerosol particles ($C_{gen}^{N}(tot)$) and number concentrations of particle of various sizes ($C_{gen}^{N}(d)$) at the lowest point in the Postojna Cave in summer 2009

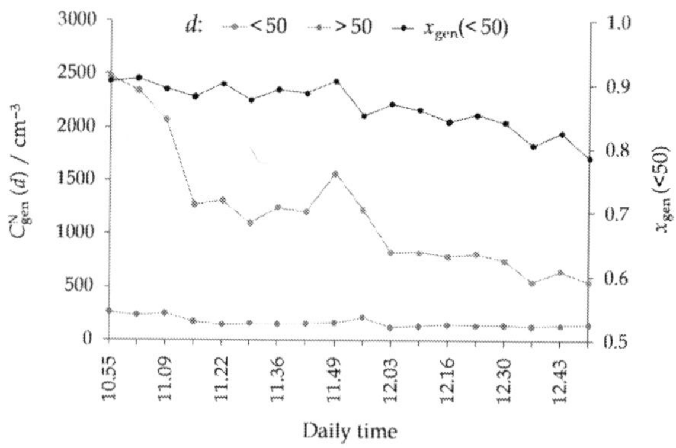

Fig. 3. Time run of number concentrations of general aerosol particles smaller ($C_{gen}^{N}(<50)$) and larger ($C_{gen}^{N}(>50)$) than 50 nm, and the fraction of the smaller particles ($x_{gen}(<50)$) at the lowest point in the Postojna Cave in summer 2009

Figure 4a shows the radon activity concentration (C_{Rn}^{A}) and equilibrium factor (F), and Fig. 4b, the equivalent equilibrium activity concentration of RnDP (C_{RnDP}^{A}) and fraction of unattached RnDP (f_{un}). As expected from previous monitoring in summer (Vaupotič, 2008b), C_{Rn}^{A} and f_{un} are high and F is low. During the period when the general aerosols were also monitored, f_{un} values were around 0.75 (Fig. 4b), resulting in a D_{CFD} value of 43.6 mSv

WLM^{-1} according to Eq. 6 or 77.4 mSv WLM^{-1} according to Eq. 7 (as compared with D_{CFE} = 5 mSv WLM^{-1} for workplaces). For this period, Fig. 5a shows the individual number concentrations of the unattached, and Fig. 5b, of the attached RnDP atoms. The measurement frequency of the EQF device is much lower (once in two hours) than that of the Grimm device (once in seven minutes) and therefore time variations of the parameters measured with the two cannot be adequately compared. The number concentration of the unattached ^{218}Po atoms is lowest because of the shorter half-life in comparison to other RnDP. On the other hand, the ratio of the numbers of its unattached and attached atoms was initially 26, far larger than those of ^{214}Pb and ^{214}Bi, being 5.2 and 0.65, respectively. There are several reasons for this high fraction for ^{218}Po. Because of its short half-life in comparison with other RnDP, it has less time available to become attached to atmospheric particles. The second reason is based on the diffusion coefficients. They are in the ranges of (0.024–0.039) cm^2 s^{-1} for the charged RnDP species and (0.068–0.085) cm^2 s^{-1} for the neutral species (Porstendörfer & Mercer, 1979; Tokonami, 1999). At the radon activity concentration of about 5500 Bq m^{-3}, as in our case (Fig. 4a), 13 % of ^{218}Po species are estimated to be charged (Chu & Hopke, 1988). Because of their lower mobility, attachment of the charged species is hindered. This effect cannot be estimated for ^{214}Pb and ^{214}Bi, because data are not available. At the end of our measurement, the ratios of the numbers of unattached and attached ^{218}Po, ^{214}Pb and ^{214}Bi atoms were 6.6, 5.0 and 0.86, respectively. The value for ^{218}Po decreased drastically, concomitantly with lowering concentration of the <50 nm particles, while the other two did not change significantly.

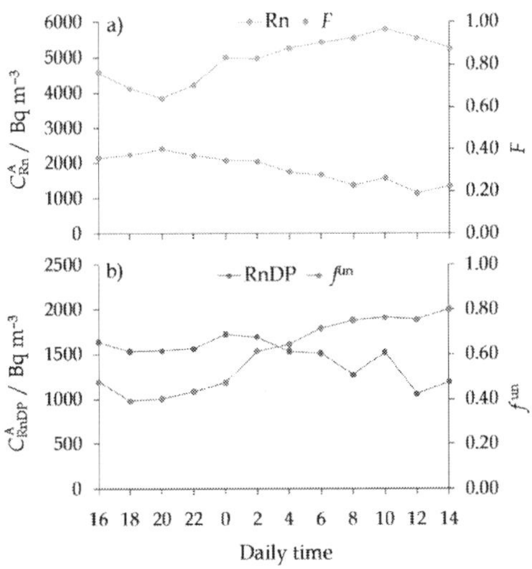

Fig. 4. Results of a continuous radon measurement at the lowest point in the Postojna Cave in summer 2009: a) radon activity concentration (C^A_{Rn}) and equilibrium factor between radon and its decay products (*F*), and b) equivalent equilibrium activity concentration of radon decay products (C^A_{RnDP}) and fraction of unattached decay products (f_{un})

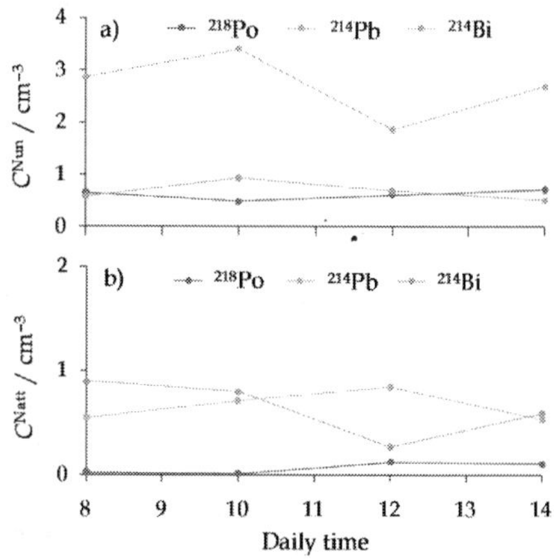

Fig. 5. Number concentrations of ^{218}Po, ^{214}Pb and ^{214}Bi atoms for the period in Fig. 4 when general aerosols were also monitored: a) in the unattached form (C^{Nun}_{214Po}, C^{Nun}_{214Pb} and C^{Nun}_{214Bi}), and b) attached form (C^{Natt}_{214Po}, C^{Natt}_{214Pb} and C^{Natt}_{214Bi})

Number concentrations of RnDP atoms are several orders of magnitude lower than the total number concentration of general aerosols (Fig. 2). Therefore, we do not believe that the reason for high f_{un} values in the cave air is only a low concentration of general aerosols (Butterweck et al., 1992; Cheng et al., 1997; Meisenberg & Tschiersch, 2009), but rather the dominating contribution of their <50 nm fraction to which the unattached RnDP are associated. This assumption is further supported by the data in Fig. 6: in a volume unit, the total surface area of all <50 nm particles, to which the unattached RnDP are related, is two to eight times lower than that of the larger particles but, on the other hand, the number of the

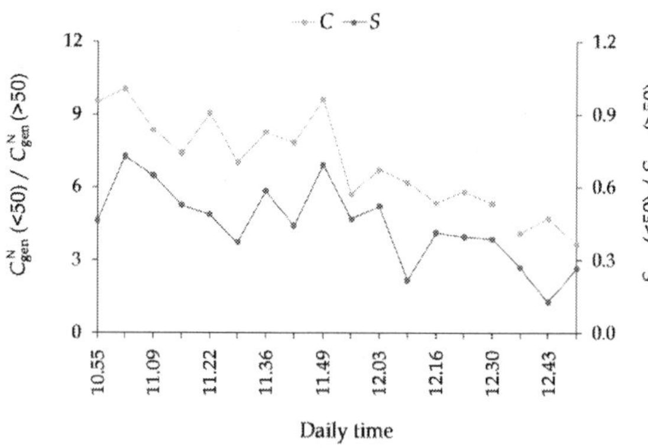

Fig. 6. Ratio of the number concentration of general aerosol particles smaller than 50 nm ($C^{N}_{gen}(\langle 50)$) to that of larger ones ($C^{N}_{gen}(\rangle 50)$), and ratio of the total surface area of all particles smaller than 50 nm ($S_{gen}(\langle 50)$) to that of all particles larger than 50 nm ($S_{gen}(\rangle 50)$) in a volume unit for the measurement in summer 2009

smaller ones is four to ten times higher than that of the larger ones. Thus, the probability of RnDP atoms being associated with smaller particle-clusters is favoured, even without the fact that the ratio of the number of surface atoms, as adsorption sites, to the total number of atoms is substantially greater on the smaller particles than on the larger ones (Oberdörster et al., 2005).

3.1.2 Winter results

The number size distribution of general aerosols, (a) before visits started and (b) during regular visits, was measured in winter (Fig. 7). The related GM_{pd} values, i. e., 110 nm and 113 nm, respectively, are substantially higher than in summer (Fig. 1). They are similar to those in the Bozokov Dolomite Cave in the Czech Republic (Thinová et al., 2005; Rovenská et al., 2008) but smaller than in a limestone cave in Australia (Solomon et al., 1992). The total concentration of general aerosols and concentrations of aerosol particles of various diameters is displayed in Fig. 8. In contrast to the situation in summer (Fig. 2), the total concentration gradually increased during visits, from 2000 cm^{-3} to 2800 cm^{-3} and is similar to the initial value in summer of (2700 cm^{-3}), but was much higher than during visits in summer (700 cm^{-3}). Higher concentration in winter cannot be understood as a result of higher air ionisation due to the radioactivity present (Pashenko & Dublyansky, 1997), because both C^{A}_{Rn} and C^{A}_{RnDP} are lower, and thus ionization is also lower in winter than in summer. Because the number size distribution in winter is similar to that observed in outdoor air in a suburban area ($C^{N}_{gen}(tot)$ = 7155 cm^{-3}, GM_{pd} = 110 nm) (Smerajec et al., 2010), although with lower total concentration, these high values are most probably reached as a result of the aerosol particles introduced in winter by the inflow of outside air. This inflow is assumed to decrease slightly during morning hours due to an increase in the outside air temperature, and thus decreased difference between temperatures in the cave and outside. Therefore, the increase in the total aerosol concentration in morning hours is not caused by the inflow of outside air but from the particles resuspended and brought by visitors walking on the paved path, which is less wet in winter than in summer. As a result of larger particles dominating in winter, $x_{gen}(\langle 50)$ values are substantially lower in winter (Fig. 9) than in summer (Fig. 3).

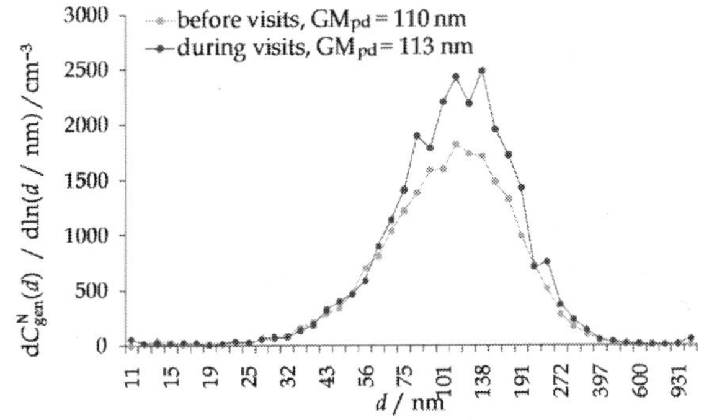

Fig. 7. Number size distribution ($dC^{N}_{gen}(d) / d\ln(d / nm)$) of general aerosol particles at the lowest point in the Postojna Cave in winter 2010: a) before visits started and b) during regular visits

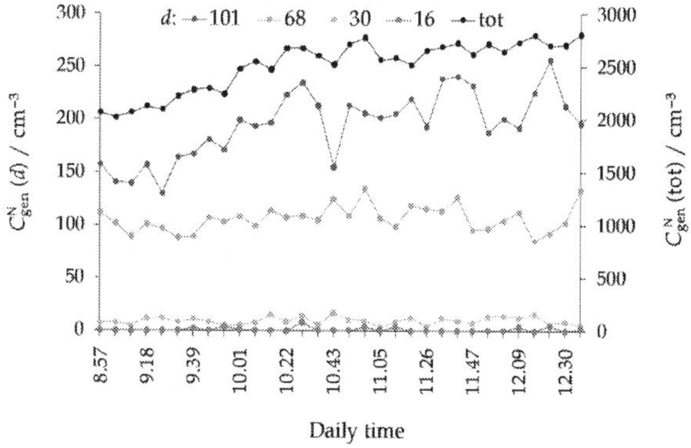

Fig. 8. Time run of the total number concentration of general aerosol particles ($C^N_{gen}(\text{tot})$) and number concentrations of particle of various sizes ($C^N_{gen}(d)$) at the lowest point in the Postojna Cave in winter 2010

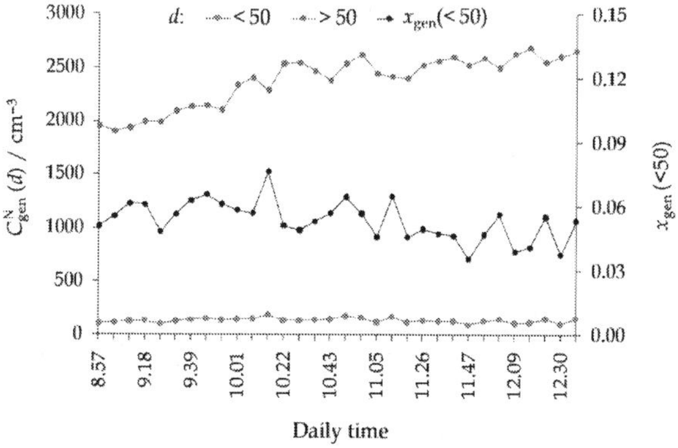

Fig. 9. Time run of number concentrations of general aerosol particles smaller than 50 nm ($C^N_{gen}(\langle 50)$) and larger than that ($C^N_{gen}(\rangle 50)$), and the fraction of the smaller ones ($x_{gen}(\langle 50)$) at the lowest point in the Postojna Cave in winter 2010

Figure 10a shows C^A_{Rn} and F, and Fig. 10b, C^A_{RnDP} and f_{un}. As expected from previous measurements in winter (Vaupotič, 2008b), C^A_{Rn} and f_{un} are lower and F is higher than in summer. The cave air is less stagnant in winter than in summer, because of the inflow of fresh air, and therefore lower F values would be expected than in summer. The opposite is true due to the higher concentration of general aerosols in winter than in summer and consequently higher availability for RnDP to be present in air. During the period when general aerosols were also monitored, f_{un} values were around 0.05 (Fig. 10b), resulting in a D_{CFD} value of 13.5 mSv WLM^{-1} according to Eq. 6 (as compared with 43.6 mSv WLM^{-1} in summer) or 11.4 mSv WLM^{-1} according to Eq. 7 (as compared with 77.4 mSv WLM^{-1} in summer, and compared with D_{CFE} =5 mSv WLM^{-1} for workplaces). Fig. 11a shows for this period the individual number concentrations of the unattached RnDP and Fig. 11b, of the attached RnDP atoms. The ratios of the numbers of unattached and attached ^{218}Po, ^{214}Pb and ^{214}Bi atoms are 0.22, 0.04 and 0.018, respectively, in the beginning, and 0.27, 0.04 and 0.021 at the end of measurement, being practically constant during measurement, in contrast to the summer situation, when the value for ^{218}Po decreased markedly. This ratio is also by far the highest for ^{218}Po in winter.

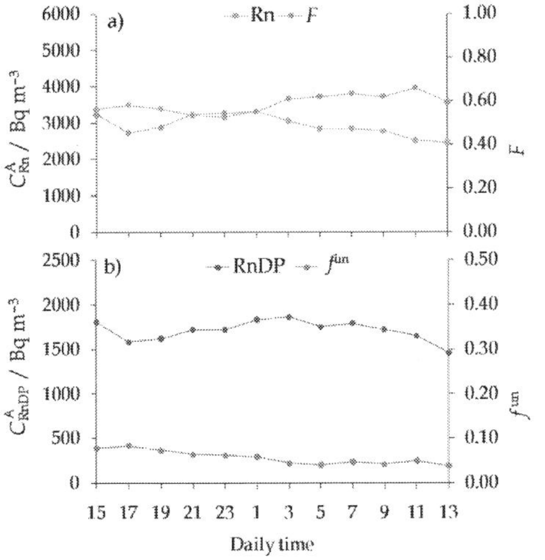

Fig. 10. Results of a continuous radon measurement at the lowest point in the Postojna Cave in winter 2010: a) radon activity concentration (C^A_{Rn}) and equilibrium factor between radon and its decay products (*F*), and b) equivalent equilibrium activity concentration of radon decay products (C^A_{RnDP}) and fraction of unattached decay products (f^{un})

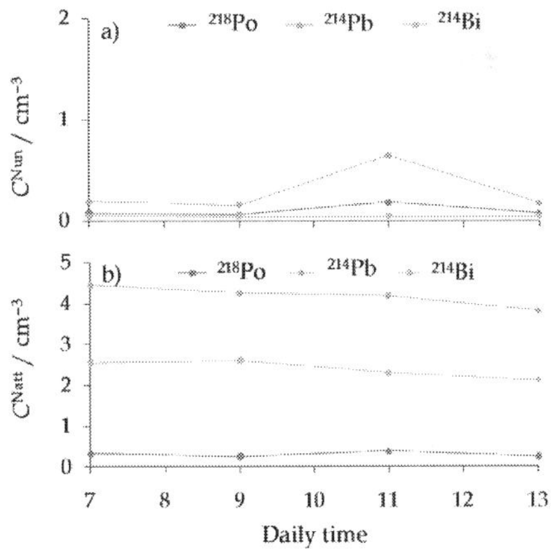

Fig. 11. Number concentrations of ^{218}Po, ^{214}Pb and ^{214}Bi atoms for the period in Fig. 10 when general aerosols were also monitored: a) in the unattached form (C^{Nun}_{214Po}, C^{Nun}_{214Pb} and C^{Nun}_{214Bi}), and b) attached form (C^{Natt}_{214Po}, C^{Natt}_{214Pb} and C^{Natt}_{214Bi})

Under winter conditions (Fig. 12), the number concentration of the <50 nm particles is about twenty times lower than that of larger ones, and their total surface area is about four hundred times smaller than that of larger ones. The probability for RnDP atoms to be associated with smaller particles is thus substantially reduced as compared with the summer conditions, and consequently, the majority of RnDP atoms appear in the attached form.

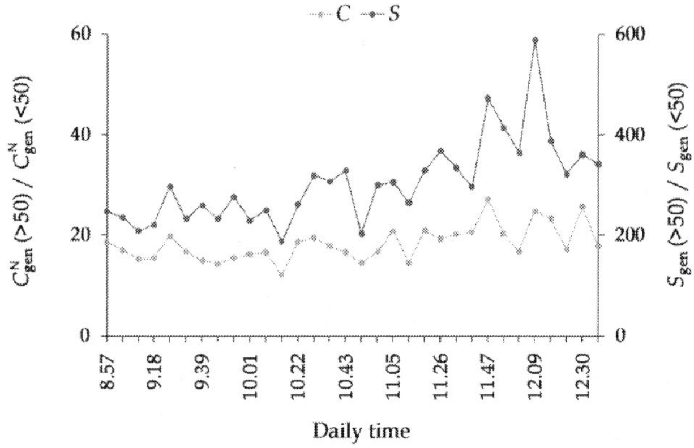

Fig. 12. Ratio of the number concentration of general aerosol particles larger than 50 nm ($C^{N}_{gen}(>50)$) to that of smaller ones ($C^{N}_{gen}(<50)$), and ratio of the total surface area of all particles larger than 50 nm ($S_{gen}(>50)$) to that of all particles smaller than 50 nm ($S_{gen}(<50)$) in a volume unit for the measurement in winter 2010

These results show why f_{un} values are higher in summer than in winter. The reason also originates from the size distribution of general aerosols in the cave, and not merely from their total number concentration. In summer, the contribution of particles smaller than 50 nm, to which the unattached RnDP are associated, is predominant, while in winter it is minor in the total aerosol concentration. When applying the dosimetric approach, the calculated dose conversion factor, and hence the dose a tourist guide or visitor receives during a visit (without mentioning that C^{A}_{RnDP} is higher in summer than in winter) is more than three times greater in summer than in winter.

3.2 Dwelling

Only results for selected period are presented here. Fig. 13a shows the total number concentration (2-h average values) of general aerosols ($C^{N}_{gen}(tot)$) and geometric means of particle diameter (GM_{pd}), and Fig. 13b, the fractions of the <10 nm particles ($x_{gen}(<10)$) and <20 nm particles ($x_{gen}(<20)$) during the period indicated, in outdoor air in the courtyard in front of the kitchen. For the entire period of measurement the following arithmetic means of the measured parameters were obtained: $C^{N}_{gen}(tot) = 6900 \pm 3200$ cm^{-3}, $GM_{pd} = 63 \pm 19$ nm, $x_{gen}(<10) = 0.03 \pm 0.04$, $x_{gen}(<20) = 0.12 \pm 0.09$, $C^{A}_{RnDP} = 10 \pm 9$ Bq m^{-3} and $f_{un} = 0.11 \pm 0.08$. $C^{N}_{gen}(tot)$ varies from about 2000 to 20,000 cm^{-3} (Fig. 13a). In both urban (Noble et al., 2003; Young & Keeler, 2007; Y. Wang et al., 2010) and semi-rural (F. Wang et al., 2010) areas, two daily

peaks were found, one in the morning and the other in late afternoon, coinciding with the traffic rush hours. In our case diurnal variation shows two maxima, one at around midnight and the other between 8h and 16h. The morning growth of $C_{gen}^{N}(tot)$, with a concomitant decrease in GM_{pd}, is caused by activities at the farm, e. g., running cars, tractors and other farming equipment, these being lowest at noon. The afternoon simultaneous increase of both $C_{gen}^{N}(tot)$ and GM_{pd} is presumably related to nucleation caused by solar radiation (Noble et al., 2003; Minoura & Takekawa, 2005; Smerajec et al., 2010) and particle growth by coagulation (Kulmala et al., 2004; Kumar et al., 2010b). Decrease in $C_{gen}^{N}(tot)$ after its midnight maximum as ascribed to the deposition of smaller particles, resulting in increasing GM_{pd}. Fig. 14 shows the particle size distribution at the GM_{pd} maximum and minimum on October 9 (at points 1 and 2 in Fig. 13) and October 10 (at points 3 and 4 in Fig. 13).
For the above period, Fig. 15 shows diurnal variation in C_{RnDP}^{A}, with maxima overnight and minima at noon. An expected (Butterweck et al., 1992; Cheng et al., 1997; Meisenberg & Tschiersch, 2009), though only approximate, coincidence of f_{un} minima and $C_{gen}^{N}(tot)$ maxima (Fig. 13a) is seen.

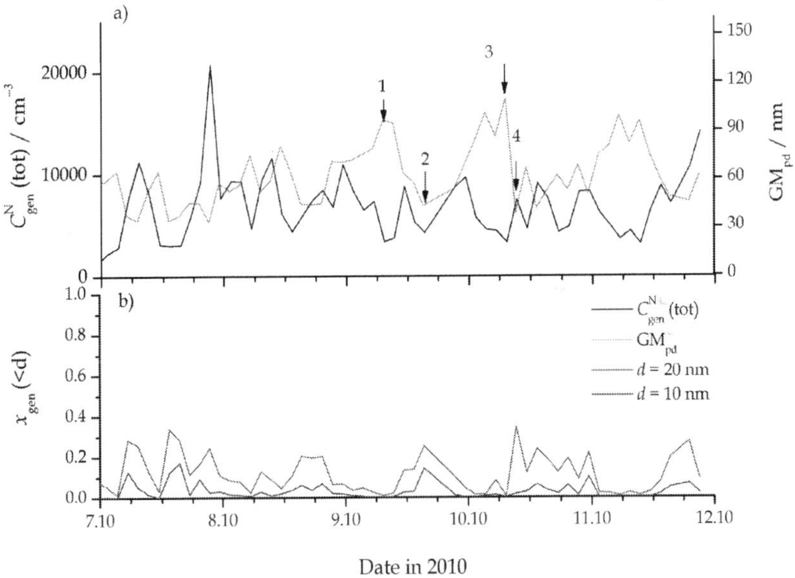

Fig. 13. Time run of: a) the total number concentration of general aerosol particles ($C_{gen}^{N}(tot)$) and geometric mean values of their diameters (GM_{pd}), b) number fraction of particles smaller than 10 nm ($x_{gen}(\langle 10)$) and smaller than 20 nm ($x_{gen}(\langle 20)$), in outdoor air in front of the dwelling in the period October 7-10

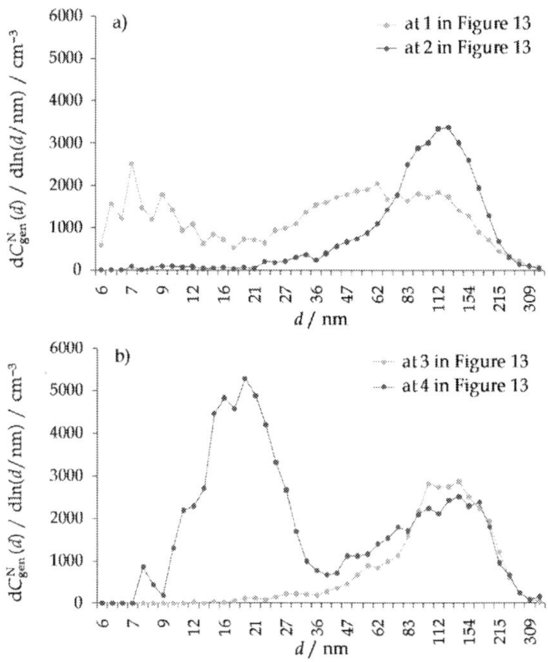

Fig. 14. Number size distributions ($dC^{N}_{gen}(d) / d\ln(d / nm)$) of general aerosol particles in outdoor air in front of the dwelling at the maximum and minimum GM_{pd} values in the time run of C^{N}_{gen}(tot) in Fig. 13 a) on October 9 (marked by arrows 1 and 2) and b) on October 10 (marked by arrows 3 and 4)

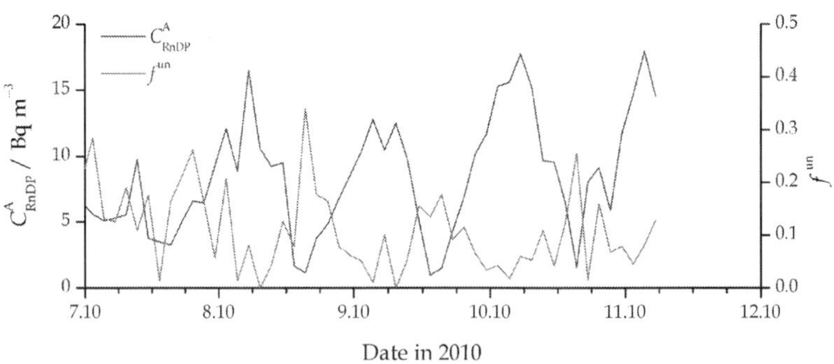

Fig. 15. Diurnal variations of the equivalent equilibrium activity concentration of radon decay products (C^{A}_{RnDP}) and fraction of the unattached decay products (f^{un}), in outdoor air in front of the dwelling in the period October 7–10

Figure 16 shows the time series of a) $C^{N}_{gen}(tot)$ and GM_{pd} and b), $x_{gen}(\langle 10)$ and $x_{gen}(\langle 20)$, and Fig. 17, of a) C^{A}_{Rn} and *F*, and b), C^{A}_{RnDP} and f^{un}, in indoor air of the basement kitchen for the period indicated. During this period, the window was open from 16.50 on October 28 to 23.30 on October and was kept closed otherwise. On October 29 at 20.05 and 20.45 a toaster was used for 10 min each time. The following arithmetic means of the measured parameters were obtained for a) the period of closed window, excluding the period of using the toaster: $C^{N}_{gen}(tot) = 5100\pm1700$ cm^{-3}, $GM_{pd} = 60\pm12$ nm, $x_{gen}(\langle 10) = 0.02\pm0.04$, $x_{gen}(\langle 20) = 0.10\pm0.09$, $C^{A}_{RnDP} = 116\pm35$ Bq m^{-3} and $f^{un} = 0.18\pm0.06$, b) the period with the window open: $C^{N}_{gen}(tot) =$ 11,500±6600 cm^{-3}, $GM_{pd} = 52\pm11$ nm, $x_{gen}(\langle 10) = 0.03\pm0.02$, $x_{gen}(\langle 20) = 0.14\pm0.08$, $C^{A}_{RnDP} = 22\pm5$ Bq m^{-3} and $f^{un} = 0.22\pm0.06$, and c) the period while using the toaster: $C^{N}_{gen}(tot) =$ 207,000±51,000 cm^{-3}, $GM_{pd} = 61\pm6$ nm, $x_{gen}(\langle 10) = 0.01\pm0.001$, $x_{gen}(\langle 20) = 0.03\pm0.01$, $C^{A}_{RnDP} =$ 89±22 Bq m^{-3} and $f^{un} = 0.09\pm0.02$.

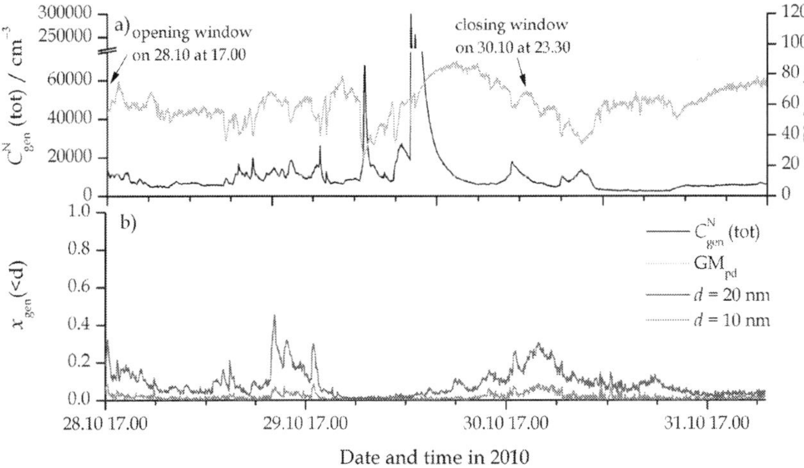

Fig. 16. Time run of: a) total number concentration of general aerosol particles ($C^{N}_{gen}(tot)$) and geometric mean values of their diameters (GM_{pd}), b) number fraction of particles smaller than 10 nm ($x_{gen}(\langle 10)$) and smaller than 20 nm ($x_{gen}(\langle 20)$), in indoor air in the kitchen in the period October 28–31

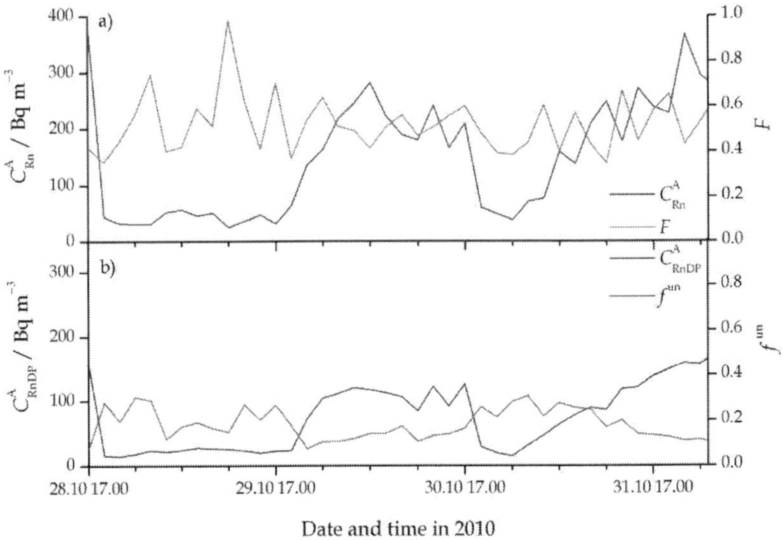

Fig. 17. Diurnal variations of: a) radon activity concentration (C^{A}_{Rn}) and equilibrium factor between radon and its decay products (F), and b) equivalent equilibrium activity concentration of radon decay products (C^{A}_{RnDP}) and fraction of unattached decay products (f^{un}), in indoor air in the kitchen in the period October 28–31

As expected, opening the window, both C^{A}_{Rn} and C^{A}_{RnDP} suddenly dropped and remained low until the window was closed at 18.50 on October 29, when they started to increase (Fig. 17). The effect of closing window is presented in Fig. 18. Because GM_{pd} in outdoor air is low at 16.50 (Fig. 13a), the inflow of outdoor air shifted the size distribution towards lower values (Fig. 19), thus reducing GM_{pd} and increasing $x_{gen}(\langle 20)$ indoors (Fig. 18b). The decrease in number concentrations of all three RnDP in the attached form is a consequence of the decrease in C^{A}_{RnDP} (Fig. 18d). The frequency of EQF measurement is too low to follow such abrupt changes and therefore the f^{un} response (Fig. 18c) is not necessarily correct. A general trend of GM_{pd} increase is seen. Neither C^{N}_{gen}(tot) nor GM_{pd} reflect the diurnal variations observed outdoors (Fig. 13a), thus proving good air tightness of the door and window.

On closing the window, only a slight increase in GM_{pd} was observed (Fig. 20a), as shown also in Fig. 21. GM_{pd} steadily increases and thus $x_{gen}(\langle 20)$ decreases (Fig. 20b). As expected, decreasing C^{N}_{gen}(tot) (Fig. 20a) is accompanied by increasing f^{un} (Fig. 20c), and increasing C^{A}_{RnDP} (Fig. 17), by increasing number concentrations of RnDP (Figs. 20d, e).

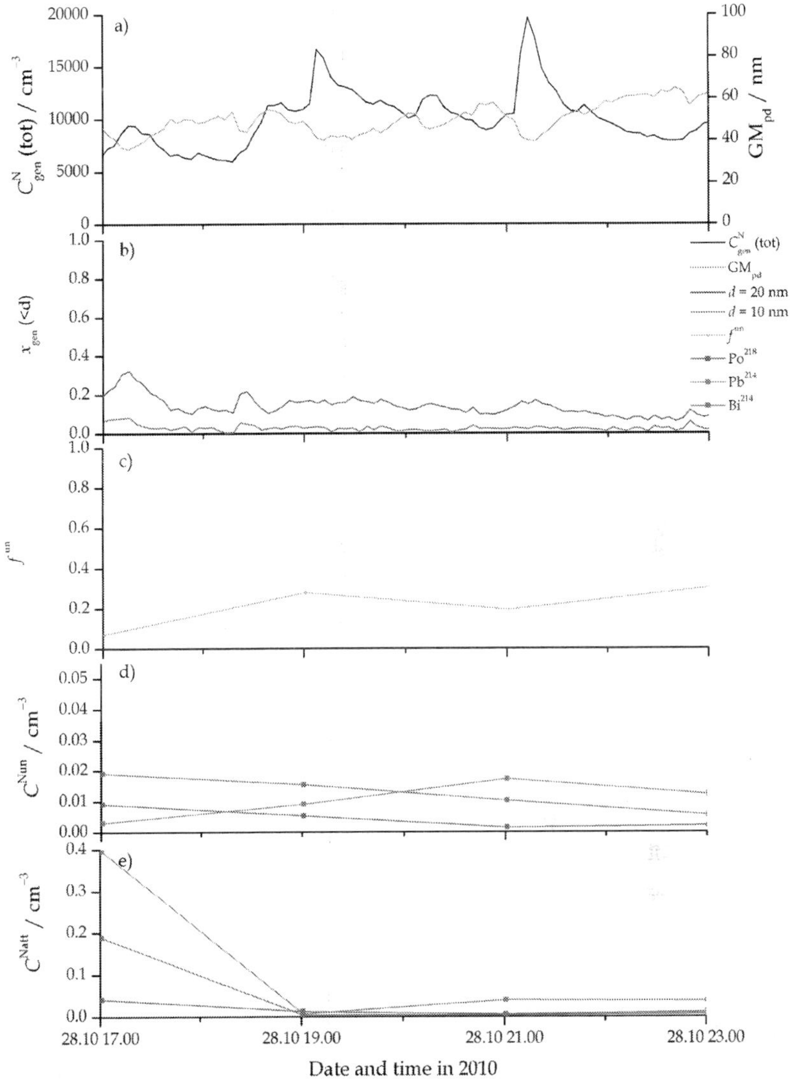

Fig. 18. Time run of: a) total number concentration of general aerosol particles ($C^{N}_{gen}(tot)$) and geometric mean values of their diameters (GM$_{pd}$), b) number fraction of particles smaller than 10 nm ($x_{gen}(\langle 10)$) and smaller than 20 nm ($x_{gen}(\langle 20)$), c) the unattached fraction of RnDP (f^{un}), d) number concentrations of the unattached RnDP atoms (C^{Nun}_{218Po}, C^{Nun}_{214Pb} and C^{Nun}_{214Bi}), and e) number concentrations of the attached RnDP atoms (C^{Natt}_{218Po}, C^{Natt}_{214Pb} and C^{Natt}_{214Bi}), in indoor air in the kitchen in the period October 28–29 (period of opening the window in the kitchen at 16.50 on October 28)

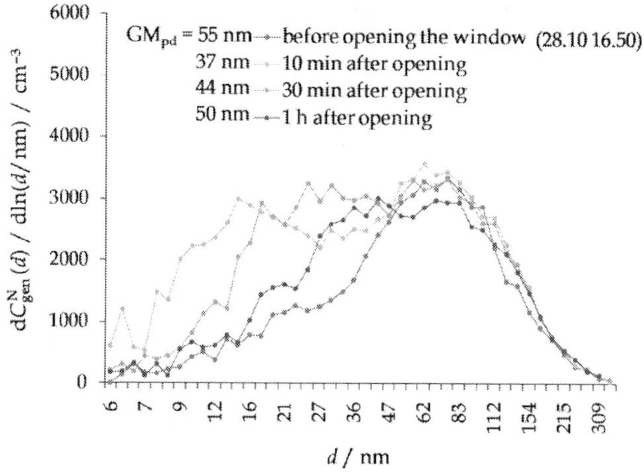

Fig. 19. Number size distributions ($dC^{N}_{gen}(d) / d\ln(d / \mathrm{nm})$) of general aerosol particles in indoor air before and during opening the window in the kitchen

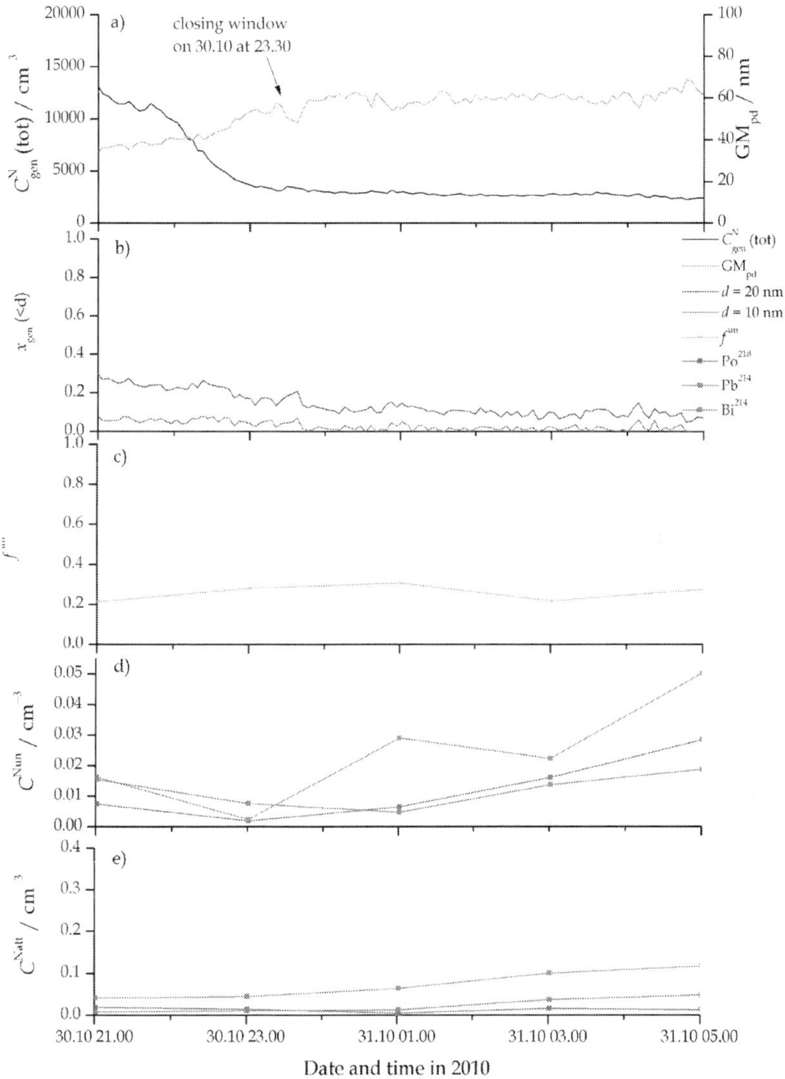

Fig. 20. Time run of: a) total number concentration of general aerosol particles ($C_{gen}^{N}(tot)$) and geometric mean values of their diameters (GM_{pd}), b) number fraction of particles smaller than 10 nm ($x_{gen}(\langle 10)$) and smaller than 20 nm ($x_{gen}(\langle 20)$),c) the fraction of the unattached RnDP (f^{un}), d) number concentrations of the unattached RnDP atoms (C_{218Po}^{Nun}, C_{214Pb}^{Nun} and C_{214Bi}^{Nun}), and e) number concentrations of attached RnDP atoms (C_{218Po}^{Natt}, C_{214Pb}^{Natt} and C_{214Bi}^{Natt}), in indoor air in the kitchen in the period October 30–31 (period of closing the window in the kitchen at 23.30 on October 30)

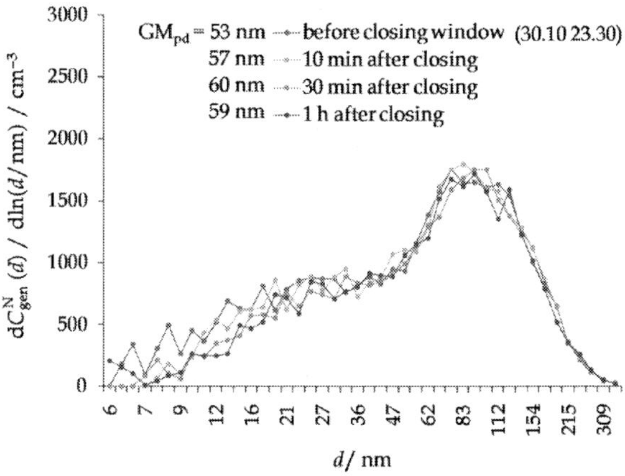

Fig. 21. Number size distributions ($dC^{N}_{gen}(d) / d\ln(d / nm)$) of general aerosol particles in indoor air before and after closing the window in the kitchen

An abrupt and very large increase in C^{N}_{gen}(tot) (reaching 300,000 cm^{-3}) on October 29 (Fig. 16a) was caused by using the toaster at 20.05. A closer look at the situation is presented in Fig. 22 in which the C^{N}_{gen}(tot) peak is split into two because of two consecutive toasting of 10 min reveals that the size distribution of particles was not significantly changed during toasting, but was only slightly shifted towards higher sizes afterwards (Fig. 23). After toasting was finished, GM$_{pd}$ started to increase steadily, presumably because of the particle growth, estimated to be 2–3 nm h^{-1} for urban area in October (Birmili et al., 2003; Kulmala et al., 2004).

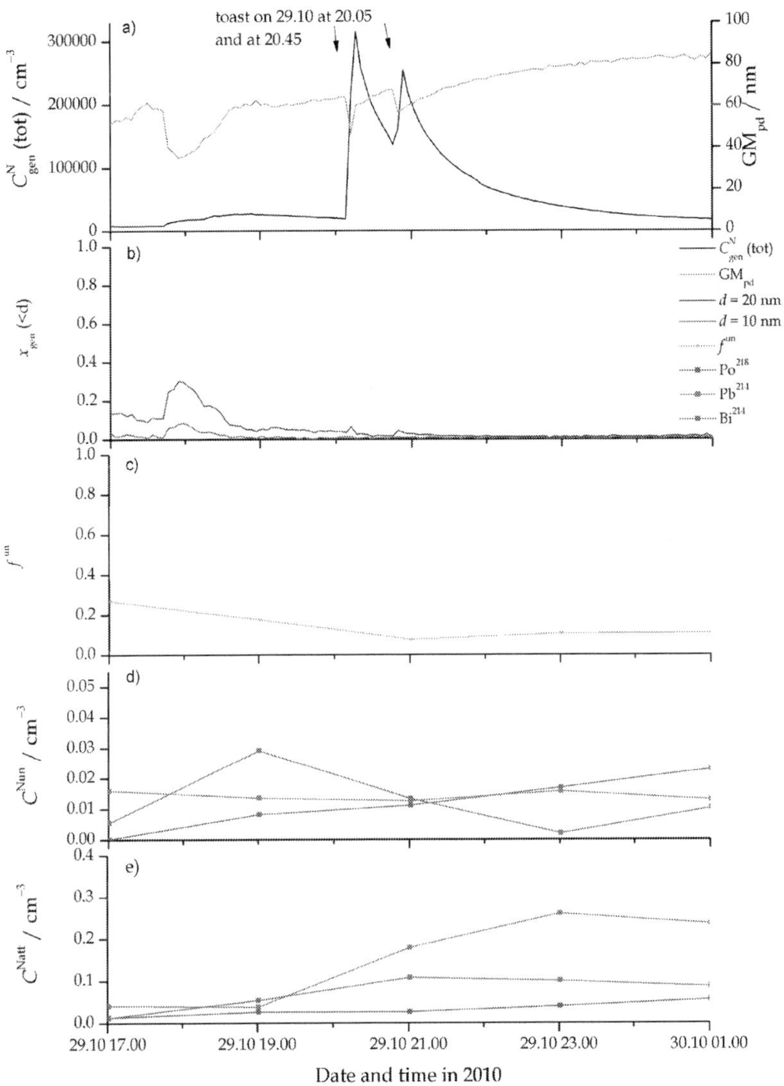

Fig. 22. Time run of: a) total number concentration of general aerosol particles ($C^{N}_{gen}(\mathrm{tot})$) and geometric mean values of their diameters (GM_{pd}), b) number fraction of particles smaller than 10 nm ($x_{gen}(\langle 10)$) and smaller than 20 nm ($x_{gen}(\langle 20)$),c) the fraction of the unattached RnDP (f^{un}), d) number concentrations of the unattached RnDP atoms (C^{Nun}_{218Po}, C^{Nun}_{214Pb} and C^{Nun}_{214Bi}), and e) number concentrations of attached RnDP atoms (C^{Natt}_{218Po}, C^{Natt}_{214Pb} and C^{Natt}_{214Bi}), in indoor air in the kitchen in the period October 30–31 (periods of using a toaster in the kitchen from 20.05 to 20.15 and from 20.45 to 20.55 on October 29)

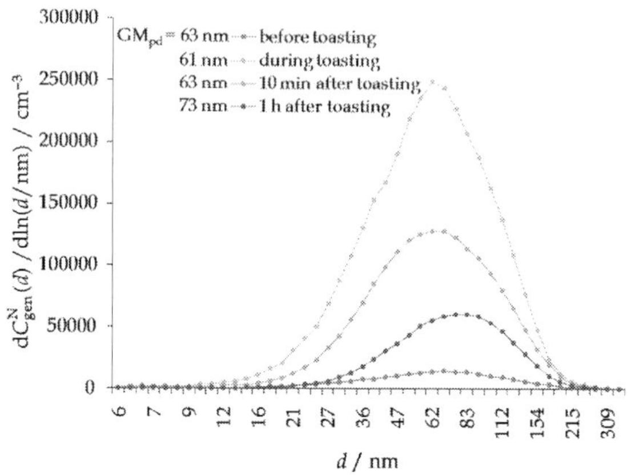

Fig. 23. Number size distributions ($dC^{N}_{gen}(d)/d\ln(d/nm)$) of general aerosol particles in indoor air in the kitchen during using a toaster

Figure 24a shows diurnal variations of $C^{N}_{gen}(tot)$ and GM_{pd}, and Fig. 24b, $x_{gen}(<10)$ and $x_{gen}(<20)$, and Fig. 25a, C^{A}_{Rn} and F, and Fig. 25b, C^{A}_{RnDP} and f_{un}, in indoor air of the basement kitchen for the period indicated. During this period, the window was kept closed and a candle was burning from 22.20 on January 5 to 1.50 on January 6. Except for the period of burning the candle, the following arithmetic means of the measured parameters were obtained: $C^{N}_{gen}(tot) =$ 6900±4000 cm^{-3}, GM_{pd} = 55±10 nm, $x_{gen}(<10) =$ 0.02±0.03, $x_{gen}(<20) =$ 0.11±0.08, C^{A}_{RnDP} = 78±18 Bq m^{-3} and f_{un} = 0.16±0.04, and during the candle period: $C^{N}_{gen}(tot) =$ 897,000±186,000 cm^{-3}, GM_{pd} = 10±1 nm, $x_{gen}(<10) =$ 0.64±0.08, $x_{gen}(<20) =$ 0.91±0.03, C^{A}_{RnDP} = 127±27 Bq m^{-3} and f_{un} = 0.10±0.01. An enormous $C^{N}_{gen}(tot)$ peak (reaching 1,200,000 cm^{-3}), accompanied by a sudden decrease in GM_{pd}, appeared during burning the candle (Fig. 24a). A closer look is presented in Fig. 26. The abrupt decrease in GM_{pd} (Fig. 26a), shown also in Fig 27, is concomitant with a substantial increase in both $x_{gen}(<10)$ and $x_{gen}(<20)$, even exceeding 0.60 and 0.90, respectively. This high fraction of small particles to which the unattached RnDP are related should result in high f_{un} values, based on our experience gained in the Postojna Cave. In contrast, f_{un} was even slightly lowered, as predicted by the inverse f_{un} - $C^{N}_{gen}(tot)$ relationship, although not as pronounced as expected (Porstendörfer, 1996; Cheng et al., 1997). Although in the period without toasting an inverse f_{un} - $C^{N}_{gen}(tot)$ correlation is observed (Fig. 28b), this is not true for the entire period shown in Fig. 28a where similar f_{un} values are seen at both low and very high $C^{N}_{gen}(tot)$ values. Neither f_{un} - $x_{gen}(<10)$ nor f_{un} - $x_{gen}(<20)$ correlations can be confirmed (Fig. 29). Generally, higher f_{un} values are associated with higher values of GM_{pd} as expected, but several low f_{un} values are found also in the high GM_{pd} region, similar to those during the candle burning with GM_{pd} of about 10 nm (Fig. 30). The explanation may be based on the plots in Fig. 31. During the candle burning $C^{N}_{gen}(<10)$ is about 2 fold higher than $C^{N}_{gen}(>10)$ and, on the other hand, the total surface area of smaller particles in a volume unit ($S^{N}_{gen}(<10)$) is about 5 fold lower than that of the larger ones ($S^{N}_{gen}(>10)$)(Fig. 31a). In the case of <20 nm particles, the following situation is evident (Fig. 31b): $C^{N}_{gen}(<20)$ is about 10 fold higher than $C^{N}_{gen}(>20)$, but on the other hand the total surface area of smaller particles in a volume unit ($S^{N}_{gen}(<20)$) is about 2

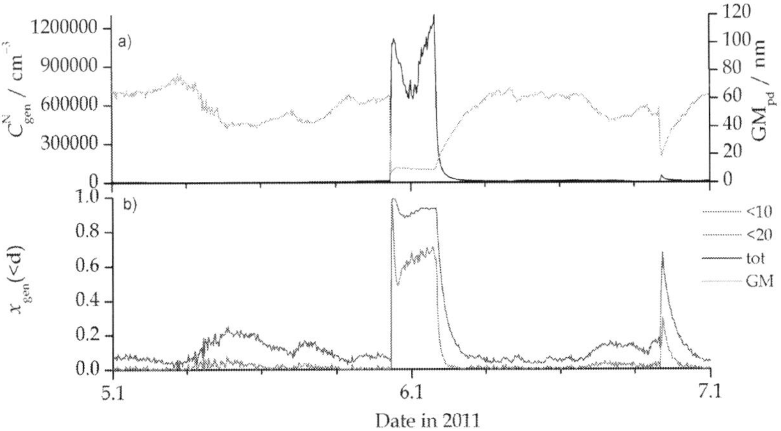

Fig. 24. Time run of: a) total number concentration of general aerosol particles ($C^{N}_{gen}(tot)$) and geometric mean values of their diameters (GM_{pd}), b) number fraction of particles smaller than 10 nm ($x_{gen}(\langle 10)$) and smaller than 20 nm ($x_{gen}(\langle 20)$), in indoor air in the kitchen in the period January 5-7

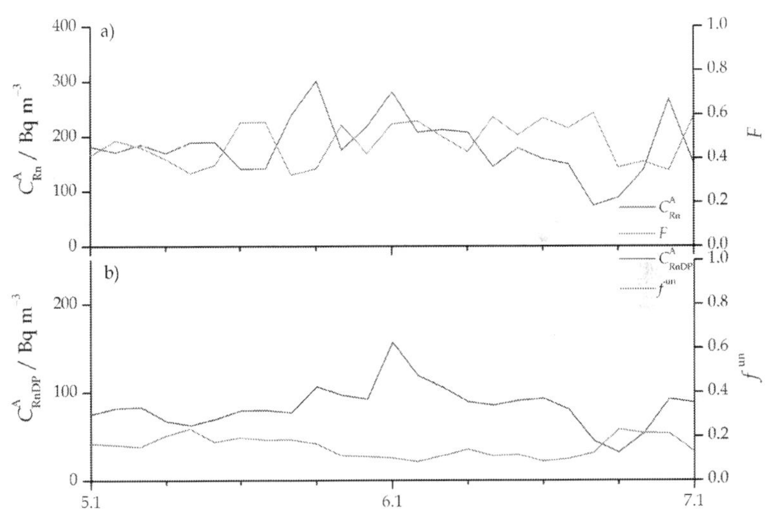

Fig. 25. Diurnal variations of: a) radon activity concentration (C^{A}_{Rn}) and equilibrium factor between radon and its decay products (F), and b) equivalent equilibrium activity concentration of radon decay products (C^{A}_{RnDP}) and fraction of unattached decay products (f^{un}), in indoor air in the kitchen in the period January 5-7

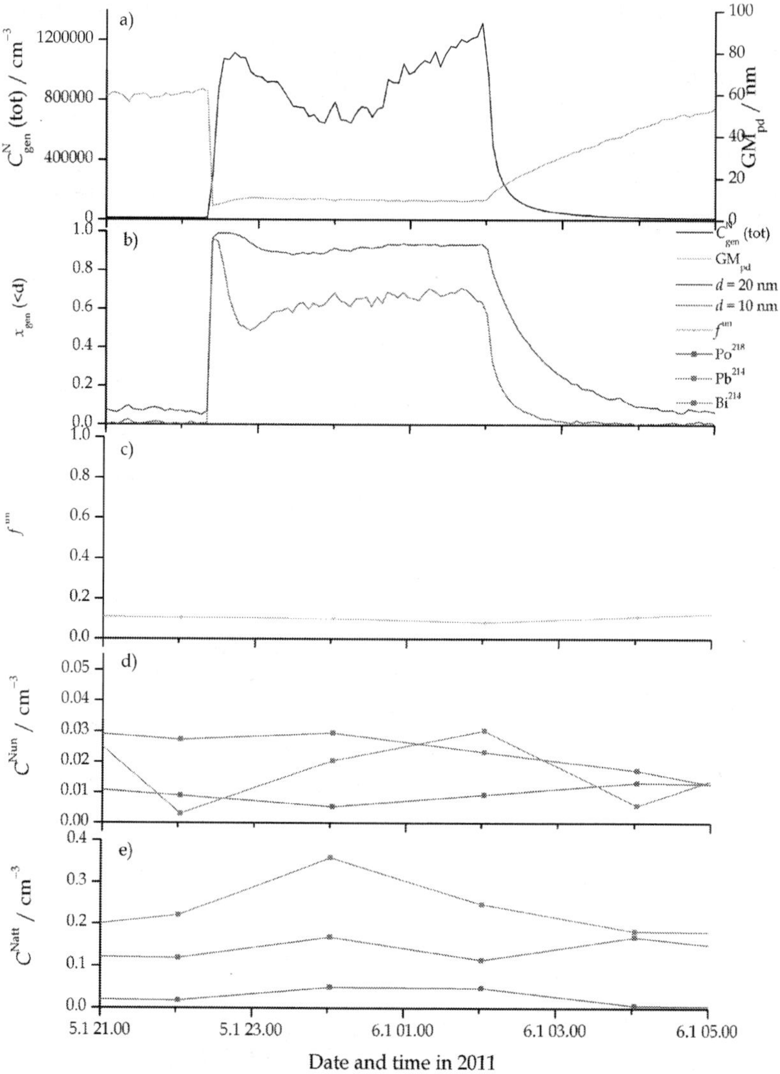

Fig. 26. Time run of: a) total number concentration of general aerosol particles ($C^N_{gen}(tot)$) and geometric mean values of their diameters (GM$_{pd}$), b) number fraction of particles smaller than 10 nm ($x_{gen}(\langle 10)$) and smaller than 20 nm ($x_{gen}(\langle 20)$), c) the unattached decay products (f^{un}), d) number concentrations of the unattached RnDP atoms (C^{Nun}_{218Po}, C^{Nun}_{214Pb} and C^{Nun}_{214Bi}), and e) number concentrations of attached RnDP atoms (C^{Natt}_{218Po}, C^{Natt}_{214Pb} and C^{Natt}_{214Bi}), in indoor air in the kitchen in the period January 5–6 (period of burning a candle in the kitchen from 22.20 on January 5 to 1.50 on January 6)

fold lower than that of the larger ones ($S^{N}_{gen}(\rangle 20)$). Therefore, a preference of RnDP atoms to be associated with particles smaller than 20 nm and smaller than 10 nm, with resulting lower f_{un} values, cannot be expected. Another reason for this may also be a very small number concentration of the particles of about 1 nm size to which most unattached RnDP are related, but whose concentration can be only guessed from Fig. 27a, but not measured with our device. In addition, process of the RnDP formation takes time and therefore f_{un} cannot follow fast changes in aerosol composition adequately.

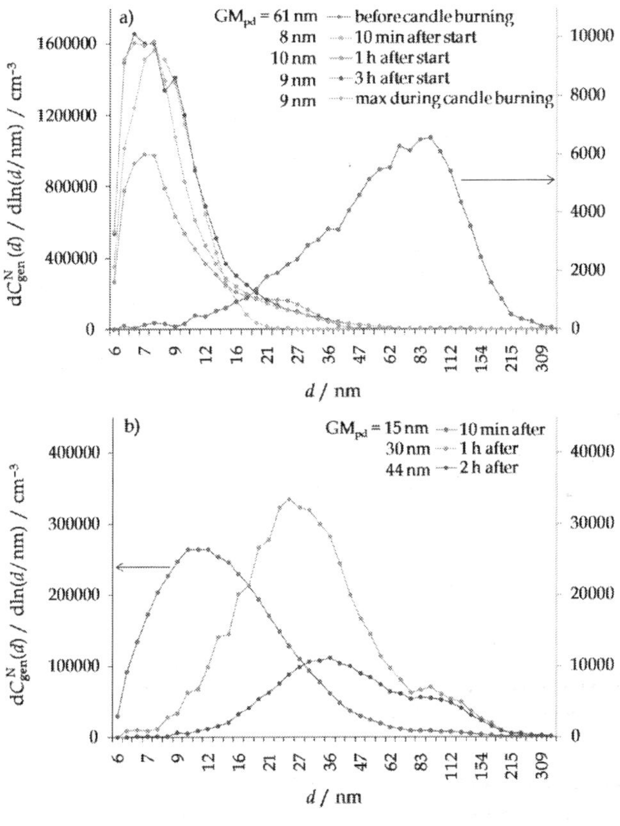

Fig. 27. Number size distributions ($dC^{N}_{gen}(d) / d\ln(d / nm)$) of general aerosol particles in indoor air a) before and during burning a candle and b) after burning a candle

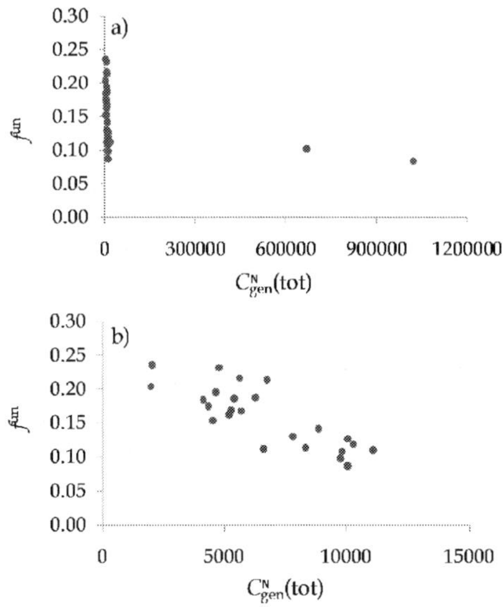

Fig. 28. Relationship between the fraction (f_{un}) of the unattached RnDP and total number concentration of general aerosols ($C_{gen}^{N}(tot)$) in indoor air in the kitchen: a) for the entire period January 5-7 and b) for the period January 5-7, except during the candle burning

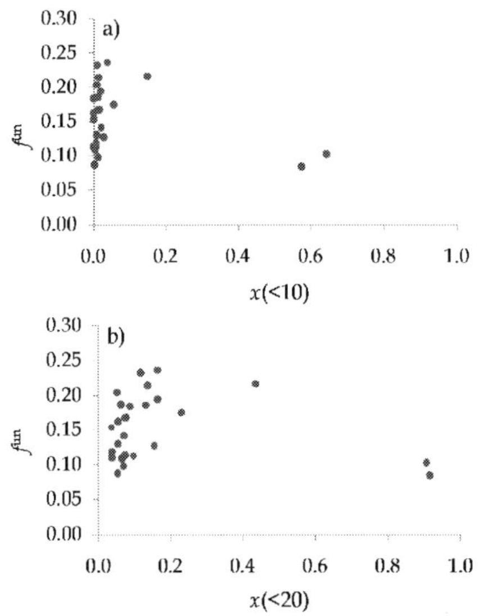

Fig. 29. Relationship between the fraction (f_{un}) of the unattached RnDP, in indoor air in the kitchen for the entire period January 5-7, and number fraction of general aerosol particles: a) smaller than 10 nm ($x(<10)$) and b) smaller than 20 nm ($x(<20)$)

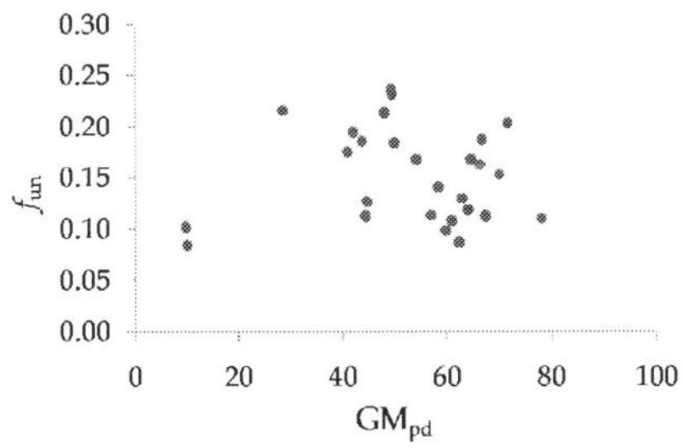

Fig. 30. Relationship between the fraction (f_{un}) of the unattached RnDP and the geometric mean of general aerosol particles (GM_{pd}), in indoor air in the kitchen for the entire period January 5-7

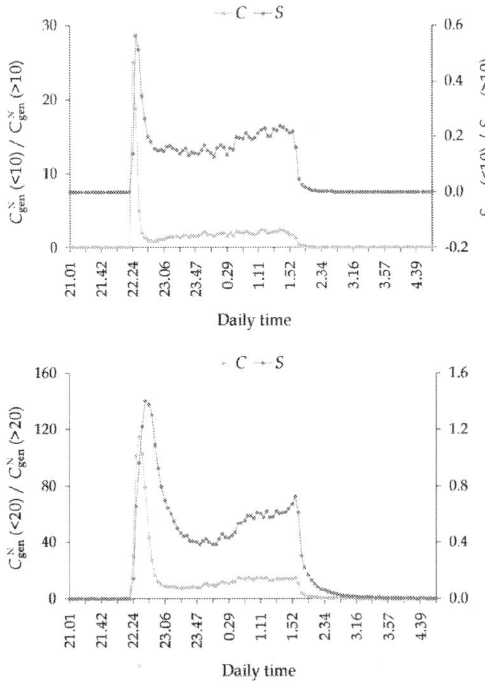

Fig. 31. Indoor air in the kitchen, entire period January 5-7: a) ratio of the number concentration of aerosol particles smaller than 10 nm ($C_{gen}^{N}(\langle 10)$) to that of larger ones ($C_{gen}^{N}(\rangle 10)$), and ratio of the total surface area of all particles smaller than 10 nm ($S_{gen}(\langle 10)$) to the total surface area of all particles larger than 10 nm ($S_{gen}(\rangle 10)$) in a volume unit, and b) ratio of the number concentration of aerosol particles smaller than 20 nm ($C_{gen}^{N}(\langle 20)$) to that of larger ones ($C_{gen}^{N}(\rangle 20)$), and ratio of the total surface area of all particles smaller than 20 nm ($S_{gen}(\langle 20)$) to the total surface area of all particles larger than 20 nm ($S_{gen}(\rangle 20)$) in a volume unit

4. Conclusion

Radon decay products (RnDP) and general aerosols were monitored in parallel in the air of the Postojna Cave (size range 10–1100 nm) in summer and in winter and in a dwelling (size range 5–350 nm) in a suburban area in an autumn - winter period, focused on the unattached fraction of RnDP (f_{un}), the key parameter in radon dosimetry. In the cave, the total number concentration of the aerosol particles ($C_{gen}^{N}(tot)$) during visits in summer was lower (700 cm^{-3}) than in winter (2800 cm^{-3}), and was dominated by the <50 nm particles (related to the unattached RnDP) in summer and by the >50 nm particles (related to the attached RnDP) in winter. This explains why f_{un} values (and thus the calculated dose conversion factors) are several times higher in summer than in winter. The difference is caused by an enhanced inflow of fresh air, driven in winter by higher air temperature in the cave than outdoors, introducing outdoor larger aerosols particles into the cave.
In the outdoor air at the dwelling, $C_{gen}^{N}(tot)$ varies in the range from 2000 cm^{-3} to 20,000 cm^{-3}, with two daily maxima, one at around midnight and the other between 8h and 16h. The geometric mean of particle diameter is highest, at 100–120 nm, at the $C_{gen}^{N}(tot)$ minimum in early morning, while it is lowest at the $C_{gen}^{N}(tot)$ minima, 30–40 nm.
In the dwelling, $C_{gen}^{N}(tot)$ during periods without any human activity was on average 4000-7000 cm^{-3} and increased markedly while toasting bread and burning a candle, reaching 300,000 cm^{-3} and 1,200,000 cm^{-3}, respectively. Even at these high $C_{gen}^{N}(tot)$ values, f_{un} did not fall below its value at $C_{gen}^{N}(tot)$ of about 10,000 cm^{-3}. Although during the candle burning more than 60 % of particles were smaller than 10 nm and more than 90 %, smaller than 20 nm, their total surface areas were three to two times, respectively, lower than those of the larger ones. Therefore, a significant enhancement of preferential association of RnDP atoms to smaller particles, and thus decreasing f_{un}, can hardly be expected.

5. Acknowledgement

The study was financed by the Slovenian Research Agency within the project contract no. J1-0745. The author thanks Ms. Mateja Smerajec for her help in operating the Grimm instrument and for data evaluations.

6. References

Agus, E. L., Young D. T., Lingard, J. J. N., Smalley, R. J., Tate, J. E., Goodman, P. S. & Tomlin, A. S. (2007). Factors influencing particle number concentrations, size distributions and modal parameters at a roof-level and roadside site in Leicester, UK. *Science of the Total Environment*, 386, pp. 65–82.

Alföldy, B., Giechaskiel, B., Hofmann, W. & Drossinos, Y. (2009). Size-distribution dependent lung deposition of diesel exhaust particles. *Journal of Aerosol Science*, 40, pp. 652–663.

Amman, M. & Butscher, H. (1990). Characterization of ultrafine aerosol particles in Mt. Etna emissions. *Bulletin of Volcanology*, 52, pp. 577–583.

Anastasio, M. & Martin, C. T. (2001). Atmospheric nanoparticles, In: *Nanoparticles and the Environment*, J. F. Bafield & A. Navrotsky, (Eds.), pp. 293–349, Mineralogical Society of America, vol. 44.

Balcázar, M., Chávez, A., Piña-Villalpando, G., Alfaro, M. & Mendoza, D. (1999). Radon decay products attached to clothes in a nuclear laboratory. *Radiation Measurements*, 31, pp. 337-324.

Baltensperger, U. & Nyeki, S. (1998). Atmospheric aerosols, In: *Physical and Chemical Properties of Aerosols*, I. Colbeck, (Ed.), pp. 280-330, Blackie Academic & Professional, an imprint of Thomas Science, ISBN 0 7514 0402 0, Bury St Edmunds, Great Britain.

Birchall, A. & James, A. C. (1994). Uncertainty analysis of the effective dose per unit exposure from radon progeny and implications for ICRP risk-weighting factors. *Radiation Protection Dosimetry*, 53, pp. 133-140.

Birmili, W., Berrensheim, H., Plass-Dülmer, C., Eslte, T., Gilge, S., Wiedensohlner, A. & Uhrner, U. (2003). The Hohenpeissenberg aerosol formation experiment (HAFEX): A long-term study including size-resolved aerosol, H_2SO_4, OH, and monoterpens measurements. *Atmospheric Chemistry and Physics*, 3, pp. 361-376.

Busek, P. R. & Adachi, K. (2008). Nanoparticles in the atmosphere. *Elements*, 4, pp. 398-394.

Butterweck, G., Porstendörfer, J., Reineking, A. & Kesten, J. (1992). Unattached fraction and the aerosol size distribution of the radon progeny in a natural cave and mine atmosphere. *Radiation Protection Dosimetry*, 45, pp. 167-170.

Cheng, Y. S., Chen, T. R., Wasiolek, P. T. & Van Engen, A. (1997). Radon and radon progeny in the Carlsbad Caverns. *Aerosol Science and Technology*, 26, pp. 74-92.

Cheng, Y. S., Chen, T. R., Yeh, H. C., Bigu, J., Holub, R., Tu, K., Knutson, E. O. & Falk, R. (2000). Intercomparison of activity size distribution of thoron progeny and a mixture of radon and thoron progeny. *Journal of Environmental Radioactivity*, 51, pp. 59-78.

Chu, K.-D. & Hopke, P. K. (1988). Neutralization kinetics for polonium-218. *Environmental Science and Technology*, 22, pp. 711-717.

Cigna, A. A. (2005). Radon in caves. *International Journal of Speleology*, 34, pp. 1-8.

Colbeck, I. (1998). Introduction to aerosol science, In: *Physical and Chemical Properties of Aerosols*, I. Colbeck, (Ed.), pp. 1-30, Blackie Academic & Professional, an imprint of Thomas Science, ISBN 0 7514 0402 0, Bury St Edmunds, Great Britain.

Dankelmann, V., Reineking, A. & Porstendörfer, J. (2001). Determination of neutralization rate of ^{218}Po ions in air. *Radiation Protection Dosimetry*, 94, pp. 353-357.

Darby, S., Hill, D., Auvinen, A., Barros-Dios, J. M., Baysson, H., Bochicchio, F., Deo, H., Falk, R., Forastiere, F., Hakama, M., Heid, I., Kreienbrock, R., Kreuzer, M., Lagarde, F., Mäkeläinen, I., Muirhead, C., Oberaigner, W. Pershagen, G., Ruano-Ravina, A., Ruosteenoja, E., Schaffrath Rosario, A., Tirmarche, M., Tomášek, A., Whitley, E., Wichmann, H.-E. & Doll, R. (2005). Radon in homes and risk of lung cancer: collaborative analysis of individual data for 13 European case-control studies. *British Medical Journal*, 330, pp. 223-227.

Dobos, E., Borbérly-Kiss, I., Kertész, Zs., Szabó, Gy. & Koltay, E. (2009). Debrecen, Hungary on the fine-fraction aerosol map of Europe. *Journal of Radioanalytical and Nuclear Chemistry*, 279, pp. 143-157.

Etiope, G. & Martinelli, G. (2002). Migration of carrier and trace gases in the geosphere: an overview. *Physics of the Earth and Planetary Interiors*, 129, pp. 185-204.

Field, S. M. (2007). Risks to cavers and cave workers from exposure to low-level ionizing a radiation from ^{222}Rn decay in caves. *Journal of Cave and Karst Studies*, 69, pp. 207–228.

Fukutsu, K., Yamada, Y. & Tokonami, S. (2004). Newly designed graded screen array for particle size measurements of unattached radon progeny products. *Review of Scientific Instruments*, 75, pp. 783–787.

Goldstein, S. D. & Hopke, P. K. (1985). Environmental neutralization of polonium-218. *Environmental Science and Technology*, 19, pp. 146–150.

Gregorič, A., Zidanšek, A. & Vaupotič, J. (2011). Dependence of radon levels in the Postojna Cave on outdoor air temperature. *Natural Hazards and Earth System Sciences* (in press).

Hakl, J., Hunyadi, I. & Varhégyi, A. (1997). Radon monitoring in caves In: *Radon Measurements by Etched Track Detectors*, S. A. Durrani & R. Ilić, (Eds.), pp. 261–283, World scientific Publishing Co. Pte. Ltd., ISBN 9810226667, Singapore.

Han, L., Zhuang, G., Sun, Y. & Wang, Z. (2005). Local and non-local sources of airborne particulate pollution in Beijing. *Science in China, Series B Chemistry*, 48, pp. 253–264.

Hofmann, W. & Koblinger, L. (1990). Monte Carlo modeling of aerosol deposition in human lungs. Part II: Deposition fractions and their sensitivity to parameter variation. *Journal of Aerosol Science*, 21, pp. 675–688.

Holmes, N. (2007). A review of particles formation events and growth in the atmosphere in the various environments and discussion of mechanistic implications. *Atmospheric Environment*, 41, pp. 2183–2201.

Holub, R. F., Raes, F., Van Dingenen, R. & Vanmarcke, H. (1988). Deposition of aerosols and unattached radon daughters in different chambers; theory and experiment. *Radiation Protection Dosimetry*, 24, pp. 217–220.

Hopke, P. K. (1996). The initial atmospheric behaviour of radon decay products. *Journal of Radioanalytical and Nuclear Chemistry*, 203, pp. 533–375.

Hopke, P.K., Wasiolek, P, Montassier, N., Cavallo, A., Gadsby, K. & Socolow, R. (1992). Measurement of activity-weighted size distributions of radon decay products in a normally occupied home. *Radiation Protection Dosimetry*, 45, pp. 329–331.

Humar, M., Šutej, T., Skvarč, J., Mljač, L., Radež, M. & Ilić, R. (1992). Indoor and outdoor radon survey in Slovenia by etched track detectors. *Radiation Protection Dosimetry*, 45, pp. 549–552.

International Commission on Radiological Protection (ICRP). (1994). *Protection against radon-222 at home and at work*. ICRP Publication 65. Pergamon Press, Oxford.

Iskra, I., Kávási, N. & Vaupotič, J. (2010). Nano aerosols in the Postojna Cave. *Acta Carsologica*, 39, pp. 523–528.

Kávási, N., Somlai, J., Szeiler, G., Szabó, B., Schafer, I. & Kovács, T. (2010). Estimation of effective doses to cavers based on radon measurements carried out in seven caves of the Bakony Mountains in Hungary. *Radiation Measurements*, 45, pp. 1068–1071.

Kertész, Zs., Birbély-Kiss, I. & Hunyadi, I. (1999). Study of aerosols collected in a speleotherapeutic cave situated below Budapest, Hungary, *Nuclear Instruments and Methods in Physics Research B*, 150, pp. 384–391.

Kulmala, M., Vehkamäki, H., Petäjä, T., Dal Maso, M., Lauri, A., Kerminen, V.-M., Birmili, W. & McMurry, P. H. (2004). Formation and growth rates of ultrafine atmospheric particles: a review of observations. *Aerosol Science*, 35, pp. 142–176.

Kumar, P., Fennell, P., Hayhurst, A. N. & Britter, R. E. (2009). Street versus rooftop level concentrations of fine particles in a Cambridge street canyon. *Boundary-Layer Meteorology*, 131, pp. 3-18.

Kumar, P., Fennell, P. & Robins, A. (2010a). Comparison of the behaviour of manufactured and other airborne nanoparticles and the consequences for prioritising research and regulation activities. *Journal of Nanoparticle Research*, 12, pp. 1523-1530.

Kumar, P., Robins, A., Vardoulakis, S. & Britter, R. (2010b). A review of the characteristics of nanoparticles in the urban atmosphere and the prospects for developing regulatory controls. *Atmospheric Environment*, 44, 5035-5052.

Mackay, C. E. & Henry, K. H. (2009). Environmental fate and transport, In: *Nanotechnology and the Environment*, pp. 123-154, CRC Press, Taylor & Francis Group, ISBN 978-1-4200-6019-5, Boca Raton, USA.

Makkonen, U., Hellén, H., Anttila, P. & Ferm, M. (2010). Size distribution and chemical composition of airborne particles in south-eastern Finland during different seasons and wildfire episodes in 2006. *Science of the Total Environment*, 408, pp. 644-651.

Markov, K. R. (1962). A rapid method to estimate radiation hazards related to radon decay products in air (in Russian). *Atomnaya Energiya*, 87, pp. 167-178.

Marsh, J.W., Birchall, A., Butterweck, G., Dorrian, M.-D., Huet, C., Ortega, X., Reineking, A., Tymen, G., Schuler, Ch., Vargas, A., Vessu, G. & Wendt, J. (2002). Uncertainty analysis of the weighted equivalent lung dose per unit exposure to radon progeny in the home. *Radiation Protection Dosimetry*, 102, pp. 229-248.

Meisenberg, O. & Tschiersch, J. (2009). Online measurement of unattached and total radon and thoron decay products. *Applied Radiation and Isotopes*, 67, pp. 843-848.

Mercer, T. T. (1976). The effect of particle size on the escape of recoiling RaB atoms from particulate surfaces. *Health Physics*, 31, pp. 173-175.

Minoura, H. & Takeakwa, H. (2005). Observation of number concentrations of atmospheric aerosols and analysis of nanoparticles behaviour at an urban background area in Japan. *Atmospheric Environment*, 39, pp. 5806-5816.

Morawska, L, & Jamriska M. (1996). Deposition of radon progeny on indoor surfaces. *Journal of Aerosol Science*, 27, pp. 305-312.

Morawska, L., Ristovski, Z., Jayarante E. R., Keogh, D. U. & Ling, X. (2008). Ambient nano and ultrafine particles from motor vehicle emissions: characteristics, ambient processing and implications. *Atmospheric Environment*, 42, pp. 8113-8138.

Nel, A., Xia, T., Mädler, L. & Li, N. (2006). Toxic potential of materials at the nanolevel. Science, 311, pp. 622-627.

Nero, Jr., A. V. (1988). Radon and its decay products in indoor air: an overview. In: *Radon and its Decay Products in Indoor Air*, W. W. Nazaroff, & A. V. Nero, Jr., (Eds.), pp. 1–53, John Wiley & Sons, New York; 1988. pp. 1–53.

Noble, A., Mukerjee, S., Gonzales, M., Rodes, C. E., Lawless, P. A., Natarajan, S., Myers, E. A., Norris, G. A., Smith, L. H., Ozkayanak, H. & Neas, L. M. (2003). Continuous measurement of fine and ultrafine particulate matter, criteria pollutants and meteorological conditions in urban El Paso, Texas. *Atmospheric Environment*, 37, pp. 827–840.

Nowack, B. & Bucheli, T. D. (2001). Occurrence, behaviour and effects of nanoparticles in the environment. *Environment Pollution*, 150, pp. 5-22.

O'Dowd, C. D. & Smith, M. H. (1997). Marine aerosol, sea-salt and the marine sulphur cycle: a short review. *Atmospheric Environment*, 31, pp. 73–80.

O'Dowd, C. D., Aalto, P., Hameri, K., Kulmala, M. & Hoffmann, T. (2002). Aerosol formation: atmospheric particles from organic vapours. *Nature*, 416, pp. 497–498.

Oberdörster, G., Oberdörster, E. & Oberdörster, J. (2005). Nanotoxicology: An emerging discipline evolving from studies of ultrafine particles. *Environmental Health Perspectives*, 113, pp. 823–839.

Pagelkopf, P. & Porstendörfer, J. (2003). Neutralisation rate and the fraction of positive ^{218}Po-clusters in air. *Atmospheric Environment*, 37, pp. 1057–1064.

Papastefanou, C. (2008). *Radioactive Aerosols* (1st Edition), Vol. 12 of the series Radioactivity in the Environment, M. S. Baxter (Ed.), Elsevier, ISBN 978-0-08-044075-0, Amsterdam.

Pashenko, S. E. & Dublyansky, J. V. (1997). Generation of cave aerosols by alpha particles: critical evaluation of the hypothesis. *Journal of Cave and Karst Studies*, 59, pp. 103–105.

Porstendörfer, J. & Mercer, T. T. (1979). Influence of electric charge and humidity upon the diffusion coefficient of radon decay products. *Health Physics*, 37, pp. 191–199.

Porstendörfer, J. (1984). Behaviour of radon daughter products in indoor air. *Radiation Protection Dosimetry*, 7, 107–113.

Porstendörfer, J. & Reineking, A. (1992). Indoor behaviour and characteristics of radon progeny. *Radiation Protection Dosimetry*, 45, pp. 303–311.

Porstendörfer, J. (1996). Radon: measurements related to dose. *Environment International*, 22 (Suppl. 1), pp. S563–S583.

Porstendörfer, J., Ch. Zock, Ch. & Reineking, A. (2000). Aerosol size distribution of the radon progeny in outdoor air. *Journal of Environmental Radioactivity*, 51, pp. 37–48.

Porstendörfer, J. (2001). Physical parameters and dose factors of the radon and thoron decay products. *Radiation Protection Dosimetry*, 94, pp. 365–373.

Reineking, A., Becker, K. H. & Porstendörfer, J. (1985). Measurements of the unattached fraction of radon daughters in homes. *Science of the Total Environment*, 45, pp. 261–270.

Rovenská, K., Thinová, L., & Ždímal, V. (2008). Assessment of the dose from radon and its decay products in the Bozkov dolomite cave. *Radiation Protection Dosimetry*, 130, pp. 34–37.

Sainz, C., Quindós, L. S., Fuente, I., Nicolás, J. & Quindós, L. (2007). Analysis of the main factors affecting the evaluation of the radon dose in workplaces: The case of tourist caves. *Journal of Hazardous Materials*, 145, pp. 368–371.

Schery, S. D. (2001). *Understanding Radioactive Aerosols and Their Measurement* (1st Edition), Kluwer Academic Publishers, ISBN 0-7923-7068-6, Dordrecht, The Netherlands.

Schwikowski, M., Seibert, P., Baltensperger, U. & Gaggeler, H. W. (1995). A study of an outstanding Sahara dust event at the high-alpine site Jungfraujoch, Switzerland. *Atmospheric Environment*, 29, pp. 1829–1842.

Shi, J. P., Evans, D. E., Khan, A. A. & Harrison, R. M. (2001). Sources and concentration of nanoparticles (<10 nm diameter) in urban atmosphere. *Atmospheric Environment*, 35, pp. 1193–1202.

Shimo, M. & Saito, H. (2000). Size distribution of radon progeny aerosols in indoor and outdoor air. *Journal of Environmental Radioactivity*, 51, pp. 49–57.

Smerajec, M., Gregorič, A., Džeroski, S. & Vaupotič, J. (2010). Continuous measurement of ultrafine particles at urban background site of Ljubljana. *Book of abstracts* of the 11th European Meeting on Environmental Chemistry - EMEC 11, Portorož, Slovenia, December 8-11, 2010.

Solomon, S. B., Cooper, M. B., O'Brien, R. S. & Wilkinson, L. (1992). Radon exposure in a limestone cave. *Radiation Protection Dosimetry*, 45, pp. 171–174.

Thinová, L., Berka, Z., Brandejsova, E., Ždímal, V. & Milka, D. (2005). Effective dose calculation using radon daughters and aerosol particles measurement in Bozkov Dolomite Cave. *International Congress Series*, 1276, pp. 381–382.

Thinová, L., & Burian, I. (2008). Effective dose assessment for workers in caves in the Czech Republic: experiments with passive radon detectors. *Radiation Protection Dosimetry*, 130, pp. 48–51.

Tokonami, S. (1999). Determination of the diffusion coefficient of unattached radon progeny with a graded screen array at the EML Radon/Aerosol chamber. *Radiation Protection Dosimetry*, 81, pp. 285–290.

Tokonami, S. (2000). Experimental verification of the attachment theory of radon progeny onto ambient aerosols. *Health Physics*, 78, pp. 74–79.

Tu, K. W., Knutson E. O. & George, A. C. (1991). Indoor radon progeny aerosol size measurements in urban, suburban, and rural regions. *Aerosol Science and Technology*, 15, pp. 170–178.

Tu, K. W., Knutson E. O. & George, A. C. (1994). Thoron versus radon: comparison of measured progeny aerosol size distributions. *Aerosol Science and Technology*, 20, pp. 266–274.

Turner, J. & Colbeck, I. (2008). Physical and chemical properties of atmospheric aerosols, In: *Environmental Chemistry of Aerosols*, I. Colbeck, (Ed.), pp. 1–29, Blackwell Publishing Ltd., ISBN 978-1-4051-3919-9, Singapore.

United Nations Scientific Committee on the Effects of Atomic Radiation (UNSCEAR). (2000). *Sources and Effects of Ionizing Radiation, UNSCEAR 2000 Report to the General Assembly, with Scientific Annexes*, vol. 1, pp. 83–156, United Nations, New York.

Vakeva, M., Hameri, K., Kulmala, M., Lahdes, R., Ruuskanen, J. & Laitinen, T. (1999). Street levels versus rooftop concentrations of submicron aerosol particles and gaseous pollutants in an urban street canyon. *Atmospheric Environment*, 33, pp. 1385–1397.

Vanmarcke, H., Landsheere, C., Van Dingenen, R. & Poffijn, A. (1991). Influence of turbulence on the deposition rate constant of the unattached radon decay products. *Aerosol Science and Technology*, 14, pp. 257–265.

Vaupotič, J., Csige, I., Radolić, V., Hunyadi, I., Planinić, J., & Kobal, I. (2001). Methodology of radon monitoring and dose estimates in Postojna Cave, Slovenia. *Health Physics*, 80, pp. 142–147.

Vaupotič, J. & Kobal, I. (2006). Effective doses in schools based on nanosize radon progeny. *Atmospheric Environment*, 40, pp. 7494–7507.

Vaupotič, J. (2007). Nano-size radon short-lived progeny aerosols in Slovenian kindergartens in wintertime. *Chemosphere*, 2007, 69, pp. 856–863.

Vaupotič, J. (2008a). Comparison of various methods of estimating radon dose in underground workplaces in wineries. *Radiation and Environmental Biophysics*, 47, pp. 527–534.

Vaupotič, J. (2008b). Nanosize radon short-lived decay products in the air of the Postojna Cave. *Science of the Total Environment*, 393, pp. 27–38.

Vaupotič, J. (2010). Radon levels in karst caves in Slovenia. *Acta Carsologica*, 39, pp. 503–512.

Wang, F., Costabileb, F., Li, H., Fang, D. & Alligrini, I. (2010). Measurements of ultrafine particle size distribution near Rome. *Atmospheric Research*, 98, pp. 69–77.

Wang, Y., Hopke, P. K., Chalupa, D. C. & Utell, M. J. (2010). Long-term study of urban ultrafine particles and other pollutants. *Atmospheric Environment*, doi: 10.1016/j.atmosenv.2010.08.022.

Wen, C. S. (1996). *The Fundamentals of Aerosol Dynamics*, (1st Edition), World Scientific Publishing Co. Pte. Ltd., ISBN 9810226616, Singapore.

Young, L.-H. & Keeler, G. J. (2007). Summertime ultrafine particles in urban and industrial air: Aitken and nucleation mode particle events. *Aerosol and Air Quality Research*, 7, pp. 379–402.

7

Effect of Updating Meteorological Data on Assessment Modeling Using VENTSAR XL©

Eduardo B. Farfán
Savannah River National Laboratory
United States of America

1. Introduction

The Savannah River Site (SRS), one of the facilities in the United States Department of Energy (DOE) complex, is located in south central South Carolina and has an area of about 800 km^2. The SRS mission was the production of special nuclear materials (such as ^{239}Pu and ^3H) for national defense from the 1950s to the early 1990s. This mission was accomplished through the operation of five nuclear reactors, chemical processing and separation plants, and various support facilities. Even though the production at most SRS facilities has been terminated, substantial quantities of high-level nuclear and mixed hazardous waste remain stored at SRS. SRS's current mission includes the stewardship of the nation's nuclear weapons stockpile, nuclear materials, and the environment. SRS will continue addressing environmental quality and managing any radioactive waste from current and future operations. Managing waste involves working with DOE, the State of South Carolina, the Environmental Protection Agency (EPA), and the Nuclear Regulatory Commission (NRC) (Mamatey, 2009).

Savannah River National Laboratory (SRNL) periodically generates an updated meteorological database in order to facilitate dosimetric calculations of accident and routine release scenarios for onsite and offsite populations. This meteorological database includes wind speed, direction, temperature, dew point, and horizontal and vertical turbulence intensities from a height of 61-m above ground-level. This information becomes the input of various environmental dosimetry codes run by the Environmental Dosimetry Group (EDG) at SRNL. The three most recent databases prior to the current one were completed for the time periods 1987-1991, 1992-1996, and 1997-2001. The current database covers the period 2002-2006 (Kabela & Weber, 2007). The advantage of updating the database at regular intervals is that meteorological observations are steadily growing more complete and reliable with the implementation of better electronic data archiving software and hardware, and improved data quality assurance procedures. Additionally, changes in the region's climate may be noticeable (Kabela & Weber, 2007).

The updated meteorological data is applied in various dosimetry models approved for risk and dose assessment at SRS. One of these models is VENTSAR XL©, which is an upgraded and improved spreadsheet version of the FORTRAN-based program named VENTSAR, which originated from the code VENTX (Smith & Weber, 1983) on the SRS IBM Mainframe**. It is a dose assessment model used to calculate dose following short-term atmospheric releases and concentrations of chemical or radiological pollutants. The user may include building effects and plume rise near a release point. The dose to individuals in or near the building can also be calculated if the pollutant is radiological. VENTSAR XL© "calculates the concentrations for a given meteorological exceedance probability or for a given stability and wind speed combination" (Simpkins, 1997). The switch to a spreadsheet-based code from Mainframe-based FORTRAN program makes it more user-friendly, allowing the user to have a better access to the code and run it without any knowledge of Mainframe commands.

VENTSAR XL© is a Gaussian Plume model that includes building effects and plume rise. The building being modeled can be a simple structure with or without a penthouse on its top. The model considers recirculation cavities, high turbulence zones, wakes beyond the building, and plume rise caused by buoyancy and momentum, and downwash (Simpkins, 1997). Doses are calculated at up to 200 user-specified increments and effective dose equivalents are estimated for plume shine and inhalation exposure pathways (Simpkins, 1997). VENTSAR XL© has been developed through the use of Macros, which are a group of coded instructions under Microsoft® Visual Basic® applications (e.g., Microsoft® Excel®) that are used to automate routine tasks and the resolution of complex mathematical calculations. VENTSAR XL© can be run on any computer that supports Microsoft® Excel® 4.0 or later (Simpkins, 1997). The code is exceptionally user-friendly and the user-input template is easy to comprehend. In addition, the code contains a number of checks to prevent the user from entering the wrong input; for example, a parameter value that is beyond the parameter range indicated on the VENTSAR XL© template. The typical input involves the location of the release, building dimensions, distance to the building, release height, vent diameter, vent gas temperature, gas molecular weight, ambient air temperature, breathing rate, meteorological conditions, radionuclides and their amount released. The output is easily converted into tables and graphs for further analysis and shows the concentrations and pathway doses for each of the incremental downwind distances (Simpkins, 1995, 1997). VENTSAR XL© has been used at SRS to investigate building effects such as reactor cooling towers in support of safety analyses. VENTSAR XL© has also been applied to Good Engineering Practice (GEP) stack height evaluations for various projects at SRS.

VENTSAR XL© test cases are executed each time for a new set of meteorological data by the EDG. Subsequently, the results from the code are compared to the test cases of the previous time period to ensure that there are no abnormalities in the new meteorological data. Lines of code in the program must be changed for VENTSAR XL© to access the new meteorological data. Executing the test cases also provides the means to verify that these changes have been properly made. This study presents test cases for four periods (1987-1991, 1992-1996, 1997-2001, and 2002-2006) and wind frequency comparisons among these four periods for various locations at SRS.

2. VENTSAR XL© methodology

The VENTSAR XL© methodology and data are described in detail in the following sections, which represent a summary of Simpkins' report (1997).

2.1 Gaussian plume model

The pollutant dispersion calculations in the VENTSAR XL© code are based on the Gaussian plume model (Hanna et al., 1982). Along the plume centerline, the dispersion factor or relative air concentration, defined as the ratio of the pollutant concentration χ (kg m^{-3} or Ci m^{-3}) to the source strength Q (kg s^{-1} or Ci s^{-1}), is given by the equation:

$$\frac{\chi}{Q}=\frac{1}{2\pi\sigma_y\sigma_z U_S}\left[e^{-\left(\frac{(z-h_e)^2}{2\sigma_z^2}\right)}+e^{-\left(\frac{(z+h_e)^2}{2\sigma_z^2}\right)}\right] \qquad (1)$$

where,

χ/Q = the dispersion factor (s m^{-3})

z = height above the ground surface (m)

h_e = effective release height (m)

U_s = wind speed at the release height (m s^{-1})

σ_y = the standard deviation of the concentration distribution in the horizontal cross-plume direction (m)

σ_z = the standard deviation of the concentration distribution in the vertical direction (m)

The annual average values of χ/Q are calculated as:

$$annual\overline{(\chi/Q)}=\sum_{i,j}^{6,7}P_{ij}\left(\frac{\chi}{Q}\right)_{ij} \qquad (2)$$

where,

i = wind speed category

j = stability class

$(\chi/Q)_i$ = relative air concentration for meteorological condition (i,j)

P_i = the probability of a particular meteorological condition (i,j) occurring within a five-year time period

2.1.1 Meteorological data

VENTSAR XL© accesses a meteorological joint frequency distribution containing six wind speed classes (Table 1) and 7 stability categories (Pasquill, 1976). VENTSAR XL© contains meteorological data files already available for used at SRS, but the user may add data files of his or her selection from any location.

Speed Category	Range (m s^{-1})
1	$0<U\leq2$
2	$2<U\leq4$
3	$4<U\leq6$
4	$6<U\leq8$
5	$8<U\leq12$
6	$U\geq12$

Table 1. Wind speed category ranges for SRS Files

Atmospheric stability is classified by standard deviations of the lateral or azimuthal wind direction. SRS meteorological towers contain instrumentation at 61 m (200 ft) that measures

horizontal (azimuth) and vertical (elevation) wind directions. In addition, direct measurements of turbulence, expressed as standard deviations of fluctuations about mean azimuth (noted either as σ_a or σ_θ) and elevation (σ_e) angles, are made at 61 m.
For calculational purposes within the spreadsheet, an assumed average value of σ_θ is selected for the atmospheric stability class of interest. Ranges for σ_θ and the values that are used within VENTSAR XL© are shown in Table 2.

Pasquill Category	Range for σ_θ (degrees)	σ_θ Used in VENTSAR XL© (degrees)
A	$23 \le \sigma_\theta$	27.5
B	$18 \le \sigma_\theta < 23$	22.5
C	$13 \le \sigma_\theta < 18$	17.5
D	$8 \le \sigma_\theta < 13$	12.5
E	$4 \le \sigma_\theta < 8$	7.5
F	$2 \le \sigma_\theta < 4$	3.75
G	$\sigma_\theta < 2$	2.00

Table 2. Classification of atmospheric stability

2.1.2 Pasquill-Briggs diffusion coefficients

The lateral and vertical diffusion coefficients within VENTSAR XL© are those derived by Pasquill (1976) and Briggs (1973), respectively. The equation representing Pasquill's lateral diffusion coefficients is

$$\sigma_y = \sigma_\theta X f(X) \tag{3}$$

where,

σ_θ = standard deviation of lateral wind direction in radians (Table 2)
X = downwind distance (km)
$f(X)$ = function of distance, X (km), as discussed below

Pasquill developed formulations for $f(X)$ with a table of values for distances less than 10 km and the following equation for distances greater than 10 km:

$$f(X) = 0.33\left[\frac{10}{X}\right]^{0.5} \tag{4}$$

For distances less than 10 km, the following equation was derived from the table of values with X in km:

$$f(X) = \frac{1}{1 + 0.031(1000X)^{0.46}} \tag{5}$$

Pasquill (1976) gives a detailed description on how the coefficients were developed using data from experiments at various sites. The vertical diffusion coefficients defined by Briggs (1973) and then refined by Briggs and published in Hanna et al. (1982) for open-country conditions are represented in Table 3 as a function of Pasquill's atmospheric stability classes.

Pasquill Stability Category	σ_z (X in m)
A	$0.20X$
B	$0.12X$
C	$0.08X(1 + 0.0002X)^{-0.5}$
D	$0.06X(1 + 0.0015X)^{-0.5}$
E	$0.03X(1 + 0.0003X)^{-1}$
F	$0.02X(1 + 0.0003X)^{-1}$
G	$0.01X(1 + 0.0003X)^{-1}$

Table 3. Brigg's vertical diffusion coefficient formulas.

2.2 Plume rise

Plume rise models are based on fundamental laws of fluid mechanics, conservation of mass, potential density, and momentum. VENTSAR XL© considers plume rise due to both buoyancy and momentum effects. Several different mechanisms can increase or decrease the height of the plume at downwind distances. Plume rise due to momentum and buoyancy effects can increase the height of the plume while downwash can decrease the height of the plume. The effective plume height at a given distance, X, downwind is

$$h(X) = h_s - \Delta h_D + \Delta h_B(X) + \Delta h_M(X) \tag{6}$$

where,

h_s = initial height of the source
Δh_D = source height change due to downwash
Δh_B = source height change due to buoyancy effects
Δh_M = source height change due to momentum effects

Downwash, buoyancy, momentum, and building wake effects considered in VENTSAR XL© are described in detail by Simpkins (1997).

2.3 Dose estimation

VENTSAR XL© can calculate inhalation and plume shine dose using dose factors provided in the United States Department of Energy (DOE) documents (1988a and 1988b). Inhalation dose is estimated by the product of the radionuclide concentration in the air that is breathed, the rate at which the air is breathed, and a factor to convert intake quantities to dose. The inhalation dose to a given individual, assuming exposure during the entire plume passage, is calculated using the following general equation:

$$D_{inh} = 3.17E-08(Q_n)(\frac{\chi}{Q})(DFI_n)(B)e^{-\lambda_n t} \tag{7}$$

where,

$3.17\text{x}10^{-8}$ =conversion factor (years per second)
Q_n = total release (Ci)
χ/Q = relative concentration at receptor (s m^{-3})
DFI_n = effective dose equivalent factor for inhalation (rem Ci^{-1})
B = adult maximum breathing rate (m3 yr^{-1})

λ_n = decay constant (s^{-1})
t = travel time from release to receptor (s)

The uniform plume model assumes that the exposed individual is located in a time integrated uniform concentration of a given nuclide throughout the infinite hemisphere above ground level. The gamma-shine external dose is therefore directly proportional to the integral air concentration and is determined by multiplying the integral concentration by an infinite-plume shine dose factor. The external dose for a given nuclide, n, is expressed as:

$$D_{PS} = (\frac{\chi}{Q})(Q_n)(DFS_n)e^{-\lambda_n t} \quad (8)$$

where,

χ/Q = relative air concentration at the receptor (s m^{-3})
Q_n = total release of nuclide n (Ci)
$DFSn$ = shine dose factor for nuclide n (mrem s^{-1} per Ci m^{-3})
λ_n = decay constant for nuclide n (s^{-1})
t = transit time between release and exposure (s)

A library of dose factors for about 500 radionuclides is contained under the file name "Dose Factor." Doses are calculated only for the radionuclides that are entered. No ingrowth is considered, but the user can enter the associated progeny as appropriate.

2.4 Relative concentration

In accordance with the Clean Air Act Amendments of 1977, GEP must be used in determining the height of any stack that will be used to disperse routine emissions (United States Environmental Protection Agency [EPA], 1981a, 1981b). With respect to stack heights, the GEP height is "the height necessary to ensure that emissions from the stack do not result in excessive concentrations of any air pollutant in the immediate vicinity of the source as a result of atmospheric downwash, eddies and wakes which may be created by the source itself, nearby structures or nearby terrain obstacles" (EPA, 1981a). The EPA has set specific criteria to determine if a stack is of the acceptable height. These criteria have been used to determine a GEP stack height for several emission assessments at SRS.

The EPA documents (1981a, 1981b) contain detailed information on how the height of a stack is determined. The general rule for stack height determination is to make the stack at least 2½ times the height of nearby buildings. This estimated height can be increased or decreased based on other factors such as plume rise, downwash, and building wake effects. According to U.S. Nuclear Regulatory Commission Guide 1.145 (NRC, 1982) ground level releases should be considered for "all release points or areas that are effectively lower than two and one-half times the height of adjacent solid structures." This regulation is applied when performing calculations for Emergency Preparedness Hazard Assessments.

Sometimes it is necessary to construct a stack with a lower height than the one required by federal or state regulations. Employing VENTSAR XL©, detailed analyses of air concentrations within the vicinity of the building can be performed to justify using a lower stack by ensuring that the maximum downwind concentration in the presence of the building is not more than 40% greater than the maximum downwind concentration without the building (EPA 1981a, 1981b). Therefore, an acceptable stack height can be demonstrated using the following equation:

$$\frac{\left(\frac{\chi}{Q_{\max}}\right)_{building}}{\left(\frac{\chi}{Q_{\max}}\right)_{\text{no building}}} \leq 1.4, \tag{9}$$

where,

$\left(\frac{\chi}{Q_{\max}}\right)_{building}$ = maximum ground level concentration with building present;

$\left(\frac{\chi}{Q_{\max}}\right)_{\text{no building}}$ = maximum ground level concentration with no building present;

χ = air concentration (Bq m^{-3}); and

Q = amount released (Bq s^{-1}).

To pinpoint the maximum ground level concentrations with a building present and without the building for Eq. (9), the annual average air concentrations (s m^{-3}) versus downwind distances are plotted from the VENTSAR XL© output.

3. Meteorological data update

Every five years meteorological data are updated for use in environmental dosimetry codes at SRNL. Data for the period 2002-2006 are available and are tested in this study using test cases and approved environmental dosimetry procedures for VENTSAR XL©. Test cases consider seven areas of SRS where radionuclide releases are possible (A, C, D, F, H, K, and P) and an alternate release location. The various SRS areas where the possibility of radionuclide releases exist are shown in Fig. 1. The meteorological data, used for dosimetry purposes at SRS, consists of hourly averages of wind speed and direction at the various SRS meteorological towers for a 5-year period. The SRS meteorological data for the periods 1987-1991, 1992-1996, 1997-2001, and 2002-2006 were reported by Kabela and Weber (2007), Weber (2002), Weber (1998), and Parker et al. (1992), respectively. The frequency at which the wind blows from the various sectors for these periods and SRS areas including L-Area is illustrated in Fig. 2.

Test cases were executed using identical parameters with the exception of the new meteorological data. The parameter input values for the test cases are presented in Table 4. The Eastern and Northern grid coordinates in VENTSAR XL© are SRS site-specific and must be included when a meteorological file name is not entered by the user. Cases 1, 2, 4 and 7 consider plume rise, Cases 1, 2, 6, and 7 involve meteorological averaging, Cases 3, 5, 6, and 8 include radioactive releases. The parameters for building and penthouse in Case 2 are set to zero; to model the absence of a building. When the VENTSAR XL© plume rise option is selected, the vent diameter, vent gas temperature, ambient air temperature, and molecular weight of the gas released must be entered. The averaging option is used when a meteorological exceedance probability is specified. For meteorological conditions that do not exceed 99.5% of the time, a value of 0.005 (0.5%) would be used (Simpkins, 1997). If the averaging option is not selected, the wind speed and stability class must be entered as VENTSAR XL© input. If the release is considered to be radioactive, the user must select the breathing rate of the individual for VENTSAR XL© to calculate dose (Simpkins, 1997).

Depending on the release (radioactive or chemical), the output and its headings vary. When a specific wind speed and stability class is selected, the annual average dilution factor (χ/Q, s m^{-3}) is not estimated. However, instead of calculating effluent concentrations (χ, Bq m^{-3}) at specific distances from the release point, dilution factors are generally estimated (by selecting the VENTSAR XL© averaging option) since the dilution factors are independent from the source strength (Q, Bq s^{-1}) (Faw & Shultis, 1999).

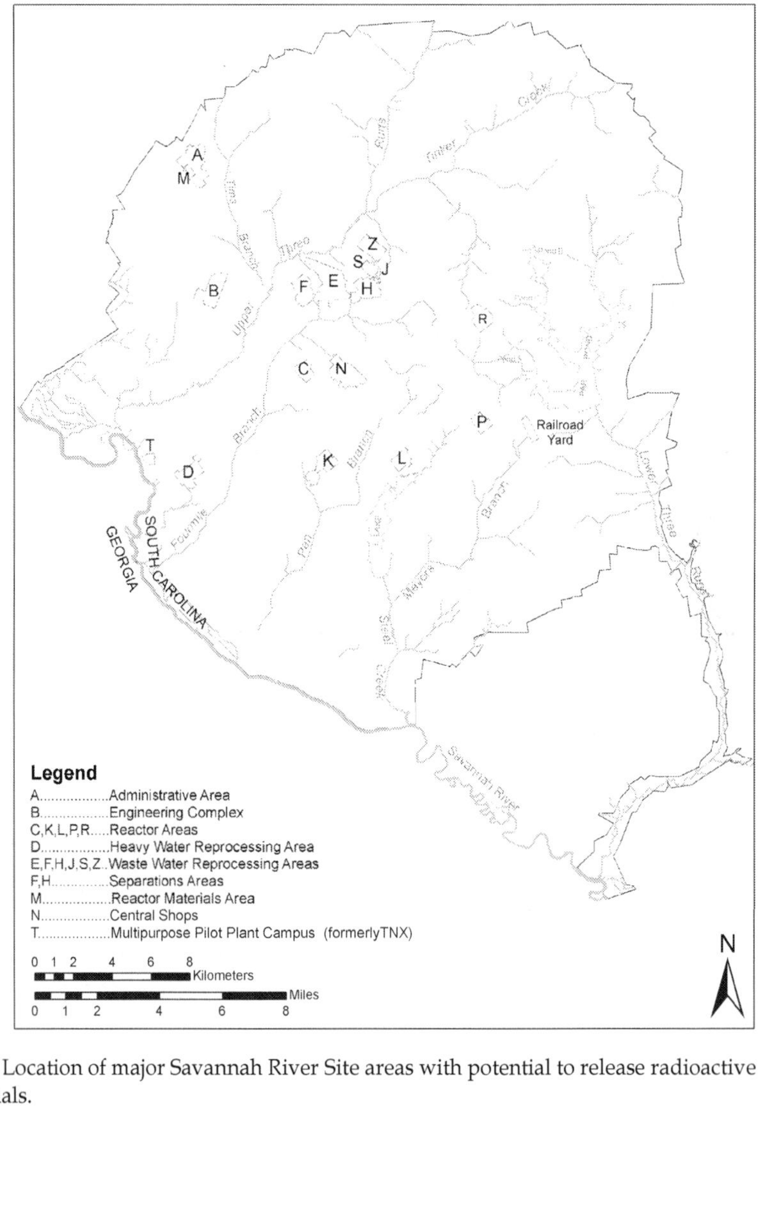

Fig. 1. Location of major Savannah River Site areas with potential to release radioactive materials.

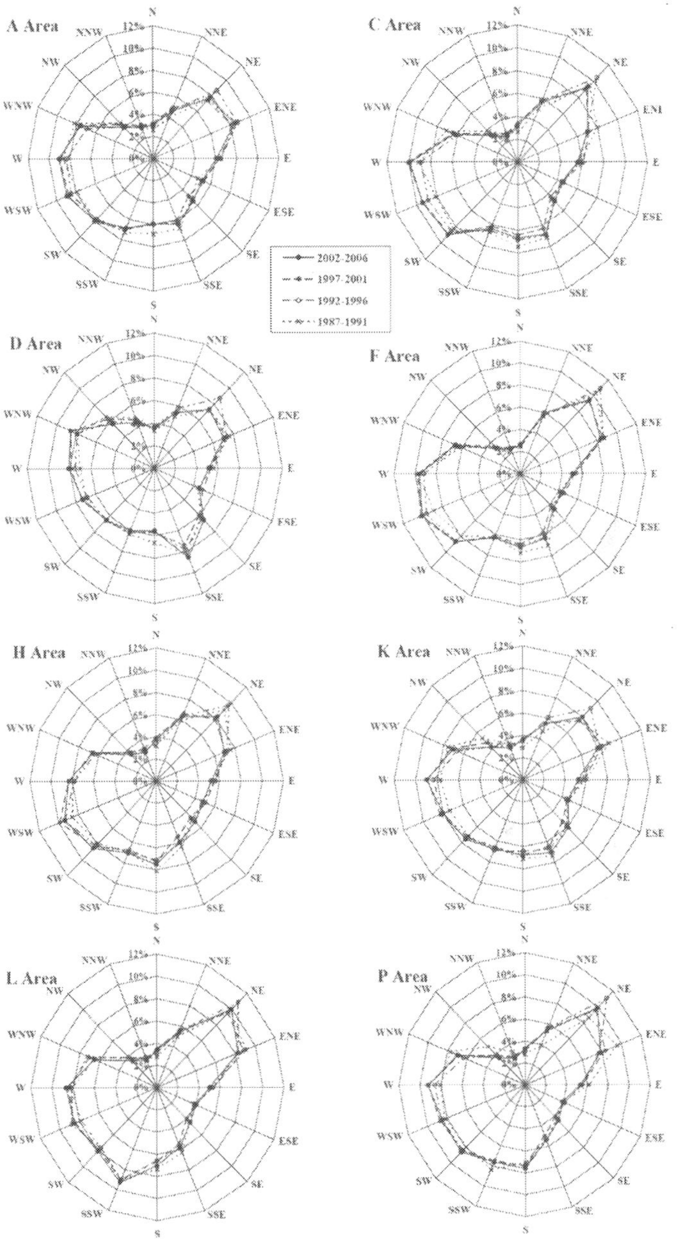

Fig. 2. Wind frequencies for 2002-2006, 1997-2001, 1992-1996, and 1987-1991 SRS meteorological data.

Parameter	Case							
	1	2	3	4	5	6	7	8
Consider Plume Rise	YES	YES	NO	YES	NO	NO	YES	NO
Area of Release	P	D	H	C	A	K	F	OTHER[a]
Building Height[b]	10	0	5	20	15	8	10	12
Building Width[b]	20	0	30	100	200	200	10	30
Building Length[b]	30	0	100	30	200	10	10	15
Penthouse Height[b]	1	0	3	6	7	3	5	6
Penthouse Width[b]	2	0	5	10	200	150	5	20
Penthouse Length[b]	3	0	5	10	100	8	5	10
Bldg. to Penthouse[b]	5	0	20	0	100	2	0	1
Min. Vent to Receptor[b]	10	10	10	10	10	10	10	10
Max. Vent to Receptor[b]	1000	1000	1000	1000	1000	1000	1000	1000
Compass Sector	NNW	S	N	NE	WNW	SSE	ENE	E
Vent to Roof Edge[b]	-500	0	-100	10	-10	500	100	-30
Vent Height[b]	50	25	5	0	100	50	5	0
Radioactive Release?	NO	NO	YES	NO	YES	YES	NO	YES
Release Rate (GBq min^{-1})	-	-	37	-	37	37	-	37
Pollutant Mole Fraction	1×10^{-6}	1×10^{-6}	-	1×10^{-6}	-	-	1×10^{-6}	-
Vent-Gas Flow Rate ($m^3 s^{-1}$)	500	750	1000	100	500	750	1000	100
Meteorological Averaging?	YES	YES	NO	NO	NO	YES	YES	NO
Probability Level	0.005	0.005	-	-	-	0.005	0.005	-
Wind Speed ($m s^{-1}$)	-	-	2	1	4	-	-	3
Stability Class	-	-	D	F	B	-	-	A
Vent Diameter[b]	3	2	3	1	3	2	2	1
Vent-Gas Molecular Weight	210	200	200	190	200	180	230	220
Vent-Gas Temp(°C)	20	17	17	18	17	17	19	14
Ambient Air Temp(°C)	15	17	17	12	17	17	16	13
Calculate Dose	N	N	Y	N	N	Y	N	N
Breathing Rate ($m^3 y^{-1}$)	-	-	12,000	-	-	12,000	-	-
Radionuclide, Source Term	^{3}H, 0.11 TBq ^{137}Cs, 74 GBq ^{137m}Ba, 74 GBq	^{3}H, 0.11 TBq ^{137}Cs, 74 GBq ^{137m}Ba, 74 GBq	^{3}H, 0.11 TBq ^{137}Cs, 74 GBq ^{137m}Ba, 74 GBq	^{3}H, 0.11 TBq ^{137}Cs, 74 GBq ^{137m}Ba, 74 GBq	^{3}H, 0.11 TBq ^{95}Zr, 0.19 TBq ^{95m}Nb, 0.19 TBq	^{3}H, 0.11 TBq ^{95}Zr, 0.19 TBq ^{95m}Nb, 0.19 TBq	^{3}H, 0.11 TBq ^{95}Zr, 0.19 TBq ^{95m}Nb, 0.19 TBq	^{3}H, 0.11 TBq ^{95}Zr, 0.19 TBq ^{95m}Nb, 0.19 TBq

[a] Release at SRS's Center of Site. [b] Units in meters (m).

Table 4. Input for VENTSAR XL tests cases.

4. Results

For most SRNL environmental studies, the airborne concentrations under a given meteorological condition are not the main focus. The center of attention is the concentrations averaged over a year and over all meteorological conditions that occur during the typical averaging period of one year, during which the weather conditions fluctuate to a great extent. Consequently, averaging the concentrations becomes necessary by measuring the wind and atmospheric stability on a daily basis for various averaging periods. From these measurements, the distribution of wind velocities are calculated and usually presented as a *wind rose*. The wind rose for the H-Area meteorological tower at SRS for the 2002-2006 period is presented in Fig. 3. H-Area is located near to the SRS center of site, which is used for risk and dose assessments. The wind class frequency graph for H-Area shown in Fig. 4 illustrates the frequency of each wind speed class.

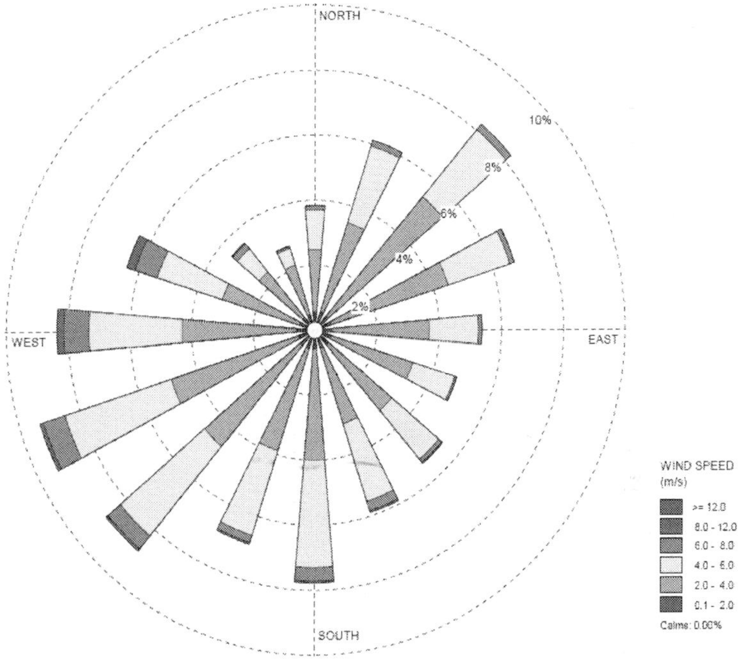

Fig. 3. Wind Rose for the SRS's H-Area meteorological tower for the 2002-2006 period.

The SRS wind direction percent differences between the meteorological periods 2002-2006 and 1997-2001 are shown in Table 5. The maximum increase (15.1%) in wind frequency is for the southeast cardinal direction for the K-Area. The maximum decrease (-12.1%) is for the northwest direction for the L-Area. The average percent difference is -0.3% for the periods 2002-2006 and 1997-2001 (Table 5). However, the average percent difference is 0.04% for the four periods and all the areas considered in this study. The minimum and maximum SRS wind direction percent differences are -30.7% (NW) and 38.1% (NNE), respectively. These considerable differences occur for P-Area when comparing the wind direction percents for the periods of 1987-1991 and 2002-2006 (Fig. 2).

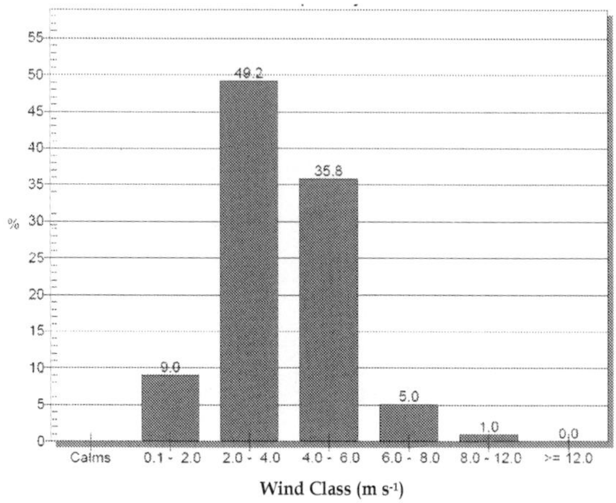

Fig. 4. Wind class frequency distribution for the SRS's H-Area meteorological tower for the 2002-2006 period.

Wind Direction	SRS Areas A	C	D	F	H	K	L	P
N	-4.8	-3.0	4.4	-3.1	-3.8	0	-1.6	0
NNE	-7.4	-3.4	-1.5	2.8	-4.5	0	-4.3	-1.8
NE	4.0	2.2	1.0	0.8	4.4	6.7	4.4	3.1
ENE	3.8	-1.4	-3.7	-3.1	-1.9	-6.2	-5.9	-6.0
E	-6.1	-8.2	-4.3	-5.4	-5.5	-10.3	-6.5	-3.6
ESE	-0.1	-2.2	-6.7	-2.7	-5.1	2.2	3.8	-4.8
SE	12.7	11.6	11.3	6.8	12.1	15.1	11.3	12.2
SSE	5.8	7.9	14.9	9.2	9.5	9.4	5.2	8.0
S	-1.2	4.7	-1.2	5.2	3.6	0	6.1	4.1
SSW	-2.1	3.3	-3.3	2.4	4.6	1.5	3.0	1.3
SW	-1.3	1.1	0.0	-0.1	3.0	2.7	1.1	2.4
WSW	1.9	0.0	-1.0	-1.1	-4.7	-3.6	0.5	-2.3
W	-0.1	-2.0	2.6	-1.0	-0.1	1.1	2.4	0.0
WNW	-0.9	-1.6	1.0	-1.9	-0.3	-2.8	-8.2	2.9
NW	-9.1	-8.6	-9.7	-8.8	-6.3	-10.6	-12.1	-10.5
NNW	-6.4	-11.1	-5.2	-4.4	-11.7	-8.8	-3.7	-3.7

Table 5. SRS wind direction percent differences between the periods 2002 2006 and 1997-2001.

Table 6 shows the VENTSAR XL© input and output for Case 6, which involves a radioactive release, 2002-2006 meteorological data for K-Area, meteorological averaging with a probability level of 0.5%, and dose calculations at various distances from the release point. The first three columns present a list of all the input parameters and their values with units. The next two columns display the valid range of the parameters. The last five columns show the output, which includes the distance (m), 99.50% dilution factor χ/Q (s m^{-3}), annual average dilution factor χ/Q (s m^{-3}), inhalation dose (mrem), plume shine dose (mrem), and total dose (mrem).

CALCULATE	VENTSAR XL ©				Distance (m)	99.50% x/Q s/m^3	Annual Average s/m^3	Inhalation Dose (mrem)	Plume Shine Dose (mrem)	Total Dose (mrem)
Test Case 6 2002-2006 met data					10.0	0.00E+00	0.00E+00	0.00E+00	0.00E+00	0.00E+00
TQ Foley (5-5184) & EB Farfan (5-2257)					15.0	0.00E+00	0.00E+00	0.00E+00	0.00E+00	0.00E+00
SRNL Environmental Dosimetry Group					19.9	3.58E-37	2.53E-39	1.45E-32	2.34E-34	1.47E-32
		Units	VALID RANGE		24.9	4.61E-25	3.26E-27	1.86E-20	3.01E-22	1.89E-20
Meteorological File Name	Kmet0206				29.8	1.61E-18	1.14E-20	6.52E-14	1.05E-15	6.62E-14
OR enter "SRS" and coordinates					34.8	1.33E-14	9.40E-17	5.37E-10	8.67E-12	5.45E-10
SRS Grid Coordinates Easting	0	ft	0	120000	39.7	4.36E-12	3.09E-14	1.76E-07	2.85E-09	1.79E-07
Northing	0	ft	0	120000	44.7	2.22E-10	1.57E-12	8.96E-06	1.45E-07	9.10E-06
Building Dimensions					49.6	3.54E-09	2.50E-11	1.43E-04	2.31E-06	1.45E-04
Building Height	8	m	0	1000	54.6	2.66E-08	1.88E-10	1.08E-03	1.74E-05	1.09E-03
Building Width	200	m	0	1000	59.5	1.20E-07	8.51E-10	4.86E-03	7.86E-05	4.94E-03
Building Length	10	m	0	1000	64.5	3.80E-07	2.69E-09	1.54E-02	2.48E-04	1.56E-02
Penthouse Height	3	m	0	1000	69.4	9.29E-07	6.58E-09	3.76E-02	6.07E-04	3.82E-02
Penthouse Width	150	m	0	1000	74.4	1.88E-06	1.33E-08	7.59E-02	1.23E-03	7.71E-02
Penthouse Length	8	m	0	1000	79.3	3.29E-06	2.33E-08	1.33E-01	2.15E-03	1.35E-01
Distance to Penthouse on Rooftop	2	m	0	1000	84.3	5.16E-06	3.66E-08	2.09E-01	3.37E-03	2.12E-01
					89.2	7.45E-06	5.27E-08	3.01E-01	4.86E-03	3.06E-01
Minimum Distance of Interest	10	m	10	99999	94.2	1.00E-05	7.12E-08	4.06E-01	6.56E-03	4.13E-01
Maximum Distance of Interest	1000	m	10	100000	99.1	1.28E-05	9.10E-08	5.19E-01	8.39E-03	5.28E-01
Number of Increments	200		1	200	104.1	1.57E-05	1.12E-07	6.36E-01	1.03E-02	6.46E-01
Compass Sector of Building	8		1	16	109.0	1.86E-05	1.32E-07	7.52E-01	1.22E-02	7.64E-01
(1-N, 2-NNE, 3-NE ...)					114.0	2.14E-05	1.52E-07	8.65E-01	1.40E-02	8.79E-01
Distance of Vent from Roof Edge	500		-1000	1000	118.9	2.40E-05	1.71E-07	9.70E-01	1.57E-02	9.86E-01
(negative is upwind)					123.9	2.64E-05	1.89E-07	1.07E+00	1.73E-02	1.09E+00
Vent Height	50	m	0	500	128.8	2.86E-05	2.06E-07	1.16E+00	1.87E-02	1.17E+00
Radioactive Calculations(Y or N)	y				133.8	3.05E-05	2.21E-07	1.23E+00	1.99E-02	1.25E+00
					138.7	3.22E-05	2.34E-07	1.30E+00	2.10E-02	1.32E+00
If No Mole Fraction of Vent Gas	0.00E+00				143.7	3.37E-05	2.46E-07	1.36E+00	2.20E-02	1.38E+00
Vent Gas Flow Rate	750	m3/s	0	1000	148.6	3.49E-05	2.56E-07	1.41E+00	2.28E-02	1.43E+00
Averaging Option (Y or N)	y				153.6	3.58E-05	2.65E-07	1.45E+00	2.34E-02	1.47E+00
If YES Specify Probability Level	0.005		0.001	0.5	158.5	3.66E-05	2.73E-07	1.48E+00	2.39E-02	1.50E+00
If NO Windspeed at Vent Height	0	m/s	0.1	15	163.5	3.72E-05	2.79E-07	1.50E+00	2.43E-02	1.53E+00
Stability Class(1-7 to A-G)	0		1	7	168.4	3.76E-05	2.84E-07	1.52E+00	2.46E-02	1.55E+00
Plume Rise(Y or N)	n				173.4	3.79E-05	2.89E-07	1.53E+00	2.48E-02	1.56E+00
Vent Diameter	2	m	0.001	100	178.3	3.81E-05	2.92E-07	1.54E+00	2.49E-02	1.56E+00
Gas Molecular Weight	180		0	400	183.3	3.81E-05	2.95E-07	1.54E+00	2.49E-02	1.57E+00
Vent Gas Temperature	17	C			188.2	3.80E-05	2.97E-07	1.54E+00	2.48E-02	1.56E+00
Ambient Air Temperature	17	C			193.2	3.79E-05	2.99E-07	1.53E+00	2.47E-02	1.56E+00
Calculate Dose?(Y or N)	y				198.1	3.76E-05	3.00E-07	1.52E+00	2.46E-02	1.55E+00
Inhalation Rate	12000	m^3/yr	8000	20000	203.1	2.01E-05	3.01E-07	8.14E-01	1.32E-02	8.27E-01
Radionuclide	**Curies**				208.0	2.09E-05	3.01E-07	8.44E-01	1.36E-02	8.58E-01
H-3	3.00E+00				213.0	2.16E-05	3.01E-07	8.72E-01	1.41E-02	8.86E-01
Zr-95	5.00E+00				217.9	2.22E-05	3.00E-07	8.98E-01	1.45E-02	9.12E-01
Nb-95m	5.00E+00				222.9	2.28E-05	3.00E-07	9.21E-01	1.49E-02	9.36E-01

Table 6. VENTSAR XL© input and output for Case 6 based on Simpkins' template (1997).

The results of the test cases were compared graphically (Fig. 5). Cases 3, 4, 5, and 8 do not use the averaged meteorological data; therefore, a wind speed, stability class, and vent diameter for each of these cases were entered as shown in Table 4 (Foley, 2008). Since these cases are independent from the meteorological data, the dilution factors for the four periods

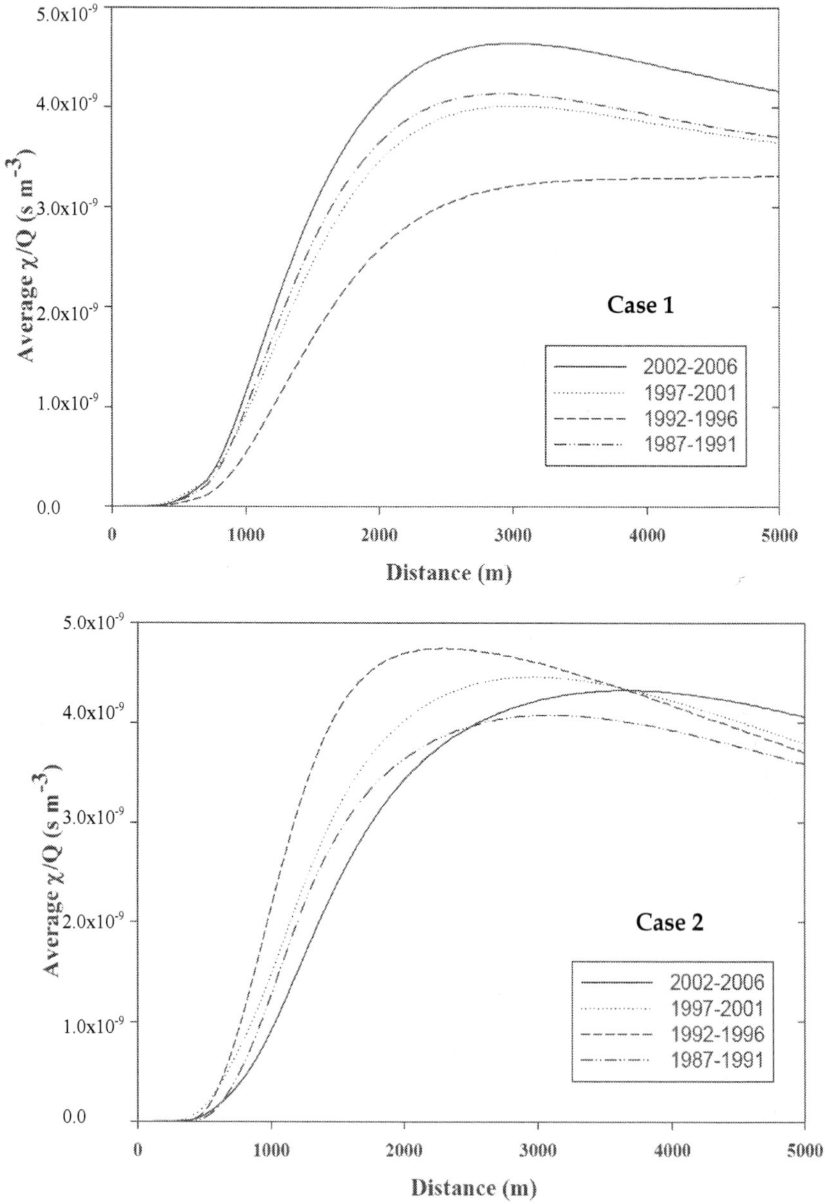

Fig. 5. Test Case comparison for periods 2002-2006, 1997-2001, 1992-1996, and 1987-1991. Cases 1, 2, 6, and 7 represent cases with average meteorological data.

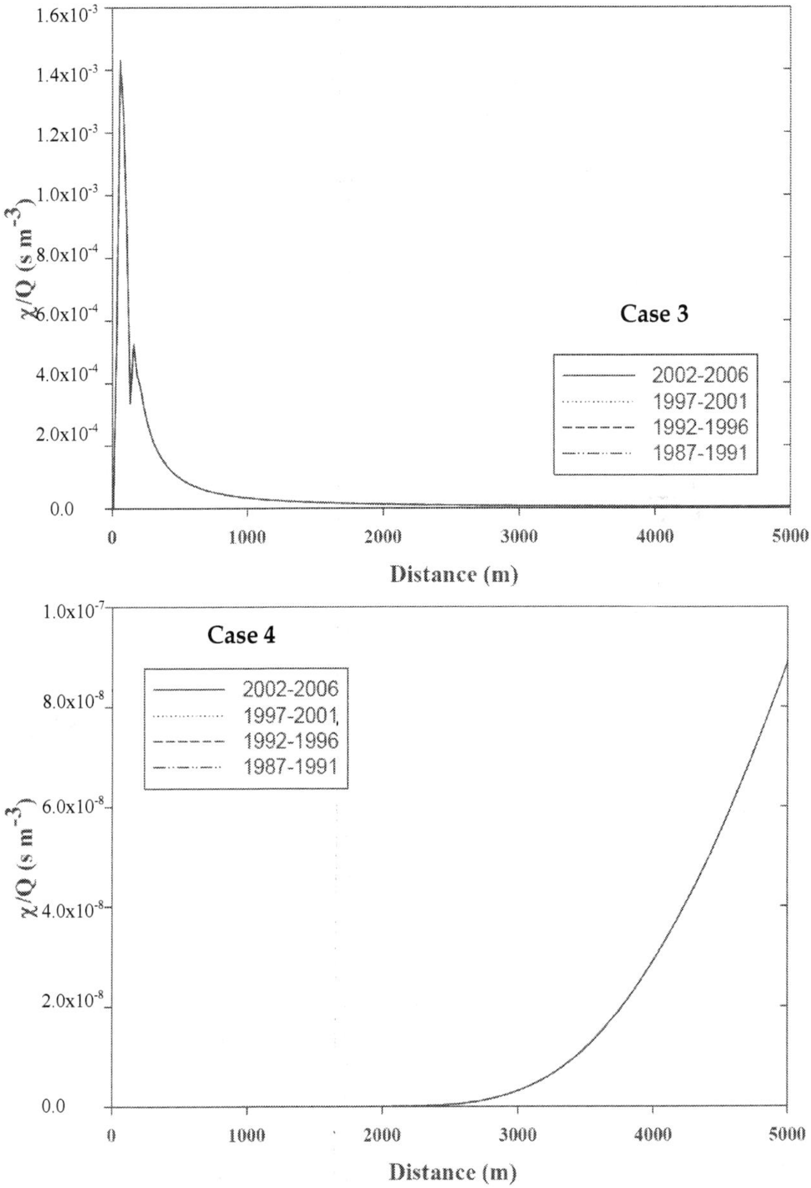

Fig. 5. Cont. Test Case comparison for periods 2002-2006, 1997-2001, 1992-1996, and 1987-1991. Cases 1, 2, 6, and 7 represent cases with average meteorological data.

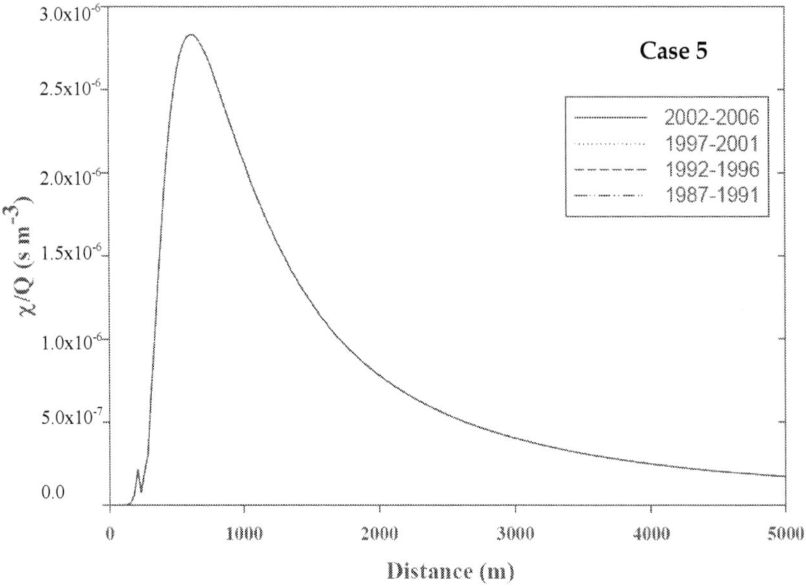

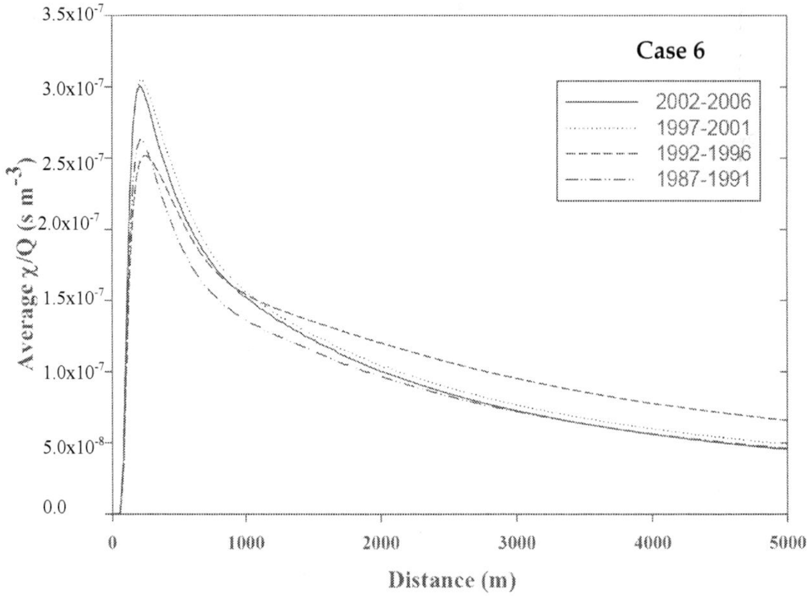

Fig. 5. Cont. Test Case comparison for periods 2002-2006, 1997-2001, 1992-1996, and 1987-1991. Cases 1, 2, 6, and 7 represent cases with average meteorological data.

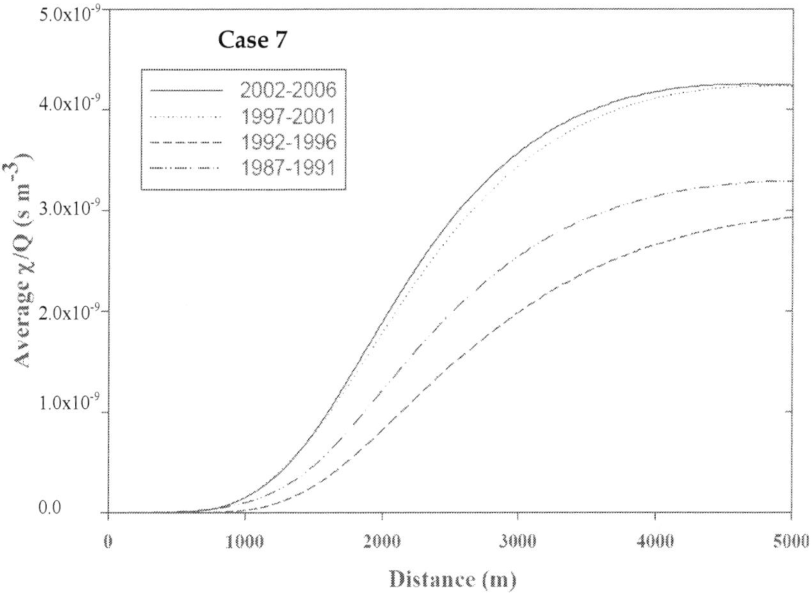

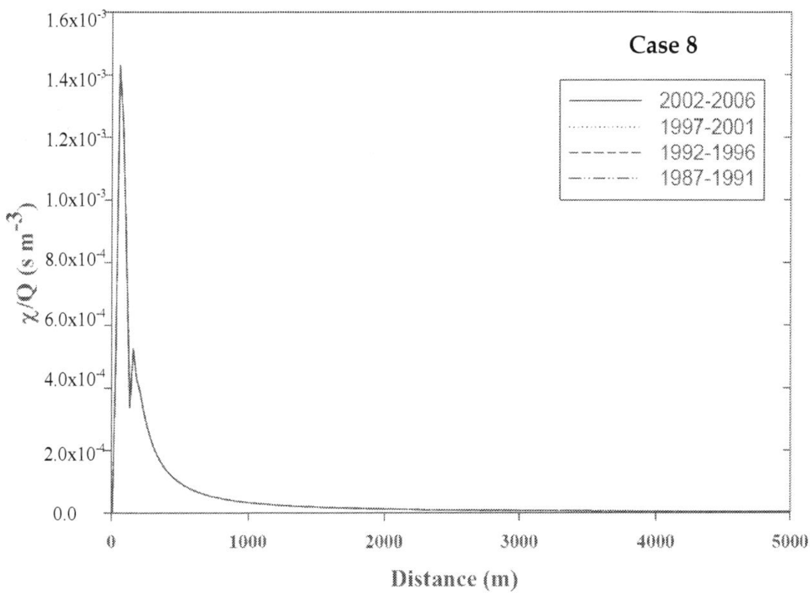

Fig. 5. Cont. Test Case comparison for periods 2002-2006, 1997-2001, 1992-1996, and 1987-1991. Cases 1, 2, 6, and 7 represent cases with average meteorological data.

of time are the same for these cases (Fig. 5). On the other hand, the average meteorological data is applied for Cases 1, 2, 6, and 7 with a probability level of 0.5%. The differences observed for these cases in Fig. 5 are caused only by the change in the meteorological data for all the periods. Two plots can be obtained for cases considering average meteorological data: annual average concentrations (χ/Q, s m^{-3}) and concentrations for meteorological conditions not exceeded 99.5% of the time (99.5% χ/Q, s m^{-3}). The curves for these cases follow a similar trend with various peaks and dips. However, the main sections of interest on these plots are the maximum values. To visually identify the maximum ground level concentrations with a building present and without the building for Eq. (9), the annual average air concentrations (s m^{-3}) versus downwind distances are plotted from the VENTSAR XL© output. The maximum concentrations are easily determined from these plots as shown in Fig. 5.

5. Conclusion

The past and current SRS missions involve dealing with significant quantities of nuclear and mixed hazardous wastes. The EDG at SRNL assesses the potential risk and doses to individuals and surrounding populations from atmospheric releases of radionuclides, using various approved computer models with SRS site-specific data. Every five years, SRNL generates a meteorological database to perform dosimetric calculations of accident or routine release scenarios for onsite and offsite populations. This information becomes the input of various environmental dosimetry codes used by the EDG. This study presents comparisons of wind frequencies among four five-year periods for various locations where the possibility of radionuclide releases exist at SRS and the comparison among test cases for these periods involving the computer model VENTSAR XL©, which is a dose assessment model used to estimate dose following short-term atmospheric releases involving GEP stack height evaluation and building effects caused by reactor cooling towers.

6. Acknowledgment

The author would like to thank the United States Department of Energy for the funding associated with the development, testing, maintenance and utilization of the dosimetry codes administered by the SRNL EDG (Contract No. DE-AC09-08SR22470). The author would also like to thanks Ali Simpkins for her efforts involving the development of VENTSAR XL© and Erik Kabela for providing Figs. 3 and 4.

7. Disclaimer

8. References

Briggs, G.A. *Diffusion Estimation for Small Emissions in Environmental Research Laboratory*, Air Resources Atmospheric Turbulence and Diffusion Laboratory, Annual Report, ATDL-106, USDOC-NOAA; 1973.

Department of Energy (DOE). *Internal Dose Conversion Factors for Calculation of Dose to the Public*, DOE/EH-0071, Washington, DC; 1988a.

Department of Energy (DOE) *External Dose-Rate Conversion Factors for Calculation of Dose to the Public*, DOE/EH-0071, Washington, DC; 1988b.

Environmental Protection Agency (EPA). *Guideline for Use of Fluid Modeling to Determine Good Engineering Practice Stack Height*, Environmental Protection Agency, EPA-450/4-81-003; 1981a.

Environmental Protection Agency (EPA). *Guideline for Use of Fluid Modeling to Determine Good Engineering Practice Stack Height (Technical Support Document for the Stack Height Regulations)*, Environmental Protection Agency, EPA-450/4-81-023; 1981b.

Faw, R.E. & Shultis, J.K. *Radiological Assessment: Sources and Doses*, American Nuclear Society, Inc. La Grange Park, Illinois. 1999.

Foley, T.Q. *2002-2006 Meteorological Data for Use with VENTSAR XL*, Savannah River National Laboratory Memo, SRNL-L6100-2008-00005. 2008.

Hanna, S.R.; Briggs, G.A.; & Hosker, R.P. *Handbook on Atmospheric Diffusion*, DOE/TIC-11223 (DE82002045), Technical Information Center, USDOE; 1982.

Kabela, E.D. & Weber, A.H. *Summary of Data and Steps for Processing the 2002-2006 SRS Meteorological Database (U)*, Washington Savannah River Company Report: WSRC-STI-2007-00613, Aiken, SC; 2007.

Mamatey, A.R. *Savannah River Site Environmental Report 2008*, Savannah River Nuclear Solutions, LLC Report: SRNS-STI-2009-00190, Aiken, SC; 2009. Available at: http://www.srs.gov/general/pubs/ERsum/er09/er2008.htm. Accessed 4 March 2011.

Parker, M.J.; Addis, R.P.; Hunter, C.H.; Kurzeja, R.J.; Tatum, C.P. & Weber, A.H. *The 1987-1991 Savannah River Site Meteorological Data Base (U)*, WSRCRP-92-598. WSRC, 9/15/92, Savannah River Technology Center, Aiken SC 29808; 1992.

Pasquill, F. *Atmospheric Dispersion Parameters in Gaussian Plume Modeling: Part II. Possible Requirements for Change in the Turner Workbook Values*, EPA Report EPA-600/4-76306, United States Environmental Protection Agency; 1976.

Simpkins, A.A.; *Verification of VENTSAR*. Westinghouse Savannah River Company Report: WSRC-RP-95-89, Aiken, SC; 1995.

Simpkins, A.A.; *VENTSAR XL - A Spreadsheet for Analyzing Building Effects and Plume Rise*. Westinghouse Savannah River Company Report: WSRC-RP-97-37, Aiken, SC; 1997.

Smith, F.G. & Weber, A.H. *A Computer Code for the Estimation of Pollutant Concentrations on or Near Buildings*, National Technical Information Service, DP-1668, Springfield, VA. 1983.

USNRC 1982, *Atmospheric Dispersion Models for Potential Accident Consequence Assessments at Nuclear Power Plants*, United States Nuclear Regulatory Commission Guide 1.145, 11/1982.

Weber, A.H. *Documentation for (92-96) Five-Year Database Software,* Inter-Office Memorandum, March 12, 1998, SRT-ETS-930073, WSRC, Aiken, SC 29808; 1998.

Weber, A.H. *Summary of data and steps for processing the CAP88 1997-2001 SRS Meteorological Database (U),* Westinghouse Savannah River Company Report: WSRC-TR-2002-00445. Aiken, SC; 2002.

Part 3

Indoor Air Quality

8

Sensing a Historic Low-CO_2 Future

Colin D A Porteous
Mackintosh Environmental Architecture Research Unit, The Glasgow School of Art
Scotland, UK

1. Introduction

The title of this chapter is intended not only to flag up the longstanding role of carbon dioxide (CO_2) as an indicator of air quality inside buildings, but also to imply an altogether different need to curb its presence, this time in our upper atmosphere and with a shorter history of awareness of the global warming phenomenon. The first word also suggests a role for human perception. In relation to air quality, we sense humidity and odour, and often have a fairly accurate idea as to causes. For example, our washing hanging out in a room can make it feel too humid, and fabric softener can emit an identifiable chemical smell. But CO_2 itself is odourless and its concentration above a certain level is normally an indicator of the 'bad company' that it keeps when air is changed too infrequently. This means that we can wrongly interpret a perception of stuffiness as being due to an inadequate supply of fresh air, when the actual cause might be that the room has become warmer than necessary in the quest for comfort. In this case, the incorrect diagnosis could lead to throwing open a window, rather than adjusting a thermostatic heating control. When direct or indirect burning of a fossil fuel provides the heating, repeated 'corrective' responses of this kind will add to global CO_2 emissions. In short, our sensing, valuable though it is, may lead us to take an expeditious action, but one that adds to both the energy and carbon emissions burden. On the other hand, if the temperature of a space is appropriate to a particular activity, we can sense freshness or stuffiness quite correctly. It is also true to say that the early, and valuable, research into control of air quality in buildings predates the means to measure CO_2. In other words it was largely reliant on perception and observation, and even after the ability to measure CO_2 was achieved, the values had to be systematically cross-checked with perceptual responses from volunteer cohorts.

This is where history is relevant to the future, all the more so at this temporal watershed when increasing demands to make buildings more airtight raise the stakes relative to the risk of poor air quality. Although we have approximately 175 years of developmental experience of mechanical control of ventilation, and the same length of scientific awareness regarding natural thermo-circulation, we have yet to fully resolve the ideal interface between electronic automation and manual intervention. The focus on air quality has a new urgency, embraced by the broader area of inquiry into 'sick building syndrome', in parallel with, and meshing with, wider research in the field of public health, particularly in the area of microbiology. Recent specialisations into health psychology and environmental psychology again spring from the public health concerns, but aligned with the specific strand of work on the human senses with its origins in the late Georgian period in the first part of the 19th C. By the mid-20th

C, architectural interest in perception psychology by Sven Hesselgren was realised in his 1954 doctoral thesis in Sweden, subsequently published in English in a more concise form (Hesselgren, 1975). The same period saw a substantial body of architecturally oriented work on the environmental forces that shape buildings (Fitch, 1972). In parallel, that in the field of human comfort (Fanger, 1970) presaged his introduction of units to measure air quality (Fanger, 1988). A strand into 'adaptive opportunity' then follows, which layers a psychological understanding of comfort on to the established physiological one (Baker & Standeven, 1994). Thereafter, the present author adopts the term 'adaptive control' (Porteous & MacGregor, 2005), to be applied to environmental comfort in a less serendipitous manner, and with more pre-emptive, design-led consequences, than 'adaptive opportunity'. This then sets the scene for a historical recapping as part of the process of establishing where we are, and where we should go, with air quality.

2. Holistic review of modern 'air quality' history – early 19th C to present

Architects of today, with their team of specialist consultants, owe a particular debt to the discipline of medicine with regard to the pioneering work on air quality. As indicated above, this started in the 19th century based on a combination of consequences, ranging from perceived discomfort to transmission of disease, and scientific advance. In the UK, the fire of 1834 that consumed a large part of the Houses of Parliament at Westminster in London, presented a rare opportunity for a Scot, Dr David Boswell Reid. He was appointed by a select committee as a consultant to use temporary accommodation as a test bed with respect to heating, ventilation, lighting and acoustics. This led to the 1836 refurbishment of the former House of Peers, described by Reid in 1837 (Sturrock & Lawson-Smith, 2006). This had been less badly damaged by the fire than other parts of the complex, and, as the temporary and experimental Commons, it was then monitored over a considerable number of years (Reid, 1844). Reid's book includes as examples tables incorporating key environmental data for two days - one on May 22nd 1837, when the population in the House varied from 40-800; and the other on May 12th 1843, when the population varied between 130-640 (Tables 1 and 2).

Each original table gives hourly values for air temperature outside, in the supply chamber and the exhaust shaft, and in three different inside locations, as well as the extent to which a valve is open, the number present, and brief comments as to weather. The 1837 readings confirm an external range of some 8 K from a high of 12.8°C at 4.0 p.m. to a low of 4.9°C at 3.0 a.m. Equivalent readings internally at the Chair (T2 below) range slowly up from 16.9°C at 4.0 p.m. to 18.9°C at 3 a.m. This modest rise of 2K reflects the increase of the population in the chamber from a rather low start, and indicates that the increasing aperture of the open valve kept the air from becoming over-warm. We may also note that clothing at this time would have provided greater personal insulation than is the case today. In the 1843 set of data the internal temperature at the Chair does get as high as 20°C in the early hours of the morning, but here the outdoor range from 16.1-11.7°C is significantly greater than the 1837 set. Despite the maximum opening being adjusted 2′ 3″ (69 cm) wider in 1843, and despite the average hourly population of the chamber being lower (370 cf. 443), the warmer ambient conditions proved dominant. Even so, 20-22°C when the chamber has been subjected to some 480 kWh of incidental heat gain from the people inside, without taking account of that from the new gas lighting, signifies rather effective environmental control for such an early experiment. Moreover, Reid's description of the filtering and washing before heating, including a secondary lime-charged phase to remove small particulates, provides a degree of confidence that the air entering the chamber was a much healthier and less malodorous than the ambient air in early Victorian London.

State of the Atmosphere	Hour	To	T1	T2	T3	Valve opening	Members+Strangers
Very dull	1600	12.8	15.8	16.9	16.4	380 mm	120
	1700	12.2	17.5	17.5	17.5	380 mm	280
	1800	11.5	18.1	18.1	17.8	305 mm	330
	1900	10.6	17.8	18.1	17.5	305 mm	200
Clear starlight	2000	10.0	18.3	18.3	18.3	460 mm	270
	2100	8.9	18.6	18.6	18.9	610 mm	348
	2200	7.2	18.6	18.6	18.9	760 mm	678
	2300	6.7	18.9	18.9	18.9	840 mm	720
	2400	6.1	18.9	18.9	18.9	840 mm	760
	0100	5.6	18.9	18.9	18.9	840 mm	760
	0200	5.0	18.9	18.9	18.9	530 mm	800
	0300	4.9	18.6	18.9	18.6	305 mm	40

Table 1. Dr D B Reid's readings for 22nd May 1837, translated to SI units, and with temperatures in supply chamber and exhaust shaft omitted. To = temperature outside; T1 = temperature East Gallery; T2 = temperature Chair; T3 = temperature West Gallery (all air temperature)

State of the Atmosphere	Hour	To	T1	T2	T3	Valve opening	Members+ Strangers
Clear & fine	1600	16.1	17.8	17.2	16.7	305 mm	130
	1700	15.6	18.9	18.3	18.3	915 mm	350
Rather dull	1800	15.0	19.4	18.9	18.9	915 mm	431
Rain	1900	13.9	19.4	18.3	18.9	760 mm	160
	2000	13.3	19.4	18.3	18.9	760 mm	170
	2100	12.8	20.0	18.3	18.9	760 mm	200
Showery	2200	11.7	20.0	18.3	19.4	1070 mm	320
	2300	11.7	20.0	18.9	20.0	1370 mm	520
	2400	11.7	20.6	19.4	20.6	1525 mm	600
	0100	11.7	21.7	20.0	21.1	1525 mm	640
	0200	11.7	22.2	20.0	21.1	1525 mm	600
	0300	11.7	21.7	19.4	20.6	1525 mm	450
	0400	11.7	20.6	18.9	20.0	1070 mm	450

Table 2. Dr D B Reid's readings for 12th May 1843, translated to SI units, and with temperatures in supply chamber and exhaust shaft omitted. To = temperature outside; T1 = temperature East Gallery; T2 = temperature Chair; T3 = temperature West Gallery (all air temperature)

The system also addressed the exhaust of combustion gases from new gas lighting, which assisted natural thermal buoyancy. It is possible that the low-level mode of delivery, effectively displacement ventilation via apertures in the floor, might have been compromised by less savoury particles from members' footwear (bear in mind horse-drawn traffic and so on). This may have been part of the reason underlying Charles Barry's opposition to Reid's displacement method when it came to the permanent replacement building still in existence. Barry favoured a plenum system, with air entering the chamber at a higher velocity from above, and this led to an insurmountable conflict for Reid that resulted in him being superseded in 1853 (Port, 1976).

Nevertheless, by 1843 we had a well-monitored model for mechanical ventilation with the science and the results disseminated via Reid's book the following year. 1844 also coincides with the year that Berlin-based Heinrich Gustav Magnus published a formula relating temperature and moisture in the air and establishing the dewpoint curve (Magnus, 1844, as cited in Lawrence, 2005). However, Mark Lawrence, in summarising the history of this scientific landmark, tells us that the formula credited to Magnus is based on earlier work by John Dalton of Manchester who started experiments to measure dewpoint at the beginning of the 19th C. Lawrence informs us that, in turn, Dalton's reasoning was used in a precursor of the Magnus formula by E F August (August, 1828, as cited in Lawrence, 2005), and argues that it should properly be called the August-Roche or August-Roche-Magnus formula. Professor James (John) Apjohn, 1796-1886, closely associated with the Royal Irish Academy, is another scientist credited with devising a formula calculating dew point around the same period. A record notes "August independently derived a similar formula at about the same time, but Apjohn's treatment was generally preferred." (Dixon, 1969)

Also in the 1840s, Walter Bernan, a civil engineer, published the methodology for determining the required supply of heated air per minute inside buildings (Bernan, 1845). This was based on a notional room with an air temperature of 17.8°C and dew point temperature of 10°C, indicating a relative humidity (RH) close to 35%. This took account of the occupants in terms of supply of air to the lungs and insensible perspiration, presence of candles or other means of lighting, area of glass, cracks around windows and doors and area of solid bounding surfaces, excluding the windows. Besides the heat exchange by convection through cracks his method indicates some knowledge of heat exchange by conduction with surfaces such as glass; no doubt based on Fourier's Law of Heat Conduction (Fourier, 1822). However, since accurate prediction in this regard was out of reach at that time (further history of measurement of thermal conductivity and vapour permeability through materials given later in this chapter), and since there was no reliable method as yet to measure CO_2 as an indicator of air quality, it is reasonable to assume that this formula was derived empirically. A starting point was the knowledge of the relative volume of the constituents of air breathed in and out - i.e. with oxygen roughly halved and CO_2 increased 5-6 times. Bernan also made reference to Reid's findings, noting that as moisture was added to the supply air, considerably more was required for fresh air supply per person - at least 30 ft^3/minute (14 l/s) per person and sometimes as much as 60 ft^3/minute - important in Reid's case at the temporary House of Commons due to the rigorous filtration and washing process, which was fundamental to improving the quality of the air supply.

The scene is thus set for a reliable instrument to measure CO_2. Jan Sundell implies the 1,000 parts per million (ppm) CO_2 air quality standard of today was established by Max von Pettenkofer in 1858 (Sundell, 2004). In fact, Pettenkofer's publication of 1872 confirmed 1,000 ppm CO_2 as a recommended maximum (Pettenkofer, 1872, as cited in Locher, 2007); and it would seem that it was his 'respiration apparatus' using the 'Pettenkofer titrimetric method' that was erected and described in 1858 (Evans, 1973; Pettenkofer, 1858, as cited in Beck, 2007). A Pettenkofer obituary claims that Dalton (see above regarding dewpoint) and Watson developed a similar method of CO_2 measurement in England (Haldane, 1901). In any event, the Pettenkofer maximum of 1,000 ppm is still effectively in use today, having been enshrined in a significant health guide for architects (Appleby, 1990).

Work aligning with Pettenkofer's 1872 paper, by Dr Francis S B Francois de Chaumont, an expert in military hygiene, was published four years later (Francois de Chaumont, 1876). He gave two sets of figures, the second corrected relative to the first as "a previously unobserved error was found in one of the constants employed"; but he points out that the error was of little consequence as the corrected value for 'fresh' is still below the value of 0.2 parts CO_2 above that outside. His methodology relied on relating measurement of CO_2 to the sensory responses of a reliably large number of participants, here 458 'fully recorded'. A CO_2 concentration of 1,000 ppm was found to be very close to the perception "Close, organic matter disagreeable"; while 700 ppm was some 100 ppm below "Rather close, organic matter becoming perceptible", allowing for an ambient level of 400 ppm. His colleague, Dr Edmund A Parkes (Parkes, 1878), also notes that Pettenkofer suggested an optimum of circa 700 ppm, slightly above that of another researcher in this field, Degen: "Pettenkofer has now adopted the limit of .7 measures of CO_2, and Degen .66 measures per 1000, as the amount when the organic matter simultaneously present becomes perceptible." This recommendation also survives to this day. It may be noted that the original figures are given as parts of CO_2 (carbonic acid) per 1,000 above the outside level, and the translated figures in ppm include 0.4 parts per 1,000 as the "average ratio of CO_2 in external air" (as cited by Parkes, where in each case it is the level above that measured outside that is recorded as due to 'respiratory impurity'). The respective findings are given in Table 3.

	Respiratory Impurity as CO_2 (i.e. parts/1,000 above that outside)	
Classes	Original figures	Corrected figures
1. Fresh	0.1830	0.1943
2. Rather close	0.3894	0.4132
3. Close	0.6322	0.6708
4. Extremely close	0.8533	0.9504

Table 3. Results of Dr Francis S B Francois de Chaumont survey published 1876

Thus we can see from Table 3, classification 3 'Close' that 0.6708 + 0.4 = 1.0708 parts CO_2/1,000 or 1,071 ppm; and for classification 2 'Rather close' 0.4132 + 0.4 = 0.8132 parts CO_2/1,000 or 813 ppm.

The Francois De Chaumont values of 1876 also accord with the recommendations of Dr John S Billings (Billings, 1889), who refers to parts of CO_2/10,000 and refers to the values in a room assuming a normal amount of 4.0 outside. For example: "between 6 and 7 parts in excess of 10,000, a faint, musty, unpleasant odor is usually perceptible to one entering from the fresh air. If the proportion reaches 8 parts, the room is said to be close." Thus 8 parts, 'close', equates with Francois de Chaumont's No. 2 'rather close'; or 800 ppm. Billings also points out that it is not CO_2 itself that is the problem: "It is because carbonic acid is usually found in very bad company, and that variations in its amount to the extent of three or four parts in ten thousand indicate corresponding variations in the amount of those gases, vapors, and suspended particles which are really offensive and dangerous, ...". Of course the nature of the bad company has changed over time, but the principle still holds. Very similar perceptive values to those of Francois De Chaumont are also used in an early 20th C manual (International Textbook Co., 1909). In this case 4-5 parts above that outside is considered 'rather close', 7-8 parts 'very close' and 12 parts 'very bad'. Although sources are not referenced, it seems likely that the values originate from Francois de Chaumont. This text includes a formula for calculating the appropriate rate of air change, taking account of CO_2 exhaled by occupants, gas lighting and appliances. Another such book refers to Sir Douglas Galton in stipulating a minimum ventilation rate per person of 1,000 ft^3/hour, or circa 8 l/s (B. F. Fletcher & H. P. Fletcher, 1907).

Returning to the 1880s, an extensive publication in the British Medical Journal (BMJ) that predates that of Billings by two years is that carried out in the east of Scotland from 1885-86 by a team comprising Prof. Thomas Carnelley and John Scott Haldane from University College, Dundee, with Dr A M Anderson, Medical Officer of Health for Dundee (Carnelley et al., 1887). Its importance lies in relating health to air quality by making use not only of Pettenkofer's invention of a reliable instrument to measure CO_2, but also of microbiological techniques to sample and analyse air. Their 1887 paper describes in some detail the method of Walter Hesse that was used in conjunction with a jelly attributed to Heinrich Koch (Carnelley et al., 1887:62). Although the date of the Hesse technique is absent, it is known that Koch's postulates concerning the use of this jelly were formalised in 1882 (Gardner, undated). It is further reported that Hesse's method followed that of Koch (Robertson, 1888). Importantly, the same address by Robertson on the study of micro-organisms in air made the point that the Carnelley-led survey had found no relationship between the amount of CO_2 and the number of micro-organisms. This was a key association for air quality, by then well-defined, and the potential impact of bacteria in particular on health, but Robertson's comment belies the complexity relative to the published findings.

In broad terms what the Scottish survey demonstrates is that although respired air, including CO_2, does not give off micro-organisms to any appreciable extent (Carnelley et al., 1887:93), the 'bad company' principle of Billings holds good, even if interpreted in a broader manner than he intended. The published paper goes on to say that micro-organisms arise from clothes and skin: "Hence, if we take the carbonic acid as a rough measure of the total impurities arising from the persons of those present in a room, it should be a rough measure of the micro-organism from the clothes and skin." (Carnelley et al., 1887:95) The authors make the further point that as the cubic space per person increases other sanitary conditions improve. This is particularly marked for single-roomed dwellings, allowing for the small sample totalling 27, where the average space per person in a single 'clean' home was 38% greater than in the average of 'dirty', 'dirtier' and 'very dirty' homes. The 'clean' home had 24% lower CO_2 (800 ppm cf. 1,055 ppm) and 70% fewer micro-organisms than the dirty set. In two-roomed dwellings, those that were considered to be 'very clean' and 'clean' had 77% fewer micro-organisms than the dirty category, although in this case the CO_2 level in the cleaner dwellings averaged 9% greater (1027 ppm cf 940 ppm). Thus this particular part of the analysis indicates that there is no strong influence of CO_2, when the values are within or not far above the Pettenkofer standard; but that the issue of cleanliness is critical for micro-organisms, and hence potentially to health. In other words, lack of hygiene constitutes injurious 'bad company'. A significantly larger cohort of cases makes it evident that space per person is critical to health outcomes - see Table 4 below. However, in this case with its reasonably large cohort of 59, the concentration of CO_2 also corresponds with the space available and the key health outcomes, noting in particular the increasing death rate according to paucity of space for children less than 5 years of age. We may also note that when the cubic space per person drops to 6.0 m^3, all four categories of disease listed exceed the average for the whole population. In terms of CO_2 from combustion, and as an indicator of further 'bad company' in this regard, it is also relevant to note that the poorest households in the most overcrowded conditions are also most likely to have the cheapest candles and other means of artificial lighting, as well as the cheapest solid fuel. The harmful by-products of smoking, in its various forms, would also be more intensive in smaller spaces. Not listed in Table 4 below are the comparative values for mould, which constitute most of the balance of micro-organisms taken together with bacteria. Subtracting respective values in row 4 from those in row 3 shows that the mould does not conform to the trend of increasing with reduced space or increased CO_2.

Description	No. of cases	4-room	3-room	2-room	1-room	Whole population
1. Space/person (m^3)	69	37.8		7.05	6.0	
2. CO_2 (ppm)	59	770		990	1,120	
3. All micro-organisms	59	9.0/1		46.0/1	60.0/1	
4. Bacteria	46	8.5/1		43.0/1	58.0/1	
5. General death rate	3,110	12.3	17.2	18.8	21.4	20.7
6. Children <5 deaths	1,347	3.3	5.8	9.8	12.3	9.0
7. Deaths - diarrhoea	253	6.1	11.3	17.4	26.4	16.9
8. Deaths - bronchitis & broncho-pneumonia	224	7.8	9.5	13.4	26.7	14.9

Table 4. Extract from Carnelley, Haldane and Anderson survey published 1887, p 74

A final component of the samples analysed is that of 'oxidisable organic matter'. This must have included combustion particles from heating, cooking and lighting, skin scales, the main food of dust mites, and dust mite faeces and corpses, as well as other matter such as pollen grains. All but the first are of great interest today in terms of the association with asthma, in particular childhood asthma. The survey also notes respiration, physical exercise and cleanliness as being influences, and provides some detailed data from tests for combustion of coal, respiration, dust and the stagnation of air, and physical exercise. However, it notes: "Cleanliness has little or no apparent influence on the quantity of organic matter in air." (Carnelley et al., 1887:86-89). In the outside air within urban contexts, it is probable that 'organic matter' was dominated by particles from combustion. Records of deposition of tar, carbonaceousless tar and ash in Glasgow over a period of several years in the 20th C prior to the Clean Air Acts of the 1950s and 1960s was very much greater than at Loch Katrine (Gilfillan, 1958). In Dundee, in the winter of 1885-86, the mean CO_2 for the town was found to be 390 ppm (i.e. below the norm of 400 ppm), and organic matter 8.9 (representing the volume of oxygen required to oxidise it in 1 million volumes of air). In its suburbs the comparative figures were 280 ppm and 2.8. To allow for such differences, analysis of air samples from inside dwellings in a variety of locations was expressed both in absolute terms and as the amount above that outside. The same pattern of decreasing values with increasing space occurs in each case; and, like CO_2, the total for micro-organisms and bacteria, the organic matter also conforms to this trend. In order to show the relativity of the various indicators clearly (in excess of outside air), the published paper includes a table where the larger dwellings of four and more rooms are reduced to unity for all indicators, with two- and one-roomed dwellings expressed proportionally - see Table 5 below.

Description	4-rooms+	2-rooms	1-room
1. Space/person	1.0	0.13	0.11
2. CO_2	1.0	1.50	2.00
3. Organic matter	1.0	1.60	4.40
4. All micro-organisms	1.0	5.10	6.70
5. Bacteria	1.0	5.10	6.90
6. Moulds	1.0	5.50	3.00

Table 5. Extract from Carnelley, Haldane and Anderson survey published 1887, p 71

Although incidence of organic matter increases with decreasing space, it did not, as indicated above, particularly relate to the judgments made as to cleanliness. For example the 'very clean' 2-roomed dwellings had more organic matter than the 'clean' ones (Carnelley et al., 1887:96).

Of at least equal importance to the housing studies were the surveys of two sets of schools, again from 1885-86. The sample in this case included 42 that were ventilated naturally, primarily via windows and fireplaces, and 26 that were mechanically ventilated. The latter achieved this by means of a plenum system with fans delivering filtered and heated air through registers located some 1.5 m from floor level, and with the vitiated air removed by natural thermal buoyancy (stack effect) via low-level outlet grilles (Carnelley et al., 1887:78). Given the current 21st C eco-publicity favouring natural ventilation of non-domestic buildings, it may seem surprising that there was clear evidence in 1886 that mechanical ventilation provided the better air quality, with marked improvements for every indicator. For example, even though the average space per person in the mechanically ventilated schools was 2% less than that in the naturally ventilated ones, the mean CO_2 in the former was 1,230 ppm compared to 1,830 ppm in the latter. Similarly organic matter was 10.1 to 16.2 and bacteria 16 per litre to 151 comparing mechanical to natural ventilation (Carnelley et al., 1887:79). Indeed it was reported: "... that of the mechanically ventilated schools only two contained more than 26 micro-organisms per litre, whereas of the naturally ventilated schools only three contained less than 26 per litre." Again the differences were clearly demonstrated by taking the indicators for the mechanically ventilated schools to unity, and expressing those of the naturally ventilated ones proportionally - see Table 6 below.

Description re Schools	Mechanical ventilation	Natural ventilation
1. Space/person	1.0	1.00
2. Temperature > outside air	1.0	0.66
3. CO_2	1.0	1.70
4. Organic matter	1.0	7.00
5. All micro-organisms	1.0	9.20
6. Bacteria	1.0	9.40
7. Moulds	1.0	2.00

Table 6. Extract from Carnelley, Haldane and Anderson survey published 1887, p 79

Here we may also note the negative impact on indoor temperature in the case of the naturally ventilated schools, the mean only 13.1°C compared with 16.7°C; and even the latter representing thermal austerity compared with today's shirtsleeve expectations.

Experiments were also reported showing that respiration was not associated with micro-organisms, and that it was clothes and skin that were primarily responsible for the correspondence between CO_2, as an indicator of occupant intensity, and the number of micro-organisms (Carnelley et al., 1887:95). In terms of CO_2's 'bad company', the report also acknowledges "other sanitary conditions improve as the cubic space increases." In the concluding comments, the report takes us back to the evidence associating poor air quality with higher death rates (see Table 4 above): "Hence we may take it as quite certain that the above differences in the death rates in Dundee are largely due to the difference in the quality of the air habitually breathed." (Carnelley et al., 1887:105) 124 years on, we are concerned on the one hand with achieving highly airtight buildings in order to conserve energy, while on the other our systems of controlling ventilation have to contend with different chemical 'cocktails'. It is to be hoped that with the economy of energy as the key driver, that we do not become overly complacent in terms of the rigour we apply to high standards of air quality.

Another research finding of the 19th C, with influence on the 20th C and potential for further relevance in the 21st C, concerned sunlight inside buildings. Arthur Downes and Thomas P Blunt found that sunlight continued to function as a natural disinfectant when transmitted through glass, particularly in the violet and blue part of the spectrum (Downes & Blunt, 1877). This research finding, published in the same year as the Scottish survey, was arguably fundamental to the 'healthy light and air' of the early 20th C modern movement, which has since been well documented (Overy, 2007). However, it was not until the 1940s that much more thorough studies, initially led by Leon Buchbinder, augmented the 19th C knowledge with respect to the bacteria-killing power of sunlight through glazing (Buchbinder et al., 1941). In hindsight, it is possible that a greater area of fenestration may have been partly responsible for the superior performance of the mechanically ventilated schools compared with their more traditional naturally ventilated counterparts.

Notwithstanding such conjecture, it seems likely that the increasing use of antibiotics from the 1950s onwards gradually displaced architectural interest in the health-giving powers of sunlight, including their contribution to enhanced air quality inside buildings. It also seems ironic that many practitioners of 'environmental architecture' in the 21st C have a bias for natural over mechanical ventilation, when evidence suggests that the latter may well be the healthier option - used in auxiliary mode at least. In this case, provided the power for fans is minimised and active cooling is not required, it should also be the more energy-efficient option.

The 19th C also saw some advances in knowledge with regard to the human senses. It was not only sense of smell that had relevance to judgements as to when air was fresh, fusty, overly dry or humid, but also the human skin receptors in terms of sensing temperature and air movement. Recapping, Pettenkofer introduced his CO_2 standard in 1872. Shortly after this Francois de Chaumont backed up his findings in 1875-76, as did the Scottish survey, published a decade later in 1887; and the latter's analysis of key organic matter and micro-organisms in air moved knowledge forward. Finally the parallel new knowledge about the disinfectant properties of sunlight inside buildings introduced a new health dimension to air quality.

The question then arises as to whether further scientific and technical advances affecting the chemical and physical nature of indoor environments, as well as greater insights with regard to our sensory stimuli, would affect future judgement regarding the original 1872 Pettenkofer standard. For example, would such changes be influential in terms of the judgements of Povl Ole Fanger's 168 subjects used to determine the decipol units to "quantify the concentration of air pollution as perceived by humans", this work being accepted for publication (Fanger, 1988) exactly a century after the Scottish survey? The corollary to this question is would such changes justify a change to air quality standards? The answer to both would seem to be 'no' or at least 'not by much'. As mentioned above, the maximum of 1,000 ppm CO_2 is still present in the 1990 RIBA publication, the Rosehaugh Guide (Appleby, 1990), and this value is usefully matched with the introduction of fresh air at the rate of 8 l/s for each person present. Eight years earlier, Neville Billington had reminded a London audience that Roscoe in 1857 showed that 4.7 l/s was not enough to clear odours in Wellington Barracks, and that 6.9 to 9.7 l/s ought to be provided (Billington, 1982). The 8.3 l/s mean average of this range is remarkably close to the Rosehaugh Guide's 8 l/s, corresponding with the Pettenkofer 1,000 ppm CO_2.

Fanger uses the standard of 10 l/s relative to his decipol, in turn based on the unit olf. To clarify this point: "One olf is the emission rate of pollutants (bioeffluents), by a standard person." In this regard, we may note that Fanger's 'bioeffluents' are part of what the 1887 Scottish survey termed 'organic matter'. One decipol is the pollution caused by one person (one olf) ventilated by 10 l/s of unpolluted air. Thus the key difference seems to be the use of 10 l/s, 25% greater than 8 l/s, and non-linearly corresponding with 874 ppm CO_2 with an ambient level of 360 ppm. In other words it is closer to the lower Pettenkofer desired value of 700 ppm. However, this is simply a matter of decimal convenience in terms of the Fanger study. What is of significance is what a rate of 10 l/s signifies in terms of 'predicted dissatisfied' (PD), and this was found to be approximately 15% for 10 l/s or 1 decipol; whereas 8 l/s or 1.25 decipols would have increased PD to approximately 17.5%. However, the above is predicated on Fanger's assertion that "the human senses are usually superior to chemical analysis of the air." This begs the question as to relevance of chemical analysis and how techniques would have advanced since the time of the 1887 Scottish survey. In that instance, perception, based on previous studies and through the medium of CO_2 as an indicator, is useful as a comparator alongside the microbiological analysis. Another issue is whether psychological and physiological understanding and knowledge regarding sensory perception has advanced so as to represent a challenge to 19th century findings, now applied in an entirely different physical and social context. The hygienic standard for Fanger's occupants assumed 0.7 baths/day, changed underwear every day, and 80% use of deodorants. Also, compared to the 19th C, clothing was of an entirely different nature, as well as furniture, furnishings and bounding materials. In fact it seems remarkable, given all these differences, that the Pettenkofer recommendations for optimum and maximum CO_2 levels are still credible as indicators of air quality.

In terms of the posited superiority, or at least validity, of the human senses to judge air quality, a historical overview of experimental psychology relative to sensation and perception is useful. In a significant mid-20th C publication, Edwin Boring complains about the paucity of knowledge at that time, 1942, with respect to the stimuli for our sense of smell, as opposed to stimulus-objects (Boring, 1942a). However he cites one early 19th C work prior to Pettenkofer's 1872 standard and De Chaumont's 1876 paper: H Cloquet in his book of 1821 (Boring, 1942a:439) with: "… the classification of odours, the psychological seat of olfaction, its mechanism, its pathology, its practical uses and individual differences in sensitivity." On the same page Boring mentions that Johannes Muller in 1838 could "find only seven pages of really solid fact on smell for his handbook" and similarly Bidder with eleven pages in Wagner's 1844 'Handworterbuch der Physiologie'. We are told that in 1847 Ernst Heinrich Weber (Boring, 1942a:440) "had found not only that cold and warm odorous liquids cannot be smelled when poured into the nostrils of the inverted head, but that eau de Cologne and acetic acid are not sensed at body temperature." We also learn that in 1862 Max Schultze "localized the olfactory membrane high up in the nasal cavity" and found the olfactory sensors there in a small area - "long cells with hair-like processes".

Boring also discusses cutaneous sensibility, and again remarks on the relative paucity of our knowledge of receptors even though we know something about the characteristics of stimuli (Boring, 1942b). The sensitivity of our skin in terms of warmth and air movement certainly relate to thermal comfort, and is known to psychologically influence our perception of air quality - e.g. too warm when stuffy; too cold when draughty. Boring discusses the issue of liminal points or limen early in his overview, citing Weber's Law of 1834 (Boring, 1942c); apparently named as such by Gustav Theodor Fechner after E H Weber. This states "two sensations are just noticeably different as long as a given constant ratio obtains between the intensities of their stimuli." For example, Weber asserts that the skin can appreciate a difference of 1:30. From this Fechner developed the Method of Limits (minimal changes) (Boring, 1942c:37-38): "the procedure in which the stimulus is changed by successive discrete serial steps until a critical point is reached, a point at which judgement is changed." This body of theory from the first half of the 19th C appears highly relevant to the work of Francois de Chaumont for example in the second half. However, although his paper explains that the methodology was based on a system of differences, it is not clear if there was an awareness of the Weber-Fechner theory. In relation to smells, Boring tells us that later researchers towards the end of the 19th C concluded that Weber's Law applies approximately, mean average ratios varying in the range 1: 3.125 to 1:2.63, with two thirds of determinations lying between 1:4 and 1:3 (Boring, 1942a:443). These findings came after H. Zwaardemaker's 1888 Olfactometer to measure the intensity of the odorous stimulus, and even a year or so after the publication of his book in 1895.

Inevitably science related to the sense of smell advanced in the next eleven decades. However, just how much of that is relevant to air quality is a moot point. Although the 19th air quality surveys such as that by De Chaumont were predicated largely on olfaction, as was that of Fanger in the 1980s in his introduction of the units olf and decipol, the reality is that human perceptions were bound to be instinctively more holistic. Indeed the descriptive terms in the former case suggest this with 'close' and 'very close', the more extreme version further defined by the adjectives like 'oppressive'. In the latter case, the term 'percentage dissatisfied' in the context of modern auditoria also invites a wider perception of atmosphere apropos air quality. In a modern auditorium, apart from the paramount issues of acoustics and sightlines, the environmental concerns that one might anticipate would relate to air quality are tangible convection and adequate, but not excessive, warmth. The fundamental attribute expected of the atmosphere would be to stimulate rather than enervate, and subjects may subliminally tend to rank between two such opposing characteristics. The field of environmental psychology gained traction in the 1960s (Gifford, 2002), some two decades before Fanger's study. The 'semantic differential' introduced by Charles E Osgood (Osgood et al., 1957) is one of the techniques used.

James J Gibson not long afterwards introduced a new way of considering the senses (Gibson, 1966): "The observer who is awake and alert does not wait passively for stimuli to impinge on his receptors, he seeks them. He explores the available fields of light, sound, odor and contact, selecting what is relevant and extracting the information." This was a potent hypothesis embodied in his book's title 'The Senses Considered as Perceptual Systems', and one of considerable interest to architects who became aware of it. 'Body, Memory and Architecture' by Kent Bloomer and Charles Moore (Bloomer & Moore, 1977), following two and five years after two USA publications already mentioned (Hesselgren, 1975; Fitch, 1972), cites J J Gibson strongly and succinctly: "His strategy was to regard the senses as aggressive, seeking mechanisms and not merely as passive sensation receivers." They also cite his fourfold categorization of 'actively detecting' senses or 'perceptual systems': the auditory system, the taste-smell system, the basic-orienting system and the haptic system. Gibson follows up his initial polemic with a passage entitled 'The Fallacy of Ascribing Proprioception to Proprioceptors' (Gibson, 1966:33-34). This is a reference back 60 years to 'The Integrative Action of the Nervous System' by Charles S Sherrington (Sherrington, 1906), whose opus seems to be the original source of three fields of sensory reception: extero-sensors, intero-sensors and proprio-sensors (introduced in Lecture IV). The first are on the outside of the body - ears, eyes, nose, mouth and skin (cutaneous); the second relate to the visceral organs; and the third to the joints and muscle tissue etc. Boring also confirms that the word 'proprioception' is Sherrington's from 1906 (Boring, 1942d).

With Gibson's thesis in mind and broadening indoor air quality to embrace the overall perceived quality of the atmosphere, it is also of historic interest therefore that Boring cites the 1909 work of Ernst Meumann (Boring, 1942d:562), who posited "... oppression as a cardiac sensation and suffocation as a pulmonary sensation, ...". While cardiac oppression, linked to anxiety and hypotension, as well as pulmonary suffocation might seem somewhat hyperbolic in relation to air quality, one can reasonably argue that a stuffy atmosphere is oppressive, on a scale that has 'fresh' at the other end. Stuffiness linked to warmth might also tempt one to take action to make breathing seem more pleasant. Such action might include wafting a programme in an auditorium or opening a window in a domestic room.

More than half a century after Meumann's publication, Victor Olgyay disseminated his 'bioclimatic chart' (Olgyay, 1963), whereby above 15 mmHg or 2 kPa of vapour pressure was considered uncomfortable in terms subjects noticing a close or depressed feeling. However, Olgyay cites climatologist Dr Paul Siple's proposition that one mph (0.45 m/s) of air movement will counteract every additional 1 mmHg (0.133 kPa) of vapour pressure (0.34 m/s for 0.1 kPa). Thus we have an architect interpreting 'air quality' in terms of thermal comfort, and crucially bringing humidity and air movement into the sensory mix. T A E Platts-Mills and A L De Weck have more recently posited vapour pressure, or the mixing ratio of moist to dry air, as a measure to limit excess dust mite population growth (Platts-Mills & De Weck, 1989). They recommend an upper threshold of 7 g/kg (1.13 kPa). More recently in the US, 'Clearing the Air' (Institute of Medicine, 2000) had a focus on asthma. It summarized: "There is sufficient evidence of a causal relationship between dust mite allergen exposure and exacerbations of asthma individuals specifically sensitized to dust mites. Continual exposure to dust mite allergens is also a contributing cause of chronic bronchial activity." The summary went further: "There is sufficient evidence of a causal relationship between dust mite allergen exposure and the development of asthma in susceptible children." The report also concludes in relation to the presence of mould or fungal spores: "There is sufficient evidence of an association between fungal exposure and symptom exacerbation in sensitized asthmatics. Exposure may also be related to non-specific chest symptoms." The second conclusion here carries a specific caveat: "There is inadequate or insufficient evidence to determine whether or not there is an association between fungal exposure and the development of asthma." The key terms 'causal relationship' and 'association' used in these conclusions have earlier been clearly defined in the Executive Summary.

Both the above findings in 'Clearing the Air' highlight the importance of controlling levels of humidity, as did the earlier 1989 paper of Platts-Mills and De Weck. Although reliable statistics on asthma are elusive, and complicated by increasing awareness and reporting of conditions, those for the city of Glasgow in the west of Scotland indicate an incremental increase from 1950-2000, with doubling in the decade prior to 1989. Industrial activity and its carbon deposition post-WW2 also accelerated prior to the introduction of the UK Clean Air Acts of 1956 and 1968. However, the period immediately following the 1968 Act was marked by a concomitant shift away from solid fuel as a dominant source of domestic heating. It also unfortunately coincided with two periods of fuel price hikes in the UK. The first followed the Yom Kippur war of 1973, and the second followed the Iranian revolution of 1979. This is apparent in Fig. 1 (left) from Brenda Boardman's seminal publication of 'Fuel Poverty' (Boardman, 1991); while the second fuel price hike, especially that for gas, coincided with a parallel surge in UK unemployment in the first half of the 1980s - see also Fig. 1 (right), which shows the specific trends for Strathclyde Region (Keating, 1988).

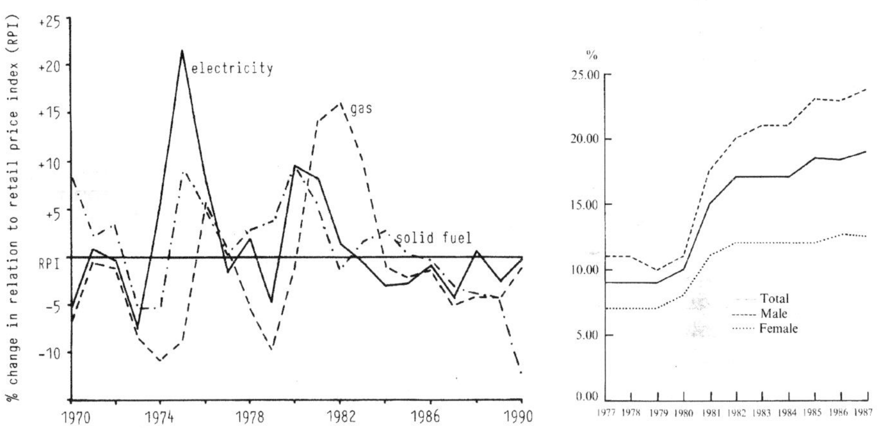

Fig. 1. (left) Annual changes in the real price of gas, electricity and solid fuel, UK (1970-90); DUKES (right) Percentage unemployment, Strathclyde Region at April, 1977-87; DoE

Thus the ironic unintended consequence of the long-needed Clean Air legislation was to move the problems of air quality from outside homes to inside them - as noted in the 1990s by the present author (Porteous, 1996). This paper also registered concern that a 1980 British Standard, BS 5295, gave a safe limit for dilution of CO_2 as 0.5% or 5,000 ppm. Although 'safe' does not signify desirable in terms of air quality and its 'bad company', this limit could send the wrong message to architects and their engineering consultants. Non-linear proportionality with the 1,000 ppm to 8 l/s relationship would permit air supply for each occupant of a room (average for men and women) as low as 1.1 l/s or 4 m^3/h. If we take a 48

m^3 living room occupied by four people, this gives a rate of air change of 0.33 ac/h. If we compare this with the 8 l/s standard of the 1990 Rosehaugh Guide, the same scenario would imply a rate of 2.4 ac/h for that period of occupation. Herein lies the dilemma for aligning energy efficiency with air quality, potentially risking conditions with high moisture levels. It also helps to explain why research exploring the potential for passive solar techniques to preheat supply air occurred in 1980s and 1990s (Ho, 1995; Porteous & Ho, 1997; Saluja et al., 1987), and why mechanical heat recovery is now in the ascendancy (Porteous, 2010).

Such work was additionally motivated by dampness from condensation in housing emerging as a major social concern in the 1980s. Black mould and mildew with its insidious smell and poverty stigma were the physical manifestations, made all the worse by the poorly insulated post-war building boom. This was the reality of 'air quality' for the low-income sector. The dense concrete construction of 1960s and 1970s systems used for towers and low-rise blocks of housing were particularly notorious, with instances of mould becoming so severe that it formed a black slime on the coldest walls. It was apparent that diagnostic processes at design stage to avoid condensation on surfaces or interstitially had not been applied. The combination of high thermal transmittance, poor and expensive means of heating, and poor control of ventilation, all contributed to this. The thermal efficiency of the building envelope needed to improve significantly along with suitable means of heating and ventilating. When sequential thermal conductivity relative to vapour permeability was analysed in multi-layer construction, it also became evident that future construction required much more careful consideration than had been the case hitherto.

We have already established above that the essential science of temperature relative to moisture had been mathematically described by 1844 by Magnus and his predecessors working in this field. Also, Joseph Fourier's Law of Heat Conduction (Fourier, 1822) stated: "the rate of heat transfer through a material is proportional to the negative gradient in the temperature, and to the area at right angles to that gradient through which the heat is flowing." Thomas Graham's Law of Diffusion (Akrill, et al., 1979), first stated in 1831, posited that the rate of effusion of a gas is inversely proportional to its molecular mass; and Adolf Fick's Laws of Diffusion of 1855 (Philibert, 2006) may be paraphrased as: a) diffusive flux travels from regions of high to low concentrations; b) the prediction of how diffusion causes the concentration field to change with time. However, in order to predict likelihood of condensation two other key advances were required. Firstly, reliable methods of measuring the thermal conductivity of materials had to be established; and secondly, similar methods were required to measure vapour permeability.

Taking conductivity first, Robert Zarr (Zarr, 2001) tells us that refrigeration industries pump-primed research from the early 1910s at The National Institute of Standards (NBS), later changing its name to the National Institute of Building and Technology (NIST). Zarr also tells us that the early experiments were conducted at an NBS laboratory in Washington DC, initially under the direction of Hobart C Dickinson and later Milton S Van Dusen. He cites several significant papers concerning experiments into thermal conductivity: one by Dickinson and Van Dusen M S in 1916, 'The testing of thermal insulators'; one by Van Dusen in 1920, 'The thermal conductivity of heat insulators'; one by Van Dusen and J L Finck in 1928, 'Heat transfer through insulating materials'; and another by the same pair in 1931, 'Heat transfer through building walls'. A post-WW2 book on insulation by Paul D Close (Close, 1947) refers to three methods of testing building materials: a hot plate apparatus by Lees in 1898 (Brown, 2006); a hot box apparatus; and the Nicholls heat meter (Nicholls, 1924). In the UK, George F C Searle published details of an apparatus to measure the conductivity of high conductors such as copper, and another apparatus for low conductors such as rubber (Searle, 1934). Thus it seems that full tables of thermal conductivity of common building materials, such as those in the book by Close, were not available until the 1940s. This signifies a remarkable period of about a century after the key knowledge defining moisture relative to temperature.

Modern methods for measuring vapour permeability started somewhat later than those for thermal conductivity. Paul Close published a paper (Close, 1930) on prevention of condensation on interior building surfaces, which only required knowledge of the thermal conductivity of bounding elements. Mark Bomberg cites J D Babbitt for his 1939 paper (Bomberg, 1989), which seems to be in the vanguard of measuring permeability through various building materials. At any rate by 1947 Close's book contains three tables, one of which is attributed to L V Teesdale, another to J D Babbitt and a third to the University of Minnesota (Close, 1947:250-252). However, the materials for which values are given reflect the typical American timber-frame construction, and so are limited in terms of a full palette of materials for architects. An article in Architects' Year Book 4 (Varming, 1952) indicates a comprehensive knowledge of vapour transmission through masonry construction of various types in Europe. However, the gap is considerable between the early work by 19th C scientists and a 20th C capability to track dew-point and structural temperature through multi-layer construction. Given that a mix of dense materials and aerated insulating products were common from the 1920s, one can only conclude that architects during this period operated on a basis of broad principles or ignorance.

In other words, the risk of interstitial condensation would be high. Nevertheless, particular buildings appeared to show a good grasp of the theory involved - for example the roof of the 1930 seaside house by Alfred Roth and Ingrid Walberg, or the wall of the 1932 house by Adolf Bens in Prague (Porteous, 2002a). Architects did not necessarily need to know exact values for vapour and thermal conductivity. If they understood four principles, they could design reasonably condensation-proof constructions: firstly, vapour travels from high to low concentrations; secondly, this will be relatively rapid through a porous material and less so through a dense material; thirdly, air movement helps to disperse water vapour; and fourthly, knowing that low conductivity relies on trapping still air, an insulating material placed on the cold side of a dense structural material will help to keep it warm.

Alvar Aalto was another architect who concerned himself with such issues of performance. In 1938 (Aalto, 1979a), in a lecture given at the Nordic Building Congress, Oslo, Aalto related flat roof construction to liberation of plan forms, and insulation on flat roofs to geographical transferability. In the same lecture Aalto prefaces his comment on roofs by stating: "Water pressure insulation has given us the opportunity to penetrate deeper into the earth." This could well have been an oblique reference to his stepped terrace of dwellings at Kauttua, constructed 1938-40, which involves significant cuts into the ground. More contentiously on the issue of air quality, in 1955 (Aalto, 1979b) in a lecture to the Central Union of Architects in Vienna, he claims that metal ducting has a deleterious effect, claiming evidence from laboratory tests.

Although this assertion is not referenced and although contemporaneous papers to this effect have not been traced, it is likely that Aalto was referring to galvanized metal ducting boosting the positive ions of the air that passed through it, the existence of ions having been known since the end of the 19th C. The longevity of this effect by such metals is also hard to track down, but it is also known that dust and other airborne particles such as bacteria and fungal spores are positively charged, and hence increase the incidence of positive ions (the 'bad' ones). Relatively recent research (Sippola & Nazaroff, 2002) shows that particle deposition is significant, especially on the floor of horizontal duct runs and considering particles larger than 40 microns; noting that experimental data of particle deposition was fitted to computer models. Production of positive ions in ducts also relates to previous microbiological work on fungal spores (Pasanen et al., 1993), which indicates the possibility of mould propagation in sections of unheated or poorly insulated ducts in certain weather conditions. On the other hand since plants emit water vapour through transpiration they produce negative ions, which also reduce dust levels (Lohr & Pearson-Mims, 1996); and, as mentioned above, sunlight in rooms will reduce bacterial activity, and hence boost negative ions, as will natural materials such as clay plaster (White et al., 1992). In another branch of science, there are concerns about chemical reactions in ventilation ducts (Andersson et al., 1996); whereby research indicates possible production of aldehydes (e.g. formaldehyde and benzaldehyde) from a reaction inside ducts between volatile organic compounds (VOCs), styrene for example, and inorganic gases such as ozone found in the ambient air. Clearly, length of ducts may be a factor, the experiments reporting use of 6.0 m tubes.

Hence, at a time when mechanical heat recovery ventilation (MHRV) is gaining traction for housing and other buildings, in part to comply with standards like PassivHaus and Minergie-P, there are notes of caution from the research community. There are some examples of buildings that have used MHRV, but deliberately avoided metal ducting - e.g. the Hockerton Housing Project near Newark in Nottinghamshire, UK, by architects Brenda and Robert Vale, using fireclay pipes (Anon., 2006); the Marché headquarters building between Zurich and Winterthur by architect Beat Kampfen, using hollow timber columns and wall-integrated ducts for vertical supply and exhaust (Anon., 2009). Since MVHR has been shown to save energy, and noting that the Marché building is designed as 'net annual zero energy' and also has excellent air quality, it would seem that further research should be carried out with regard to ducting materiality and cleanliness relative to a healthy ion balance in the served spaces. Similar additional research could be carried out to identify any perceived effects in terms of air quality in served spaces, which are attributable to the presence of verdure, moving water and natural linings such as clay plaster and untreated timber.

The capability is certainly there for such investigations in terms of advanced modelling techniques validated by field measurement. One such study, led by John F Straube, sets a context of tackling indoor pollutants, described as "a chemical soup of VOCs", by means of avoidance, removal and exclusion (Straube & deGraauw, 2001). For removal he cites a role for plants and running water, as well as for hygroscopic materials, which constitutes the core of his findings. A renowned example of vegetation and running water as the ambient context for a home is Frank Lloyd Wright's famous Fallingwater. One could argue that Jay Appleton's 'prospect-refuge' theory (Appleton, 1996) and the 'hazard symbolism' expounded by Grant Hildebrand when analysing this building (Hildebrand, 1991) has relevance from a holistic view of indoor air quality. The psychology of perception here favours generous fenestration both to enhance the drama of prospect and to admit winter sunshine to a secluded refuge deep within the room - fire of the sun to fire of the hearth.

There are other significant buildings of that ilk, such as the 1988 ING Bank building by Ton Alberts, or Joachim Eble's 1992 Oko-haus in Frankfurt visited by the present author (Porteous, 2002b). In each case atria with planting and running water contribute to indoor air quality as well as to prospect and refuge for the occupants, if not hazard symbolism. There have also been several papers published on the air-cleansing qualities of certain indoor plants. For example, that led by Martina Giese examines detoxification of formaldehyde by the spider plant (Giese et al., 1994). Another led by Bill Wolverton explores the ability of plants to remove a range of VOCs from the indoor environment (B. C. Wolverton & J. D. Wolverton, 1993); and one three years later examines the influence of indoor plants on the reduction of airborne microbes and influence inside energy-efficient buildings (B. C. Wolverton & J. D. Wolverton, 1996). In the same year, Virginia Lohr and colleagues demonstrated that interior plants could improve worker productivity and reduce stress in a windowless environment (Lohr et al., 1996) - i.e. creating an internal prospect of verdure. J R Girman and a team at Lawrence Berkeley Laboratory in California also investigated emission rates from various indoor combustion sources, including cigarettes, this still being relevant in private buildings at least (Girman, 1982). More recently, a useful online design guide 'Design and Detailing for Toxic Chemical Reduction in Building' by the Scottish Ecological Design Association (SEDA) has been available (Liddell et al., 2008). However, this has proved controversial for some manufacturers and SEDA's response to complaints thus far has been online redaction of some passages.

Returning to Straube's own analytical findings, which include results from modelling hygric response as well as field measurements, a strong case is made for moisture buffering by using hygroscopic lining materials. A mathematical model for non-isothermal vapour transport in hygroscopic building materials, validated by an experimental study and led by Menghau Qin was recently published (Qin et al., 2008). Further work on determining moisture buffering potential in situ by Evy Vereecken and colleagues (Vereecken et al., 2011), adds to that of Straube and introduces the units HIR (Hygric Inertia for a Room, $g/m^3.\%RH$) and MBV (Moisture Buffer Value, $g/m^2.\%RH$). These units express the mass of moisture taken up by materials in field experiments over different time intervals, and validated by a test room in the VLIET test-building in KU Leuven in Belgium. They also include the potential contents of rooms, such as piles of newspapers and books, as well as the lining materials like wood-wool cement board.

There has been parallel work targeted at reducing airborne pathogens in buildings. For example that by a team led by Peng Xu investigated the efficacy of ultraviolet irradiation of upper room air in this regard (Xu et al., 2003); and there has been more recent work reviewing various means of protecting occupants from airborne pathogens (Balashikov & Melikov, 2009). The latter notes that knowledge of the influence of RH on pathogenic bacteria is scant, while 40-60% RH is accepted as more lethal to non-pathogenic bacteria. Both papers are concerned with 'active' methods of intervention in air handling systems, rather than the passive solar methods explored in the 1940s by Leon Buchbinder and colleagues, which were mentioned earlier. Arguably, a case can be made for hybrid techniques. Enhancing the ability of mechanical ventilation to improve air quality in partnership with effective passive solar environmental design seems a workable strategy. However, in terms of health, it would mean avoiding re-circulation in favour of MHRV. An earlier investigation led by John F Brundage, proved that mechanically ventilated barracks with 95% recirculated air, increased incidence of respiratory infection compared with older naturally ventilated barracks (Brundage et al., 1988); an irony given all the 19th C research work on ventilation by army surgeons.

Recirculation is a 20th C phenomenon, aimed at saving energy. In the 19th C the norm was a once-through system with the filtered, washed and heated air introduced mechanically and exhausted by wind-assisted thermal buoyancy through terminals at the top of buildings. One of the earliest 20th C buildings that was until recently of this type did introduce a modicum of recirculation - the 1901 Kelvingrove Art Gallery and Museum by architects John W Simpson and E J Milner Allen. Technical information about this building was presented at the Twenty-Second Congress of the Sanitary Institute, and published in a book 'Municipal Enterprises - Glasgow' (Anon., 1904). Two electrical enclosed fans, designed to deliver a total of 5 million ft^3/h or 140,000 m^3/h, powered the plenum system in this building. This design rate corresponds to approximately 1.1 ac/h. The system was also designed to cool the ambient air by at least 4°F (2.2 K) in summer, while the embedded masonry ducting provided a similar preheating effect in winter. Some recirculation was enabled via doors to the main supply shafts off the two main glazed atria either side of the main entrance concourse, but this would have been insignificant compared to the Brundage barracks case studies. With only one minor upgrade to replace worn out parts in the 1950s, Kelvingrove provided excellent air quality and comfort up to its conversion to full air conditioning in 2003-06. Now in the refurbished basement at least, there is a tendency to over-chill the air supply. In the restaurant the distribution is particularly poor, and diners below the ceiling registers are liable to be subjected to uncomfortably cool draughts. Even in the main gallery spaces on the ground and first floor, the flow rate from supply registers in walls is such that movement of air is more evident than formerly. Also, daylight has been banished from all the peripheral galleries, and, given the psychological twinning of 'light and airy', the overall ambience is now perceptually less pleasant than formerly.

In a later 20th C example from the 1970s, the Bourdon Building at The Glasgow School of Art housing the Mackintosh School of Architecture, the mechanical ventilation supply with a high proportion of recirculation proved problematic. This was evidenced by the rapidity with which smoke emanating from a single inside source circulated around the entire building. A 'black-box' lecture theatre was particularly poor in terms of its air quality and was eventually fitted with additional freestanding air conditioning units. However, a monitoring study (Fung, 2008a) found a significant proportion of dissatisfaction among the students during a one-hour lecture. The highest level of satisfaction occurred some 15 minutes into the lecture after an initial surge in CO_2 up to 1,380 ppm started to drop down to 1,080 ppm. As the level steadily increased up to 1,170 during the remainder of the time, the sum of 'very dissatisfied and 'somewhat dissatisfied' also gradually increased up to 31% of those present. Again, this study tends to support the Pettenkofer limit of 1,000 ppm. This finding also corresponded with a similar survey of dissatisfaction within offices on the campus.

A visit by the present author to the Whatcom Museum in Bellingham, Washington, USA, completed in 2009 by architect Jim Olson, confirmed a similar tendency to that at Kelvingrove in terms of rather cool temperatures, especially when passing in the 'line of fire' of plenum supply registers delivering at relatively high velocity. Temperatures varied between 18.3°C and 19.4°C in lower and upper galleries respectively, while RH was in the unexpectedly high range 62-68%. Since the weather outside was sunny and warm, this proved chilly for lightly clad visitors. While such a building may be supplying perfectly healthy, clean air, the perception of air quality is compromised by lack of thermal comfort.

A domestic example that has issues of both low temperature of air delivery and recirculation is that of a 2003 rural housing project in the Scottish village of Ballantrae by architects ARP Lorimer Associates for Ayrshire Housing Association. This employs a proprietary positive-pressure system supplying air continuously through a single register on the landing at the top of the staircase. The double-height circulation space thus acts as a plenum delivering fresh air to all rooms, with exhaust via windows. A manifold in the attic seeks air from a number of sources depending on the setting by the occupant. When set to 'warm' this could be from the apex of the attic or from a double-height passive solar buffer, located between the main rooms. Unfortunately this meant that cooking odours, smoke and so forth from the kitchen or the living room could form part of the 'fresh' supply. When set to 'cool', the supply air is taken from a vent in the eaves at the north side of the roof, which is useful when there is a need for summer cooling. However, when set to 'warm' on a cold evening the inflow will come from either the attic or the sunspace, and may then be perceived as an uncomfortably cool draught by residents.

3. Experience from recent Scottish housing research projects

The experience at Ballantrae suggests that if solar buffers are to be a successful provider of preheated fresh air, the control of supply must be more fully in the hands of occupants. On the other hand, in order to preheat, the direction of flow must be from outside to inside. An urban retrofit project in Glasgow that adopts such a strategy was analysed by Janice Fung, and found to have satisfactory levels of air quality (Fung, 2008b). Spot measurements taken in twelve flats in winter gave a maximum in the living room of about one third over the Pettenkofer limit, and a mean comfortably below it. In spring all measurements taken in ten flats were below 1,000 ppm, indicating a tendency to ventilate more freely with increasing solar irradiation. In the case of the single aspect and corner flats in these towers, the continuous extract from bathrooms will tend to encourage ingress of air from the outer edge, via the glazed buffer spaces into the living rooms and towards the interior circulation space. However, negative pressure on the leeward side of the towers may compete with this tendency, and indicative measurements may also reflect the overall improvement in energy efficiency, as well as the presence of glazed spaces. Equivalent values in nearby towers with smaller single-glazed windows and no other thermal improvements were above 1,000 ppm in December, and March. In a third type of tower, again in the same neighbourhood and with some thermal improvements (double-glazing and 'heat with rent' tariff), respective comparative values were in between the first and second examples. Table 7 below summarises these comparative data in ranked order, with least energy-efficient first. Note also that the air quality is inversely proportional to level of airtightness afforded by respective types of fenestration - metal frames, single-glazed, to plastic double-glazed, to plastic double-glazed plus buffer spaces.

Description of towers	Time of year	Maximum living rm	Mean living rm
1. SG, high U's	20/12/05	2,110 ppm	1,274 ppm
2. Ditto 1	01/03/06	1,600 ppm	1,020 ppm
3. DG, high U's, HWR	20/12/05	1,670 ppm	1,066 ppm
4. Ditto 3	01/03/06	1,380 ppm	904 ppm
5. DG, low U's, SB	21/12/05	1,370 ppm	852 ppm
6. Ditto 5	08/03/06	950 ppm	804 ppm

Table 7. Comparative CO_2 in three types of tower block from Fung, pp (6)-32 & (6)-34
Legend: SG, DG = single, double-glazing; U's = U-values; HWR = heat with rent; SB = solar buffer

Durational data from individual flats in the second and third of these tower-block sets, also confirmed a significantly lower incidence of the mixing ratio exceeding 7 g/kg - the threshold of Platts-Mills and De Weck cited above to limit dust mite propagation - for the most energy-efficient case with the glazed buffer spaces. Table 8 summarises this.

Description of towers	Time of year	Living rm	Bedroom	Kitchen
1. DG, high U's, HWR	23/10/06 - 18/12/06	34.9	10.4	8.1
2. Ditto 1	30/11/06 - 04/12/06	97.6	98.0	87.9
3. DG, low U's, SB	24/10/06 - 23/11/06	13.5	5.7	2.6

Table 8. Comparative % mixing ratio >7g/kg in two types of tower block, Fung, p (6)-48

Even though one should be cautious about individual comparisons, born out by the range between two flats in the HWR tower, the findings align with CO_2 levels from a larger number of cases in Table 7. Moreover, a 42% incidence of mould on walls was recorded in the entire set of HWR flats, compared with none for the SB towers and 67% for the least thermally efficient (rows 2 and 3 in table 7 above). Fung was only able to include two flats with MHRV in her study (Fung, 2008b), this a block designed by architect Ian Ritchie for Scotland's Home of the Future in 1999. In one case, it was a non-smoking household that made use of the MHRV, but also frequently overrode it by opening windows. Thus although the CO_2 was generally below the 1,000 ppm level, it was not used optimally in terms of energy efficiency. A smoking household occupied the other flat and decided not to use the MVHR due to a fear of "creepy crawlers coming through the vents"; and, in this case, the levels exceeded 1,000 ppm. The non-smoking household in this case also achieved similar percentages above the mixing ratio of 7 g/kg to the SB flat in Table 8 for the living room, although somewhat higher for the kitchen and bedroom. More recent work by the Mackintosh Environmental Architecture Research Unit (MEARU) has found CO_2 levels in the living room of new houses in Glasgow with MVHR going up to circa 2,000 ppm during periods of occupation by four persons, with a maximum of over 2,500 ppm, while a gathering of six persons took the level above 3,500 ppm.

Olaf Adan showed in his 1990s doctoral thesis that short-term peaks of high humidity can support fungal growth (Adan, 1994, as cited in: Ginkel & Hasselaar, 2005; Straube & deGraauw, 2001; Viitanen & Ojanen, T, 2007). This characteristic has been shown to be common in a series of 2-week surveys carried out as part of a Glasgow project 'Environmental Assessment of Domestic Laundering' led by the present author. In most cases CO_2 peaks coincide with vapour pressure peaks, indicating that both are due to relatively intensive occupation with closed windows, but not with particularly airtight construction. Typically in this situation vapour pressure rises steeply to over 2.0 kPa (nearly 100% above the 7 g/kg threshold) from less than 1.0 kPa (well below the 7 g/kg threshold). This contrasts with a situation whereby drying a washing load overnight gives a gradual increase of vapour pressure, while CO_2 levels fall in the absence of occupants. In the second case, however, the mixing ratio still rises above 7 g/kg to approximately 11 g/kg (1.78 kPa) by morning; while CO_2 falls from a peak of 2,000 ppm in late evening. Therefore, although CO_2 and vapour pressure maxima are more severe in the first case, they are a cause for concern in both. Neither are these particular two situations unusual. Table 9 below summarises key data seasonally for 23 households, all monitored for approximately two weeks at different times of the year.

Description: CO_2 ppm & RH%	Spring (10)	Summer (5)	Autumn (5)	Winter (3)
1. CO_2 ppm: mean	953	762	1,112	1,178
2. CO_2 ppm: mean maximum	2,630	1,448	2,267	3,844
3. CO_2 ppm: maximum	4,983	1,896	5,000*	5,000*
4. CO_2 ppm: mean minimum	448	495	505	560
5. CO_2 ppm: minimum	299	431	434	380
6. RH: mean	47	51	64	41
7. RH: mean maximum	65	67	78	70
8. RH: maximum	83	76	86	83
9. RH: mean minimum	32	29	47	26
10. RH: minimum	20	21	36	22

Note*: 5,000 ppm is the maximum possible reading on the instrument used

Table 9. Comparative CO_2 and RH for living rooms in domestic laundering study, Glasgow

Table 9 indicates a distinct tendency for the lowest CO_2 levels in summer and highest in winter, suggesting linkage with ventilation regimes relative to intensity of occupation. The considerable range of RH values reflect a similar range of temperatures. In this regard some aberrant values due to the location of sensors in direct radiant view of the sun have been omitted from the analysis. Even so, passive solar gain in the absence of occupants during the daytime has given some very high absolute values - see Table 10 below.

Description (°C)	Spring (10)	Summer (5)	Autumn (5)	Winter (3)
1. Temp: mean	20.7	21.0	18.6	21.0
2. Temp: mean maximum	24.3	23.2	22.8	24.8
3. Temp: maximum	34.6	33.3	27.8	26.6
4. Temp: mean minimum	16.2	18.2	15.7	12.7
5. Temp: minimum	12.1	16.2	13.1	14.0

Table 10. Range of temperatures for living rooms in domestic laundering study, Glasgow

In general, the values in Table 9 are concerning. For example, the mean maximum in winter of 3,844 ppm is 20% greater than the absolute maximum of the 19th C Scottish survey for single-room dwellings of 3,210 ppm, the smaller figure at a time when overcrowding and fetid air was a major health concern. Given some of the high CO_2 and RH values, one might reasonably anticipate problems with condensation and mould growth. However, although the colony forming units (CFUs) vary considerably there is no strong seasonal effect - see Table 11 below. Nor is there any consistency when taking individual cases and comparing the CFU value with CO_2, vapour pressure and presence of mould itself - see Table 12 below - although a breakdown of mould types might be relevant here. Nevertheless, the CFU level is generally of some concern. Vivienne Ryan of Belfast City Council has usefully categorized approximate low, moderate and high levels, respectively <500, from 500-1,300 and from 1,300-5,000+ (Ryan, 2002). In the 22 case studies where CFU/m^3 values were obtained from air samples, none were in the 'low' category as defined above; 19 were in the 'moderate' range, but with 4 of these close to the upper limit; and 3 were in the 'high' range.

Also some cases of the 'moderate' category that are not listed in Table 12 above had mould reported - see Table 13 below. More detailed analysis of mould types and sampling dates is ongoing.

Description (CFU/m³)	Spring (10)	Summer (5)	Autumn (5)	Winter (2)
1. Mould count: mean	1,352	809	672	1,115
2. Mould count: maximum	2,975	1,275	1,045	1,265
3. Mould count: minimum	545	595	548	965

Table 11. Comparative CFU/m³ for living rooms in domestic laundering study, Glasgow

Case study (date)	CFU/m³	CO_2 peaks	VP	Notes
No. 2 (14/04/09)	Liv: 2,960	996 (1212)	1.13 (1210)	no mould
No. 7 (14/04/09)	Liv: 735	948 (1811)	1.04 (1812)	no mould
No. 8 (13/04/09)	Liv: 1,565	1,139 (1104)	1.24 (1105)	no mould
		1,068 (1117)	1.37 (1117)	
		1,278 (2245)	1.12 (2245)	
No. 11 (07/04/09)	Liv: 2,975	1,239 (1824)	1.79 (1824)	no mould
	Br2: 2,900	2,990 (0724)	1.64 (0934)	mould + K & Ba
No. 12 (26/05/09)	Liv: 545	1,323 (1810)	1.42 (1820)	no mould
No. 15 (11/06/09)	Liv: 1,275	640 (1234)	1.04 (1214)	no mould, Ba only
No. 22 (06/01/10)	Liv: 1,265	4,209 (2113)	2.42 (2053)	no mould

Table 12. CFU/m³ cf. CO_2 and vapour pressure in domestic laundering study, Glasgow
Legend: CO_2 peaks in ppm; VP = vapour pressure in kPa; K = Kitchen; Ba = Bathroom

Case study (date)	CFU/m³	CO_2 peaks	Notes
No. 6 (17/04/09)	K: 1,110	1,093 (1800)	1.32 kPa, mould K + drying pulley
No. 13 (27/05/09)	Br1: 595	1,646 (2316)	16.4°C, 62.8% RH, mould; drying?
No. 14 (29/05/09)	Br1: 875	1,157 (1408)	20.5°C, 75.5% RH, mould; drying?
No. 18 (15/10/09)	Br2: 515	4,217 (2355)	1.47 kPa, 23.8°C, mould; ironing?
No. 19 (19/10/09)	K: 510	723 (2233)	1.47 kPa, 19.4°C, mould; drying?
No. 21 (06/01/10)	Br1: 1,025	4,031 (1853)	23.8°C, 84.8% RH, mould; drying?

Table 13. CFU/m³ cf. CO_2 and mould presence in domestic laundering study, Glasgow

The association with some aspect of domestic laundering, usually drying and/or ironing, is somewhat tentative, although there are notes in diaries kept by occupants to support this contention. The CO_2 peaks are also variable, and this may suggest a liberal or frugal ventilation regime. But it is possible that peaks of vapour from occupants play a role in tandem with a laundering activity such as ironing. Although the mean maximum of vapour pressures in each of the six cases in Table 13 is 1.69 KPa on the date when the air sampling was carried out, the mean maximum for the whole period of measurement is 2.07; and absolute maxima for Case 21, the highest, are respectively 2.46 kPa and 2.69 kPa. In this instance the high RH value of 84.8% in Table 13 for the first day of measurement is not typical of the monitored period, the average dropping to 45.5%. Therefore, one has to be careful about expectations with regard to cause and effect. The overall lack of consistent association between mould indicators (CFUs), presence of mould, and CO_2 or humidity is in accord with the findings of the Scottish survey of the 1880s. In any case, given that mould was a relatively frequent occurrence, the guideline of keeping RH below 70%, which corresponds approximately with 1.4 kPa at 20°C air temperature and dew point of 12°C, is generally accepted as 'good practice'.

In summary, the above findings suggest a case both for better control of ventilation, and for improved drying facilities for domestic laundering. In the first regard, MHRV will undoubtedly have a role to play. However, there is presently an attitude of undue complacence in the ability of MHRV to resolve all conditions of ventilation control satisfactorily. For example, automated changes in volume flow rate switched by moisture level and CO_2 should become the norm in domestic systems. Also filters should always be readily accessible, materiality of ducts should receive greater consideration, and thermal insulation of ducts, as well as sound reduction through ducts, should be more effective. Since even a low background noise at the lowest flow rate can be disturbing in a bedroom, consideration could be given for a facility to isolate parts of a system manually. Bedrooms should not require to be heated overnight in a well-insulated envelope, so that open windows rather than MHRV overnight should be feasible without compromising energy efficiency. With regard to dedicated drying facilities, both passive and active solar systems have been shown to be capable of playing a part (Porteous & Menon, 2010). However, it seems a dubious tactic to employ photovoltaic (PV) arrays to displace thermal energy for tumble dryers. For example, small-scale renewable combined heat and power (RCHP), with waste heat from electrical generation used directly for communal dryers could be more effective; and solar upgrading of passive drying facilities should be economically viable.

4. The holstic value of indoor sunlight and light today for air quality

Another issue relating solar energy to air quality is the perceived conflation of 'light and airy', already mentioned. The theorists Colin Rowe and Robert Slutzky claim that dictionaries define 'transparent' as "a material condition - pervious to light and air" (Rowe & Slutzky, 1963). Notwithstanding whether this assertion is wishful relative to the pedantry of dictionaries, a fenestrated façade that is 'pervious to light and air' seems a perfectly reasonable literal proposition, since, even in the case of a sealed curtain wall, there will be some air leakage. Another theorist, David Leatherbarrow talks about 'translucency' and 'transparency' as 'indicative terms' in relation to Rudolph Schindler's work: "His usage invoked the customary sense of the words, meaning optical or visual conditions, the state of some material being impervious to or penetrable by light or air, ..." (Leatherbarrow, 2009). Again, one might relate such a claim to fenestration but not simply to transparency or a transparent material. It has also been stated above that the psychology of 'prospect-refuge' theory sits well the physics of sunlight's bacteria-killing power through glazing and that indoor planting further enhances air quality. All this goes together with surveys carried out several decades ago to establish the popularity and social need for sunlight opportunity indoors.

A 1965 survey in the Netherlands (Bitter & van Ierland, 1965) found sunshine indoors valued over view, especially in living rooms where afternoon sunlight was preferred to that of morning for bedrooms; and the combined effect of "the warmth, the light and the atmosphere" synergistically valued significantly more than the sum of these individual factors. A UK housing study (Markus & Gray, 1973), concerning "the psychological significance of sunshine, daylight, view and visual privacy", employed semantic scales as psychological measures. These explored reactions to outlook from windows relative to sunshine, rather than reactions to sunlight within rooms. Another in Zurich (Gilgen & Barrier, 1976), found that enough sunlight ranked as important as plenty of space indoors, and sunlight is desired in children's rooms as much as living rooms. Also perception of sunniness and measured sunniness closely matched. In the same year a UK social survey (Ne'eman et al., 1976) found the visual experience of sunshine to be beneficial in terms of psychological well-being, enhanced colours of interior surfaces and contact with the exterior; while thermal benefit is most noticeable in absence of adequate heating, whether perceived or actual. In the 35 years since that work, psychological metrics have become more sophisticated, while housing providers have arguably become less concerned with the social impact of access to sunshine. However, a social survey by Robert Day and Chris Creed in the 1990s, which analyzed responses to seven questions, made a case for further research to establish: "a sunlight index, using an appropriate reference quantity to calibrate a scale of subjective benefit/loss" (Day & Creed, 1996). Thereafter, Fung's study in Glasgow, cited above, indicated an association between the size of windows to living rooms and positive affectivity or mood (Fung, 2008c). This tentative finding makes a case for further work by environmental architects in collaboration with health or environmental psychologists and microbiologists; this to test the hypothesis that 'sunny and airy' dwellings can be energy-efficient and also enhance mental and physical health. Moreover the strand focused on psychological well-being could be supported by clinical analysis of bio-markers; while microbiological analysis concerning mould, bacteria and other organisms or particles, could be augmented by further chemical research relating materials to pollutants. Such composite investigations would also help to resolve differences of attitude within the community advocating adoption of the PassivHaus standard (applicable to domestic and non-domestic sectors); on the one hand Minergie-P in Switzerland aligning with PassivHaus (Mennel et al. 2007), while explicitly using passive solar techniques, and on the other hand a view that orientation is irrelevant and window sizes should be curtailed.

This then provides a strategic aim for a future whereby housing, workplaces and other buildings enjoy improved air quality, considered within a holistic spectrum of indicators, and securely based on precedent from the near and more distant past. For non-domestic buildings the 19th C provided a template for what we now call 'mixed mode' ventilation. An early modern building that took this forward for commerce was Frank Lloyd Wright's Larkin Building (Porteous, 2002c). Although built shortly after Glasgow's Kelvingrove Museum and Art Gallery, Wright used the technology to promote a paradigm for the workplace where an atrium, a conservatory and the space between its horizontal laylight and pitched glass roof, play an essential role in terms of ventilation cycle along with a system of ducts integrated in structural walls, columns and beams. Interpretations or applications of the Larkin prototype began to appear in the late part of the 20th C, for example, architect Mick Pearce's Eastgate Centre in Harare, 1993-96 (Baird, 2001). Moving into the 21st C, Mick Pearce uses very much the same method, but without the atrium, in a tighter urban context in Melbourne - a corner block, with nine floors of offices above a double-height floor of retail units, with servicing below (Slessor, 2007). There are other good illustrations of mixed-mode ventilation in educational and cultural buildings, where plan and section revolve round a sunlit atrium, such as Gunter Behnisch and Partner's Deutsches Postmuseum in Frankfurt, 1984-1990, and BDP's Saltire Centre at Glasgow Caledonian

University, 2004-06 (Porteous, 2008). There are also projects where major spaces are naturally lit and ventilated with minimal active back up, for example 'The Bridge', a new arts centre and library in Glasgow completed in 2006 by Gareth Hoskins Architects (Porteous, 2008).

Chris Twinn of Arup helps to bring this technology up to date with recent exemplars, where the emphasis is on natural ventilation, but there is active assistance available for extreme conditions (Twinn, 2011). He cites Paignton Library by Kensington Taylor, 2007-09, where the air supply is via acoustically attenuated damper boxes located close to floor level, and Lauriston Primary School in Hackney by Meadowcroft Griffin, completed 2009, which uses similar devices below windows. In each of these buildings high-level exhaust can be augmented actively using motorised dampers and fans. Twinn also illustrates the Robert Burns Birthplace Museum by Simpson & Brown Architects, completed 2010. In this case a ground source heat pump tempers the air supply, mixing fresh air entering via and underground labyrinth with recirculated air from the gallery, before it is exhausted at high level via controllable louvres. Twinn highlight areas or continuing research interest as "heat recovery with natural ventilation" (HRNV) and a "better understanding of how single-sided ventilation (SSV) works under fluctuating wind pressures through a single window."

In the case of HRNV, another interesting technology that merges air with light is that of 'air supply' windows. For example, these have been used passively in the Orchard Building, completed 2006 at the Mater Hospital in Dublin, Ireland, as described by Jay Stuart, the sustainable building consultant (Stuart, 2006). In order to recover heat lost through the inner layer of glazing and to control the air supply passively, 'Dwell-Vent' pressure-sensitive trickle vents were used. An example of SSV, Christ's College Secondary School in Guildford, UK, 2007-09 by architects DSDHA, uses MHRV units below wide internal sill-boards. The fresh air supply enters through a narrow perforated plate directly below external preformed metal sills, while the exhaust is via widened butt joints in the brickwork below sill level (Anon., 2010; McGuirk, 2010).

Twinn's article does reveal cause for some concern relating to ventilation standards for schools. Building Bulletin 101 (Department for Education and Skills, 2006) only requires "the capability of achieving a minimum of 8l/s per person at any one time." It accepts "a minimum daily average of 5 l/s per person" and "a minimum of 3 l/s per person". These last two lower standards of supply compared with the 8 l/s or 1,000 ppm Pettenkofer upper limit perhaps recognize the lower CO_2 expiration rate from younger children. If we assume a mean male/female adult rate of respiration of CO_2 of 0.0185 l/s, and the outdoor air averages 360 ppm, the above values correspond respectively with 1,000, 1,399 and 2,073 ppm (i.e. 640, 1,028 and 1,713 ppm added inside). However, if we assume an average expiration rate of 0.012 l/s per young child, the respective values drop significantly to 777, 1,077, and 1,110 ppm. Even so, the standard allows the maximum concentration during the teaching day to reach 5,000 ppm. If this were calculated on the basis of an entire population of young children, it would imply a meagre supply of fresh air considerably under the 3 l/s minimum - some 0.72 l/s per child. In other words, Building Bulletin 101 expresses a minimum supply in l/s that appears to be at odds with the maximum CO_2 concentration. The other concern at low rates of ventilation is that the 'bad company' of CO_2, such as pollutants from internal finishes or from the ambient air, will become increasingly significant, as will the concentration of bacteria.

A 2008 report on primary schools led by Karl Wall cites the earlier Building Bulletin 87, 2003b:16 (Wall et al., 2008), which describes "acceptable" CO_2 rates as 1,000 ppm or 8 l/s per person. In other words it is using the adult equivalence in terms of CO_2 respiration, and here the value signifies acceptability, not attainability. This 2008 report also cites a paper by Coley and Greaves: "Elevated carbon dioxide levels, above 2,000 ppm, were found to be associated with impaired cognitive function expressed in measures of task attention" (Coley & Greaves, 2004, as cited in Wall et al., 2008:12). The Wall-led report itself also alludes to the synergy of light and air, specifically in terms of sunlight by citing the 1931 Hadow Report, which advocated "plenty of sunlight with fresh air" (Wall et al., 2008:2). However its own findings, while only stressing the pejorative aspect of "thermal discomfort linked to sunlight entering un-shaded windows" (Wall et al., 2008:15), does recognize the psychological importance of the internal environment of schools: "Whether physiological or psychological, a person's health or productivity may be affected." (Wall et al., 2008:19). Building Bulletin 90 (Department for Education and Employment, 1999) tackles 'lighting' in terms of daylight displacing artificial light. It defines Uniform Ratio (UR) as the ratio of minimum to average Daylight Factors (DF). It recommends a UR of 0.3-0.4 in side-lit spaces and 0.7 in top-lit spaces. Notwithstanding the large body of published research into passive solar gain, which can displace CO_2 emissions from heating (an assumed given for this chapter), sunlight in Building Bulletin 90 is only mentioned in terms of adequate shading control. There is indeed a paucity of research into the composite value of sunlight in schools, although E. Ne'eman carried out work on sunlight and thermal comfort in non-specific buildings in the 1970s (Ne'eman, 1977). For the workplace, where solar overheating is often a concern, research led by Mohamed Boubekri found in 1991 that a 25-40% ratio of sunlight patch to floor area was optimal for a feeling of excitement and cheerfulness (Boubekri et al., 1991, cited in Boubekri, 2008). However, although there is some evidence that non-daylit classrooms affect concentration negatively due to disturbance of pattern of stress hormone (Kuller & Lindsten, 1992), evidence of the positive effects of sunlight is harder to find.

5. Conclusion

Such varying standards therefore epitomise the environmental dilemma of today. On the one hand, they are aimed at maintaining minimum, rather than optimum, levels for health. On the other hand, they are aimed at maximising energy efficiency. CO_2 is the common denominator critical to both, coming into play via the different, but related, media of respiration and combustion. The psychological twinning of light and air clearly affects both aspects of CO_2 as well as having physical and mental impacts on the occupants of buildings. However, research projects undertaken by environmental architects in partnership with psychologists and other specialists in medicine (e.g. apropos well-being biomarkers) and science (e.g. microbiological analysis) remain scarce. The above review serves to strengthen the case for an upgrade of regulations pertaining to air quality, which would require both consistent design standards and a new model for post occupancy evaluation (POE) or building performance evaluation (BPE).

The issue of ownership of environmental control relative to building type would be central to this model. In some public buildings the user does not wish to be concerned with such matters. There is simply an expectation to enjoy the interior of the building for what it offers, be it gallery, museum, concert hall, theatre, velodrome and so on. But in other buildings, the issue of personal control becomes critical, and has significant economic impact - for example, a range from lack of vitality within, to absenteeism from, the workplace. It was superior control that gave the mechanically ventilated schools in Scotland of the 1880s better air quality than the naturally ventilated ones; in this case almost certainly with considerable manual caretaking involved by a member of servicing staff. Similarly, it is control that makes natural ventilation feasible in the recent buildings cited by Twinn above, this time with 'smart' electronic operation of vents as well as automated mechanical backup when conditions require it. Thus the apparent paradox between 21st C paradigms for good air quality predicated on natural ventilation, as opposed to 19th C models based on mechanical ventilation is not really so. The presentational 'spin' may suggest a paradox, but the reality is that both are reliant on a combination of automated and manual control. The challenge is to get this balance right, as at the Marché headquarters building near Zurich, where CO_2 sensors play a part on a hot summer's day alongside workers opening windows for a short period in the morning. Similarly there are opportunities for entirely passive, pressure-sensitive trickle vents as one means of control for ingress of fresh air, while electronically-switched valves are another.

Getting this balance right is key to resolving the apparent tension between energy efficiency and air quality. It is also the nub of the case for holistic, adaptive and perception-oriented evaluation of air quality, and one that embraces the synergetic sensory roles of light, prospect and spatial atmosphere.

6. Acknowledgment

I would like to acknowledge the valuable support I have received from David Buri, the Architectural Librarian at The Glasgow School of Art. I would also like to thank my long-time and long-suffering partner, Mary Patrick, for coming to my aid with regard to typesetting and scanning.

7. References

Aalto, A. (trans S. Wrede) (1979a). The Influence of Construction and Materials on Modern Architecture (1938), In: *Sketches, Alvar Aalto*, ed. G Schildt, pp. 60-63, MIT Press, ISBN 0-262-01053-4, Cambridge, USA.

Aalto, A. (1979b). Between Humanism and Materialism (1955), op. cit., pp. 130-133.

Akrill, T. B., Bennet, G. A. G. & Millar, C. J. (1979). Ch 10 Thermometers and the Ideal Gas, In: *Physics*, p. 130, Edward Arnold, ISBN 0-7131-0297-7, London, UK.

Andersson, B., Andersson, K., Nillson, C-A, Sandstrom, M., Stjerrnberg, M. & Sundell, J. (1996). Chemical Reactions in Ventilation Ducts, *Proceedings of Indoor Air '96*, Copenhagen, Vol.3, pp. 947-952.

Anon. (1904). 2 Art Galleries and Museums, In: *Municipal Enterprises - Glasgow*, pp. 15-18, Souvenir Handbook of Glasgow issued by The Corporation on the occasion of The Twenty-Second Congress of The Sanitary Institute, Glasgow, 25-30 July 1904.

Anon. (2006). Technical Fact Sheet 6, Ventilation & Airtightness, Hockerton Housing Project Trading Ltd., Hockerton, Southwell, Notts, UK, NG25 0QU.

Anon. (2009) Marché International Support Office, Zero-Energy Architecture in Switzerland, *Detail Green*, May 2009, p 36, ISSN 1868-3843

Anon. (2010). College in Guildford, *Detail, Review of Architecture*, Vol.5, Facades, pp. 480-485.

Appleby, P. (1990). Ch C1 Indoor Air Quality and Ventilation Requirements, In: *Buildings and Health, The Rosehaugh Guide To the Design, Construction, Use and Management of*

Buildings, S Curwell, C March and R Venables, pp. 176-177, RIBA Publications, ISBN 0 947877 78 9, London, UK.

Appleton, J. (1996). Ch 3 Behaviour and Environment. In: *The Experience of Landscape, revised edition*, pp 66, 250-254, John Wiley & Sons Ltd, ISBN 0-471-96233-3 (hbk), 0-471-96236-X (pbk) Chichester, UK.

Baird, G. (2001). Ch 13 Eastgate Centre, Harare, Zimbabwe, In: *The Architectural Expression of Environmental Control Systems*, pp. 164-180, Spon Press, ISBN 0-419-24430-1, London and New York.

Baker, N. & Standeven, M. (1994). Comfort criteria for passively cooled buildings: A PASCOOL task, *Renewable Energy*, Vol.5, No.5, pp 977-984.

Balashikov, Z. D. & Melikov, A. K. (2009). Methods for air cleaning and protection of building occupants from airborne pathogens, *Building and Environment*, Vol.44, pp. 1378-1385.

Beck, E-G. (2007). 180 Years of Atmospheric CO_2 Gas Analysis by Chemical Methods, In: *Energy & Environment*, Vol.18, No.2, pp. 259, 263 in 259-282.

Bernan, W. (1845). Essay XVII, In: *On the History & Art of Warming & Ventilation, Vol II*, pp 313-335 (p 319), George Bell, Fleet St., London, UK.

Billings, J. S. (1889). Ch 1, Introductory - Expense of Ventilation–Difference between "Perfect" and Ordinary Ventilation - Relations of Carbonic Acid to the Subject - Methods of Testing Ventilation, In *The Principles of Ventilation and Heating and their Practical Application*, 2nd Ed., pp 13-26 (pp 20-21), The Engineering & Building Record, New York, USA.

Billington, N. S (1982). Energy Efficient Domestic Ventilation Systems for Achieving Acceptable Indoor Air Quality, *Proceedings of 3rd AIC Conference*, London, UK, September 20-23, 1982.

Bitter, C. & van Ierland, F. F. A. A. (1965). Appreciation of sunlight in the home, *Proceedings of CIE Conference Sunlight in Buildings*, Newcastle, 1965; *Bouwcentrum International*, Rotterdam, 1967, pp 27-38.

Bloomer, K. C. & Moore, C. W. (1977). Ch 4 Some Twentieth-Century Models of Sense Perception, In: *Body, Memory, and Architecture*, p. 33, Yale University Press, ISBN 0-300-0231-9, New Haven and London.

Boardman, B. (1991). Ch 2 Emergence of Fuel Poverty, In: *Fuel Poverty, from cold homes to affordable warmth*, p. 19, Belhaven Press, ISBN 1-85293-139-6, London and New York.

Bomberg, M. (1989). Testing Water Vapor Transmission: Unresolved issues. *National Research Council Canada, Institute for Research in Construction*, NRCC 3091; reprinted from: Water Vapor Transmission Through Building Materials and Systems: Mechanisms and Measurement, *American Society for Testing and Materials, Standard Technical Publication*, 1039, pp 157-167 (IRC Paper No. 1615).

Boring, E. G. (1942a). Ch 12 Smell and Taste, In: *Sensation and Perception in the History of Experimental Psychology*, pp. 437-462, Appleton-Century-Crofts Inc., 660-6, New York, USA.

Boring, E. G. (1942b). Ch 13 Tactual Sensibility, op. cit., p. 487.

Boring, E. G. (1942c). Ch 1 Sensation and Perception, op. cit., p. 35.

Boring, E. G. (1942d). Ch 14 Organic Sensibility, op. cit., p 525, 534.

Boubekri, M. (2008). Ch 5 Psychology of Lighting, In: *Daylighting, Architecture and Health: Building Design Strategies*, p. 94, Architectural Press, ISBN 978-0-7506-6724-1, Oxford, UK.

Brown, R. (2006). Ch 14 Thermal Properties, In: *Physical Testing of Rubber*, 4th Ed., p. 279, Springer, ISBN 10 0-387-28286-6, New York, USA.

Brundage, J. F., Scott, R. McN., Lednar, W. M., Smith, D. W. & Miller, R. N. (1988). Building-Associated Risk of Febrile Acute Respiratory Diseases in Army Trainees, *Journal of the American Medical Association*, April 8, Vol.259, No.14, pp 2108-2112.

Buchbinder, L., Soloway, M. & Phelps, E. B., (1941). Studies on microorganisms in simulated room environments III: The survival rates of streptococci in the presence of natural daylight, sunlight and artificial illumination, *Journal of Bacteriology*, Vol.42, No.3, pp 353-366.

Carnelley, T., Haldane, J. S. & Anderson, A. M. (1887). IV The Carbonic Acid, Organic Matter, and Micro-organisms in Air, more especially of Dwellings and Schools, *Philosophical Transactions of the Royal Society*, Series B, 178, pp. 71-72 in 61-111.

Close, P. D. (1930). Preventing Condensation on Internal Building Surfaces, *ASH & VE Transactions* Vol.36, No.854, New York, USA.

Close, P. D. (1947). Transmission Coefficients and Tables, In: *Building Insulation*, 3rd Ed., p. 117, American Technical Society, Chicago, USA.

Day, R & Creed, C. (1996). Sunlight in Dwellings: Validation of Insolation criteria, Lighting Research Technology, Vol.28, No.1, pp.43-51.

Department for Education and Skills (2006). *Building Bulletin 101: Ventilation of School Buildings*, Version 1.4, 5th July 2006, p. 6.

Dixon, F. E. (1969). Some Irish Meteorologists, *Irish Astronomical Journal*, Vol.9, No.4, Irish Astronomical Society, Armagh, Northern Ireland, pp 116-117.

Downes, A. & Blunt, T. P. (1877). Researches on the effect of light upon bacteria and other organisms, *Proceedings of the Royal Society*, Vol.26, pp. 488-500.

Evans A. S. (1973). Pettenkofer Revisited, The Life and Contributions of Max von Pettenkofer (1818-1901), *Yale Journal of Biology and Medicine* Vol.46, pp 167 in 161-176.

Fanger, P. O. (1970). Thermal Comfort, Danish Technical Press, Copenhagen (Doctoral dissertation reprinted several times, inter alia McGraw Hill Book Company, New York).

Fanger, P. O. (1988). Introduction of the olf and the decipol Units to Quantify Air Pollution Perceived by Humans Indoors and Outdoors, *Energy and Buildings*, Vol. 12, pp 1-6.

Fitch J. M. (1972). *American Building, The Environmental Forces that Shape It*, Houghton Mifflin Co., Boston, Mass., ISBN 0-39512667-3, USA.

Fletcher, B. F. & Fletcher H. P. (1907). Ch XIII Ventilation, In: *Architectural Hygiene*, 3 Ed. Rev., pp. 144 in 137-158, Whittaker & Co., London.

Fourier, J. (1822). *Théorie Analytique de la Chaleur*, Firmin Didot Pere et Fils, Paris.

François de Chaumont, F. S. B. (1876). Supplementary Note on the Theory of Ventilation, *Proceedings of The Royal Society*, May 4, 1876, Vol XXV, pp 1-4.

Fung, J. W. (2008a). Ch 3 Methodology, In: *The Unintended Consequences of Decision-making in Glasgow's Social Housing Sector*, PhD Thesis, pp. (3)-6-(3)-8, Mackintosh School of Architecture, The Glasgow School of Art, University of Glasgow, UK.

Fung, J. W. (2008b). Ch 6 Decisions, Consequences and/or Conditions and Correlations, op. cit., pp. (6)-34 and (6)-82.

Fung, J. W. (2008c). Ch 7 Discussion of UNCs with Further Comparisons between Case Studies, op. cit., p (7)-20.

Gardner, B. (undated) Robert Koch, In: *Science and Technology, Famous People of Science*, http://web.ukonline.co.uk/b.gardner/Koch/htm.

Gibson, J. J. (1966). Ch II The Obtaining of Stimulation, In: *The Senses Considered as Perceptual Systems*, pp. 32 in 31-46, Houghton Mifflin, ISBN 0-313-23961-4, Boston, USA (Reprinted 1983, Greenwood Press, Westport, Connecticut, USA).

Giese, M., Bauer-Doranth, U., Langebartels, C. & Sandermann, H. (1994). Detoxification of Formaldehyde by the Spider Plant, *Plant Physiology*, Vol.104, pp 1301-1309.

Gifford, R. (2002). Ch 1 The Nature and Scope of Environmental Psychology, In: *Environmental Psychology, Principles and Practice*, 3rd ed., pp. 1-19, Optimal Books, ISBN 0-9688543-0-3, Colville, Washington, USA.

Gilfillan, J. B. S. (1958). Ch 1 The Site and its Development, In: *The Third Statistical Account of Glasgow*, pp. 30 in 17-47, William Collins and Sons C. Ltd., Glasgow, UK.

Gilgen, A. & Barrier A (1976). Part 4 Sunlight and natural lighting indoors, In: *Environmental factors in urban planning*, Grandjean, E. & Gilgen, A., pp 155-198, Taylor & Francis, ISBN 0 85066 084 X, London.

Ginkel, J. T. V. & Hasselaar, E. (2005). Housing characteristics predicting mould growth in bathrooms, *Proceedings of Indoor Air 2005*, 10th International Conference on Indoor Air, Sep. 4-9, Beijing, China, pp. 2425-2429.

Girman, J. R, Apte, M. G, Traynor, G.W. Allen, J. R. & Hollowell, C. D. (1982), Pollutant Emission rates from Indoor Combustion Appliances and Sidestream Cigarette Smoke, *Environment International*, Vol.8, pp. 213-221.

Haldane, J. S. (1901). The Work of Max von Pettenkofer, *Journal of Hygiene*, Vol.1, No.3, pp 293 in 289-294.

Hesselgren, S. (1975) *Man's Perception of Man-made Environment*, Dowden, Hutchison & Ross Inc, Stroudsburg, Pennsylvania, USA. (Note previous English translation of 1954 PhD thesis: 1967, 'The Language of Architecture', Studentlitteratur, Lund, Sweden.)

Hildebrand, G. (1991). Ch 6 Fallingwater, In: *The Wright Space: Pattern and Meaning in Frank Lloyd Wright's Houses*, pp. 104 in 93-105, University of Washington Press, ISBN 0-295-97108-8, Seattle, USA.

Ho, H. M. (1995). *User-Performance Sensitivity of Small Sunspaces in a Scottish Housing Context*, PhD Thesis, October 1995, Mackintosh School of Architecture, Glasgow, UK.

Institute of Medicine (2000). Executive Summary and Ch 5 Indoor Biologic Exposures, In: *Clearing the Air, Asthma and Indoor Exposures*, pp. 5 in 1-18 & 144, 173 in 105-222, National Academic Press, ISBN 0-309-06496-1, Washington, DC, USA.

International Textbook Co. (1909). Heating and Ventilation, In: *ICS International Reference Library*, pp 28-29, International Correspondence Schools Ltd., London, UK.

Keating, M. (1988). Ch 7 Jobs and Industry, In: *The City that Refused to Die, Glasgow: the politics of urban regeneration*, pp. 171 in 168-192, Aberdeen University Press, ISBN 0 08 034612-8, Aberdeen, UK.

Kuller R. & Lindsten C. (1992). Health and behaviour of children in classrooms with and without windows, *Journal of Environmental Psychology*, Vol.12, pp. 305-317.

Lawrence, M. G. (2005). The Relationship between Relative Humidity and the Dewpoint Temperature in Moist Air, *American Meteorological Society*, Feb. 2005, pp 225-232.

Leatherbarrow, D. (2009). Ch 6 Sitting in the City, In: *Architecture Oriented Otherwise*, pp. 164 in 143-173, Princeton Architectural Press, ISBN 978-1-56898-811-5, New York, USA.

Liddell, H., Gilbert, J. & Halliday, S. (2008). *Design and Detailing for Toxic Chemical Reduction in Buildings*, SEDA Design Guides for Scotland: No. 3.

Locher, W. G. (2007). Max von Pettenkofer (1818-1901) as a Pioneer of Modern Hygiene and Preventive Medicine, *Environmental Health and Preventive Medicine,* Vol.12, pp 238-245.

Lohr, V. I. & Pearson-Mims, C. H. (1996). Particulate Matter Accumulation on Horizontal Surfaces in Interiors: Influence of Foliage Plants, *Atmospheric Environment*, Vol.30, pp 2565-2568.

Lohr, V. I., Pearson-Mims, C. & Goodwin, G. (1996). Interior Plants May Improve Worker Productivity and Reduce Stress in a Windowless Environment, *Journal of Environmental Horticulture*, Vol.14, pp 97-100.

Markus T. A. & Gray A. (1973). Windows in low rise, high density housing - the psychological significance of sunshine, daylight, view and visual privacy, *Proceedings of Windows and their Functions in Architectural Design*, Technical University, Istanbul 22-27/10/73, pp. 1.1-1.32.

McGuirk, T. (2010). Christ's College, Guildford, The Architects' Journal, Vol.232, No.12, pp. 46-52.

Mennel, S., Menti, U-P. & Notter, G. (2007). Minergie-P® - a building standard for the future, *Proceedings of Clima 2007 WellBeing Indoors*, 10-14th June, Helsinki, Finland.

Ne'eman, E., Craddock, J. & Hopkinson R. G. (1976). Sunlight requirements in buildings: 1 Social survey, *Building and Environment*, Vol.11, pp. 217-238.

Ne'eman, E. (1977). Sunlight Requirements in Buildings - II. Visits of an Assessment Team and Experiments in a Controlled Room, *Buildings & Environment*, Vol.12, pp. 147-152.

Nicholls, P. (1924). A Heat Transmission Meter, *Industrial & Engineering Chemistry*, Vol.16, No.5, pp. 490-493.

Olgyay, V. (1963). Ch II The Bioclimatic Approach, In: *Design with Climate, a bioclimatic approach to architectural regionalism*, pp. 20 in 14-23, ISBN 0-442-01110-5, Princeton University Press, New Jersey, USA.

Osgood C. E., Suci, G. J. & Tannenbaum, P. H. (1957). Ch 3 The Semantic Differential as a Measuring Instrument, In: *The Measurement of Meaning*, pp. 76-124, University of Illinois Press, ISBN 0-252-74539-6, Urbana and Chicago, USA.

Overy, P. (2007). *Light, Air and Openness, Modern Architecture Between the Wars*, Thames & Hudson, ISBN 978-0-500-34242-8, London, UK.

Parkes, E. A. (1878). Ch 3 Ventilation, In: *A Manual of Practical Hygiene*, 5th Edition (edited F. S. B. Francois de Chaumont), pp. 140-147, J & H Churchill, London, UK.

Pasanen, P., Pasanen, A-L. & Jantunen, M. (1993). Water Condensation promotes Fungal Growth in Ventilation Ducts, *Indoor Air*, 1993, Vol.3, pp 106-113.

Philibert, J. (2006). Adolf Fick and Diffusion Equations, *Defect and Diffusion Forum*, Vol.249 (Diffusion in Solids - Past, Present and Future), pp. 1-6.

Platts-Mills, T. A. E. & De Weck, A. L. (1989). Dust mite allergens and asthma - A worldwide problem, *Journal of Allergy & Clinical Immunology*, Vol.83, pp. 422 in 416-427.

Port, M. H. (1976). Ch XI The Building Services, In: *The Houses of Parliament*, pp. 218-231, Yale University Press, ISBN 0-300-02022-8, New Haven and London.

Porteous, C. (1996). Airing energy efficiency: home truths, *alt'ing, The Scottish Journal of Architectural Research*, The Rutland Press, ISSN 1363-075x, Vol.1, No.1, pp. 17 in 17-28.

Porteous, C. D. A. & Ho, H. M. (1997). Do sunspaces work in Scotland? Lessons learnt from a CEC solar energy demonstration project in Glasgow, *The International Journal of Ambient Energy*, Vol.18, No.1, January 1997, pp. 23-35, ISSN 0143-0750.

Porteous, C. (2002a). Ch 2 Protagonists - aesthetic and tectonic attitude, Ch 3 Condensation factor, In: *The New eco-Architecture, alternatives from the modern movement*, pp. 9-10, 29-30, Spon Press, ISBN 0-415-25624-0, London and New York.

Porteous, C. (2002b). op. cit. Ch 7 Eco-footprint, Ch 11 Spirit of the Age, pp. 102, 177-178.

Porteous, C. (2002c). op. cit. Ch 7 Eco-footprint, pp 100-101.

Porteous C. & MacGregor K. (2005). Ch 4 Adaptive Control, In: Solar Architecture in Cool Climates, pp. 81-126, Earthscan, ISBN-10: 1-84407-281-9, London, UK.

Porteous, C. (2008). Shaping light and air, *Engineering & Technology*, Vol.3, Issue 9, pp. 52-55.

Porteous, C. (2010). Passive solar PassivHaus paradigm for Scotland in zero-carbon quest? Lessons from a study tour in Switzerland and Germany, August 2009, *Innovation Review* (CIC Start Online), Issue 2, March 2010, pp 32-46.

Porteous, C. D. A. & Menon R. A. (2010). Displacing electrical energy for drying domestic laundry by practical solar upgrades - proposed Glasgow housing case studies, *Proceedings Eurosun 2010 International Conference on Solar Heating, Cooling and Buildings*, Topic 1, 28 Sep - 1 Oct, Graz, Austria, 6 pages.

Qin, M., Belarbi, R., Ait-Mokhtar, A. & Nilsson, L-O. (2008). Nonisothermal moisture transport in hygroscopic building materials: modelling for the determination of moisture transport coefficients, *Transport in Porous Media*, March 72, No.2, pp 255-271.

Reid, D. B. (1844). CH IX On Vitiated Air, Pt IV: Ventilation of the Present Houses of Parliament, In: *Illustrations on the Theory and Practice of Ventilation with Remarks on Warming, Exclusive Lighting and the Communication of Sound*, pp 270-299, Longman, Brown, Green & Longmans, London, UK.

Robertson, J. D. (1888). A Study of Micro Organisms in Air, Especially those in Sewer Air, and a New Method of Demonstrating Them, *Abstract of Presidential Address, The British Medical Journal*, Dec 15, Vol.2, No.1459, pp. 1330-1334.

Rowe, C. & Slutzky, R. (1963). Transparency: Literal and Phenomenal, *Perspecta*, Vol.8, pp. 45-54.

Ryan, V. (2002). Condensation in Dwellings, *Journal of Environmental Health Research*, Vol.1, Issue 1, 7 pages.

Saluja, G. S., Porteous, C. D. A. & Holling, A. (1987). Experience of Passive Solar Heating in a Scottish Housing Scheme at 58°N Latitude, *Proceedings of Advances in Solar Energy Technology*, Eds. W H Bloss and F Pfisterer, 13-18 September, Hamburg, Germany, Pergamon Press, ISBN 0-08-034315-5, Vol.4, pp. 3686-3691.

Searle, G. F. C. (1934). Ch IX Experiments in Heat, In: *Experimental Physics, A Selection of Experiments*, pp. 273-284, Cambridge University Press, London, UK.

Sherrington, C. S. (1906). Lecture IV Interaction Between Reflexes, In: *The Integrative Action of the Nervous System*, pp. 114-149, Archibald Constable & Co. Ltd., London, UK.

Sippola, M. R. & Nazaroff, W. W. (2002). Modelling particle deposition in ventilation ducts, *Proceedings of Indoor Air*, 2002, pp. 515-520.

Slessor, C. (2007). Green Gauge, *The Architectural Review*, Vol.CCXXI, No.1322, pp. 72-75.

Straube, J. F. & deGraauw, J. P. (2001). Indoor Air Quality and Hygroscopically Active Materials, *ASHRAE Transactions*, Vol.107, Pt.1, 19 pages.

Stuart, J. (2006). Mind Over Mater, *Construct Ireland*, Issue 14, April, 2006.

Sturrock, N. & Lawson-Smith, P. (2006). The Grandfather of Air-Conditioning - The Work and Influence of David Boswell Reid, Physician, Chemist, Engineer (1805-63), *Proceedings of the Second International Conference on Construction History*, 29th March - 2nd April, Queens College, Cambridge, UK.

Sundell, J. (2004). On the History of Indoor Air Quality and Health, *Indoor Air 2004*; 14 (Suppl. 7), pp. 52 in 51-58.

Twinn, C. (2011). Natural Ventilation, *Ecotech*, 25, February 2011, pp. 10-25.

Varming, J. (1952). Vapour Transmission and Condensation, In: Architects' Year Book 4, J B Drew and T Dannatt: Eds, pp. 113-118, Paul Elek, London.

Vereecken, E., Roels, S. & Janssen, H. (2011). In situ determination of the moisture buffering potential of room enclosures, *Journal of Building Physics*, Vol.34, Issue 3, pp. 223-246.

Viitanen, H. & Ojanen, T. (2007). Improved Model to Predict Mold Growth in Building Materials, *Proceedings of Thermal Performance of the Exterior Envelopes of Buildings, ASHRAE THERM X*, Clearwater, Florida, USA, Dec. 2007, 5th of 8 pages.

Wall, K., Dockrell, J. & Peacey, N. (2008). Primary Schools: The Built Environment, *Research Survey 6/1, Primary Review*, Faculty of Education, University of Cambridge, p.11.

White, I. D., Mottershead, D. N. & Harrison, S. J, (1992). Ch 11 The Weathering System, In: *Environmental Systems, An Introductory Text*, 2nd Ed., p. 274. Chapman & Hall, London.

Wolverton, B. C. & Wolverton, J. D, (1993). Plants and Soil Microorganisms - Removal of Formaldehyde, Xylene and Ammonia from the Indoor Environment, *Journal of the Mississippi Academy of Sciences*, Vol.38, No.2, pp. 11-15.

Wolverton, B. C. & Wolverton, J, D. (1996). Interior Plants: Their Influence on Airborne Microbes Inside Energy-Efficient Buildings, *Journal of the Mississippi Academy of Sciences*, Vol.41, No.2, pp. 99-105.

Xu, P., Peccia, J., Fabian P., Martyny, J. W., Fennelly, K. P., Hernandez, M. & Miller, S. L. (2003). Efficacy of ultraviolet germicidal irradiation of upper-room air in inactivating airborne bacterial spores and mycobacteria in full-scale studies, *Atmospheric Environment*, Vol.37, pp 405-419.

Zarr, R. R. (2001). A History of Testing Heat Insulators at the National Institute of Standards and Technology, *ASHRAE Transactions 2001*, Vol.107, Pt 2.

9

One-Way ANOVA Method to Relate Microbial Air Content and Environmental Conditions

José A. Orosa
University of A Coruña
Spain

1. Introduction

During the past few years, people passed most of their lives indoors than ever before. While one part was spent in their homes, the other part was spent in their working environments, such as factories and offices.
As a consequence of spending their lives in these two environments, we found that people developed certain health-related symptoms, such as headache, fatigue, nausea and getting irritated with other people. When these symptoms detected in indoor environments, it is called sick building syndrome (SBS). On the other hand, when these symptoms are stronger and related to their workplace, it is called work risk hazard.
To identify these symptoms, sampling apparatus is employed. There is a growing interest in employing bioindicators that, after tests in the laboratory, can now be employed in real case studies.
In this chapter, a new methodology based on the statistical study of One-Way ANOVA was developed to test bioindicators, such as fungi, in real case studies. With this statistical study it was possible to relate bioindicators with indoor conditions like pets' presence, limited space and presence of localized humidity problems.

2. Problematic indoor ambiences

2.1 Sick building syndrome

The SBS is defined by the World Health Organization (WHO, 1983) as the occurrence of an increased prevalence of no specific symptoms among populations in determined buildings (Thörn, 1998).
Between its more important and common symptoms, we find eyes, nose and throat irritation, mental fatigue, headaches, nausea, skin irritation, irritability and lack of concentration (Hedge *et al.*, 1996; Raw *et al.*, 1996; Gupta *et al.*, 2007).
It is difficult to detect SBS due to its non-specific symptoms that appear with a higher prevalence than the expected value. Consequently, there is no clear consensus at the moment to define SBS and to determine if it is diagnosed by the exclusion of other causes.
On the other hand, subjective detection of SBS is being analyzed in recent research works. This detection is based on surveys and questionnaires about the perception of occupants of an indoor environment, such as that developed previously to analyze indoor thermal comfort (ASHRAE, 2003). All the subjective and objective parameters are analyzed in depth below.

As stated above, there are some parameters that can define SBS. To control these parameters like indoor air temperature, relative humidity and dust, amongst others, are some of the questions that must be asked.

The first question is if there is a need to employ any natural or mechanical ventilation. After a review of recent research works on this topic, it was concluded that there is a prevalence of some symptoms when the mechanical ventilation is working.

These symptoms are, for example, some upper respiratory problems and fatigue. The results showed that these effects can be related to the dust deposition in ducts, humidifiers and chillers of the HVAC system. On the other hand, the air remains clean when it is introduced into the house through natural ventilation, such as doors and windows.

The corrective actuation way to relax these symptoms is by HVAC system corrections and building design improvements. For example, recent research works like that of Kolari *et al.* (2005) showed that duct cleaning will not imply an improvement of indoor air quality, but is related to an improvement of indoor perception of air quality and its relation to nasal symptoms. Furthermore, a clear reduction, in the building, of the volatile organic compounds (VOC), carbon dioxide (CO_2) and fungal spore concentrations reach values below the actual after a duct cleaning process.

At the same time, the percentage of outdoor air that can be introduced into the indoor ambience must be considered. In this sense, it is very important to remember that, in the last few years, to reduce the energy consumption in HVAC system the air changes of indoor air was reduced (Osayintola & Simonson, 2006); consequently, a percentage of indoor air returns to the same indoor ambience mixed with outdoor air with a corresponding increment in CO_2 and moist air humidity.

We must consider that air changes are due to mechanical ventilation and infiltrations of outdoor air. In this sense, the mechanical ventilation must be improved and buildings must be less airtight. What is more, an increment in air changes of indoor air implies an improvement of indoor air quality, as reported by Haghighatt & Donnini, 1999.

Finally, outdoor air is not the only source of contaminants of indoor environments. Other parameters, such as indoor materials contamination must be considered as a source of pollution.

Once the objective parameters to detect the SBS (WHO, 1989) are defined, it is the right moment to define the subjective parameters. Among the main subjective parameters, we must consider job stress and perception of the indoor air quality.

Studies by Hedge *et al.* (1996) showed a clear relationship between SBS and working conditions than with environmental parameters. Furthermore, parameters such as type of job can influence the perception of SBS.

The other parameter, which is perception of indoor air quality, must be considered. In this sense, we find that the perception of indoor air quality depends not only on combustion gases, VOC, dust, particles and but also on parameters like mucous membranes temperature and humidity that can alter these perceptions (Salonvaara & Simonson, 2000 and Simonson *et al.*, 2001). Other parameters such as air movement perception can alter this situation too.

Finally, to control the subjective parameters, we can employ questionnaires that are recognized as a more important tool to define the relationship between objective and subjective perception.

These surveys must cover an area for objective parameters that must be sampled, such as indoor air temperature and relative humidity and, at the same time, the same survey must

be present in another region for the subjective parameters that must question the occupants of a building, such as indoor air perception and job stress.
The SBS detected by symptoms or by some bioindicators must be corrected. In other words, we must consider that SBS is related with building construction and not necessarily with the occupants. Consequently, parameters like air temperature, relative humidity and dust are sampled to define the SBS.

2.2 Objective detection of indoor air quality

Despite the fact that indoor air quality can be sampled with different apparatus like multi-gas samplers, it was learned that, in the past few years, there is an increasing interest in employing some natural indicators and monitors because they present some advantages with respect to most of loggers.
The main advantages of these indicators are based on the fact that they do not need any kind of calibration or energy source. It is the work of fungi to evaluate indoor environments and mosses and lichens to evaluate outdoor ambiences.
From these concepts, we do an initial definition of bioindicator and biomonitor. A bioindicator is an organism that can be used for identification and qualitative determination of human-generated environmental factors.
At the same time, we can define accumulative bioindicators which have the ability to store contaminants in their tissues and are used for the integrated measurement of the concentration of such contaminants in the environment, as a result of the equilibrium process of biota compound intake/discharge from and into the surrounding environment.
Biomonitors are organisms used for the quantity determination of contaminants and can be sub-classified as sensitive and accumulative. The methodology based in biomonitors present the problem of the need of a background level of this contaminant in the environment objective of study.

3. Detection of indoor and outdoor air quality

3.1 Fungi

Owing to its feasibility to be employed in a real case study, once defined, a few examples of biomonitors and bioindicators are explained below. To sample outdoor ambiences, we can employ mosses and lichens, and to sample indoor ambiences we can employ fungi.
Despite the fact that fungi develop in nature some functions, such as recycled energy and nutrients, most of these tasks are not adequate if it is to be developed in indoor environments.
It is due to fungi being related to spores, fungal fragments, mycotoxins and VOCs emissions is the reason for the failing health of the occupants. In this sense, nowadays we find indoor environments that present higher VOCs concentration than in years before and related to the fact that building designs were modified in accordance with energy saving, and consequently, buildings are more airtight and present a low number of air changes.
When most researchers tried to find where fungi were located in indoor environments, the results showed that it developed on walls, roofs and in materials wherever it can be find dust to develop a growing media.
This fungi development increased with higher air temperatures and relative humidity values of over 75%. This is the higher value of relative humidity that must never be passed to prevent fungi developing in indoor environments.

At the same time, fungi emit mycotoxins that are low molecular weight compounds and toxic for animals and men (Cabral , 2010).

3.2 Mosses and lichens

To analyze outdoor air quality, bioindicators, such as lichens and mosses, are employed due to they present some advantages respect traditional sampling methods. For example, these bioindicators were selected as they do not present any seasonal variation and their longevity. Consequently, these bioindicators let us sample indoor conditions for long periods without calibration and without any kind of energy source, which is a clear advantage respect traditional loggers.
Lichens are defined as a symbiotic association (Newbound *et al.*, 2010) of a fungus and an alga, and can be employed to develop a map of all the species detected in a sampling area. Another method is basically in the sampling process of pollutants in the thallium of the lichen.
Other methods are based in the transplantation of native lichens to a place where it will be killed by pollutants after a reduced period of time. Finally, new methods are being developed to define the climate change with different lichens sampling processes over calcareous rocks.
As a function of previous sampling processes, two indexes can be defined based in lichens measurements. The first index is called the index of atmospheric purity and the second is the index of poleotolerance.
Mosses are employed to define, at the same time as lichens, outdoor air quality (Szczepaniak & Biziuk, 2003). Mosses present some advantages as bioindicator as, for example, it can be employed in different regions due to their growth in different environmental conditions, such as industrial and urban areas.
Another advantage of mosses is based on the fact that sampling process is cheap and simple. It will allow a very large number of sites to be sampled, obtaining a better sampling map. At the same time, these natural indicators present some disadvantages. For example, to develop a comparative study with mosses and lichens, the same species of lichens and trees are needed to obtain adequate results to do a comparative study.
On the other hand, fungi growth in indoor environments depends on indoor temperature and relative humidity in all zones of the building. Consequently, one of the methods to control fungi growth in indoor ambiences is to control occupants' habits. For example, most researchers proposed an increment in the air changes during cooking in the kitchens, in the bathroom during mornings and in bedrooms during the nights. This increment in air changes can be obtained by natural ventilation through open windows.

4. Practical case study in flats

Recent research works is related to airborne microorganisms with some infections or allergic disorders (Parat *et al.*, 1997), and some epidemiologic studies have been related to dust mite exposure with some degree of asthma. In general, however, we can say that in ambiences with higher allergen exposure we find higher asthma prevalence (Liu, 2004).
Despite this, there is not always a relationship between fungi development and asthma in children. Consequently, some researchers reached the conclusion that, with actual methodologies, it cannot be done (Jovanovic *et al.*, 2004).
Nowadays, there is very little information on how to prevent allergies, such as environmental hygiene, avoidance of some foods and prevention of contact with some kind of pets.

However, things are more complicated and not the same for all allergens. Although increased exposure to house dust mite allergen is paralleled by increased sensitisation rates, the same is not true for cat allergen.

One possible explanation for the different effects of different allergens may be their biochemical properties: mite allergens, in contrast to cat and dog allergens, contain proteolytic enzymes. It has been shown that, concerning house dust mites, a low-allergen environment can be achieved (Lauener, 2003).

To summarize, we can say that new standards are needed to show the better methodology to sample indoor ambiences and define the effect of indoor pollutants over health to reduce sensitization to these parameters (Lauener, 2003).

Other factors like distance of the building from the source (a nearby park) and supermicrometre particle concentrations will be associated with the concentration levels of fungi in indoor ambiences of occupied buildings (Hargreaves *et al.*, 2003).

At the present time, the only way to guarantee lower mite allergen levels in modern homes in the western world is to remove the carpets and to encase the mattresses and beddings. Furthermore, to reduce this exposure, we must improve IAQ.

The three primary considerations in improving IAQ are (1) evaluation of construction failures that allow moisture into the walls and roofs, (2) poor ventilation, causing excessive humidity and accumulation of gaseous and/or chemical exposure from materials in the living space, and (3) poorly designed or failing HVAC systems that contribute to poor air calculation.

About the two last points, some authors (Parat *et al.*, 1997) have analyzed that massive proliferation of microorganisms may take place in HVAC unit with certain risk factors, such as low efficiency filters, cold mist humidifiers using water recycling, areas in which condensation water remains stagnant, large recirculation of air and faulty or deficient maintenance conditions.

They demonstrated that compared to a naturally ventilated building, a HVAC system which is well designed and well maintained improves the microbiological quality of indoor air and, in consequence, can reduce health hazards for its occupants.

Finally, not all indoor allergens are necessarily equal in their propensity to cause asthma and its related health effects like shortness of breath and coughing. For example, using dust allergen concentration as a proxy for exposure, recent studies have revealed that indoor cockroach allergen exposure, but not mite or cat allergen exposure, is a significant risk factor for asthma (Hens, 2007).

Despite the fact that the data collected on household characteristics varied greatly between the studies and that building materials techniques are very different in different parts of the world, some common themes have emerged.

For example, concentrations of moulds varied hardly between areas (Jovanovic *et al.*, 2004), and neither climatologically conditions nor differences between urban and rural regions exhibited a systematic influence.

In another example (Perfetti *et al.*, 2004), no association was found between the concentration of mite allergens and the environmental characteristics (geographic location, floor above ground, type of ventilation) and no correlation was found between indoor humidity and allergen levels.

We must consider the fact that house dust mites live in an environment where there is no liquid water, and they are dependent on the ambient humidity to absorb water from the atmosphere. To get this, water dust mites can gain it by diffusion through the body or extract the water vapor from air via hygroscopic crystals in their supracoxal glands, located at the base of their first pair of legs.

The optimum relative humidity for mite growth is 75-95%, at temperatures of 15-30ºC, whereas above 70% relative humidity conditions may be optimal for fungal growth (Liao *et al.*, 2004). Relative humidity has a major influence on the survival of mite colonies and therefore levels of mite allergens.

Although laboratory and early field studies suggested that there was a strong relationship between relative humidity and mite allergen levels, this had not been conformed by more recent large-scale studies when other factors have been considered in a multivariate analysis.

For example, freezing and/or dry weather can damage fungi and reduce the spore counts on outdoor samples, but the conditions indoors may be very hospitable to fungal growth non-seasonally (Zhou *et al.*, 2000).

From these studies, we can conclude that outdoor relative humidity influences indoor relative humidity, but other household factors can influence mite allergen levels and that means allergen levels in different geographical areas tend to be influenced by the local climate.

As a result of this, novel techniques have been developed recently, which allows measurements of relative humidity to be made within the mite microhabitat, that is, where it matters, in the depth of the carpet or mattress.

This has revealed that the relative humidity in the carpet may be higher than that in the room air, and that with different types of construction, the differences between room RH and floor RH will vary. This suggests that the relative humidity in the room air does not necessarily reflect the RH in the micro-habitat of the mite—in the depth of the carpet pile.

The object of this chapter was to get information about Spanish apartments' microbial levels and relate it with their characteristics. Results will be useful to get a healthy home, taking into account costs versus energy saving, and improve health outcomes (Bernstein *et al.*, 2008).

4.1 Materials and methods

4.1.1 Apartments

In our case study, different apartments located in the northwest of Spain were selected, in accordance with different criteria defined to obtain realistic comparisons between indoor ambiences.

The first criterion is that, in all apartments, the residents presented some kind of health problems related with the relative humidity, which is typical in this area. The reason for this work is to relate health problems with indoor conditions, in building constructions' humid areas and occupants' habits. In particular, fungi and bacterial growth were sampled in these indoor environments.

All apartments present natural ventilation to remove all indoor air. Despite this, other mechanical ventilation system was located in toilets to reduce the humidity released during bath.

The heating system consists of heat water radiators and is employed only for a few months in the winter season. On the other hand, there is no cooling system, as the temperatures during the summer season are not too high. Consequently, during summer ventilation is enough to reduce indoor temperature.

To obtain adequate comparative results in apartments, sampling process was developed in accordance with the daily life conditions indicated by occupants. Furthermore, all buildings present the same construction and located in the same city. Consequently, outdoor weather conditions were the same for all the buildings.

Outdoor weather conditions were sampled by some weather stations located in a representative zone of the area where the buildings are located. This sampling process was developed by weather stations from MeteoGalicia (2002), with a sampling and frequency of 10 minutes.

On the other hand, humidity and temperature were measured by a 1221 Datalogger, with sensors of temperature and relative humidity, and tinytag Plus 2 dataloggers were employed.

These loggers were located in each apartment, in accordance with the ISO Standard indications. In particular, each sampling point was separated from heat sources conditions and as near as possible of center of gravity each room to obtain representative values of indoor

As explained earlier, a microbiological analysis off indoor ambiences was done. Consequently, two culture media were employed: Trypticase Soy Agar to find the total number of bacteria and Malt Agar was used to define fungi growth.

4.1.2 One-way ANOVA

To compare sampled mean values of temperature and relative humidity for fungi and bacteria, a statistical study of one-way ANOVA was done. This statistical study consists in an analysis of the variance of one factor for a significance level of 0.05.

Furthermore, different statistical studies, such as Duncan and Student-Newman-Keuls *post hoc* analyses, let us define groups of apartments that present the same condition for this level of significance.

In this study, two assumptions were defined. The first is based on the fact that the dependent variable is normally distributed. The second is that the two groups have approximately equal variance on the dependent variable.

In the same study, two hypotheses were considered. The hypothesis null is that there are no significant differences between the groups' mean scores. The alternate hypothesis is that there is a significant difference between the groups' mean scores. Finally, to develop this task, the statistical software SPSS 11.0 was employed. More information on how to employ this software SPSS can be found in their website (SPSS).

5. Results and discussion

At this point, the main results obtained are shown. In particular, Table 1 shows us the main indoor and outdoor air temperature and relative humidity of 25 apartments during the sampling process.

Despite this, we must consider the fact that this table shows the main value of sampled data, and consequently, conclusions about instantaneous values at different hours cannot be obtained from this table.

From Figs. 1, 2 and 3, we can conclude that Coruña, located in the northwest coast of Spain, presents a mild climate. In this sense, we can see that outdoor temperature is not too high in summer and too low in winter season. Mean temperature values of 11°C during winter and 16°C during summer, respectively, can be expected.

Flat	tindoor (°C)	RHindoor	toutdoor (°C)	RHoutdoor
A	20.97	63.45	16.72	74.6
B	24.09	65.09	17.76	88.8
C	19.42	62.1	14.12	82
D	20.38	64.87	17.7	59.6
E	21.46	63.59	17.7	62
F	23.43	65.03	22.28	66.8
G	22.2	70.92	16.24	94
H	19.92	63.73	15.64	75
I	21.24	49.62	18.08	45
J	23.78	55.7	17.76	47
K	25.11	48.09	17.6	58.8
L	23.63	65.58	20.04	73.2
M	22.37	67.19	19.28	78.4
N	21.74	63.65	15.6	74.76
O	24.05	50.31	15.2	74.2
P	20.29	59.22	12.08	88
Q	20.32	62.23	11.84	88
R	20.1	62.66	14.4	89.2
S	17.4	69.88	14.4	87.4
T	19.42	64.79	16.6	72
U	21.13	61.1	15.8	76
V	22.63	65.22	21.4	59.4
W	25.12	61.18	21.64	64.4
X	19.14	64.56	14.84	78
Y	23.83	65.65	21.5	60.8

Table 1. Indoor/outdoor sampled variables.

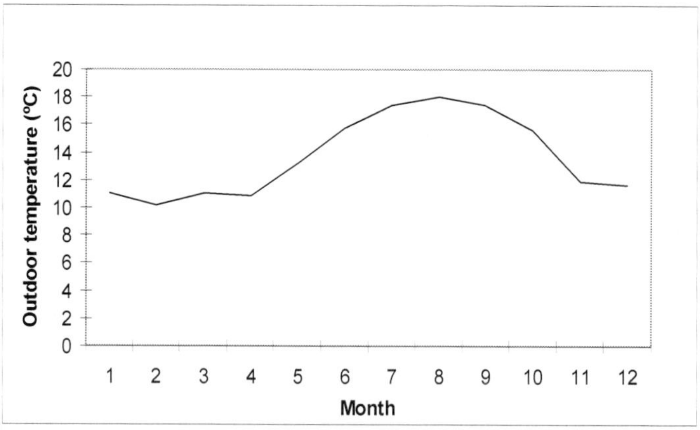

Fig. 1. Outdoor temperature.

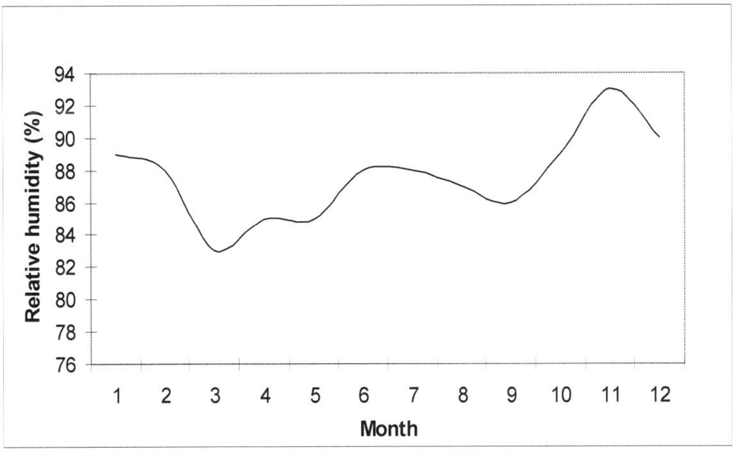

Fig. 2. Outdoor relative humidity.

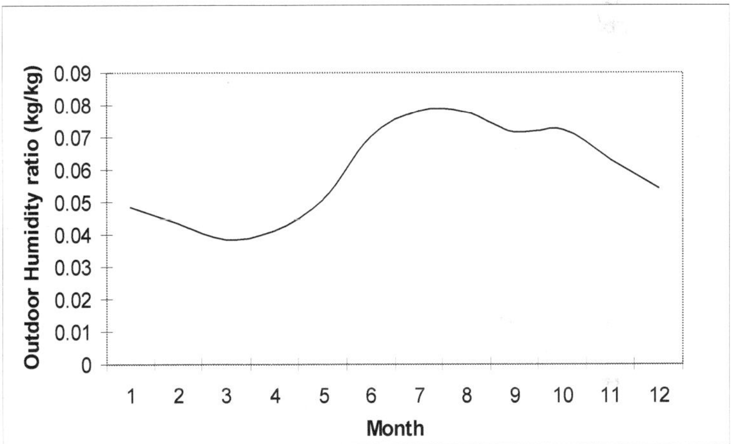

Fig. 3. Outdoor humidity ratio.

On the other hand, outdoor relative humidity showed mean values between 83% in March and 93% in November. Consequently, mean outdoor relative humidity of 86% is obtained throughout the year, as we can see in Fig. 2.

Finally, another way to show the relationship between temperature and relative humidity was to express the outdoor air humidity ratio, as we can see in Fig. 3. This humidity has shown yearly values between 0.04 and 0.08 kg of water per kilogram of dry air.

This high humidity is incremented with different indoor moisture sources. This increment of humidity ratio under temperatures of 21°C will imply relative humidity values over 75% in some daily life periods, as we can see in Fig. 3.

From Table 1, we can conclude that the northwest of Spain present apartments with an indoor mean relative humidity about 62% and temperature of about 21.7°C. Thus, value is relatively high, but not excessively high for a coastal area. However, due to an indoor relative humidity of not more than 75% with an adequate cleaning procedure, development of fungi can be prevented.

Flat	Characteristics
A	Pet
B	Normal
C	Normal
D	Limited space
E	Normal
F	Humidity problems
G	Humidity problems
H	Limited space
I	Normal
J	Pet
K	Humidity problems
L	Limited space
M	Normal
N	Humidity problems
O	Normal
P	Normal
Q	Pet
R	Normal
S	Normal
T	Humidity problems
U	Normal
V	Normal
W	Humidity problems
X	Normal
Y	Normal

Table 2. Observed characteristic.

At the same time, temperature and relative humidity were sampled and different characteristics of each apartment flood were considered (Table 2). This Table shows us parameters like pets' presence, limited space and presence of localized humidity problems in the walls and roofs that were considered.

Finally, if none of previous commented parameter was detected, then the apartment was considered normal.

Once Tables 1 and 2 have shown us the main value of temperature, relative humidity and apartment characteristics, it is the right moment to show the results of fungi and bacteria growth, as we can see in Figs. 4 and 5.

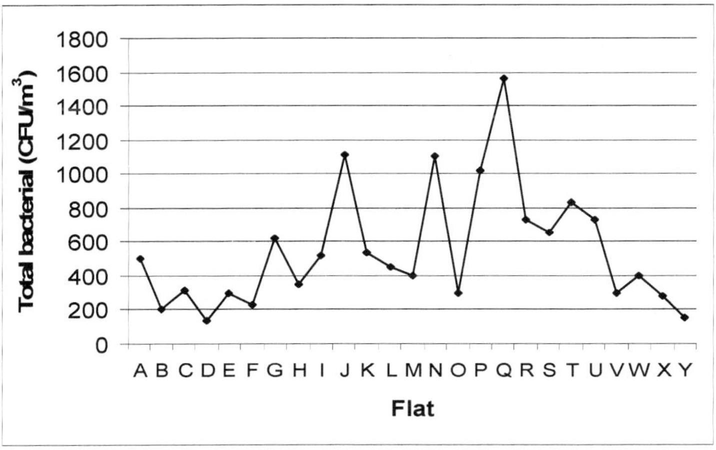

Fig. 4. Total bacteria sampled (CFU/m^3).

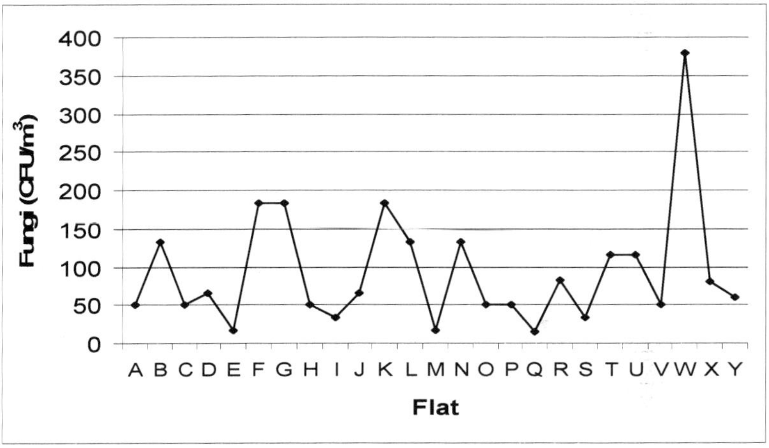

Fig. 5. Total fungi sampled (CFU/m^3).

After fungi and bacteria growth in these indoor environments are sampled to relate these pollutants with indoor relative humidity, like in most of laboratory studies (NTP 335, 2008), it is necessary that this is applied to real case studies. In particular, in this chapter a relationship between fungi and bacteria with the particular parameters detected in each building, reflected in Table 2, was proposed.

From this study, we see that there is no possibility to obtain an adequate linear regression between humidity and fungi reflected by a correlation factor below 0.9, see Figure 6.

It is related with the fact that, in real buildings, there are other parameters that does not influence in most laboratory studies, but in real case studies it can alter the situation. It is the case of the presence of pets, moisture sources and moisture-damaged walls and roofs.

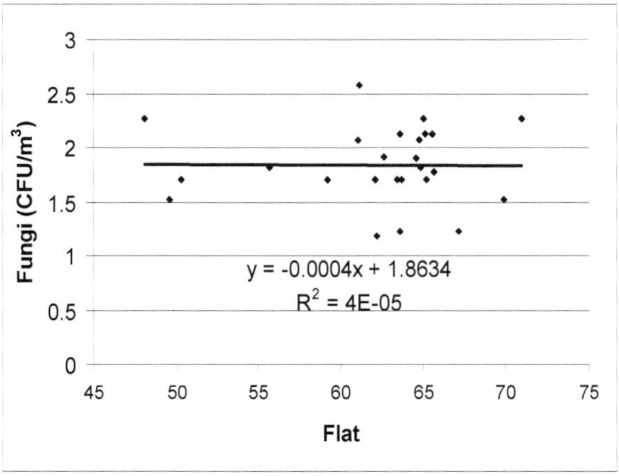

Fig. 6. Fungi linear regression.

To relate these indoor conditions with sampled indoor parameters, it was proposed to develop one-way ANOVA analyses with different *post hoc* studies to define which groups of indoor environments experience the same evolution with time.

In particular, the Duncan *Post hoc* analysis with mean total bacteria and fungi was developed as we can see in Tables 3 and 4.

From the one-way ANOVA analysis, we can concluded that apartments having pets showed the same indoor air bacteria evolution with time and, consequently, can be separated as an independent group, as we can see in Table 3.

A	J	Q	B	C	E	I	M	O	P	R	S	U	V	X	Y	D	H	L	F	G	K	N	T	W
Group 1: Pets			Group 2: Normal, limited space and humidity problems																					

Table 3. One-way ANOVA and Duncan *post hoc* with mean total bacterial (CFU/m^3).

A	J	Q	B	C	E	I	M	O	P	R	S	U	V	X	Y	D	H	L	F	G	K	N	T	W
Group 1: Normal, limited space and pets																			Group 2: Humidity problems					

Table 4. One-way ANOVA and Duncan *post hoc* with mean fungi (CFU/m^3).

On the other hand, it was concluded that there exists a clear different indoor air fungi developed in apartments that present some humidity problems on walls and roofs with respect to others, as we can see in Table 4.

6. Conclusions and future research works

This research work tried to relate indoor air conditions with fungi and bacteria growth. In this sense, objective and subjective parameters were considered. So, parameters like indoor and outdoor temperature and relative humidity were sampled in relation to bacteria and fungi growth. At the same time, parameters such as presence of pets and humidity problems in walls and roofs were considered too.

The results showed us that it is not easy to relate fungi growth with indoor air relative humidity like in laboratory studies. It is owing to the fact that there are some factors that can alter this situation. Furthermore, pets' presence was related to the increment in bacteria in indoor air, and humidity problems were related with fungi developed in a statistical way.

In conclusion, we can say that one-way ANOVA is an interesting tool to be employed by engineers to approach real case studies with laboratory conclusions.

7. Acknowledgement

I express my gratitude to INEGA, the University of A Coruña, Xunta de Galicia and all individuals and institutions that collaborated during the writing of this chapter.

8. References

ASHRAE Standard 55P -*Thermal Environmental Conditions for Human Occupancy*, (2003). American Society of Heating, Refrigerating and Air-conditioning Engineers, Atlanta, USA.

Bernstein, JA, Alexis, N., Bacchus, H., Bernstein, L., Fritz, p., Horner, E., Li, N., Mason, S., Nel, A., Oullette, J., Reijula, K., Reponen, T., Seltzer, J., Smith, A. & Tarlo, S. (2008). The health effects of nonindustrial indoor air pollution. *American Academy of Allergy, Asthma & Immunology*. Article in press.

Cabral, JPS. (2010). Can we use indoor fungi as bioindicator of indoor air quality? Historical perspectives and open questions. *Science of the Total Environment*. Vol.408, pp.4285-4295.

Gupta, S, Khare M & Goyal R (2007) Sick building syndrome- A case study in a multi-storey centrally air-conditioned building in the Delhi City. *Building and Environment*, Vol.42, pp. 2797-2809.

Haghighat, F & Donnini, G (1999) Impact of psycho-social factors on perception of the indoor air environment studies in 12 office buildings. *Building and Environment* Vol.34, pp. 479-503.

Hargreaves, M., Parappukkaran, S., Morawska, L., Hitchins, J., He, C. & Gilbert, D. (2003) A pilot investigation into associations between indoor airborne fungal and non-biological particle concentrations in residential houses in Brisbane, Australia. *The Sci Total Environ*. Vol. 312. pp.89-101.

Hedge A, Erickson WA & Rubin G (1996) Predicting sick building syndrome at the individual and aggregate levels. *Environmental International* Vol.22, No.1, pp.3-19.

Hens H. (2007). Indoor climate in student rooms: measured values. IEAEXCO energy conservation in buildings and community systems annex 41 "moist-eng" Glasgow meeting.

Jovanovic, S., Felder-Kennel, A., Gabrio, T., Kouros, B., Link, B., Maisner, V., Piechotowski, I., Schick, K., Schrimpf, M., Weidner, U., Zöllner, I. & Schwenk, M. (2004). Indoor fungi levels in homes of children with and without allergy history. *International Journal of Hygiene and environmental Health*. Vol.207.pp.369-378.

Kolari S, Heikkilä-Kallio U, Luoma M, Pasanen P, Coronen P, Nykyri E & Reijula K (2005) The effect of Duct clearing on perceived work environment and symptoms of office employees in non-problem buildings. *Building and Environment* Vol.40, pp.1665-1671.

Lauener R.P. (2003). Primary prevention of allergies. Revue française d`allergologie et d`immunologie clinique. Vol.43. pp.423-426.

Liao C.M., Luo W.C., Chen S.C., Chen J.W. & Liang H.M. (2004).Temporal/seasonal variation of size-dependent airborne fungi indoor/outdoor relationship for a wind-induced naturally ventilated airspace. *Atmos Environ.* Vol. 38 pp.4415-4419.

Liu A.H. (2004) Something Old, something New: Indoor endotoxin, allergens and asthma. *Paedriatic Respiratory Reviews.* Vol 5 (Suppl A), pp.65-71.

MeteoGalicia. Anuario climatolóxico de Galicia (2002). Consellería de Medio Ambiente. *Xunta de Galicia.* ISBN: 84-453-3520-0.

National Institute of Safety and Hygiene at Work. (2008). NTP 335: Indoor air quality: pollen grains and fungi spores evaluation. http://www.mtas.es/insht/ntp/ntp_335.htm.

Newbound, M., Mccarthy, MA & Lebel, T. (2010). Fungi and the urban environment: a review. *Landscape and urban planning,* Vol. 96, pp.138-145.

Osayintola, O. F. & Simonson, C.J. (2006) Moisture buffering capacity of hygroscopic buildings materials: Experimental facilities and energy impact. *Energy and Buildings.* Vol. 38. pp. 1270-1282.

Parat, S., Perdrix, A., Fricker-hidalgo, H., Saude, I., Grillot, R. & Baconniers, P. (1997) Multivariate analysis comparing microbial air content of an air-conditioned building and a naturally ventilated building over one year. *Atmospheric environment.* Vol. 31. pp. 441-449.

Perfetti, L., Ferrari, M., Galdi, E., Pozzi, V., Cottica, D., Grignani, E., Minoia, C. & Moscato C. (2004) House dust mites (Der p 1, Der f 1), cat (Fel d 1) and cockroach (Bla g 2) allergens in indoor work-places (offices and archives). *Science of the Total Environment.* Vol. 328. pp.15-21.

Raw, GJ, Roys, M AS & Whitehead, C and Tong, D (1996) Questionnaire design for sick building syndrome: an empirical comparison of options. *Environmental International.* Vol.22, No.1, pp.61-72.

Salonvaara, MH & Simonson, CJ. (2000). Mass transfer between indoor air and a porous building envelope: Part II. Validation and numerical studies. *Proceedings of Healthy Buildings,* Vol. 3.

Simonson, CJ, Salonvaara, M & Ojalen, T. (2001) Improving indoor climate and comfort with wooden structures. Espoo 2001.Technical Research Centre of Finland, VTT Publications 431.200p.+ app 91 p.

Szczepaniak, K. & Biziuk, M. (2003) Aspects of biomonitoring studies using mosses and lichens as indicators of metal pollution. *Environmental Research.* Vol.93, pp.221-230.

Thörn, A (1998) The sick building syndrome: A diagnostic dilemma. *Social Science & Medicine* vol.47 No.9, pp.1307-1312.

World Health Organization (1983). Indoor air pollutants: exposure and health effects: report on a WHO meeting. WHO Regional Office for Europe. (EURO reports and studies 78). Copenhagen:

World Health Organization. (1989). Indoor air quality: organic pollutants. WHO Regional Office for Europe (EURO Report and Studies I 111), Copenhagen.

Zhou, G., Whong, W.Z., Ong, T. & Chen, B. (2000).Development of a fungus-specific PCR assay for detecting low-level fungi in a indoor environment. *Molecular and Cellular Probes.* Vol 14. pp.339-348.

10

Indoor Air Quality - Volatile Organic Compounds: Sources, Sampling and Analysis

Alessandro Bacaloni, Susanna Insogna and Lelio Zoccolillo
Department of Chemistry, University of Rome "La Sapienza"
Italy

1. Introduction

Since the 70s, research has found in Europe and in the United States that individuals spend between 70 and 90% of their time indoors. Health studies have found that exposures to a variety of air pollutants indoors can be substantially higher than outdoors, even in urban environment. Volatile Organic Compounds (VOCs) are often considered among the more important indoor pollutants, because of by their continue emission from many sources and their diffusion properties. With the aim to evaluate the occupants' discomfort and health effects and in order to develop guidelines and standards, Indoor Air Quality (IAQ) assessment and control is an essential step. IAQ assessment will complain:

- Sources: Identification and characterization of sources, as emissions from materials, products or activities, is best done under laboratory conditions; so it is possible measuring rates of emissions (especially chemicals such as VOCs). Exposure characterization is the second level of source identification; after the measurement of contaminants' concentrations in controlled environment, characterised by known sources, adsorbing and absorbing surfaces, these data can be used in validation of current exposure models.
- Sampling methods: In order to determine concentrations of VOCs and exposures of building occupants via inhalation, field studies can be carried out by sampling methods (and analysis) in accordance with existing official methods (EC, NIOSH, OSHA, ACGIH, etc.). This way may be expensive and cumbersome; in addition, it can be not exhaustive in predicting the discomfort and health impact. A greater number of perspectives are offered by using some "descriptors" that can be more adequate in characterizing anthropogenic pollution. Specific sampling methods may be reserved for contaminants with specific toxic effects (e.g., formaldehyde, benzene, monomers, etc.). Measurements of specific contaminants' can be necessary for sources that cause high room concentrations for relatively short periods (e.g., in case of freshly applied coatings on walls, etc.). Currently, diffusion (passive) samplers are mainly used in order to evaluate long term exposures (days to weeks and more)
- Analysis and data meaning: existing analytical methods are validated and generally show adequate limits of detection (LODs) and limits of quantification (LOQs) even in measuring subtoxic contaminants' concentrations. Analytical methods for the more

frequent indoor contaminants are presented and commented. More difficulties lie in the interpretation of results, due to the limited indications suggested by European and International Standards (such as EN, EN ISO, etc.) that will be discussed. The difficulty in data evaluation is growing depending on the simultaneous occurrence of contaminants at subtoxic concentration.

In this chapter, there will be presented the way to handle the problem in IAQ assessment, and some practical applications, in order to provide the logical pathway to face the majority of actual cases.

2. Indoor air quality and VOCs

Research on pollutant sources is needed to identify pollutants and emission levels from building materials and other products. Work in this area will serve two purposes: (1) providing measurement protocols and data to employ in exposure reduction actions, and (2) providing correlations between research on health effects and pollutant sources of contaminant critical levels. Finally, research is required to determine definition, causes of, and solutions to multiple chemical sensitivity (MCS). Identifying the physiologic nature of MCS is the first step in understanding whether and how IAQ eventually contributes to the syndrome. Maroni et al. (1995) recommended the following definitions that should be used in the description of indoor related complaints and illnesses:

- Building-Related Environmental Complaints (BREC): Complaints of poor IAQ or poor indoor air environment. BREC are usually registered in the complaint (annoyance) part of questionnaire studies.
- Building-Related Symptoms (BRS) or Building Related Health Complaints (BRC): The health complaints (subjective symptoms) reported by the single individual as occurring inside a building and usually subsiding shortly after leaving it.
- Sick Building Syndrome (SBS): denotes a situation where a significant number of the occupants of a building complain of a typical group of general, unspecific, and irritating symptoms, including particularly headache, lethargy, dry eyes, blocked nose, and sore throat. The symptoms usually fade after the person has left the indoor environment but the specific cause is unidentified.
- Building-Related Illness (BRI): clinical condition with defined symptoms and signs in which the cause (aetiology) is building-related and identifiable.

The difference between BRI and SBS is that the building problems are identified in the former and undiagnosed in the latter. The term "sick building" probably should not be used on its own, but it might be replaced with the expression "building with indoor climate problems" or "problem building."

VOCs are often more important in assessing IAQ because of their ubiquity; VOCs may be used like “descriptors” that can be more adequate in characterizing anthropogenic pollution. Specific sampling methods may be reserved for contaminants with specific toxic effects (e.g., formaldehyde, benzene, monomers, etc.). In many cases, investigations of the indoor air are carried out because of complaints about poor IAQ, which are made by persons living or working indoors. Such complaints may be perception of unknown or unpleasant odours, headache, irritation of the eyes, nose and throat, dryness of the skin or symptoms like tiredness, lack of concentration and unspecific hypersensitivity reactions. Investigations resulting from observed or suspected health problems of occupants are quite similar and require the same sampling strategy.

3. Sources

Identification of indoor pollutant sources may be very important; emission from point (or surface) sources generates a pronounced concentration gradient near the source. The existence of such concentration gradients can be used to trace the source in an indoor environment by a careful selection of sampling sites. This strategy can lead to recommendations on how to improve indoor air quality, e.g. removing identified materials and strong emitters.

Maroni et al (1995) proposed these recommendations concerning source control:

Material Selection

- Manufacturers should be required to provide sufficient information for the evaluation of their products' safety. Protection of legitimate confidentiality must not impede evaluation.
- Standardization and harmonization of emission testing procedures must continue at international level. Meanwhile, research and application of different techniques and methodological intercomparison should be encouraged.
- Labelling of products and their ranking for safety and emission properties should be encouraged to facilitate appropriate selection and use by consumers. The choice of process may vary according to the nature of the risk involved and the local priorities.

Biological Contaminants

- The growth of moulds and other microorganisms in the indoor environment should be avoided. This can be achieved effectively by elimination of moisture.
- Good housekeeping and sanitation rather than the use of biocides are appropriate control measures also for mites.

Volatile Organic Compounds (VOCs)

- Overall exposure to VOCs should be kept at the lowest possible level primarily by proper control of the emission sources.
- Pending improved methods, the total volatile organic compounds concept is a useful tool for practical assessment of the overall quality of materials and of indoor air. However, at least the type of major VOC species must be known.
- More research is recommended for the sensory effects of VOC mixtures and for some individual compounds typically present in indoor air mixtures for which little toxicological knowledge is available; this recommendation is strongly current, due to the growing interest in persistent organic pollutants (POPs) as endocrine disruptors, also in classic subtoxic concentrations.

4. Sampling methods

The primary objective of taking indoor air is to determine the quality of indoor air with the aim of assessing any risk to the health of the population and of individuals due to indoor air pollution. Monitoring is useful also in testing effectiveness of remedial actions, such as modifications to a building, its systems or equipment, aimed at reducing indoor air pollution. The measurements for testing their effectiveness are comparison measurements before and after the remedial actions. It is also important to record intermediate and long-term trends of indoor air pollution concentration. Respective analysis of trends will help to maintain, improve and establish abatement or risk management procedures. Finally, beside screening measurements, it can be very helpful to evaluate indoor air concentrations at abnormal or "worst-case" conditions of the indoor climate (temperature, humidity, ventilation etc.) and during particular activities. Sampling and analytical methods may be also used in validation of indoor pollution models.

To determine reference values of indoor air pollutants, the following elements of sampling strategy shall be used in addition to the basic information characterising every sampling method (EN 14412, 2004):

- Time of sampling: to rule out any seasonal effects the individual sampling events shall be evenly distributed over the year. The time of individual sampling shall be fixed in a way that representative concentration values can be assessed. However, for sampling for ≥ one week, the choice of sampling time tends to be less important.
- Sampling duration and sampling frequency: the duration of sampling has to be set in a way that the limits of determination for all compounds of interest are exceeded and that representative readings are obtained. It is necessary to maintain a sampling duration of one week or a multiple to achieve a maximum of representativity per sampling site. This is to assess all different pollution loads that occur on the different days of the week at least once. One sampling per site is adequate.
- Sampling site and determinations with resolution in space. The sampling site shall be representative, e.g. the centre of the room most inhabited by all occupants shall be used.

General methods (like NIOSH, OSHA or similar) developed for industrial workplaces may be adapted for VOC determination in air; as an example, we report the synthesis of NIOSH Method 2549 (Table 1.1).

APPLICABILITY: This method has been used for the characterization of environments containing mixtures of volatile organic compounds (see Table 1.2). The sampling has been conducted using multi-bed thermal desorption tubes. The analysis procedure has been able to identify a wide range of organic compounds, based on operator expertise and library searching.

INTERFERENCES: Compounds which coelute on the chromatographic column may present an interference in the identification of each compound. By appropriate use of background subtraction, the mass spectrometrist may be able to obtain more representative spectra of each compound and provide a tentative identity (see Table 1.2).

OTHER METHODS: Other methods have been published for the determination of specific compounds in air by thermal desorption/gas chromatography [1-3]. One of the primary differences in these methods is the sorbents used in the thermal desorption tubes.

REAGENTS:

1. Air, dry
2. Helium, high purity
3. Organic compounds of interest for mass spectra verification (See Table 1.2).*
4. Solvents for preparing spiking solutions: carbon disulfide (low benzene chromatographic grade), methanol, etc. (99+% purity)

   * See SPECIAL PRECAUTIONS

EQUIPMENT:

1. Sampler: Thermal sampling tube, ¼″ s.s. tube, multi-bed sorbents capable of trapping organic compounds in the C_3-C_{16} range. Exact sampler configuration depends on thermal desorber system used. See Figure 1 as an example.

SAMPLING AND MEASUREMENT		
SAMPLER:		THERMAL DESORPTION TUBE (multi-bed sorbent tubes containing graphitized carbons and carbon molecular sieve sorbents [See Appendix])
FLOW RATE:		0.01 to 0.05 L/min
VOL	-MIN:	1L
	-MAX:	6L *(even more for non industrial environments)*
SHIPMENT:		Ambient in storage containers
SAMPLE STABILITY:		Compound dependent (store @ -10 °C)
BLANKS:		1 to 3 per set
TECHNIQUE:		THERMAL DESORPTION, GAS CHROMATOGRAPHY, MASS SPECTROMETRY
ANALYTE:		See Table 1.1
DESORPTION:		Thermal desorption
INJECTION VOLUME:		Defined by desorption split flows (See Appendix)
TEMPERATURE	-DESORPTION:	300 °C for 10 min
	-DETECTOR (MS):	280 °C
	-COLUMN:	35 °C for 4 min; 8 °C/min to 150 °C, 15 °C/min to 300 °C
CARRIER GAS:		Helium
COLUMN:		30 meter DB-1, 0.25-mm ID, 1.0-µm film *(or equivalent)*
CALIBRATION:		Identification based on mass spectra interpretation and computerized library searches.
RANGE:		not applicable
ESTIMATED LOD:		100 ng per tube or less
PRECISION (S_r):		not applicable

Table 1.1. NIOSH method 2549: volatile organic compounds (modified).

2. Personal sampling pump, 0.01 to 0.05 L/min, with flexible tubing.
3. Shipping containers for thermal desorber sampling tubes.
4. Instrumentation: thermal desorption system, focusing capability, desorption temperature appropriate to sorbents in tube (~300°C), and interfaced directly to a GC-MS system.
5. Gas chromatograph with injector fitted with 1/4" column adapter, 1/4" Swagelok nuts and Teflon ferrules (or equivalent).
6. Syringes: 1-µL, 10-µL (liquid); 100-µL, 500-µL (gas tight).
7. Volumetric Flasks, 10-mL.
8. Gas bulb, 2 L.

SPECIAL PRECAUTIONS: Some solvents are flammable and should be handled with caution in a fume hood. Precautions should be taken to avoid inhalation of the vapours from solvents as well. Skin contact should be avoided.

SAMPLING:

NOTE: Prior to field use, clean all thermal desorption tubes thoroughly by heating at or above the intended tube desorption temperature for 1-2 hours with carrier gas flowing at a rate of at least 50 mL/min. Always store tubes with long-term storage caps attached, or in containers that prevent contamination. Identify each tube uniquely with a permanent number on either the tube or tube container. Under no circumstances should tape or labels be applied directly to the thermal desorption tubes.

1. Calibrate each personal sampling pump with a representative sampler in line.
2. Remove the caps of the sampler immediately before sampling. Attach sampler to personal sampling pump with flexible tubing.
   NOTE: With a multi-bed sorbent tube, it is extremely important to sample in the correct direction, from least to maximum strength sorbent.
3. For general screening, sample at 0.01 to 0.05 L/min for a maximum sample volume of 6 L. Replace caps immediately after sampling. Keep field blanks capped at all times. Tubes can act as diffusive samplers if left uncapped in a contaminated environment.
4. Collect a "humidity test" sample to determine if the thermal adsorption tubes have a high water background.
   NOTE: At higher sample volumes, additional analyte and water (from humidity) may be collected on the sampling tube. At sufficiently high levels of analyte or water in the sample, the mass spectrometer may malfunction during analysis resulting in loss of data for a given sample.
5. Collect a "control" sample. For indoor air samples this could be either an outside sample at the same location or an indoor sample taken in a non-complaint area.
6. Ship in sample storage containers at ambient temperature. Store at -10°C.

SAMPLE PREPARATION:

1. Allow samples to equilibrate to room temperature prior to analysis. Remove each sampler from its storage container.
2. Analyze "humidity test" sampler first to determine if humidity was high during sampling (step 10).
3. If high humidity, dry purge the tubes with purified helium at 50 to 100 mL/min for a maximum of 3 L at ambient temperature prior to analysis. .
4. Place the sampler into the thermal desorber. Desorb in reverse direction to sampling flow.

CALIBRATION AND QUALITY CONTROL:

1. Tune the mass spectrometer according to manufacturer's directions to calibrate.
2. Make at least one blank run prior to analyzing any field samples to ensure that the TD-GC-MS system produces a clean chromatographic background. Also make a blank run after analysis of heavily concentrated samples to prevent any carryover in the system. If carryover is observed, make additional blank runs until the contamination is flushed from the thermal desorber system.
3. Maintain a log of thermal desorber tube use to record the number of times used and compounds found. If unexpected analytes are found in samples, the log can be checked to verify if the tube may have been exposed to these analytes during a previous sampling use.
4. Run spiked samples along with the screening samples to confirm the compounds of interest. To prepare spiked samples, use the procedure outlined in the Appendix .

MEASUREMENT:

1. See Appendix for conditions. MS scan range should cover the ions of interest, typically from 20 to 300 atomic mass units (amu). Mass spectra can either be identified by library searching or by manual interpretation (see Table 1.2). In all cases, library matches should also be checked for accurate identification and verified with standard spikes if necessary.

EVALUATION OF METHOD:

The method has been used for a number of field screening evaluations to detect volatile organic compounds. Estimate of the limit of detection for the method is based on the analysis of spiked samples for a number of different types of organic compounds. For the compounds studied, reliable mass spectra were collected at a level of 100 ng per compound or less. In situations where high levels of humidity may be present on the sample, some of the polar volatile compounds may not be efficiently collected on the internal trap of the thermal desorber. In these situations, purging of the samples with 3 L of helium at 100 mL/min removed the excess water and did not appreciably affect the recovery of the analytes on the sample.

Compound/ Synonyms	**CAS# RTECS**	**Empirical Formula**	**MW[a]**	**BP[b] (°C)**	**VP[c] @ 25°C mm Hg**	**kPa**	**Characteristic Ions, m/z**
Aromatic Hydrocarbons							
Benzene /benzol	71-43-2 CY1400000	C_6H_6	78.11	80.1	95.2	12.7	78*
Xylene /dimethyl benzene	1330-20-7 ZE2100000	C_8H_{10}	106.7				91, 106*, 105
o-xylene				144.4	6.7	0.9	
m-xylene				139.1	8.4	1.1	
p-xylene				138.4	8.8	1.2	
Toluene /toluol	108-88-3 XS5250000	C_7H_8	92.14	110.6	28.4	3.8	91, 92*

Compound/ Synonyms	CAS# RTECS	Empirical Formula	MW[a]	BP[b] (°C)	VP[c] @ 25°C mm Hg	kPa	Characteristic Ions, m/z
Aliphatic Hydrocarbons							
n-Pentane	109-66-0 RZ9450000	C_5H_{12}	72.15	36.1	512.5	68.3	43, 72*, 57
n-Hexane /hexyl-hydride	110-54-3 MN9275000	C_6H_{14}	86.18	68.7	151.3	20.2	57, 43, 86*, 41
n-Heptane	142-82-5 MI7700000	C_7H_{16}	100.21	98.4	45.8	6.1	43, 71, 57, 100*,41
n-Octane	111-65-9 RG8400000	C_8H_{18}	114.23	125.7	14.0	1.9	43, 85, 114*, 57
n-Decane /decyl hydride	124-18-5 HD6500000	$C_{10}H_{22}$	142.29	174	1.4	0.2	43, 57, 71, 41, 142*
Ketones							
Acetone /2-propanone	67-64-1 AL3150000	C_3H_6O	58.08	56	266	35.5	43, 58*
2-Butanone /methyl ethyl ketone	78-93-3 EL6475000	C_4H_8O	72.11	79.6	100	13	43, 72*
Methyl isobutyl ketone /MIBK, hexone	108-10-1 SA9275000	$C_6H_{12}O$	100.16	117	15	2	43, 100*, 58
Cyclohexanone /cyclohexyl ketone	108-94-1 GW1050000	$C_6H_{10}O$	98.15	155	2	0.3	55, 42, 98*, 69
Alcohols							
Methanol /methyl alcohol	67-56-1 PC1400000	CH_3OH	32.04	64.5	115	15.3	31, 29, 32*
Ethanol /ethyl alcohol	64-17-5 KQ6300000	C_2H_50H	46.07	78.5	42	5.6	31, 45, 46*
Isopropanol /1-methyl ethanol	67-63-0 NT8050000	C_3H_7OH	60.09	82.5	33	4.4	45, 59, 43
Butanol /butyl alcohol	71-36-3 E01400000	C_4H_9OH	74.12	117	4.2	0.56	56, 31, 41, 43
Glycol Ethers							
Butyl cellosolve /2-butoxyethanol	111-76-2 KJ8575000	$C_6H_{14}O_2$	118.17	171	0.8	0.11	57, 41, 45, 75, 87
Diethylene glycol ethyl ether /Carbitol	111-90-0 KK8750000	$C_6H_{14}O_3$	134.17	202	0.08	0.01	45, 59, 72, 73, 75, 104
Phenolics							
Phenol /hydroxybenzene	108-95-2 SJ3325000	C_6H_5OH	94.11	182	47	0.35	94*, 65, 66, 39
Cresol	1319-77-3 G05950000	C_7H_7OH	108.14				108*, 107, 77, 79
2-methylphenol	95-48-7			190.9	1.9	0.25	
3-methylphenol	108-39-4			202.2	1.0	0.15	
4-methylphenol	106-44-5			201.9	0.8	0.11	

Compound/ Synonyms	CAS# RTECS	Empirical Formula	MW[a]	BP[b] (°C)	VP[c] @ 25°C mm Hg	kPa	Characteristic Ions, m/z
Chlorinated Hydrocarbons							
Methylene chloride /dichloromethane	75-09-2 PA8050000	CH_2Cl_2	84.94	40	349	47	86*, 84, 49, 51
1,1,1-Trichloroethane /methyl chloroform	71-55-6 KJ2975000	CCl_3CH_3	133.42	75	100	13.5	97, 99, 117, 119
Perchloroethylene /hexachloroethane	127-18-4 KX3850000	CCl_3CCl_3	236.74	187 (subl)	0.2	<0.1	164*, 166, 168, 129, 131, 133, 94, 96
o-,p-Dichlorobenzenes		$C_6H_4Cl_2$	147.0				146*, 148, 111, 113, 75
/1,2-dichlorobenzene	95-50-1 CZ4500000			172-9	1.2	0.2	
/1,4-dichlorobenzene	106-46-7 CZ4550000			173.7	1.7	0.2	
1,1,2-Trichloro-1,2,2-trifluoroethane /Freon 113	76-13-1 KJ4000000	CCl_2FCClF_2	187.38	47.6	384	38	101, 103, 151, 153, 85, 87
Terpenes							
d-Limonene	5989-27-5 OS8100000	$C_{10}H_{16}$	136.23	176	1.2		68, 67, 93, 121, 136*
Turpentine (Pinenes)	8006-64-2	$C_{10}H_{16}$	136.23	156 to 170	4 @ 20"		93, 121, 136*, 91
α-pinene	80-56-8			156			
β-pinene	127-91-3			165			
Aldehydes							
Hexanal /caproaldehyde	66-25-1 MN7175000	$C_6H_{12}O$	100.16	131	10	1.3	44, 56, 72, 82, 41
Benzaldehyde /benzoic aldehyde	100-52-7 CU4375000	$C_7H_{12}O$	106.12	179	1.0	0.1	
Nonanal /pelargonic aldehyde	124-19-6 RA5700000	$C_9H_{18}O$	142.24	93	23	3	43, 44, 57, 98, 114
Acetates							
Ethyl acetate /acetic ether	141-78-6 AH5425000	$C_4H_8O_2$	88.1	77	73	9.7	43, 88*, 61, 70, 73, 45
Butyl acetate /acetic acid butyl ester	123-86-4 AF7350000	$C_6H_{12}O_2$	116.16	126	10	1.3	43, 56, 73, 61
Amyl acetate /banana oil	628-63-7 AJ1925000	$C_7H_{14}O_2$	130.18	149	4	0.5	43, 70, 55, 61
Other							
Octamethylcyclotetra-siloxane	556-67-2 GZ4397000	$C_8H_{24}O_4Si_4$	296.62	175			281, 282, 283

[a] Molecular Weight
[b] Boiling Point
[c] Vapour Pressure
*Indicates molecular ion

Table 1.2. Common volatile organic compounds with mass spectral data.

Appendix

Multi-bed sorbent tubes: Other sorbent combinations and instrumentation/conditions shown to be equivalent may be substituted for those listed. In particular, if the compounds of interest are known, specific sorbents and conditions can be chosen that work best for that particular compound(s). The tubes that have been used in NIOSH studies with the Perkin Elmer ATD system are ¼ " stainless steel tubes, and are shown in the figure in the next page:

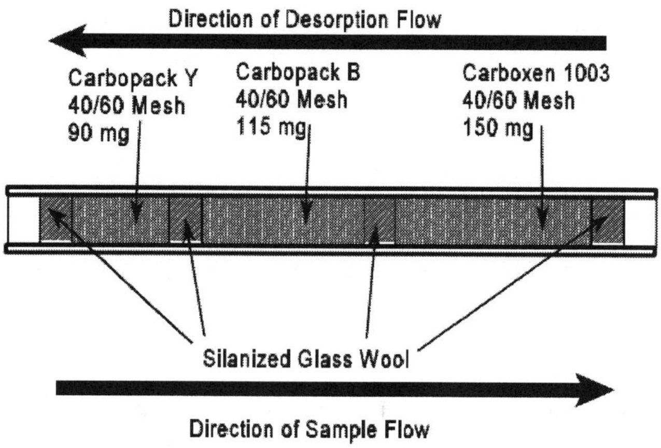

Fig. 1. Carbopack™ and Carboxen™ adsorbents are available from Supelco, Inc.

Preparation of spiked samples: Spiked tubes can be prepared from either liquid or gas bulb standards.

Liquid standards: Stock solutions are prepared by adding known amounts of analytes to 10-mL volumetric flasks containing high purity solvent (carbon disulfide, methanol, toluene). Solvents are chosen based on solubility for the analytes of interest and ability to be separated from the analytes when chromatographed. Highly volatile compounds should be dissolved in a less volatile solvent. For most compounds, carbon disulfide is a good general purpose solvent, although this will interfere with early eluting compounds.

Gas bulb standards: Inject known amounts of organic analytes of interest into a gas bulb of known volume filled with clean air. Prior to closing the bulb, a magnetic stirrer and several glass beads are placed in the bulb to assist in agitation after introduction of the analytes. After injection of all of the analytes of interest into the bulb, warm the bulb to 50 °C and place it on a magnetic stirring plate and stir for several minutes to ensure complete vaporization of the analytes. After the bulb has been stirred and cooled to room temperature, remove aliquots from the bulb with a gas syringe and inject into a sample tube as described below.

Tube spiking Fit a GC injector with a ¼" column adapter. Maintain the injector at 120 °C to assist in vaporization of the injected sample. Attach cleaned thermal desorption tubes to injector with ¼" Swagelok nuts and Teflon ferrules, and adjust helium flow though the injector to 50 mL/min. Attach the sampling tube so that flow direction is the same as for sampling. Take an aliquot of standard solution (gas standards 100 to 500 µL; liquid standards, 0.1 to 2 µL) and inject into the GC injector. Allow to equilibrate for 10 minutes.

Remove tube and analyze by thermal desorption using the same conditions as for field samples.

Instrumentation: Actual media, instrumentation, and conditions used for general screening of unknown environments are as follows: Perkin-Elmer ATD 400 (automated thermal desorption system) interfaced directly to a Hewlett-Packard 5980 gas chromatograph/HP5970 mass selective detector and data system.

ATD conditions:

Tube desorption temperature: 300°C
Tube desorption time: 10 min.
Valve/transfer line temperatures: 150°C
Focusing trap: Carbopack B/Carboxen 1000, 60/80 mesh, held at 27°C during tube desorption
Focusing trap desorption temperature: 300°C
Desorption flow: 50-60 mL/min.
Inlet split: off
Outlet split: 20 mL/min.
Helium: 10 PSI

GC conditions:

DB-1 fused silica capillary column, 30 meter, 1-µm film thickness, 0.25-mm I.D. Temperature program: Initial 35°C for 4 min, ramp to 100°C at 8°/min, then ramp to 300°C at 15°/min, hold 1-5 min.
Run time: 27 min.

MSD conditions:

Transfer line: 280°C
Scan 20-300 amus, EI mode
EMV: set at tuning value
Solvent delay: 0 min for field samples; if a solvent-spiked tube is analyzed, a solvent delay may be necessary to prevent MS shutdown caused by excessive pressure.

Alternative methods may be obtained adapting, for instance, NIOSH Method 1500 (Hydrocarbons b.p. 36 - 216 °C) or NIOSH Method 1501 (Hydrocarbons, aromatic).
Main differences - not fundamentals - may consist in: sorbent choice (charcoal tube), desorption technique (elution with CS_2), sampling time (longer than in workplaces) and other. More important is the possibility of using diffusive (passive) samplers.

5. Passive sampling

The original purpose of the development on passive sampling (based on Fick's laws) is to provide technology at low cost, enabling air quality surveys to be routinely executed at multiple locations within urban and rural areas, industrial sites and forests. This requires the examination of the performance characteristics of the diffusive sampler over long sampling periods, in comparison with established methods, and in practical applications of urban monitoring (Brown et al., 1994; Brown et al., 1999). Several authors proposed the use of different models of diffusive (passive) devices (Brown, 1993; Harper, 2000; Bertoni et al.,

2001; Brown, 2002; Kot-Wasik et al., 2007) conceived for the determination of long-term averaged concentration of some airborne volatile and semi-volatile organic compounds, relevant on the human health protection (VOCs, polycyclic aromatic hydrocarbons and nicotine). The diffusive sampling technique is known to be the cheapest and easiest way to perform extensive sampling campaigns, both in temporal and spatial terms. Moreover, this is the only collection technique allowing the true separation of the vapour phase species from the particle bound fraction (Bertoni et al., 2004; Namiesik et al., 2005).
A lot of different models of passive sampler have been proposed by many researchers; in this chapter we can show only a "classic" sampler (Analyst, developed by the Italian National Research Council, IIA-CNR, Figg. 2 and 3) and a more recent sampler, based on radial diffusion principle (Ring, IIA-CNR, Fig. 4).
All the diffusive samplers have to respect the requirements and test methods according to EN 13528-1, EN 13528-2, EN 13528-3, EN 14412 and EN-ISO 16017-2.

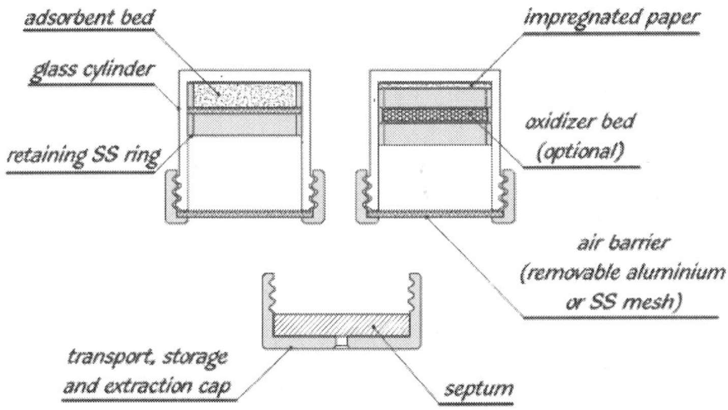

Fig. 2. Analyst sampler scheme

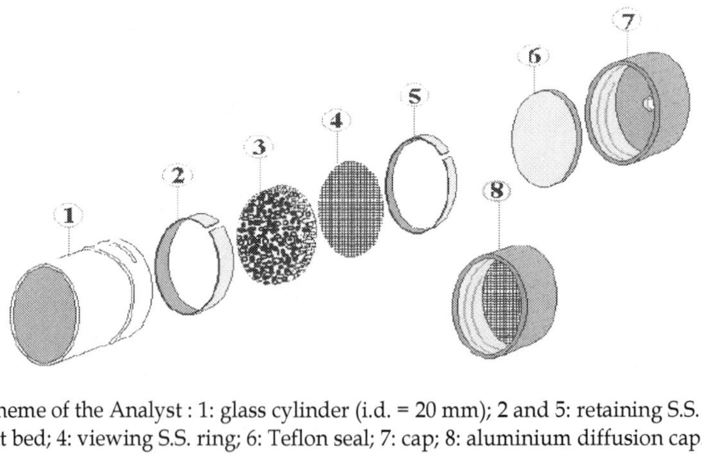

Fig. 3. Scheme of the Analyst : 1: glass cylinder (i.d. = 20 mm); 2 and 5: retaining S.S. rings; 3 adsorbent bed; 4: viewing S.S. ring; 6: Teflon seal; 7: cap; 8: aluminium diffusion cap.

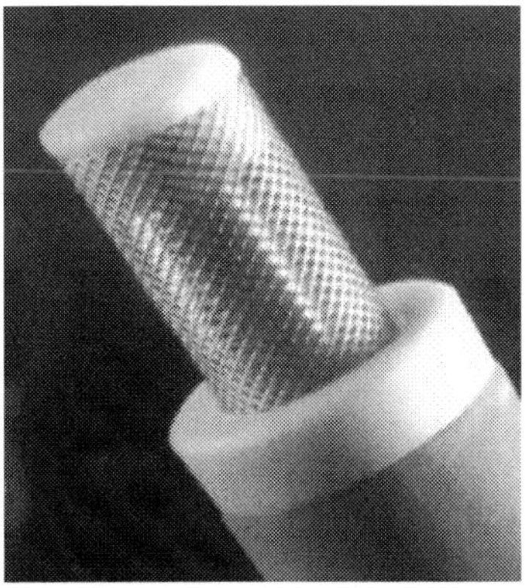

Fig. 4. Diffusive sampler "Ring" (IIA CNR).

6. Data evaluation and exposure risk assessment

Studies have found that home levels of several organics average 2 to 5 times higher indoors than outdoors. During and for several hours immediately after certain activities, such as paint stripping, levels may be 1,000 folds background outdoor levels. Despite direct reading instruments can be used (like the IR used in indoor monitoring shown in Figure 5), as above mentioned, currently it is better to try to obtain very low LODs and LOQs by employing sampling methods as described.

Sometimes they are used Threshold Limit Values (TLVs) that are the guideline values set by the American Conference of Governmental Industrial Hygienists (ACGIH) to minimize workers exposure to hazardous concentrations as much as possible. The TLVs are published yearly for more than 700 chemical substances and physical agents (ACGIH, 2010). This use is incorrect and unauthorized by ACGIH, because the TLVs respect is not a warranty of no health effects, that is on the contrary a fundamental requirement in nonindustrial environment.

No standards have been set for VOCs in non-industrial settings. OSHA regulates formaldehyde, a specific VOC, as a carcinogen. OSHA has adopted a Permissible Exposure Level (PEL) of 0.75 ppm, and an action level of 0.5 ppm. Based upon current information, it is advisable to mitigate formaldehyde that is present at levels higher than 0.1 ppm. Recently the World Health Organization (WHO, 2010) provided health-based guidelines for 55 airborne inorganic and organic compounds for carcinogenic and non-carcinogenic health endpoints. The non-carcinogenic endpoints include development toxicity, reproduction toxicity, respiratory toxicity, neurotoxicy, hepatoxicity, hematoxicity, eye/nose/throat irritation, and odor annoyance. The lowest concentration at which effects are observed in humans, animals, and plants was used as a staring point for the non-carcinogenic endpoints. Uncertainty factors determined through scientific judgment in consensus and averaging time were also taken into account in determining the health endpoint for non-carcinogenic compounds. The classification by the international Agency for Research on Cancer (IARC) was used to determine a chemical as a carcinogen. The endpoint of carcinogen was determined by linear extrapolation from the high dose level, which is characteristic of animal experiments or occupation exposure with cancer responses.

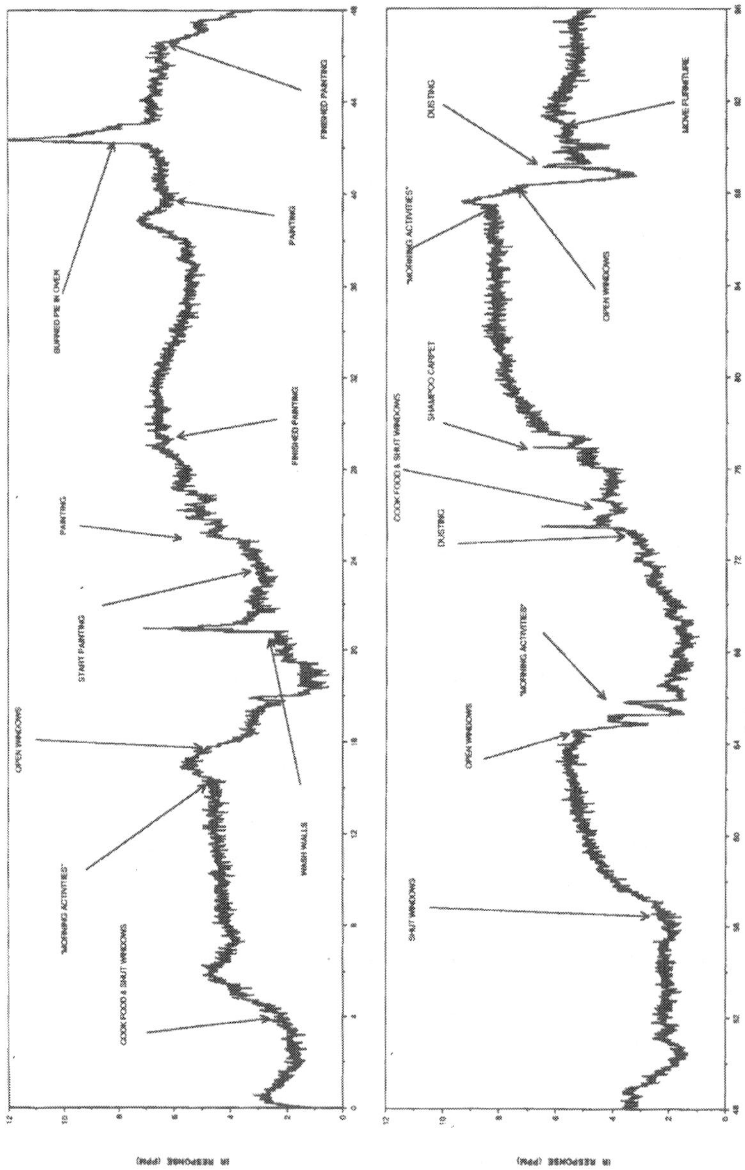

Fig. 5. Variation in indoor VOC levels in house house as detected by an IR instrument. VOC levels are expressed in ppm of equivalent 33% propane/67% butane. Human activities at different times are shown (Clobes et al., 1992).

This objective requires measurements for checking whether specified limit or guideline values are being exceeded. Examples of limit and guideline values (except workplace atmospheres and ambient air) for indoor environments are given in Table 2.1 and 2.2, established by the ad hoc committee of IRK/AOLG of Germany for organic chemicals.

Pollutant	Country	Limit value	Pollutant
Tetrachloroethene	Germany	0.1 mg/m^3	7 days

Table 2.1. Examples of limit values.

Pollutant	Country/Organisation	Guideline Value	Averaging Time
Formaldehyde	Germany	0.12 mg/m^3	not specified
Nitrogen dioxide	Germany	60 $\mu g/m^3$ [a]	7 days
Toluene	Germany	0.3 mg/m^3 [a]	not specified
Styrene	Germany	0.03 mg/m^3 [a]	7 days
Dichloromethane	Germany	0.2 mg/m^3 [a]	24 h
TVOC	Germany	0.2 - 0.3 mg/m^3	not specified

[a] Guideline value (RW I-value, "Richtwert I"= guideline value I) aimed at hygienic prevention.

Table 2.2. Examples of guideline values.

7. References

American Conference of Governmental Industrial Hygienists (ACGIH) 2010 – Threshold Limit Values and Biological Exposure Indices, Cincinnati OH - USA.

Bertoni G., Tappa R., Allegrini I., (2001) "The internal consistency of the 'Analyst' Diffusive sampler - a long-term field test" - Chromatographia 54, November (No. 9/10) 653 - 657

Bertoni G., Tappa R., Cecinato A. (2001) - "Environmental Monitoring of semi-volatile polyciclic aromatic hydrocarbons by means of diffusive sampling devices and GC-MS analysis" Chromatographia, 53, Suppl., S312 - S316.

Bertoni G., Ciuchini C., Tappa R. (2004) "Long-term diffusive samplers for the indoor air quality evaluation", - Ann. Chim. 94 (Rome), 637 - 644.

Brown R.H. (2002) Monitoring volatile organic compounds in air – the development of ISO standards and a critical appraisal of the methods - J Environ Mon 4, 112N-118N

Brown R. H., Wright M. D. and Plant N. T. (1999) The Use of Diffusive Sampling for Monitoring of Benzene, Toluene and Xylene in Ambient Air (IUPAC Technical Report) Pure Appl. Chem., Vol. 71, No. 10, pp. 1993-2008.

Brown R.H. (1993) The use of diffusive samplers for monitoring of ambient air - Pure & Appl. Chem. 65, 1859-1874

Brown R.H., Harvey R.P., Purnall C.J., and Saunders K.J., (1984) A Diffusive Sampler Evaluation Protocol - Am. Ind. Hyg. Assoc. J. 45, 67.

Clobes A.L., Ananth G.P., Hood A.L., Schroeder J.A. and Lee K.A. (1992) Human Activities as Source of Volatile Organic Compounds in Residential Environments; in "Sources of Indoor Air Contaminants", Tucker W.G., Leaderer B.P., Molhave L. and Cain W.S. editors. Annals of the New York Academy of Sciences, vol 641, 79-86.

Harper M. Review - Sorbent trapping of volatile organic compounds from air - J Chrom A 885 (2000) 129-151

Kot-Wasik A., Zabiegala B., Urbanovicz M., Dominiak E., Wasik A. and Namiesnik J. (2007) Advances in passive sampling in environmental studies - Anal Chim Acta 602, 141-164

Maroni M., Axelrad R., Bacaloni A. (1995) "NATO's Efforts to set IAQ Standards" Am. Ind. Hygiene Ass. J., 56, 499-508

Namiesnik J., Bozena Z., Kot-Wasik A., Partyka M., Wasik (2005) A. Passive sampling and/or extraction techniques in environmental analysis: a review - Anal Bioanal Chem 381, 279-301

EN 13528 - 1 (Sept. 2002) Ambient air quality: Diffusive samplers for the determination of concentrations of gases and vapours - Requirements and test methods. Part 1: General requirements

EN 13528 - 2 (Sept. 2002) Ambient air quality: Diffusive samplers for the determination of concentrations of gases and vapours - Requirements and test methods. Part 3: Specific requirements and test methods

EN 13528 - 3 (Dec. 2003) Ambient air quality: Diffusive samplers for the determination of concentrations of gases and vapours - Requirements and test methods. Part 3: Guide to selection, use and maintenance

EN 14412 (Sept. 2004) Indoor air quality - Diffusive samplers for the determination of concentrations of gases and vapours - Guide for selection, use and maintenance

EN ISO 16017-2 (May 2003) Indoor, ambient and workplace air - Sampling and analysis of volatile organic compounds by sorbent tube/thermal desorption/capillary gas chromatography. Part 2: Diffusive sampling

NIOSH Manual of Analytical Methods, Method 2549 Volatile Organic Compounds (screening) 4th rev. (1996), issue 1- ed. Cincinnati, OH - USA.

NIOSH Manual of Analytical Methods, Method 1500 Hydrocarbons b.p. 36 - 216 °C (2003), issue 3- ed. Cincinnati, OH - USA.

NIOSH Manual of Analytical Methods, Method 1501 Hydrocarbons, aromatic (2003), issue 3- ed. Cincinnati, OH - USA.

WHO guidelines for indoor air quality: selected pollutants (2010) ISBN 978 92 890 0213 4

http://www.epa.gov/iaq/atozindex.html

http://www.umweltbundesamt.de/uba-info-daten-e/daten-e/irk.htm#4

http://www.umweltbundesamt.de/uba-info-daten/daten/irk.htm

http://www.osha.gov/dts/osta/otm/otm_iii/otm_iii_2.html

11

Statistical Considerations for Bioaerosol Health-Risk Exposure Analysis

M.D. Larrañaga[1], E. Karunasena[2], H.W. Holder[3],
E.D. Althouse[4] and D.C. Straus[5]
[1]*Oklahoma State University*
[2]*Texas Tech University*
[3]*SWK, LLC*
[4]*Air Intellect, LLC*
[5]*Texas Tech University Health Sciences Center*
USA

1. Introduction

Air and surface sampling was conducted to confirm the types of microbiological contamination within a hospital facility in the southern United States, identify indicators of indoor microbiological contamination, and profile the aero-biological makeup of the inside air (ISA) for comparisons to outside air (OSA) and reference concentrations, where applicable. The investigation strategy recommended by the American Conference of Governmental Industrial Hygienists (ACGIH) was utilized to assess indoor environmental quality and conditions found within the hospital facility. The ACGIH methodology is guided by the text Bioaerosols: Assessment and Control. (Macher 1999) Biological contamination within a hospital environment is of great concern as bacteria and fungi are important causes of nosocomial infection (NI). It is estimated that the overall hospital-acquired infection rate in Europe and North America is between 5 and 10%. (Kalliokoski 2003) A substantial number of bacteria and fungi are capable of spreading via the airborne route in hospitals, and airborne transmission accounts for approximately 10% of all NI. (Eickhoff 1994; Kalliokoski 2003) Contaminated Heating, Ventilating, and Air Conditioning (HVAC) systems and infiltration of unfiltered outside air have been implicated in airborne outbreaks of NI via infective aerosols, dust, and contaminated filters. (Lentino, Rosenkranz et al. 1982; Rhame 1991; Eickhoff 1994) Certain underlying diseases, procedures, hospital services, and categories of age, sex, race, and urgency of admission have been shown to be significant risk factors for nosocomial infection. (Freeman and McGowan 1978) The day-specific incidence of nosocomial infection rises from near zero on the first hospital day to maximal during the fourth through seventh weeks of hospital stay. (Freeman and McGowan 1981) Nosocomial infections can affect patients in any location within a hospital. (Boss and Day 2003)

The Centers for Disease Control (CDC) estimates that 2 million patients develop hospital-acquired infections annually and as many as 88,000 die as a result. (CDC 1992) Hospitals typically maintain a patient population with increased susceptibility to infection and factors inherent to the healthcare environment contribute to the risk associated with acquiring an infection during a hospital admission. (Dulworth and Pyenson 2004) Environmental control and high efficiency filtration are critical to preventing person-to-person and environmentally related infections in hospitals. (Wenzel 1997; Boss and Day 2003) Current evidence indicates that excessive moisture indoors promotes microbial growth and is associated with an increased prevalence of symptoms due to irritation, allergy, and infection. It is widely accepted in various scientific communities that indoor microbiological contamination presents unacceptable conditions for the preservation of human health, and that removal and prevention of microbial contamination is necessary and prudent. (Pope, Patterson et al. 1993; Macher 1999; Agency 2001; ACOEM 2002; Fung and Hughson 2002; Redd 2002; CDC 2003; Fung and Hughson 2003)

The inherent variability of microbiological organisms in air presents a challenge for conducting air sampling that provides meaningful results for the evaluation of human exposure and health risk. In general, bioaerosol air sampling is highly variable and prone to error. Multiple and replicate samples over subsequent days are necessary to characterize exposure and multiple samples per sample location are required to evaluate human exposure in that particular location. An air and surface sampling plan was designed to address the inherent variability and error associated with air sampling and to evaluate the exposure and subsequent health risk to patients, visitors, and staff in a hospital facility. The objectives of this chapter are to highlight the necessity for multiple (replicate) air samples per sample location to conduct valid assessments of the airborne concentrations of bioaerosols.

2. Materials and methods

Sampling was conducted to provide a bioaerosol profile of the air within the spaces under the control of 9 separate HVAC systems over a two-year period. The spaces under the environmental control or each HVAC system are identified as air handling units (AHUs) 10, 11 (no final filters present), 13 (90% final filters), 15, 16, 17, 18, and 19 during the summer of 2005. In early 2006, a follow-up investigation was conducted within the spaces controlled by AHUs 17, 19, and 21. Except where indicated, AHU final filters had a filtration efficiency of 95%. Air and surface sampling data and analysis, observations, and the collective experience of a team experienced in moisture intrusion in hospital facilities along with input from professionals in medical microbiology, industrial hygiene, medicine, engineering, and public health were utilized to interpret data for hypothesis testing. The hypotheses were:

a. Hypothesis A: The 90-95% final filters control particulate matter generated by the AHUs and preventing contamination downstream of the filters. Note: AHU 11 does not have final filters and therefore this hypothesis does not apply to AHU 11.
b. Hypothesis B: The 90-95% final filters prevent microbial contamination downstream of the filters. Note: AHU 11 does not have final filters and therefore this hypothesis does not apply to AHU 11.
c. Hypothesis D: Staff is being exposed to potentially harmful concentrations of biological contaminants.
d. Hypothesis E: Patients are being exposed to harmful quantities of biological contaminants.

Air sampling was conducted to establish mean airborne concentrations of fungal and biological aerosols indoors and outdoors to characterize the fungal and bacterial aerobiological profiles of the areas controlled by each AHU. Culturable air samples were analyzed at the species level as information on species is crucial for determining whether indicator organisms are present. Indicator species of fungi whose presence indicates excessive indoor moisture or a health hazard were evaluated. Indicator organisms identified via air sampling in the hospital were *Aspergillus versicolor, A. flavus, A. fumigatus, Fusarium* species, yeasts, and species of *Penicillium.* (Macher 1999) In addition, the American Industrial Hygiene Association (AIHA) has consistently recommended urgent risk management decisions be made when the confirmed presence of these indicator organisms are identified indoors These indicator organisms include those listed above in addition to *Stachybotrys chartarum.* The confirmed presence is defined as colonies in several samples, many colonies in any sample, or, where a single colony was found in a single sample, evidence of growth of these fungi on building materials by visual inspection or source sampling. (Macher 1999) As early as 1996, AIHA stated that urgent risk management decisions are required of the industrial hygienist in the following conditions: a) the confirmed presence of facultative pathogens (fungi capable of inducing pulmonary infections in humans) such as *A. versicolor and Fusarium moniliforme,* and b) The presence of fungi, such as *S. chartarum* and *F. moniliforme,* known to result in occupational diseases in part due to their potent toxins. AIHA recommended that these urgent risk management decisions be made promptly as opposed to weeks or months later. (Dillon, Heinsohn et al. 1996; Prezant, Weekes et al. 2008) Based on the literature and in consideration of the recommendations made elsewhere by governmental and nongovernmental entities and other professional societies, the 1996 and 2008 AIHA recommendations and the 1999 ACGIH recommendations (Macher 1999) continue to be appropriate risk management guidance for the industrial hygienist and indoor environmental professional.

The estimated cumulative sampling and analytical error for each air sample is defined as $E_C = (P^2 + T^2 + A^2 + O^2)^{1/2}$ where E_C is the cumulative error, P is the pump error (estimated at ± 5%), T is the time error (estimated at ± 2.5%), A is the analytical error (estimated to be ± 25%), and O is other error associated with calibration and technician variability (estimated to be ± 25%). (Macher 1999; Burton 2006) Solving for E_C, the cumulative sampling and analytical error for each sample is estimated to be approximately ± 36%. Therefore, data derived from individual samples should be viewed as qualitative. The inherent variability of the air concentrations of bioaerosols over time or within a space far outweighs any errors associated with measurement of airborne microbiological concentrations. Thus, the interpretation of a single sample is difficult without information on the variability of the concentrations of biological agents identified in the environment because the variability in the measurement is almost always large. (Macher 1999)

Duplicate, side-by-side air samples were taken at each location to address the error associated with individual samples. Duplicate side-by-side air sampling is considered adequate to define the mean and the random sampling error given the high temporal and spatial variability of bioaerosol concentrations in air. (Dillon, Heinsohn et al. 1996) Multiple samples from multiple random and non-random locations were taken on separate days to characterize exposures within the space under the environmental control of each air handling unit. (Macher 1999) Replicate samples within the space controlled by each air handling unit were taken to address sampling variability (Weber and Page 2001) and allow for the estimation of the sampling data's variances so that differences between two environments (e.g. ISA v. OSA concentrations) could be identified. (Macher 1999) A minimum of 6 replicate samples were taken at 10 indoor locations ($6 \leq n \leq 48$ samples per location) and 2 outdoor locations ($12 \leq n \leq 72$ total outdoor reference samples per indoor location). An ANOVA was utilized to compare indoor and outdoor concentrations of biological agents using SAS Statistical Software. The level of significance was prescribed as $\alpha = 0.05$. A significant difference identified by the ANOVA indicates that the difference is unlikely to have occurred by chance and that there is statistical evidence that there is a difference. At the $\alpha = 0.05$ (5%) level of significance, the result could have occurred by chance one time in 20. Statistical techniques evaluate an observed difference in view of its precision to determine with what probability it might have arisen by chance (the level of significance). Values with a low probability of occurring by chance are called statistically significant and are considered to represent a real effect (e.g. difference between means). (Conover 1999; Montgomery 2001)

Ideally, human respiratory exposure is measured using air samples taken near the breathing zone, or within 12 inches of the mouth. This was not feasible considering the large size and weight of bioaerosol air sampling equipment. Most bioaerosol sampling is done to characterize ambient aerosols and the ambient conditions are utilized as quantitative estimates of bioaerosol exposure. Although the characterization of the ambient environment is not the ideal exposure sampling scheme, when low flow rate suction impactors (e.g. Andersen type) are utilized, the error introduced is small. (Pope, Patterson et al. 1993) Low flow rate suction impactors were utilized to take bioaerosol samples.

A comparison of total fungal or bacterial concentrations may be used as a preliminary indicator of a difference in two environments, but not as evidence of similarity or dissimilarity. Indoor/outdoor comparisons are used to document the presence or infer the absence of indoor, biologically derived contamination. These comparisons cannot be made unless the genera and species found indoors and outdoors have been identified. (Macher 1999) Air sampling mean total counts were utilized as a preliminary indicator of a difference in two environments (e.g. indoor vs. outdoor air concentrations) to determine the effectiveness of the filters in removing aerobiological particulates from the air stream. Air sampling indicators of indoor microbiological contamination were identified from sampling results where both the genera and species are identified (culturable fungal and bacterial samples incubated at 25°C and 37°C). Culturable air samples were taken for incubation at two separate temperatures (25°C and 37°C) to enhance the detection of both environmental and pathogenic microorganisms in the air. (Dillon, Heinsohn et al. 1996; Macher 1999)

Air sampling indicators of indoor contamination were compared to both surface sampling results and outdoor air sampling concentrations of like-organisms for interpretation. Note that monitoring for allergens can help characterize environments with respect to specific allergens (e.g., fungi and/or bacteria), and measurements can be semi-quantitative (e.g., "presence or absence" or "low, medium, or high"). (Pope, Patterson et al. 1993) Airborne bioaerosol sampling was conducted so that comparisons and interpretations could be made between ISA and OSA culturable and non-culturable bioaerosols, mean concentrations and variability of concentrations, and species could be compared. Each set or type of sampling results should be viewed in consideration with the other sampling results and not independently. For example, one should not consider spore trap (total and non-culturable fungal samples) alone, as spore trap sampling does not identify fungi at the species level and may mask important differences in species present between test and reference locations. Analysis of spore traps alone or total fungal or bacterial concentrations could lead to incorrect conclusions. The sampling interpretation considered all sampling results for the development of interpretations and conclusions.

The American Society for Heating, Refrigerating, and Air Conditioning Engineers (ASHRAE) defines critical care areas as the following functional spaces within a hospital: 1) Surgery and Critical Care, 2) Nursing, 3) Ancillary, 4) Diagnostic and Treatment, and 5) Sterilization and Supply. Critical care areas include but are not limited to intensive care units, coronary care units, angiography laboratories, cardiac catheterization laboratories, delivery rooms, operating rooms, recovery rooms, emergency departments, and other special care units where enhanced engineering controls are required for the protection of patients and staff. Although patients spend most of their time within a specific area, they may be exposed to bioaerosols in non-critical locations of a hospital where engineering controls are not as stringent. Non-critical functional areas are administration and service locations within the Hospital. (ASHRAE 2003) Table 1 lists the critical and non critical locations within the areas investigated.

AHU #	Area Served	Critical Area
10	Endoscopy, EKG, EGG, PFT, Pharmacy, Sterile Preparation, Recovery	Yes
11	1st floor PBX, Waiting Room, HR, Medical Records, Security	No
13	Patient Rooms	Yes
15	Intermediate Nursery, Ambulatory Surgical Center, Endoscopy, Operating Rooms, Sterile Supply, Pre- and Post- Op areas	Yes
16	Patient Rooms, Wound Therapy, Lactation Center	Yes
17	Emergency Room, Radiology and CT-Scan	Yes
18	Radiology, Ultrasound, Mammography, Laboratory, Laboratory Biohazard, Histology, Pathology	Yes
19	1st Floor, Administration Area, Labor and Delivery, Recovery, C-Section Room, Pre-Op Area, Intensive Care, Rehabilitation	Yes
21	Radiology, Purchasing, Shipping and Receiving, Plant Operations	Yes

Table 1. Classification of Hospital areas by critical or non-critical area.

3. Results

3.1 Total outside vs. total indoor concentrations (2005 data)

Figure 1 depicts the percent difference of bioaerosols from the OSA reference concentrations to the ISA concentrations for the space controlled by each AHU. The results show that the existing filters within the facility were removing bioaerosols from the air within the Hospital.

95% final filters were installed in the AHUs investigated, with the exception of AHUs 11 and 13. AHU 11 did not have final filters and AHU 13 had 90% final filters installed. Ninety-five percent final filters are rated to remove greater than 90% of particles between 1.0 and 10 micrometers in size and 85-95% of particles between 0.3 and 1.0 micrometers in size. Ninety percent final filters are rated to remove 90% of particulate matter between 1 and 10 micrometers in size and 75-85% of particles between 0.3 and 1.0 micrometers in size.

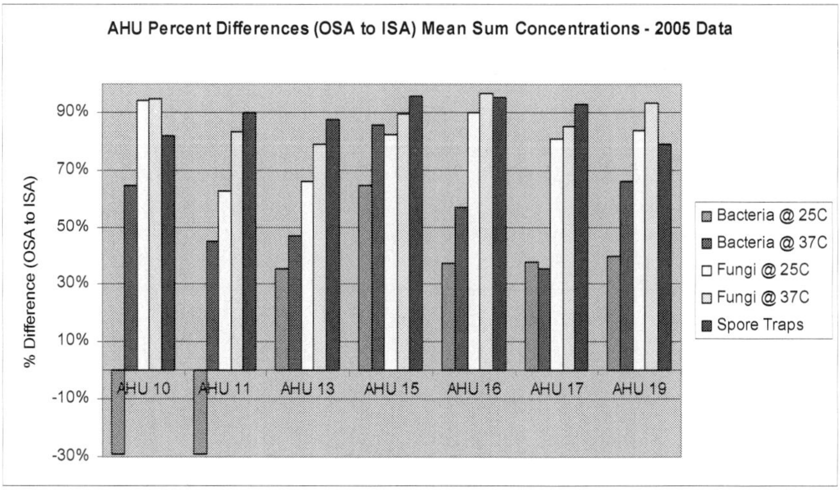

Fig. 1. AHU percent differences from outside air to inside air.
Note: A negative percent difference indicates that the indoor concentration was higher than outdoors.

(ASHRAE 1992; ASHRAE 1999) Therefore, with the exception of AHU 11, the percent differences for fungal particulates between OSA and ISA should approach 90% for organic (e.g. fungi and bacteria) particulate matter ranging from 1.0 to 10 micrometers in aerodynamic diameter and 78-85% of organic matter for particles ranging from 0.3-1.0 micrometers in aerodynamic diameter.

Non-culturable fungal air sampling via spore trap measures the airborne fungal particle concentrations in spores/m^3 of air. The minimum percent difference (reduction) of 79% (AHU 19) for total spore count (spore trap) results indicates that at least 79% of the non-culturable total spore concentrations from the outside air are being removed by the HVAC systems prior to entering the building. The minimum percent difference of all fungal air sampling results is 63% for the fungal samples incubated at 25°C in the space controlled by AHU 11. Excluding AHU 11 because it does not have final filters, the minimum percent difference of all fungal sampling results is 66% for AHU 13, which has 90% final filters. This indicates that at least 66% of the fungal bioaerosols are being removed by the final filters of the AHUs investigated, with the exception of AHU 11. The concentrations of total and culturable fungal bioaerosols within the spaces controlled by the AHUs with 95% final filters (all except AHUs 11 and 13) are at least 75% lower than outside air concentrations, indicating that the filters are performing and removing particulates from the air.

The bacterial concentrations (incubated at 25°C) within the space controlled by AHUs 10 and 11 were higher indoors than outdoors. Unlike fungi, bacteria have natural reservoirs indoors (including humans), and total bacterial concentrations are often higher indoors than outdoors. (Macher 1999) The bacterial organisms identified (incubated at 25°C) as indicators of an indoor source were *Micrococcus* species, *Micrococcus luteus*, *Staphylococcus capitus*, and *Staphylococcus hyicus*, which are human-shed bacteria. (Wilson 2005) Because these organisms are human-shed and were not identified as indoor contaminants via surface sampling, it cannot be concluded that these higher concentrations of bacteria detected via air sampling were the result of building-related sources of bacterial contamination. (Macher 1999) Higher indoor concentrations of human-shed bacteria are an anticipated condition within a building. The bacteria *Bordetella bronchiseptica* was identified inside AHUs 10, 11, 13, 16, 18, 19 and was not identified indoors via air sampling. Viridans streptococci was identified inside AHU 18 and not identified in indoor air samples. *Staphylococcus aureus* was identified inside AHU 21 and not identified in indoor air samples. This is a strong indication that the filters may prevent the transmission of *Bordetella bronchiseptica,* Viridans streptococci, and *Staphylococcus aureus* through the filters and into the occupied space of the hospital.

With the exception of the bacterial (incubated at 25°C) air samples in AHUs 10 and 11, the percent differences for bacterial sample sets comparing OSA to ISA were at least 35%. This is an indication that the filters were removing bacteria from both the outside and re-circulated air of the Hospital.

3.2 Total outside vs. total indoor concentrations (2006 data)

The 2006 sampling data indicate that, in general, the total air concentrations are lower indoors than outdoors. However, indicators of indoor contamination were identified. The 2006 sampling strategy was to compare outdoor fungal and bacterial airborne concentrations with the concentrations identified 1) within each AHU before the filter, 2) within each AHU after the filter, and 3) within a room controlled by the AHU. AHUs 17, 19, and 21 were tested in early 2006.

3.2.1 AHU 17

AHU 17 serves the first floor emergency room and radiology, which are critical areas. Figure 2 summarizes the 2006 data for AHU 17.

The AHU 17 data indicate fungal and bacterial percent differences from OSA to ISA greater than 74%, with the exception of total fungi (spore traps), which is shown as a difference of 1%. Percent differences between before filter and after filter concentrations also show minimum percent differences of 58%, which indicate bioaerosol removal from the air stream after it passes through the filter. The low percent differences for spore traps and culturable air sampling (25°C and 37°C samples) between OSA and ISA (room) indicate indoor contamination as shown by the negative percent difference shown for the after filter to room samples. This is not an indication that the filters are inadequately filtering particles, but rather an indication of an indoor source and/or infiltration due to negative pressure contributing to the indoor concentrations of bioaerosols.

Percent differences show an increase in airborne fungal and bacterial concentrations between the locations after the filter and within the room, indicating an increase in airborne concentrations in the air for both culturable bacteria and fungi and total fungi after it leaves the AHU. Negative percent differences are identified for spore trap samples from before the filter to room and after the filter to room. The increase in concentrations as the air moves from the AHU to the room indicates the presence of an indoor source of fungi and bacteria, specifically yeasts, fungal species of *Cladosporium, Penicillium, Aspergillus,* and *Fusarium,* and bacterial species of *Staphylococcus, Micrococcus,* and *Sphingomonas paucimobilis.*

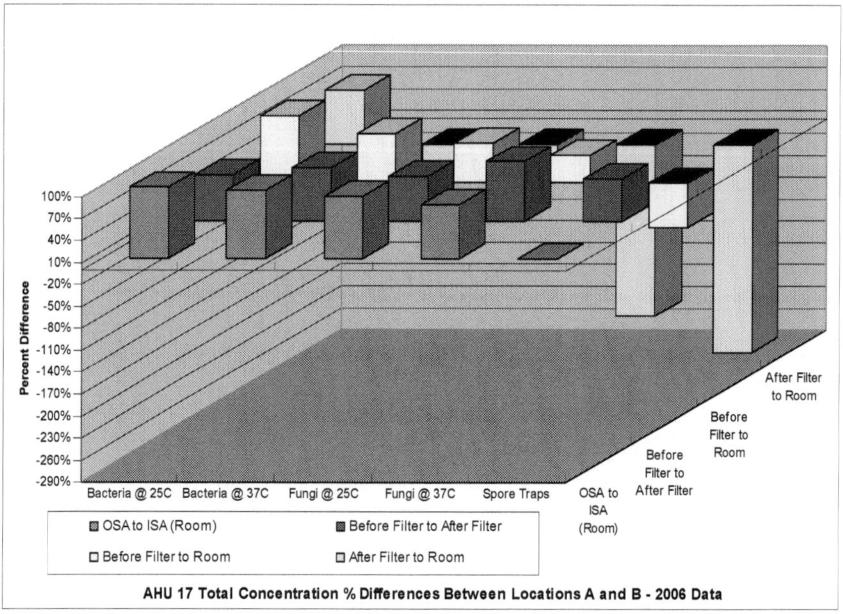

Fig. 2. AHU 17 percent differences in mean total concentrations between spaces (2006 data).
Note: A negative percent difference indicates that the indoor concentration was higher than outdoors for OSA to ISA comparisons.
Note: A black top of a bar in the graph in the 0% plane indicates a negative percent difference indicating an increase in concentration from one space to another when a decrease is expected.

3.2.2 AHU 19

AHU 19 serves the 1st Floor Administration Area, Labor and Delivery, Recovery, the Cesarean Section room, Pre-Op Area, the Intensive Care Unit, and a Rehabilitation area. All areas supplied by AHU 19 are considered critical care areas except the Administration Area. The AHU 19 data indicate percent differences that approach or are greater than 90% for fungi and bacteria between OSA and ISA. This indicates that greater than 90% of the bioaerosols in the OSA are being removed from the air stream. The AHU 19 data indicate positive percent differences signifying decreasing concentrations from before the filter to after the filter, before the filter to room, and after the filter to room. Figure 3 summarizes the 2006 data for AHU 19.

3.2.3 AHU 21

AHU 21 serves Radiology, Purchasing, Shipping and Receiving, and Plant Operations. Radiology is a critical area. The AHU 21 data indicate positive percent differences for airborne concentrations comparisons of OSA to ISA. The minimum percent difference is 54% (spore traps) for OSA to ISA comparisons and show more than 50% of the mean total concentrations of fungi and bacteria were being removed from OSA. Percent differences for bacteria incubated at 25°C were negative for the before filter to after filter, before filter to

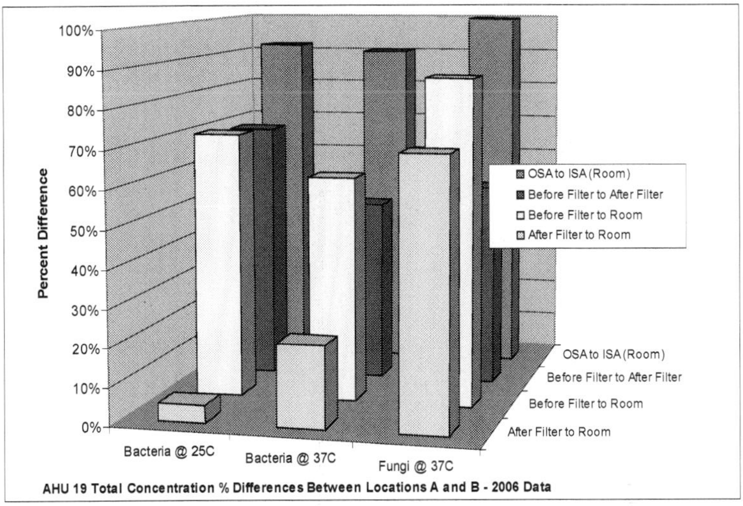

Fig. 3. AHU 19 percent differences in mean total concentrations between spaces (2006 data).
Note: A negative percent difference indicates that the indoor concentration was higher than outdoors for OSA to ISA comparisons.
Note: A black top of a bar in the graph in the 0% plane indicates a negative percent difference indicating an increase in concentration from one space to another when a decrease was expected.
Note: The 2006 AHU 19 laboratory results were not received from the analytical laboratory for spore trap and fungal samples incubated at 25°C.

room, and after filter to room sample sets, indicating an indoor source of bacteria. Percent differences for bacteria incubated at 37°C were negative for the before filter to room and after filter to room sample sets, indicating an indoor source of bacteria. Percent differences were negative for all sampling sets comparing after filter to room, indicating indoor sources of bacteria and fungi. Percent differences show an increase in airborne fungal and bacterial concentrations between air concentrations after the filter and air concentrations within the room, indicating an increase in airborne concentrations in the air for both culturable bacteria and fungi and total fungi after the air leaves the AHU.

The low percent differences for spore traps and culturable air sampling (25°C and 37°C samples) between OSA and ISA (room) indicate indoor contamination as shown by the negative percent difference shown for the after filter to room samples. This indicates an indoor source or infiltration due to negative building pressure as opposed to inadequate particle filtration by the AHU filters. The increase in concentrations as the air moves from the AHU to the room indicates the presence of an indoor source of fungi and bacteria, specifically fungal species of *Rhodotorula, Penicillium, Aspergillus, Verticillium, Nigrospora, Paecilomyces, Cladosporium, Engyodontium, Rhizopus,* and *Scytalidium,* and bacterial species of *Acinetobacter, Chryseomonas, Pseudomonas, Tatumella,* and *Staphylococcus*. Table 4 summarizes the 2006 data for AHU 21.

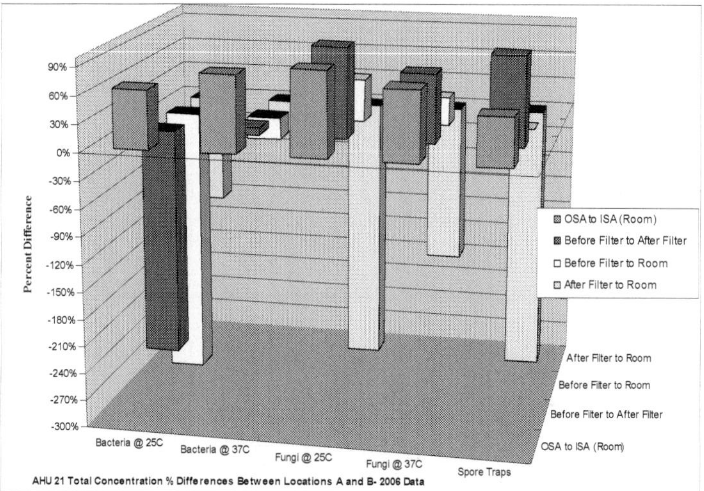

Fig. 4. AHU 21 percent differences in mean total concentrations between spaces (2006 data).
Note: Negative scale was truncated at -300% for simplicity in graphical presentation.
Note: A negative percent difference indicates that the indoor concentration was higher than outdoors for OSA to ISA comparisons.
Note: A black top of a bar in the graph in the 0% plane indicates a negative percent difference indicating an increase in concentration from one space to another when a decrease was expected.

3.3 Air sampling indicators of indoor iontamination

Indoor vs. outdoor comparisons of fungi and bacteria are used to document the presence or infer the absence of indoor, biologically derived contamination. (Macher 1999) A substantial number of bacteria and fungi are capable of spreading via the airborne route in hospitals. (Eickhoff 1994) The presence of contamination in dust or on surfaces or water is often considered de facto evidence of human exposure to fungal aerosols. (Burge 2000)

When evaluated as total concentrations, the air sampling results generally indicate lower concentrations of fungi and bacteria indoors compared to outdoors. Unless the genera and species have been identified total counts merely indicate gross numbers and indoor vs. outdoor comparisons are not meaningful. (Macher 1999; Weber and Page 2001) An investigator cannot make meaningful indoor vs. outdoor comparisons unless the genera and species found indoors and outdoors have been identified. (Macher 1999) For this project, culturable air samples were analyzed at the species level so that meaningful comparisons could be made.

When evaluated according to genera, air sampling results indicate the potential presence of indoor contamination sources for both bacteria and fungi. Air sampling indicators of an indoor source were identified in the space controlled by each AHU. Indicators of indoor contamination were defined as:

1. Potential indoor biological source (indoor mean concentration > outdoor mean concentration [tested by the ANOVA, $\alpha = 0.05$, with differences identified by SNK Post Hoc Grouping] or the organism was identified in the indoor air samples but not identified in the outdoor reference samples ($n \geq 12$) for the space controlled by each AHU),

2. Meets the criteria of 1 above (a potential indoor biological source) and is a confirmed indoor contamination source via surface sampling in the space controlled by each AHU,
3. Meets the criteria of 1 above (a potential indoor biological source) and the organism was not identified in all outdoor air samples (n ≥ 48), indicating an indoor source of air contamination, and
4. Meets the criteria of both 2 and 3 above, confirming an indoor source of air contamination.

Air sampling identified indicators of indoor contamination within each space investigated in both the 2005 and 2006 data. While indoor bacterial sources are expected within occupied buildings, the species of fungi found in indoor and outdoor air should be similar. (Weber and Page 2001) Outdoor air samples serve as the primary comparison to indoor bioaerosol samples and the types of fungi present indoors should not be significantly different from the outdoor environment. (Spicer and Gangloff 2000) To determine if the indoor and outdoor bioaerosol profiles were similar, the means of the sampling data were tested for biodiversity. The numbers of air sampling indicators identified per AHU from the 2005 air sampling data are shown in Figure 5.

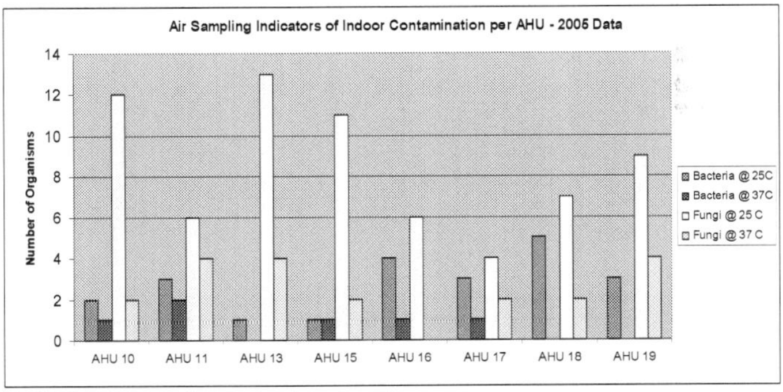

Fig. 5. Number of Air Sampling Indicators of Indoor Contamination per AHU.

Air sampling indicators identified per AHU from the 2006 air sampling data are shown in Figure 6.

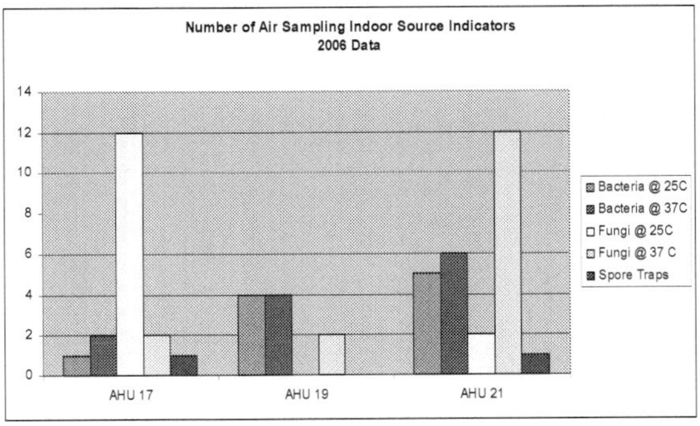

Fig. 6. Number of air sampling indicators identified via air sampling (2006 data).

Figures 7 and 8 illustrate the percentage of air sampling indicators discussed in the **Air Sampling Indicators of Indoor Contamination section** that were confirmed via surface sampling or not detected in the outside air reference samples, indicating an indoor source of contamination.

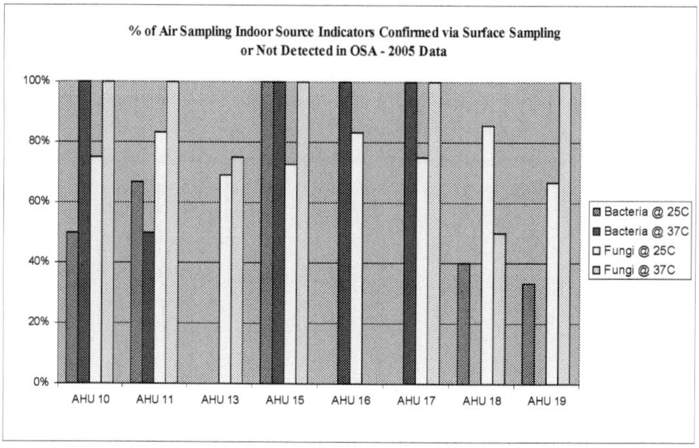

Fig. 7. Percentage of air sampling indoor source indicators confirmed via surface sampling or not detected in the OSA.

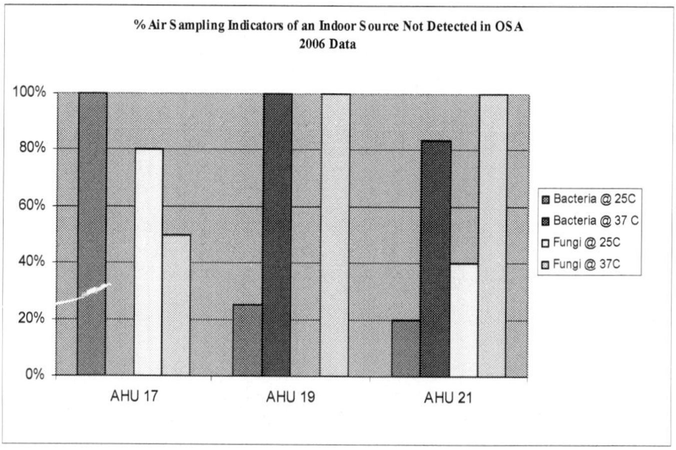

Fig. 8. Percentage of air sampling indoor source indicators not detected in the OSA.

Indicators of indoor contamination that were not identified in outdoor air samples or were identified in indoor samples were found within the space controlled by each AHU investigated. Indicators of indoor contamination identified on surface sampling are associated with the building while indicators not identified in the outdoor air are likely to be associated with the building, building occupants, or another indoor source.

3.4 Tests for biodiversity

The Spearman's Rank Correlation (SRC) is a non-parametric statistical test for comparing bioaerosol samples from separate environments. SRC is used to assess the similarity of the genera and species of culturable fungi and bacteria in air or source samples and the types of fungal spores in spore-trap samples. Data from a reference site (OSA) can be compared to data from a test site (ISA). Mean concentrations of multiple samples from each sampling location are preferred for use in the calculations for biodiversity. (Macher 1999) Indoor air sampling locations were identified broadly as the space under the environmental control of each AHU investigated. Several samples were taken from each sampling area (space under the environmental control of the AHU) so that mean concentrations could be analyzed statistically. The SRC test for biodiversity allows inferences to be made based upon whether the bioaerosol profile of one location is statistically similar to the bioaerosol profile of another. For fungi, indoor and outdoor profiles should be statistically similar. (Dillon, Heinsohn et al. 1996; Macher 1999; Spicer and Gangloff 2000) For bacteria, however, it is not unusual to have differences in biodiversity, as many bacteria are human shed. (Macher 1999)

The mix of airborne fungal species indoors should be similar to that found in the outdoor air. (Weber and Page 2001) In assessing indoor air quality with regards to airborne fungi, the types of fungi present in the indoor environment should not be significantly different from the outdoor environment. Similarity indicates that the building is not promoting or amplifying the growth of microorganisms. (Spicer and Gangloff 2000) A zero value was used as a replacement for microorganisms that were not detected in a sample. (Spicer and Gangloff 2000)

The biodiversity can be measured either as a combined aerobiological profile of fungi and bacteria, or separately. Here, bacteria and fungi are considered separately. (Macher 1999) Indoor source aerosols tend to be dominated by the readily released spores of *Aspergillus* and *Penicillium* species. (Burge 2000) The culturable air sampling results of *Aspergillus* and *Penicillium* species (at both 25°C and 37°C) were tested to determine if indoor vs. outdoor species identified were similar. Table 2 summarizes the SRC test results for biodiversity. Tests for biodiversity were conducted with SPSS 14 Statistical Software. The level of significance was prescribed as $\alpha = 0.05$. A significant correlation (p-value $\leq \alpha \leq 0.5$) indicates that there is statistical evidence that the biodiversity of the indoor air is similar to the biodiversity of the OSA. Similarity indicates that the building is not promoting or amplifying the growth of microorganisms, while dissimilarity indicates that the biodiversity of indoor and OSA are independent and that the indoor environment is amplifying the growth of microorganisms. (Dillon, Heinsohn et al. 1996; Macher 1999; Spicer and Gangloff 2000) Similarities in the biodiversity of indoor and outdoor air are unlikely to have occurred by chance, and at the $\alpha = 0.05$ (5%) level of significance, the result could have occurred by chance one time in 20. Statistical techniques evaluate an observed difference in view of its precision to determine with what probability it might have arisen by chance (the level of significance). Values with a low probability of occurring by chance are called statistically significant and are considered to represent a real effect (e.g. difference or similarity in biodiversity). (Dillon, Heinsohn et al. 1996; Conover 1999; Macher 1999) The SRC applied to bioaerosol samples in the tests for differences between OSA and ISA biodiversity at the species level for culturable bacteria and fungi and at the genus level for spore traps.

The 2005 data show consistent dissimilarity between the bioaerosol profiles of ISA and OSA for species of *Penicillium* and *Aspergillus* at both 25°C and 37°C (86% dissimilar). For AHUs 16, 17, 18, and 19, bacterial and fungal biodiversities are similar to OSA. Yet, the biodiversity of *Penicillium* species and *Aspergillus* species analyzed independently are consistently dissimilar (86% dissimilar) indicating indoor amplification. Spore traps consistently showed 100% similarity because only non-culturable fungi were analyzed at the genus level, masking differences in species present in OSA and ISA. This is an expected occurrence with spore trap analysis because spore traps analysis is not sufficient for identification at the species level. For the 2005 data, the biodiversity of bacteria between OSA and ISA were consistently similar (75% similar). The tests for biodiversity indicate that the diversity of bacterial taxa identified indoors were consistently dissimilar to the bacterial taxa identified outdoors in the space controlled by AHUs 10, 11, 13, and 15.

The 2006 data indicate that the biodiversity between OSA and ISA was dissimilar for the space under the environmental control of AHUs 17, 19, and 21. For the 2006 data, the biodiversity of bacterial tests were consistently dissimilar (83% dissimilar). This may be due to differences in seasons that the tests were conducted. The 2005 samples were taken in summer and the 2006 samples were taken in winter. During winter, OSA bioaerosol concentrations may be lower due to less than ideal growth conditions. Conversely, indoor bioaerosol concentrations of bacteria may be similar year-round due to the presence of human shed sources of bacteria.

4. Sampling interpretation summary

Allergic reactions to indoor allergens [including fungi and bacteria] can produce inflammatory conditions of the eyes, nose, throat, and bronchi. A substantial number of bacteria and fungi are capable of spreading via the airborne route in hospitals. (Eickhoff 1994) Contaminated ventilation or air conditioning systems have been implicated in nosocomial outbreaks, via infective aerosols, dust, and colonized filters. (Eickhoff 1994) Biofilms or slime on HVAC system pan or coil surfaces is an indicator of microbiological amplification. (Pope, Patterson et al. 1993) The growth of microorganisms downstream from the cooling coils can be promoted by water droplets being blown off coil surfaces and into the air. (Pope, Patterson et al. 1993; CDC 2003) Poorly designed HVAC systems may provide for amplification of fungi and *Actinomycetes* in wet niches of the system. (Pope, Patterson et al. 1993; CDC 2003) In a hospital with high efficiency filters, airborne fungal spores reflect incomplete filtration, infiltration of outside air, and shedding of adherent spores from indoor growth. (Rhame 1991) Nosocomial aspergillosis occurs in direct proportion to the mean ambient hospital airborne spore content. (Rhame 1991) Mini-bursts of spores occur from disturbance of settled spores in dust, shedding spores from clothes or sources such as indoor growth. (Rhame 1991) Immunocompromised persons, children and the elderly are at risk of from exposure to infectious microorganisms. (Boss and Day 2003) In general, the filters show that total quantities of aerobiological contaminants are being removed from the air stream. See the filter discussion in the **Total Outside Concentrations vs. Total Indoor Concentrations (2005 Data)** section. However, both surface and air samples indicate the presence of indoor microbiological amplification, which is problematic in the indoor environment.

Results for Spearman's Rank Correlation Testing for Biodiversity									
AHU	**10**	**11**	**13**	**15**	**16**	**17**	**18**	**19**	**21**
2005 Data									
Bacteria @ 25°C	D	D	S	D	S	S	S	S	Not tested in year 1.
Bacteria @ 37°C	S	S	D	S	S	S	S	S	
Fungi @ 25°C	D	S	D	D	S	S	S	S	
Fungi @ 37°C	D	S	D	D	S	S	S	S	
Spore Traps	S	S	S	S	S	S	S	S	
Aspergillus @ 25°C	D	D	D	D	D	D	D	D	
Penicillium @ 25°C	D	D	D	D	D	S	D	D	
Aspergillus @ 37°C	D	S	D	D	*	D	S	S	
Penicillium @ 37°C	*	D	D	D	*	D	D	D	
2006 Data									
Bacteria @ 25°C	Not tested in year 2.					D	Not tested in year 2.	D	D
Bacteria @ 37°C						D		D	S
Fungi @ 25°C						S		**	S
Fungi @ 37°C						D		*	D
Spore Traps						S		**	S
Aspergillus @ 25°C						*		*	*
Penicillium @ 25°C						S		*	*
Aspergillus @ 37°C						*		*	*
Penicillium @ 37°C						*		*	*

S indicates that the aerobiological profiles between indoor and outdoor air are "similar" and not independent (i.e. the populations appear to be related or the samples could have been drawn from the same environment), ($\alpha \leq 0.05$, Spearman's Rank Correlation). (Dillon, Heinsohn et al. 1996; Macher 1999)

D indicates that the aerobiological profiles between indoor and outdoor air are "dissimilar" and independent (ie. The populations appear to be unrelated and drawn from separate environments). (Macher 1999; Dillon, Heinsohn et al. 2005)

* indicates that the Spearman's Rank Correlation test could not be performed due to indoor concentration data that were constant (zero) or the number of species identified for each test was not sufficient to warrant testing via Spearman's Rank Correlation. (Spicer and Gangloff 2000)

** Note: Sample results were not received from the analytical laboratory.

Note: One tailed test for correlation. Significant correlation indicates that the indoor and outdoor biological profiles are similar. (Conover 1999; Macher 1999)

Table 2. Spearman's Rank Correlation Test for Biodiversity between outdoor and indoor sampling locations. (Significance level $\alpha \leq 0.05$)

4.1 Fungal samples

The mix of airborne fungal species indoors should be similar to that found in the outdoor air. Fungal organisms identified indoors, that are not present in the outdoor air or control locations, suggests the presence of an amplifier (growth site) for that species in the building. (Macher 1999; Weber and Page 2001) Most fungi can become opportunistic pathogens in a severely immunocompromised patient. (Weber and Page 2001) Airborne fungi within the hospital setting are especially dangerous because antifungal therapy is still rather ineffective. (Kalliokoski 2003) Although average air concentrations of total culturable fungi indoors were consistently lower than those found outdoors, many types of fungi identified indoors were not found in outdoor reference samples; especially within the species of *Aspergillus* and *Penicillium*. (Weber and Page 2001) The biodiversity of species of *Penicillium* and *Aspergillus* between indoor and outdoor air was consistently dissimilar for sampling results within the sampling space controlled by each air handling unit. Species of fungi identified within the hospital in air and on surfaces that are associated with airborne transmission or NI are *Penicillium, Aspergillus, Rhizopus, Acremonium,* and *Fusarium*. (CDC 2003)

Indicator organisms identified via air sampling in the hospital are *Aspergillus versicolor, A. flavus, A. fumigatus,* species of *Fusarium* and *Penicillium,* and yeasts. (Macher 1999) Some indicator organisms identified on indoor surfaces within the Hospital are species of *Aspergillus, Chaetomium, Stachybotrys, Penicillium, Fusarium, Acremonium, Trichoderma,* and yeasts. The presence of indicator organism contamination on surfaces indicates long-term or severe moisture problems. (Boss and Day 2003) Indicator organisms were identified on surfaces within the space controlled by each AHU investigated in 2005.

Exposure to fungi actively growing indoors may present unusual health risks even when total fungal concentrations are higher outdoors. (Macher 1999) Exposure to damp indoor environments and the presence of molds in damp indoor environments are associated with asthma symptoms in sensitized asthmatic persons. (Institute of Medicine Committee on Damp Indoor Spaces and Health 2004) Serious respiratory infections resulting from exposure to *Aspergillus* species and *Fusarium* species are common in persons who are immunocompromised. It is likely that many of these fungal infections are contracted through contact with fungi in indoor environments, because poor health conditions limit people with severely impaired immune systems to spend most of their time indoors. (Institute of Medicine Committee on Damp Indoor Spaces and Health 2004) The lungs of persons with chronic pulmonary disorders such as cystic fibrosis, asthma, and chronic obstructive pulmonary disorder may become colonized and potentially infected with *Aspergillus* species. (Institute of Medicine Committee on Damp Indoor Spaces and Health 2004) Healthy persons exposed to damp or moldy indoor environments report that they are more prone to respiratory infections, including the common cold, sinusitis, tonsillitis, otitis, and bronchitis. (Institute of Medicine Committee on Damp Indoor Spaces and Health 2004) Fungi have become the largest cause of occupational diseases among healthcare workers in Finland. (Kalliokoski 2003)

Indoor source aerosols tend to be dominated by the readily released spores of *Aspergillus* and *Penicillium* species (Burge 2000), which produce large numbers of spores that are easily released into the air. (Institute of Medicine Committee on Damp Indoor Spaces and Health 2004) Indoor exposure to *Aspergillus* and *Penicillium* species spores has been shown to be associated with an increased risk of allergic sensitization in children (Wilson, Holder et al. 2004) and are chiefly involved in the genesis of asthma and allergic alveolitis (pulmonitis due

to hypersensitivity). (Perdelli, Christina et al. 2006) Bronchial asthma is frequently provoked by airborne fungal spores belonging to the genera *Aspergillus* and *Penicillium*. (Smith 1990) Species of *Aspergillus* and *Penicillium* are considered indicator organisms that may signal unwanted moisture intrusion and/or a potential for health problems. (Macher 1999) As such, the biodiversity of both *Penicillium* and *Aspergillus* species was tested to determine whether the outdoor air was the primary source for the fungi identified in the indoor air. (Macher 1999)
Aspergillus and Penicillium species are two of the most ubiquitous fungi known. Large quantities of fungal spores are produced when these fungi are actively growing. During germination, large quantities of spores are produced, and when sporulation occurs, several thousand spores may be disseminated per cubic meter of air. (Wenzel 1997) *Penicillium* and *Aspergillus* spores are sphere-like, measuring from approximately 2-5 micrometers in diameter and can be suspended very easily in the air. (Wenzel 1997; Straus 2004) Spores of *Penicillium* species often cannot be distinguished via microscopic examination from spores of *Aspergillus* species and vice versa. (Stetzenbach and Yates 2003) Once suspended, they may remain suspended for prolonged periods, and when those spores settle, they can contaminate any surface in contact with air. (Wenzel 1997) Once inhaled, these spores travel through the airways into the lower regions of the lungs, leading to the potential development of respiratory symptoms. (Straus 2004)

4.1.1 Aspergillus species

The presence of *Aspergillus* species in health-care environments is a substantial extrinsic risk factor for opportunistic invasive aspergillosis (invasive aspergillosis being the most serious form of the aspergillosis). (CDC 2003) The presence of *Aspergillus* contamination and/or growth within a health care facility is of particular concern due to the presence of immunocompromised persons. Causative agents of aspergillosis are *Aspergillus flavus, Aspergillus fumigatus, Aspergillus niger, Aspergillus terreus,* and *Aspergillus nidulans.* (CDC 2003) Airborne concentrations of *Aspergillus* species at or below 0.1 cfu/m^3 have been recommended for the prevention of nosocomial aspergillosis. (Weber and Page 2001) Low concentrations of *A. fumigatus* and *A. flavus* have been associated with nosocomial aspergillosis in immunocompromised patients. (Arnow, Sadigh et al. 1991) Among immunosuppressed patients in general, invasive aspergillosis remains a serious complication and may be lethal. (Perdelli, Christina et al. 2006)
The genus *Aspergillus* is one of the most ubiquitous fungi known. Large quantities of fungal spores are produced when actively growing. The *Aspergillus* spores are inhaled easily because of their small aerodynamic size and can easily reach and colonize the upper respiratory tract, including paranasal sinuses and terminal airways. (Wenzel 1997) Nosocomial pulmonary and disseminated aspergillosis arises from inhalation of fungal spores. (Rhame, Streifel et al. 1984) The use of powerful new chemotherapy protocols for malignancies and certain immunologic disorders and the increasing use of organ transplantation are risk factors for nosocomial aspergillosis. Patients with acute or chronic myelogenous leukemia and AIDS are particularly susceptible to nosocomial aspergillosis. In the transplant population and in patients with aplastic anemia, *Aspergillus* has emerged as a major cause of death. (Wenzel 1997)
Nosocomial aspergillosis is primarily established when an immunocompromised host inhales fungal spores present in the air. (Rhame, Streifel et al. 1984; Wenzel 1997) Any dust-generating activity, such as maintenance of ventilation systems, cleaning, vacuuming, and dry mopping, can render *Aspergillus* spores airborne and potentially cause outbreaks of

nosocomial aspergillosis. The achievement of a spore-free air within an area or ward of a hospital may not be sufficient to eradicate nosocomial aspergillosis because of "non-ward" sources of *Aspergillus* within the hospital, such as radiology, radiation therapy units, and other areas visited by patients where engineering and environmental controls may not be as stringent. (Wenzel 1997) Fungi actively growing indoors compounds the problem associated with the prevention of nosocomial aspergillosis.

Species of *Aspergillus* found indoors should be similar to species identified outdoors. With the exception of AHU 11, all AHUs have filters of 90% efficiency or greater. 90% and 95% filters are considered high efficiency (Boss and Day 2003) and rated to remove 90% of particles from 1-10 micrometers and almost all particles greater than 10 micrometers in size. (ASHRAE 1992; ASHRAE 1999) Due to the presence of high efficiency filters in the AHUs investigated (except AHU 11), indoor concentrations should be statistically lower than outdoor concentrations and organisms not detected outdoors should not be detected indoors (Weber and Page 2001) (with the exception of indoor-source bacteria (Macher 1999)). However, many species of *Aspergillus* were detected indoors and not outdoors or indoor concentrations were greater than or not statistically different than outdoor concentrations, indicating an indoor source or infiltration.

The average concentration of *A. fumigatus* for culturable fungal air samples at 25°C was 0.50 cfu/m3 (n=14 indoors, n=38 outdoors) in the space controlled by AHU 15. The biodiversity of species of *Aspergillus* between indoor and outdoor air was consistently dissimilar for sampling results (incubated at 25°C) within the sampling space controlled by each air handling unit, indicating the presence of indoor fungal amplifiers; the building appears to be promoting or amplifying the growth of species of *Aspergillus*. (Macher 1999; Spicer and Gangloff 2000; Weber and Page 2001)

Of particular concern in a hospital setting are the presence of thermotolerant fungi, including *A. fumigatus* and *A. flavus*. Thermotolerant fungi are of primary concern in healthcare facilities, since they can cause infection in at-risk patients, even when concentrations are very low. (Page and Trout 2001) Concentrations of thermotolerant *A. fumigatus* (incubated at 37°C) were 1.17 (n=12 outdoors, n=6 indoors) and 3.50 cfu/m3 (n=12 indoors, n=22 outdoors) in the spaces controlled by AHUs 13 and 15, respectively. Indoor concentrations of *A. flavus* were not statistically different than outdoor concentrations in the spaces controlled by AHUs 11 and 17 for samples incubated at 25°C and AHUs 11, 17, and 18, for samples incubated at 37°C. This is an expected condition within the space controlled by AHU 11 due to the absence of final filters. This suggests the presence of an indoor source of *A. flavus* or outdoor air infiltration within the spaces controlled by AHUs 17 and 18. Indoor *A. flavus* concentrations detected in the space controlled by AHUs 10 and 19 were lower than OSA concentrations (statistically significant difference at the α = 0.05 level of significance). The biodiversity of species of thermotolerant *Aspergillus* between indoor and outdoor air was consistently dissimilar for sampling results within the sampling space controlled by AHUs 10, 13, 15, and 17, indicating the presence of indoor fungal amplifiers; the building appears to be promoting or amplifying the growth of species of *Aspergillus*. (Macher 1999; Spicer and Gangloff 2000; Weber and Page 2001)

Indoor concentrations for culturable *Aspergillus* species incubated at 25°C (*A. flavus, A. sydowii, A. nidulans, A. niger, A. fumigatus, A. flavipes , A. sclerotiorum, A. terreus, A. versicolor*) that exceeded indoor concentrations, were not statistically different to outdoor concentrations, or not detected in the outdoor reference samples were identified in the space controlled by AHUs 11, 13, 15, 16, 17, 18, and 19. Indoor concentrations of thermotolerant species of *Aspergillus* (*A. flavus, A. sydowii, A. niger, A. fumigatus, A. terreus*) that exceeded indoor concentrations, were not statistically different to outdoor concentrations, or not detected in the outdoor reference samples were identified in the space controlled by AHUs 10, 11, 13, 15, 17, and 18.

Potentially hazardous concentrations of *A. fumigatus* were identified in the spaces controlled by AHUs 13 and 15. (Weber and Page 2001) Moisture intrusion via infiltration, leaks, inadequate HVAC control, etc. has provided a chronically moist indoor environment ideal for fungal growth. Air infiltration due to negative building pressurization allows the introduction of unfiltered air into the hospital. Air samples indicated the presence of indoor fungal reservoirs/amplifiers within the spaces controlled by all AHUs investigated. Air and surface sampling indicated microbial growth, dissemination and, hence, occupant exposure, from the indoor microbial reservoirs. (Weber and Page 2001) An indoor environment has been created in which immunosuppressed or allergic patients within the Hospital are not fully protected against the risk of infection and the allergenic effects of *Aspergillus* species (Perdelli, Christina et al. 2006) and otherwise healthy persons may suffer exacerbation of allergies and be prone to increased incidences of respiratory infections, including the common cold, sinusitis, tonsillitis, otitis, and bronchitis. (Institute of Medicine Committee on Damp Indoor Spaces and Health 2004)

General controls for prevention of nosocomial aspergillosis are: air filtration, positive pressurization, avoidance of dust-generating activities, attention to non-filtered air infiltration, protection of immunocompromised patients who enter areas without highly filtered air, and isolation of hospital construction. (Wenzel 1997) Highly filtered air is essential to preventing person-to-person and environmentally related infections. (Boss and Day 2003) However, it is important to note that the use of highly filtered air conditioning systems does not provide complete protection against fungi actively growing or infiltrating into a facility. (Perdelli, Christina et al. 2006) Nosocomial aspergillosis has been associated with poorly maintained and/or malfunctioning HVAC systems. (CDC 2003)

4.1.2 Penicillium species

As with *Aspergillus* species, the genus *Penicillium* is one of the most ubiquitous fungi known and one of the most commonly isolated molds from contaminated buildings. With *Penicillium* species the main cause for concern is allergic disease, as infections due to *Penicillium* species are rare. Indoor concentrations of *Penicillium* species greater than outdoor concentrations are associated with negative health effects in humans. *Penicillium* species has been correlated with allergic asthma, allergic alveolitis, atopy, increased lower respiratory infections in children during the first year of life, and wheezing. (Straus 2004)

Inhalation of *Penicillium* species spores has been shown to provoke immediate and delayed-type asthma in individuals already sensitized to *Penicillium*. (Straus 2004) Infants with high risk for the development of asthma (e.g. due to premature birth and/or ethnicity) may be at significant risk for persistent cough and wheeze when exposed to *Penicillium* species spores. *Penicillium* species have been shown to cause allergic alveolitis due to exposure from a faulty installation of a HVAC system. (Straus 2004) *Penicillium* is a large group of fungi valued as producers of antibiotics. *Penicillium* may cause allergic reactions, exacerbate asthma, and cause other adverse health effects when dispersed through air. (Boss and Day 2003; Institute of Medicine Committee on Damp Indoor Spaces and Health 2004) The blue-green molds of *Penicillium* are common contaminants of indoor environments. Inhalation of spores is the major route of entry. *Penicillium* species have been associated with asthma and hypersensitivity pneumonitis (Weber and Page 2001) and can cause NI in the immunocompromised host. (Fox, Chamberlin et al. 1990; Walsh and Groll 1999; CDC 2003) *Penicillium* species spores have been shown to be associated with an increased risk of allergic sensitization in children (Straus 2004) and are chiefly involved in the genesis of asthma and allergic alveolitis (pulmonitis due to hypersensitivity). (Perdelli, Christina et al. 2006) One 2.5 cm diameter colony of *Penicillium* species can produce 400,000,000 spores that can become airborne and generate increased concentrations of airborne spores. (Hitchcock, Mair et al. 2006)

The biodiversity of *Penicillium* species between ISA and OSA was consistently dissimilar for sampling results (incubated at 25°C) within the sampling space controlled by each AHU (except AHU 17), indicating the presence of indoor fungal amplifiers; the building appears to be promoting the growth of species of *Penicillium*. (Spicer and Gangloff 2000) Thermotolerant fungi are of primary concern in healthcare facilities, since they can cause infection in at-risk patients, even when concentrations are very low. (Weber and Page 2001) The biodiversity of species of thermotolerant *Penicillium* between indoor and outdoor air was consistently dissimilar for sampling results within the sampling space controlled by each AHU (except AHUs 10 and 16 because tests could not be performed), indicating the presence of indoor fungal amplifiers; the building appears to be promoting or amplifying the growth of species of *Penicillium*. (Macher 1999; Spicer and Gangloff 2000; Weber and Page 2001)

Species of *Penicillium* found indoors should be similar to species identified outdoors. With the exception of AHU 11, all AHUs have filters of 90% efficiency or greater and should remove greater than 90% of particles between 1 and 10 micrometers in size. Due to the presence of high efficiency filters in the AHUs investigated (except AHU 11), indoor concentrations should be statistically lower than outdoor concentrations and organisms not detected outdoors should not be detected indoors (with the exception of indoor-source bacteria). (Weber and Page 2001) However, many species of *Penicillium* were detected indoors and not outdoors or indoor concentrations were greater than or not statistically different than outdoor concentrations. Indicators of indoor contamination of *Penicillium* species were identified in the space controlled by each AHU.

Indoor airborne concentrations of culturable Penicillium species incubated at 25°C (P. citrinum, P. chrysogenum, P. corylophilum, P. decumbens, P. duclauxii, P. funiculosum, P. glabrum, P. implicatum, P. janthinellum, P. oxalicum, P. pinophilum, P. purporogenum, P. sclerotiorum, P. variabile, P. waksmani) that exceeded or were similar to outdoor concentrations or not detected in the outdoor reference samples were identified in the space controlled by each AHU. Indoor airborne concentrations of thermotolerant (incubated at 37°C) species of Penicillium (P. citrinum, P. chrysogenum, P. decumbens, P. funiculosum, P. janthinellum, P. oxalicum, P. pinophilum, P. simplicissimum) that exceeded or were similar to outdoor concentrations or not detected in the outdoor reference samples were identified in the space controlled by each AHU, with the exception of AHU 10 where no thermotolerant species of Penicillium were identified.

4.1.3 Yeasts

Yeasts are found in a variety of natural habitats or organic substrates such as plant leaves, flowers, soil, and salt water. Some yeasts are part of the normal human flora. Although yeasts may be part of the normal human flora, they were detected on indoor surfaces.

Therefore, it can be concluded that the yeasts identified as indicators of indoor contamination were most likely from an indoor contamination source. (Macher 1999) The presence of yeasts actively growing indoors is of concern, as yeasts are considered an indicator organism and can cause infections in the immunocompromised host. Some yeasts are reported to be allergenic, and may cause problems in individuals with previous exposure and developed hypersensitivities. Yeast infections are among the most common fungal infections in humans. Their form ranges from localized cutaneous or mucocutaneous lesions, to fungemia or disseminated systemic mycoses. (AerotechP&K 2006) Indoor concentrations of yeasts exceeded outdoor concentrations or were detected indoors and not outdoors in the spaces controlled by AHUs 10, 13, 15, and 19. See indicators of indoor contamination for both 2005 and 2006 data. Yeasts were identified via surface sampling within the spaces controlled by AHUs 10, 11, 13, 15, 16, 17, 18, and 19.

4.2 Bacterial samples

Like fungi, bacteria actively growing indoors can release spores into the air (Institute of Medicine Committee on Damp Indoor Spaces and Health 2004). Additionally, bacteria secrete enzymes that can act as allergens. Enzymes and spores from Gram-positive bacilli and thermophilic *Actinomycetes* have been implicated in epidemics of hypersensitivity pneumonitis and work-related asthma. Concentrations of bacteria associated with sensitization or provoking human allergic reactions are unknown. (Pope, Patterson et al. 1993) Bacteria are known to cause diseases either as pathogens or as opportunistic pathogens in the immunocompromised host. (Boss and Day 2003) Environmental bacteria also grow in all wet spaces and are found in most cases where there is fungal growth. (Institute of Medicine Committee on Damp Indoor Spaces and Health 2004) Some bacteria that are common in outdoor air may penetrate to building interiors and may also grow indoors. (Macher 1999) Unlike fungi, bacteria have natural reservoirs indoors (primarily humans), and total bacterial concentrations are often higher indoors than outdoors. (Macher 1999)

The bacterial organisms identified (incubated at 25°C) as indicators of an indoor source were Acinetobacter lwoffi, Gram (+) cocci, Micrococcus luteus, Micrococcus species, Staphylococcus species (S. auricularis, S. capitus, S. epidermis, S. hominis, S. hyicus, S. warneri, and S. xylosus) which are human-shed bacteria. (Wilson 2005) Because these organisms are human-shed and were not identified as indoor contaminants via surface sampling, it cannot be concluded that these higher concentrations of indoor-source bacteria detected via air sampling were the result of building-related sources of bacterial contamination. (Macher 1999)

The bacterial organisms identified (incubated at 37°C) as indicators of an indoor source were Gram (-) cocci, *Pseudomonas aeruginosa, Pseudomonas stutzeri, Rhizobium radiobacter, and Tatumella ptyseos*. Gram (-) bacteria are usually not present on the skin (with the exception of *Acinetobacter* species). *Pseudomonas aeruginosa* can be found on skin, but is also considered an environmental organism and may be associated with building contamination and/or infiltration. (Wilson 2005) *Pseudomonas stutzeri and Rhizobium radiobacter* are considered environmental source organisms. The presence of airborne concentrations of *Pseudomonas stutzeri* (identified via surface sampling on a ceiling tile in the space controlled by AHU 19), and *Rhizobium radiobacter* indicates the presence of an indoor environmental source (amplification).

Acinetobacter species may cause NI and death in infants during periods of airborne dissemination. Environmental conditions leading to an increase in air conditioner condensate in HVAC systems may increase the risk of nosocomial infection with *Acinetobacter* species. (McDonald, Walker et al. 1998) *Acinetobacter* species have been cultured from air conditioners in a hospital nursery. (CDC 2003) *Acinetobacter lwoffi* species was identified as an indicator of an indoor source via air sampling in the space controlled by AHUs 13 and 21. *Acinetobacter* species are widely distributed in the environment and are frequently found on human skin. (Wilson 2005) *Acinetobacter* sp. are a main causative agent of pneumonia, which is a leading cause of morbidity and mortality and is the sixth most common cause of death in the United Kingdom and the Untied States. (Wilson 2005) The presence of *Acinetobacter* sp. in the air is of concern in the hospital environment. There is no indication that the indoor source of *Acinetobacter* species was associated with building contamination, but is likely associated with infiltration.

Pseudomonas species are common in the outdoor air, but rarely occur in indoor air. (Macher 1999) *P. aeruginosa* was identified as an indicator of indoor contamination in the space controlled by AHU 11. This may be due to the absence of final filters in AHU 11. Airborne infections by *P. aeruginosa* have been reported. (Kalliokoski 2003) Transmission of *P. aeruginosa* may occur through direct patient-to-patient contact, environmental contamination, or via the hands of health care workers. (Kerr, Moore et al. 1995; Beggs and Kerr 2000) Airborne *P. aeruginosa* was detected within the hospital in 2005 during a pilot study of proposed sampling equipment in the basement of the Hospital. These sampling results must be viewed as a qualitative indication of an indoor source of *P. aeruginosa*, as the equipment utilized in the 2005 pilot study was determined to be inaccurate for the determination of airborne concentrations but sufficient for qualitative identification of the bacterium at high concentrations within the space tested. *P. aeruginosa* is an environmental organism typically found on skin of people and has been isolated environmentally from soil, manure, canal water, and straw. *P. aeruginosa* is a ubiquitous soil organism that proliferates in standing water and wet and warm materials such as leaking hot water pipe insulation and showers. (Boss and Day 2003; CDC 2003; Stetzenbach and Yates 2003) *P. aeruginosa* is an opportunistic pathogen and can be especially problematic for those with cystic fibrosis and burn victims. (Boss and Day 2003; CDC 2003; Gaynes and Edwards 2005) Nosocomial infections due to *P. aeruginosa* have been associated with poorly maintained and/or malfunctioning HVAC systems. (CDC 2003) *P. aeruginosa* causes urinary tract and skin infections, septicemia (blood infections), and meningitis. (Boss and Day 2003; Wilson 2005) *P. aeruginosa* is a main causative agent of pneumonia, which is a leading cause of morbidity and mortality and is the sixth most common cause of death in the United Kingdom and the Untied States. (Wilson 2005) The presence of *P. aeruginosa* in the air is of concern in the hospital environment.

Pseudomonas oryzihabitans was identified as an indicator of indoor contamination within the space controlled by AHU 21 (2006 data). *P. oryzihabitans* is an opportunistic pathogen and is a common soil bacterium. (Freney, Hansen et al. 1988; Bendig, Mayes et al. 1989; Munro, Buckland et al. 1990; Podbielski, Mertens et al. 1990; Reina, Odgardd et al. 1990; Esteban, Valero-Moratalla et al. 1993; Lam, Isenberg et al. 1994; Lucas, Kiehn et al. 1994; Rahav, Simhon et al. 1995; Romanyk, Gonzalez-Palacios et al. 1995; Liu, Shi et al. 1996; Anzai, Kudo et al. 1997; Kiris, Over et al. 1997; Lin, Hsueh et al. 1997; Marin, Garcia de Viedma et al. 2000) *Pseudomonas stutzeri* was identified as an indicator of an indoor source in the space controlled by AHU 15 (2005 data) and AHU 19 (2006 Data). It is considered an environmental organism and has been isolated from soil, manure, canal water, and straw. It has been isolated from the respiratory tract, wounds, blood, urogenital tract, spinal and joint fluid of humans and is associated with NI. (Palleroni, Doudoroff et al. 1970; Reisler and Blumberg 1999; Taneja, Meharwal et al. 2004; Lalucat, Bennasar et al. 2006; Yee-Guardino, Danziger-Isakov et al. 2006) *P. stutzeri* surface contamination was confirmed from a ceiling tile within the space controlled by AHU 19. The presence of indoor contamination of *P. stutzeri* is of concern in the hospital environment.

Pseudomonas species are one of the most antibiotic resistant bacteria and resistant to antiseptics such as quaternary ammonium compounds. (Georgiev 1998; Boss and Day 2003) This property allows them to survive environmental conditions which are lethal to many other bacteria. (Georgiev 1998) *Pseudomonas* species in general are among the most clinically relevant healthcare associated pathogens. (CDC 2003) In general, the presence of *Pseudomonas* in a hospital setting is problematic. (Boss and Day 2003) *P. fluorscens* was identified as a contaminant in the drain pan water of AHU 19 and *P. Stutzeri* was identified as surface contamination in the space controlled by AHU 19. The presence of *Pseudomonas* species actively growing in the indoor healthcare environment is especially problematic. (Boss and Day 2003)

Rhizobium species are environmental source fungi typically found in the roots of plants and in soils. (Stetzenbach, Buttner et al. 2004) *Rhizobium radiobacter* was identified as an indicator of indoor contamination in the space controlled by AHU 11, and was not detected in the outdoor air (n = 118 outdoor air samples). This indicates the presence of an indoor source. *Rhizobium* sp. is recognized as an opportunistic human pathogen associated with NI in the immunocompromised host. (Lai, Teng et al. 2004) It is likely that the absence of final filters in the space controlled by AHU 11 prevented indoor concentrations of *Rhizobium* sp. from being removed from the air stream. It is likely that the *R. radiobacter* contamination originated from within the Hospital from an environmental source.

Staphylococcus aureus was identified in the condensate water of AHU 21 (2006 data). The presence of *S. aureus* actively growing indoors should be considered a health risk in the hospital environment. *S. aureus* produces toxins and can infect surgical wounds, develop resistance to antibiotics, and is the agent of toxic shock syndrome. *S. aureus* also produces toxins that cause food poisoning. (Boss and Day 2003) *S. aureus* was not identified in the indoor air via air sampling.

Tatumella sp. is a member of the family *Enterobacteriaceae*, which are widely distributed in soil, water, plants, and animals. (Hollis, Hickman et al. 1981; Georgiev 1998) *Enterobacteriaceae* are responsible for over half of the NI in the United States. (Hollis, Hickman et al. 1981) *Tatumella ptyseos* was identified as an indicator of an indoor source in the space controlled by AHUs 10 (2005 data) and 21 (2006 data). *T. ptyseos* is associated with NI, but there is no indication that the indoor source of *Tatumella* species was associated with building contamination.

4.2.1 Actinomycetes

Actinomycetes were once considered fungi because of their resemblance to fungi. However, these organisms are not fungi, but are bacteria. (Georgiev 1998; AerotechP&K 2006) The presence of *Actinomycetes* is rare in buildings and outdoors. The presence of *Actinomycetes* indoors may be considered an indication of an indoor environmental source (Macher 1999) and may add to the complexity of the environmental problem. (Straus 2004) The

Actinomycetes have the potential to become opportunistic, especially in immunocompromised hosts. (McNeil and Brown 1994; Georgiev 1998)

Thermophilic *Actinomycetes* are usually found in closed barns, silos, grain mills, and bagasse (sugar cane waste). *Actinomycetes* have been found in problematic or poorly maintained air conditioning ducts. (AerotechP&K 2006) Allergic respiratory disease caused by the *Actinomycetes* is referred to as farmer's lung, a hypersensitivity reaction from repeated exposure to antigens produced by the *Actinomycetes*. *Actinomycetes* may also cause other diseases such as ocular infections, periodontal disease, and abscess formations, which can infect humans. (Georgiev 1998) Several reports indicate that infections by these bacteria are not rare (especially from the *Actinomycete* genus *Nocardia*), are frequently misdiagnosed, or are under diagnosed, and that the incidence of infection is increasing. The spectrum of disease caused by *Nocardia* is broad and varies from a self-limited, asymptomatic infection to an aggressive, destructive disease resulting in death. *Nocardial* infections are commonly diagnosed in previously healthy adults with no predisposing factors. (McNeil and Brown 1994) The *Nocardiae* are frequently being recognized as emerging opportunistic pathogens; the most common underlying predispositions include organ transplantation, malignancies, use of corticosteroids, alcohol abuse, diabetes, or other debilitating factors. (McNeil and Brown 1994; AerotechP&K 2006)

Nocardioform bacilli and/or presumptive *Nocardioforms* were identified as indicators of an indoor microbiological source in the space controlled by AHUs 17, 18, and 19. *Nocardioforms* include the genus *Nocardia* and the *Nocardioform Actinomycetes*. (Georgiev 1998; Boss and Day 2003; Gibson, Gilleron et al. 2003; Stetzenbach and Yates 2003) *Nocardia* species are the *Nocardioform* most often isolated from NI. (Georgiev 1998) *Nocardia* are found in soil around the world, and the indoor concentrations of *Nocardioforms* were not identified via surface sampling and could not be associated with building contamination. (Macher 1999) *Nocardia* morphologically resembles *Actinomyces* species, and both are bacteria that are often pathogenic and opportunistic. (McNeil and Brown 1994; Georgiev 1998; Boss and Day 2003; AerotechP&K 2006)

Actinomycetes were detected indoors via air sampling and not outdoors (n=94 outdoor air samples) within the space controlled by AHU 10 and AHU 17 for the fungal samples incubated at 25°C. *Actinomycetes* were detected indoors and not outdoors (n=48 outdoor air samples) via air sampling within the space controlled by AHU 19 for the fungal samples incubated at 37°C. *Actinomycetes* were confirmed via surface sampling within the spaces controlled by AHUs 13, 15 (*Actinomyces* sp.), 16, and 17 (*Actinomycetes*-like). Therefore, it can be concluded that the airborne concentrations of *Actinomycetes* identified within the Hospital were due to an indoor source of surface contamination. (Macher 1999) The presence of *Actinomycetes* indoors is of concern in the indoor hospital environment.

5. Surface sampling interpretation and health risk model

Allergic reactions to indoor allergens can produce inflammatory diseases of the eyes, nose, throat, and bronchi, which are medical problems that come under the headings of allergic conjunctivitis, allergic rhinitis, allergic asthma, and hypersensitivity pneumonitis (extrinsic allergic alveolitis) respectively. (Pope, Patterson et al. 1993) The Health Risk Model (HRM) considers the type of microbial contamination and the type of person expected to be within a specific Hospital location. Critical care areas are areas of the Hospital where it is expected that immunocompromised persons will be present and therefore contamination within a critical care area is given a higher weight in the overall determination of health risk.

Risk assessment is a process designed to evaluate the potential relationship that may exist between exposure to aeroallergens and a particular effect (e.g. toxic effect, allergic sensitization, infection, allergic disease). (Pope, Patterson et al. 1993) A HRM was utilized to semi-quantitatively identify the health risk associated with fungal and bacterial surface contamination within the hospital. Monitoring for allergens can help characterize environments with respect to specific allergens (e.g., fungi and/or bacteria). Both fungi and bacteria secrete enzymes that act as allergens. (Pope, Patterson et al. 1993) Source or reservoir samples have been used as indicators of exposure to indoor allergens and measurement interpretations can be semi-quantitative (e.g., "presence or absence" or "low, medium, or high). (Pope, Patterson et al. 1993) Environmental bacteria also grow in all wet spaces and are found in most cases where there is mold growth. (Institute of Medicine Committee on Damp Indoor Spaces and Health 2004)

The American Industrial Hygiene Association's consensus document *A Strategy for Assessing and Managing Occupational Exposures* (Mulhausen and Damiano 1998) served as the basis for the HRM. The HRM utilized criteria and recommendations of the Centers for Disease Control and Prevention (CDC 2003), US Environmental Protection Agency (USEPA 2001), American Conference of Governmental Industrial Hygienists (ACGIH 1999), Institute of Medicine (Pope, Patterson et al. 1993), the New York City Department of Health and Mental Hygiene (NYCDHMH 2006), the American Society of Heating, Refrigerating, and Air Conditioning Engineers (ASHRAE 2003) and the Vanderbilt University Medical Center (VUMC 2006) in establishing the risk factors for the model. A literature search was conducted to determine if the organisms identified via surface sampling within the Hospital were allergenic, pathogenic or opportunistic, and capable of producing fungal or bacterial toxin. The HRM resulted in a Health Risk classification of the space controlled by each AHU.

Health Risk was classified as High, Medium, Low, and de Minimis. The risk classifications were determined with input from experts in medical microbiology, industrial hygiene, public health, engineering controls, infection control, and medicine. A de Minimis risk score means that no indoor environmental contamination was found. A low risk score means the environmental conditions present do not indicate extensive biological contamination and/or the risk associated with adverse health affects to building occupants is low. A medium risk score indicates that environmental conditions present an increased risk for adverse health effects to building occupants due to environmental contamination and remediation is necessary. A high risk score indicates that conditions exist for adverse health effects due to exposure to biological contaminants and immediate intervention is necessary. Figure 9 below displays the HRM scores for the indoor space controlled by each AHU.

Indoor surface fungal and bacterial surface contamination was identified in every area of the hospital investigated. Air sampling confirmed the presence of indicators of indoor contamination in each of the spaces investigated. See **Section 4. Sampling Interpretation Summary** above. The spaces under the control of every AHU placed within the medium risk category. The environmental conditions are present such that immunocompromised or allergic patients are not fully protected against the risk of NI due to environmental bioaerosols. (Perdelli, Christina et al. 2006) Healthy hospital workers are not protected

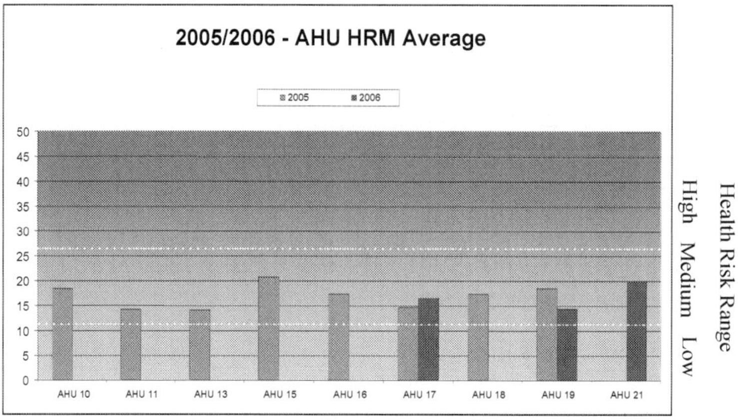

Fig. 9. Health Risk Model Scores for the space controlled by each AHU.

against allergic reactions to indoor bioaerosols growing within the facility and are at an increased risk of respiratory infections, including the common cold, sinusitis, tonsillitis, otitis, and bronchitis. (Institute of Medicine Committee on Damp Indoor Spaces and Health 2004) Hospital workers who are immunocompromised (e.g., diabetics, asthmatics, those undergoing cancer therapy or who have recent invasive surgery) are more susceptible to allergic reactions and the risk of work-related infections. The results of the HRM indicate that patients and staff are being exposed to microorganisms that are actively growing within the hospital which present a risk higher than what is expected in a hospital without water damage, microbial contamination, moisture infiltration, and OSA infiltration.

Periods of maintenance and non-routine operation of HVAC systems within the hospital can result in filter bypass, dissemination of biological contamination, and the infiltration of unfiltered OSA into the hospital, placing the hospital within the High Risk category due the creation of an exposure pathway during these times. Hence, times during and immediately after maintenance and non-routine operation of the HVAC systems present a high risk for health effects due to bioaerosols in the indoor environment. (CDC 2003)

6. Discussion and hypothesis testing

Indoor microbial contaminants and infectious agents are closely related to water and moisture-related conditions. (Bartley 2000) The scenario that has emerged within the hospital is one in which immunosuppressed or allergic patients are not fully protected against the risk of NI or allergies due to environmental bioaerosols. (Perdelli, Christina et al. 2006) Where an indoor environment is exhibiting growth or airborne suspension of bioaerosols, risk may be determined to exist even if levels are less than outdoor baseline and control levels. (Boss and Day 2003) While indoor bioaerosol concentrations were consistently lower than OSA, the biodiversity was consistently different. Air sampling indicators of an indoor source and microbial surface contamination were identified in each of the areas investigated.

The objectives of environmental control in buildings are to prevent or minimize occupant exposures that can be deleterious to human health and to provide for the comfort and well-being of the occupants. (Pope, Patterson et al. 1993) The inability of the hospital's environmental systems in the areas investigated to manage moisture has resulted in a situation where patients, visitors, and staff are exposed to airborne microorganisms from indoor surfaces and OSA infiltration that normally would not be present within the building. This has resulted in an indoor environment that is not hygienic from the perspective of environmental contamination associated with the building and building systems.

Infection prevention in the hospital environment is one of the goals of healthcare workers and facilities. (Boss and Day 2003) Visitors, volunteers, and staff can both be infected and infectious. Highly filtered air is essential to preventing infection in healthcare facilities. (Boss and Day 2003) With the exception of AHU 11, all areas investigated were under the environmental control of an air conditioning system with >90% efficiency filters. The presence of the high efficiency final filters was largely responsible for the relatively low concentrations of total bioaerosols identified in the hospital ISA. Of particular concern is that a breach in the integrity of the final filters or dissemination of contamination during maintenance activities or non-routine operation could place the areas of the hospital which now fall into the medium risk category into the high risk category by creating an exposure pathway from contamination within the air handlers, OSA, and re-circulated air of the hospital to the occupant. The high efficiency filters within the AHUs are the hospital's main defense against nosocomial infection associated with indoor environmental contamination.

Although HVAC systems equipped with high efficiency filters can significantly reduce indoor concentrations of bioaerosols (Perdelli, Christina et al. 2006), the areas of the hospital investigated were under negative pressure, allowing the infiltration of unfiltered OSA, and had visible and confirmed microbial contamination on indoor surfaces and within each AHU. The presence of unfiltered air entering the Hospital via infiltration due to negative building pressurization is of concern (CDC 2003) and is significant risk factor for NI. (Streifel, Lauer et al. 1983; Rhame, Streifel et al. 1984; Rhame 1991; Nolard 1994) Biofilms (communities of bacteria in a matrix) and standing water were present in each of the AHUs investigated in 2005 and indicate the presence of excess moisture and poorly draining drain pans. Biofilms are conglomerates of microorganisms and indicate microbiological amplification within the AHUs. Visible surface contamination was identified within all the areas investigated.

Note that a breach in filter media or during times of HVAC system maintenance such as changing filters or starting and stopping the units, will cause an increase in airborne microorganisms or infiltration of unwanted OSA into the ISA of the Hospital. A failure or malfunction of any HVAC system component may subject patients to discomfort and exposure to airborne contaminants. Accumulation of dust and moisture in HVAC systems increases the risk for spread of health-care-associated environmental fungi and bacteria. If moisture is present in the HVAC system, periods of stagnation should be avoided. Bursts of organisms can be released upon system start-up, increasing the risk of airborne infection. If the ventilation system is out of service, rendering indoor air stagnant, sufficient time must be allowed to clean the air and re-establish the appropriate number of air changes once the HVAC system begins to function again. Reactivation of HVAC systems after shutdown can dislodge substantial amounts of dust and create a transient airborne increase of fungal spores. (CDC 2003) Therefore, during and after non-routine operation of the HVAC systems, the hospital is at high risk for nosocomial infection from environmental microorganisms.

The hypotheses testing results are:

Hypothesis A: The 90-95% final filters are controlling particulate matter generated by the AHUs and preventing contamination downstream of the filters. Note: AHU 11 does not have final filters.

Test: OSA versus ISA comparisons of total bioaerosols and spore traps.

Method: Investigator observations, interpretations, and literature review.

Test Result: Accept Hypothesis A—The final filters are controlling the dissemination of particulate matter and preventing the dissemination of particles downstream of the filters during routine operation. Statistical comparisons showed that indoor concentrations of non-culturable (spore traps) fungal bioaerosols were significantly lower than OSA concentrations. At least 79% (2005 data) of the non-culturable fungal bioaerosols are being removed by the final filters of the AHUs investigated, with the exception of AHU 11, which does not have final filters. At least 66% (2005 data) of the culturable fungal bioaerosols are being removed by the final filters of the AHUs investigated, with the exception of AHU 11, which does not have filters. With the exception of the bacterial (25°C) air samples in AHUs 10 and 11, the percent differences from outside air to indoor air were at least 35%. This is an indication that the filters were removing bacteria from both the outside and re-circulated air of the Hospital. The 2006 data showed a minimum percent reduction of 54% of all fungal samples (culturable and non-culturable) for the before filter to after filter comparisons. Therefore, particulate matter of similar aerodynamic diameters to fungi and bacteria are being removed from the air stream by the final filters.

Hypothesis B: The 90-95% final filters are preventing microbial contamination downstream of the filters.

Test: OSA versus ISA comparisons of bioaerosols.

Method: Investigator observations, interpretations, and literature review.

Test Result: Accept Hypothesis B—The final filters are preventing fungal particulate dissemination downstream of the filters by filtering the particles from the airstreams during routine operation. See Test Result for Hypothesis A. Indoor concentrations of non-culturable fungal bioaerosols were consistently lower and spore trap sampling results indicated that indoor concentrations of fungal bioaerosols were significantly lower than indoor concentrations. At least 66% (2005 data) of the culturable fungal bioaerosols are being removed by the final filters of the AHUs investigated, with the exception of AHU 11, which does not have filters. The 2006 data showed a minimum percent reduction of 54% of all fungal samples (culturable and non-culturable) for the before filter to after filter comparisons. The final filters are preventing the dissemination of bacterial contamination downstream of the filters during routine operation. Species of bacteria were identified within the AHUs and not identified post-filter, indicating that the filters were preventing the dissemination of contamination downstream. Indoor bacterial concentrations were consistently lower than outdoor concentrations. With the exception of the bacterial (25°C) air samples in AHUs 10 and 11, the percent differences from outside air to indoor air were at least 35% for the 2005 data. For the 2006 data, the minimum percent difference from OSA to ISA for bacteria was 64%. This is an indication that the filters were removing bacteria from the air streams. Note that indoor concentrations of bacteria are expected to be higher than outdoors but were consistently lower in the Hospital.

Hypothesis C: Staff is being exposed to potentially harmful concentrations of biological contaminants.
Test: OSA versus ISA comparisons of bioaerosols.
Method: Air samples, indoor surface sampling, HRM, Investigator observations, interpretations, and literature review.
Test Result: Accept Hypothesis D—Both the 2005 and 2006 data show indicators of indoor contamination and confirmed microbial surface contamination in each area investigated. Analysis for biodiversity showed that the bioaerosol profiles of ISA and OSA were consistently different. The HRM places all of the areas investigated in the medium risk category, indicating that staff are at risk for adverse health effects due to environmental contamination.
Hypothesis D: Patients are being exposed to harmful quantities of biological contaminants.
Test: OSA versus ISA comparisons of bioaerosols.
Method: Air samples, indoor surface sampling, HRM, Investigator observations, interpretations, and literature review.
Test Result: Accept Hypothesis E—Both the 2005 and 2006 data show indicators of indoor contamination and confirmed microbial surface contamination in each area investigated. Analysis for biodiversity showed that the bioaerosol profiles of ISA and OSA were consistently different, especially for species of *Penicillium* and *Aspergillus*. The HRM places all of the areas investigated in the medium risk category, indicating that patients are at risk for adverse health effects due to environmental contamination.

7. Conclusion

This 2-year investigation of bioaerosol concentrations in a hospital facility demonstrates that consideration of error, sampling variability, identification of microorganisms at the species level, and at least 6 replicate samples per location are necessary to detect significant differences in bioaerosol concentrations. Furthermore, spore trap samples were not sufficient to detect these differences regardless of the number of samples taken. Environmental investigators must consider this in their investigation strategy. Of particular importance is the failure of spore trap sampling to detect the same differences in airborne fungal bioaerosol species identified by culturable air sampling. Spore trap sampling should be considered a tool to determine a gross estimation of the quality of the indoor environment with regards to fungal bioaerosols and should not be used to make estimations on health or bioaerosol exposure. In all cases, spore trap sampling failed to detect differences in the aerobiological profile, masking significant concentrations of airborne fungal bioaerosols, including species of *Aspergillus* and *Penicillium*.
Statistical validity, analytical/sampling error, and species diversity should be considered for any bioaerol sampling intended to make comparisons on airborne concentrations of bioaersols between two or more environments. Identification of surface contamination should be considered de facto evidence of exposure to potentially harmful biological compounds. This study shows that high efficiency filtration can result in the environmental control of airborne bioaerosols. This is especially important in the hospital setting. PCR sampling shows promise because longer air sampling times can be used, surface samples can detect the presence of organisms on surfaces for an indication of a building's microbial burden, and because of relatively simple PCR sampling methods.

8. References

ACGIH (1999). Bioaerosols: Assessment and Control. Cincinnati, OH, American Conference of Governmental Industrial Hygienists.

ACOEM (2002). Adverse Human Health Effects Associated with Molds in the Indoor Environment, American College of Occupational and Environmental Medicine: 1-10.

AerotechP&K. (2006). "Microbial and IAQ Glossaries." Retrieved August, 2006, from www.aerotechpk.com.

Agency, U. S. E. P. (2001). "Mold remediation in schools and commercial buildings." USEPA EPA 402-K-01-001.

Anzai, Y., Y. Kudo, et al. (1997). "The phylogeny of the genera Chryseomonas, Flavimonas, and Pseudomonas supports synonymy of these three genera." Int J Syst Bacteriol 47(2): 249-251.

Arnow, P. M., M. Sadigh, et al. (1991). "Endemic and epidemic aspergillosis associated with in-hospital replication of Aspergillus organisms." J Infect Dis 164(5): 998-1002.

ASHRAE (1992). Gravimetric and Dust-Spot Procedures for Testing Air-Cleaning Devices Used in General Ventilation for Removing Particulate Matter Atlanta, GA, American Society of Heating, Refrigerating, and Air-Conditioning Engineers: 32.

ASHRAE (1999). Method of Testing General Ventilation Air-Cleaning Devices for Removal Efficiency by Particle Size Atlanta, GA, ASHRAE: 41.

ASHRAE (2003). 2003 ASHRAE handbook : heating, ventilating, and air-conditioning applications. Atlanta, Ga., ASHRAE.

Bartley, J. M. (2000). "APIC state-of-the-Art report: the role of infection control during construction in health care facilities." Am J Infect Control 28(2): 156-169.

Beggs, C. and K. Kerr (2000). "The Threat Posed by Airborne Micro-Organisms." Indoor Built Environmnet 9: 241-245.

Bendig, J. W., P. J. Mayes, et al. (1989). "Flavimonas oryzihabitans (Pseudomonas oryzihabitans; CDC group Ve-2): an emerging pathogen in peritonitis related to continuous ambulatory peritoneal dialysis?" J Clin Microbiol 27(1): 217-218.

Boss, M. J. and D. W. Day (2003). Biological risk engineering handbook : infection control and decontamination. Boca Raton, Fla., Lewis Pub.

Burge, H. A. (2000). The Fungi. Indoor Air Quality Handbook. J. D. Spengler, J. M. Samet and J. McCarthy, McGraw-Hill: 45.41-45.33.

Burton, D. J. (2006). Industrial Hygiene Workbook: the Occupational Health Sciences. Bountiful, UT, IVE, Inc.

CDC (1992). "Public health focus: surveillance, prevention, and control of nosocomial infections." Morb Mort Wkly Rep 42(42): 783-787.

CDC (2003). Guidelines for environmental infection control in health-care facilities: Recommendations of CDC and the HealthCare Infection Control Practices Advisory Committee. Atlanta, GA.

Conover, W. J. (1999). Practical nonparametric statistics. New York, Wiley.

Dillon, H. K., P. A. Heinsohn, et al. (2005). Field guide for the determination of biological contaminants in environmental samples. Fairfax, VA, American Industrial Hygiene Association.

Dillon, H. K., P. A. Heinsohn, et al. (1996). Field guide for the determination of biological contaminants in environmental samples. Fairfax, VA, American Industrial Hygiene Association.

Dulworth, S. and B. Pyenson (2004). "Healthcare-associated infections and length of hospital stay in the medicare population." American Journal of Medical Quality 19(3): 121-127.

Eickhoff, T. C. (1994). "Airborne nosocomial infection: a contemporary perspective." Infection Control and Hospital Epidemiology 15(10): 663-672.

Esteban, J., M. L. Valero-Moratalla, et al. (1993). "Infections due to Flavimonas oryzihabitans: case report and literature review." Eur J Clin Microbiol Infect Dis 12(10): 797-800.

Fox, B. C., L. Chamberlin, et al. (1990). "Heavy contamination of operating room air by Penicillium species: identification of the source and attempts at decontamination." Am J Infect Control 18(5): 300-306.

Freeman, J. and J. E. McGowan, Jr. (1978). "Risk factors for nosocomial infection." J Infect Dis 138(6): 811-819.

Freeman, J. and J. E. McGowan, Jr. (1981). "Differential risk of nosocomial infection." Am J Med 70(4): 915-918.

Freney, J., W. Hansen, et al. (1988). "Postoperative infant septicemia caused by Pseudomonas luteola (CDC group Ve-1) and Pseudomonas oryzihabitans (CDC group Ve-2)." J Clin Microbiol 26(6): 1241-1243.

Fung, F. and W. G. Hughson (2002). Health effects of indoor bioaerosol exposure. Indoor Air.

Fung, F. and W. G. Hughson (2003). "Health effects of indoor fungal bioaerosol exposure." Appl Occup Environ Hyg 18(7): 535-544.

Gaynes, R. and J. R. Edwards (2005). "Overview of nosocomial infections caused by gram-negative bacilli." Clin Infect Dis 41(6): 848-854.

Georgiev, V. S. (1998). Infectious diseases in immunocompromised hosts. Boca Raton, CRC Press.

Gibson, K. J. C., M. Gilleron, et al. (2003). "Structural and functional features of Rhodococcus ruber lipoarabinomannan." Microbiology 149: 1437-1445.

Hitchcock, P. J., M. Mair, et al. (2006). "Improving performance of HVAC systems to reduce exposure to aerosolized infectious agents in buildings; recommendations to reduce risks posed by biological attacks." Biosecur Bioterror 4(1): 41-54.

Hollis, D. G., F. W. Hickman, et al. (1981). "Tatumella ptyseos gen. nov., sp. nov., a member of the family Enterobacteriaceae found in clinical specimens." J Clin Microbiol 14(1): 79-88.

Institute of Medicine Committee on Damp Indoor Spaces and Health (2004). Damp indoor spaces and health. Washington, DC, National Academies Press.

Kalliokoski, P. (2003). "Risks caused by airborne microbes in hospitals - source control is important." Indoor and Built Environment 12: 41-46.

Kerr, J. R., J. E. Moore, et al. (1995). "Investigation of a nosocomial outbreak of Pseudomonas aeruginosa pneumonia in an intensive care unit by random amplification of polymorphic DNA assay." Journal of Hospital Infection 30: 125-131.

Kiris, S., U. Over, et al. (1997). "Disseminated Flavimonas oryzihabitans infection in a diabetic patient who presented with suspected multiple splenic abscesses." Clin Infect Dis 25(2): 324-325.

Lai, C. C., L. J. Teng, et al. (2004). "Clinical and microbiological characteristics of Rhizobium radiobacter infections." Clin Infect Dis 38(1): 149-153.

Lalucat, J., A. Bennasar, et al. (2006). "Biology of Pseudomonas stutzeri." Microbiol Mol Biol Rev 70(2): 510-547.

Lam, S., H. D. Isenberg, et al. (1994). "Community-acquired soft-tissue infections caused by Flavimonas oryzihabitans." Clin Infect Dis 18(5): 808-809.

Lentino, J. R., M. A. Rosenkranz, et al. (1982). "Nosocomial aspergillosis: a retrospective review of airborne disease secondary to road construction and contaminated air conditioners." Am J Epidemiol 116(3): 430-437.

Lin, R. D., P. R. Hsueh, et al. (1997). "Flavimonas oryzihabitans bacteremia: clinical features and microbiological characteristics of isolates." Clin Infect Dis 24(5): 867-873.

Liu, P. Y., Z. Y. Shi, et al. (1996). "Epidemiological typing of Flavimonas oryzihabitans by PCR and pulsed-field gel electrophoresis." J Clin Microbiol 34(1): 68-70.

Lucas, K. G., T. E. Kiehn, et al. (1994). "Sepsis caused by Flavimonas oryzihabitans." Medicine (Baltimore) 73(4): 209-214.

Macher, J. (1999). Bioaerosols: assessment and control. Cincinnati, Ohio, American Conference of Governmental Industrial Hygienists.

Marin, M., D. Garcia de Viedma, et al. (2000). "Infection of hickman catheter by Pseudomonas (formerly flavimonas) oryzihabitans traced to a synthetic bath sponge." J Clin Microbiol 38(12): 4577-4579.

McDonald, L. C., M. Walker, et al. (1998). "Outbreak of Acinetobacter spp. bloodstream infections in a nursery associated with contaminated aerosols and air conditioners." Pediatr Infect Dis J 17(8): 716-722.

McNeil, M. M. and J. M. Brown (1994). "The medically important aerobic actinomycetes: epidemiology and microbiology." Clinical Microbiology Reviews: 357-417.

Montgomery, D. C. (2001). Design and analysis of experiments, John Wiley and Sons, Inc.

Mulhausen, J. R. and J. Damiano (1998). A strategy for assessing and managing occupational exposures. Fairfax, VA, AIHA Press.

Munro, R., G. Buckland, et al. (1990). "Flavimonas oryzihabitans infection of a surgical wound." Pathology 22(4): 230-231.

Nolard, N. (1994). "[Links between risks of aspergillosis and environmental contamination. Review of the literature]." Pathol Biol (Paris) 42(7): 706-710.

NYCDHMH (2006). "Guidelines on Assessment and Remediation of Fungi in Indoor Environments." The New York City Department of Health and Mental Hygiene.

Page, E. H. and D. B. Trout (2001). "The role of *Stachybotrys* mycotoxins in building related illness." AIHA J 62: 644-648.

Palleroni, N. J., M. Doudoroff, et al. (1970). "Taxonomy of the aerobic pseudomonads: the properties of the Pseudomonas stutzeri group." J Gen Microbiol 60(2): 215-231.

Perdelli, F., M. L. Christina, et al. (2006). "Fungal contamination in hospital environments." Infection Control and Hospital Epidemiology 27(1): 44-47.

Podbielski, A., R. Mertens, et al. (1990). "Flavimonas oryzihabitans septicaemia in a T-cell leukaemic child: a case report and review of the literature." J Infect 20(2): 135-141.

Pope, A. M., R. Patterson, et al. (1993). Indoor allergens: assessing and controlling adverse health effects. Washington, D.C., National Academy Press.

Pope, A. M., Patterson, R., Burge, H., Ed. (1993). Indoor Allergens. Assessing and controlling adverse health effects. Washington, D.C., National Academy Press.

Prezant, B., D. M. Weekes, et al., Eds. (2008). Recognition, Evaluation, and Control of Indoor Mold, American Industrial Hygiene Association.

Rahav, G., A. Simhon, et al. (1995). "Infections with Chryseomonas luteola (CDC group Ve-1) and flavimonas oryzihabitans (CDC group Ve-2)." Medicine (Baltimore) 74(2): 83-88.

Redd, S. C. (2002). "State of the Science on Molds and Human Health." Center for Disease Control and Prevention, U.S. Department of Health and Human Services.: 1-11.

Reina, J., J. Odgardd, et al. (1990). "Flavimonas oryzihabitans (formerly CDC group Ve-2) bacteremia in a pediatric patient on assisted ventilation." Eur J Clin Microbiol Infect Dis 9(10): 786-788.

Reisler, R. B. and H. Blumberg (1999). "Community-acquired Pseudomonas stutzeri vertebral osteomyelitis in a previously healthy patient: case report and review." Clin Infect Dis 29(3): 667-669.

Rhame, F. S. (1991). "Prevention of nosocomial aspergillosis." J Hosp Infect 18 Suppl A: 466-472.

Rhame, F. S., A. J. Streifel, et al. (1984). "Extrinsic risk factors for pneumonia in the patient at high risk of infection." Am J Med 76(5A): 42-52.

Romanyk, J., R. Gonzalez-Palacios, et al. (1995). "A new case of bacteraemia due to Flavimonas oryzihabitans." J Hosp Infect 29(3): 236-237.

Smith, G. E. (1990). Sampling and identifying allergenic pollens and molds. San Antonio, TX, Blewstone Press.

Spicer, R. C. and H. J. Gangloff (2000). "Limitations in Application of Spearman's Rank Correlation to Bioaerosol Sampling Data." AIHA J 61(May/June): 362-366.

Stetzenbach, L. D., M. P. Buttner, et al. (2004). "Detection and enumeration of airborne biocontaminants." Curr Opin Biotechnol 15(3): 170-174.

Stetzenbach, L. D. and M. V. Yates (2003). The dictionary of environmental microbiology. San Diego, Calif., Academic Press.

Straus, D. C., Ed. (2004). Sick Building Syndrome. Advances in Applied Microbiology. London, Elsevier Academic Press.

Streifel, A. J., J. L. Lauer, et al. (1983). "Aspergillus fumigatus and other thermotolerant fungi generated by hospital building demolition." Appl Environ Microbiol 46(2): 375-378.

Taneja, N., S. K. Meharwal, et al. (2004). "Significance and characterisation of pseudomonads from urinary tract specimens." J Commun Dis 36(1): 27-34.

USEPA (2001). Mold remediation in schools and commercial buildings. United States Environmental Protection Agency. EPA 402-K-01-001.

VUMC (2006). Infection Control Interventions During Construction in Patient Care Areas. Procedure 10-10.17. Nashville, TN, Vanderbilt University Medical Center. 10-10.17.

Walsh, T. J. and A. H. Groll (1999). "Emerging fungal pathogens: evolving challenges to immunocompromised patients for the twenty-first century." Transpl Infect Dis 1(4): 247-261.

Weber, A. and E. Page (2001). "Renovation of contaminated building materials at a facility serving pediatric cancer outpatients." Applied Occupational and Environmental Hygiene 16(1): 2-31.

Wenzel, R. P. (1997). Prevention and control of nosocomial infections. Baltimore, William & Wilkins.

Wilson, M. (2005). Microbial inhabitants of humans : their ecology and role in health and disease. New York, Cambridge University Press.

Wilson, S. C., W. H. Holder, et al. (2004). "Identification, remediation, and monitoring processes used in a mold-contaminated high school." Adv Appl Microbiol 55: 409-423.

Yee-Guardino, S., L. Danziger-Isakov, et al. (2006). "Nosocomially acquired Pseudomonas stutzeri brain abscess in a child: case report and review." Infect Control Hosp Epidemiol 27(6): 630-632.

12

Distributed Smart Sensing Systems for Indoor Monitoring of Respiratory Distress Triggering Factors

Octavian Postolache, José Miguel Pereira,
Pedro Silva Girão and Gabriela Postolache
*Instituto de Telecomunicações ESTSetubal (LabIM),
Polytechnic Institute of Setúbal Escola de Saúde,
Universidade Atlântica
Portugal*

1. Introduction

Indoor air quality pollution [1][2] represents one of the factors associated with the etiology of chronic obstructive pulmonary disease and also plays an important role in respiratory distress, the second most common symptom of adults that request emergency transportation to the hospital and is associated with a relatively high overall mortality before hospital discharge [3][4][5][6]. The prevention of acute respiratory distress or asthma attacks can be possible by monitoring the air quality conditions using distributed smart sensing systems characterized by accuracy, short time response, and robustness as well as by data processing, data logging and data communication capabilities.

Considering the importance of indoor air quality monitoring, different distributed measuring system architectures and associated calibration methods and systems are presented in the literature [7][8][9]. The main elements of these kind of systems are not only temperature and relative humidity sensors, but also gas detectors and gas concentration sensors whose metrological characteristics, such as accuracy and linearity are very limited, which implies the design and implementation of signal processing algorithms namely for numerical linearization and common factors correction [10][11][12].

Taking into account the indoor spatial distribution of the temperature and relative humidity values as well as the concentration values of pollutants (e.g CO, CO_2 resulting of combustion), the development of distributed measuring systems [13][14] that can include personal computers (PCs) or mobile devices (e.g. PDAs [15] or smart phones [16]) based human-sensing system interface represents an important requirement for optimal indoor air quality monitoring.

This chapter presents a practical approach concerning distributed smart sensing solutions for air quality monitoring, highlighting the original contributions of the authors in this area. The first part of the chapter deals with the relation between the subject's health status, respiratory distress condition and air quality conditions. The second part contains a brief presentation of solid state sensors [17] that materialize the sensing component of air quality monitoring systems and the third part presents a distributed architecture based on an embedded Web server for air quality monitoring including elements of data processing. In the fourth part, a Bluetooth wireless distributed system including smart sensing nodes and a smart phone programmed as assisted human - distributed air quality monitoring system interfacing device is presented.

Referring to the distributed air quality monitoring system based on embedded Web server nodes, the sensing part of each node is expressed by a thick film metal oxide semi-conductor sensor array that includes general air contaminant, alcohol and organic solvent detection, and CO sensing. The measurement data and pollution alarms from the nodes, which are parts of a wired or wireless network, are obtained through the browser that accesses the nodes' Web pages. A set of temperature and relative humidity sensors are included in the node's hardware in order to increase the gas sensor accuracy through the correction of temperature and humidity influences. This chapter also includes a brief description of the multiple-input–single-output neural network design and implementation [18] that is used to obtain temperature and humidity compensated gas concentration values on the client software side. A Bluetooth enabled wireless sensing network designed and implemented for continuous monitoring of indoor humidity and temperature conditions as well as to detect general air contaminants is described in the chapter. Bluetooth compatible nodes, characterized by data acquisition capabilities, are connected to a mobile device expressed by a smart phone programmed using Java2ME to perform different tasks including data communication, data logging, data processing, alarm generation and graphical user interfacing with the indoor air quality monitoring system. Elements regarding the smart phone embedded software configuration and logged data transfer according to the network architecture and air quality monitoring tasks are discussed and an example of particular implementation is also presented. Using the distributed measurement system, an intelligent assessment of air conditions for risk factor reduction of asthma or chronic obstructive pulmonary disease is proposed.

2. Air quality and its impact on respiratory diseases

Air conditions and respiratory assessment represent an important challenge taking into account that distress is the second most common symptom of adults transported by ambulance and is associated with a relatively high overall mortality before hospital discharge [3]. Among the most common causes of respiratory distress in this setting are congestive heart failure, pneumonia, chronic obstructive pulmonary disease and asthma [4]. It is projected that chronic obstructive pulmonary disease (COPD) will be the third leading cause of death worldwide by 2020, due to an increase in smoking rates and demographic changes in many countries [5]. Worldwide, some 300 million people currently suffer from asthma. It is the most common chronic disease among children [6]. The economic burden of COPD in the US in 2007 was 42.6 billion in health care costs and lost productivity [19]. The indoor air pollution is one of the factors associated with etiology of chronic obstructive pulmonary disease and asthma. There are evidences that the environmental factors acting during early life and interacting with specific "asthma genes" are crucial for the development of chronic, persistent form of disease [20][21]. The identification of the indoor air associated with pathophysiology of COPD and asthma disease will thus be crucial for the primary-prevention strategy.

Poor indoor air quality is becoming an increasing problem around the world because, in general, people are spending more time indoors. This problem is greater in infants who now spend less time playing outside. Reduction of indoor air quality - produced by mould growth, smoke exposure, cooking fire smoke (often using biomass fuels such as wood and animal dung), house dust mites in bedding, carpets and stuffed furniture, chemical irritants (i.e. perfumes), pet dander - may adversely affect the health of building occupants and exacerbate asthma and COPD attacks. Asthma attacks are mainly related to mould growth that is enabled by relative humidity high values for different temperature conditions. Mould spores, bacteria, and mildew thrive in dampened towels, washcloths, and moist or humid areas. Additionally, people with immune or respiratory system problems may more easily succumb to poor health caused by mould growth at home, which is mainly associated with humidity and temperature values. Improved heating systems and less ventilation from outside has also provided more suitable conditions for mould growth. Using air conditions sensing components as parts of an air quality measuring system, high risk disease conditions for indoor occupants can be avoided. Several solutions have been presented in the literature [9][13][22]. In order to assure mobility and flexibility, a wireless network for air quality is an interesting solution considering that the measuring nodes can be distributed in different regions of the house according with different monitoring scenarios. As the interface between a user and the network (human machine interface, HMI), can be used a low cost smart phone (Bluetooth enabled), a PDA (personal digital assistant), or situated displays with interaction capabilities (touch screen enabled).

3. Air quality sensing and data processing

This section contains the description of the main components of a distributed smart sensing system that can be used for air quality assessment. Particular attention is dedicated to the implementation of the sensing nodes, to signal conditioning, and to signal processing of measurement data.

A. Sensing nodes

The sensing nodes are designed and implemented to perform the air quality (AirQ) monitoring using low cost gas sensors and, at the same time, to get additional information about the temperature (T) and relative humidity (RH). This information is used to increase gas concentration measurement accuracy, performing the error compensation caused by temperature and humidity influence.

The gas sensors can be sintered SnO_2 semiconductor heated sensors, as those provided by Figaro [23], that assure pollution event detection (TGS800 - general air contaminant sensor - AC), methane detection (TGS842-M), alcohol and organic solvent detection (TGS822-SV) and carbon monoxide detection (TGS203-CO). Information about temperature and relative humidity are obtained using Smartec SMT160-30 [24] and Humirel HM1500 [25] temperature and relative humidity transducers, respectively.

The gas sensors, connected to proper conditioning circuits, are devices that produce voltages whose values depend on the concentrations of gas expressed in ppm. The used conditioning circuit for the air pollution sensor TGS800, solvent vapors (TGS822) and methane sensor (TGS842) are presented in figure 1.

Electrochemical cells can also be used to implement the sensing units. The NAP-505 [17] is a typical example of this kind of implementation. In this cased, the 3 terminals measuring cell consists of 3 porous noble metal electrodes separated by an acidic aqueous electrolyte, housed within a plastic enclosure. The working principle of the sensing unit is based on chemical reactions between gas and other elements. From the electrical charges that are involved in those reactions it is possible to measure an electrical current that is proportional to gas concentration. Using multiple cells it is possible to measure the concentration of different gas types.

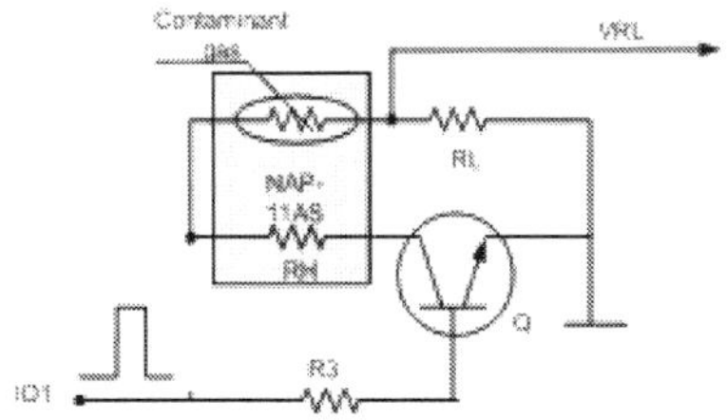

Fig. 1. Gas sensing unit based on semiconductor heated sensors (Vc - circuit voltage, VH - heater voltage, VGS - gas sensor output voltage, RL - load resistance)

Figure 2 represents the main elements of a gas sensing unit based on an electrochemical cell.

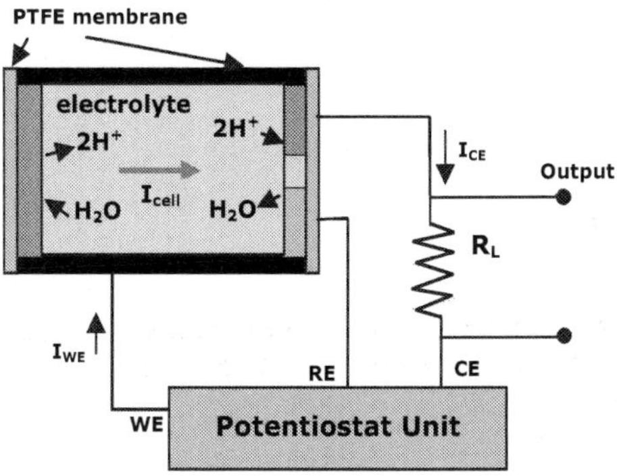

Fig. 2. Gas sensing unit based on an electrochemical cell (RE - reference electrode, CE - counting electrode, WE - working electrode, RL - load resistance)

The measuring cell includes a working electrode (WE), a counter electrode (CE) and a reference electrode (RE) [26]. The conditioning circuit is basically a potentiostat unit that measures the gas dependent current amplitude (I_{cell}) that flows between the CE and WE through cell's electrolyte. The current amplitude is directly proportional to the gas concentration but its value is usually very low, about a few tens of nA. For this reason a careful design of the potentiostat is crucial to obtain an acceptable measurement. Figure 3 represents the electrical diagram of a typical potentiostat conditioning circuit [27][28]. The negative feedback loop, provided by operational amplifiers (OA1 and OA2) and the electrical connection that exists between CE and RE electrodes through the sensing element, assures that the operational amplifiers are working in their linear zones. Since the current between the working and the reference electrodes is very low, the differential voltage between working and counter electrodes is equal to VRE and the output voltage (VADC) from the current to voltage converter implemented by sub-circuit 2 is given by

$$V_{ADC} = -R_F \cdot [f_{sol.}(V_{DAC}) - I_B] \tag{1}$$

where RF represents the feedback resistor of the current to voltage converter, IB represents the polarization current of OA2, VDAC is the output voltage of the D/A converter and fsol is generally a non-linear function that depends on solution characteristics and applied voltage (VWE).

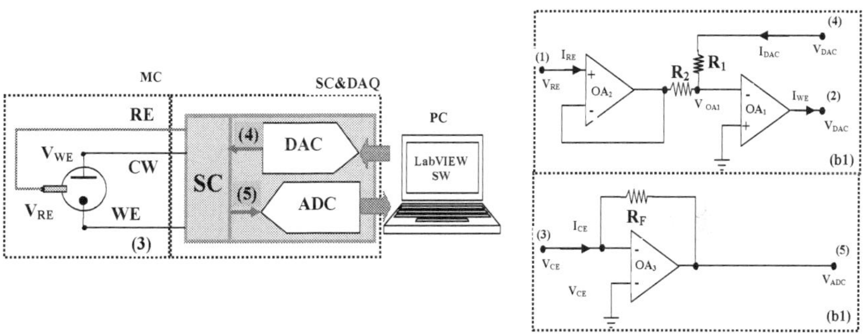

(a) (b)

Fig. 3. Electrical circuit of a voltammetry measuring system (MS- measuring cell, RE-reference electrode, CE- counting electrode, WE- working electrode, SC&DAQ- signal conditioning and data acquisition, ADC- analogue to digital converter, DAC- digital to analog converter, OA- operational amplifier)

Another attractive solution that can be used to implement the sensing nodes is based on surface acoustic wave (SAW) devices [29][30]. The sensor consists of an interdigitated transducer etched onto a piezoelectric substrate, covered with a thin film. The mass of the film increases as its material selectively adsorbs a chemical substance from the air. This causes a shift in resonance to a slightly lower frequency giving information about the amount of gas species in the air.

B. Measurement data interpolation

To perform the interpolation of the calibration data in order to obtain the inverse characteristic of the measurement data, two methods are usually considered, namely, polynomial interpolation and artificial neural networks (ANNs).
Assuming, for simplicity, a single variable function (f) and a LMS polynomial interpolation function defined by [31]

$$P_n(x) = \sum_{k=0}^{p} a_k \cdot x^k \tag{2}$$

where p represents the degree of the polynomial curve fitting function and x represents the independent variable - measured quantity - it is possible to demonstrate that the LMS deviation between calibration and curve fitting data is obtained when the coefficients of the curve fitting polynomial function are given by

$$[\alpha] = \left[X_C^T \cdot X_C\right]^{-1} \cdot \left[X_C^T \cdot Y\right] \tag{3}$$

being vector Y and matrix X_C defined, for a set or n calibration points, by

$$Y = \begin{bmatrix} y_1 \\ y_2 \\ \vdots \\ y_n \end{bmatrix} \quad X_C = \begin{bmatrix} 1 & x_1 & x_1^2 & \cdots x_1^p \\ 1 & x_2 & x_2^2 & \cdots x_2^p \\ \vdots & \vdots & \vdots & \cdots \vdots \\ 1 & x_n & x_n^2 & \cdots x_n^p \end{bmatrix} \tag{4}$$

The concerns related with polynomial interpolation are mainly associated with the choice of the polynomial degree. If a low polynomial degree is used, the interpolation error is generally high because the polynomial function can not fit correctly a large number of calibration points. Conversely, if an excessive polynomial degree is used, the LMS deviation between calibration data and the values obtained from the polynomial interpolation function may be very low, but the interpolation errors of points between calibration data are usually very high. This problem is usually known as overfitting and the previous one as underfitting.

Regarding ANN [18][32][33], the curve fitting function can be computed using the following expression:

$$F_{ANN}(x_i) = F_N\left(W_N{}^*\left(F_{N-1}\left(\ldots F_2\left(W_2{}^*F_1(W_1{}^*x_i + B_1) + B_2\right)\ldots + B_{N-1}\right) + B_N\right)\right. \tag{5}$$

where N represents the number of neural network (NN) layers, Bi the bias vectors, Wi the weight vectors and Fi the activation transfer function of each layer.

The most common ANN structure for measurement applications contains a hidden layer of neurons with sigmoidal activation functions whose input is the measured data, and an output layer of neurons with linear activation functions. This ANN structure calculates an output vector given by

$$F_{ANN}(x_i) = \text{purelin}\left(W_2{}^*\text{tansig}(W_1{}^*x_i + B_1) + B_2\right) \tag{6}$$

where purelin() and tansig() are linear and hyperbolic tangent sigmoidal activation transfer functions, respectively.

This architecture has proved capable of approximating any function with a finite number of discontinuities and with arbitrary accuracy. Generally a more complex function, such as transducer characteristics that are strongly non-linear, requires more sigmoidal neurons in the hidden layer.

To evaluate the capability of a given solution to generalize the learned function, a second more dense set of data points - testing set - is used and the correspondent interpolated errors are evaluated. The best values of [B] and [W] matrices, associated with the bias and weights of each neuron, can be computed by minimizing the mean square error

$$MSE(x) = \frac{1}{n}\sum_{i=1}^{n}(F_{ANN}(x_i) - y_i)^2 \tag{7}$$

Several gradient methods [34][35], like back propagation (generalized Δ rule), can minimize the error function during the ANN training phase. During training, a set of input values corresponding to the calibration points is used to adjust the weights and biases of the neurons by minimizing the difference between the ANN output and the calibration values.
Even if there is no general rule to choose polynomial or ANN based curve fitting methods for a given application, when a reduced number of calibration points are available, and especially when extrapolation capabilities are desired, ANN can usually give better results in terms of measurement accuracy. This is particularly true for non-linear and non-deterministic sensors' characteristics and, if the number of calibration points is small, there is not an excessive penalty in terms of the computational load caused by an higher number of mathematical operations, usually caused by the need to evaluate non-linear transfer functions (tanh()) [36][37][38][39].

C. Data processing: an application example

In order to take advantage of the joint use of polynomial and artificial neural network (ANN) curve fitting techniques [12][40], this section describes a hybrid solution based on polynomial modelling (PM) and artificial neural networks modelling (ANN-M) that can be used to estimate the values of air quality parameters, such as, temperature, relative humidity, and polluting gases concentration.
For the particular case of broadband gas sensors, different methods can be used to convert the measured data into concentration of possible gas contaminants, such as, methane, carbon monoxide, isobutane, hydrogen, ethanol or cigarette smoke. Considering the voltage generated by a gas sensing unit based on a semiconductor heated sensor (TGS800 from Figaro), an air quality index ζ, is defined using the following relation

$$\zeta = \frac{R_S}{R_{S0}} = \left(\frac{V_C}{V_{RL}} - 1\right) \cdot \frac{1}{\left(\frac{V_C}{V_{RL0}} - 1\right)} \tag{8}$$

where R_{S0} represents the sensor resistance for a clean air condition, R_S represents the sensor resistance for the tested air, V_C is the circuit power supply voltage, V_{RL} is the load resistor voltage and V_{RL0} is the load resistor voltage for clean air.
Sensor's characteristic is non-linear and monotonic, decreasing sensor's resistance ratio with contaminant gas concentration. Higher concentrations of contaminants originate lower values of resistance ratios. Moreover, since the sensor is designed for general contaminants detection, it is not possible to identify specific contaminants. So, according to the application requirements in terms of the maximum acceptable level of contamination, a coefficient (ζ) value equal to 0.3 is considered for air pollution alarm. Considering that the used sensor has not good selectivity for each potential air contaminant, a look-up table, a polynomial, and a multilayer perceptron single-input single output neural network were designed and implemented to convert the value of ζ into air contaminants' concentrations expressed in parts per million (ppm). The measurement data processing scheme that was implemented is represented in figure 4.

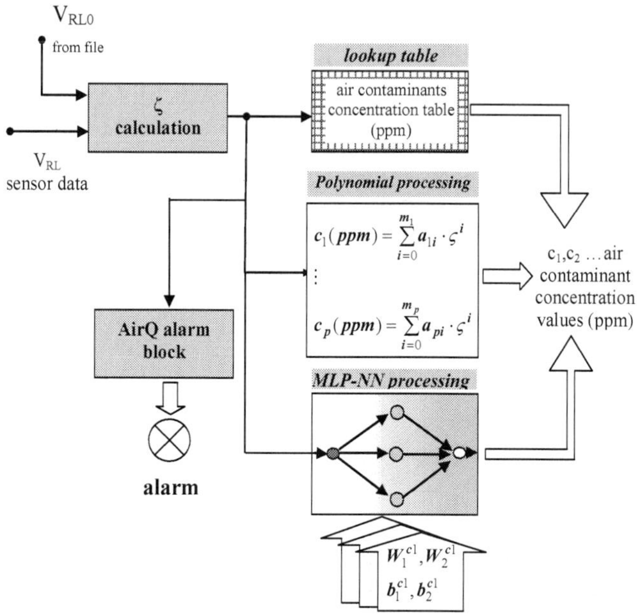

Fig. 4. Block diagram of the hybrid data processing scheme that was used to evaluate contaminants' air concentrations.

To test the performance of the proposed modeling scheme, a set of coefficient values (ζ), contained in the interval between 0.15 and 1 (no pollution), and the correspondent values of air contaminants' concentrations obtained from TGS800 sensitivity curves for methane, carbon monoxide, isobutane, hydrogen and ethanol, were considered. The calculation of polynomial coefficients, a_{1i}, a_{2i}, ... a_{pi}, is based on LS linear fit function (Givens method) that is implemented in LabVIEW. The calculated polynomial coefficients values that correspond to TGS800 sensitivity curves, such as the ones represented in figure 5, are stored in a memory and then used to perform the evaluation of air contaminants' concentrations.

The used neural processing blocks (NPB_i) is related with the inverse modeling [41] of gas sensor multivariable nonlinear characteristics, which are strongly dependent on temperature and humidity but also influenced by the concentration of other gases of the analyzed gas mixture. Based on the designed NPB_i, a digital read-out of the gases concentration with temperature and compensation [16] is obtained.

Regarding the NPB_i, two inputs one output multilayer perceptron neural networks were considered. Figure 6 represents the NPBi architecture including the normalization blocks and denormalization blocks used for ANN input and output data, respectively.

The NPB_i's internal parameters (weights and biases) are off-line calculated using the MATLAB program. The neural network training data were obtained in the system calibration phase. They are voltage values (V_{GSi}) acquired from the gas concentration measurement channel for different values of the gas concentration (C_{Gi}), and different temperature (T_p) and relative humidities (RH_i) values.

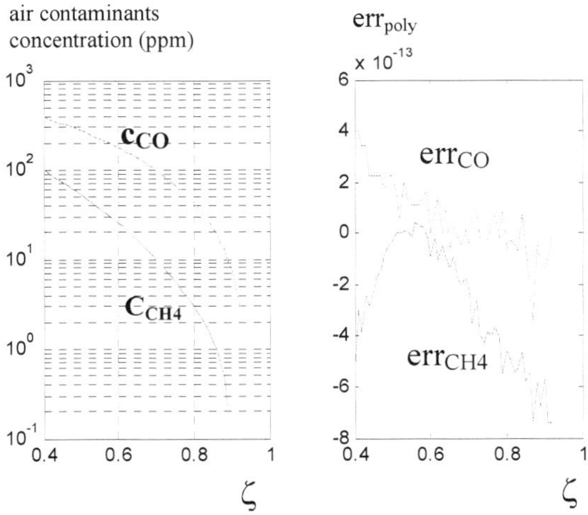

Fig. 5. Polynomial approximation of air contaminants curves (CO and methane case) and polynomial approximation error (err_{CO}, err_{CH4})

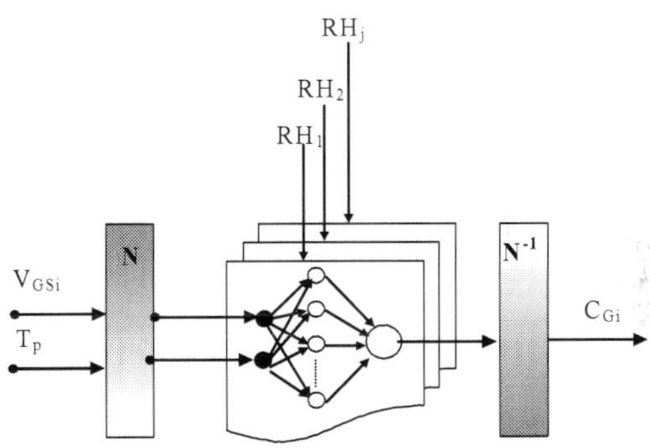

Fig. 6. NPB_i architecture (N, N^{-1}: normalization and denormalization blocks; RH_j: humidity selector; C_{Gi}: temperature and humidity compensated values of G_i gas concentration; T_P: temperature input value, V_{GSi}: input voltage value on the GS_i channel).

The neural network algorithm developed in MATLAB software calculates different sets of weights and biases for each RH_i experimental value (e.g. RH={45%, 55%, 65%}. The NPB_i input is the normalized voltage associated with each gas sensor's channel and normalized temperature, while the NPBi's output is the temperature compensated gas concentration (C_{Gi}). The NPBi normalized inputs are defined by:

$$V_{GSi}^{N}=\frac{V_{GS_i}}{V1_S},\qquad T^{N}=\frac{T}{\max(T)} \tag{9}$$

where $V1_S$ represents the gas sensor normalization factor (GS_i voltage supply=+10V in the present case).

Because GSi characteristics depend on humidity, an accurate measurement of the gas concentration is provided using different $NPB_i|_{RH}$ whose weights and biases are calculated using the data obtained for predefined relative humidity conditions (RH=45%, 55% and 65%) and by the interpolation method presented in [42].

The number of NPBi's layers is three. The hidden layers have 2 to 5 tansignoid (tansig(x)) neurons, and the output layer has 1 linear (l(x)) neuron. The implemented tansig(x) calculates its output according to

$$\text{tansig}(x)=\frac{2}{1+\exp(-2x)}-1 \tag{10}$$

which leads to a reduction of the computational load.

Two criteria for NPBi design were considered, the type and the number of neurons on the hidden layer, both determining the capabilities of the NPBi to adapt to a given characteristic. Different neuron nonlinear activation functions require different memory space and processing capabilities from the hardware platform.

To reduce the weights and biases in vector sizes, several simulation tests concerning the number of neurons for a required NPB_i performance, expressed by a modeling error, were performed. ANNs with a higher number of neurons increase processing load and, moreover, require larger memories to store weights and biases matrices. The results of these simulations are particularly important when embedded systems are used to implement the neural processing architecture (e.g. 512k EEPROM in the IPμ8930 case).

For the particular case of the CO measuring channel, the training set includes, as target, fifteen CO concentration values uniformly distributed in the 30 to 300ppm interval. The input values are the voltage values acquired from the TGS203 CO concentration measuring channel corresponding to the above-mentioned concentrations. The measured temperature in the testing chamber was T_p[°C]=10×p, p={1,2,3,4,5} and the relative humidity RH=35%. The Levenberg Marquardt algorithm [43] was used to calculate the weights and biases (W_{NPBi}, B_{NPBi}) of the neural network. Imposing a sum square error stop condition SSE=0.01, and for neural networks characterized by 4, 5 or 6 hidden neurons, different measuring channel modeling error characteristics (e_{CGsi}) were obtained (figure 7). The modeling error is defined by:

$$e_{CGsi}=\frac{C_{CGsi}-C_{CGsi}^{NPB}}{FS}\times 100 \tag{11}$$

where FS represents the measurement range, CCGsi is the experimental used gas concentration (e.g. carbon monoxide concentration) expressed in ppm, and C_{CGsi}^{NPB} the concentration of gas calculated by the corresponding neural processing module.

Since the used gas sensors characteristic depends on temperature, a study related with the CO channel modeling error (e_{CO}) versus temperature was carried out (figure 8).

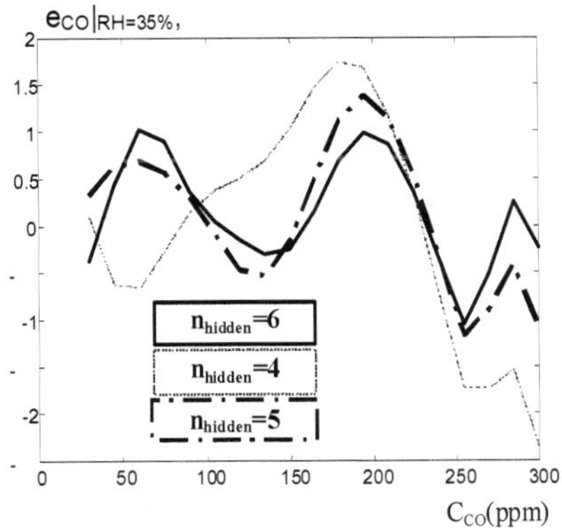

Fig. 7. The modeling error versus concentration for different NPB_{CO} architectures (T=10°C)

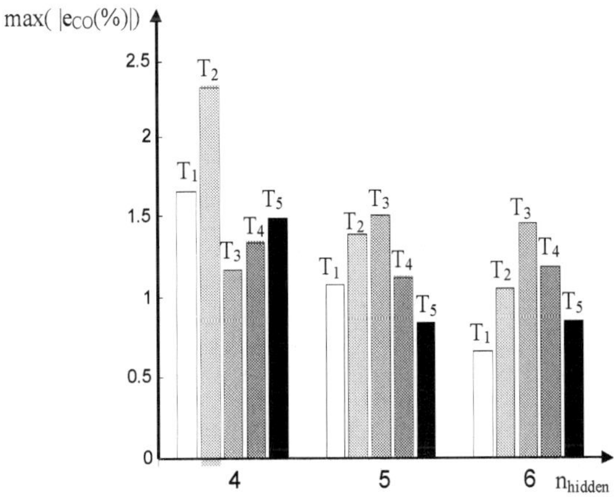

Fig. 8. The maximum inverse modeling error for different NPB_{CO} architectures (n_{hidden}={4, 5, 6}) and different temperatures $T_p=10 \cdot p$ °C

Being humidity an influence quantity, different values of the relative humidity lead to different primary gas selectivity characteristics and hence to different gas concentration measurement accuracies. Thus, experimental data obtained for three different values of relative humidity, RH_1=35%, RH_2=65% and RH_3=95%, and five values of temperatures included in the I_T=[10;50]°C were considered. The imposed gas concentrations for measurement system testing were: 10 values of methane concentration distributed in the I_{CM}=[500;5000] ppm interval, 15 values of carbon monoxide concentration I_{CCO}=[30;300] ppm, and 15 values of solvent vapors (Ethanol vapors) concentration, C_{SV}=[50;5000] ppm.

Based on the GS_i voltages for the considered gases concentrations, and taking into account temperature and humidity, three sets of weights and biases (35%, 65% and 95% relative humidity) were calculated for carbon monoxide, methane and solvent vapor measurement channels.

4. Smart sensing networks for air quality assessment

Gas sensors networks provide a promising mechanism for mining information from the monitored areas. Point-to-point and multipoint wireless network architectures, including sensing nodes, materialize the implementations in the air quality monitoring for indoor and outdoor conditions.

A. Point-to-point network architecture

Different architectures were developed by the authors, one of them based on a Bluetooth PDA [15]. In this case, the air quality measuring system is a virtual one (AIR-Q VMS) that joins hardware and software components to assure higher flexibility, mobility, data processing and data transmission. The block diagram of the mobile indoor air quality monitor system is presented in figure 9.

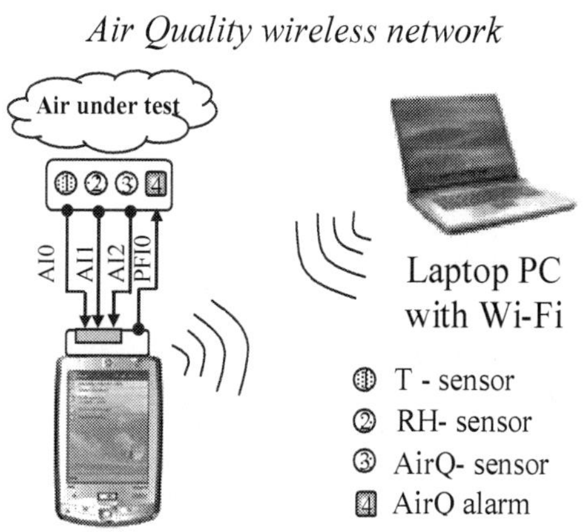

Fig. 9. Mobile Air Quality system based on a PDA with a compact flash (CF) multifunction I/O board

The sensing node includes sensors (temperature, relative humidity, and air quality), conditioning circuits, a compact flash data acquisition device DAQ (NI CF-6004) and a PDA with wireless communication capabilities (Wi-Fi or Bluetooth). A point-to-point connection between the measurement node and an advanced processing and communication unit (a PC) permits to deliver the air quality data from the sensing node to the PC and to receive information, such as alarm thresholds, that is used to implement alarm mechanisms in the PDA. The acquired data is processed by the PDA and the results are displayed by the PDA GUI.

Considering the cost of the implementation of the air quality sensing node based on a DAQ board plugged to a PDA, and also taking into account the evolution of the area of pervasive computing,, the authors decided to develop air quality monitoring systems based on smart phones and Bluetooth enabled smart sensors. The implemented architecture is presented in figure 10.

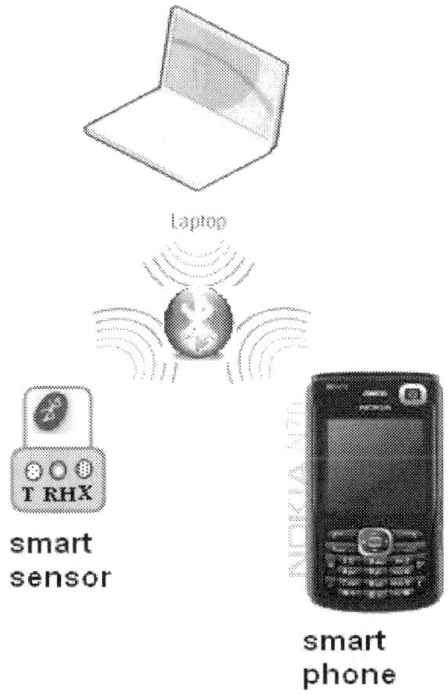

Fig. 10. Air quality virtual measuring system's architecture based on a smart sensing node (SN) with Bluetooth communication capabilities, and on a smart phone

The smart sensors indicated in figure 10 are specialized for temperature, air quality and air quality index measurement [16]. When monitoring large spaces, the number of sensing nodes increases, which means that point-to-multipoint architectures must be considered.

B. Point-to-multipoint Bluetooth architecture and embedded smart phone software

An implementation of a point-to-multipoint network architecture that uses Bluetooth compatible smart sensing nodes is presented in figure 11. The sensing nodes provide information about the level of relative humidity, temperature, and air contaminants (e.g. undesired odours that can trigger respiratory disorders). As computation units and human machine interface are included a laptop PC that works as the system server, a touch panel computer (TPC) and a smart phone (SP).

The implemented Bluetooth scatter net architecture assures the remote monitoring of the sensing nodes and data communication between the mobile device and smart sensor nodes. The hardware component of the system includes: sensors and conditioning circuits, a data acquisition device Bluetooth enabled (e.g. BlueSentry from Grid Connect), a smart phone with Bluetooth interface (e.g. N70 from Nokia), a situated display (NI TPC2106) Bluetooth compatible through a RS232-to-Bluetooth bridge, and a data communication, data processing and data storage unit (laptop PC).

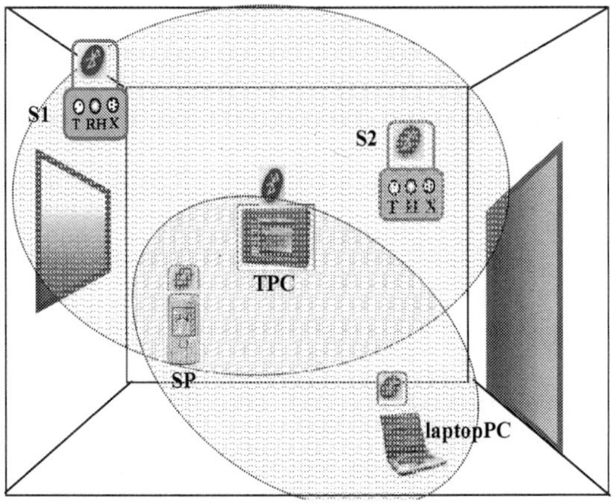

Fig. 11. Distributed air quality measurement architecture associated with respiratory distress triggering factors monitoring based on Bluetooth networking protocol (S1 and S2 are the sensing nodes characterized by T-temperature, H – relative humidity and X- air quality index measurement channel, TPC- touch panel computer, SP- smart phone)

The software technologies used to develop the applications for the smart phone running Symbian OS and for the TPC running Windows CE OS, were Java2Me and LabVIEW. The application embedded in the smart phone was named SmartSense Mobile. AirQUbicomp is the application developed using LabVIEW 8.6 Touch Panel Module for the TPC. This application provides the information about indoor air quality.

The SmartSense application has the ability to identify the active smart sensing nodes, to establish a connection via Bluetooth with the nodes, to control the on/off state of the air quality index sensor (XairQ-sensor), and to collect voltage samples from relative humidity, temperature and air quality measuring channels of each node in single-shot mode or in continuous mode.

SmartSense also assures the transfer of the indoor air quality values calculated and stored in the smart phone memory extension to the laptop PC through Bluetooth synchronization.

After node(s) selection, the operator can choose between the "one sample" acquisition and continuous acquisition. The sample acquisition is triggered by the user in order to test the normal functioning of the sensing node or to verify the measurement accuracy of the considered air parameters (humidity, temperature or broadband pollution) during the system setup.

Working in continuous acquisition mode, the smart phone application permits to prevent the asthma or COPD attacks through warnings issued when imposed thresholds previously stored in the SmartSense Mobile configuration file are exceeded. The sampling rate of the continuous acquisition mode is set using the text files received through Bluetooth from the PC that runs a SmartAdmin application [16]. Values of time intervals between two successive acquisitions in the 0.5 min to 60 min interval were considered. These values are adapted to the smart phone's available memory and also to the time constants of the system associated with temperature, humidity and XairQ index measurement. During continuous monitoring of the indoor air quality, the voltages received from the sensing nodes are converted into physical values by the SmartSense Mobile application and stored in this format. The acquired data are saved in a file or can be sent as an SMS to the phone whose number was written in the SmartSense configuration file.

The continuous acquisition and data conversion software modules work together with an implemented alarm module that permits to generate acoustic alarms to inform that indoor air conditions are critical. The used threshold values (th) are included in table 1.

Th	Measured factors		
	RH[%]	T[°C]	XairQ[%]
thmin	30	15	50
thmax	50	30	-

Table 1. Threshold values for relative humidity, temperature and air quality index that was used to signalize a likelihood of asthma attack.

During visual or acoustic signalling, a set of useful recommendations related to indoor air factors values and the actions necessary to change the indoor air conditions from critical to normal are available through the smart phone GUI.

The AirQUbicomp application is designed to continuously monitor the air quality, generating visual and acoustic alarms according to the imposed thresholds. Active interaction with the touch panel computer is permitted after identification of the user through a numeric password. After identification, the user can modify the thresholds related to asthma or can define data logging elements such as the time between readings and the monitoring period (DLog TAB in Fig. 12). The values of temperature, air quality and air quality index as well as the alarms LEDs (AirQ Alarm) are part of the T-RH-XAirQ software TAB. In figure 12 the GUI associated with AirQUbicomp is presented.

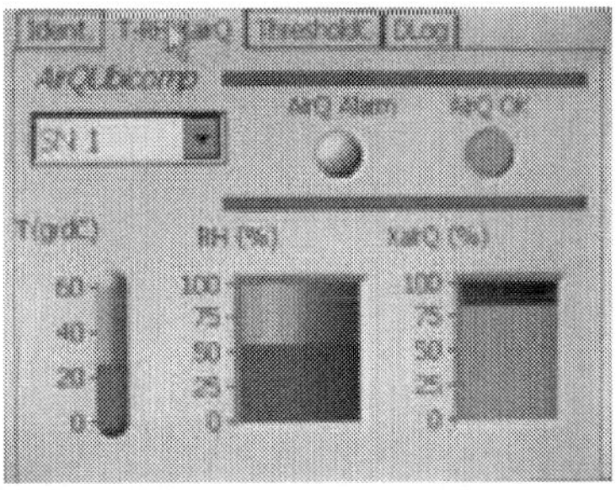

Fig. 12. AirQUbicomp GUI

Using the developed SmartSense Mobile application different tests associated with indoor air quality monitoring were carried out. The data stored in the Nokia N70, is wireless transferred to the database implemented in the laptop PC. Some data related with continuous measurement of the asthma or COPD attack triggering factors are presented in figure 13.

In figure 13 (a) the relative humidity is in the limit of automatic alarm generation (RH>50%) while temperature and air quality are inside the interval values associated with "no asthma or COPD attack conditions".

In figure 13 (b) can be observed low levels of the XairQ index when the measurement session started. Based on the information displayed, the user acted to improve the air quality (e.g. by opening the window). The air quality started to improve and, at the same time, room's temperature and humidity change significantly.

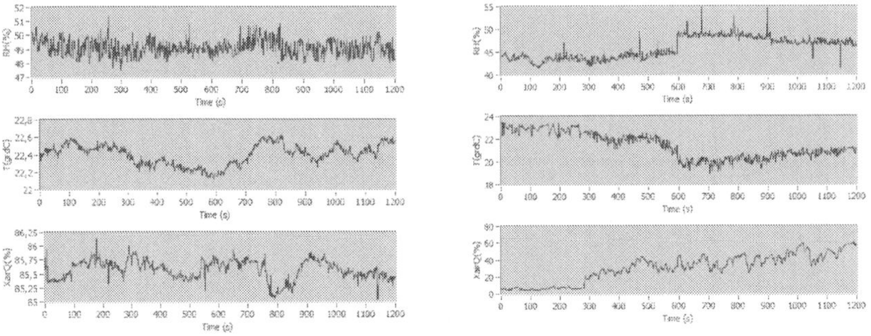

a) b)

Fig. 13. S1 node monitoring of respiratory distress triggering factors

In order to find correlations between the air quality and the values of physiological parameters, such as oxygen saturation (SpO2) and heart rate (HR), a digital pulse oxymeter and electrocardiograph apparatus ECG Medlab P-OX 100 was used for testing purposes. Table 2 presents the results of SpO2 and HR for two volunteers, with and without respiratory distress history (RD-N, RD-Y). The physiological values were measured in the same room and for the volunteers seated on a chair.

Analyzing the data from table 2, one can notice that in case of the healthy individual (RD-N), values of XairQ lower than 80% and of RH near 50% do not induce changes in HR and SpO2, while a significant increase in the HR of RD-Y is felt.

Sensor node	T (°C)	RH (%)	XairQ (%)	RD-N		RD-Y	
				HR	SpO2	HR	SpO2
S1	17.2	47.4	62.7	72	98	96	92
S2	17.8	44.2	69.6				

Table 2. S1 node: air quality and physiological parameter values for two volunteers, with and without respiratory distress history

Nowadays, smart phones are provided with operating systems, such as Android OS and iOS, which make the implementation of complex software modules easier and faster. The authors have been working to develop an AirQ Android OS application for a multichannel sensing node. The graphical interface of the implemented application is presented in figure 14.

The AirQ dashboard includes elements related with respiration activity (respiration rate). The data logging procedure is done using a smart phone embedded database that can synchronize with Web-based information system database through Wi-Fi or 3G/UMTS communication protocol.

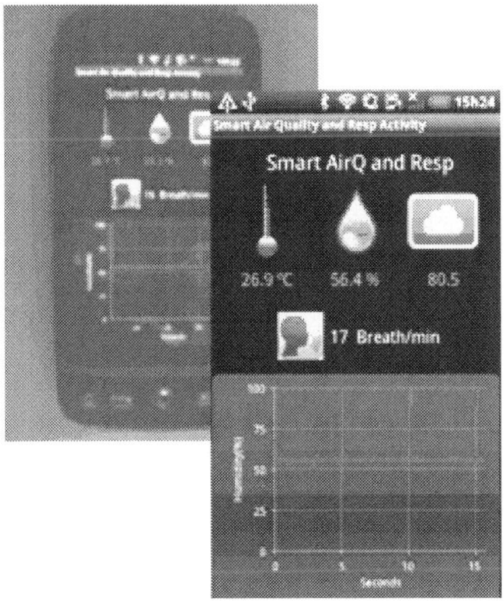

Fig. 14. AirQ graphical interface implemented in the AndroidOS smart phone

5. Conclusion

The quality of life of pulmonary patients greatly depends on the quality of the air they breathe. The identification of the indoor air associated with pathophysiology of COPD and asthma disease is crucial for the primary-prevention strategy. In the preceding paragraphs the authors summarize the main elements of a distributed smart sensing network for indoor air quality assessment. Regarding sensing nodes and signal conditioning, two possible solutions were presented. One based on semiconductor heated sensors and another based on three electrodes' cells. For data processing purposes, a hybrid solution based on polynomial and artificial neural networks modelling is presented. The last part of the chapter includes possible solutions for indoor air quality smart sensing networks.

6. References

[1] USA Environment Protection Agency, "Indoor Air Quality", on-line at http://www.epa.gov/iaq/

[2] European Environment Agency, "Every breath you take – air quality in Europe", on-line at http://www.eea.europa.eu/articles/air-quality-in-europe

[3] I.G. Stiel, D.W. Spaite, B. Field, L.P. Nesbitt, D. Munkley, J. Maloney et al, "Advanced life support for out-of hospital respiratory distress", NEJM, vol. 356, issue 24, pp. 2156-64, 2007

[4] K.F. Rabe, S. Hurd, A. Anzuento, P.J. Barnis, S.A. Buist, P. Calverley, et al, "Global strategy for the diagnosis management, and prevention of chronic obstructive pulmonary disease: GOLD executive summary", Am J Respir Crit Care Med, vol. 176, issue 6, pp. 532-55, 2007

[5] J.C. Hogg, F. Chu, S. Utokaparch S et al, "The nature of small-airway obstruction in chronic obstructive pulmonary disease", NEJM, vol. 350, issue 26, pp. 2645-53, 2004
[6] WHO, 2007, Asthma: http://www.who.int/topics/asthma/en/2007
[7] J. Bartolomeo, "Detecting CO in the home", Home Automation and Building Control, pp. 51–55, Oct. 1995.
[8] F. Sarry and M. Lumbreras, "Gas discrimination in an air-conditioned system", IEEE Trans. Instr. Meas., vol. 49, no. 4, pp. 809–812, Aug. 2000
[9] O. Postolache, J.M. Dias Pereira, P. Girão, "Smart sensor network for air quality monitoring", Proc. IEEE Instrumentation and Measurement Technology Conf., Ottawa, Canada, vol. 1, pp. 537 - 542, 2005
[10] S. Marco, A. Ortega, A. Pardo, and J. Samitier, "Gas identification with tin oxide sensor array and self-organizing maps: Adaptive correction of sensor drifts", IEEE Trans. Instr. Meas., vol. 47, no. 1, pp. 316–320, Feb. 1998
[11] F. Ménil, M. Susbielles, H. Debéda, C. Lucat, and P. Tardy, "Evidence of a correlation between the non-linearity of chemical sensors and the asymmetry of their response and recovery curves," Sens. Actuators B, Chem., vol. 106, no. 1, pp. 407–423, Apr. 29, 2005
[12] J. M. Dias Pereira, P. M. B. Silva Girão, and O. Postolache, "Fitting transducer characteristics to measured data", IEEE Signal Process. Mag., vol. 4, no. 4, pp. 26–39, Dec. 2001
[13] N. Ulivieri, C. Distante, T. Luca, S. Rocchi, P. Siciliano, "IEEE1451.4: A way to standardize gas sensor", Sens. Actuators B: Chemical vol. 114, issue 1, pp. 141-151, 2006
[14] Octavian A. Postolache, J. M. Dias Pereira, P. M. B. Silva Girão "Smart sensors Network for Air Quality Monitoring Applications", IEEE Trans. Instr. Meas., Vol. 58, No. 9, Sept. 2009
[15] O. Postolache, P. Girão, M. Pereira, "PDA based virtual measuring system for broadband air quality monitoring", Proc. IMEKO World Congress, on CD, September, Brasil, 2006
[16] O. Postolache, P. Girão, J. M. Dias Pereira, N. Barroso, G. Ferreira, G. Postolache, "Indoor monitoring of respiratory distress triggering factors using a wireless sensing network and a smart phone", Proc. of I2MTC 2009, Singapore, May 2009
[17] Nemoto Chemical Sensors Catalogue, Nemoto & Co., Tokyo, Japan, on line at http://www.nemoto.co.jp/product
[18] S. Haykin, "Neural Networks: A Comprehensive foundation (2nd Edition)", Wiley, 1998.
[19] NHLBI Morbidity and Mortality Chart Book, on line at http://www.nhlbi.nih.gov/resources/docs/cht-book.htm
[20] C.M. Patino, F.D. Martinez, "Interaction between genes and environment in the development of asthma", Allergy, vol. 56, pp.279-86, 2001
[21] Indoor Air Quality Impacts Child Asthma, Humidex Reduces Contaminates that Trigger Asthma Attacks, on line at http://www.naturalnews.com/010255.html
[22] M. Tomizuka, C. Bang Yun, V. Giurgiutiu, "A smart indoor air quality sensor network", Proc. of SPIE, Smart Structures and Materials, vol. 6174, pp. 1277-1290, 2006
[23] Figaro, "Gas sensors catalog", on line at http://www.figaro.co.jp/en/make_html/item_1.html.

[24] Smartec, "Smartec- Sensors Catalog", on line at http://www.mmselectronics.co.uk/smt160-30.htm.

[25] Humirel, "Relative Humidity Sensor-HS1100", Humirel Inc, 1999

[26] J.M. Dias Pereira, O. Postolache, Ricardo Salgado, P. Silva Girão, "Voltammetry based automated instrument for in-situ and on-line measurement of heavy metals concentration in water", 1st IMEKO TC19 Symposium on Measurement and Instrumentation for Environmental Monitoring, pp. 85-90, Iasi, Romania, September 2007

[27] J.M. Dias Pereira, O. Postolache, P. Silva Girão, "A smart and portable solution for heavy metals concentration measurements", 24th IEEE Instrumentation and Measurement Technology Conference (IMTC/2007), Warsaw, Poland, May 2007

[28] M. D. Pereira, O. Postolache, P. S. Girão, "Improving celerity of heavy metals measurements", International Instrumentation & Measurement Technology Conference 2010 (I2MTC'2010), pp. 1073-1077, Austin, May 2010

[29] R.M. White, F.W. Voltmer, "Direct piezoelectrical coupling to surface elastic waves", Applied Physics Letters, (7), pp. 314-316, 1965

[30] S.Y. Yurish and M.T. Gomes, "Smart Sensors and MEMS", Proc. of the NATO Advanced Study Institute, II-Mathematics, Physics and Chemistry, Vol. 181, Kluwer Academic Publisher, September 2003

[31] S.D. Conte, C. de Boor, "Elementary numerical analysis: An algorithmic approach", 2nd edition, New York, McGraw-Hill, 1972

[32] H. Demuth, M. Baele, "Neural Network Toolbox for Use with MATLAB - User's Guide", The MathWorks Inc., September 1993

[33] J. Hertz, A. Krogh, R.G. Palmer, "Introduction to the theory of neural computation", Addison Wesley, 1991

[34] F. M. Silva, L. B. Almeida, "Speeding up backpropagation", Advanced Neural Computers, R. Eckmiller edition, pp. 151-160, North Holland, 1990

[35] P.L.C. Simon, P.H.S. de Vries and S. Middelhoek, "Autocalibration of silicon hall devices", Sens. Actuators A, Vol. 52, pp. 203-207, 1996

[36] J.M. Dias Pereira, O. Postolache, P. Silva Girão, M. Cretu, "Minimising temperature drift errors of conditioning circuits using Artificial Neural Networks", IEEE Trans. Instr. Meas., Vol.49, No,5, October 2000

[37] J.M. Dias Pereira, O. Postolache, P. Silva Girão, J.A. Brandão Faria, M. Cretu, "An optical temperature transducer based on a bimetallic sensor", 10th International Symposium on Development in Digital Measuring Instrumentation (ISDDMI'98), Vol.2, pp. 661-664, Nápoles, Itália, September 1998

[38] O. Postolache, P. Silva Girão, J.M. Dias Pereira, "Neural Network Application in a carbon monoxide measurement system", IJCNN'01, Washington DC. USA, 2001.

[39] O. Postolache, J.M. Dias Pereira, P. Silva Girão, M. Cretu, "A virtual magnetometer based on Artificial Neural Networks", 10th International Symposium on Development in Digital Measuring Instrumentation (ISDDMI'98), Vol.2, pp. 580-584, Naples, Italy, September 1998.

[40] O. Postolache, J.M. Dias Pereira, M. Cretu, P. Silva Girão, "An ANN Fault Detection procedure applied in virtual measurement systems case", Proc. of IMTC/98", Vol. 1, pp. 257-260, St. Paul, Minnesota, USA, May 1998

[41] J. Patra, A. Kot, G. Panda, "An intelligent pressure sensor using Neural Netwoks", IEEE Trans. Instr. Meas., Vol. 49, No. 4, pp. 829-834, August 2000
[42] J. Pereira, O. Postolache, P. Girão, "A temperature-compensated system for magnetic field measurements based on Artificial Neural Networks", IEEE Trans. Instr. Meas., Vol 47, No.2., pp.494-498, April 1998
[43] M.T. Hagan and M.B. Menhaj, "Training feedforward networks with the Marquardt Algorithm", IEEE Trans. Neural Networks, Vol. 5, No. 6, pp. 989-993, 1994

13

An Exposure Model for Identifying Health Risk due to Environmental Microbial Contamination in the Healthcare Setting

Michael D. Larrañaga[1], Enusha Karunasena[2], H.W. Holder[3], Eric D. Althouse[4] and David C. Straus[5]

[1]*Oklahoma State University*
[2]*Texas Tech University*
[3]*SWK, LLC*
[4]*Air Intellect, LLC*
[5]*Texas Tech University Health Sciences Center*
USA

1. Introduction

It is widely accepted in various scientific communities that indoor microbiological contamination presents unacceptable conditions for the preservation of human health, and that removal and prevention of microbial contamination is necessary and prudent. (Pope, Patterson et al. 1993; Macher 1999; EPA 2001; ACOEM 2002; Redd 2002) Additionally, it is well established that fungal and bacterial bioaerosols cause infections and hypersensitivity diseases and that bioaerosols in the indoor environment can cause toxic effects (Karunasena E, Larrañaga MD et al. 2010) and nosocomial infections to immunocompromised individuals, it is reasonable to use indicators of environmental contamination for evaluating the need for remediation in a preventative context. (Macher 1999) The presence of fungi on indoor surfaces is often considered de facto evidence of human exposure to fungal aerosols, and the apparent absence of visible or measurable indoor growth does not ensure the absence of exposure. (Burge 2000) Fungi are designed for airborne dispersal from surface growth, and for many fungi, air movement is sufficient to produce spore aerosols. (Burge 2000) The objective of this chapter is to provide a mechanism for the indoor environmental professional to describe the health risk of the indoor environment with a single unit of measurement, providing decision makers a useful evaluation of the risk presented by the growth of microorganisms indoors.

2. Assessing health risk as a function of environmental contamination

To assess the health risk associated with the environmental microbial contamination within a hospital facility so that administrators could use this information in their decision-making processes, a health risk model (HRM) was utilized based on the text A Strategy for Assessing and Managing Occupational Exposures, Second Edition, a consensus document published by the American Industrial Hygiene Association (AIHA). (Mulhausen and

Damiano 1998) Industrial Hygiene is the science and art devoted to the recognition, evaluation, and control of environmental factors or stresses, arising in or from the work place, which may cause sickness, impaired health and well-being, or significant discomfort and inefficiency among workers or among the citizens of the community. (ABIH 2006)

For the industrial hygienist, exposure assessment and risk assessment are inextricably mixed such that they cannot be separated. Consider the following relationship between health risk and exposure:

Health Risk = (Exposure)(Toxicity [and/or Pathogenicity])

In the world of industrial hygiene, evaluation of exposure is half the assessment of health risk. The other half is evaluation of the health effects per unit exposure, or the toxicity and/or pathogenicity of the agent to which the person is exposed. Any exposure in an industrial hygiene sense is only meaningful in its relationship to the health effects the exposure might cause...The industrial hygienists' ultimate goal is to provide reasonable assurance of occupant health. (Mulhausen and Damiano 1998) In the case of exposure to biological contaminants, toxicity also has a pathogenicity component, as biological contaminants can be pathogenic and/or toxic.

The role of the industrial hygienist is to direct the health assessment so that he or she can make professional judgments on the acceptability of exposure and the associated health risks. The participation of other technical professionals such as engineers, environmental scientists, physicians, toxicologists, safety professionals, etc. is a proven way to streamline the exposure assessment process and improve the quality of assessments. (Mulhausen and Damiano 1998) For the hospital project application of the exposure assessment strategy, an inter-disciplinary team of professionals participated in preparing the model and validating its effectiveness per the interpretations of the investigators and characterization of the overall condition within the facility. Professionals from the following areas participated in the modification, interpretation, and validation of the HRM: medical microbiology, infection control, public health, medicine, engineering, mechanical contracting, medicine, mold assessment consulting, and statistics.

For the HRM, total surface areas for both fungal and bacterial contamination were quantified and sampled to confirm the presence of microbial contamination. Contaminated surfaces were then prescribed a toxicity/pathogenicity score based on the type of microbial contamination identified by sampling. Exposure scores were calculated and multiplied by the toxicity/pathogenicity scores to provide an indication of health risk.

3. Hospital HRM application

The HRM for the hospital project utilized the AIHA Exposure assessment Strategy as a framework for computing health risk. Health Risk is defined as:

Health Risk = (Exposure Score)(Toxicity/Pathogenicity Score)/(Exposure Pathway Score)

In the case of exposure to biological contaminants, toxicity also has a pathogenicity component, as it is well established that bioaerosols can cause infections and hypersensitivity disease and that bioaerosols in the indoor environment may cause toxic effects and nosocomial infections in immunocompromised individuals. Indicators of environmental contamination may be considered for prescribing preventative methods of control and for making decisions regarding building-related illness and building-related symptoms. (Macher and American Conference of Governmental Industrial Hygienists. 1999)

3.1 Exposure Score (ES) modeling

The ACGIH, EPA, IOM, and CDC recommendations emphasize that active fungal growth in indoor environments is inappropriate and may lead to adverse health effects. The confirmed presence of fungal growth is strong evidence that exposure may occur, and the conditions leading to this should be corrected and the growth removed under appropriate conditions. (Macher and American Conference of Governmental Industrial Hygienists. 1999) This is the premise behind the establishment of increasing levels of protection and containment necessary for remediation of increasing surface areas of contaminated surfaces prescribed by the United States Environmental Protection Agency document "Mold Remediation in Schools and Commercial Buildings" (EPA 2001), the "New York City Guidelines on Assessment and Remediation of Fungi in Indoor Environments" (NYCDHMH 2006), and the ACGIH text Bioaerosols: Assessment and Control. (Macher 1999)

Hence, the Exposure Score modeling is based on the same premise that increasing surface areas of contamination dictate an increased potential of exposure. For a detailed summary of recommendations associated with the above references, see Damp Indoor Spaces and Health published by the Institute of Medicine, 2004. (Institute of Medicine Committee on Damp Indoor Spaces and Health 2004) The ES determination differentiates between critical and non-critical areas within the hospital. Critical areas were defined as the following functional spaces within a hospital: 1) Surgery and Critical Care, 2) Nursing, 3) Ancillary, 4) Diagnostic and Treatment, and 5) Sterilization and Supply; non-critical functional areas were defined as Administration and Service. (ASHRAE 2003) The exposure score, then, is based on the location of exposure, type of procedures conducted in the location, and persons expected to be exposed in those locations.

It is expected that immunocompromised persons, the elderly, newborns, and sick children will be present in the hospital. Protecting children from indoor pollutants is particularly important because 1) children are still developing physically and affected by pollutants to a greater degree than adults, 2) the number of children with asthma has risen approximately 49 percent since 1982, 3) children below the age of 10 have three times as many colds as adults, 4) and children have a higher rate of metabolism than adults and may ingest or inhale more air and surface contaminants than adults (Bayer 2000). Allergic disease (nasal allergy, asthma, and other allergies) is also the number one chronic childhood illness. (Richards 1986) To fully estimate the risk associated with exposure to the immunocompromised and sick children, the HRM was employed by utilizing the maximum score of the tape sample score or the swab sample score and the maximum toxicity/pathogenicity score for each sample in the health risk calculation.

In cases where the exposure pathway was impeded, the exposure score was decreased by one half. The exposure pathway was considered impeded when contamination or growth was identified behind intact vinyl wallpaper or an air handling unit was post-filtered with 90% or 95% final filters, as mandated by the Texas Department of State Health Services. (TDSHS 1994) The exposure score modeling was adapted from the AIHA Exposure Assessment Strategy (Mulhausen and Damiano 1998) to associate increasing amounts of contamination with increased surface area of contamination. Microbial contamination above the false ceiling was not considered impeded because microbial contamination above ceiling tiles has been shown to move through pores in ceiling tiles and cause nosocomial infections in the space below the false ceilings (Arnow, Andersen et al. 1978) and positive pressure above the false ceilings allows the exchange of air between the space indoors and the space above the false ceilings. See the Assured HVAC report for a detailed description of pressure differentials within the Hospital.

3.1.1 Determination of the exposure score

The exposure score is the maximum of the tape sample and swab sample scores divided by the Exposure Pathway Score:

ES = (MAX[Tape Sample Score, Swab Sample Score])/(Exposure Pathway Score))

Determination of Tape Sample Score:

Tape Sample Score = MAX[Growth Score, Tape Contamination Score]

The information necessary to determine the growth and contamination scores are identified in the Center for Indoor Air Research's Standard Operating Procedures laboratory result sheets for each sample. The laboratory sheets for each sample specify the following: 1) the presence of a fungal growth site for determination of the Growth Score (Table 1), and 2) the laboratory defined level of contamination identified by the tape sample for determination of the Tape Sample Score (Table 2) and 3) the laboratory defined level of contamination identified by the swab sample for determination of the swab sample score. Note: Utilize the maximum value of the scores from Tables 1-2 when determining the Tape Sample Score.
The Growth Score (Table 1) and Tape Contamination Score (Table 2) can be determined by utilizing information presented in the lab sheet for each sample and the estimated surface area of growth or contamination to calculate a value.

3.1.1.1 Determination of the growth score

Table 1 is utilized to determine the Growth Score. If the location is not identified as a growth site, then the growth score is zero.
To determine the Growth Score, multiply the Location Multiplier (values of 1 or 2) by the Growth Multiplier (values of 0 to ≥ 4) found in Table 1. Two Examples are provided:

1. For 35 square feet of fungal growth in a critical area, then the Growth Score is found by multiplying the Location Multiplier which is 2 for a critical area by the Growth Multiplier represented by 35 square feet of growth.
   a. The Location Multiplier for a critical Area is 2 and the Growth Multiplier representing 35 square feet of growth is 2.5. Therefore, the Growth Score would be 2 x 2.5 = 5.
2. For 225 square feet of growth in a non-critical area, the Growth Score is found by multiplying the Location Multiplier which is 1 for a non-critical area and the Growth Multiplier represented by 225 square feet of growth, which would be 4 + 1 (for 100 square feet of additional growth) = 5. Therefore, the Growth Score would be 5 x 1 = 5.

To determine the Tape Contamination Score, multiply the Location Multiplier (values of 1 or 2) by the Tape Multiplier (values of 0 to ≥ 4) found in Table 2. Two Examples are provided:

3. For 35 square feet of Moderate or Medium Contamination in a critical area, then the Tape Contamination Score is found by multiplying the Location Multiplier which is 2

Location Multiplier	**Growth Score Determination Matrix (Growth Site Identified) (Ranges of Possible Growth Score Values)**				
Critical Area*=2	0	2-3.6	4-5.5	6-7.6	≥ 8
Non-Critical Area**=1	0	1-1.8	2-2.75	3-3.8	≥ 4
Growth Multiplier (determined by amount of contaminated surface area***)	0 (No Growth)	>0-2 ft^2 = 1 >2-4 ft^2 = 1.2 >4-6 ft^2 = 1.4 >6-8 ft^2 =1.6 >8-10 ft^2 =1.8	>10-20 ft^2 =2 >20-30 ft^2 =2.25 >30-40 ft^2 =2.5 >40-50 ft^2 =2.75	>50-60 ft^2 = 3 >60-70 ft^2 =3.2 >70-80 ft^2 =3.4 >80-90 ft^2 =3.6 >90-100 ft^2 = 3.8	4 for >100 ft (Add 1 to Growth Multiplier for every additional 100 ft^2 of contamination)

Note: If a growth site is not identified, then utilize Table 2 to score the matrix.
* A critical area is defined as areas within the following healthcare function spaces:
Surgery and Critical Care, Ancillary, Nursing, Diagnostic and Treatment, and Sterilizing and Supply. (ASHRAE 2003)
** A non-critical care area is defined as areas within the following healthcare function spaces:
Administration and Service. (ASHRAE 2003)
*** See the following references for protection levels associated with surface areas of contamination: (Macher 1999; EPA 2001; NYCDHMH 2006; VUMC 2006)

Table 1. Fungal Growth Scoring Matrix - Growth Score

Location Multiplier		**Tape Contamination Score Determination Matrix (Ranges of Possible Tape Contamination Score Values)**			
Critical Area*=2		0-5.6	0-7.5	0-7.6	≥ 0
Non-Critical Area**=1		0-2.8	0-3.75	0-3.8	≥ 0
Tape Multiplier (See Lab Contamination Description)	No Growth	0	0	0	0
	Very Light or Few, Light	1	1	1	1
	Moderate or Medium	1	2	2	≥ 3
	Heavy, Very Heavy	2	3	3	≥ 4
Area of Surface Contamination Modifier* (add to Tape Multiplier value)**		>0-2 ft^2 = add 0 >2-4 ft^2 = add 0.2 >4-6 ft^2 = add 0.4 >6-8 ft^2 = add 0.6 >8-10 ft^2 = add 0.8	>10-20 ft^2 = add 0 >20-30 ft^2 = add 0.25 >30-40 ft^2 = add 0.5 >40-50 ft^2 = add 0.75	>50-60 ft^2 = add 0 >60-70 ft^2 = add 0.2 >70-80 ft^2 = add 0.4 >80-90 ft^2 = add 0.6 >90-100 ft^2 = add 0.8	>100 ft^2 (Add 1 to Tape Multiplier for every additional 100 ft^2 of contamination)

* A critical area is defined as areas within the following healthcare function spaces:
Surgery and Critical Care, Ancillary, Nursing, Diagnostic and Treatment, and Sterilizing and Supply. (ASHRAE 2003)
** A non-critical care area is defined as areas within the following healthcare function spaces:
Administration and Service. (ASHRAE 2003)
*** See the following references for protection levels associated with surface areas of contamination: (Macher 1999; USEPA 2001; NYCDHMH 2006; VUMC 2006)

Table 2. Fungal Contamination Scoring Matrix - Tape Contamination Score for a critical area by the Tape Multiplier represented by 35 square feet of Moderate or Medium Contamination is 2 + 0.5 (modifier to account for 35 square feet of contamination) = 2.5.

   a. The Location Multiplier for a critical Area is 2 and the Tape Multiplier representing 35 square feet of moderate or medium contamination is 2.5. Therefore, the Tape Contamination Score is 2 x 2.5 = 5.

4. For 225 square feet of moderate or medium contamination in a non-critical area, the Tape Contamination Score is found by multiplying the Location Multiplier which is 1 for a non-critical area and the Tape Multiplier represented by 225 square feet of moderate or medium contamination, which would be 3 + 1 (for 100 square feet of additional contamination) = 4. Therefore, the Tape Contamination Score would be 4 x 1 = 4.

3.1.1.2 Determination of Swab Sample Score (SS): accounts for swab contamination results

The laboratory sheets for each sample specify the laboratory defined level of contamination identified by the swab sample for determination of the swab sample score.
To determine the Swab Sample Score, multiply the Location Multiplier (values of 1 or 2) by the Tape Multiplier (values of 0 to ≥4) found in Table 3.

Location Multiplier	**Swab Sample Score Determination Matrix (bacteria and/or fungi) (Ranges of Possible Swab Sample Score Values)**			
Critical Area*=2	0-5.6	0-7.5	0-7.6	≥ 0
Non-Critical Area**=1	0-2.8	0-3.75	0-3.8	≥ 0
Swab Multiplier (See Laboratory Contamination Description) – No Growth	0	0	0	0
Very Light or Few	1	1	1	1
Light or Medium	1	2	2	≥ 3
Heavy, Very Heavy	2	3	3	≥ 4
Area of Surface Contamination Modifier* (add to Tape Multiplier value)**	>0-2 ft² = add 0 >2-4 ft² = add 0.2 >4-6 ft² = add 0.4 >6-8 ft² = add 0.6 >8-10 ft² = add 0.8	>10-20 ft² = add 0 >20-30 ft² = add 0.25 >30-40 ft² = add 0.5 >40-50 ft² = add 0.75	>50-60 ft² = add 0 >60-70 ft² = add 0.2 >70-80 ft² = add 0.4 >80-90 ft² = add 0.6 >90-100 ft² = add 0.8	>100 ft² Add 1 to Swab Multiplier for every additional 100 ft²

* A critical area is defined as areas within the following healthcare function spaces: Surgery and Critical Care, Ancillary, Nursing, Diagnostic and Treatment, and Sterilizing and Supply. (ASHRAE 2003)

** A non-critical care area is defined as areas within the following healthcare function spaces: Administration and Service. (ASHRAE 2003)

*** See the following references for protection levels associated with surface areas of contamination: (Macher 1999; USEPA 2001; NYCDHMH 2006; VUMC 2006)

Table 3. Microbial Contamination Scoring Matrix - Swab Sample Score

Two Examples are provided:

1. For 35 square feet of Light or Medium Contamination in a critical area, then the Swab Sample Score is found by multiplying the Location Multiplier which is 2 for a critical area by the Swab Multiplier represented by 35 square feet of Light or Medium Contamination is 2 + 0.5 (modifier to account for 35 square feet of contamination) = 2.5.
   a. The Location Multiplier for a critical Area is 2 and the Swab Multiplier representing 35 square feet of light or medium contamination is 2.5. Therefore, the Swab Sample Score is 2 x 2.5 = 5.
2. For 225 square feet of growth in a non-critical area, the Swab Sample Score is found by multiplying the Location Multiplier which is 1 for a non-critical area and the Swab Multiplier represented by 225 square feet of growth, which would be 3 + 1 (for 100 square feet of additional growth) = 4. Therefore, the Swab Contamination Score would be 4 x 1 = 4.

3.2 Exposure Pathway score (EP)

The Health Risk equation was modified to include an Exposure Pathway Score that compensated for microbial contamination or growth that was likely impeded from reaching a building occupant and causing pathogenic effects. See Table 4.

EP	Interpretation
1	Exposure pathway present.
2	Exposure pathway impeded (e.g. contamination is behind in-tact vinyl wallpaper or air is filtered through 90% (or higher) filters prior to entering the space per State of Texas Regulations). (TDSHS 1994)

Table 4. Exposure Pathway Score Scoring Matrix

3.3 Determination of Toxicity/Pathogenicity score (TP)

The literature was reviewed to determine if the organisms identified inside the hospital were associated with pathogenic or opportunistic infections in humans. See Tables 6 and 7 below. Organisms were identified as opportunistic/pathogenic, allergenic, and/or toxigenic if the organism identified was identified as potentially capable of producing a toxin (e.g. aflatoxin, endotoxin, satratoxin). In cases where multiple organisms were identified on a sample, the highest toxicity/pathogenicity score was assigned to the health risk calculation.

TP = sum of individual components: Not Identifiable, Allergenic (A), Toxigenic (T), and/or Opportunistic/Pathogenic based on the organisms identified.

Note: Where multiple organisms are identified on the same sample, the highest Toxicity/Pathogenicity score of the identified organisms is assigned to the calculation.

4. Calculation of health risk score

The Health Risk Score is calculated by multiplying the Exposure Score by the Toxicity/Pathogenicity Score. (Mulhausen and Damiano 1998)

Health Risk = (Exposure Score)(TP Score)

Toxicity/Pathogenicity Identifier	Contamination Score	Interpretation
No Organism Identified	0	No organism was identified.
Not identifiable as A, T, or O/P	1	The organism is not known to be infectious, toxic, or allergenic to humans.
Allergenic (A)	2	The organism has been shown to induce allergy in some individuals (Pope, Patterson et al. 1993; W.B. Saunders 2000)
Toxigenic (T)	2	The organism produces one or more toxins (W.B. Saunders 2000)
Opportunistic/ Pathogenic (FOP) fungi identified	3	The identified organism is either a microorganism that does not ordinarily cause disease but that may cause disease in immunocompromised hosts (opportunistic) and/or any disease producing organism (pathogenic). (W.B. Saunders 2000)
Opportunistic/ Pathogenic (BOP) bacteria identified	3	The identified organism is either a microorganism that does not ordinarily cause disease but that may cause disease in immunocompromised hosts (opportunistic) and/or any disease producing organism (pathogenic). (W.B. Saunders 2000)

Table 5. Determination of Toxicity/Pathogenicity Score

4.1 Health risk scoring interpretation

The criteria for determining the health risk ratings of de minimis, low, medium, and high risk were determined by input from the investigators, peer reviewers, and specialists. Since no guidelines or limits of exposure exist, the expert input was utilized to create estimates of risk based on professional judgment and experience in the fields of medicine, engineering, infection control nursing, industrial hygiene, public health, and medical microbiology. The risk score interpretations and defining criteria are defined as:

de Minimis: No indication of environmental contamination was identified and therefore the risks associated with indoor microbiological contamination are negligible. No remediation is necessary. No further action is warranted. (Spengler, Samet et al. 2001)

Health Risk Score = 0

Low: The environmental conditions present do not indicate extensive biological contamination and/or the risk associated with adverse health affects to building occupants is low. Remediation may be necessary. Containment may be necessary. Remediation may necessitate increased levels of protection (e.g. High Efficiency Particulate Air (HEPA) filtration, full containment). If remediation is warranted, all persons must be removed from the immediate work area, and vacating people from spaces adjacent to the work area is not necessary but is recommended in the presence of infants (less than 12 months old), persons recovering from recent surgery, immune suppressed people, or people with chronic inflammatory lung diseases (e.g., asthma, hypersensitivity pneumonitis, and severe allergies). (NYCDHMH 2006) Containment may be limited or no containment may be required. See Table 2 of the document Mold Remediation in Schools and Commercial Buildings, the New York Guidelines and Guidelines for Environmental Infection Control in Health-Care Facilities. (EPA 2001; CDC 2003; NYCDHMH 2006)

Defining Criteria: Non-Critical Care Area with <10 square feet of mold growth and heavy or very heavy contamination on either the swab or tape. The Toxicity/Pathogenicity component is equal to Allergenic + Toxigenic, and the exposure pathway is not impeded.

Low Risk Range = 1-11 (rounded down)

Medium: The environmental conditions present an increased risk for adverse health effects to building occupants due to environmental contamination. The indoor environment suggests that immunosuppressed or allergic patients within the hospital are not fully protected against the risk of infection and the allergenic effects due to exposure to environmental-source fungi and bacteria. (Pope, Patterson et al. 1993; Perdelli, Christina et al. 2006) Remediation is necessary. Containment is necessary. Persons within the remediation area must be vacated. Further vacating of people from spaces near the work area is recommended in the presence of infants (less than 12 months old), persons having undergone recent surgery, immune suppressed people, or people with chronic inflammatory lung diseases (e.g., asthma, hypersensitivity pneumonitis, and severe allergies). Containment may be limited or full, with negative air pressure and HEPA filtration exhausted outdoors. Containment may necessitate increased environmental monitoring to establish the effectiveness of containment. See Table 2 of the document Mold Remediation in Schools and Commercial Buildings, the New York Guidelines, and Guidelines for Environmental Infection Control in Health-Care Facilities. (Agency 2001; CDC 2003; NYCDHMH 2006)

Defining Criteria: Non-Critical Care Area, 10-100 square feet of growth with heavy to very heavy contamination on the swab or tape. The Toxicity/Pathogenicity component is equal to Opportunistic/Pathogenic + Toxigenic + Allergenic, and the exposure pathway is not impeded.

Medium Risk Range = >11-26 (rounded down)

High: An indoor environment has been created in which immunosuppressed or allergic patients within the hospital are not fully protected against the risk of infection and the allergenic effects of exposure to environmental-source fungi and bacteria. (Pope, Patterson et al. 1993; Perdelli, Christina et al. 2006) The environmental conditions present a high risk for building occupants and intervention is necessary. The conditions exist for adverse health effects due to exposure to biological contaminants. Remediation is necessary, and during remediation, persons within the remediation area must be vacated. Vacating people from spaces adjacent to the work area is not necessary but is recommended in the presence of infants (less than 12 months old), persons having undergone recent surgery, immune suppressed people, or people with chronic inflammatory lung diseases (e.g., asthma, hypersensitivity pneumonitis, and severe allergies). (NYCDHMH 2006) Full containment is warranted with negative air pressure and HEPA filtration exhausted outdoors. Environmental Monitoring is warranted to establish the effectiveness of containment. See Table 2 of the document Mold Remediation in Schools and Commercial Buildings, the New York Guidelines, and Guidelines for Environmental Infection Control in Health-Care Facilities. (EPA 2001; CDC 2003; NYCDHMH 2006)

High Risk Range = >26

5. Assumptions and limitations of the determination of the exposure score

Assumptions

The following assumptions were made for the HRM:

1. Increasing surface area of microbial contamination represents an increased potential for exposure to fungal or bacterial environmental contaminants.
2. The maximum surface contamination score (from Tables 1-3) is utilized in determining the Health Risk Score as a marker of exposure.
3. The presence of an impeded exposure does not eliminate the risk associated with biological contamination within the hospital, and therefore, the EP was limited to reducing the exposure score by one-half (Table 4). Environmental disturbances, routine maintenance, climate change, etc. can disturb bioaerosols that may be impeded and release them into the building. (Arnow, Andersen et al. 1978; Loo, Bertrand et al. 1996; Pegues, Daar et al. 2001; CDC 2003) Therefore, a residual risk of exposure remains, even if the exposure is considered impeded.
4. The risk levels are based on the input of experts in related fields to the hospital project. The HRM does not set exposure limits but presents semi-quantitative risk levels based on exposure to microbiological contamination for the estimation of health risk in a hospital setting, where there is no doubt that persons who are ill will be present. There is also no doubt that microbial contamination is present indoors as confirmed by source sampling. See Assured report.

Limitations

The following limitations were identified when applying the HRM to the data:

1. The HRM may underestimate the health risk associated with small areas of growth or contamination in critical areas. For example, the investigators felt that the HR associated with a small amount microbial growth/contamination in a trauma room or surgical suite was significant and presented a high risk. The HRM, however could return a Health Risk Score falling in the low risk range for a small area of fungal growth or heavy contamination in a critical area. The HRM was designed to assess the risk of the entire facility in a broad sense and should not be utilized to assess risk based on one or a few samples. The samples should be of a sufficient number to characterize contaminated surfaces in the space under the control of each air handling unit.
2. The HRM is not sensitive to health risk associated with hidden microbial contamination, as invasive testing was not conducted. A large proportion of contaminated surfaces within buildings may remain hidden and are not visible without invasive investigation. (Dillon, Heinsohn et al. 1996) Therefore, the HRM may underestimate the health risk associated with hidden contamination.
3. The HRM may overestimate the health risk associated with large areas of contamination that are common to most buildings. Specifically, the investigators felt that the

contamination identified in the return air ducts of the HVAC systems was unavoidable and not inconsistent with contamination that could be expected in a return air duct. The investigators felt that the contamination within return air ducts that did not have mold growth did not eliminate the risk associated with contamination, but was not represented by the HRM. Therefore, when calculating the ES for contamination within the return air ducts, the maximum square footage utilized in the calculations was 100 ft^2. After reviewing the values of the HRM associated with the return air ducts, the investigators agreed that the Health Risk Score utilizing a maximum contamination surface area of 100 ft^2 adequately represented the health risk associated with contamination.

4. It is unlikely that adverse conditions and exposure to microbial contamination present within the hospital will affect building occupants equally and there are no exposure limits that would allow the calculation of an uncertainty rating to compare identified conditions with published exposure limits. Dose-response relationships are not available for comparison to environmental levels of indoor bioaerosols. There is no doubt, however that building occupants are being exposed to biological contaminants (allergens, opportunistic pathogens, and biological contaminants that can produce toxic metabolites) that have proliferated on indoor surfaces within the areas of the hospital investigated. The HRM prescribes semi-quantitative estimates of risk based on input from a multi-disciplinary team of professionals whose areas of specialization include microbial contamination in the indoor environment and indoor environmental control. The HRM provides the hospital administration with a method to quantify the risk associated with indoor environmental contamination based on the conditions within the hospital.
5. The HRM does not consider additive or synergistic effects of exposure to multiple organisms and/or toxins/metabolites.
6. The HRM does not represent the indoor conditions of the facility during and immediately after maintenance activities, disruption in electrical service, or the start-up and shut down of the HVAC systems.

6. Conclusion

Allergic reactions to indoor allergens can produce inflammatory diseases of the eyes, nose, throat, and bronchi, which are medical problems that come under the headings of allergic conjunctivitis, allergic rhinitis, allergic asthma, and hypersensitivity pneumonitis (extrinsic allergic alveolitis) respectively. (Pope, Patterson et al. 1993) The Health Risk Model (HRM) considers the type of microbial contamination and the type of person expected to be within a specific Hospital location. Critical care areas are areas of the Hospital where it is expected that immunocompromised persons will be present and therefore contamination within a critical care area is given a higher weight in the overall determination of health risk.

Risk assessment is a process designed to evaluate the potential relationship that may exist between exposure to aeroallergens and a particular effect (e.g. toxic effect, allergic sensitization, infection, allergic disease). (Pope, Patterson et al. 1993) A HRM was utilized to semi-quantitatively identify the health risk associated with fungal and bacterial surface contamination within the hospital. Monitoring for allergens can help characterize environments with respect to specific allergens (e.g., fungi and/or bacteria). Both fungi and bacteria secrete enzymes that act as allergens. (Pope, Patterson et al. 1993) Source or reservoir samples have been used as indicators of exposure to indoor allergens and measurement interpretations can be semi-quantitative (e.g., "presence or absence" or "low, medium, or high). (Pope, Patterson et al. 1993) Environmental bacteria also grow in all wet spaces and are found in most cases where there is mold growth. (Institute of Medicine Committee on Damp Indoor Spaces and Health 2004)

The American Industrial Hygiene Association's consensus document *A Strategy for Assessing and Managing Occupational Exposures* (Mulhausen and Damiano 1998) served as the basis for the HRM. The HRM utilized criteria and recommendations of the Centers for Disease Control and Prevention (CDC 2003), US Environmental Protection Agency (USEPA 2001), American Conference of Governmental Industrial Hygienists (ACGIH 1999), Institute of Medicine (Pope, Patterson et al. 1993), the New York City Department of Health and Mental Hygiene (NYCDHMH 2006), the American Society of Heating, Refrigerating, and Air Conditioning Engineers (ASHRAE 2003) and the Vanderbilt University Medical Center (VUMC 2006) in establishing the risk factors for the model. A literature search was conducted to determine if the organisms identified via surface sampling within the hospital were allergenic, pathogenic or opportunistic, and capable of producing fungal or bacterial toxin. The HRM resulted in a Health Risk classification of the space controlled by each AHU.

Health Risk was classified as High, Medium, Low, and de Minimis. The risk classifications were determined with input from experts in medical microbiology, industrial hygiene, public health, engineering controls, infection control, and medicine. A de Minimis risk score means that no indoor environmental contamination was found. A low risk score means the environmental conditions present do not indicate extensive biological contamination and/or the risk associated with adverse health affects to building occupants is low. A medium risk score indicates that environmental conditions present an increased risk for adverse health effects to building occupants due to environmental contamination and remediation is necessary. A high risk score indicates that conditions exist for adverse health effects due to exposure to biological contaminants and immediate intervention is necessary. Figure 9 below displays the HRM scores for the indoor space controlled by each AHU.

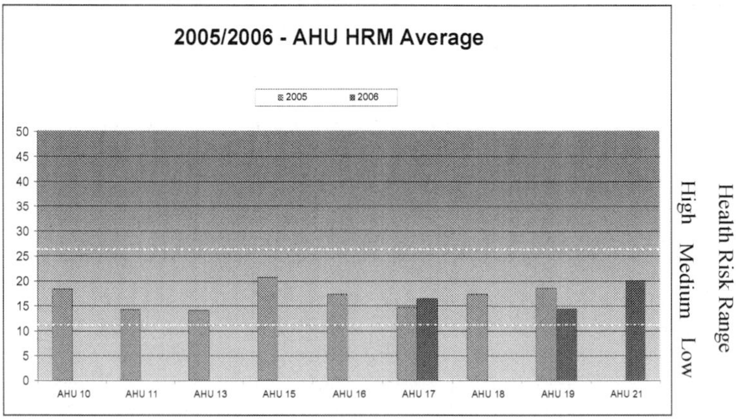

Fig. 9. Health Risk Model Scores for the space controlled by each AHU.

Indoor surface fungal and bacterial surface contamination was identified in every area of the hospital investigated. Air sampling confirmed the presence of indicators of indoor contamination in each of the spaces investigated. See **Section 4. Sampling Interpretation Summary** above. The spaces under the control of every AHU placed within the medium risk category. The environmental conditions are present such that immunocompromised or allergic patients are not fully protected against the risk of NI due to environmental bioaerosols. (Perdelli, Christina et al. 2006) Healthy hospital workers are not protected against allergic reactions to indoor bioaerosols growing within the facility and are at an increased risk of respiratory infections, including the common cold, sinusitis, tonsillitis, otitis, and bronchitis. (Institute of Medicine Committee on Damp Indoor Spaces and Health 2004) Hospital workers who are immunocompromised (e.g., diabetics, asthmatics, those undergoing cancer therapy or who have recent invasive surgery) are more susceptible to allergic reactions and the risk of work-related infections. The results of the HRM indicate that patients and staff are being exposed to microorganisms that are actively growing within the hospital which present a risk higher than what is expected in a hospital without water damage, microbial contamination, moisture infiltration, and OSA infiltration.

Periods of maintenance and non-routine operation of HVAC systems within the hospital can result in filter bypass, dissemination of biological contamination, and the infiltration of unfiltered OSA into the hospital, placing the hospital within the High Risk category due the creation of an exposure pathway during these times. Hence, times during and immediately after maintenance and non-routine operation of the HVAC systems present a high risk for health effects due to bioaerosols in the indoor environment. (CDC 2003)

Organisms Identified (fungal)	Allergen*	Opportunistic / Pathogenic*	Toxin Producer*	References:
Absidia species	1	1		(Holzberg and Artis 1983; Gonzalez, del Palacio et al. 1996; Tomsikova 2002; AerotechP&K 2006)
Acremonium species	1	1	1	(Kwon-Chung and Bennett 1992; Macher 1999; Walsh and Groll 1999; Groll and Walsh 2001; Fleming, Walsh et al. 2002; Tomsikova 2002; CDC 2003; Hilmioglu-Polat, Metin et al. 2005; Robles Garcia, Dierssen Sotos et al. 2005; AerotechP&K 2006)
Acrodontium species				
Actinomyces	1	1		(Schaal and Lee 1992; Macher 1999; AerotechP&K 2006)
Actinomycetes	1	1		(Schaal and Lee 1992; Macher 1999; AerotechP&K 2006)
Alternaria sparsus	1	1		(Tomsikova 2002; Ramphal 2006)
Alternaria species	1	1		(Botticher 1966; Aloi, Cervetti et al. 1987; Body, Sabio et al. 1987; Wiest, Wiese et al. 1987; Anaissie, Bodey et al. 1989; Kwon-Chung and Bennett 1992; Niedoszytko,

Organisms Identified (fungal)	Allergen*	Opportunistic / Pathogenic*	Toxin Producer*	References:
				Chelminska et al. 2002; Tomsikova 2002; Wheat, Goldman et al. 2002; Robles Garcia, Dierssen Sotos et al. 2005; AerotechP&K 2006)
Alternaria terreus	1	1		(Venugopal, Venugopal et al. 1989; Hilmioglu-Polat, Metin et al. 2005; AerotechP&K 2006)
Aphanocladium japonicus				
Aphanocladium species				
Arthrinium species	1			(AerotechP&K 2006)
Arthrographis species	1	1		(Degavre, Joujoux et al. 1997; Perlman and Binns 1997; Chin-Hong, Sutton et al. 2001; Biser, Perry et al. 2004; Xi, Fukushima et al. 2004; AerotechP&K 2006)
Ascospores	1	1	1	(AerotechP&K 2006)
Aspergillus alliaceus	1	1	1	(AerotechP&K 2006; Aspergillus.org 2006)
Aspergillus clavato-nanicus	1	1		(AerotechP&K 2006)
Aspergillus clavatus	1	1	1	(Macher 1999; AerotechP&K 2006; Aspergillus.org 2006)
Aspergillus flavipes	1	1	1	(AerotechP&K 2006; Aspergillus.org 2006)
Aspergillus flavus	1	1	1	(Macher 1999; CDC 2003; AerotechP&K 2006; Aspergillus.org 2006; Ramphal 2006)
Aspergillus fumigatus	1	1	1	(Kwon-Chung and Bennett 1992; Macher 1999; AerotechP&K 2006; Aspergillus.org 2006)
Aspergillus japonicus	1	1		(AerotechP&K 2006)
Aspergillus nidulans	1	1	1	(Kwon-Chung and Bennett 1992; Macher 1999; CDC 2003; AerotechP&K 2006; Aspergillus.org 2006)
Aspergillus niger	1	1	1	(Kwon-Chung and Bennett 1992; Macher 1999; CDC 2003; AerotechP&K 2006; Aspergillus.org 2006)
Aspergillus niveus	1	1		(AerotechP&K 2006)
Aspergillus oryzae	1	1	1	(Kwon-Chung and Bennett 1992; AerotechP&K 2006; Aspergillus.org 2006)
Aspergillus parasiticus	1		1	(Macher 1999; AerotechP&K 2006; Aspergillus.org 2006)
Aspergillus sclerotiorum	1		1	(AerotechP&K 2006; Aspergillus.org 2006)

Organisms Identified (fungal)	Allergen*	Opportunistic / Pathogenic*	Toxin Producer*	References:
Aspergillus sojae	1			(AerotechP&K 2006)
Aspergillus sydowii	1	1		(AerotechP&K 2006)
Aspergillus terreus	1	1	1	(Kwon-Chung and Bennett 1992; CDC 2003; AerotechP&K 2006; Aspergillus.org 2006)
Aspergillus ustus	1	1	1	(Kwon-Chung and Bennett 1992; AerotechP&K 2006; Aspergillus.org 2006)
Aspergillus versicolor	1	1	1	(Kwon-Chung and Bennett 1992; Macher 1999; AerotechP&K 2006; Aspergillus.org 2006)
Aureobasidium species		1		(Venugopal, Venugopal et al. 1989; Trupl, Minarik et al. 1995; Huttova, Kralinsky et al. 1998; Tomsikova 2002)
Basidiomycetes		1		(Bartz-Schmidt, Tintelnot et al. 1996; Nenoff, Horn et al. 1996; Rihs, Padhye et al. 1996; Nenoff, Friedrich et al. 1997; Sigler, Estrada et al. 1997; Verweij, van Kasteren et al. 1997)
Basidiospores	1			(AerotechP&K 2006)
Beauveria species	1			(AerotechP&K 2006)
Bipolaris species	1	1	1	(Rao, Forgan-Smith et al. 1989; Walsh, Gonzalez et al. 1995; Walsh and Groll 1999; Groll and Walsh 2001; Fleming, Walsh et al. 2002; Tomsikova 2002; Robb, Malouf et al. 2003; AerotechP&K 2006; Toul, Castillo et al. 2006)
Chaetomium species	1	1	1	(Kwon-Chung and Bennett 1992; Naidu 1993; Tomsikova 2002; AerotechP&K 2006)
Chrysosporium species	1	1	1	(Kwon-Chung and Bennett 1992; AerotechP&K 2006)
Circinella species				
Cladosporium cladosporioides	1	1		(Kwon-Chung and Bennett 1992; AerotechP&K 2006)
Cladosporium herbarum	1	1		(AerotechP&K 2006)
Cladosporium-like	1	1		(AerotechP&K 2006)
Cladosporium macrocarpum	1	1		(AerotechP&K 2006)
Cladosporium oxysporum	1	1		(AerotechP&K 2006)
Cladosporium species	1	1		(CDC 2003; AerotechP&K 2006)
Cladosporium sphaerospermum	1			(AerotechP&K 2006)
Coelomycetes species				
Corynespora species				(AerotechP&K 2006)
Curvularia species	1	1		(Loveless, Winn et al. 1981; Anaissie, Bodey et al. 1989; Venugopal, Venugopal et al. 1989; Naidu 1993; AerotechP&K 2006)

Organisms Identified (fungal)	Allergen*	Opportunistic / Pathogenic*	Toxin Producer*	References:
Emericella nidulans		1	1	(AerotechP&K 2006)
Engyodontium species		1		(Abarca 2000)
Epicoccum species	1	1		(AerotechP&K 2006)
Eupenicillium species				
Exophiala species	1	1	1	(AerotechP&K 2006)
Exserohilum species	1	1		(Tomsikova 2002)
Fonsecaea species	1	1		
Fusarium species	1	1	1	[3-12, 16, 20, 28, 45-47]
Geotrichum species	1	1		(AerotechP&K 2006)
Gliocladium species	1			(AerotechP&K 2006)
Hormographiella species		1		(Verweij, van Kasteren et al. 1997; AerotechP&K 2006)
Malbranchea species		1		
Mucor species	1	1		(Mikat 1980; Holzberg and Artis 1983; Fotedar and Banerjee 1992; Kwon-Chung and Bennett 1992; Gonzalez, del Palacio et al. 1996; Muhm, Zuckermann et al. 1996; AerotechP&K 2006)
Myxomycetes	1			(AerotechP&K 2006)
Nigrospora species	1			(AerotechP&K 2006)
Ochroconis species		1		(Tomsikova 2002)
Paecilomyces marquandii	1	1		(Kwon-Chung and Bennett 1992; Naldi, Lovati et al. 2000; AerotechP&K 2006)
Paecilomyces species	1	1		(Kwon-Chung and Bennett 1992; Walsh and Groll 1999; Naldi, Lovati et al. 2000; Groll and Walsh 2001; Fleming, Walsh et al. 2002; Tomsikova 2002; AerotechP&K 2006)
Paecilomyces variabile	1	1		(AerotechP&K 2006)
Paecilomyces variotii	1	1		(Akhunova and Shustova 1989; Akhunova 1991; Naidu 1993; Dhindsa, Naidu et al. 1995; Young, Hertl et al. 1995; Athar, Sekhon et al. 1996; AerotechP&K 2006)
Penicillium/Aspergillus-like	1			(AerotechP&K 2006)
Penicillium aurantiogriseum	1		1	(Frisvad and Filtenborg 1983; Yeulet, Mantle et al. 1988; AerotechP&K 2006)
Penicillium brevicaulis	1			(AerotechP&K 2006)
Penicillium brevicompactum	1	1	1	(Frisvad and Filtenborg 1983; AerotechP&K 2006)
Penicillium chrysogenum	1	1	1	(Frisvad and Filtenborg 1983; Kwon-Chung and Bennett 1992; Macher 1999; Lyratzopoulos, Ellis et al. 2002; AerotechP&K 2006)

Organisms Identified (fungal)	Allergen*	Opportunistic / Pathogenic*	Toxin Producer*	References:
Penicillium citrinum	1		1	(Vujanovic, Smoragiewicz et al. 2001; AerotechP&K 2006)
Penicillium commune	1			(AerotechP&K 2006)
Penicillium corylophilum	1			(AerotechP&K 2006)
Penicillium decumbens	1	1		(Kwon-Chung and Bennett 1992; Lyratzopoulos, Ellis et al. 2002; AerotechP&K 2006)
Penicillium duclauxii	1			(AerotechP&K 2006)
Penicillium fellutanum	1			(AerotechP&K 2006)
Penicillium funiculosum	1			(AerotechP&K 2006)
Penicillin griseofulvum	1		1	(Macher 1999; AerotechP&K 2006)
Penicillium implicatum	1			(AerotechP&K 2006)
Penicillium janthinellum	1			(AerotechP&K 2006)
Penicillium miczynskii	1			(AerotechP&K 2006)
Penicillium minioluteum	1			(AerotechP&K 2006)
Penicillium oxalicum	1		1	(Macher 1999; AerotechP&K 2006)
Penicillium oxysporum	1			(AerotechP&K 2006)
Penicillium pinophilum	1			(AerotechP&K 2006)
Penicillium purpurogenum	1	1		(Breton, Germaud et al. 1998; AerotechP&K 2006)
Penicillium rugulosa	1			(AerotechP&K 2006)
Penicillium sclerotiorum	1			(AerotechP&K 2006)
Penicillium simplicissimum	1			(AerotechP&K 2006)
Penicillium species	1	1	1	(Frisvad and Filtenborg 1983; Streifel, Stevens et al. 1987; Yeulet, Mantle et al. 1988; Fox, Chamberlin et al. 1990; Chakrabarti, Nayak et al. 1992; Gaye, Samb et al. 1992; Kwon-Chung and Bennett 1992; Walsh and Groll 1999; Lyratzopoulos, Ellis et al. 2002; CDC 2003; Robles Garcia, Dierssen Sotos et al. 2005; AerotechP&K 2006)
Penicillium thomii	1			(AerotechP&K 2006)
Penicillium variabile	1			(AerotechP&K 2006)
Penicillium waksmanii	1			(AerotechP&K 2006)
Periconia species				(AerotechP&K 2006)
Peronospora species				(AerotechP&K 2006)
Pithomyces species			1	(Macher 1999; AerotechP&K 2006)
Ramichloridium species		1		(Naim ur, Mahgoub et al. 1988; Jamjoom, al-Hedaithy et al. 1995; Sutton, Slifkin et al. 1998; Podnos, Anastasio et al. 1999; De Hoog, Queiroz-Telles et al. 2000; Kanj, Amr et al. 2001; Brandt and

Organisms Identified (fungal)	Allergen*	Opportunistic / Pathogenic*	Toxin Producer*	References:
				Warnock 2003; Kantarcioglu and de Hoog 2004; Al-Abdely, Alkhunaizi et al. 2005)
Rhizomucor species	1	1		(del Palacio Hernanz, Fereres et al. 1983; Severo, Job et al. 1991; Gonzalez, del Palacio et al. 1996; AerotechP&K 2006)
Rhizopus species	1	1		[1-4, 11, 49, 75-84]
Rhizopus oryzae	1	1		(Bottone, Weitzman et al. 1979; Telles Filho Fde, Coelho et al. 1985; Kwon-Chung and Bennett 1992; Gonzalez, del Palacio et al. 1996; Muhm, Zuckermann et al. 1996; Linder, Keller et al. 1998; AerotechP&K 2006; Lai, Liaw et al. 2006)
Rhodotorula species		1		(Walsh, Gonzalez et al. 1995; Huttova, Kralinsky et al. 1998; Costa, Marinho et al. 2000; Groll and Walsh 2001; Tomsikova 2002; Centeno and Machado 2004; AerotechP&K 2006)
Scedosporium species		1		[3-6, 11, 34, 60, 65, 87-90]
Scytalidium species		1		(Summerbell, Kane et al. 1989; Gaye, Samb et al. 1992; Brandt and Warnock 2003; AerotechP&K 2006)
Stachybotrys species	1		1	(Sudakin 1998; AerotechP&K 2006; Solomon, Hjelmroos-Koski et al. 2006)
Stemphylium species	1	1		(AerotechP&K 2006)
Sterile mycelia	1			(AerotechP&K 2006)
Syncephalastrum racemosus	1	1		(AerotechP&K 2006)
Tetraploa species				(AerotechP&K 2006)
Torula species	1			(Walsh, Gonzalez et al. 1995; Huttova, Kralinsky et al. 1998; Costa, Marinho et al. 2000; Groll and Walsh 2001; Tomsikova 2002; Centeno and Machado 2004; AerotechP&K 2006)
Trichoderma species	1	1	1	(Guarro, Antolin-Ayala et al. 1999; Richter, Cormican et al. 1999; Walsh and Groll 1999; Fleming, Walsh et al. 2002; Myoken, Sugata et al. 2002; Kredics, Antal et al. 2003; De Miguel, Gomez et al. 2005; Hilmioglu-Polat, Metin et al. 2005; AerotechP&K 2006)

Organisms Identified (fungal)	Allergen*	Opportunistic / Pathogenic*	Toxin Producer*	References:
Trichoderma viride	1	1	1	(Kwon-Chung and Bennett 1992; De Miguel, Gomez et al. 2005; AerotechP&K 2006)
Tritirachium species		1		(Rodrigues and Laibson 1975; AerotechP&K 2006)
Ulocladium species	1	1		(Gaye, Samb et al. 1992; Duran, Del Pozo et al. 2003; Hilmioglu-Polat, Metin et al. 2005; AerotechP&K 2006)
Verticillium species		1		(Amici, Grandesso et al. 1994; Shin, Kim et al. 2002; AerotechP&K 2006)
yeast		1		(AerotechP&K 2006; Ramphal 2006)

*A 1 signifies that the organism is opportunistic or pathogenic and or capable of producing a toxin.

Table 6. Pathogenicity and Toxicity Potential of Fungal Organisms

Organisms Identified (Bacteria)	Pathogenicity*	Toxin Producer (endotoxin)*	References
Acinetobacter lwoffi	1	1	(Crawford, Conway et al. 1997; Rathinavelu, Zavros et al. 2003; Larson, Cimiotti et al. 2005; Mathews, Mathews et al. 2005)
Acinetobacter species	1	1	(Bergogne-Berezin 2001; Alvarez-Lerma, Palomar et al. 2005; Benitez and Ricart 2005; Agodi, Zarrilli et al. 2006)
Actinomycetes	1		(Schaal and Lee 1992)
Aerococcus viridans			
Aeromonas hydrophila (Cheng, Horng et al. 2004)	1		(NNISR 1979; Poirier, Laurens et al. 1993; Davin-Regli, Bollet et al. 1998; Cheng, Horng et al. 2004)
Bacillus species	1		(Richard, Van der Auwera et al. 1988; Matsumoto, Suenaga et al. 2000; Yang, Xu et al. 2000; Newman 2002)
Bordetella bronchiseptica	1	1	(Bizet and Bizet 1995; Stevens-Krebbers, Schouten et al. 1999; Huebner, Christman et al. 2006)
Burkholderia cepacia	1	1	(Jang, Kuo et al. 1999; Belchis, Simpson et al. 2000; Matrician, Ange et al. 2000; Bureau-Chalot, Piednoir et al. 2003; Shehabi, Abu-Al-Soud et al. 2004; Balkhy, Cunningham et al. 2005; Berthelot, Grattard et al. 2005)

Organisms Identified (Bacteria)	Pathogenicity*	Toxin Producer (endotoxin)*	References
Burkholderia gladioli	1	1	(Wilsher, Kolbe et al. 1997; Otterbein, Splettstoesser et al. 1998; Clode, Metherell et al. 1999; Segonds, Heulin et al. 1999; Segonds and Chabanon 2001)
Burkholderia species	1	1	(Otterbein, Splettstoesser et al. 1998; Segonds, Heulin et al. 1999; Segonds and Chabanon 2001)
Chryseomonas luteola	1		(Hawkins, Moriarty et al. 1991; Ndugulile, Jureen et al. 2005)
Citrobacter freundii	1		(Hodges, Degener et al. 1978; Tejada Artigas, Bello Dronda et al. 2001; Fiorio, Marroni et al. 2004; Ndugulile, Jureen et al. 2005)
Comamonas acidovrans			
Diptheroids	1		(Schofferman, Zucherman et al. 1991)
Enterobacter agglomerans	1		(Geere 1977; Goldmann, Dixon et al. 1978; Maki 1981; Astagneau, Gottot et al. 1994)
Escherichia coli	1	1	(Hoogkamp-Korstanje, Cats et al. 1982; Raymond 2000; Newman 2002; Larson, Cimiotti et al. 2005; Kramer, Schwebke et al. 2006; Toniolo, Endimiani et al. 2006)
Flavimonas oryzihabitans	1	1	(Hawkins, Moriarty et al. 1991)
Flavobacterium meningosepticum	1	1	(Abrahamsen, Finne et al. 1989; Liu, Wong et al. 1999; Bellais, Girlich et al. 2002; Seetha, Bairy et al. 2002)
Flavobacterium breve	1	1	(Bellais, Girlich et al. 2002)
Gram (+) cocci	1		(Peter, Jehl et al. 1988; Rosina 1991; Zhang 1991; Pechere 1993; Astagneau 1998; Gayvallet-Montredon, Sauvestre et al. 1998; Raymond 2000)
Gram (+) cocci in clumps	1		(Peter, Jehl et al. 1988; Rosina 1991; Zhang 1991; Pechere 1993; Astagneau 1998; Gayvallet-Montredon, Sauvestre et al. 1998; Raymond 2000)
Gram (-) cocci	1		(Berk and Verghese 1989; Donowitz 1989; Zhang 1991;

Organisms Identified (Bacteria)	Pathogenicity*	Toxin Producer (endotoxin)*	References
			Carlisle, Gucalp et al. 1993; Pechere 1993; Du, Chen et al. 1996; Astagneau 1998; Gayvallet-Montredon, Sauvestre et al. 1998; McEachern and Campbell 1998; Jones, Low et al. 1999; Lang, Livesley et al. 1999; Karchmer 2000; Raymond 2000; Raymond and Aujard 2000; Chang, Carvalho et al. 2003; Palabiyikoglu, Tekeli et al. 2006)
Gram Negative Rod Non-Fer	1	1	(Berthelot, Grattard et al. 2005)
Gram Negative Rods	1	1	(LaForce 1981; Carlisle, Gucalp et al. 1993; Pechere 1993; McEachern and Campbell 1998; Berthelot, Grattard et al. 2005; Toniolo, Endimiani et al. 2006)
Micrococcus luteus	1		(Marinella, Pierson et al. 1997)
Micrococcus sp.	1		(Meyer, Eitzen et al. 1981; Hughes, Williams et al. 1986; Marinella, Pierson et al. 1997; Davies, Mehr et al. 2000)
Micrococcus species	1		(Meyer, Eitzen et al. 1981; Hughes, Williams et al. 1986; Marinella, Pierson et al. 1997; Davies, Mehr et al. 2000)
Myroides odoratum	1	1	(Mammeri, Bellais et al. 2002)
Nocardia sp.	1		(Simpson, Stinson et al. 1981; Schaal and Lee 1992)
Nocardioform	1		(Poonwan, Kusum et al. 1995; Votava, Skalka et al. 1997)
Nocardioform bacilli	1		(Poonwan, Kusum et al. 1995; Votava, Skalka et al. 1997)
Nocardioform bacilli Cog.	1		(Poonwan, Kusum et al. 1995; Votava, Skalka et al. 1997)
Presumptive *Nocardioform*	1		(Poonwan, Kusum et al. 1995; Votava, Skalka et al. 1997)
Pseudomonas aeruginosa	1	1	(Hoogkamp-Korstanje, Cats et al. 1982; Celis, Torres et al. 1988; Zhang 1991; Du, Chen et al. 1996; Hijazi and MacIntyre 2000; Yang, Xu et al. 2000; Esen and Leblebicioglu 2004; Fiorio, Marroni et al. 2004; Berthelot, Grattard et al. 2005; Branger 2005; Crnich, Safdar et al. 2005; Wang, Chang et al. 2005; Toniolo, Endimiani et al. 2006)

Organisms Identified (Bacteria)	Pathogenicity*	Toxin Producer (endotoxin)*	References
Pseudomonas diminuta	1	1	(Forbes, Sahm et al. 1998)
Pseudomonas fluorescens	1	1	(Franzetti, Cernuschi et al. 1992; Burgos, Torres et al. 1996; Forbes, Sahm et al. 1998; Hsueh, Teng et al. 1998; Forbes, Sahm et al. 2002)
Pseudomonas orizihabitans		1	
Pseudomonas stutzeri	1	1	(Forbes, Sahm et al. 1998)
Ralstonia picketti	1	1	(Adiloglu, Ayata et al. 2004)
Rhizobium radiobacter	1	1	(Potvliege, Vanhuynegem et al. 1989; Lai, Teng et al. 2004)
Sphingomonas paucimobilis	1	1	(de Otero, Masip et al. 1998; Hsueh, Teng et al. 1998; Perola, Nousiainen et al. 2002)
Staphylococcus aureus	1	1	(McGowan 1988; Berk and Verghese 1989; Peters 1991; Astagneau 1998; Barie 1998; Raymond 2000; Yang, Xu et al. 2000; Fiorio, Marroni et al. 2004; Branger 2005; Lee, Hua et al. 2005; Jerassy, Yinnon et al. 2006; Toniolo, Endimiani et al. 2006)
Staphylococcus auricularis			
Staphylococcus capitis	1		(Wang, Liu et al. 1999; Van Der Zwet, Debets-Ossenkopp et al. 2002)
Staphylococcus cohnii	1		(Narayani, Naseema et al. 1990; Szewczyk, Piotrowski et al. 2000)
Staphylococcus capitos			
Staphylococcus cohnii			
Staphylococcus epidermis	1		(Peters 1991; Perez Monras, Azahares Romero et al. 1992; Branger 2005; Larson and Dinulos 2005)
Staphylococcus hominis	1		(Ponce de Leon, Guenthner et al. 1986; Narayani, Naseema et al. 1990; Lang, Livesley et al. 1999; Szewczyk, Piotrowski et al. 2000; Basaglia, Moras et al. 2003)
Staphylococcus hyicus			
Staphylococcus haemolyticus	1		(Ponce de Leon, Guenthner et al. 1986; Narayani, Naseema et al. 1990)
Staphylococcus hominis	1		(Ponce de Leon, Guenthner et al. 1986; Narayani, Naseema et

Organisms Identified (Bacteria)	Pathogenicity*	Toxin Producer (endotoxin)*	References
			al. 1990; Lang, Livesley et al. 1999; Szewczyk, Piotrowski et al. 2000)
Staphylococcus salvarius			
Staphylococcus saprophyticus	1		(Hoogkamp-Korstanje, Cats et al. 1982; Cohen 1986; Narayani, Naseema et al. 1990; Hell, Kern et al. 1999; Lang, Livesley et al. 1999; Szewczyk, Piotrowski et al. 2000; von Eiff, Proctor et al. 2001; von Eiff, Peters et al. 2002)
Staphylococcus sciuri	1		(Lang, Livesley et al. 1999; Stepanovic, Dakic et al. 2002)
Staphylococcus sp. coag neg	1		(Maki 1981)
Staphylococcus warneri	1		(Ponce de Leon, Guenthner et al. 1986; Buttery, Easton et al. 1997)
Staphylococcus xylosus	1		(Narayani, Naseema et al. 1990; Won, Kwon et al. 2002)
Staphylococcus sp. Cog.	1		(Maki 1981)
Stenotroph maltophilia	1		(Berthelot, Grattard et al. 2005)
Suggestive *Diptheroids*			
Tatumella ptyseos	1		(Hollis, Hickman et al. 1981)

*A 1 signifies that the organism is opportunistic or pathogenic and/or gram negative.

Table 7. Pathogenicity and Toxicity Potential of Bacterial Organisms

7. References

Abarca, M. L. (2000). "[Taxonomy and identification of the species involved in nosocomial aspergillosis]." Rev Iberoam Micol 17(3): S79-84.

ABIH. (2006). "Definition of industrial hygiene." Retrieved August, 2006, from http://www.abih.org/general/history.html.

Abrahamsen, T. G., P. H. Finne, et al. (1989). "Flavobacterium meningosepticum infections in a neonatal intensive care unit." Acta Paediatr Scand 78(1): 51-55.

ACGIH (1999). Bioaerosols: Assessment and Control. Cincinnati, OH, American Conference of Governmental Industrial Hygienists.

ACOEM (2002). Adverse Human Health Effects Associated with Molds in the Indoor Environment, American College of Occupational and Environmental Medicine: 1-10.

Adiloglu, A. K., A. Ayata, et al. (2004). "Case report: nosocomial Ralstonia pickettii infection in neonatal intensive care unit." Mikrobiyol Bul 38(3): 257-260.

AerotechP&K. (2006). "Microbial and IAQ Glossaries." Retrieved August, 2006, from www.aerotechpk.com.

Agency, U. S. E. P. (2001). "Mold remediation in schools and commercial buildings." USEPA EPA 402-K-01-001.

Agodi, A., R. Zarrilli, et al. (2006). "Alert surveillance of intensive care unit-acquired Acinetobacter infections in a Sicilian hospital." Clin Microbiol Infect 12(3): 241-247.
Akhunova, A. M. (1991). "[Infectious-allergic bronchopulmonary paecilomycosis]." Ter Arkh 63(10): 19-24.
Akhunova, A. M. and V. I. Shustova (1989). "[Paecilomyces infection]." Probl Tuberk(8): 38-42.
Al-Abdely, H. M., A. M. Alkhunaizi, et al. (2005). "Successful therapy of cerebral phaeohyphomycosis due to Ramichloridium mackenziei with the new triazole posaconazole." Med Mycol 43(1): 91-95.
Aloi, F. G., O. Cervetti, et al. (1987). "[Alternaria mycosis in a kidney transplant patient]." G Ital Dermatol Venereol 122(1-2): 35-38.
Alvarez-Lerma, F., M. Palomar, et al. (2005). "Infections caused by Acinetobacter spp. in critically ill ICU patients." Enferm Infecc Microbiol Clin 23(9): 533-539.
Amici, G., S. Grandesso, et al. (1994). "Verticillium peritonitis in a patient on peritoneal dialysis." Am J Nephrol 14(3): 216-219.
Anaissie, E. J., G. P. Bodey, et al. (1989). "Emerging fungal pathogens." Eur J Clin Microbiol Infect Dis 8(4): 323-330.
Arnow, P. M., R. L. Andersen, et al. (1978). "Pumonary aspergillosis during hospital renovation." Am Rev Respir Dis 118(1): 49-53.
ASHRAE (2003). 2003 ASHRAE handbook : heating, ventilating, and air-conditioning applications. Atlanta, Ga., ASHRAE.
Aspergillus.org. (2006). "Toxic Metabolites of Aspergillus." Retrieved August, 2006, from http://www.aspergillus.org.uk/indexhome.htm?secure/mycotoxin/index.php~main.
Astagneau, P. (1998). "[Epidemiology of nosocomial infections]." Rev Prat 48(14): 1525-1529.
Astagneau, P., S. Gottot, et al. (1994). "Nosocomial outbreak of Enterobacter agglomerans pseudobacteraemia associated with non-sterile blood collection tubes." J Hosp Infect 27(1): 73-75.
Athar, M. A., A. S. Sekhon, et al. (1996). "Hyalohyphomycosis caused by Paecilomyces variotii in an obstetrical patient." Eur J Epidemiol 12(1): 33-35.
Balkhy, H. H., G. Cunningham, et al. (2005). "A National Guard outbreak of Burkholderia cepacia infection and colonization secondary to intrinsic contamination of albuterol nebulization solution." Am J Infect Control 33(3): 182-188.
Barie, P. S. (1998). "Antibiotic-resistant gram-positive cocci: implications for surgical practice." World J Surg 22(2): 118-126.
Bartz-Schmidt, K. U., K. Tintelnot, et al. (1996). "Chronic basidiomycetous endophthalmitis after extracapsular cataract extraction and intraocular lens implantation." Graefes Arch Clin Exp Ophthalmol 234(9): 591-593.
Basaglia, G., L. Moras, et al. (2003). "Staphylococcus cohnii septicaemia in a patient with colon cancer." J Med Microbiol 52(Pt 1): 101-102.
Bayer, C. W. (2000). "Humidity control and ventilation in schools." ASHRAE Journal(Summer).
Belchis, D. A., E. Simpson, et al. (2000). "Histopathologic features of Burkholderia cepacia pneumonia in patients without cystic fibrosis." Mod Pathol 13(4): 369-372.
Bellais, S., D. Girlich, et al. (2002). "EBR-1, a novel Ambler subclass B1 beta-lactamase from Empedobacter brevis." Antimicrob Agents Chemother 46(10): 3223-3227.

Benitez, L. and M. Ricart (2005). "Pathogenesis and environmental factors in ventilator-associated pneumonia." Enferm Infecc Microbiol Clin 23 Suppl 3: 10-17.
Bergogne-Berezin, E. (2001). "The Increasing Role of Acinetobacter Species As Nosocomial Pathogens." Curr Infect Dis Rep 3(5): 440-444.
Berk, S. L. and A. Verghese (1989). "Emerging pathogens in nosocomial pneumonia." Eur J Clin Microbiol Infect Dis 8(1): 11-14.
Berthelot, P., F. Grattard, et al. (2005). "Epidemiology of nosocomial infections due to Pseudomonas aeruginosa, Burkholderia cepacia and Stenotrophomonas maltophilia." Pathol Biol (Paris) 53(6): 341-348.
Biser, S. A., H. D. Perry, et al. (2004). "Arthrographis keratitis mimicking acanthamoeba keratitis." Cornea 23(3): 314-317.
Bizet, C. and J. Bizet (1995). "[Comparative susceptibility of Ochrobactrum anthropi, Agrobacterium tumefaciens, Alcaligenes faecalis, Alcaligenes denitrificans subsp. denitrificans, Alcaligenes denitrificans subsp. xylosidans and Bordetella bronchiseptica against 35 antibiotics including 17 beta-lactams]." Pathol Biol (Paris) 43(4): 258-263.
Body, B. A., H. Sabio, et al. (1987). "Alternaria infection in a patient with acute lymphocytic leukemia." Pediatr Infect Dis J 6(4): 418-420.
Botticher, W. W. (1966). "Alternaria as a possible human pathogen." Sabouraudia 4(4): 256-258.
Bottone, E. J., I. Weitzman, et al. (1979). "Rhizopus rhizopodiformis: emerging etiological agent of mucormycosis." J Clin Microbiol 9(4): 530-537.
Brandt, M. E. and D. W. Warnock (2003). "Epidemiology, clinical manifestations, and therapy of infections caused by dematiaceous fungi." J Chemother 15 Suppl 2: 36-47.
Branger, B. (2005). "2001 national survey of nosocomial infection prevalence among newborns and under-eighteen children and adolescents in France." Arch Pediatr 12(7): 1085-1093.
Breton, P., P. Germaud, et al. (1998). "[Rare pulmonary mycoses in patients with hematologic diseases]." Rev Pneumol Clin 54(5): 253-257.
Bureau-Chalot, F., E. Piednoir, et al. (2003). "Nosocomial Burkholderia cepacia outbreak in an intensive pediatric care unit." Arch Pediatr 10(10): 882-886.
Burge, H. A. (2000). The Fungi. Indoor Air Quality Handbook. J. D. Spengler, J. M. Samet and J. McCarthy, McGraw-Hill: 45.41-45.33.
Burgos, F., A. Torres, et al. (1996). "Bacterial colonization as a potential source of nosocomial respiratory infections in two types of spirometer." Eur Respir J 9(12): 2612-2617.
Buttery, J. P., M. Easton, et al. (1997). "Pediatric bacteremia due to Staphylococcus warneri: microbiological, epidemiological, and clinical features." J Clin Microbiol 35(8): 2174-2177.
Carlisle, P. S., R. Gucalp, et al. (1993). "Nosocomial infections in neutropenic cancer patients." Infect Control Hosp Epidemiol 14(6): 320-324.
CDC (2003). Guidelines for environmental infection control in health-care facilities: Recommendations of CDC and the HealthCare Infection Control Practices Advisory Committee. Atlanta, GA.
Celis, R., A. Torres, et al. (1988). "Nosocomial pneumonia. A multivariate analysis of risk and prognosis." Chest 93(2): 318-324.

Centeno, S. and S. Machado (2004). "Assessment of airborne mycoflora in critical areas of the Principal Hospital of Cumana, state of Sucre, Venezuela." Invest Clin 45(2): 137-144.
Chakrabarti, A., N. Nayak, et al. (1992). "Surveillance of nosocomial fungal infections in a burn care unit." Infection 20(3): 132-135.
Chang, M. R., N. C. Carvalho, et al. (2003). "Surveillance of pediatric infections in a teaching hospital in Mato Grosso do Sul, Brazil." Braz J Infect Dis 7(2): 149-160.
Cheng, N. C., S. Y. Horng, et al. (2004). "Nosocomial infection of Aeromonas hydrophila presenting as necrotizing fasciitis." J Formos Med Assoc 103(1): 53-57.
Chin-Hong, P. V., D. A. Sutton, et al. (2001). "Invasive fungal sinusitis and meningitis due to Arthrographis kalrae in a patient with AIDS." J Clin Microbiol 39(2): 804-807.
Clode, F. E., L. A. Metherell, et al. (1999). "Nosocomial Acquisition of Burkholderia gladioli in patients with cystic fibrosis." Am J Respir Crit Care Med 160(1): 374-375.
Cohen, M. L. (1986). "Staphylococcus aureus: biology, mechanisms of virulence, epidemiology." J Pediatr 108(5 Pt 2): 796-799.
Costa, S. F., I. Marinho, et al. (2000). "Nosocomial fungaemia: a 2-year prospective study." J Hosp Infect 45(1): 69-72.
Crawford, P. M., Jr., M. D. Conway, et al. (1997). "Trauma-induced Acinetobacter lwoffi endophthalmitis with multi-organism recurrence: strategies with intravitreal treatment." Eye 11 (Pt 6): 863-864.
Crnich, C. J., N. Safdar, et al. (2005). "The role of the intensive care unit environment in the pathogenesis and prevention of ventilator-associated pneumonia." Respir Care 50(6): 813-836; discussion 836-818.
Davies, M. W., S. Mehr, et al. (2000). "Bacterial colonization of toys in neonatal intensive care cots." Pediatrics 106(2): E18.
Davin-Regli, A., C. Bollet, et al. (1998). "A cluster of cases of infections due to Aeromonas hydrophila revealed by combined RAPD and ERIC-PCR." J Med Microbiol 47(6): 499-504.
De Hoog, G. S., F. Queiroz-Telles, et al. (2000). "Black fungi: clinical and pathogenic approaches." Med Mycol 38 Suppl 1: 243-250.
De Miguel, D., P. Gomez, et al. (2005). "Nonfatal pulmonary Trichoderma viride infection in an adult patient with acute myeloid leukemia: report of one case and review of the literature." Diagn Microbiol Infect Dis 53(1): 33-37.
de Otero, J., J. Masip, et al. (1998). "[Bacteremia caused by Sphingomonas (Pseudomonas) paucimobilis]." Enferm Infecc Microbiol Clin 16(8): 388-389.
Degavre, B., J. M. Joujoux, et al. (1997). "First report of mycetoma caused by Arthrographis kalrae: successful treatment with itraconazole." J Am Acad Dermatol 37(2 Pt 2): 318-320.
del Palacio Hernanz, A., J. Fereres, et al. (1983). "Nosocomial infection by Rhizomucor pusillus in a clinical haematology unit." J Hosp Infect 4(1): 45-49.
Dhindsa, M. K., J. Naidu, et al. (1995). "Chronic suppurative otitis media caused by Paecilomyces variotii." J Med Vet Mycol 33(1): 59-61.
Dillon, H. K., P. A. Heinsohn, et al. (1996). Field guide for the determination of biological contaminants in environmental samples. Fairfax, VA, American Industrial Hygiene Association.
Donowitz, L. G. (1989). "Nosocomial infection in neonatal intensive care units." Am J Infect Control 17(5): 250-257.

Du, B., D. Chen, et al. (1996). "[Nosocomial bacterial infection in comprehensive intensive care unit]." Zhonghua Yi Xue Za Zhi 76(4): 262-266.

Duran, M. T., J. Del Pozo, et al. (2003). "Cutaneous infection caused by Ulocladium chartarum in a heart transplant recipient: case report and review." Acta Derm Venereol 83(3): 218-221.

EPA (2001). Mold remediation in schools and commercial buildings. United States Environmental Protection Agency. EPA 402-K-01-001.

Esen, S. and H. Leblebicioglu (2004). "Prevalence of nosocomial infections at intensive care units in Turkey: a multicentre 1-day point prevalence study." Scand J Infect Dis 36(2): 144-148.

Fiorio, M., M. Marroni, et al. (2004). "Nosocomial infections in a general surgical ward." Recenti Prog Med 95(1): 11-14.

Fleming, R. V., T. J. Walsh, et al. (2002). "Emerging and less common fungal pathogens." Infect Dis Clin North Am 16(4): 915-933, vi-vii.

Forbes, B. A., D. F. Sahm, et al. (2002). Bailey & Scott's diagnostic microbiology. St. Louis, Mosby.

Forbes, B. A., D. F. Sahm, et al. (1998). Bailey & Scott's diagnostic microbiology. St. Louis, Mosby.

Fotedar, R. and U. Banerjee (1992). "Nosocomial fungal infections--study of the possible role of cockroaches (Blattella germanica) as vectors." Acta Trop 50(4): 339-343.

Fox, B. C., L. Chamberlin, et al. (1990). "Heavy contamination of operating room air by Penicillium species: identification of the source and attempts at decontamination." Am J Infect Control 18(5): 300-306.

Franzetti, F., M. Cernuschi, et al. (1992). "Pseudomonas infections in patients with AIDS and AIDS-related complex." J Intern Med 231(4): 437-443.

Frisvad, J. C. and O. Filtenborg (1983). "Classification of terverticillate penicillia based on profiles of mycotoxins and other secondary metabolites." Appl Environ Microbiol 46(6): 1301-1310.

Gaye, O., K. Samb, et al. (1992). "[Fungi in the hospital environment and infectious risk]." Dakar Med 37(1): 11-14.

Gayvallet-Montredon, N., C. Sauvestre, et al. (1998). "[Bacteriologic surveillance of nosocomial septicemia and bacteremia in a pediatric hospital]." Arch Pediatr 5(11): 1216-1220.

Geere, I. W. (1977). "Enterobacter agglomerans: the clinically important plant pathogen." Can Med Assoc J 116(5): 517-519.

Goldmann, D. A., R. E. Dixon, et al. (1978). "The role of nationwide nosocomial infection surveillance in detecting epidemic bacteremia due to contaminated intravenous fluids." Am J Epidemiol 108(3): 207-213.

Gonzalez, A., A. del Palacio, et al. (1996). "[Zygomycosis: review of 16 cases]." Enferm Infecc Microbiol Clin 14(4): 233-239.

Groll, A. H. and T. J. Walsh (2001). "Uncommon opportunistic fungi: new nosocomial threats." Clin Microbiol Infect 7 Suppl 2: 8-24.

Guarro, J., M. I. Antolin-Ayala, et al. (1999). "Fatal case of Trichoderma harzianum infection in a renal transplant recipient." J Clin Microbiol 37(11): 3751-3755.

Hawkins, R. E., R. A. Moriarty, et al. (1991). "Serious infections involving the CDC group Ve bacteria Chryseomonas luteola and Flavimonas oryzihabitans." Rev Infect Dis 13(2): 257-260.

Hell, W., T. Kern, et al. (1999). "Staphylococcus saprophyticus as an unusual agent of nosocomial pneumonia." Clin Infect Dis 29(3): 685-686.

Hijazi, M. H. and N. R. MacIntyre (2000). "Advances in infection control: ventilator-associated pneumonia." Semin Respir Crit Care Med 21(3): 245-262.

Hilmioglu-Polat, S., D. Y. Metin, et al. (2005). "Non-dermatophytic molds as agents of onychomycosis in Izmir, Turkey - a prospective study." Mycopathologia 160(2): 125-128.

Hodges, G. R., C. E. Degener, et al. (1978). "Clinical significance of Citrobacter isolates." Am J Clin Pathol 70(1): 37-40.

Hollis, D. G., F. W. Hickman, et al. (1981). "Tatumella ptyseos gen. nov., sp. nov., a member of the family Enterobacteriaceae found in clinical specimens." J Clin Microbiol 14(1): 79-88.

Holzberg, M. and W. M. Artis (1983). "Hydroxamate siderophore production by opportunistic and systemic fungal pathogens." Infect Immun 40(3): 1134-1139.

Hoogkamp-Korstanje, J. A., B. Cats, et al. (1982). "Analysis of bacterial infections in a neonatal intensive care unit." J Hosp Infect 3(3): 275-284.

Hsueh, P. R., L. J. Teng, et al. (1998). "Outbreak of Pseudomonas fluorescens bacteremia among oncology patients." J Clin Microbiol 36(10): 2914-2917.

Hsueh, P. R., L. J. Teng, et al. (1998). "Nosocomial infections caused by Sphingomonas paucimobilis: clinical features and microbiological characteristics." Clin Infect Dis 26(3): 676-681.

Huebner, E. S., B. Christman, et al. (2006). "Hospital-acquired Bordetella bronchiseptica infection following hematopoietic stem cell transplantation." J Clin Microbiol 44(7): 2581-2583.

Hughes, W. T., B. Williams, et al. (1986). "The nosocomial colonization of T. Bear." Infect Control 7(10): 495-500.

Huttova, M., K. Kralinsky, et al. (1998). "Prospective study of nosocomial fungal meningitis in children--report of 10 cases." Scand J Infect Dis 30(5): 485-487.

Institute of Medicine Committee on Damp Indoor Spaces and Health (2004). Damp indoor spaces and health. Washington, DC, National Academies Press.

Jamjoom, A. B., S. A. al-Hedaithy, et al. (1995). "Intracranial mycotic infections in neurosurgical practice." Acta Neurochir (Wien) 137(1-2): 78-84.

Jang, T. N., B. I. Kuo, et al. (1999). "Nosocomial gram-negative bacteremia in critically ill patients: epidemiologic characteristics and prognostic factors in 147 episodes." J Formos Med Assoc 98(7): 465-473.

Jerassy, Z., A. M. Yinnon, et al. (2006). "Prospective hospital-wide studies of 505 patients with nosocomial bacteraemia in 1997 and 2002." J Hosp Infect 62(2): 230-236.

Jones, R. N., D. E. Low, et al. (1999). "Epidemiologic trends in nosocomial and community-acquired infections due to antibiotic-resistant gram-positive bacteria: the role of streptogramins and other newer compounds." Diagn Microbiol Infect Dis 33(2): 101-112.

Kanj, S. S., S. S. Amr, et al. (2001). "Ramichloridium mackenziei brain abscess: report of two cases and review of the literature." Med Mycol 39(1): 97-102.

Kantarcioglu, A. S. and G. S. de Hoog (2004). "Infections of the central nervous system by melanized fungi: a review of cases presented between 1999 and 2004." Mycoses 47(1-2): 4-13.

Karchmer, A. W. (2000). "Nosocomial bloodstream infections: organisms, risk factors, and implications." Clin Infect Dis 31 Suppl 4: S139-143.

Karunasena E, Larrañaga MD, et al. (2010). "Building-associated neurological damaged modeled in human cells: a mechanism of neurotoxic effects by exposure to mycotoxins in the indoor environment." Mycopathalogia Dec(6): 377-390.

Kramer, A., I. Schwebke, et al. (2006). "How long do nosocomial pathogens persist on inanimate surfaces? A systematic review." BMC Infect Dis 6: 130.

Kredics, L., Z. Antal, et al. (2003). "Clinical importance of the genus Trichoderma. A review." Acta Microbiol Immunol Hung 50(2-3): 105-117.

Kwon-Chung, K. J. and J. E. Bennett (1992). Medical mycology. Philadelphia, Lea & Febiger.

LaForce, F. M. (1981). "Hospital-acquired gram-negative rod pneumonias: an overview." Am J Med 70(3): 664-669.

Lai, C. C., S. J. Liaw, et al. (2006). "Empyema thoracis due to Rhizopus oryzae in an allogenic bone marrow transplant recipient." Med Mycol 44(1): 75-78.

Lai, C. C., L. J. Teng, et al. (2004). "Clinical and microbiological characteristics of Rhizobium radiobacter infections." Clin Infect Dis 38(1): 149-153.

Lang, S., M. A. Livesley, et al. (1999). "The genomic diversity of coagulase-negative staphylococci associated with nosocomial infections." J Hosp Infect 43(3): 187-193.

Larson, A. A. and J. G. Dinulos (2005). "Cutaneous bacterial infections in the newborn." Curr Opin Pediatr 17(4): 481-485.

Larson, E. L., J. P. Cimiotti, et al. (2005). "Gram-negative bacilli associated with catheter-associated and non-catheter-associated bloodstream infections and hand carriage by healthcare workers in neonatal intensive care units." Pediatr Crit Care Med 6(4): 457-461.

Lee, S. C., C. C. Hua, et al. (2005). "Risk factors of mortality for nosocomial pneumonia: importance of initial anti-microbial therapy." Int J Clin Pract 59(1): 39-45.

Linder, N., N. Keller, et al. (1998). "Primary cutaneous mucormycosis in a premature infant: case report and review of the literature." Am J Perinatol 15(1): 35-38.

Liu, C. E., W. W. Wong, et al. (1999). "Flavobacterium meningosepticum bacteremia: an analysis of 16 cases." Zhonghua Yi Xue Za Zhi (Taipei) 62(3): 125-132.

Loo, V. G., C. Bertrand, et al. (1996). "Control of construction-associated nosocomial aspergillosis in an antiquated hematology unit." Infect Control Hosp Epidemiol 17(6): 360-364.

Loveless, M. O., R. E. Winn, et al. (1981). "Mixed invasive infection with Alternaria species and Curvularia species." Am J Clin Pathol 76(4): 491-493.

Lyratzopoulos, G., M. Ellis, et al. (2002). "Invasive infection due to Penicillium species other than P. marneffei." J Infect 45(3): 184-195.

Macher, J. (1999). Bioaerosols: assessment and control. Cincinnati, Ohio, American Conference of Governmental Industrial Hygienists.

Macher, J. and American Conference of Governmental Industrial Hygienists. (1999). Bioaerosols : assessment and control. Cincinnati, Ohio, ACGIH.

Maki, D. G. (1981). "Nosocomial bacteremia. An epidemiologic overview." Am J Med 70(3): 719-732.

Mammeri, H., S. Bellais, et al. (2002). "Chromosome-encoded beta-lactamases TUS-1 and MUS-1 from Myroides odoratus and Myroides odoratimimus (formerly Flavobacterium odoratum), new members of the lineage of molecular subclass B1 metalloenzymes." Antimicrob Agents Chemother 46(11): 3561-3567.

Marinella, M. A., C. Pierson, et al. (1997). "The stethoscope. A potential source of nosocomial infection?" Arch Intern Med 157(7): 786-790.
Mathews, D., J. P. Mathews, et al. (2005). "Preseptal cellulitis caused by Acinetobacter lwoffi." Indian J Ophthalmol 53(3): 213-214.
Matrician, L., G. Ange, et al. (2000). "Outbreak of nosocomial Burkholderia cepacia infection and colonization associated with intrinsically contaminated mouthwash." Infect Control Hosp Epidemiol 21(11): 739-741.
Matsumoto, S., H. Suenaga, et al. (2000). "Management of suspected nosocomial infection: an audit of 19 hospitalized patients with septicemia caused by Bacillus species." Jpn J Infect Dis 53(5): 196-202.
McEachern, R. and G. D. Campbell, Jr. (1998). "Hospital-acquired pneumonia: epidemiology, etiology, and treatment." Infect Dis Clin North Am 12(3): 761-779, x.
McGowan, J. E., Jr. (1988). "Gram-positive bacteria: spread and antimicrobial resistance in university and community hospitals in the USA." J Antimicrob Chemother 21 Suppl C: 49-55.
Meyer, C. L., H. E. Eitzen, et al. (1981). "Should linen in newborn intensive care units be autoclaved?" Pediatrics 67(3): 362-364.
Mikat, D. M. (1980). "Unusual fungal conditions of the skin." Int J Dermatol 19(1): 18-23.
Muhm, M., A. Zuckermann, et al. (1996). "Early onset of pulmonary mucormycosis with pulmonary vein thrombosis in a heart transplant recipient." Transplantation 62(8): 1185-1187.
Mulhausen, J. R. and J. Damiano (1998). A strategy for assessing and managing occupational exposures. Fairfax, VA, AIHA Press.
Myoken, Y., T. Sugata, et al. (2002). "Fatal necrotizing stomatitis due to Trichoderma longibrachiatum in a neutropenic patient with malignant lymphoma: a case report." Int J Oral Maxillofac Surg 31(6): 688-691.
Naidu, J. (1993). "Growing incidence of cutaneous and ungual infections by non-dermatophyte fungi at Jabalpur (M.P.)." Indian J Pathol Microbiol 36(2): 113-118.
Naim ur, R., E. S. Mahgoub, et al. (1988). "Fatal brain abscesses caused by Ramichloridium obovoideum: report of three cases." Acta Neurochir (Wien) 93(3-4): 92-95.
Naldi, L., S. Lovati, et al. (2000). "Paecilomyces marquandii cellulitis in a kidney transplant patient." Br J Dermatol 143(3): 647-649.
Narayani, T. V., K. Naseema, et al. (1990). "Prevalence of coagulase negative Staphylococcus species among hospital personnel and surgical patients." Indian J Pathol Microbiol 33(3): 258-262.
Ndugulile, F., R. Jureen, et al. (2005). "Extended spectrum beta-lactamases among Gram-negative bacteria of nosocomial origin from an intensive care unit of a tertiary health facility in Tanzania." BMC Infect Dis 5: 86.
Nenoff, P., T. Friedrich, et al. (1997). "Rare fatal simultaneous mould infection of the lung caused by Aspergillus flavus and the basidiomycete Coprinus sp. in a leukemic patient." J Med Vet Mycol 35(1): 65-69.
Nenoff, P., L. C. Horn, et al. (1996). "[Invasive mold infections in the university clinics of Leipzig in the period from 1992-1994]." Mycoses 39 Suppl 1: 107-112.
Newman, M. J. (2002). "Neonatal intensive care unit: reservoirs of nosocomial pathogens." West Afr J Med 21(4): 310-312.
Niedoszytko, M., M. Chelminska, et al. (2002). "[Fungal allergy--part II]." Pol Merkuriusz Lek 12(70): 314-317.

NNISR (1979). "Nosocomial acquisition of aeromonas hydrophila." Natl Nosocomial Infect Study Rep: 23-25.

NYCDHMH (2006). "Guidelines on Assessment and Remediation of Fungi in Indoor Environments." The New York City Department of Health and Mental Hygiene.

Otterbein, C. K., W. D. Splettstoesser, et al. (1998). "Development and characterization of a murine monoclonal antibody reactive with a 64 kDa somatic antigen of Burkholderia cepacia." Hybridoma 17(2): 143-150.

Palabiyikoglu, I., E. Tekeli, et al. (2006). "Nosocomial meningitis in a university hospital between 1993 and 2002." J Hosp Infect 62(1): 94-97.

Pechere, J. C. (1993). "[Microbiology of nosocomial infections]." Bull Acad Natl Med 177(5): 705-717; discussion 717-708.

Pegues, C. F., E. S. Daar, et al. (2001). "The epidemiology of invasive pulmonary aspergillosis at a large teaching hospital." Infect Control Hosp Epidemiol 22(6): 370-374.

Perdelli, F., M. L. Christina, et al. (2006). "Fungal contamination in hospital environments." Infection Control and Hospital Epidemiology 27(1): 44-47.

Perez Monras, M., L. Azahares Romero, et al. (1992). "[The surveillance of nosocomial bacteremia in the Intensive Care Unit of the Hospital Pediatrico Docente Centro Habana]." Rev Cubana Med Trop 44(1): 25-28.

Perlman, E. M. and L. Binns (1997). "Intense photophobia caused by Arthrographis kalrae in a contact lens-wearing patient." Am J Ophthalmol 123(4): 547-549.

Perola, O., T. Nousiainen, et al. (2002). "Recurrent Sphingomonas paucimobilis -bacteraemia associated with a multi-bacterial water-borne epidemic among neutropenic patients." J Hosp Infect 50(3): 196-201.

Peter, J. D., F. Jehl, et al. (1988). "[Aztreonam treatment of severe infections caused by gram-negative aerobic bacilli]." Pathol Biol (Paris) 36(5): 525-530.

Peters, G. (1991). "[Infections caused by staphylococci. The human as a source of infection for S. aureus and coagulase negative staphylococci]." Fortschr Med 109(22): 437-440.

Podnos, Y. D., P. Anastasio, et al. (1999). "Cerebral phaeohyphomycosis caused by Ramichloridium obovoideum (Ramichloridium mackenziei): case report." Neurosurgery 45(2): 372-375.

Poirier, T., E. Laurens, et al. (1993). "[Nosocomial Aeromonas hydrophila pneumonia complicating toxic coma]." Ann Fr Anesth Reanim 12(1): 72-74.

Ponce de Leon, S., S. H. Guenthner, et al. (1986). "Microbiologic studies of coagulase-negative staphylococci isolated from patients with nosocomial bacteraemias." J Hosp Infect 7(2): 121-129.

Poonwan, N., M. Kusum, et al. (1995). "Pathogenic Nocardia isolated from clinical specimens including those of AIDS patients in Thailand." Eur J Epidemiol 11(5): 507-512.

Pope, A. M., R. Patterson, et al. (1993). Indoor allergens: assessing and controlling adverse health effects. Washington, D.C., National Academy Press.

Potvliege, C., L. Vanhuynegem, et al. (1989). "Catheter infection caused by an unusual pathogen, Agrobacterium radiobacter." J Clin Microbiol 27(9): 2120-2122.

Ramphal, L. (2006). Personal communication between Michael Larranaga and Dr. Ramphal, MD, MPH.

Rao, A., R. Forgan-Smith, et al. (1989). "Phaeohyphomycosis of the nasal sinuses caused by Bipolaris species." Pathology 21(4): 280-281.

Rathinavelu, S., Y. Zavros, et al. (2003). "Acinetobacter lwoffii infection and gastritis." Microbes Infect 5(7): 651-657.

Raymond, J. (2000). "[Epidemiology of nosocomial infections in pediatrics]." Pathol Biol (Paris) 48(10): 879-884.

Raymond, J. and Y. Aujard (2000). "Nosocomial infections in pediatric patients: a European, multicenter prospective study. European Study Group." Infect Control Hosp Epidemiol 21(4): 260-263.

Redd, S. C. (2002). "State of the Science on Molds and Human Health." Center for Disease Control and Prevention, U.S. Department of Health and Human Services.: 1-11.

Richard, V., P. Van der Auwera, et al. (1988). "Nosocomial bacteremia caused by Bacillus species." Eur J Clin Microbiol Infect Dis 7(6): 783-785.

Richards, W. (1986). "Allergy, asthma, ans school problems." Journal of School Health 56(4): 151-152.

Richter, S., M. G. Cormican, et al. (1999). "Fatal disseminated Trichoderma longibrachiatum infection in an adult bone marrow transplant patient: species identification and review of the literature." J Clin Microbiol 37(4): 1154-1160.

Rihs, J. D., A. A. Padhye, et al. (1996). "Brain abscess caused by Schizophyllum commune: an emerging basidiomycete pathogen." J Clin Microbiol 34(7): 1628-1632.

Robb, C. W., P. J. Malouf, et al. (2003). "Four cases of dermatomycosis: superficial cutaneous infection by Alternaria or Bipolaris." Cutis 72(4): 313-316, 319.

Robles Garcia, M., T. Dierssen Sotos, et al. (2005). "Prevention of nosocomial infection of fungal origin: verification of the environmental biosafety in surgery rooms." Rev Clin Esp 205(12): 601-606.

Rodrigues, M. M. and P. Laibson (1975). "Exogenous corneal ulcer caused by Tritirachium roseum." Am J Ophthalmol 80(5): 804-806.

Rosina, M. (1991). "[Current microbiological aspects of nosocomial infections]." Recenti Prog Med 82(2): 100-103.

Schaal, K. P. and H. J. Lee (1992). "Actinomycete infections in humans--a review." Gene 115(1-2): 201-211.

Schofferman, L., J. Zucherman, et al. (1991). "Diptheroids and associated infections as a cause of failed instrument stabilization procedures in the lumbar spine." Spine 16(3): 356-358.

Seetha, K. S., I. Bairy, et al. (2002). "Bacteraemia in high-risk patients." Indian J Med Sci 56(8): 391-396.

Segonds, C. and G. Chabanon (2001). "Burkholderia cepacia: dangers of a phytopathogen organism for patients with cystic fibrosis." Ann Biol Clin (Paris) 59(3): 259-269.

Segonds, C., T. Heulin, et al. (1999). "Differentiation of Burkholderia species by PCR-restriction fragment length polymorphism analysis of the 16S rRNA gene and application to cystic fibrosis isolates." J Clin Microbiol 37(7): 2201-2208.

Severo, L. C., F. Job, et al. (1991). "Systemic zygomycosis: nosocomial infection by Rhizomucor pusillus." Mycopathologia 113(2): 79-80.

Shehabi, A. A., W. Abu-Al-Soud, et al. (2004). "Investigation of Burkholderia cepacia nosocomial outbreak with high fatality in patients suffering from diseases other than cystic fibrosis." Scand J Infect Dis 36(3): 174-178.

Shin, J. Y., H. M. Kim, et al. (2002). "Keratitis caused by Verticillium species." Cornea 21(2): 240-242.

Sigler, L., S. Estrada, et al. (1997). "Maxillary sinusitis caused by Schizophyllum commune and experience with treatment." J Med Vet Mycol 35(5): 365-370.

Simpson, G. L., E. B. Stinson, et al. (1981). "Nocardial infections in the immunocompromised host: A detailed study in a defined population." Rev Infect Dis 3(3): 492-507.

Solomon, G. M., M. Hjelmroos-Koski, et al. (2006). "Airborne mold and endotoxin concentrations in New Orleans, Louisiana, after flooding, October through November 2005." Environ Health Perspect 114(9): 1381-1386.

Spengler, J. D., J. M. Samet, et al. (2001). Indoor air quality handbook. New York, McGraw-Hill.

Stepanovic, S., I. Dakic, et al. (2002). "Surgical wound infection associated with Staphylococcus sciuri." Scand J Infect Dis 34(9): 685-686.

Stevens-Krebbers, A. H., M. A. Schouten, et al. (1999). "Nosocomial transmission of Bordetella bronchiseptica." J Hosp Infect 43(4): 323-324.

Streifel, A. J., P. P. Stevens, et al. (1987). "In-hospital source of airborne Penicillium species spores." J Clin Microbiol 25(1): 1-4.

Sudakin, D. L. (1998). "Toxigenic fungi in a water-damaged building: an intervention study." Am J Ind Med 34(2): 183-190.

Summerbell, R. C., J. Kane, et al. (1989). "Onychomycosis, tinea pedis and tinea manuum caused by non-dermatophytic filamentous fungi." Mycoses 32(12): 609-619.

Sutton, D. A., M. Slifkin, et al. (1998). "U.S. case report of cerebral phaeohyphomycosis caused by Ramichloridium obovoideum (R. mackenziei): criteria for identification, therapy, and review of other known dematiaceous neurotropic taxa." J Clin Microbiol 36(3): 708-715.

Szewczyk, E. M., A. Piotrowski, et al. (2000). "Predominant staphylococci in the intensive care unit of a paediatric hospital." J Hosp Infect 45(2): 145-154.

TDSHS. (1994). "PHYSICAL PLANT AND CONSTRUCTION REQUIREMENTS FOR NEW AND EXISTING AMBULATORY SURGICAL CENTERS." Title 25, Part 1, Chapter 135, Sub Chapter C, 2007, from http://www.sos.state.tx.us/tac/

Tejada Artigas, A., S. Bello Dronda, et al. (2001). "Risk factors for nosocomial pneumonia in critically ill trauma patients." Crit Care Med 29(2): 304-309.

Telles Filho Fde, Q., A. Coelho, et al. (1985). "Subcutaneous mucormycosis caused by Rhizopus oryzae probable nosocomial acquired infection." Rev Inst Med Trop Sao Paulo 27(4): 201-206.

Tomsikova, A. (2002). "Causative agents of nosocomial mycoses." Folia Microbiol (Praha) 47(2): 105-112.

Tomsikova, A. (2002). "[Risk of fungal infection from foods, particularly in immunocompromised patients]." Epidemiol Mikrobiol Imunol 51(2): 78-81.

Toniolo, A., A. Endimiani, et al. (2006). "Microbiology of postoperative infections." Surg Infect (Larchmt) 7 Suppl 2: S13-16.

Toul, P., L. Castillo, et al. (2006). "A pseudo tumoral sinusitis caused by Bipolaris sp." J Infect.

Trupl, J., T. Minarik, et al. (1995). "Nosocomial bacterial and fungal meningitis in cancer patients." Support Care Cancer 3(6): 425-427.

USEPA (2001). Mold remediation in schools and commercial buildings. United States Environmental Protection Agency. EPA 402-K-01-001.

Van Der Zwet, W. C., Y. J. Debets-Ossenkopp, et al. (2002). "Nosocomial spread of a Staphylococcus capitis strain with heteroresistance to vancomycin in a neonatal intensive care unit." J Clin Microbiol 40(7): 2520-2525.

Venugopal, P. L., T. L. Venugopal, et al. (1989). "Mycotic keratitis in Madras." Indian J Pathol Microbiol 32(3): 190-197.

Verweij, P. E., M. van Kasteren, et al. (1997). "Fatal pulmonary infection caused by the basidiomycete Hormographiella aspergillata." J Clin Microbiol 35(10): 2675-2678.
von Eiff, C., G. Peters, et al. (2002). "Pathogenesis of infections due to coagulase-negative staphylococci." Lancet Infect Dis 2(11): 677-685.
von Eiff, C., R. A. Proctor, et al. (2001). "Coagulase-negative staphylococci. Pathogens have major role in nosocomial infections." Postgrad Med 110(4): 63-64, 69-70, 73-66.
Votava, M., B. Skalka, et al. (1997). "[Rhodococcus equi--a newly recognized opportunistic pathogen in man]." Epidemiol Mikrobiol Imunol 46(2): 58-66.
Vujanovic, V., W. Smoragiewicz, et al. (2001). "Airborne fungal ecological niche determination as one of the possibilities for indirect mycotoxin risk assessment in indoor air." Environ Toxicol 16(1): 1-8.
VUMC (2006). Infection Control Interventions During Construction in Patient Care Areas. Procedure 10-10.17. Nashville, TN, Vanderbilt University Medical Center. 10-10.17.
W.B. Saunders (2000). Dorland's illustrated medical dictionary. London, W.B. Saunders.
Walsh, T. J., C. Gonzalez, et al. (1995). "Fungemia in children infected with the human immunodeficiency virus: new epidemiologic patterns, emerging pathogens, and improved outcome with antifungal therapy." Clin Infect Dis 20(4): 900-906.
Walsh, T. J. and A. H. Groll (1999). "Emerging fungal pathogens: evolving challenges to immunocompromised patients for the twenty-first century." Transpl Infect Dis 1(4): 247-261.
Wang, K. W., W. N. Chang, et al. (2005). "Post-neurosurgical nosocomial bacterial meningitis in adults: microbiology, clinical features, and outcomes." J Clin Neurosci 12(6): 647-650.
Wang, S. M., C. C. Liu, et al. (1999). "Staphylococcus capitis bacteremia of very low birth weight premature infants at neonatal intensive care units: clinical significance and antimicrobial susceptibility." J Microbiol Immunol Infect 32(1): 26-32.
Wheat, L. J., M. Goldman, et al. (2002). "State-of-the-art review of pulmonary fungal infections." Semin Respir Infect 17(2): 158-181.
Wiest, P. M., K. Wiese, et al. (1987). "Alternaria infection in a patient with acquired immunodeficiency syndrome: case report and review of invasive alternaria infections." Rev Infect Dis 9(4): 799-803.
Wilsher, M. L., J. Kolbe, et al. (1997). "Nosocomial acquisition of Burkholderia gladioli in patients with cystic fibrosis." Am J Respir Crit Care Med 155(4): 1436-1440.
Won, Y. S., H. J. Kwon, et al. (2002). "Identification of Staphylococcus xylosus isolated from C57BL/6J-Nos2(tm1Lau) mice with dermatitis." Microbiol Immunol 46(9): 629-632.
Xi, L., K. Fukushima, et al. (2004). "First case of Arthrographis kalrae ethmoid sinusitis and ophthalmitis in the People's Republic of China." J Clin Microbiol 42(10): 4828-4831.
Yang, X. H., X. H. Xu, et al. (2000). "Clinical study on nosocomial infection in patients with burns." Hunan Yi Ke Da Xue Xue Bao 25(4): 388-390.
Yeulet, S. E., P. G. Mantle, et al. (1988). "Nephrotoxicity of Penicillium aurantiogriseum, a possible factor in the aetiology of Balkan endemic nephropathy." Mycopathologia 102(1): 21-30.
Young, V. L., M. C. Hertl, et al. (1995). "Paecilomyces variotii contamination in the lumen of a saline-filled breast implant." Plast Reconstr Surg 96(6): 1430-1434.
Zhang, Y. (1991). "[A two-year prospective survey on nosocomial infections]." Zhonghua Yi Xue Za Zhi 71(5): 253-256, 218.

14

Air Change Measurements Using Tracer Gases

Detlef Laussmann and Dieter Helm
Robert Koch Institute
Germany

1. Introduction

Both comfortable and healthy indoor climate conditions can only be achieved by constant fresh air supply. However, the minimum of air change required to reach this goal is depending upon different perspectives. Measures for thermal insulation and energy savings are difficult to bring in line with air quality requirements that result from findings of epidemiological studies (Seppänen et al., 1999; Seppänen & Fisk, 2004; Wargocki et al., 2002), and supply air facilities considering the technical construction of buildings and safety aspects (Erhorn & Gertis, 1986). Moreover, construction deficits impairing the integrity of the building envelope, meteorological conditions (thermal and flow induced pressure differences) and, not least, the behaviour of the residents, affect the air change in a variety of ways (Heidt, 1987).

For a number of different reasons it is necessary and desirable to examine the real fresh air flow between indoor and outdoor climate under given circumstances. Air change processes are of particular importance in studies focusing on their relationship to indoor air pollutants.

For manifold reasons the air quality inside buildings has been intensively investigated since quite a long time. Of particular relevance are the following aspects (Seifert & Salthammer, 2003):

- In countries with a cold or temperate climate, inhabitants spend more than 50% of the time in their homes. In certain population groups (e.g. infants and the elderly), this proportion is even exceeding 90%.
- Indoor air contains a wide range of different organic and inorganic components. Therefore, the carbon dioxide concentration alone cannot always be regarded as an indicator of air quality (Fanger, 1988; Persily, 1997). Above all, organic compounds which are released from the building, furnishings, household and hobby devices, as well as by daily activities of the inhabitants, such as cooking, baking or frying, and especially smoking, altogether contribute to air pollution caused by volatile organic compounds.

The energy crisis in the 1970s caused an increase in energy costs, evoking an urgent need to reduce the consumption of heat energy. The prevention of heat loss in homes is a very effective way to save energy and related costs. For climate protection and the reduction of global CO_2 emissions the economical use of energy resources is of outstanding significance. Thermal insulation of the building envelope protects efficiently against heat loss, but enhanced tightness includes the disadvantage of reduced air change, thereby increasing the indoor pollutant concentration. To evaluate the indoor air quality under conditions of natural ventilation, the air change rate (ACR) can be determined through standard tracer gas measurement.

For this purpose, a small amount of the tracer gas is released in the room (or building) under study and its concentration is then recorded as a function of time. Subsequently, using appropriate evaluation algorithms the ACR can be calculated from the data obtained.

The first section is introducing basic physical principles of air change processes between indoor and outdoor environment, followed by a description of tracer gases which are frequently used in daily routine, and methods determining the ACR. Moreover, the applicability of carbon dioxide as a tracer gas has been compared with results obtained by hexafluorobenzene or sulphur-hexafluoride. In addition, we are discussing the impact of weather conditions on ACR data obtained under natural ventilation. This section is followed by a review on ACR due to passive ventilation through facades in selected residences in Berlin. Finally, the relationship between air change and the concentration of selected volatile compounds under worst case conditions will be discussed.

2. Modelling air change in indoor rooms

According to VDI (2001) 4300, part 7, air change is defined as the ratio of air supply $Q(t)$ into a zone (i.e. a room or space) in relation to the volume of this zone V_R (room volume) and is generally expressed as air change per hour [h^{-1}] or [ACH]. The following equation expresses this definition:

$$\lambda(t) = Q(t)/V_R \tag{1}$$

$\lambda(t)$ is the ventilation rate or air change rate [h^{-1}],
$Q(t)$ is the air supply into a room [m^3/h],
V_R is the room's volume [m^3], and t = time [h].

2.1 Model assumptions

For the model described here, which is designed to calculate the time course of the tracer gas concentration, the following simplifying assumptions has been made:

- The tracer gas is considered to be chemically stable and inert; i.e. there will be no chemical reactions capable to alter the concentration of the tracer gas in the room.
- There will be no adsorption processes on walls, ceiling or furnishings of the room that may lower the concentration of the tracer gas in the room.
- The air is considered to be completely mixed throughout the measurements. Inside the room there are no concentration gradients, i.e. the concentration of the tracer gas at a given time is the same for the whole room.
- An exchange of tracer gas-containing air with ambient air only occur in those areas that are in contact with the outside, i.e. air change with other interior spaces is considered to be negligible. The room in which the tracer gas was released is considered a single zone system.
- The exchange processes that take place during the measurement period are assumed to be temporally invariant. The air supply rate $Q(t)$ and, thus, the air change rate $\lambda(t)$ are constant. $Q(t)$ and $\lambda(t)$ can be replaced by Q and λ.

2.2 Model equations

The basis for the description of the relationship between the mass or concentration of a gaseous substance in a space as a function of time is the mass balance equation. This equation expresses that the mass and - in a fixed volume - the concentration of a tracer gas can only change when either more tracer gas is added to the original amount or tracer gas is removed by elimination processes. Considering the above assumptions, the following supply and removal processes are significant for the tracer gas concentration in the room air (physical dimensions of these variables are given in brackets):

- Transport of tracer gas from the room air to the outside: Q^*C_i [mass per time unit]
- Transport of tracer gas from the outside air into the room air Q^*C_a [mass per time unit]
- (Constant) emission E of tracer gas into the space by a tracer gas source [mass per time unit]

Thus, the mass balance equation can be formulated as the following differential equation (Heidt & Werner, 1986):

$$V_R^*dC_i(t)/dt = -C_i^*Q + C_a^*Q + E \quad \text{or}$$

$$V_R^*dC_i(t)/dt = -(C_i - C_a)^*Q + E \tag{2a}$$

After dividing both sides of the equation by the volume V_R, we obtain an ordinary differential equation which describes the concentration change of a tracer gas in the room per time unit:

$$dC_i(t)/dt = -(C_i - C_a)^*Q/V_R + E/V_R \tag{2b}$$

C_a: tracer gas concentration in outside air [mass / volume]
C_i: tracer gas concentration in the indoor air [mass / volume]
Q: exchange air flow between room and outside [volume / time unit]
E: amount of tracer gas emitted per unit time [mass / time unit]
V_R: room volume
t: time

As stated above, Q/V_R is defined as the air change rate. When Q/V_R in Eq. 2b is replaced by λ (Eq. 1) we obtain Eq. 2c:

$$dC_i(t)/dt = -(C_i - C_a)^*\lambda + E/V_R \tag{2c}$$

Equation 2c expresses that under constant homogeneous mixing the concentration change of the tracer gas is proportional to the concentration difference between indoor and outdoor spaces ($C_i - C_a$) at time t, the air change rate λ, and the amount of tracer gas emitted per time unit (emission rate E).

By integration we obtain the starting conditions at time 0 ($C(t=0) = C_0$):

$$C_i(t) = C_a + E/(\lambda^*V_R) + [C_0 - C_a - E/(\lambda^*V_R)] \exp(-\lambda t_i) \tag{3a}$$

Expanding and transposing yields:

$$C_i(t) = (C_0 - C_a) \exp(-\lambda t_i) + C_a + E/(\lambda^*V_R) [1 - \exp(-\lambda t_i)] \tag{3b}$$

Integration of the differential equation thus leads to an exponential function with the air change rate λ in the exponent. Equation 3b is the basis for the mathematical analysis of tracer gas measurements that are recorded as concentration-time curves. This function is characterised by the following features: If no tracer gas is emitted into the room (i.e. E = 0), and there is already a non-zero tracer gas concentration C_0 present at the time t = 0 which is higher than the outdoor air concentration C_a, then the expression $(C_0 - C_a)\exp(-\lambda t) + C_a$ describes the elimination of the tracer gas out of the room. The curve starts with the initial concentration C_0 and decays exponentially until the ambient tracer gas concentration C_a or any other constant background concentration is reached. If a tracer gas is used, which does not occur in the outside air (i.e. $C_a = 0$) the concentration decreases over time to the value zero. Equation 3b can be simplified to:

$$C_i(t) = C_0 \exp(-\lambda t_i) \tag{3c}$$

If a source in the room is emitting tracer gas with a constant rate E and if the initial tracer gas concentration in the room at the beginning of the measurement C_0 is equal to the outside concentration C_a ($C_0 = C_a$), then the tracer gas will accumulate within the room. Starting with the concentration C_a, the mathematical function grows with increasing t asymptotically towards C_{eq}, which reaches the final value of $C_a + E/(\lambda^* V_R)$ at infinity.

$$C(t) = C_a + [E/(\lambda^* V_R)][1-\exp(-\lambda t)] \tag{4a}$$

When the outdoor air concentration can be neglected, then:

$$C(t) = [E/(\lambda^* V_R)][1-\exp(-\lambda t)] \tag{4b}$$

For $t >> 1/\lambda$ Equation 4b can be simplified to:

$$C_{eq} = E/(\lambda^* V_R) \tag{4c}$$

C_{eq}: equilibrium concentration

Equation 4c reflects the fact that under equilibrium conditions (i.e. emission equals elimination) the tracer gas concentration C_{eq} is proportionally dependent on the emission rate E but inversely proportionally dependent on both the air change rate λ and the volume of the room V_R. This relationship forms the basis for the determination of the air change rate with tracer gas measurements made under equilibrium conditions.

3. Determination of the air change rate (ACR) using tracer gases

In addition to the physical properties discussed in Section 2.1 (model assumptions) tracer gases should fulfil some other requirements in regard to their practical suitability. These include health safety aspects, low environmental burden, high availability and good handling in practical use at the lowest costs possible and, not least, tracer gases should be well recordable with established measurement techniques over a wide concentration range and with high selectivity (Raatschen, 1995).

3.1 Requirements for ideal and commonly used tracer gases

In the past, a number of gaseous substances such as helium, hydrogen, oxygen, carbon monoxide, methane, acetone, and the radioactive noble gases argon-41 and krypton-85, have been studied and tested for the determination of ACR, mostly in comparison with other tracer gases. Reviews are given by Grimsrud et al. (1980), Shaw (1984), and Sherman (1990). Until the early 1990s, krypton-85 was still used for air change measurement (Schulze &

Schuschke, 1990), but later skipped for safety reasons (radiation protection). Nowadays, the following tracer gases are mainly used in practice (Raatschen, 1995):

- Nitrous oxide (N_2O)
- Sulphur hexafluoride (SF_6)
- Halogenated hydrocarbons, such as hexafluorobenzene (C_6F_6) and perfluorocarbons (PFC).

3.1.1 Nitrous oxide

Formerly, nitrous oxide (N_2O) was widely used for air change measurement in buildings, primarily in Europe (Heidt & Werner, 1986; Keller & Beckert, 1994; Salthammer, 1994; Wegner, 1983, 1984). In the U.S. it was rarely used because of its low TLV (threshold limit value) of 50 ppm (Lagus & Grot, 1997). Germany's equivalent of TLV, the MAK, is however 100 ppm. For precautionary reasons it should not be used in occupied buildings (Raatschen, 1995). Other disadvantages of N_2O are the ease of adsorption on surfaces at concentrations below 1000 ppm, and its high solubility in water, which means that the air change rate can be substantially overestimated in very airtight rooms (Schulze & Schuschke, 1990).

3.1.2 Sulphur hexafluoride (SF_6)

After Gregory's observation in 1962, that sulphur hexafluoride (SF_6) can be measured reliably on the nanogram scale with electron capture detection - ECD (Gregory, 1962), it has been widely used for air infiltration measurement in buildings since the early 1970s (Drivas et al., 1972; Hunt & Burch, 1975). Of all candidates, the characteristics of SF_6 are nearest to the ideal of a tracer gas. Today, SF_6 is the most frequently used tracer gas worldwide, which is confirmed by the number of relevant publications. In this chapter, only on a very limited selection of publications could be considered (Chuah et al., 1997; Howard-Reed et al., 2002; Kumar et al., 1979; Lagus & Grot, 1997; Raatschen, 1995; Shaw, 1984; Walker & Forest, 1995; Wilson et al., 1996). SF_6 is very stable and only decomposes above 550 °C. Background concentration of SF_6 in ambient air is ≈1 ppt (6 ng/m^3) (Raatschen, 1995). Although its density is about five times higher than that of air, this difference causes no systematically distorting effects on the results of air change measurements with concentrations usually applied in practice (Niemelä et al., 1991; Shaw, 1984). SF_6 can be used in occupied buildings. It can be recorded with high accuracy within a wide concentration range. According to Raatschen (1995) the concentrations commonly used for indoor air change measurement are not exceeding one hundredth of the German MAK value which was defined to be 1000 ppm or 6100 mg/m^3 (TRGS 900, 1999). It should be noted that this value is not a toxicity limit. It is simply defined as the still manageable analytical upper limit for gases which are not imminently toxic (BIA-Report, 2001). Due to its high stability SF_6 is only very slowly degraded in the atmosphere and it belongs, like the perfluorinated hydrocarbons, to the climatic relevant greenhouse gases. Therefore, to avoid unnecessary environment hazard it should be used carefully and sparingly in concentrations as low as possible, like other tracer gases.

3.1.3 Hexafluorobenzene (C_6F_6) and perfluorocarbon-hydrocarbon tracers (PFT)

Hexafluorobenzene, and perfluorocarbon-hydrocarbon tracers (PFT) such as perfluorodimethylcyclobutane, perfluoromethylcyclobutane, perfluorodimethylcyclo-hexane, and perfluoromethylcyclohexane are also appropriate as tracer gases and are preferably used in the determination of ACR with the constant injection method applying the passive sampler technique. Enrichment for active sampling is also possible (Cheong & Riffat, 1995; Dietz & Cote, 1982; Dietz et al., 1986; Krooß et al., 1997; Mailahn et al., 1989; Salmon et al., 2000). Since these compounds, unlike SF_6, adsorb well on activated carbon or Tenax they are frequently used in field studies (surveys) to determine indoor air change rates (Andersen et al., 1997; Bornehag et al., 2005; Hirsch et al., 2000; Lembrechts et al., 2001; Øie et al., 1997, 1998; Parker, 1986; Pandian et al., 1993; Ruotsalainen et al., 1992; Sakaguchi & Akabayashi, 2003). The disadvantage of these compounds, however, is that they tend to attach to room surfaces and the emission rates of these gases are strongly temperature dependent (Hill et al., 2000). Thus, a sufficiently long conditioning period is necessary as well as an accurate temperature control of the storage vessels.

3.2 Methods for the determination of ACR

Basically, three appropriate methods exist for the determination of ACR using tracer gases. According to VDI (2001) 4300, part 7, these are the concentration decay method, the constant injection method, and the constant concentration method.

3.2.1 Concentration decay method

Tracer gas is injected into the room for a short period of time, either from a gas bottle with pressure reducer or manually from filled gas tanks. After mixing with the room air the tracer gas concentration is measured at regular time intervals. Because the decay curve of the tracer gas concentration C follows an exponential course when completely mixed with the room air (see Eq. 3c) each sampling will not only remove old air but also a certain amount of fresh air supply as well. This essentially means that there will be still ≈37% ($100/e^1$) of the originally added tracer gas (37% old room air), after a complete air change cycle has occurred. The time ($1/\lambda$), after which the air change cycle is completed is known as the nominal time constant τ (Maas, 1997; Sherman, 1990). After 3τ ($3/\lambda$), 4τ ($4/\lambda$) and 4.6τ ($4.6/\lambda$), the tracer gas concentration in the room volume under study is 5%, 2% and 1% of its initial value, respectively (Fig. 1a).
By using the concentration decay function

$$C(t) = C_0 * e^{-\lambda * t} \qquad (5)$$

and applying non linear regression analysis we can determine the ACR λ (C_0: tracer gas concentration at time t=0 (Sherman, 1990)). However, in most cases logarithmic concentration values are used to obtain a linear relationship between the logarithm of the tracer gas concentration $C_i(t)$ and the time t (Fig. 1b):

$$\ln C_i(t) = \ln C_0 - \lambda * (t_i) \qquad (5a)$$

The ACR λ is then calculated via linear regression analysis according to Eq. 5a. Both evaluation options - the linear and the non-linear regression analysis - are particularly well suited for the examination of concentration-time curves recorded over a longer period of time. If only a few measurements are available, e.g. when sampling is performed with syringes or other appropriate devices, then the ACR can be determined using the following relationship

$$[\ln C_i(t=t_i) - \ln C_i(t=t_{i+1})] \;/\; (t_{i+1} - t_i) = \lambda. \qquad (5b)$$

$C_i(t_i)$ is the tracer gas concentration at time t_i
$t_{i+1} - t_i$ is the time interval between two measurements

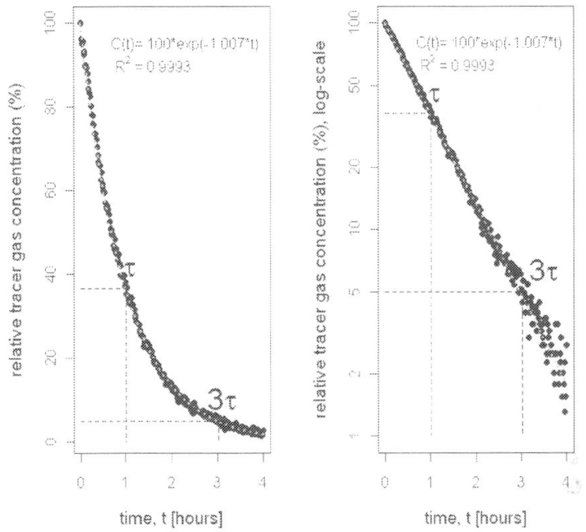

Fig. 1. Relationship between tracer gas concentration and time; a, untransformed ordinate; b, after logarithmic transformation.

An illustration of the procedure is given in Fig. 2a. Sulphur hexafluoride (SF_6) was released in an office space (volume 48 m^3, equipped with two double glazed box windows). Initial concentration of the tracer gas was adjusted to ≈5 mg/m^3. The concentration was measured for 7 hours in 15 minutes intervals using a photo-acoustic infrared detector with selective filter (Bruel & Kjaer Single Gas Monitor). During the measurement the office was unoccupied; door and windows were kept closed. Tracer gas and room air were not continuously mixed with a fan or similar device. For the analysis, SF_6 concentration values were chosen in such a way that each concentration-time pair - $c(t_{i+1})$, $c(t_i)$ - covered exactly a time interval of one hour, meaning that the denominator was equal to 1 when determining the air change rate (ACR). After choosing the six measurement points depicted in Fig. 2a we calculated an average ACR of 0.327 h^{-1}. The standard deviation was 0.038 h^{-1}, which corresponds to a coefficient of variation of ≈12%. When the same measuring points were analysed by linear regression analysis we obtained an ACR of 0.323 h^{-1} (Fig. 2b). Both values are almost equal. The determination of ACR using concentration-time pairs is also known as the two-point method (Sherman, 1990). For this method, a number of at least five concentration-time pairs are recommended which should be evenly distributed over the entire measurement period (ASTM E 741 - 00; VDI , 2001). In order to reduce the statistical error to ≤10% the time interval between the first and the last measuring point should be in the order of magnitude of the nominal time constant (ASTM E 741 - 00; Heidt & Werner, 1986; Maas, 1997; Sherman, 1990). This prerequisite is, however, hardly fulfilled for ACR between 0.1 and 0.2 h^{-1} or even lower, since the required minimum decay time would be between 5 and 10 hours or more. Most probably, in the case of very low ACR a statistical error of more than 10% must be accepted. Recommended minimum time intervals between the first and the last measurement and the corresponding measurement intervals are summarised in Table 1, according to ASTM E 741 - 00 and VDI (2001) .

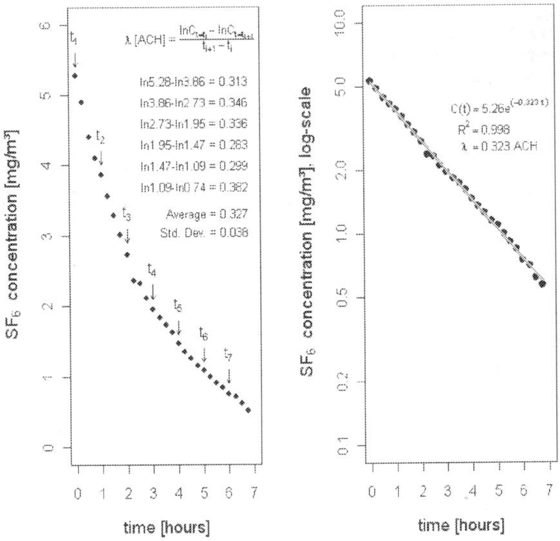

Fig. 2. ACR determination from concentration decay with the two-point method, time interval $t_{i+1} - t_i$ between two measurements: one hour; a, linear scale; b, logarithmic scale, linear regression.

Air change rate [ACH]	Minimum time span of tracer gas measurement [hours]	Air change rate [ACH]	Sampling interval [min]
0.05	20		
0.125	8		
0.25	4	< 0.5	30 - 40
1	1	0.5 to 1	20 - 30
2	0.5	1 to 2	10
4	0.25	2 to 5	5
10	0.1	> 10	< 2

Table 1. Examples for minimum time spans between first and last sampling based on recommendations of ASTM E 471 - 00 and recommended sampling intervals (VDI, 2001) for air change measurement with the decay method.

The concentration decay method is the most commonly used one in practice. It is particularly well suited for the determination of ACR up to 10 h^{-1} (10 ACH) in indoor rooms with a volume below 500 m^3 (VDI, 2001). If data loggers are used to record the concentration decay, the measurement intervals can be shortened down to seconds and in consequence ACR >10 h^{-1} can be determined.

3.2.2 Constant injection method

To determine ACR with this method a diffusion tube containing C_6F_6 or PFT is frequently used as tracer gas source. A defined amount of tracer gas is constantly emitted over a certain period of time. Thus, the tracer gas concentration increases with time and reaches a stable value (equilibrium concentration) which depends on the room volume V_R, the air change rate λ, and the emission rate E. At that time point, one or more air samples are taken and the tracer gas concentration is determined for each sample. The ACR can be calculated after solving Eq. 4c for λ. Fig. 3a depicts an example of the constant injection method.
Crucial for this method is, however, that sampling can only be started when the tracer gas concentration is near the equilibrium. If sampling starts too soon, an overestimation of the ACR will be the consequence. The time to reach approximate equilibrium conditions depends on λ (see Fig. 3b).

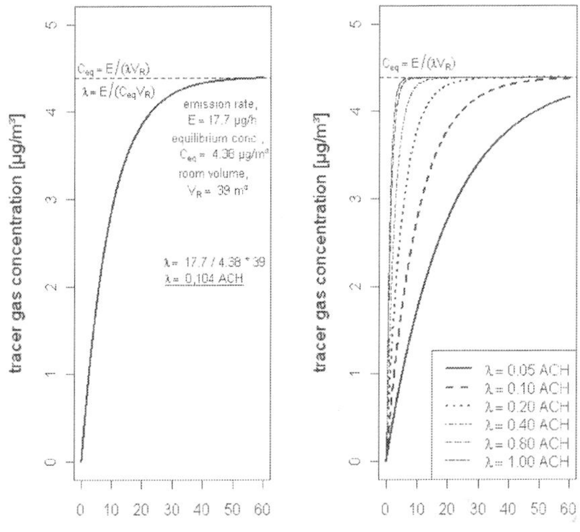

Fig. 3. Determination of ACR; a, constant injection method; b, concentrations after constant injection for different ACR.

According to ASTM E 741 - 00, the tracer gas concentration should have reached at least 95% of the equilibrium concentration before measurements can be started. The time until approximate equilibrium is reached can be estimated applying Eqs. 4b, 4c. However, under worst case conditions in rooms with very low air changes, it will be difficult to correctly measure ACRs with this method, because the build-up time to reach nearly equilibrium concentrations is too long (Table 2).
The time intervals for the tracer gas concentration to reach 95%, 98% or 99% of the target equilibrium value (C_{eq}) are $3/\lambda$, $4/\lambda$ and $4.6/\lambda$, respectively (cf. Table 2).
As Table 2 shows, the time periods needed to reach approximate equilibrium can be several days when the ACR is only 0.1 h^{-1} or even lower. For this reason, the constant injection method is suitable only for short-term measurements of ACR above 0.3 h^{-1} when diffusion

tubes are used as tracer gas source. Under usual "worst case" conditions (i.e. last active window ventilation 10-12 hrs before sampling) and ACR <0.3 h^{-1} the tracer gas concentrations are still too far away from the equilibrium. The advantage of the constant injection method is that it can be used for indoor hygiene studies addressing the relationship between pollution and air change in daily used rooms. In this case, tracer gas measurements can be performed with passive samplers and exposure times range from days to weeks.

	Time span [hours] to reach nearly equilibrium concentration C_{eq}		
Air change rate [hours]	95% C_{eq}	98% C_{eq}	99% C_{eq}
0	-	-	-
0.01	300	400	460
0.05	60	80	92
0.1	30	40	46
0.2	15	20	23
0.3	10	13	15
0.4	7.5	10	11.5
0.5	6	8	9.2
1	3	4	4.6
1.5	2	2.7	3
3	1	1.3	1.5

Table 2. Time spans [hours] needed to reach 95%, 98% or 99% of equilibrium concentration (C_{eq}) for a wide range of air change rates.

3.2.3 Constant concentration method

During constant and thorough mixing with the indoor air, tracer gas is released in the room until a predefined concentration is reached. During the entire measurement the tracer gas concentration is kept constant with an automated dosing and control system. Under the condition of constant tracer gas concentration the air supply is proportional to the tracer gas supply rate. The air supply rate can then be calculated from the ratio of the tracer gas supply to the tracer gas concentration. If the room volume is known, the ACR can be calculated from this ratio (Chao et al., 2004; Kumar et al., 1979; Maas, 1997). An advantage of the constant concentration method is that even short-term changes of air supply can be detected. Compared to the previously described methods the technical equipment required for this method is, however, rather expensive and thus this method is comparatively rarely used for indoor air quality evaluation.

3.3 The use of carbon dioxide as a tracer gas

Carbon dioxide (CO_2) is one of the gaseous organic compounds always detectable in the indoor air. Since humans exhale metabolic carbon dioxide in considerable quantities, its concentration can increase to several thousand ppm (ml/m^3 room air) within a short time. CO_2 concentration is often used to assess the air quality of occupied rooms. In this context we remind of Pettenkofer's reference concentration. Already in 1858 the German chemist and hygienist pointed out that a CO_2 concentration of 1000 ppm (0.1 vol %) is the upper tolerable limit in indoor environments. Nowadays CO_2 measurements are often used for the determination of the indoor ACR, because it can be easily quantified and the required

devices are reasonably priced and easy to operate. Moreover, CO_2 fulfils a number of the above mentioned specifications of a good tracer gas.
A huge number of studies are published testing the feasibility of exhaled human carbon dioxide as tracer gas in air change settings. Next to the already mentioned work of Pettenkofer (1858) we would like to allude to the studies done by Penman (1980), Penman & Rashid (1982), and Smith (1988) which are of special importance. Results of Dols & Persily (1992), Nabinger et al. (1994), and Persily (1997) have, however, demonstrated that ACR cannot be reliably determined from spot, peak or average values of the CO_2 concentration inside buldings, because these values are strongly influenced by the number of occupants in the rooms, their times of stay, and, hence, the incessantly changing carbon dioxide supply rates. Depending on the amount of natural ventilation the ACR is sometimes over-estimated up to 2-fold of the real value. The reason is that the air-tightness of modern buildings and the usual sojourn times of the occupants prevent in most cases that the equilibrium concentration can be approached. In practice it is much better to derive the ACR from the decay or build-up curve as was already shown (Barankova, 2005; Bekö et al., 2010; Chao et al., 1997; Chung & Hsu, 2001; Guo & Lewis, 2007; Menzies et al., 1995; Roulet & Foradini, 2002; Schulze & Schuschke, 1990; Sekhar, 2004; Shaw, 1984).

3.3.1 Statistical evaluation of CO_2 decay curves (with examples)

After CO_2 is released in a room, either as exhaled breath or via a gas container, its concentration will decay exponentially, if no further CO_2 supply occurs. To exemplify this, the CO_2 decay recorded in a bedroom of an older building (built 1908) is depicted in Fig. 4. In 1990 this bedroom was equipped with a double box window; the room volume is 30 m^3. The room was doped by a person with exhaled carbon dioxide, then the measurement was started and the room was left. During the entire procedure the room was unoccupied - door and window were kept closed. Measurement of CO_2 concentration was done by a CO_2 sensitive probe with infrared absorption. A Testo 400 device (Testo, Lenzkirch, Germany) was used for data logging and as control unit.
Unlike other tracer gases, CO_2 has the particularity that there is always a certain amount of CO_2 in the outdoor air meaning that the background concentration cannot be neglected. In general the outdoor air concentration of CO_2 is between 350 and 450 ppm or even higher, depending on the season. Thus, the CO_2 decay curve will not decline to zero. Instead, the CO_2 concentration decreases to values which are near to that of the outdoor air. This must be taken into account when analysing CO_2 decay curves. Eq. 5 cannot be applied; the air change rate (ACR) must be determined using the term derived from Eq. 3b:

$$C(t) = (C_0 - C_a)*e^{-\lambda*t} + C_a \tag{6}$$

C_0: initial concentration, C_a: concentration in the ambient air (i.e. background concentration)
It is possible to examine the linearised curve when the background concentration is subtracted before linearising. Therefore, the background concentration must be determined by an additional measurement.

$$\ln(C(t) - C_a) = \ln(C_0 - C_a) - \lambda*t \tag{6a}$$

With this relation the ACR can be determined by linear regression analysis. In practice, however, this will be difficult in most cases since it is not feasible to record the CO_2-concentration until the decay curve reaches background level. Modern buildings are in general designed to achieve high airtightness resulting in a very slow decay, and hence the linearisation of the decay curve using the logarithm of the concentration differences is not possible. Furthermore, the background concentrations are often unknown or not accessible to direct measurement. In these cases, the air change rate λ can be determined only by non-linear regression with iterative calculation methods according to the model as given in Eq. 6. Iterative calculation methods start with initial values (raw values) specified by the user for the function parameters, which are then improved iteratively by using the method of least squares until the model function is fitted optimally to the measured curve.

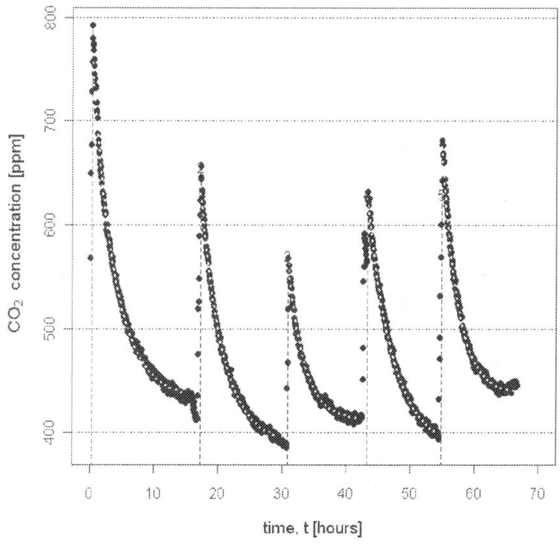

Fig. 4. Decay curves of CO_2 (black glyphs) in a bedroom equipped with double box window and located in a building built in 1908 , estimated with nonlinear regression (white curves).

curve	time [hours]	air change λ [ACH] (95% CI)	background conc. [ppm] (95% CI)	goodness of curve fit (R^2)
1	14.5	0.250 (0.246 – 0.254)	428 (427 – 429)	0.997
2	12.8	0.301 (0.295 – 0.306)	391 (390 – 392)	0.997
3	11.3	0.394 (0.383 – 0.406)	412 (411 – 413)	0.992
4	11.0	0.249 (0.243 – 0.254)	384 (382 – 385)	0.998
5	10.8	0.400 (0.393 – 0.407)	438 (437 – 439)	0.998
Avg. 95% CI		0.319 (0.226 – 0.411)		

Table 3. Air change rates in a bedroom located in an older building (built in 1908) equipped in 1990 with wooden framed double box window, decay method and CO_2 as tracer gas. Avg.: Average, CI: confidence interval

The results obtained from the CO_2 decay curves depicted in Figure 4 were evaluated using the method of nonlinear regression and are shown in Table 3. The calculated ACR ranged from 0.25 to 0.4 h^{-1} (mean 0.32 h^{-1}). The goodness of fit of the regression model is very high, since more than 99% of the variability of the CO_2 concentration can be explained by the respective regression functions. This is also reflected by high accuracy of the estimates for the various ACR. The 95% confidence intervals are very narrow and deviate only about 2 - 3% downward and upward from each individual value. These results were due to both the low variance of the CO_2 values from the fitted curves (white curves in Fig. 4) and the extensive measurement periods (several hours) which yielded high numbers of nodes (concentration-time data pairs) at measuring intervals of 3 minutes. However, the differences between day-to-day measurements are much larger. The 95% confidence interval for the mean ACR ranges from 0.226 to 0.411 h^{-1}. A possible explanation for this relatively high day-to-day variation is given below (see section 3.5). ACR, which were determined under the terms of exclusive joint ventilation (i.e. windows and doors closed), vary considerably and show extremes that span two orders of magnitude. This is shown in Figures 5a and 5b where six examples of CO_2 decay curves are depicted which can typically be recorded indoors. For illustrative reasons decay curves were selected which start at about almost the same initial concentration of 1500 - 1800 ppm and were recorded over a measurement period of more than 10 hours (Fig. 5a). Another selection criterion was the existence of stable weather conditions over the entire measurement period. Both figures demonstrate that CO_2 decay curves can decrease exponentially over a period of 50 hours and longer under stable weather conditions (Fig. 5a).

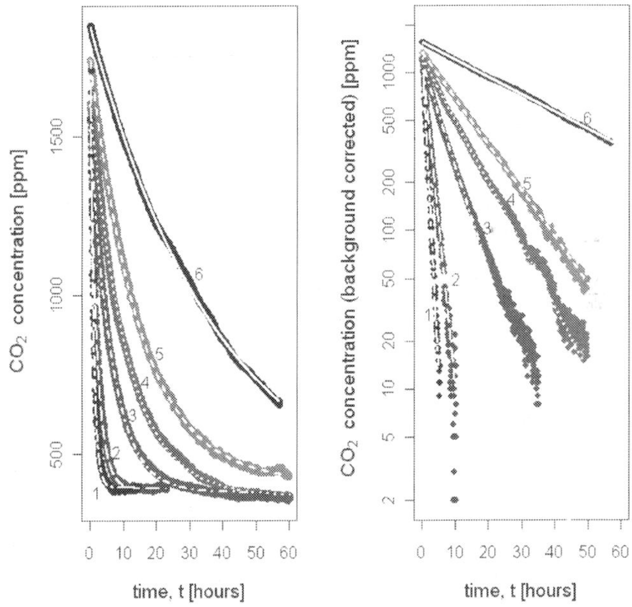

Fig. 5. CO_2 decay curves (black) in 6 rooms of different air tightness. White curves show results of regression analyses; a, nonlinear regression; b, linear regression of logarithmic values

This is an indication that tracer gas and room air can remain homogeneously mixed over such long periods without the need of a fan. The ACR estimated from the decay curves are given in Table 4, and the analysis of the curves was done with non-linear regression analysis (see Eq. 6). In addition to the air change rate λ, the background concentration C_a was also determined with the regression model and is presented in Table 4. After subtracting the respective background concentration from the measured CO_2 concentration values, the logarithmic plot of the concentration differences shows a nearly perfect linear concentration-time relationship (Fig. 5b).

curve	time [hours]	air change, [ACH] (95% CI)	background conc. [ppm]	goodness of fit (R^2)
1	14.5	0.942 (0.931 – 0.953)	393 (391 – 394)	0.998
2	23.0	0.465 (0.463 – 0.468)	392 (391 – 393)	0.999
3	48.0	0.154 0.153 – 0.155)	380 (379 – 381)	0.998
4	60.3	0.091 (0.091 – 0.092)	367 (366 – 368)	0.999
5	60.0	0.067 (0.066 – 0.067)	406 (405 – 407)	0.9996
6	57.0	0.0255 (0.0253 – 0.0257)	357 (350 – 365)	0.999

Table 4. ACR and background concentrations with their 95% confidence intervals (95% CI) measured in 6 rooms with differing natural ventilation (windows and doors closed); estimated from CO_2 decay curves with the statistical method of nonlinear regression (time interval between two concentration values: 3 min).

3.3.2 Determination of ACR from the concentration increase of carbon dioxide

When CO_2 is supplied at a constant rate, it is possible to determine the ACR from the increasing concentration values (Figure 6). Evaluation of the concentration curve is the same as with the constant injection method. The air change rate λ is calculated with the non-linear regression model approach $y = a_0 + a_1*[1\text{-}exp(\text{-}a_2*t)]$ (cf. Eq. 4a), where $a_0 = C_a$, $a_1 = E/(\lambda*V_R)$, and $a_2 = \lambda$. When the ACR is small, the build-up curve will approach the equilibrium only slowly, and this will lead to a quasi-linear concentration curve if the measurement period is too short. Therefore, the ACR calculated from this curve can be afflicted with large uncertainties. For this reason, to obtain a more reliable estimate of the ACR the concentration build-up must be recorded over a period as long as several hours, with intervals of a few minutes. When metabolic CO_2 is used, measurements should be done best during sleep (of the occupants), because this comes nearest to the requirement of a constant CO_2 supply. Figure 6 also shows the ACR determined from the analysis of the CO_2 respective build-up curves depicted here.

These CO_2 build-up curves were recorded in the same room as the decay curves in Figure 4. The measurement of the CO_2 increase occurred, however, four years before, when the room was occupied by 3 persons (two adults, one child). The ACR of both series of measurements, decay and build-up curves, differ only slightly from each other taking daily variation into account.

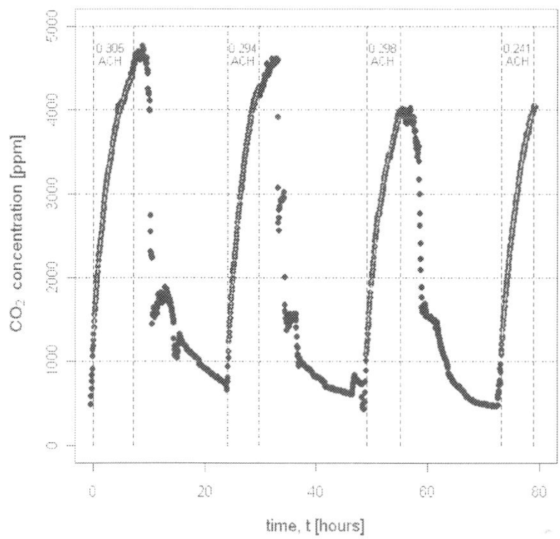

Fig. 6. CO_2 concentration build-up curves in a bedroom equipped with a double box window and located in a building built in 1908. Corresponding ACR, estimated with nonlinear regression (fitted values: white curves). Constant-injection method, room volume 30 m^3, occupancy: 2 adults, one child.

3.4 Comparison of different tracer gases for the determination air change rates

The above examples have shown the applicability of determining ACR from CO_2 concentration curves, when either the influence of the background concentration can be mathematically eliminated or a constant supply of CO_2 is provided. It can, however, not be concluded that conventional and established tracer gases are dispensable. Compared to CO_2 these have some indisputable advantages, namely a much wider range of possible applications and virtually negligible background concentrations. Both of which greatly simplifies the determination of ACR.

Under certain circumstances the use of established tracer gases is not possible, and in these cases we have to resort to the CO_2 method. This may be necessary in patients with environmentally related health problems, who will not accept the use of tracer gases in their homes because they fear health hazards. In such cases, the CO_2 method is a valuable alternative. It is therefore necessary to examine the extent to which results obtained with CO_2 differ from those obtained with conventional and established tracer gases. To achieve this we performed a number of comparative studies by which an established tracer gas like hexafluorobenzene (C_6F_6) or sulphur hexafluoride (SF_6) was used in parallel to CO_2.

3.4.1 Determination of air change rates with CO_2 and C_6F_6

The measurements described below were done in two different rooms, firstly, in a bedroom (24 m^3 effective volume) of a terraced house (built December 1997) equipped with double glazed windows and insulating rubber seals, conducting measurements over a period of 5

days, and secondly, in an office space (72 m^3 effective volume) of an old brick building (built ≈1900) equipped with two double-box windows. Windows and doors were closed during the experiment. A total of two measurement cycles were performed at intervals of two months.
At the start of the conditioning phase a diffusion tube which contained hexafluorobenzene was placed in both rooms. First sampling on Tenax-tubes occurred about 72 hours after the windows and doors were closed, further samples (two samples each) were taken every 24 hours. Room air was collected using sampling tubes (1 litre per tube) with a bellows pump, type Accuro (Dräger, Lübeck, Germany). ACR were calculated from the known emission rate of the C_6F_6-tubes, the equilibrium concentration and the effective room volume, according to the procedure for the constant injection method as described in VDI (2001), using Eq. 4c.
In the bedroom, CO_2 concentration was measured continuously over the entire study period with measuring intervals of 3 minutes using the Testo 400-device equipped with CO_2 probe, and the measured values were stored in the internal data logger. In the office space, the CO_2 concentration was measured with a second device in a 24 hour rhythm according to the American standard ASTM D 6245-98. Measurements occurred mainly at nighttime in the then unoccupied office building. Prior to the start of the measurement, the CO_2 concentration in the office space was set to ≈600 ppm by the experimenter. The calculation of the ACR was achieved with non-linear regression analysis using the statistical package SPSS, where the measured concentration values were fitted to exponential terms with consideration of the calculated background concentrations.

Results of the CO_2 measurements:

The CO_2 time-concentration curve recorded for the bedroom is depicted in Figure 7. The bold black line represents the measured concentration values, whereas the grey line describes the graph which was obtained by means of section-wise non-linear regression analysis. Above the individual curve sections, the calculated ACR are plotted. Given the high air-tightness of the bedroom ACR as low as 0.05 to 0.1 h^{-1} were determined. It is obvious that the CO_2 concentration increases very quickly during the night (one adult) to levels greater than 4000 ppm and decay exponentially during the day to values of 1500 - 2000 ppm. Due to the low ACR, the equilibrium concentration of CO_2 is not compassed during a 7 hours' sleep. The statistical analysis of the CO_2 curves obtained for the office space equipped with double box windows resulted, however, in significantly higher ACR, which ranged between 0.2 and 1.4 h^{-1} because of different window and door features.

Results of the hexafluorobenzene measurements:

In Figure 8, the values of the hexafluorobenzene concentrations from a series of measurements in both rooms are shown together with calculated curves. The solid lines represent the concentration curves which are calculated using Eq. 4b on the basis of the known emission rate E, the room volume V_R and the given air change rate λ, whereas the points represent the concentrations that have been measured (bedroom: triangles; office space: squares). It is evident that the measurements in the bedroom oscillate about values which would be expected for an ACR of about 0.05 h^{-1}. The reason for the large deviation of the two first measurements from the curve is probably due to largely differing flow resistances of the Tenax tubes used. In comparison to the bedroom, the CO_2 values in the office (lower curve) show a much lower level despite the double emission rate. On the one hand, the lower concentration increase is caused by the larger room volume, but another reason is the significantly higher ACR of 0.45 h^{-1}. The calculation of the ACR from the individual concentration values measured led to values ranging from 0.05 to 0.14 h^{-1} for the bedroom and from 0.2 to 1.7 h^{-1} for the office space, respectively.

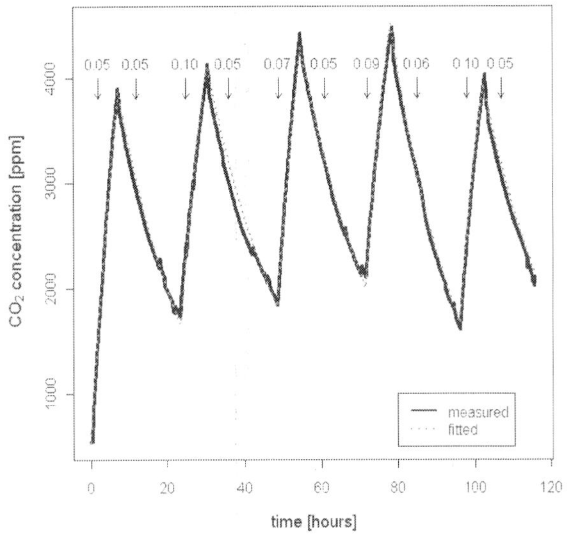

Fig. 7. Course of CO_2 concentration in a bedroom equipped with an insulating glass window with rubber sealing. Window and door closed. Measured values (black) with corresponding air change rates [in ACH], dotted curve: theoretical values. Room volume: 24 m^3, occupancy: one adult.

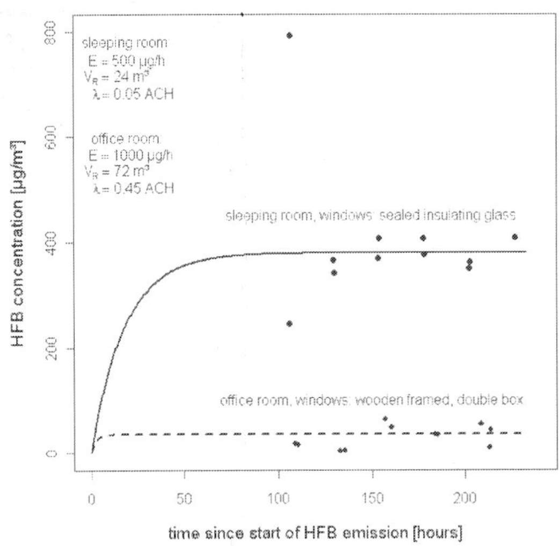

Fig. 8. Course of hexafluorobenzene concentration (HFB) in two rooms with different volumes and different window types, theoretical curves calculated from given emission rates E, air change rates λ, known room volumes V_R. Black glyphs: measured values.

The results of the tracer gas comparison are shown in Table 5 for both test series. At large, a very good agreement was found for ACR determined with the two different tracer gases.

		CO_2 (ACH)		Hexafluorobenzene (ACH)	
building	Window construction	serie 1	serie 2	serie 1	serie 2
new	Insulating glass with rubber sealing	0.05 to 0.08	0.05 to 0.1	0.05 to 0.06	0.09 to 0.14
		n = 10	n = 10	n = 5	n = 6
old	wooden framed bouble box	0.3 to 0.5	0.2 to 1.4	0.2 to 0.9	0.3 to 1.7
		n = 10	n = 10	n = 5	n = 6

Table 5. ACR results of comparative measurements with CO_2 and hexafluorobenzene as tracer gases. Windows and doors closed.

3.4.2 Parallel testing of CO_2 and SF_6

In another series of investigations an office space with a volume of 48 m^3 and two double box windows was doped with CO_2 and SF_6 in parallel on 6 consecutive days, and the ACR were determined from the decay curves. Measurement of SF_6 was done in the same way as described in the above section "decay curve". An example of the parallel concentration decay of both tracer gases is given in Figure 9; results of the 6 experiments, which differ only slightly, are compiled in Table 6.

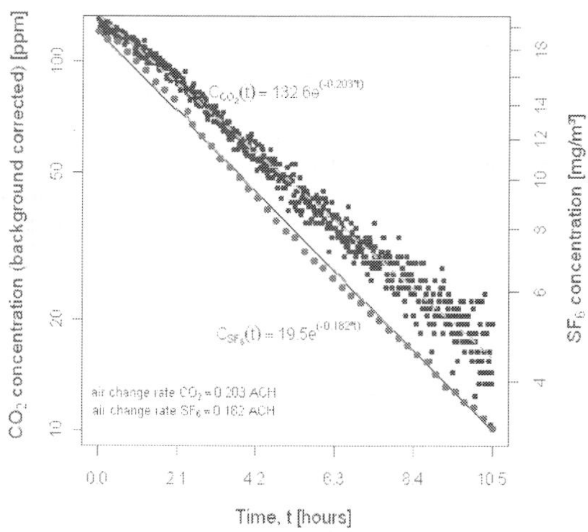

Fig. 9. Decay curves of CO_2 and sulphur hexafluoride (SF_6) and curve fits; simultaneous measurements in the same room.

measurement	air change rate [ACH]		Quotient of air changes
	CO_2	SF_6	ACR_{CO2} / ACR_{SF6}
1	0.211	0.240	0.879
2	0.447	0.438	1.021
3	0.330	0.327	1.009
4	0.203	0.182	1.115
5	0.330	0.320	1.031
6	0.300	0.280	1.071
average	0.304	0.298	1.021
std.dev.	0.090	0.087	0.080
COV (%)	29.7	29.2	8.0

Table 6. Comparative measurements with CO_2 und SF_6; office space (volume 48 m^3, two double box windows); COV (%): coefficient of variation = 100*standard deviation (std.dev.) / average.

The average ACR was 0.304 h^{-1} for CO_2 (SD ± 0.09 h^{-1}) and 0.298 h^{-1} for SF_6 (SD ± 0.09 h^{-1}). To check for systematic differences between the individual determinations we calculated the ratio of the two air change rates and got a mean ratio of 1.02 with a standard deviation of ± 0.08. Thus, no statistically significant systematic differences between the compared tracer gases could be detected. Since our sample size was rather small (only 6 determinations for each tracer gas) we included results from other studies (Guo and Lewis 2007; Roulet and Foradini 2002; Shaw, 1984) for comparison. After inclusion of these data the difference was statistically significant, albeit the systematic deviation between the ACR, determined under the same conditions with either CO_2 or SF_6, was very small (Fig. 10). A similarly good agreement was found by Stavova (Baránková) et al. (2006) comparing CO_2 and Freon 134a. The mean ratio of the two air change rates (CO_2/SF_6) is about 1.06 (95% CI: 1.01 - 1.12). Standard deviation ranged from 0.93 to 1.21. The ≈6% overestimation of the ACR, as obtained with CO_2, is yet much smaller than the 50 percent overestimation obtained by Schulze and Schuschke (1990) who used krypton 85. On the basis of existent data it is so far not possible to conclude with certainty whether or not very small ACR (i.e. 0.1 h^{-1}) are responsible for higher systematic deviations.

3.5 Wind and temperature effects on the natural air flow in case of joint ventilation

Natural ventilation is driven by the air-pressure difference between the interior and external environment which prevails on the room's opening areas like windows and doors. The air change rate is also influenced by the tightness of windows and doors versus joints and cracks in the building envelope. The air-pressure difference is predominantly caused by temperature differences between indoor and outdoor climate, and by local wind effects. On the windward side of a building, the wind is trapped reaching its highest dynamic power when the flow angle equals 90°. On the sides of a building that parallel the wind direction and on the far side, negative air-pressures occur regularly. These air-pressure differences result in raised rim hole rates, particularly in case of leaks or open windows on opposite sides of a building, due to suction and pressure cycles of the wind (i.e. cross ventilation).

The entire air-pressure onto the façade is dependent on both the wind speed and squalls, as well as the building's constructional quality, namely the tightness of the façade.

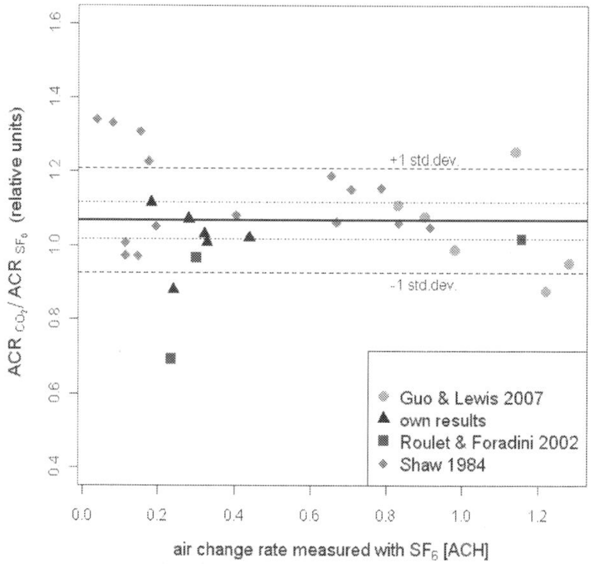

Fig. 10. Comparative air change determinations with CO_2 and SF_6 as tracer gases. Results from other authors and own results. Solid line: mean, dotted lines: 95% confidence interval of the mean, dashed lines: interval of ± 1 standard deviation.

3.5.1 Wind influences

To study the effects of weather conditions on the ACR, measurements were conducted in two different office buildings of the Robert Koch Institute, Berlin, Germany. Building #1 is an older brick building (built ca. 1900), surrounded by allotment gardens (to the north and south) and large open spaces. On the east side is a three-storey row house, and on the west side a four-storey office building. Due to the location of the building, unhindered wind flow is limited to south-east or south-west directions. All of the five rooms which have been selected in building #1 were located on the south side and equipped with double box windows (rooms A - D and F; room volumes between 30 and 90 m³).

The rooms in building #2 constructed in the late 1970s were equipped with tightly closing windows featuring single glazing and insulating rubber seals. The room selected for the measurements had a large window sill with a couple of windows in north-east direction and a room volume of 40 m³ (room E). In front of the windows was an open space (length ≈150 m), limiting wind flow to solely from north-east to south-east directions. On the south side, room E was protected by a cross-building located in a distance of ≈10 m. Likewise, on the north side, the building was protected by a north-east to south-west oriented wing of the same building (in a distance of about 70 m from room E).

Between January 1999 and December 2010 a total of 611 air change measurements were performed in the five rooms of building #1 (A - D and F) using the CO_2 decay method. The decay curves were mainly recorded at night or on weekends according to ASTM D 6245-98, when the building was unoccupied. Analogously, 388 air change measurements were carried out in room E of building #2 between July 2002 and September 2008. During the measurements all doors and windows were kept closed. Data about meteorological conditions during the tests, such as outside air temperature, wind speed and wind direction were obtained from the web sites of local Berlin weather stations. In room E of building #2 the indoor temperature was recorded during the measurements. Table 7 provides an overview of the air change rates determined. For rooms with double box windows individual values of ACR varied from <0.05 h^{-1} to 1.8 h^{-1} (i.e. a factor of ≈40), whereas the medians (50th percentile) obtained for the rooms A – D and F varied only between 0.3 and 0.6 h^{-1}. Significant differences between the individual rooms could not be detected.

room	sample size	Minimum [ACH]	Median (95% CI)	Maximum [ACH]
A	15	0.209	0.457 (0.369 - 0.715)	1.320
B	26	0.163	0.445 (0.351 - 0.552)	1.370
C	11	0.090	0.470 (0.126 - 0.696)	0.720
D	42	0.079	0.512 (0.352 - 0.639)	1.426
F	517	0.043	0.342 (0.317 - 0.374)	1.841
E	388	0.016	0.078 (0.070 - 0.085)	0.403

Table 7. Air change rates [ACH] in office spaces with windows of different construction, rooms A-D, F: wooden double box windows, room E: insulating glass window with plastic frame and rubber sealing

Significantly lower ACR were determined for room E in building #2 equipped with double glazed windows for which the median was 0.08 h^{-1}. The individual values varied between 0.016 and 0.40 h^{-1}, i.e. maximum and minimum differ more than a magnitude. Wind influences, especially wind speed, were identified as the main reason for the variability of the ACR in the two buildings (Fig. 11).

In building #1 (rooms A - D and F) the ACR ranged from <0.1 h^{-1} when wind speed was lower than 5 km/h to 1.2–1.4 h^{-1} at a wind speed up to 30–40 km/h. The strong influence of the wind caused significant and uncontrollable heat losses through leaky windows which, in turn, caused unhealthy draught effects.

Surprisingly, the results obtained for room E in building #2 equipped with insulated windows and rubber seals, were significantly dependent on wind influences, as well, although the wind effects were much smaller than those of building #1. For room E, ACR were less or equal to 0.05 h^{-1} when wind speed was lower than <5 km/h but increased up to 0.25 h^{-1} with a wind speed of more than 40 km/h. The relations described here essentially confirm the results formerly obtained by Wegner (1983). Besides the wind speed, the flow direction of the wind had a significant impact on the amount of the ACR. This is evident from the graphical presentation in Figure 12, grouping ACRs according to the wind direction prevailing in test periods.

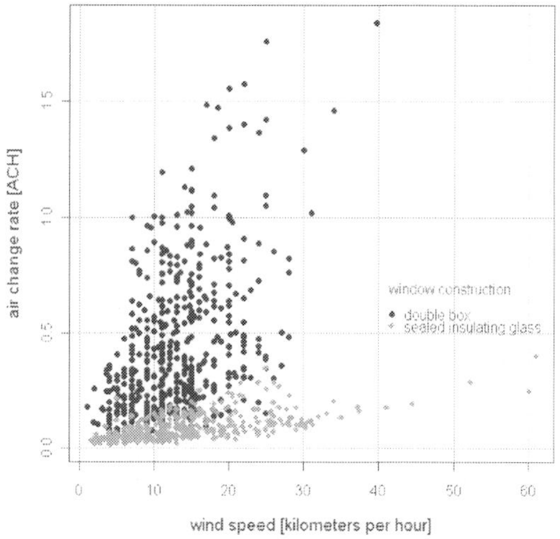

Fig. 11. Influence of wind speed on ACR in office spaces equipped with windows of different construction. Windows and doors closed. Rooms with double box windows (A-D, F: n= 611), room with insulating glass window: n= 388).

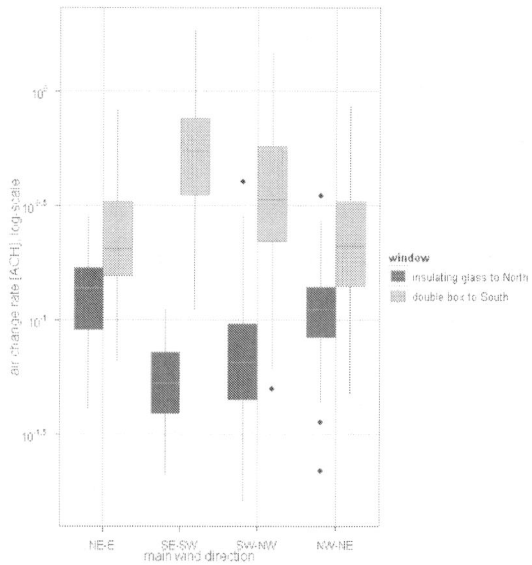

Fig. 12. Wind direction and ACR in office spaces equipped with windows of different construction. Windows and doors closed. Rooms with double box windows (A-D, F: n= 601), room with insulating glass window: n= 388).

Adverse wind blowing against the window or in parallel to the window front are of higher impact than wind flow from the backside of the building, the latter causing lower ACR than the two former (see Table 8).

room	wind direction	sample size	ACR [ACH]	$Quotient_{raw}$	$Quotient_{adj}$
			Median (95% CI)	Median (95% CI)	Median (95% CI)
A - D, F	NE-E	122	0.202 (0.171 - 0.243)	0.92 (0.779 - 1.107)	1.06 (0.941 - 1.189)
	SE-SW	242	0.55 (0.505 - 0.596)	2.50 (2.295 - 2.709)	2.89 (2.601 - 3.207)
	SW-NW	163	0.329 (0.294 - 0.395)	1.50 (1.340 - 1.801)	1.38 (1.239 - 1.547)
	NW-NE	74	0.22 (0.179 - 0.261)	1 (Ref.)	1 (Ref.)
E	NE-E	83	0.138 (0.109 - 0.153)	2.09 (1.651 - 2.317)	2.21 (2.038 - 2.399)
	SE-SW	81	0.054 (0.047 - 0.063)	0.82 (0.714 - 0.957)	1.10 (1.013 - 1.198)
	SW-NW	171	0.066 (0.056 - 0.072)	1 (Ref.)	1 (Ref.)
	NW-NE	53	0.111 (0.095 - 0.135)	1.68 (1.438 - 2.043)	1.92 (1.742 - 2.115)

Table 8. Dependance of ACR on wind direction, raw and adjusted ACR quotients, Rooms A-D, F: windows to south, room E windows to north-east.

In case of the rooms with double box windows, this led to a relative threefold increase of air change, expressed as ratio ac(luv) : ac(lee), for wind directions from south to west. In the room with double glazed windows, we measured an almost doubled increase of air change (70 - 100%) for wind directions from north to east. The overall air change ratios depending upon the relative wind direction, respectively were not remarkably influenced by the wind speed ($Quotient_{adj}$ in Table 8). Summarising wind effects we could demonstrate that regardless of wind strength and the window type an air flow directed to the window front can yield a 1.5- to more than 2-fold increase of the ACR in comparison to wind from the backside.

3.5.2 Temperature influences

The temperature difference between inside and outside is one of the major forces driving ventilation. We investigated the impact of the temperature gradient (room air temperature minus outside air temperature) on the ACR in room E and found a statistically significant effect on the extent of air change (Fig. 14a). Whereas the average air change rate λ, after correction for the wind influence, was only 0.06 h^{-1} at a temperature difference ΔT of 0 K, it increased up to 0.1 h^{-1} when the temperature gradient rose to about 30 K. However, as compared to wind effects (Fig 14b), the exponential relationship between ACR and temperature gradient (Fig. 14a), turned out to be much weaker, at least for room E with its tightly closing window.

3.6 Air change rate during window ventilation

Wind action and temperature gradient both proved to be significant factors influencing the amount of air change through joints and cracks in buildings. For indoor spaces in buildings with high leak tightness, window ventilation plays a crucial role in the removal of excess humidity, carbon dioxide, odors and other volatile compounds. The effectiveness of ventilation via opened windows is not only dependent on their number, size and arrangement in the room, but also on the modus and frequency of active window ventilation. Windows that are only opened partway (i.e. tilted) ventilate less effective than fully opened windows. Cross-ventilation can be achieved when windows on opposite sides of a room are opened simultaneously. Cross-ventilation removes moisture and contaminants most effectively from the indoor air to the outside. The effectiveness of the different modes of window ventilation expressed in terms of the ACR, was examined with a couple of tests in room E, the results of which are shown in Figure 13 in conjunction with the joint ventilation results.

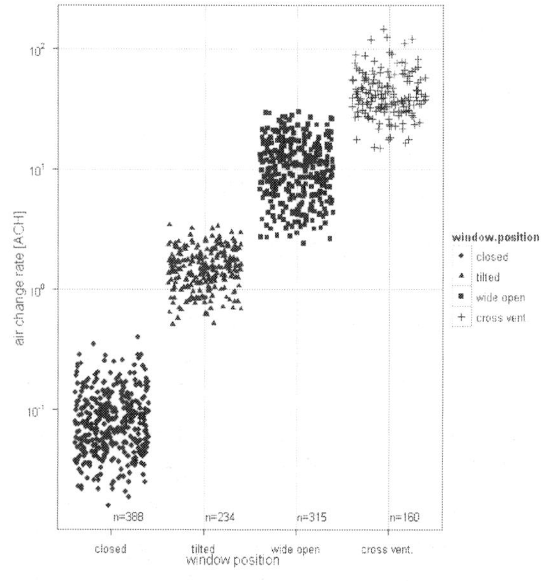

Fig. 13. ACR and window position. Office room E.

ACR determinations were carried out by the decay method; duration of the measurements varied between 0.5 h and 3 h with recording intervals of 12 s to 1 min. After doping the room with CO_2, one window in the room (dimensions: width 0.8 m and length 1.2 m) was either in tilted position (i.e. 10 cm wide upper opening) or widely open (i.e. opening angle 80 °). To achieve cross-ventilation, an opposite window was also widely open. For both ventilation modes - tilted or full open - the following ACR were determined (Table 9a): With the tilted window a total of 239 measurements was performed; the average ACR (i.e. median) was 1.5 h^{-1} with day-to-day variations that ranged from 0.5 h^{-1} to 3.4 h^{-1}. In the case of full open window, the median ACR was 10 h^{-1}, with a minimum of 2.4 h^{-1} and a maximum of 30 h^{-1} (321 measurements). For cross ventilation a number of 179 measurements were performed. The ACR varied between 15 h^{-1} and 146 h^{-1} (median: 40 h^{-1}). On the basis of these data we calculated the ventilation times which are necessary to reduce an indoor CO_2 concentration of 1500 ppm by 98% (corresponding to a CO_2 concentration that is about 30 ppm higher than outside) and yielded 2 hours for the tilted window (range 1

- 7 hours), ≈0.3 hours for the wide open window (range 0.1 - 1.5 hours) and 0.1 hours for cross-ventilation (range 1 - 14 min; see also Table 9b).

			Air change rate [ACH]					
room	window construction	window position	Min	05. P	Median	95. P	Max	sample size
A – D, F	double box	closed	0.043	0.111	0.36	1.030	1.841	602
E	insulating glass	closed	0.016	0.029	0.078	0.200	0.403	388
		tilted	0.52	0.70	1.50	2.60	3.43	239
		wide open	2.4	3.5	10.4	23.4	29.7	321
		cross vent.	15.0	22.5	40.1	104.7	146.4	179

Table 9a. Dependance of ACR on window construction and opening

			Ventilation time [hours]				
room	window construction	window position	Max	95. P	Median	05. P	Min
A – D, F	double box	closed	84	32.4	10	3.5	2.0
E	insulating glass	closed	225	124	46	18	8.9
		tilted	6.9	5.1	2.4	1.4	1.0
		wide open	1.5	1.0	0.35	0.15	0.12
		cross vent.	0.24	0.16	0.09	0.03	0.02

Table 9b. Ventilation time needed to reduce CO_2 concentration of 1500 ppm by 98%

These results clearly confirm the great importance of ventilation with fully open windows and, even more, of cross-ventilation for the efficient removal of indoor pollutants. The broad variation in ACR can be explained to a great deal by the effect of the temperature gradient between indoor and outside (Fig. 14a).

Compared to the temperature gradient, wind effects played only a minor role at least for the window openings "tilted" and "widely open" tested here (Fig. 14b). The dependence of the ACR on the temperature gradient could be estimated by non-linear exponential curve fitting for all tested window opening modes (Fig. 14a). From these findings we can conclude that in case of ventilation via tilted window, with a temperature gradient of 10 K, an ACR of 0.92 * exp (0.045 * 10) = 1.4 h^{-1} can be expected for the room examined here (95% prediction interval: 1.0 h^{-1} - 2.0 h^{-1}). With a temperature gradient of 20 K an average ACR of 2.3 h^{-1} can be expected (95% prediction interval 1.6 h^{-1} - 3.1 h^{-1}). For ventilation with fully open window an ACR of 4.5 * exp (0.071 * 10) = 9.2 h^{-1} (ΔT = 10 K) can be expected and 18.6 to h^{-1} (ΔT = 20 K) with 95% prediction intervals 5.5 h^{-1} - 15.3 h^{-1} and 11.1 h^{-1} - 31.2 h^{-1}, respectively. In the case of cross ventilation an average ACR would expected to be 39.0 h^{-1} and 50.2 h^{-1} for ΔT = 10 K and 20 K, with 95% prediction intervals of 23.0 h^{-1} - 65.8 h^{-1} and 29.6 h^{-1} - 85.1 h^{-1}, respectively.

3.7 Air change rates in Berlin's housing stock

The results presented so far do not allow general statements on the air change in buildings, since the number of rooms examined was far too small. As part of site inspections and indoor air investigations during 1999–2005, ACR were determined for a further number of different buildings and rooms applying the CO_2 concentration decay method. A total of 198 living and working spaces, situated in 152 buildings in the Berlin area and in the immediate surrounding were studied, comprising of 143 residential buildings of different

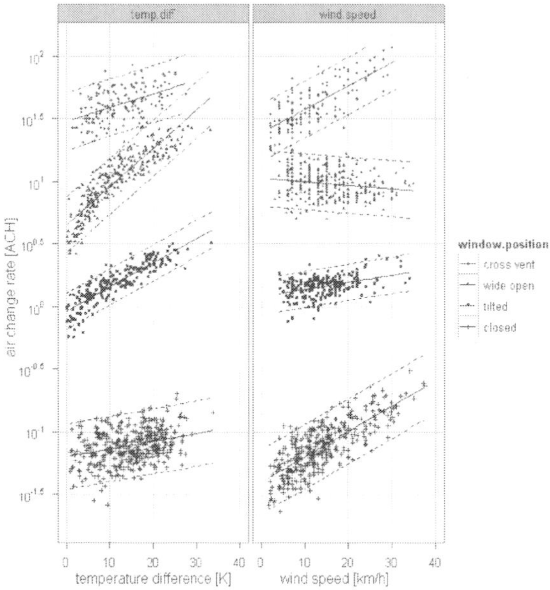

Fig. 14. a. Influence of temperature gradient (indoor to outdoor) on ACR, corrected for wind effects, b. Influence of wind speed on ACR, corrected for temperature effects, in an office space (room E) for different ventilation conditions: natural ventilation: "closed", "window tilted", "widely open" and "cross ventilation", (from bottom to top); room volume 40 m^3, solid lines: fitted curve, dashed lines: 95% prediction intervals.

age and storey numbers, and nine different office buildings. None of the rooms had an additionally installed ventilation system. Tenants were asked not to enter the rooms during the tracer gas measurements and to keep windows and doors closed. The duration of the tracer gas measurements varied between 2 hours and 24 hours. To obtain decay curves with a high number of nodes we choose time intervals of 1 to 3 minutes between CO_2 measurements. Two Testo 400 devices with CO_2 probe (infrared absorption) were used for recording and data logging. The results show that under worst case conditions the vast majority of the ACR are below the range of 0.5 h^{-1} - 1 h^{-1} which was recommended by the former German Federal Health Office (Bundesgesundheitsamt, 1993) for common living quarters. The ACR of the studied rooms ranged from minimum 0.02 h^{-1} to maximum 1.98 h^{-1}. Of the 198 naturally ventilated rooms the majority (167=84%) had ACR of 0.5 h^{-1} or below. ACR of 0.8 h^{-1} or higher could be determined for only 12 of the rooms (6%). The 50th percentile (median) of the ACR was 0.2 h^{-1} with a 95% confidence interval (CI) of 0.17 h^{-1} - 0.24 h^{-1}; arithmetic mean was 0.31 h^{-1} (95% CI 0.27 h^{-1} - 0.36 h^{-1}) and geometric mean was 0.22 h^{-1} (95% CI 0,20 h^{-1} - 0,25 h^{-1}), respectively (see Table 10). Thus, our studies essentially confirm the results obtained by Münzenberg (2004) and Salthammer et al. (1995), who conducted ACR measurements under similar methodological conditions in Germany, and received results consistent with those from surveys in Scandinavia by Andersen et al. (1997); Bornehag et al. (2005) and Harving et al. (1992).

Measurements performed in the same rooms using diffusion tubes and hexafluorobenzene (C_6F_6) or PFT as a tracer gas yielded somewhat higher ACR compared to that obtained with the CO_2 concentration decay method. This can be explained by necessarily longer exposure times of collection tubes ranging from several days to weeks, which essentially means that there have been one ore more active ventilation phases during the measurement period influencing the air change (Table 10). However, even in these studies, the proportion of rooms with ACR below 0.5 h^{-1} was remarkably high. Studies conducted in the USA also show a high proportion of buildings (50 percent and more) with ACR below 0.5 h^{-1} (Persily et al., 2010), suggesting that a basically desirable ACR of at least 0.5 h^{-1} is difficult to achieve under normal conditions. Similar conclusions have been drawn by Erhorn and Gertis (1986) and Münzenberg (2004).

Stratification for window types showed significantly higher ACR (factor ≈2) for rooms with double box windows or composite windows than for rooms with specially insulated windows (double glazing and continuous rubber seal; Table 11). However, also for rooms with double box windows or composite windows the majority of the ACR determinations yielded values <0.5 h^{-1}. The 50th percentile (median) was 0.33 h^{-1}, and the 75th percentile was about 0.56 h^{-1}. In comparison, rooms with double glazed windows had a median ACR of 0.16 h^{-1} and the 75th percentile was 0.25 h^{-1}.

study	method	tracer gas	sample size	Min [ACH]	Median [ACH	Max [ACH]	AM [ACH]	GM [ACH]	< 0.5 ACH (%)
Salthammer et al. (1995)	concentration decay	N_2O	150	< 0.1	0.25	1.7	0.36	n.a	> 70
Münzenberg (2004)		CO_2	80	< 0.05	0.18	> 1.5	0.26	n.a	90
own results		CO_2	198	0.02	0.21	1.98	0.31	0.22	84
Harving et al. (1992)		TFE	114	< 0.1	0.28	> 1.5	n.a	n.a	72
Bekö et al. (2010)	concentration decay and build-up	CO_2	300 cases		0.44		0.62	0.46	57
			200 bases		0.42		0.62	0.46	
Hirsch et al. (2000)	constant-injection (diffusion tubes)	C_6F_6	78					0.73 1)	
								0.52 2)	
Andersen et al. (1997)		PFT	117	0.16	n.a	0.96	0.37	0.34	82
Kuotsalainen et al. (1992)		PFT	242	0.07	≈ 0.5	1.55	0.52	n.a.	48
Øie et al. (1997)		PFT	38	0.15	n.a.	1.4	n.a.	n.a.	58
Øie et al. (1998)		PFT	344	n.a.	n.a.	n.a.	n.a.	n.a.	36
Bornehag et al. (2005)		PFT	320 3)				0.36 3)		80
			43 4)				0.48 4)		60

Table 10. ACR measurements in Germany and Scandinavia, results of surveys. TFE: 1.1.2-trichloro-1.2.2-trifluoroethane, 1) before reconstruction. 2) after reconstruction, 3) single family houses, 4) multi-family houses, AM: arithmetic mean,. GM: geometric mean, n.a. not available, < 0.5 h^{-1}: proportion (%) of air change rates < 0.5 h^{-1}

window construction	Sample size	Min	05. P	25. P	Median (95 % CI)	75. P	95. P	Max	AM (95 % CI)	GM (95 % CI)
Double box/ composite window	80	0.05	0.11	0.20	0.33 (0.27 – 0.44)	0.56	1.50	1.98	0.47 (0.37 – 0.57)	0.34 (0.28 – 0.41)
Insulating glass	118	0.02	0.05	0.10	0.16 (0.13 – 0.20)	0.25	0.54	1.00	0.22 (0.18 – 0.25)	0.16 (0.14 – 0.19)

Table 11. ACR in naturally ventilated rooms (windows and doors closed) equipped with different window types. Min: Minimum. Max: Maximum. 05. P: 5th percentile. 25. P: 25th percentile. 75.P: 75th percentile. 95. P: 95th percentile. 95% CI: 95% confidence interval. AM: arithmetic mean. GM: geometric mean

The empirical frequency distribution of the ACR for the total sample had the typical shape of a log-normal distribution with a steep rise on the left and a long tail on the right side (Fig. 15a). A good fit to the normal distribution could be achieved by calculating the logarithms of the ACR (Fig. 15b).

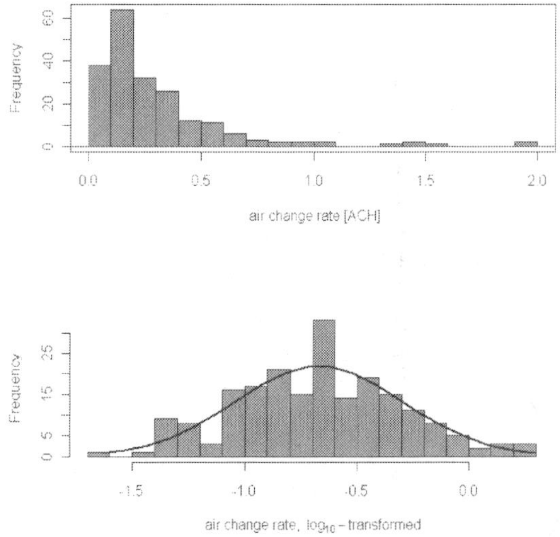

Fig. 15. Empirical frequency distribution of ACR (natural ventilation) for 198 rooms in the City of Berlin; a, values as determined; b, log-transformed values. Solid line represents normal distribution.

This is consistent with the results of Scandinavian studies by Andersen et al. (1997); Bekö et al. (2010); Harving et al. (1992); Øie et al. (1998) and Ruotsalainen et al. (1992) who found the same form of empirical frequency distributions of ACR in Danish, Finnish and Norwegian residential buildings.

3.8 Ventilation and pollutant concentrations

Adequate ventilation is the cheapest and easiest way to effectively remove indoor pollutants. Previous studies have shown that insufficient air change is mainly due to renovation and other structural changes, namely in most cases the installation of new air-tight windows (Wegner 1983, 1984). The results of our own investigations presented here have led to similar conclusions, apart from significantly lower ACR as compared to previous studies (Wegner 1983, 1984). Generally insufficient fresh air supply resulted in an considerable increase of indoor air pollutants. This is the case for formaldehyde (Li et al., 2002; Salthammer et al., 1995; Wegner, 1983) and other volatile compounds (Hodgson et al., 2003) as well as for the radioactive noble gas radon (Andersen et al., 1997; Chao et al., 1997; Sentikova, 1999). Our own studies also found a correlation between the concentration of volatile organic compounds and the ventilation rate in naturally ventilated rooms, as described below.

We have chosen some frequently found indoor air pollutants, which may be taken as typical representatives of different substance classes, namely formaldehyde, acetaldehyde, acetone, 2-ethyl-1-hexanol, hexanal, d-limonene, α-pinene, decamethylcyclopentasiloxane (D5-siloxane), and toluene. Air sampling to analyse these contaminants was done in the rooms under study (closed windows and doors), either after a CO_2 tracer gas measurement was finished or directly before the start of a CO_2 decay recording with Aircheck samplers (model 224-PCXR8) SKC Inc., Eighty Four, PA, USA. For enrichment we used the adsorption media recommended for these pollutants. Air volumes varied from 90 l to 120 l and the sampling air flow was 1.5 - 2 l/min. Formaldehyde was somewhat special since it was measured semi-quantitatively in situ in most rooms. An Interscan 4000 device (Ansyco, Lenzkirch, Germany) was used to carry out the measurements. Prior to the determination of the formaldehyde concentration, a zero balance was produced by pumping air through an upstream activated carbon filter for 10 minutes. Thereafter, the concentration value was recorded when the reading remained constant. Comparison of measurements for 24 different rooms yielded a good agreement of concentration levels obtained with the electrochemical formaldehyde method using the Interscan 4000 and the laboratory results obtained by analysis of the air samples that were collected with DNPH cartridges.

Statistics for the pollutants investigated here are listed in Table 12; if the sample size was sufficient the 95% confidence intervals were also included. Acetone, formaldehyde and acetaldehyde showed the highest concentrations, with median values of 75 $\mu g/m^3$, 23 $\mu g/m^3$ and 31 $\mu g/m^3$, respectively. Due to boiling points below 60° C, these pollutants are classified as very volatile organic compounds (VVOC). Therefore, they are - with the exception of formaldehyde - not regularly recorded indoors.

As Figure 16 shows, the frequency of formaldehyde concentrations >100 $\mu g/m^3$ seems to correlate with the ACR. In 8 of 77 category-1 rooms with ACR <0.2 h^{-1} formaldehyde amounted to ≥100 $\mu g/m^3$, i.e. the frequency of exceeding values was 10.4% (95% CI 4.6% - 19.5%). Among the category-2 und -3 rooms was only one each with exceeding formaldehyde (1.6% and 3.2% relative frequency, 95% CI 0% - 8.5% and 0% - 16.7%, respectively). According to X^2 testing these differences were close to be significant (p = 0.060), but this seems unlikely due to strong overlaps of the very broad 95% CI. Our finding thus points to successful emission reduction measures through low-emission products by which pollutant concentrations can be accomplished that are within the range of the accepted limits, even in very airtight rooms.

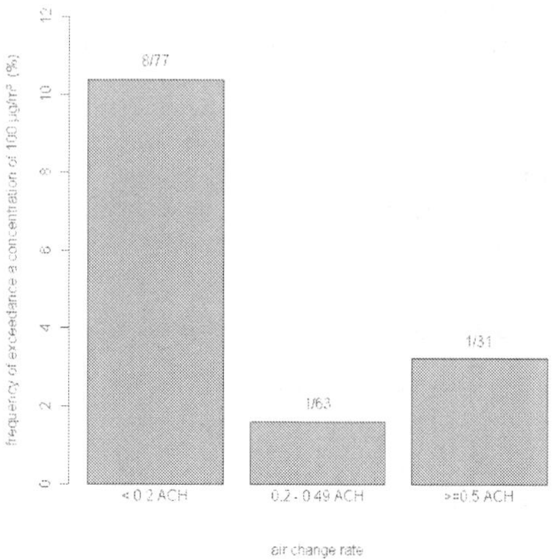

Fig. 16. Frequency of formaldehyde concentrations >100 µg/m³ and ACR (natural ventilation) in rooms of the City of Berlin.

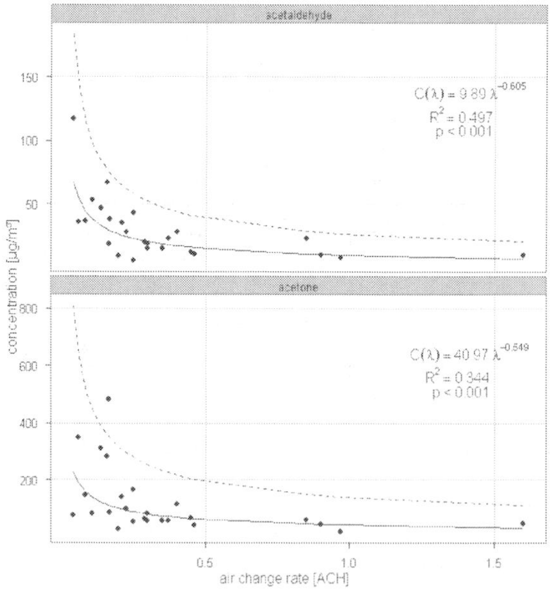

Fig. 17. a. and b. Concentrations of acetaldehyde (a) and acetone (b) and corresponding air change rates in rooms of the City of Berlin. Sampling: worst case conditions, solid line: mean concentration, dotted line: 90% prediction limit.

Unlike formaldehyde, the concentrations of acetaldehyde and acetone showed a clear dependence of the air change rate λ, which could be described by the power function ($C(\lambda) = a^*\lambda^{-k}$). The independent variable air change rate could explain about 50% ($R^2 = 0.5$) of the total variance of the acetaldehyde concentration values and about 34% ($R^2 = 0.34$) of the acetone concentration values. The results of a curve fitting are shown in Fig. 17a and b.

According to the resulting function parameters it can be estimated that an increase of the ACR from 0.1 h^{-1} to 0.5 h^{-1} would yield a reduction of the indoor concentrations from 40 μg/m³ to 15 μg/m³ for acetaldehyde, and from 145 μg/m³ to 60 μg/m³ for acetone. This is a reduction factor of 2.4 - 2.7.

adsorbent	compound	Sample size	Min	10. P (95 % CI)	Median (95 % CI)	90 P (95 % CI)	MAX	AM (95 % CI)	GM (95 % CI)
	very volatile								
In situ detection	formaldehyde	171	< 20	< 20 (< 20)	31 (25 – 38)	70 (63 – 96)	190	41 (36 – 45)	33 (30 – 37)
DNPH-Silica	acetaldehyde	25	6	9 (n.d.)	23 (11 – 37)	53 (n.d.)	117	29 (19 – 39)	22 (16 – 30)
DNPH-Silica	acetone	25	18	39 (n.d.)	75 (53 – 128)	308 (n.d.)	481	118 (71 – 166)	84 (60 – 117)
	volatile								
Anasorb 747 / Tenax	2-ethyl-1-hexanol	148	< 1	1 (1 – 1,5)	4 (3 – 6)	12 (10 – 19)	65	7 (5 – 8)	4 (4 – 5)
Anasorb 747 / DNPH	hexanal	93	< 2	4 (3 – 6)	17 (14 – 21)	37 (30 – 45)	75	20 (17 – 22)	14 (12 – 17)
charcoal (NIOSH)	limonene	133	< 1	2 (1 – 3)	10 (8 – 13)	66 (36 – 89)	256	24 (17 – 31)	10 (8 – 13)
charcoal (NIOSH)	α-pinene	133	< 1	2 (2 – 5)	15 (11 – 22)	102 (57 – 143)	623	45 (28 – 62)	16 (12 – 20)
charcoal (NIOSH)	d5-siloxane	99	< 1	< 1 (<1 - 1)	3 (2 – 4)	94 (39 – 321)	736	39 (17 – 61)	4 (3 – 6)
charcoal (NIOSH)	toluene	133	2	4 (3 – 5)	14 (10 – 21)	91 (46 – 144)	396	34 (25 – 43)	16 (13 – 20)

Table 12. Concentrations [μg/m³] and 95% confidence intervals of selected VVOC and VOC in the air of rooms in the City of Berlin; n.d.: 95% confidence interval not detected (small sample size), 10. P: 10th percentile, 90. P: 90th percentile, AM: arithmetic mean, GM: geometric mean.

Similar relations were found for other volatile organic compounds (Table 12). Again, the dependence of their indoor air concentrations on the ACR could be characterized by a power function (see Figures 17c to 17h). Regression analysis revealed that an increase of the ACR from 0.1 h^{-1} to 0.5 h^{-1} decreased the concentrations of 2-ethyl-1-hexanol and hexanal (or hexanaldehyde) by a factor of ≈2.3, since 2-ethyl-hexanol declined from 6.2 μg/m³ to 2.6 μg/m³, and hexanal declined from 21.2 μg/ m³ to 9.1 μg/m³. For limonene and α-pinene the reduction factors were 2.6 and 2.2, respectively. For these pollutants, the indoor air concentrations would be reduced from 24 μg/m³ and 30 μg/ m³ to values of 9 μg/m³ and 14 μg/m³, respectively For decamethylcyclopentasiloxane (D5) and toluene the reduction factors amounted to 2.7 and 2.4, corresponding to a decline 6.5 μg/m³ to 2.4 μg/m³ (D5) or 32 μg/m³ to 13 μg/m³ (toluene).

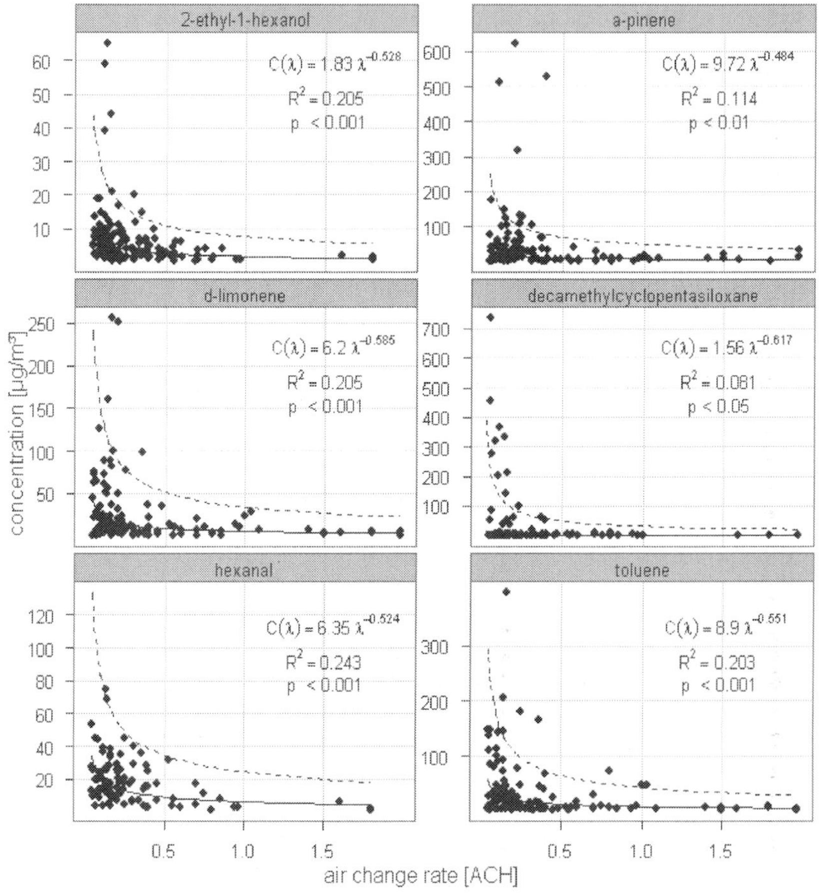

Fig. 17. c-h. Concentrations of selected volatile organic compounds (VOC) and corresponding ACR in rooms of the City of Berlin. Sampling under worst case conditions, solid line: mean concentration, dotted line: 90% prediction limit.

These results may allow to conclude that increasing the ACR to the fourfold will reduce indoor air concentration of VVOC and VOC to half. This means that the concentration is inversely proportional to the square root of the ACR, but seems to be contradictory to theoretical considerations, if one assumes a nearly constant emission rate of volatile compounds. In this case, the concentration would be inversely proportional to the ACR under equilibrium conditions, meaning that a fourfold increase in the ACR would reduce the concentration to a quarter.

One of the reasons for the deviating results is surely the fact that nowadays many buildings are constructed with such a high air-tightness that a kinetic equilibrium cannot be achieved under typical sampling conditions (i.e. time span of several hours between last window ventilation and start of the measurement; see also Figure 3b). That leads, firstly, to low ACR ($<0.2\ h^{-1}$) and, secondly, to an underestimation of the VOC concentrations. The regression analysis yields a flatter curve and an exponent $k < 1$ for the air change rate λ. Another reason is the different interior decoration (furniture, wallpapers, other equipment) of the spaces investigated here, and thus their differing pollution loads. Our VOC sample comprised rooms which differed greatly in respect to number and kind of possible sources. Therefore, we can assume that even in rooms with low air change either no VOC sources existed or only such sources with low emission rates. As a consequence, only low concentrations could have been measured in the air of these rooms. This fact would also contribute to the flattening of the air change-concentration curve.

The influence of ACR on the concentrations of VOC is also valid when we consider substance classes. The spectrum of individual compounds of these classes resembles that described by Schleibinger et al. (2001, 2002). As with formaldehyde, the analysis was performed with grouped data to reveal the dependency on the ACR, and categorisation was done as above. For determining of frequencies of exceeding concentrations the individual measurement values were summed up. Subsequently, the sum values were dichotomised on the basis of the medians of the empirical distributions. The empirical medians agreed quite well with the limits proposed by Schleibinger et al. (2002) - especially when taking into account the 95% confidence intervals of the percentiles (Table 13). Exceptions here are the siloxanes, for which the 90th percentile was clearly above Schleibinger's recommendation (Schleibiger, 2002). The marked deviation from previous studies (Schleibinger et al., 2001; Scholz, 1998) may indicate an increased (and still increasing) use of siloxanes as constituents of coating materials, paints, cosmetics etc., thus contributing more than ever to indoor air pollution.

Compound class	Sample size	Median (95% CI)	90. P (95% CI)	MAX	AM (95% CI)	GM (95% CI)	Target value	Reference value
aldehydes (C4 – C10)	93	49 (36 – 59)	88 (76 – 127)	203	51 (43 –58)	37 (30 – 45)	50	120
aromatics	133	39 (30 – 51)	206 (149 - 248)	1140	82 (60 – 103)	45 (37 – 54)	50	200
siloxanes (D3 – D5)	99	4 (2 – 6)	125 (51 - 338)	741	43 (21 – 65)	5 (4 – 8)	5	10
terpenes und sesquiterpenes	133	51 (36 – 61)	230 (150 – 297)	1008	103 (73 – 133)	47 (37 – 58)	40	150

Table 13. Indoor air concentrations and 95% confidence intervals [$\mu g/m^3$] for selected classes of chemical compounds. Results of a survey performed in the city of Berlin. Published target and reference values by Schleibinger et al. (2002) are included. 90. P: 90th percentile, AM: arithmetic mean, GM: geometric mean.

The relationship between ambient air concentrations of selected substance classes and the ACR the following picture can summarised as follows (Fig. 18): Depending on the substance class, between 50% and 70% of rooms with ACR $<0.2\ h^{-1}$ (sample size of this subgroup varied from n = 41 to n = 53) showed concentrations which were above the median of the respective substance class. For ACR between $0.2\ h^{-1}$ and $0.5\ h^{-1}$ (sample size between n = 37 and n = 52) this was true for only 40 - 50% of the rooms. This proportion was further reduced to 10 - 20% when the ACR were equal to or greater than $0.5\ h^{-1}$ (n = 15-27). This trend was statistically significant for all substance classes studied here ($\chi 2$ test for trend: aldehydes $p < 0.01$, aromatics $p < 0.001$, terpenes $p < 0.001$), Even for siloxanes the trend was significant, although not very pronounced ($p < 0.05$).

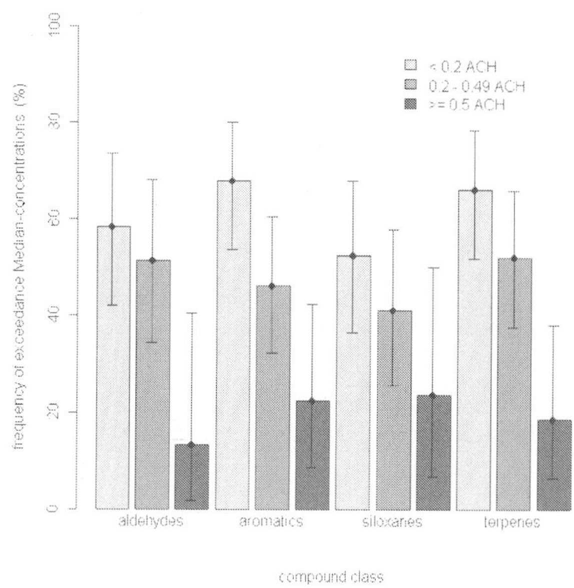

Fig. 18. Frequency of concentrations above median (50th percentile), given for selected compound classes and stratified by ACR. Active air sampling under worst case conditions, error bars: 95% confidence intervals.

Overall, it can be deduced that the probability of exceeding pollutant concentrations not only depends on the number of indoor sources and their emitting power, but also on the air-tightness of rooms. However, no exceeding concentrations can occur without the presence of indoor sources. For this reason, the use of low-emission products should given priority in order to avoid preventable future exposure to air pollutants. Whether a given indoor air concentration is above the usual burden and appropriate remedial measures are necessary, can be clearly evaluated by the ACR. It would therefore be desirable to define air-change-related reference values for common and tolerable concentrations of volatile indoor air pollutants.

4. Conclusion

Tracer gas measurements facilitate the determination of air change between indoor and outdoor environment under different ventilation conditions. In this paper methods are described which can be used to determine the air change rate with tracer gas measurements. Special emphasis was given to carbon dioxide, since this tracer gas is often used to assess indoor air quality and air tightness of rooms. We have shown ways to determine air change rates with carbon dioxide and compared the results with those obtained by other tracer gases.

The dependence of the natural air change from the prevailing weather conditions, such as the current wind and temperature conditions could be demonstrated by own investigations. Wheather effects in conjunction with different window types were leading to considerable day-to-day variability, thereby limiting the reproducibility of measurements to a large extent. However, focusing on the relationship between air change and window ventilation we could show that ventilation with wide open windows or cross-ventilation are by far the cheapest ways to remove indoor pollutants as quickly as possible from indoor air. The effectiveness of window ventilation strongly depends on the temperature difference between room air and outside air, whereas the influence of wind seems to be less important.

Studies on the air quality in Berlin's housing stock revealed a very high proportion of rooms with low air change rates. About 80% of the rooms examined by us showed air change rates below 0.5 h^{-1}, half of which were even below 0.2 h^{-1}, when determined under so-called "worst case" conditions. Especially in rooms that had newly installed windows with insulated glass and peripheral sealing strip, very low air change rates were frequently determined. Of these, about 50% (median) had air change rates below 0.16 h^{-1}. This was in significant contrast to rooms in older buildings equipped with wooden framed double box windows or composite windows, which displayed mean air change rates of 0.47 h^{-1}.

Low air change rates often contribute to elevated VOC concentrations, in rooms with insulated windows, unless emissions are not limited by appropriate low-emission products. This relationship could be confirmed by our investigations. Improvement of ventilation i.e. increasing the air change rate is an efficient measure to additionally reduce the pollutant load in indoor spaces. Therefore, to achieve limitation of unhealthy indoor-borne pollutants a combination of both measures can strongly be recommended.

5. Acknowledgement

We are grateful to Dipl. Ing. P. Tappler, Innenraum Mess- und Beratungsservice, Vienna, Federal Republic of Austria, for providing instrumentation and sulphur hexafluoride for the measurement of air change rates. We also thank Dr.-Ing. H. Schleibinger, Institute for Research in Construction, National Research Council of Canada, Ottawa, for his generous assistance in the acquisition of literature, and to the ALAB staff, Analytic Laboratory Berlin, for carefully conducting the sample analyses. We gratefully thank Dr. Liv Bode for her support in revising the manuscript.

6. References

Andersen, C.E., Bergsøe, N.C., Majborn, B. & Ulbak, K. (1997). Radon and Natural Ventilation in Newer Danish Single-Family Houses. *Indoor Air,* Vol.7, No.4, (December 1997), pp. 278-286, ISSN 0905-6947

ASTM International (1998). ASTM Standard D 6245 - 98 (Reapproved 2002): Standard guide for using Carbon dioxide concentrations to evaluate indoor air quality and ventilation. American Society for Testing and Materials. West Conshohocken, Pennsylvania

ASTM International (2001). ASTM Standard E 741-00. Standard Test Method for Determining Air Change in a Single Zone by Means of a Tracer Gas Dilution. American Society for Testing and Materials. West Conshohocken, Pennsylvania

BIA-Report (2001). Tracergasmesstechnik zur Ermittlung von Kenngrößen, *Berechnungsverfahren und Modellbildung in der Arbeitsbereichsanalys,.* U. Eickmann (Ed.), 99-114, Hauptverband der gewerblichen Berufsgenossenschaften (HVBG), ISBN 3-88383-588-9, Sankt Augustin, Federal Republic Germany

Baránková, P. (2005). Měření intenzity větrání metodou značkovacího plynu CO_2. *Vytapeni, vetrani, instalace,* Vol.14, No.1/2, (únor 2005/duben 2005), pp. 31-34/77-81, ISSN 1210-1389

Bekö, G., Lund, T., Nors, F., Toftum, J. & Clausen, G. (2010). Ventilation rates in the bedrooms of 500 Danish children, *Building and Environment,* Vol.45, No.10, (October 2010), pp. 2289–2295, ISSN 0360-1323

Bornehag, C.G., Sundell, J., Hägerhed-Engman, L. & Sigsgaard, T. (2005). Association between ventilation rates in 390 Swedish homes and allergic symptoms in children, *Indoor Air,* Vol.15, No.4, (August 2005), pp. 275–280, ISSN 0905-6947

Bundesgesundheitsamt: Kommission Innenraumlufthygiene des BGA (1993). Raumklimabedingungen in Schulen, Kindergärten und Wohnungen und ihre Bedeutung für die Bestimmung von Formaldehydkonzentrationen, *Bundesgesundheitsblatt,* Vol.36, No.2, (February 1993), pp. 76-78, ISSN 1436 -9990

Chao, C.Y.H., Tung, T.C.W. & Burnett, J. (1997). Influence of ventilation on indoor radon level. *Building and Environment,* Vol.32, No.6, (November 1997), pp. 527-534, ISSN 0360-1323

Chao, C.Y., Wan, M.P., & Law, A.K. (2004). Ventilation performance measurement using constant concentration dosing strategy. *Building and Environment,* Vol.39, No.11, (November 2004), pp. 1277-1288, ISSN 0360-1323

Cheong, K.W., & Riffat, S.B. (1995). New approach for measuring airflows in buildings using a perfluorocarbon tracer. *Applied Energy,* Vol.51, No.3, (July 1995), pp. 223-232, ISSN 0306-2619

Chuah, Y.K., Fu, Y.M., Hung, C.C. & Tseng, P.C. (1997). Concentration variations of pollutants in a work week period of an office. *Building and Environment,* Vol.32, No.6, (November 1997), pp. 535-540, ISSN 0360-1323

Chung, K.Ch. & Hsu, S.-P. (2001). Effect of ventilation pattern on room air and contaminant distribution. *Building and Environment,* Vol.36, No.9, (November 2001), pp. 989-998, ISSN 0360-1323

Dietz, R.N. & Cote, E.A. (1982). Air infiltration measurements in a home using a convenient perfluorocarbon tracer technique. *Environment International,* Vol.8, No.1-6, (July-December), pp. 419-433, ISSN 0160-4120

Dietz, R.N., Goodrich, R.W., Cote, A. & Wieser, R.F. (1986). Detailed description and performance of a passive perfluorocarbon tracer system for building ventilation and air change measurements. In: *Measured Air Leakage of Buildings,* Trechsel, H.R. & Lagus, P.M., pp. 203-264, American Society for Testing and Materials, ISBN 978-803149717, Philadelphia

Dols, W.S. & Persily, A.K. (1992). Study of Ventilation Measurement in an Office Building, In: *Airflow Performance of Building Envelopes, Components, and Systems* (ASTM STP 1255), Modera, M.P. & Persily, A.K., pp. 23-46, American Society for Testing and Materials, ISBN 978-0803120235,. West Conshohocken, Pennsylvania

Drivas, P.J., Simmonds, F.G. & Shair, F.H. (1972). Experimental characterization of ventilation systems in buildings. *Environmental Science and Technology,* Vol.6, No.7, (July 1972), pp. 577-666, ISSN 0013-936X

Erhorn, H. & Gertis, K. (1986). Mindestwärmeschutz oder/und Mindestluftwechsel? *gi Gesundheits-Ingenieur: Haustechnik - Bauphysik - Umwelttechnik,* Vol.107, No.1, (February 1986), pp. 12-14 & 71-76, ISSN 0932-6200

Fanger, P.O. (1988). Introduction of the olf and the decipol units to quantify air pollution perceived by humans indoors and outdoors. *Energy and Buildings,* Vol.12, No.1, (April 1988), pp. 1-6, ISSN 0378-7788

Gregory, N.L. (1962). Detection of nanogram quantities of Sulphur Hexafluoride by electron capture methods. *Nature,* Vol. 196, No. 4859, (October 1962), p. 162, ISSN 0028-0836

Grimsrud, D.T., Sherman, M.H., Janssen, J.E., Pearman, A.N. & Harrje, D.T. (1980). An intercomparison of tracer gases used for air infiltration measurements. *ASHRAE Transactions,* Vol.86, pp. 258-267, ISSN 0001-2505

Guo, L. & Lewis, O.J. (2007). Carbon Dioxide Concentration and its Application on Estimating the Air Change Rate in Typical Irish Houses. *International Journal of Ventilation,* Vol.6, No.3, (December 2007), pp. 235-245, ISSN 1473-3315

Harving, H., Dahl, R., Korsgaard, J. & Linde, S.A. (1992). The indoor environment in dwellings: a study of air-change, humidity and pollutants in 115 Danish residences. *Indoor Air,* Vol.2, No.2, (June 1992), pp. 121-126, ISSN 1600-0668

Heidt, F.-D. & Werner, H. (1986). Microcomputer-aided measurement of air change rates. *Energy and Buildings,* Vol.9, No.4, (December 1986), pp. 313-320, ISSN 0378-7788

Heidt, F.-D. (1987). Fortschritte bei der Luftwechselmessung durch Mikrocomputereinsatz. *Heizung - Lüftung - Haustechnik,* Vol.38, No.8, (August 1987), pp. 391-395, ISSN 0341-0285

Hill, M., Gehrig, R., Dorer, V., Weber, A. & Hofer, P. (2000). Are measurements of air change rates with the PFT-method biassed by sink and temperature effects? *Proceedings of Healthy Buildings 2000,* Vol.2, pp. 333-338, ISBN 952-5236-05-6, Espoo Finland, August 6-10

Hirsch, T., Hering, M., Burkner, K., Hirsch, D., Leupold, W., Kerkmann, M.L., Kuhlisch, E. & Jatzwauk, L. (2000). House-dust-mite allergen concentrations (Der f 1) and mold spores in apartment bedrooms before and after installation of insulated windows and central heating systems. *Allergy,* Vol.55, No.1, (January 2000), pp. 79-83, ISSN 0105-4538

Hodgson, A.T., Faulkner, D., Sullivan, D.P., DiBartolomeo, D.L., Russell, M.L. & Fisk, W.J. (2003). Effect of outside air ventilation rate on volatile organic compound concentrations in a call center. *Atmospheric Environment,* Vol.37, No.39-40, (December 2003), pp. 5517-5527, ISSN 1352-2310

Howard-Reed, C., Wallace, L.A., & Ott, W.R. (2002). The effect of opening windows on air change rates in two homes. *Journal of the Air & Waste Management Association,* Vol.52, No.2, (February 2002), pp. 147-59, ISSN 1047-3289

Hunt, C.M. & Burch, D.M. (1975). Air infiltration measurements in a four bedroom townhouse using Sulphur Hexafluoride as a tracer gas. *ASHRAE Transactions,* Vol.81, No.1, pp. 186-201, ISSN 0001-2505

Keller, R. & Beckert, J. (1994). Untersuchungen zum Vorkommen organischer flüchtiger Verbindungen in dem Musterraum eines Neubaus unter der Berücksichtigung des „natürlichen" Luftwechsels. *Zentralblatt für Hygiene und Umweltmedizin,* Vol.195, No.5-6, (June 1994), pp. 432-443, ISSN 0934-8859

Krooß, J., Siemers, U., Stolz, P., Weiss, N. & Clausnitzer, K.-D. (1997). Luftwechselraten in Wohn- und Arbeitsräumen. *Gefahrstoffe - Reinhaltung der Luft,* Vol.57, No.9, (September 1997), pp. 357-362, ISSN 0949-8036

Kumar, R., Ireson, A.D. & Orr, H.W. (1979). An automated air infiltration measuring system using SF6 tracer gas in constant concentration and decay methods. *ASHRAE Transactions,* Vol.85, No.2, pp. 385-395, ISSN 0001-2505

Lagus, P.L. & Grot, R.A. (1997). Control room envelope unfiltered air inleakage test protocols. In: *Proceedings of the 24th DOE/NRC Nuclear Air Cleaning and Treatment Conference,* First, M.W., pp. 400-427, The Havard Air Cleaning Laboratory, NUREG/CP-0153, CONF-960715, Boston, MA

Lembrechts, J., Janssen, M. & Stoop, P. (2001). Ventilation and radon transport in Dutch dwellings: computer modelling and field measurements. *The Science of the Total Environment,* Vol.272, No.1-3, (May 2001), pp. 73-78, ISSN 0048-9697

Li, Y.-Y., Wu, P.-C., Su, H.-J., Chou, P.-C. & Chiang, C.-M. (2002). Effects of HVAC ventilation efficiency on the concentrations of formaldehyde and total volatile organic compounds in office buildings. *Proceedings of the 9th International Conference on Indoor Air Quality and Climate,* pp. 376-381, ISBN 0-9721832-0-5, Monterey, CA, June 30 - July 5 2002

Maas, A. (1997). Tracergasmeßtechniken zur Ermittlung des Luftwechsels. *gi Gesundheits-Ingenieur: Haustechnik - Bauphysik - Umwelttechnik,* Vol.118, No.5, (October 1995), pp. 256- 267, ISSN 0932-6200

Mailahn, W., Seifert, B., Ullrich, D. & Morikse, H.-J. (1989). The use of a passive sampler for the simultaneous determination of long-term ventilation rates and VOC concentrations. *Environment International*, Vol.15, No.1-6, (January-December 1989), pp. 537-544, ISSN 0160-4120

Menzies, R., Schwartzman, K., Loo, V. & Pasztor, J. (1995). Measuring ventilation of patient care areas in hospitals. Description of a new protocol. *American Journal of Respiratory and Critical Care Medicine*, Vol.152, No.6, (December, 1995), pp. 1992-1999, ISSN 1073-449X

Münzenberg, U. (2004). Der natürliche Luftwechsel in Gebäuden und seine Bedeutung bei der Beurteilung von Schimmelpilzschäden. In: *Umwelt, Gebäude & Gesundheit: Innenraumhygiene, Raumluftqualität und Energieeinsparung. Ergebnisse des 7. Fachkongresses der Arbeitsgemeinschaft Ökologischer Forschungsinstitute (AGÖF) am 04. und 05. März 2004 in München*, AGÖF - Arbeitsgemeinschaft Ökologischer Forschungsinstitute, (Ed.), 263-271, AGÖF, ISBN 3-930576-05-8, Springe -Eldagsen

Nabinger, S.J., Persily, A.K. & Dols, W.S. (1994). Study of Ventilation and Carbon Dioxide in an Office Building. *ASHRAE Transactions*, Vol.100, No.2, pp. 1264-1274, ISSN 0001-2505

Niemelä, R., Lefevre, A., Muller, J.P. & Aubertin, G. (1991). Comparison of three tracer gases for determination ventilation effectiveness and capture efficiency. *Annals Occupational Hygiene*, Vol.35, No.4, (August 1991), pp. 405-417, ISSN 0003-4878

Øie, L., Magnus, P. & Johansen, B.V. (1997). Suspended particulate matter in Norwegian dwellings in relation to fleece and shelf factors, domestic smoking, air change rate, and presence of hot wire convection heaters, Environment International, Vol.23, No.4, (July-August 1997), pp. 465-473, ISSN 0160-4120

Øie, L. Stymne, H., Boman, C.-A. & Hellstrand, V. (1998). The ventilation rate of 344 Oslo residences. *Indoor Air*, Vol.8, No.3, (September 1998), pp. 190-196, ISSN 0905-6947

Pandian, M.D., Ott, W.R. & Behar, J.V. (1993). Residential air change rates for use in indoor air and exposure modelling studies. *Journal of Exposure Analysis and Environmental Epidemiology*, Vol.3, No.4, (October 1993), pp. 407-416, ISSN 1053-4245

Parker, G.B. (1986). Measured air change rates and indoor air quality in multifamily residences. *Energy and Buildings*, Vol.9, No.4, (December 1986), pp. 293-303, ISSN 0378-7788

Penman, J.N. (1980). An experimental determination of ventilation rate in occupied rooms using atmospheric carbon dioxide concentration. *Building and Environment*, Vol.15, No.1, (April-March 1980), pp. 45-47, ISSN 0360-1323

Penman, J.N. & Rashid, A.A.M. (1982). An experimental determination of air flow in a naturally ventilated room using metabolic carbon dioxide. *Building and Environment*, Vol.17, No.4, pp. 253-256, (October–December 1982), ISSN 0360-1323

Persily, A.K. (1997). Evaluating building IAQ and ventilation with indoor Carbon Dioxide. *ASHRAE Transactions*, Vol.103, No.2, pp. 193-204, ISSN 0001-2505

Persily, A.; Musser, A. & Emmerich, S.J. (2010). Modeled infiltration rate distributions for U.S. housing. Indoor Air, Vol.20, No.6, (December 2010), pp. 473-485, ISSN 1600-0668

Pettenkofer, M. (1858). *Besprechung allgemeiner auf die Ventilation bezüglicher Fragen. Über den Luftwechsel in Wohngebäuden,* Cottaische Verlagsbuchhandlung, München, pp. 69-126

Raatschen, W. (1995). Tracergasmessungen in der Gebäudetechnik: Luftaustausch - Messung und Simulation. *GI - Gesundheits-Ingenieur: Haustechnik, Bauphysik, Umwelttechnik,* Vol.116, No.2-3, (April, June 1995), pp. 78-87 and 129-138, ISSN 0932-6200

Roulet, C.A. & Foradini, F. (2002). Simple and cheap air change rate measurement using CO_2 concentration decays. *International Journal of Ventilation,* Vol.1, No.1, (June 2002), pp. 39-44, ISSN 1473-3315

Ruotsalainen, R. Rönnberg, R., Säteri, J., Majanen, A., Seppänen, O. & Jaakkola, J.J.K. (1992). Indoor climate and the performance of ventilation in Finnish residences. *Indoor Air,* Vol.2, No.3, (September1992), pp. 137-145, ISSN 0905-6947

Sakaguchi, J. & Akabayashi, S. (2003). Field survey of indoor air quality in detached houses in Niigata Prefecture. *Indoor Air,* Vol.13, No.S6, (January 2003), pp. 42-49, ISSN 0905-6947

Salmon, L.G., Cass, G.R., Bruckman, K. & Haber J. (2000). Ozone exposure inside museums in the historic central district of Krakow, Poland. *Atmospheric Environment,* Vol.34, No.22, (June 2000), pp. 3823-3832, ISSN 1352-2310

Salthammer, T. (1994). Luftverunreinigende organische Substanzen in Innenräumen. *Chemie in unserer Zeit,* Vol.28, No.6, (December 1994), pp. 280-290, ISSN 0009-2851

Salthammer, T., Fuhrmann, F., Kaufhold, S., Meyer, B. & Schwarz, A. (1995). Effects of Climatic Parameters on Formaldehyde Concentrations in Indoor Air. *Indoor Air,* Vol.5; No.2, (April 1995), pp. 120-128, ISSN 0905-6947

Schleibinger, H., Hott, U., Marchl, D., Braun, P., Plieninger, P. & Rüden, H. (2001). VOC-Konzentrationen in Innenräumen des Großraums Berlin im Zeitraum von 1988 - 1999. *Gefahrstoffe – Reinhaltung der Luft,* Vol.51, No.1/2, (January-February 2001), pp. 26-38, ISSN 0949-8036

Schleibinger, H., Hott, U., Marchl, D., Braun, P., Plieninger, P. & Rüden, H. (2002). Ziel- und Richtwerte zur Bewertung der VOC-Konzentrationen in der Innenraumluft - ein Diskussionsbeitrag. *Umweltmedizin in Forschung und Praxis,* Vol.7, No.3, (June 2002), pp. 139-147, ISSN 1430-8681

Scholz, H. (1998). Vorkommen ausgewählter VOC in Innenräumen und deren Bewertung. In: *Gebäudestandard 2000: Energie und Raumluftqualität,* Arbeitsgemeinschaft ökologischer Forschungsinstitute (Ed.), pp. 205-214, AGÖF, ISBN 3-930576-01-5, Springe-Eldagsen

Schulze, H.G. & Schuschke, G. (1990). Studie über die Notwendigkeit und Zuverlässigkeit von Luftwechsel und Luftvolumenstrommessungen. *Gesundheits-Ingenieur,* Vol.111, No.1, (February 1990), pp. 12-16, ISSN 0016-9277

Seifert, B. & Salthammer, T. (2003). Innenräume. Belastung der Umweltmedien. In: *Handbuch der Umweltmedizin,* H.-E. Wichmann, H.-W. Schlipköter & G. Fülgraff

(Ed.), pp. 26. Erg.Lfg. 4/03 - 1-34, ecomed Fachverlag, ISBN 978-3-609-71180-5, Landsberg/Lech

Sekhar, S.C. (2004). Enhancement of ventilation performance of a residential split-system air-conditioning unit. *Energy and Buildings*, Vol.36, No.3, , (March 2004), pp. 273-279, ISSN 0378-7788

Senitkova, I. (1999). Building redesign for Radon elimination. *Proceedings of the 8th International Conference on Indoor Air Quality and Climate*, Vol.2, pp. 953-958, ISBN 860812953, Edinburgh, Scotland, 8 - 13 August 1999

Seppänen, O.A. & Fisk, W.J. (2004). Summary of human responses to ventilation. *Indoor Air*, Vol.14, No.S7, (August 2004), pp. 102-118, ISSN 0905-6947

Seppänen, O.A., Fisk, W.J. & Mendell, M.J. (1999). Association of ventilation rates and CO_2 concentrations with health and other responses in commercial and institutional buildings. *Indoor Air*, Vol.9, No.4, (December 1999), pp. 226-252, ISSN 0905-6947

Shaw, C.Y. (1984). The effect of tracer gas on the accuracy of air-change measurements in buildings. *ASHRAE Transactions*, Vol.90, No.2816 (=pt 1A), pp. 212-225, ISSN 0001-2505

Sherman, M.H. (1990). Tracer-gas techniques for measurement ventilation in a single zone. *Building and Environment*, Vol.25, No.4, (April 1990), pp. 365-374, ISSN 0360-1323

Smith, P.N. (1988): Determination of ventilation rates in occupied buildings from metabolic CO_2 concentrations and production rates. *Building and Environment*, Vol.23, No.2, (January 1988), pp. 95-102, ISSN 0360-1323

Stavova (Barankova), P., Melikov, A.K., Sundell, J. & Drkal, F. (2006). Validation of a CO2 Method for Ventilation Measurement in Dwellings - Laboratory Study. *Proceedings of Healthy Buildings 2006, 8th International Conference on Indoor Air Quality and Health*, p. 375, ISBN 989-95067-1-0, Lisbon, Portugal, June 4-8, 2006

TRGS 900 (1999). Technische Regeln für Gefahrstoffe: Grenzwerte in der Luft am Arbeitsplatz - Luftgrenzwerte (TRGS 900), *Bundesarbeitsblatt*, Vol.6. p. 87, ISSN 0007-5868

VDI (2001). VDI 4300 - Messen von Innenraumluftverunreinigungen. Bestimmung der Luftwechselzahl in Innenräumen. In: *VDI/DIN-Handbuch Reinhaltung der Luft Band 5*, Kommission Reinhaltung der Luft (KRdL) im VDI und DIN - Normenausschuß (Ed.), Beuth-Verlag, ICS: 13.040.01, Berlin

Walker, I.S. & Forest, T.W. (1995). Field measurements of ventilation rates in attics. *Building and Environment*, Vol.30, No.3, (July 1995), pp. 333-347, ISSN 0360-1323

Wargocki, P., Sundell, J., Bischof, W., Brundrett, G., Fanger, O., Gyntelberg, F., Hanssen, S.O., Harrison, P., Pickering, A., Seppänen, A. & Wouters, P. (2002). Ventilation and health in non-industrial indoor environments: report from a European Multidisciplinary Scientific Consensus Meeting (EUROVEN). *Indoor Air*, Vol.12, No.2, (June 2002), pp. 113-128, ISSN 0905-6947

Wegner, J. (1983). Untersuchungen des natürlichen Luftwechsels in ausgeführten Wohnungen, die mit sehr fugendichten Fenstern ausgestattet sind. *gi - Gesundheitsingenieur*, Vol.104, No.1, (February 1983), pp. 1-56, ISSN 0932-6200

Wegner, J. (1984). Schadstoffanfall, Luftwechsel in Wohnungen, freie Lüftung. *Haustechnik - Bauphysik - Umwelttechnik - Gesundheits - Ingenieur*, Vol.105, No.3, (June 1984), pp. 117-123, ISSN 0172-8199

Wilson, A.L., Colome, S.D., Tian, Y., Becker, E.W., Baker, P.E. Behrens, D.W. & Billick, I.C.H. (1996). California residential air change rates and residence volumes. *Journal of Exposure Analysis and Environmental Epidemiology*, Vol.6, No.3, (July-September 1996), pp. 311-326, ISSN 1053-4245

15

Olfactory Comfort Assurance in Buildings

Sârbu Ioan and Sebarchievici Călin
"Politehnica" University of Timisoara
Romania

1. Introduction

Reducing energy consumption in buildings is one of the main current directions of research in building constructions. An important part of household energy consumption is necessary to achieve, in living spaces, indoor microclimate parameters. Therefore, is particularly important the achievement of structural elements, building equipments and operating modes to allow getting both adequate comfort parameters and energy saving.

The greatest majority of people carry on 80...90% of their lives inside buildings, which must satisfy the objective and subjective requests linked to vital functions of the occupants. That is why the enclosed spaces must insure the possibility for both physical and intellectual work, as well as for some recreation activities, for rest and sleep under most favourable conditions. The achievement of these conditions depends on very many factors that decisively influence the sensation of comfort perceived, the work capacity and man's regeneration capacity. The design of the rooms must take into consideration these conditions and present tendencies to reduce the energy consumption, that are decisively influencing the optimal or admissible values of comfort parameters. Thus the inside microclimate of a building must be the result of a computation of multicriterial optimization, taking into account technical and psychological comfort and the energy saving.

The concept of technical comfort comprises all parameters achieved and controlled with HVAC systems that act upon building occupants senses. This includes thermal, acoustic, olfactory and visual comfort. In accordance with the dissatisfied person percent of the ensured comfort: 10%, 20%, 30%, rooms are classified into three categories: A, B, C.

The perception and appreciation of basic comfort elements to man are influenced by some psychological factors, but at the same time evolution and man psychological equilibrium are closely linked with the environment. So, between psychological and technical comfort is a reciprocal connection. Human psyche depends also by other independent factors like: age, gender, etc., influencing the technical comfort level appreciation. So pleasant sensation may occur as a result of optimum technical and psychological comfort parameters (Fig. 1).

Subjective comfort of persons in a room depends on many factors: temperature, humidity and air circulation; smell and respiration; touch and touching; acoustic factors; sight and colours effect; building vibrations; special factors (solar-gain, ionization); safety factors; economic factors; unpredictable risks.

Because of some technical conditions the common influence of these factors can not be analysed, and the adaptation of the human body to a certain environment is a complex process, this one reacting to the common action of more parameters.

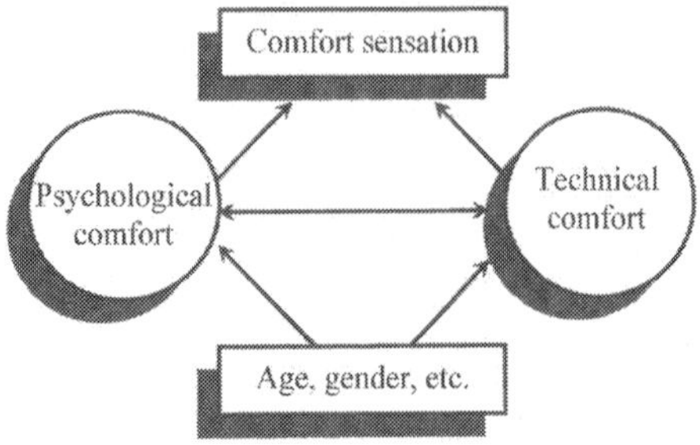

Fig. 1. Comfort sensation

In this chapter an olfactory comfort analysis in buildings is performed starting from a general description of the comfort fundamental components (thermal, olfactory, acoustic and visual comfort). It is developed a computational model for indoor air quality numerical simulation, as well a methodology to determine the outside airflow rate and to verify the indoor air quality in enclosed spaces, according to the European Standard CEN 1752 (1998). Also, it is presented the influence of carbon dioxide on human performance and productivity.

2. Generalities on fundamental components of the comfort

2.1 Thermal comfort

The subjective sensation of thermal comfort is decisively determined by the following parameters (Sârbu & Kalmar, 2002): indoor air temperature (t_i); mean radiant temperature (t_{mr}) of bordering surfaces; relative humidity of air (φ_i); partial water vapours pressure (p_a); air velocity (v_i); thermal resistance of clothing (R_{cl} or R_h), and their influence on the vaporization; heat production of human body and human thermoregulation. The first four are physical parameters, and the other two expresses the capacity of the human body to adapt itself in order to maintain the thermal equilibrium. The main factors that influence the thermal equilibrium of the human body are:

- heat production of human body, which depends on the activity level, age, sex, etc.
- body heat loss, which depends on clothing and on the other parameters mentioned previously.

In general, comfort occurs when body temperatures are held within narrow ranges, skin moisture is low, and the physiological effort of regulation is minimized. Surprisingly, although climates, living conditions, and cultures differ widely throughout the world, the temperature that people choose for comfort under similar conditions of clothing, activity, humidity, and air movement has been found to be very similar (Busch, 1992; Dear et al., 1991; Fanger, 1972).

In order to evaluate the sensation of thermal comfort we use the thermal sensation scale with seven levels (ASHRAE, 2009): +3 (hot); +2 (warm); +1 (slightly warm); 0 (neutral); −1 (slightly cool); −2 (cool); −3 (cold).

Numerical prediction of thermal comfort in a room is performed by using the PMV - PPD model, and testing is achieved at asymmetric thermal radiation, caused by building elements with a temperature much different from the mean radiant temperature. Radiant asymmetry is the difference in radiant temperatures seen by a small flat element looking in

opposite directions. Four calculation methods of radiant temperature asymmetry are available in the technical literature (Frohner et al., 2004).
PMV (predicted mean vote) index has the optimum value equal to zero, but according to the prescriptions ISO Standard 7730 (ISO, 2005) it is considered that the domain of thermal comfort corresponds to values between -0.5 and +0.5. The use of PMV index is recommended only for values between +2 and -2. In the Fig. 2 are represented the values of operative comfort temperature t_c (corresponding to index PMV = 0), correlated to thermal resistance of clothing (R_{cl} and R_h), metabolic rate i_M and metabolic heat production M.
In Table 1 is illustrated the optimal values of operative temperature t_c according to building use, summer and winter, according to European Standard CEN 1752 (CEN, 1998).

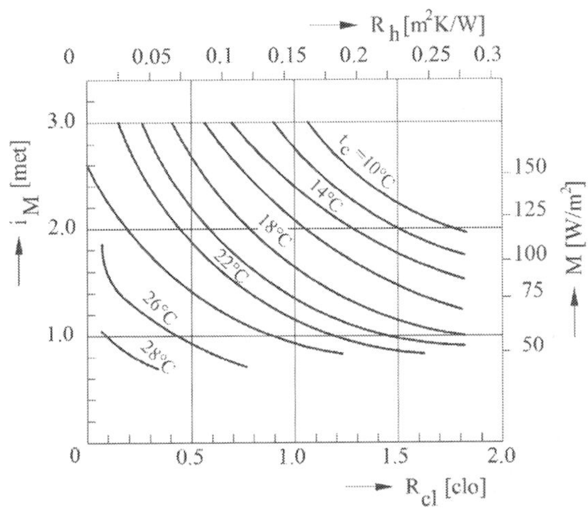

Fig. 2. Operative comfort temperature function of clothing and activity

In recent study (Sârbu & Bancea, 2009) is developed a computational and testing model of thermal comfort in buildings, according to the European Standard CEN 1752. On the bases of this model there was elaborated a computer program, implemented on compatible microsystems IBM-PC.

2.2 Olfactory comfort

Comfort and indoor air quality (IAQ) depend on many factors, including thermal regulation, control of internal and external sources of pollutants, supply of acceptable air, occupant activities and preferences, and proper operation and maintenance of building systems. Ventilation and infiltration are only part of the acceptable indoor air quality and thermal comfort problem. The condition to achieve human body metabolism in a enclosed space is oxygen (O_2) taking and carbon dioxide (CO_2) releasing. After respiration process air reaches the lungs through upper and lower airways. Upper airways filter the inspired air, while providing to it the proper temperature and humidity. The oxygen is transported from the lungs to tissues through blood that carry back the carbon dioxide. Expired CO_2 flow rate are illustrated in Table 2.

Room destination	R_{cl} [clo] Summer	R_{cl} [clo] Winter	i_M [met]	Pers./ /m² floor	Room category	t_c [°C] Summer	t_c [°C] Winter
Small offices	0.5	1.0	1.2	0.10	A	24.5±0.5	22.0±1.0
					B	24.5±1.5	22.0±2.0
					C	24.5±2.5	22.0±3.0
Large offices	0.5	1.0	1.2	0.07	A	24.5±0.5	22.0±1.0
					B	24.5±1.5	22.0±2.0
					C	24.5±2.5	22.0±3.0
Conference rooms	0.5	1.0	1.2	0.50	A	24.5±0.5	22.0±1.0
					B	24.5±1.5	22.0±2.0
					C	24.5±2.5	22.0±3.0
Study rooms	0.5	1.0	1.2	1.50	A	24.5±0.5	22.0±1.0
					B	24.5±1.5	22.0±2.0
					C	24.5±2.5	22.0±3.0
Restaurant rooms	0.5	1.0	1.4	0.70	A	23.5±1.0	20.0±1.0
					B	23.5±2.0	20.0±2.0
					C	23.5±2.5	20.0±2.5
Classrooms	0.5	1.0	1.2	0.50	A	24.5±0.5	22.0±1.0
					B	24.5±1.5	22.0±2.0
					C	24.5±2.5	22.0±3.0
Kindergartens	0.5	1.0	1.4	0.50	A	23.5±1.0	20.0±1.0
					B	23.5±2.0	20.0±2.0
					C	23.5±2.5	20.0±2.5
Deposits	0.5	1.0	1.6	0.15	A	23.0±1.0	19.0±1.5
					B	23.0±2.0	19.0±3.0
					C	23.0±3.0	19.0±4.0

Table 1. Optimum values of operative temperature t_c

Activity	M [W/pers]	Inspired air [m³/h]	Expired CO_2 [l/h]	Consumed O_2 [l/h]
Sedentary	-	0.30	12	14
Intellectual	120	0.375	15	18
Physical very easy	150	0.575	23	27
Physical easy	190	0.75	30	35
Physical hard	>270	>0.75	>30	>35

Table 2. Expired CO_2 flow rate

Air composition in living spaces differs from that of the outside air. Carbon dioxide concentration in outside air is between 300 and 400 ppm, and in living spaces is of about 900 ppm. The maximum admitted limit of CO_2 concentration in the inhaled air is of 1000 ppm (Pettenkofer's number). Table 3 presents the effect of different CO_2 concentrations on human body.

CO_2 concentration		Effect
[%]	[ppm]	
3	30000	Deep breathing, strong
4	40000	Head aches, pulse, dizziness, psychic emotions
5	50000	After 0.5…1 hours may cause death
8...10	80000...100000	Sudden death

Table 3. The effect of CO_2 concentration on human body

Air quality is prevailingly determined by people's sensations to different odorants. Because is impossible to be measured each of the air contaminants quantitatively and qualitatively (about 8000), Fanger proposed that all these compounds to be measured by one parameter: the odour.

The odours arise in inhabited areas by the release of the human body (ammonia, methan, fatty acids), emanations of the furniture, carpets, paintings and other building materials (formaldehyde), by combustion and heating processes (carbon monoxide, fuel vapour), by exhaust gas poluted air infiltration or air from industrial areas, meal preparation, toilettes areas, mold chemical reactions, mushrooms or any decomposition products. Most of these unpleasant products are made of complex organic substance.

Excitation level in confront of some odours is very low. For example, mercaptan odour is perceived starting from a concentration of 0.00000004 mg/l. The olfactory organ main feature is adaptation. After a while, due to continuous charging, sensation of smell intensity decreases (Fig. 3).

A large number of pollutants come a time in the air with tobacco smoke. This affects the eyes, the nose and it is a risk factor for different diseases. Reduce air pollution by tobacco smoke can not be done only by increasing the air exchange rate. Thus, to annihilate the negative feelings created by smoking one cigarette are requested 100 m^3 of fresh (outside) air.

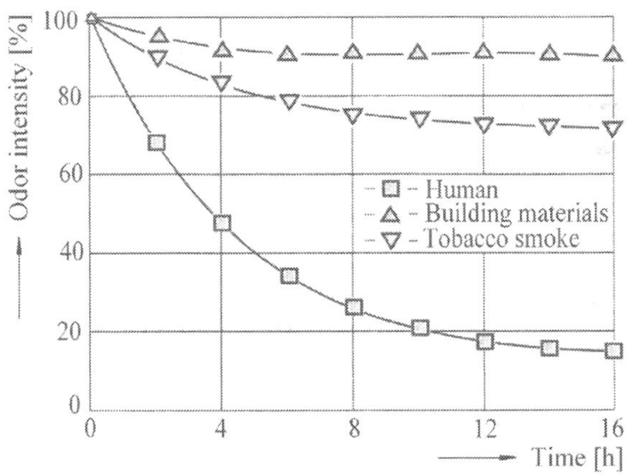

Fig. 3. Time evolution of odour intensity

Along with tobacco smoke in the air reaches carbon monoxide (CO). Maximum carbon monoxide concentration permissible values, given by international prescriptions, are 10

mg/m^3 for housing spaces and 20 mg/m^3 for kitchen and ancillary areas (3 h maximum residence time). Fig. 4 illustrates the variation of CO concentration depending of smoked cigarettes number and fresh air flow-rate introduced into the room. The smoking weighty influences the CO content of expired air, whose values are indicated in Table 4.

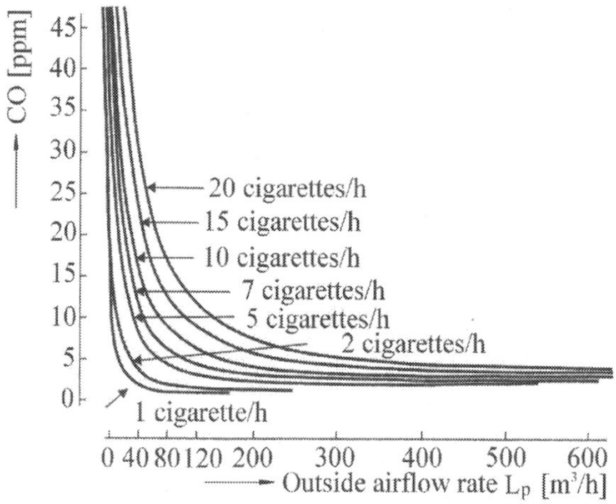

Fig. 4. Variation of CO concentration

Odorant perception depends, on one side, on objective factors: concentration and toxicity of air pollutants (bio-effluents), activity level, outside airflow rate, and on the other hand, on psychological factors with subjective character.

No.	Category	CO concentration [mg/m^3]	
		Male	Female
0	1	2	3
1	Nonsmoker	7.1	5.8
2	Former smoker	7.8	6.5
3	Cigar and pipe smoker	9.6	12.0
4	Tobacco smoker	24.3	21.1

Table 4. Average values of CO concentration in expired air

The relation between perceived odorant intensity and its concentration conforms to a power function (Stevens, 1957):

$$S=kC^{\beta} \tag{1}$$

where: S is odorant intensity (magnitude); C - odorant concentration, in ppm; β - exponent of psychophysical function; k - constant characteristic of material.

In the olfactory realm: $\beta < 1.0$. Accordingly, a given percentage change in odorant concentration causes a smaller percentage change in perceived odour intensity.

Sometimes IAQ scientists cannot successfully resolve complaints about air in offices, schools, and other nonindustrial environments. Customarily, complaints are atributed to elevated pollutant concentrations; frequently, however, such high concentrations are not found, yet complaints persist.

Assuming that the inability to find a difference between air pollutant levels in buildings with registered complaints and those without complaints is due to inadequacies of prevailing measurement techniques, Fanger and other changed the focus from chemical analysis to sensory analysis (Fanger, 1987, 1988; Fanger et al., 1988). Fanger quantified air pollution sources by comparing them with a well-known source: a sedentary person in thermal comfort. A new unit, the *olf,* was defined as the emission rate of air pollutants (bioeffluents) from a standard person. A *decipol* (*dp*) is one olf ventilated at a rate of 10 l/s of unpolluted air.

The percentage of persons dissatisfied (PPD) with air polluted by human bio-effluent (1 olf) can be calculated from the equations (Fanger et al., 1988):

$$PPD = 395\ \exp\left(-3.66 L_p^{0.36}\right) \quad \text{for } L_p \geq 0.332\ \text{l/s} \tag{2}$$

$$PPD = 100 \quad \text{for } L_p < 0.332\ \text{l/s} \tag{3}$$

where: L_p is the outside air flow-rate, in l/s.

The curve (Fig. 5) generated by equations (2) and (3) is based on experiments involving more than 1000 European subjects (Fanger & Berg-Munch, 1983). Experiments with American (Cain et al., 1983) and Japanese (Iwashita et al., 1990) subjects show very similar results.

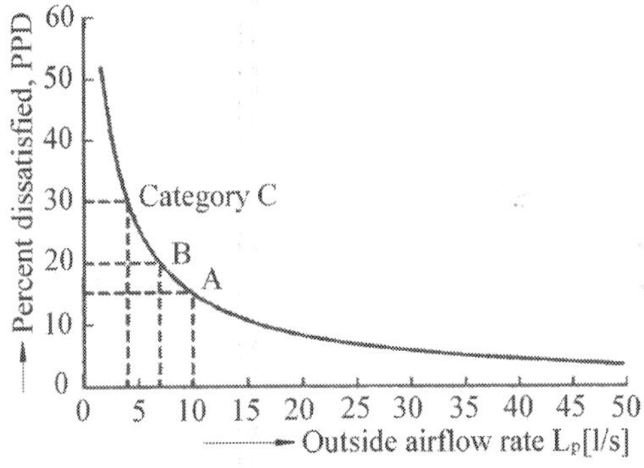

Fig. 5. PPD as a function of ventilation rate per standard person (i.e., per olf)

The idea behind the olf is to express both human and nonhuman sensory sources in a single unit: *equivalent standard persons* (i.e., in olfs). A room should therefore be ventilated to handle the total sensory load from persons and building. Table 5 shows the sensory loads from different pollution sources used in CEN.

The sensory load on the air in a space can be determined from Fig. 5 by measuring the outside air flow-rate and determining the percent dissatisfied, using an untrained panel with a minimum of 20 impartial persons (Gunnarsen & Fanger, 1992). The required outside air flow-rate depends on the desired percentage of occupant satisfaction.

No.	Source	Sensory load
0	1	2
1	Sedentary person (1... 1.5 met)	1 olf
2	Person exercising low level (3 met)	4 olf
3	Person exercising medium level (3 met)	10 olf
4	Children, kindergarten (3... 6 yrs)	1.2 olf
5	Children, school (4... 16 yrs)	1.3 olf
6	Low-polluting building	0.1 olf/m^2
7	Non-low-polluting building	0.2 olf/m^2

Table 5. Sensory pollution load from different pollution sources

Various factors make odor control an important consideration in ventilation engineering:

- contemporary construction methods result in buildings that allow less air infiltration through the building envelope;
- indoor sources of odours associated with modern building materials, furnishings, and office equipment have increased;
- outdoor air is often polluted;
- energy costs encourage lower ventilation rates at a time when requirements for a relatively odour-free environment are greater than ever.

Outdoor air requirement for acceptable indoor air quality have long been debated. Historically, the major considerations have included the amount of outdoor air required to control moisture, carbon dioxide and tobacco smoke generated by occupants. These considerations have led to prescriptions of a minimum rate of outdoor air supply per occupant.

Tables 6 and 7 present the minimum rate of outside airflow per occupant for different activities, according to European Standard CEN 1752, that also takes into consideration the smokers in ventilated rooms.

No.	Activity	Outside airflow rate [m^3/(h·pers)]
0	1	2
1	Intellectual	30
2	Physical very easy	30
3	Physical easy	40
4	Physical hard	50

Table 6. Minimum rate of outside airflow

Room category	Outside airflow rate [m^3/(h·pers)]			
	Without smokers	20% smokers	40% smokers	100% smokers
0	1	2	3	4
A	36.0	72.0	108.0	108.0
B	25.2	50.4	75.6	75.6
C	14.4	28.8	43.2	43.2

Table 7. Minimum rate of outside airflow for rooms with smokers

The perceived olfactory sensation depends not only on the pollutant source but also to a great extend, on the dilution degree with outside air. The olfactory pollution degree of a room is given by:

$$C_i = C_p + \frac{10G}{L_p} \quad (4)$$

where: C_i is the indoor air quality, in dp; C_p - outdoor air quality, in dp; G - contaminants concentration of room air.

2.3 Acoustic comfort

Another key element contributing to the overall comfort in an enclosed space is acoustic comfort. Should be made the difference between the notions of *sound* and *noise*. Thus, the notion of sound has several definitions, depending on the purpose of interpretation.

If sound is considered a physical phenomenon, then it represents the elementary vibration of elastic matter which spreads as wave in transport medium.

The concept of sound can have the sense of external excitement of creatures auditory organ, which leads to different reactions from them. Perceptible sound field to human is illustrated in Fig. 6.

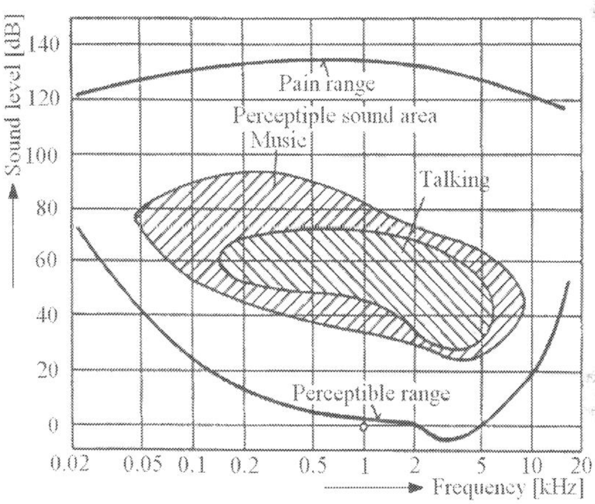

Fig. 6. Range of human audibility with normal hearing

The third sense of the sound concept is aesthetic and understanding effect. From this point of view the sound has a coded form information content, that the brain decodes it and the correct perception of sound crucial influence the human comfort.

The main element of relationship between people is the speaking and any event that disturbs the understanding it creates uncertainty and with this the discomfort. So there are sound effects, negatively perceived by human, called noises and they are indistinguishable by sound after the first two definitions. The difference between them is made by human through the given interpretation.

From researches performed in medical science it is known that the noises act on the autonomic nervous system. Nervous system reflex response is strangling noises capillaries.

Increased blood flow resistance does not involve the increase of heart beatings. This leads to a decrease blood flow rate that determines the amount reduction of oxygen transported to cells. Reduced amount of oxygen is manifested by various symptoms such as headaches, migraines, decreased concentration, blurred vision etc.
The negative effect of the arterial system bottleneck is felt especially during sleep when the body functions are reduced to a minimum. If in this state the body is subjected to the effect of noise, then reduce the amount of oxygen that feeds the cells causes lengthening the period of regeneration. It was shown that frequent sleep disturbances lead to nervousness, loss of efficiency, fatigue and nervous system degeneration. Also, intellectual activity done in terms of high noise level the yield is particularly low.
Acceptable interior equivalent noise level, according to prescriptions of the European Standard CEN 1752, is shown in Table 9.

Room destination	Cate-gory	Noise level [dB (A)]	Room destination	Cate-gory	Noise level [dB (A)]
0	1	2	3	4	5
Small offices	A B C	30 35 40	Restaurant rooms	A B C	35 45 55
Large offices	A B C	35 40 45	Classrooms	A B C	30 35 40
Conference rooms	A B C	30 35 40	Kindergaertens	A B C	30 40 45
Study rooms	A B C	30 33 35	Deposits	A B C	40 45 50

Table 9. Acceptable interior noise level, according to CEN 1752

2.4 Visual comfort

Human life is closely connected with the visual environment, because the information is collected at a rate of 90% within sight, and activity is also linked to vision in most cases.
The visual comfort is conscious state, which occurs due to physiological and psychological actions, expressing satisfaction with the environment.
Visual environment in an enclosed space appears if it is illuminated and has two components:

- the room delimited by opaque or transparent surfaces (passive component);
- the light that makes the room visible (active component).

Human perceives only the light reached in the eyes. Thus, are seen only the surfaces that send light in the observer`s eye. Usually those are opaque surfaces (walls, floors, furniture etc.). The surfaces that allow natural light (windows, skylights etc.) have a particular importance because:

- glazed and opaque surfaces do not create the same visual environment;
- usually the visual environment does not coincide with built interior space.

In function on the room destination there are requirements for visual environmental characteristics which can be divided as follows:

- the informations about certain parts of the space have to be accurate;
- the visible space must not create discomfort (visual disturbance).

Accurate vision means also the right colors perception. They are regarded as suitable if they coincide with the colors seen in natural light, hence the need for natural lighting.
For artificial lighting the source light quality can be differentiated from the point of view of body natural colors reveling. So if the bodies color lighted with artificial source corresponds to their natural color when color rendering is perfect, otherwise the color rendering is more or less good.
The visual comfort sensation represents the concordance between lighting and the quality of light, caracterised by light color or its temperature. The color pleasant effect refers to the fact that a less lighting ambient presents visual comfort if is the result of a rich warm color and an lighted ambient presents visual comfort if is the result of cold light. In terms of color rendition, the visual comfort can be analyzed in base of Kruithoff charts, that presents the variation of lighting E with light temperature T (Fig. 7).

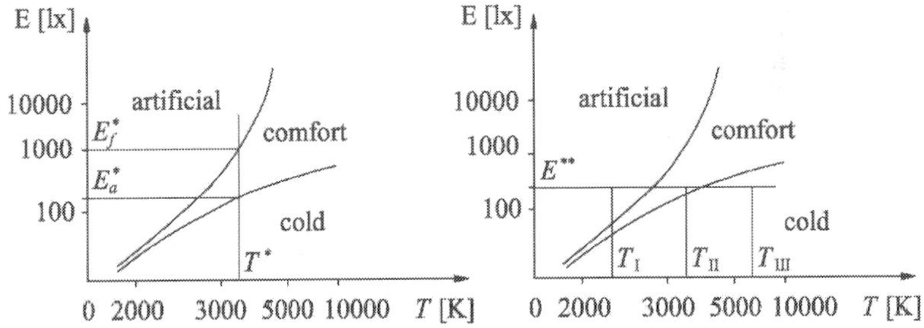

Fig. 7. Relation between lighting level, lighting temperature and visual comfort

So, a light source with temperature T^* can create different observer sensations depending of lighting value:

- for $E < E_a^*$, the lighting is perceived as cold;
- for $E_a^* < E < E_f^*$ are satisfied the visual comfort conditions;
- for $E > E_f^*$, the lighting is perceived as artificial (disturbing).

For a given value of lighting E^{**}, created subjective visual sensation depends by light temperature: artificial lighting at T_I, visual comfort at T_{II} and cold lighting at T_{III}.

2.5 Sick building syndrome

Concern about the health effects associated with indoor air dates back several hundred years, and has increased dramatically in recent decades. This attention was partially the result of increased reporting by building occupants of complaints about poor health associated with exposure to indoor air. Since then, two types of diseases associated with exposure to indoor air have been identified: *sick building syndrome* (SBS) and *building related illness* (BRI).
The people with their activity in buildings with large glazed exterior surfaces, equiped with complex building climatisation systems (commercial areas, office buildings, etc.) are affected of SBS. Sick building syndrome describes a number of adverse health symptoms related to

occupancy in a „sick" building, including mucosal irritation, fatigue, headache, and, occasionally, lower respiratory symptoms and nausea. There is no windespread agreement on an operational definition of SBS. Some authors define it as acute discomfort (e.g., eye, nose, or throat irritation; sore throat; headache; fatigue; skin irritation; mild neurotoxic symptoms; nausea; building odours) that persists for more than two weeks at frequencies significantly greater than 20%.
In Figs. 8 and 9 are presented the main symptoms of the "sick" building for their occupants and also the consequences of these ones. The most common causes of the SBS are thermal comfort and inadequate air quality.

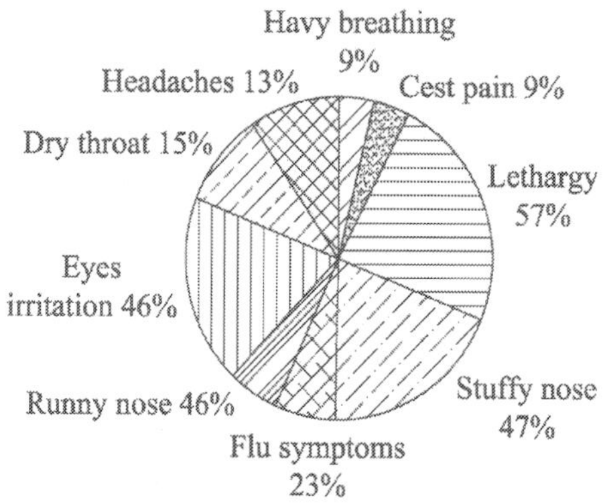

Fig. 8. The "sick" building symptoms distribution

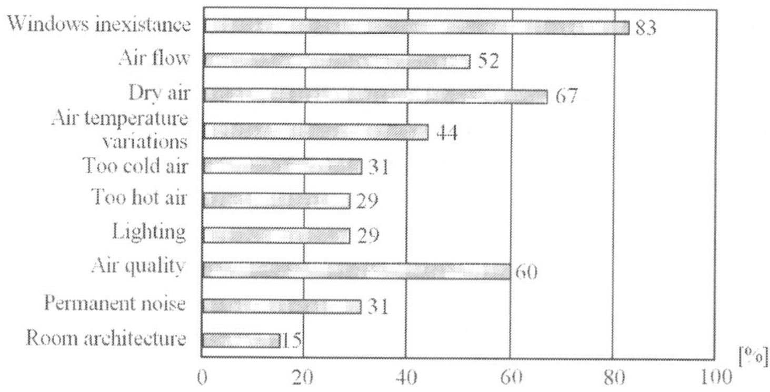

Fig. 9. Causes of sick building syndrome

The increased prelevance of health complaints among office workers is typical of sick building syndrome (Burge et al., 1987). Widespread occurrence of these symptoms has prompted the World Healt Organization to classify SBS into several categories (Morey et al., 1986):

- sensory irritation in the eyes, nose, or throat;
- skin irritation;

- neurotoxic symptoms;
- odor and taste complaints.

Some investigations have sought to correlate SBS symptoms with reduced neurological and psysiological performance. In controlled studies, SBS symptoms can reduce performance in susceptible individuals (Molhave et al., 1986).

Research performed on the basis of questionnaires on a sample of 4000 people (43.1 % men and 56.9% women) with their activity in a Frankfurt administrative building revealed the following discomfort factors (Bánhidi & Kajtár, 2000): indoor climate (65.4%); noises (32.7%); noncorresponding lighting (25.5%); tobacco smoke (24.7%); small work space (23.9%); overtime work hours (12.8%); stress caused by chief (9.7%); competition (7.1%).

3. Indoor air quality simulation model

3.1 General equation for the time evolution of a contaminant concentration

Consider a single zone compartment, where there is a source of pollution that has gas exchanges with the outside air, and where an air purifier may be used. Admitting the possibility of deposition and absorption of the pollutant on the walls and other surfaces, the temporal evolution of the concentration of a pollutant is modeled by the following differential equation (Gameiro da Silva, 2009):

$$\frac{\mathrm{d}c}{\mathrm{d}\tau} = \frac{P}{V} + nc_p - nc - \mathrm{v}_d \frac{S}{c} - \frac{L_p}{V} c\,\varepsilon_p \tag{5}$$

in which: c is the instantaneous average contaminant concentration, in mg/m^3; P - pollutant generation inside the compartment, in mg/h; V - room volume, in m^3; n - air exchange rate, in h^{-1}, i.e., the fresh air flowrate divided by the room volume; c_p - concentration of the contaminant in outside air, in mg/m^3; v_d - deposition rate of pollutant, in mg/(h·m^2); S - surface of deposition, in m^2; L_p - flow rate through the air purifier, in m^3/h; ε_p - efficiency of the air purifier.

The effects of absorption or deposition of the pollutant inside the compartment and the removal of filtering through the air purifier system may be considered in a simplified form, reducing the intensity of the sources of their value to them. Thus, for purposes or simplification, their terms in the equation may be despised, coming:

$$\frac{\mathrm{d}c}{\mathrm{d}\tau} = \frac{P}{V} + nc_p - nc(\tau) \tag{6}$$

That, integrated for a situation where V, P, c_p, L_p remain constant, since an initial time instant $\tau = 0$, where the initial concentration $c_0 = c_p$, till a generic time instant τ, will come:

$$c(\tau) = c_{eq} + \left(c_0 - c_{eq}\right) \cdot e^{-n} \tag{7}$$

in which c_{eq} is the equilibrium concentration.

3.2 Equilibrium concentration

The equilibrium concentration c_{eq}, in the equation (7), is the value that occurs when it ceases the variation in concentration. Thus, it is obtained equalizing the first member of equation (6) to zero, which gives:

$$c_{eq} = c_p + \frac{P}{Vn} \tag{8}$$

that as $n = L_p / V$, comes:

$$c_{eq} = c_p + \frac{P}{L_p} \tag{9}$$

The equations are solved numerically. The outputs of the simulation are instantaneous concentration of the pollutant and the graphical results of the time-evolution of the pollutant concentration.

3.3 Metabolic CO_2 computation

The expressions to calculate the volumes of oxygen (O_2) and carbon dioxide (CO_2) in the human respiration process are given (Emmerich & Persily, 2001), as a function of the metabolic rate and the corpulence of the studied person. The volume of consumed O_2 is given by:

$$V_{O2} = \frac{0.00276 A_D M}{0.23r + 0.77} \frac{1}{S} \tag{10}$$

where:

$$A_D = 0.202 m^{0.425} h^{0.725} \tag{11}$$

in which: A_D is the nude body surface area (DuBois & DuBois, 1916), in m^2; M - metabolic rate, in met (1 met = 58.15 W/m^2); r - ratio between the volume of released CO_2 and the consumed volume of O_2.; m - mass of human body, in kg; h - height of human body, in m.
The ratio r usually takes the value of 0.83, but that may vary till 1, for a person with a very high metabolic rate (more than 5 met).
Once the volume of O_2 has been calculated, the volume of released CO_2, for normal metabolic rate cases is computed from:

$$V_{CO2} = 0.83 \times V_{O2} \tag{12}$$

In recent study (Gameiro da Silva, 2009) is presented a software tool for indoor air quality simulation. Typical values of pollutants usually checked in IAQ audits and released by one cigarette burning are given in Table 10, which summarizes information collected in (Charles et al., 2008; REHVA, 2004).

Type	Pollutant	Unit [mg]
0	1	2
Particle	Particles suspended in air	18
Chemical	Carbon dioxide (CO_2)	160...550
	Carbon monoxide (CO)	60
	Formaldehyde (HCHO)	0.4
	Total volatile organic compunds (VOC)	3.6

Table 10. Typical emission from a smoked cigarette

4. Computation of outside air flow–rate and indoor air quality control

4.1 Mathematical model

Computation of outside air flow-rate in a room can be performed function of:

- number of occupants, keeping CO_2 concentration under the maximum admitted level, according to Romanian Norms (I 5, 1998);
- number of occupants and room surface, according to German Norms (DIN 1946, 1994);
- indoor air quality, according to European Standard CEN 1752, described below.

Ventilation efficiency ε_v is a criterion for energy and fan performances. This is used to evaluate a ventilation system and is defined by following expression:

$$\varepsilon_v = \frac{c_i - c_p}{c_{zl} - c_p} \tag{13}$$

where: c_i is the contaminants concentration in the exhausted air; c_p – contaminants concentration the supply outside air; c_{zl} – contaminants concentration in the working area.
The value of ε_v depends on the entrance place and the exhaust way of outside air, and on the difference between the outside air temperature t_e and indoor air temperature t_i (Table 11).
The outside air flow-rate L_p, in l/s, can be computed function of IAQ from equation:

$$L_p = 10\frac{G}{\left(C_i - C_p\right)\varepsilon_v} \tag{14}$$

where:

$$G = G_{oc} + G_{ob} \tag{15}$$

in which: G is the contaminants concentration of room air, in olf; C_i – indoor air quality, in dp (Table 12); C_p – outside air quality, in dp (Table 13); G_{oc} – contaminants quantity from the occupants (Table 14); G_{ob} – contaminants quantity from room objects (building elements, furniture, carpets, etc.) having the values in Table 15.

System type	t_e–t_i [°C]	ε_v
0	1	2
up–up	<0	0.9…1.0
	0…2	0.9
	2…5	0.8
	>5	0.4…0.7
up–down	<–5	0.9
	–5…0	0.9…1.0
	>0	1.0
down–up	<0	1.2…1.4
	0…2	0.7…0.9
	>2	0.2…0.7

Table 11. Ventilation efficiency, ε_v

Room category	C_i [dp]	Percent dissatisfied [%]
A	1.0	15
B	1.4	20
C	2.5	30

Table 12. Indoor air quality, C_i

No.	Air source	C, [dp]
1	Mountain, sea	0.05
2	Locality, fresh air	0.1
3	Locality, mean air	0.2
4	Locality, polluted air	0.5

Table 13. Outside air quality, C_p

No.	Contaminants source	G_{oc} [olf/pers]
0	1	2
1	Adults resting, if the percentage of smokers is:	
	0 %	1
	20 %	2
	40 %	3
	100 %	6
2	Adults, if metabolic rate is:	
	reduced (3 met)	4
	medium (6 met)	10
	high (10 met)	20
3	Children:	
	children under school age (2.7 met)	1.2
	pupils (1…1.2 met)	1.3

Table 14. Contaminants quantity from the occupants, G_{oc}

The outside air flow-rate L_p, in m³/h, can be computed and function of hygienic sanitary conditions follow as:

$$L_p = \frac{P}{\left(c_{i\max} - c_{p\max}\right)\varepsilon_v} \tag{16}$$

where: P is the power of the indoor pollutant source, in mg/h; $c_{i\ max}$ – maximum admitted concentration of the most critical contaminant of room air, in mg/m^3; $c_{p\ max}$ – maximum admitted concentration of the most critical contaminant of outside air, in mg/m^3.

No.	Building destination	G_{ob} [olf/m²]
0	1	2
1	Offices	0.02...0.95
2	Schools	0.12...0.54
3	Kindergarten	0.20...0.74
4	Meeting rooms	0.13...1.32
5	Houses	0.05...0.10

Table 15. Contaminants quantity from room objects, G_{ob}

To determine the time evolution of the contaminants concentration c_i in indoor air two hypotheses are assumed:

- constant pollution in time, where the balance equation within an infinitesimal time interval dτ is:

$$L_p c_p d\tau + P\,d\tau - L_p c_i d\tau = V dc_i \tag{17}$$

where V is the room volume.
Integrating the equation (17), with the initial condition $c_i = c_p$, for $\tau = 0$, we obtain:

$$c_i = c_p + \frac{P}{L_p}\left(1 - e^{-n\tau}\right) \tag{18}$$

where n is the air exchange rate.

- instantaneous pollution at moment $\tau = 0$; consequently, the contaminant initial concentration is given by the equation:

$$c_0 = \frac{P}{V} \tag{19}$$

The balance equation for an infinitesimal time interval dτ is:

$$L_p c_p d\tau - L_p c_i d\tau = V dc_i \tag{20}$$

Integrating equation (20) with the initial condition $c_i = c_0$, for $\tau = 0$, is obtained following expression:

$$c_i = c_p - c_0 e^{-n\tau} \tag{21}$$

4.2 Computer program COMFORT 2.0

The computer program COMFORT 2.0 allows to determine the outside air flow-rate and air exchange rate for the ventilation of a room and the variation in time of contaminants concentration of room air both on the basis of the mathematical model described above and on some national norms (I5, DIN 1946), as well as to analyse the influence of different parameters on these characteristics.

- The inputs data are: geometrical characteristics of the room, in m; number of occupants; activity type; room category; outside air quality, in dp; ventilation system type; ventilation efficiency; smokers existence in room.
- The results of program are the following: outside air flow-rate; air exchange rate; polluting substances of indoor air; time variation of CO_2 concentration from room air.

4.3 Numerical application

It is considered an A category room with geometrical dimensions 10×10×2.7 m, where there are 11 persons having an intellectual activity, and smoking is forbidden. Production rate of CO_2 for the occupants is of 15 l/(h·pers) and CO_2 concentration in outdoor air has the value of 350 ppm. The floor finishing is made of parquet floor or PVC.

A comparative study for computation of outside air flow-rate and indoor air quality control according to the European Standard CEN 1752 and national norm I5 and DIN 1946 is performed using the computer program COMFORT 2.0.

The values obtained for the outside air flow-rate and the air exchange rate are reported in Table 16. In Fig. 10 is represented the variation of outside air flow-rate function of the indoor and outdoor air quality. The Fig. 11 shows the time variation of CO_2 concentration in room air.

Computational norm	Method	L_p [m³/h]	n [h⁻¹]
0	1	2	3
CEN 1752	Air quality - parquet floor	2306	8.54
	Air quality - PVC floor	8494	31.46
I5	Number of occupants	330	1.22
DIN 1946	Surface of the room	600	2.22
	Number of occupants	660	2.44

Table 13. Outside air flow-rate L_p and air exchange rate, n

From Table 13 it is to be seen that the outside air flow-rate computed according to norm I5 has the smallest value, leading to the highest values of CO_2 concentration of room air. This CO_2 concentration determines a state of strong tiredness and head aches for the occupants.

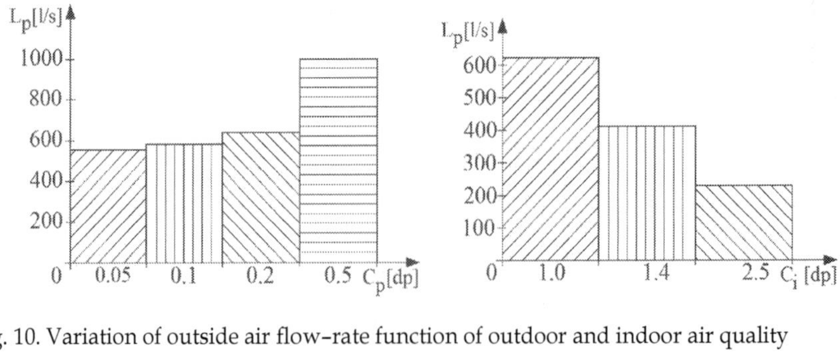

Fig. 10. Variation of outside air flow-rate function of outdoor and indoor air quality

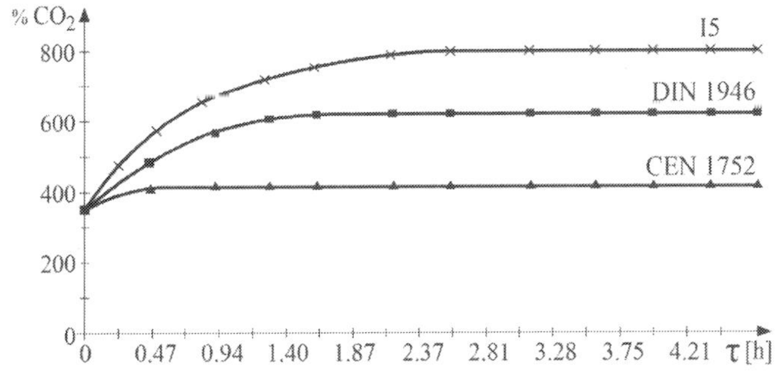

Fig. 11. Time variation of CO_2 concentration

5. Influence of carbon dioxide on human performance and productivity

Interdisciplinary efforts in the last decade of many researches converge to the necesity to find and develop the "resources" for achieving optimal environment conditions.
Although there are many weaknesses and divergents in the knowledge of environment effects on productivity and health activities, one thing is universally recognized, namely the obligativity of keeping the comfort parameters: temperature, humidity, sound level, indoor air quality, air velocity, etc. That is why VAC specialists must convince that positive effects of air-conditioning are much larger and important than negative ones.
Temperature influence is decisive both in terms of comfort and productivity. Another cause that causes discomfort associated with interior temperatures beyond the comfort limit is the dioxide carbon concentration. Manifestation of this discomfort is characteristic to densely populated rooms: theatrical and meeting rooms, public buildings (bank lobbies, counter rooms etc.), and especially to rooms from education buildings. Classrooms are usually densely populated and the length of stay is at least four hours.
It is considered that air is contaminated when the content of carbon dioxide ranges between 0.1 and 0.15%, which corresponds for a concentration of 1000 ppm and 1500 ppm. Becomes harmful to the body starting from a concentration of 2.5%. Normal air contents corbon dioxide in ranges of 0.032% and 0.035%, reaching 0.04% in urban centers.
Measurements of carbon dioxide concentration in a natural ventilated classroom through leaks had shown that in 10...15 minutes the concentration reaches values over 1000 ppm, and after 45 minutes values to 2500 ppm.
The sedentary human body approximatively releases 15 l/h of carbon dioxide. For not exceeding the maximum concentration it is has to be assured a fresh airflow rate L_p for each person:

$$L_p = \frac{Q}{c_{i\max} - c_{p\max}} = \frac{15000}{1500 - 400} = 12.5\,\text{m}^3/(\text{h} \cdot \text{pers}) \tag{22}$$

in which: Q is the CO_2 emission of human body, in cm^3/h; $c_{i\,max}$ - maximum admitted CO_2 concentration for indoor air, in ppm; $c_{p\,max}$ – maximum admitted CO_2 concentration for outside (fresh) air, in ppm.

This airflow rate can not be assured with a natural ventilation through leaks. For rooms with these destinations, the simple windows opening (intense natural ventilation) can not satisfy, because in winter time only in break time will partly solve the reduction of carbon dioxide. In summer time, even if the windows can be opened, comes out the growth temperature and noise intesity inconvenience.

In Fig. 12 is represented in a nomogram the outside air flow-rate depending by maximum admitted CO_2 concentration.

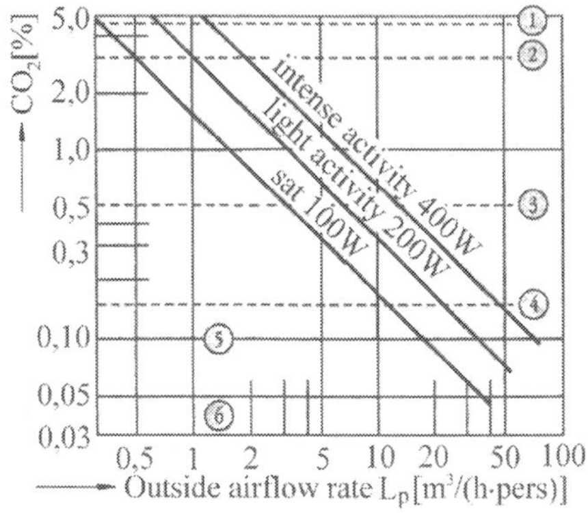

1-expired air; 2-underground rooms; 3-industrie maximum concentration; 4-offices maximum concentration; 5- Pettenkofer's number; 6-outdoor air

Fig. 12. Outside airflow rate depending by admitted CO_2 concentration

Taking into account the overlapping of contaminant, an insanitary environment, correlated with a high temperature reduces even more the human productivity.

Even if only in these two aspects (temperature and CO_2 concentration) the VAC specialists have the duty to convince over the need to assure these two parameters by using air–conditioning systems.

6. Conclusions

Computer model developed offer the possibility of detailed analyses on olfactory comfort in enclosed spaces, being of a real use for the design and research activity and in the environmental studies.

Results show that the microclimate in rooms influence not only the comfort but also the health of the occupants, and preserving the comfort parameters at the optimal values is the mission of the engineer for designing and operating of HVAC systems.

The indoor environment is influenced by the way the equipments are designed, produced and operated. That is why it is necessary to decide upon the equipments that are able to give the proper microclimate conditions, permanently observing the elimination of all secondary effects (professional illnesses) that have negative influences upon human health. It is possible at the design stage of HVAC systems and buildings to take into account most of the comfort criteria.

In addressing the relationship between building energy efficiency and IAQ, one needs to consider more than just outdoor air ventilation rates as the sole link between these critical goals. One challenge is the use of ventilation for IAQ control under conditions of poor outdoor air quality. While increased ventilation can help reduce occupant symptoms (Seppänen & Fisk, 1999), that relationship assumes that the outdoor air is clean. Ventilating with polluted outdoor air will degrade IAQ, at least in terms of contaminants that are not generated indoors.

High performance buildings should provide better IAQ conditions than exist in current buildings, and many strategies will not conflict with energy efficiency. As more experience and information is generated, the goal of truly high performance, sustainable and healthy buildings will be more fully realized in practice.

7. References

ASHRAE (2009). *Fundamentals Handbook*, American Society of Heating, Refrigerating and Air Conditioning Engineers, ISBN 978-1-933742-55-7, Atlanta, USA.

Bánhidi, L. & Kajtár, L. (2000). *Komfortelmélet*, Műegyetemi Kiadó, Budapest, Hungary.

Burge, S.; Hedge, A.; Wilson, S.; Bass, J.H. & Robertson, A. (1987). Sick building syndrome: A study of 4373 office workers, *Annals of Occupational Hygiene*, 31(4A), pp. 493-504, ISSN 0003-4878.

Busch, J.F. (1992). A tale of two populations: Thermal comfort in air–conditioned and naturally ventilated offices in Thailand, *Energy and Buildings*, vol. 18, pp. 235-249, ISSN 0378-7788.

Cain, W.S.; Leaderer, B.P.; Isseroff, R.; Berglund, L.G.; Huey, R.I.; Lipsitt, E.D. & Perlman, D. (1983). Ventilation requirements in buildings - I: Control of occupancy odor and tobacco smoke odor, *Atmospheric Environment*, vol. 17, pp. 1183-1197, ISSN 1352-2310.

CEN (1998). *Ventilation for buildings. Design criteria for the indoor environment*, Technical Report CR 1752, European Committee for Standardization, Brussels, Belgium.

Charles, S.M.; Jia, C.; Batterman, S.A. & Godwin, C. (2008). VOC and particles emissions from commercial cigarettes: Analysis of 2.5 DMF as an ETS tracer, *Environmental Science and Technology*, 42(4), pp. 1324-1331, ISSN 1520-5851.

Dear, R.; Leow, K. & Ameen, A. (1991). Thermal comfort in the humid tropics - Part I, *ASHRAE Transactions*, 97(1), pp. 874-879, ISSN 0001-2505.

DIN (1994). *Raumlufttechnik; Gesundheitstechnische Anforderungen*, VDI–Lüftungsregeln, DIN Standard 1946-2, Deutsches Institut für Normung, Berlin, Germany.

DuBois, D. & DuBois, E.F. (1916). A formula to estimate approximate surface area, if height and weight be known, *Archives of Internal Medicine*, 17(6_2), pp. 863-871, ISSN 0003-9926.

Emmerich, S.J. & Persily, A.K. (2001). *State of the Art Review of CO_2 Demand Controlled Ventilation Technology and Application*, NISTIR 6729, National Institute of Standards and Technology, USA.

Fanger, P.O. (1972). *Thermal Comfort*, McGraw-Hill, New York, USA.

Fanger, P.O. & Berg-Munch, B. (1983). Ventilation and body odor, *Proceedings of Engineering Foundation Conference on Management of Atmospheres in Tightly Enclosed Spaces*, ASHRAE, USA, pp. 45-50.

Fanger, P.O. (1987). A solution to the sick building mystery, *Proceedings of the International Conference on Indoor Air and Climate*, Institute of Water, Soil and Air Hygiene, Berlin, Germany.

Fanger, P.O. (1988). Introduction of the olf and decipol units to quantify air pollution perceived by humans indoors and outdoors, *Energy and Buildings*, vol. 12, pp. 1-6, ISSN 0378-7788.

Fanger, P.O.; Lauridsen, J.; Bluyssen, P. & Clausen, G. (1988). Air pollution sources in offices and assembly halls quantified by the olf unit, *Energy and Buildings*, vol. 12, pp. 7-19.

Frohner, I.; Bánhidi, L.; Láng, E. & Vajda J. (2004). Current questions of determining the radiation asymmetry, *Proceedings of the 13th Conference „Building Services and Ambient Comfort"*, ISBN 973-625-140-3, Timişoara, Romania, April 22-23, vol. 1, pp. 30-39.

Gameiro da Silva M. (2009). Indoor air quality simulation tool, *Rehva Journal*, 46(4), pp. 30-32, ISSN 1307-3729.

Gunnarsen, L. & Fanger, P.O. (1992). Adaptation to indoor air pollution, *Energy and Buildings*, vol. 18, pp. 43-54, ISSN 0378-7788.

I5 (1998). *Proiectarea şi executarea instalaţiilor de ventilare şi climatizare*, Normativ I5, Editura Impuls, Bucuresti, Romania.

ISO (2005). *Moderate thermal environment - Determination of the PMV and PPD indices and specification of the conditions for thermal comfort*, Standard 7730, International Organization for Standardization, Geneva.

Iwashita, G.; Kimura, K.; Tanabe, S.; Yoshizawa, S. & Ikeda, K. (1990). Indoor air quality assessment based on human olfactory sensation, *Journal of Architecture, Planning and Environmental Engineering*, vol. 410, pp. 9-19, ISSN 1340-4210.

Morey, P.R.; Hodgson, M.J.; Sorenson, W.G.; Kullman, G.J.; Rhodes, W.W. & Visvesvara, G.S. (1986). Environmental studies in moldy office buildings, *ASHRAE Transactions*, 93(1B), pp. 399-419, ISSN 0001-2505.

Molhave, L.; Bach, R. & Pederson, O.F. (1986). Human reactions to low concentrations of volatile organic compounds, *Environment International*, no. 12, pp. 167-175, ISSN 0160-4120.

REHVA (2004). Ventilation and Smoking, *Rehva Guide Book*, no. 4.

Sârbu, I. & Kalmár, F. (2002). *Energetical optimization of buildings (in Romanian)*, Matrix Rom Press, ISBN 973-685-438-8, Bucharest, Romania.

Sârbu, I. & Bancea, O. (2009). Analysis of thermal and olfactory comfort in closed spaces, *WSEAS Transaction on Heat and Mass Transfer*, 4(3), pp. 65-74, ISSN 1790-5079.

Seppänen, O.A. & Fisk, W.J. (1999). Association of ventilation rates and CO_2 concentrations with health and other responses in commercial and institutional buildings, *Indoor Air*, 9(4), pp. 226-252, ISSN 0905-6947.

Stevens, S.S. (1957). On the psychophysical law, *Psychological Review*, vol. 64, pp. 153, ISSN 1939-1471.

16

Chronic Solvent Encephalopathy in a Printing Unit for Flexible Packaging

Aida Benzarti Mezni and Abdelmajid Ben Jemâa
Department of occupational medicine Rabta hospital University of Medicine of Tunis, Tunisia

1. Introduction

The solvents are a large group of chemicals which are used industrially in the production of paints, glues, coatings, degreasing agents, dyes, polymers, pharmaceuticals and printing inks. They have three characteristics in common, namely they are volatile liquids at normal temperature, they are strongly lipophilic and most of them, if the dose is sufficiently high, affect the functioning of the central nervous system.

Exposure to organic solvents can occur by inhalation, ingestion or absorption through the skin, but inhalation is usually the main route of entry.

The health impact of workplace solvent exposure remains an issue of substantial interest and concern to occupational health professionals. In fact, there is a growing recognition that organic solvents are neurotoxic to the nervous system and its neuropsychological functions. For an agent to be neurotoxic, it must produce an adverse change in the structure and function of the nervous system following exposure to agents such as solvents.

It is generally accepted that workers exposed to solvents for a long time may develop a Chronic toxic encephalopathy (CTE). The clinical picture of CTE is dominated by vague and non specific symptoms such as irritability, headache, sleep disturbances, concentrations problems and impaired memory.

In 1985, a WHO Working Group presented diagnostic criteria and a classification for solvent-induced chronic toxic encephalopathy (CTE). In the same year, the "Workshop on neurobehavioral effects of solvents" in Raleigh, N.C., USA introduced a somewhat different classification for CTE. The objective of a systematic review of criteria for solvent-induced chronic toxic encephalopathy, was to study the diagnostic procedures that are used to establish the diagnosis of CTE, and the extent to which the diagnostic criteria and classification of the WHO, and the classification of the Raleigh Working Group, are applied. Thus a systematic search of studies on CTE was performed, and the diagnostic criteria and use of the WHO and Raleigh classifications were listed. They retrieved 30 original articles published in English from 1985 to 1998, in which CTE was diagnosed. Only two articles did not report the duration of solvent exposure. The type of solvent(s) involved was described in detail in four articles, poorly in 17 articles, and not at all in nine articles. Tests of general intelligence were used in 19 articles, and tests of both attention and mental flexibility and of learning and memory were used in 18 articles. Exclusion, by interview, of potentially confounding conditions, such as somatic diseases with central nervous effects and psychiatric diseases, was reported in 21 and 16 articles, respectively. In only six of the articles were both the WHO diagnostic criteria and the WHO or Raleigh classifications used. [1]

These organic solvents occupy an important place in most industrial sectors because of their interesting technological properties, particularly in the printing industry. In Europe, the use of solvents in printing inks represent 6% of their total industrial use.

In Tunisia, the sector of packaging and printing is an important industry. There are more than 350 companies employing about 17,000 workers.

The printing industry is one on of the sectors which expose workers to organic solvent. Some studies were performed on printing workers which are usually exposed to mixtures of organic solvents. In fact, Yu, Ignatius Tak-Sun MBBS and al showed in their study about occupational exposure to mixtures of organic solvents among printing workers in Hong Kong, that this exposure increases the risk of neurological Symptoms. It was a cross-sectional survey conducted among 762 male printing workers to assess the impacts of exposure to mixtures of *n*-hexane, toluene, isopropyl alcohol, and benzene on neurological and other symptoms. After controlling for age, smoking, alcohol drinking, past exposure history, working hours and shift work, current exposure to solvent mixtures was significantly associated with the total number of neurological symptoms and with the prevalence of specific symptoms of the nervous system and mucous membrane irritation. The adjusted odds ratio of neurovegetative lability (1.7-5.9) and memory loss (1.8) were significantly increased in the exposed group, especially when the summation index of exposure exceeded one. Finally, the health effects of low-dose occupational exposure to organic solvents remained unclear. [2]

Another study addresses the prevalence of organic brain syndrome (OBS) among long-term toluene-exposed rotagravure workers who are still working. The prevalence of OBS in 22 workers exposed to toluene for a minimum of 12 years and 19 unexposed control subjects, matched for age and employment status, was assessed with a comprehensive clinical psychiatric interview. There was a significantly greater prevalence of mild chronic encephalopathy and organic affective syndrome in the toluene-exposed group. [3]

Elsewhere, a 2-year prospective study examined the neurobehavioral effects of acute and chronic exposure to mixed solvents in workers in a screen printing business. A total of 30 subjects participated in the study in two field testings over a 12 month period. Each subject completed a detailed medical and occupational questionnaire, had a neurological examination, and underwent a battery of neuropsychological tests. Industrial hygiene investigation identified the following chemical exposures as present: toluene, methyl ethyl ketone, mineral spirits, β-ether, methylene chloride, and acetic acid. Different departments and jobs had varying degrees of exposure to these chemicals, the highest exposures being in the ink mix area and the screen washroom area. However, all exposure levels were below recommended threshold limit values. Persons categorized as having higher acute exposure demonstrated significantly impaired test performance on tasks involving manual dexterity, visual memory, and mood. Those with higher chronic exposure demonstrated significantly poorer performance on visual memory tasks and mood. Results suggest that the mixed solvents used in the screen printing industry have an effect on central nervous system functioning in the absence of obvious clinical disease. [4]

Our study consists on a descriptive epidemiological survey with 75 employees working in a printing company for flexible packaging to detect chronic neuropsychological effects secondary to the manipulation of solvents by employees.

Objectives

- Assess objectively the degree of exposure to solvents by an environmental study.
- Assess neuro-psychological effects related to chronic exposure to solvents

2. Material and methods

2.1 Material

1. **Presentation of the company**: The industrial company of packaging and derivatives is a print on flexible packaging for various food products, located in an industrial area. It has a large workshop with an area of 2000 m2 and about 8 m high divided into several units including:
   - Print Soloflex
   - Print Heliography
   - Laminating
   - Shop for storing raw materials and finished products
   - Appendices located outside the main workshop used for the preparation of a mixture of inks and accessories for cleaning printing machines.
2. **Population**: The total number of employees in this company is 75 people divided into four occupational categories: managers, technicians, skilled and versatile workers. Employees are divided into two groups: one group exposed to solvents and other non-exposed group.

2.2 Methodology

Our study consisted of two parts:

- Environmental Study
- Medical study

2.2.1 Environmental study

- The qualitative assessment: It is based on an inventory of chemicals handled. Data collection was based on canvas: Outline for the handled products and canvas for post study.
  It helped to develop a job-exposure matrix taking into account the following parameters: The exposure, the frequency of exposure and The length of the post
  This matrix has identified three risk levels (high, moderate and low).
- The Quantitative assessment:
- Sampling techniques and analysis used:
  The technical team carried out by sampling dynamic industrial pumps in the respiratory tract. The used adsorbent is activated carbon, the pumps are calibrated at a flow rate of 0.2 L / min. The average duration of each sample was 150 min. A total of 8 samples were taken at different units studied covering two days of work. The samples were analyzed by gas chromatography after desorption with carbon disulfide (CS_2). Analytes in air samples are determined by consulting the data sheets and safety are the following: ethyl acetate (TWA: 100 ppm), ethanol (TWA: 1000ppm), the methoxy propanol (TWA: 1000ppm). The values measured concentrations are compared to average values of exposure (international standards)
- Calculation of concentration index of the mixture:

In the case of exposure to several substances whose effects are the same, we must take into consideration, to assess the average exposure (TWA), the sum of the effects of each substance. For this, we must determine the concentration index of the mixture which is calculated as follows: we determine the actual atmospheric concentration of each substance divided by its average value of exposure.

To meet the standard permissible concentration index of the mixture should not exceed unity.

$$Rm = C1/N1 + C2/N2 + ... + Cn / Nn$$

Rm = concentration index of the mixture

... C1 ... Cn: actual concentration of substances N1 ... Nn ...: Standards Substance

2.2.2 Medical study

This is a transversal epidemiological study based on a set questionnaire, modelled on the Q16 usually used for the detection of neurotoxic effects in workers exposed to solvents. The questionnaire was administered by a physician to all employees. It included the following parts:

- Socio-professional employee
- The disease history and lifestyle.
- Health care consumption
- The functional neuro-psychological symptoms reported by patients

Solvent exposure was assessed by the research of direct handling of solvents or exposure to vapours from one or more positions of neighbourhoods, and our population was divided into two groups:

- *The exposed group*: those who directly handle solvents and those who do not handle solvents but are exposed to fumes from one or more positions of neighbourhoods (indirect exposure)
- *The non exposed group*: those who do not handle solvents and are not exposed to vapours from neighbourhoods posts (administrator, drivers).

The medical study included a descriptive and analytical part type exposed - unexposed.

The Statistical Analysis: Data entry and data analysis were performed by Epi Info version 6. It is essentially a comparison of actual and number of responses between the group of exposed and unexposed. For this, we used the chi 2 test with a confidence interval of 95% which corresponds to a set P (0.05).

3. Results

3.1 Qualitative assessment

- *Inventory of used products*: Solvent exposure comes primarily from printing units, laminating, preparing ink, solvent recycling and cleaning accessories for printing machines. At these units the mostly used products were: ethyl acetate, ethanol and methoxy propanol.
- *Post Study*: There are several units within the company. *(Table N° 1)*

- Flexographic Printing
- Print Heliography
- laminating
- Appendices located outside the main workshop:

- Preparation of ink mixing
- Cleaning accessories printing machines
- Solvent recycling

Product	Exposed workers	Post	Quantity	Route exposure	Conditions of work
• Buthanol • Ethanol • 1 propanol • 1 metoxy propanol • Isopropanol • Propanol • Ethoxyprop-anol	20	- Mixed ink - Flexographic and rotogravure - complexing paraffin - Flexography - heliograph	70256 Kg/ mois	- Cutaneous - Respiratory	- not wearing PPE - natural and artificial ventilation provided by windows, doors and 2 extractors
• Acétate d'isopropyle	6	- Flexography - Héliography - complexing paraffin - Mixed ink	100,148 Kg/ mois	- Cutaneous - Respiratory	- not wearing PPE - natural and artificial ventilation provided by windows, doors and 2 extractors
• Acétate d'éthyle	28	- all posts	106,148 Kg/ mois	-Cutaneous - Respiratory	-not wearing PPE unless the worker in the cleaning room accessories, wearing unsuitable masks - natural and artificial ventilation provided by windows, doors and 2 extractors above the unit flexography and laminating

Table 1. Fact handled by product

3.2 Quantitative assessment

We note that the average concentrations measured are below the requirement for the solvents tested in all units studied, but they do not reflect the reality of occupational exposure as employees of the units are exposed to a mixture of solvents, whose effects are identical, so we calculated the concentration index of the mixture. The results showed that the latter is less than 1 in all units studied. (Table N° 2, Table N° 3)

Post	Number of sampls	Average concentrations of solventsanalyzed in air (ppm)		
		Acétate d'éthyle (VME=400ppm)	Ethanol (VME=1000 ppm)	Methoxypropyl (VME=100 ppm)
Heliography Unit	2	102,3	36,6	-
flexogarphy Unit (Soloflex)	1	73,7	83,5	5,7
unit Lamination simplex	1	52,9	-	6,7
unit of ink Preparation for Heliography	1	158,6	-	-
unit of ink Preparation Soloflex	1	133,8	12,6	-
unit of accessories Cleaning	2	359,9	-	-

Table 2. Results of analysis of air samples by the workstation

Post	Index of mixture
Héliography unit	0,28
Flexography unit Soloflex	0,32
Unit of lamination simplex	0,2
Préparation unit of ink Héliogarphy	0,39
Préparation unit of ink Soloflex	0,34
Unit of accessories Cleaning	0,9

Table 3. Concentration index of the mixture per workstation

3.3 Medical study

3.3.1 Descriptive study

3.3.1.1 Socio-occupational study of population:

- **Age**: The average age is 40.6 years. More than half of the employees belong to the age group between 30 and 39 years.
- **Sex**: Sex - ratio M / F = 5.48. More than three quarters of the study population are men.
- **Grade**: Almost half the study population is not educated beyond the secondary level.
- **Marital Status**: Almost two thirds of the study population are married
- **Professional Category**: Two-thirds of the staff of this company is mainly represented by skilled and versatile workers.

- **Seniority in the workplace**: The average tenure is 13.62 years ± 9.63.
- **Exposure to solvents**: it is noted that the study population is composed of 56 persons exposed and 18 unexposed people.

3.3.1.2 Socio-occupational characteristics of exposed population

- **Average age** is 40.59 years ± 8.71.
- **Sex**: Sex - ratio M / F = 6
- **Qualification**: Most of the papers or 87.5% are skilled and versatile workers.
- **Seniority in the post**: The average tenure is 13.7 ± 9.5. Most presentations have a length exceeding 16 years
- **Exposure to solvents**: There are two modes of exposure:
- *An indirect exposure*: it is the exposure to solvent vapours from one or more neighbourhood positions
- *Direct exposure*: that is to say direct handling of solvents *(Table No. 4)*

Direct exposure relates generally employees occupying the positions of printing, laminating, preparation of inks, accessories and cleaning solvent recycling.

The distribution of exposure by frequency of exposure revealed that exposure to solvents is essentially daily in 75% of cases. Employees who are exposed for a few days a month and a few days a week does not exceed ¼ of the actual presentation is 25% *(Table No. 5)*

Exposition	Number	%
Direct exposure	38	67,9
Non direct exposure	18	32,1
Total	56	100%

Table 4. Distribution of risk by mode of exposure

Exposure frequency	Number	%
Daily	42	75,0
A few times a week	9	16,0
A few times a month	5	8,9
Total	56	100%

Table 5. Distribution of exposure by frequency of exposure

The qualitative assessment of exposure:

Matrix: Exposure Mode - Frequency: According to this matrix "exposure mode, exposure frequency" we can identify four groups of exposure. (Table No. 6)

Exposure frequency	Exposure mode	
	Direct Exposure	**Non direct Exposure**
Daily	33	9
A few times a week	3	6
A few times a month	2	3
Total	38	18

Table 6. Distribution of exposure by mode and frequency of exposure

G1: very important Exposure: 33 persons (daily direct exposure)

G2: High exposure: 12 people (direct exposure within a few times a week or daily indirect exposure)
G3: Moderate exposure: 8 persons (direct exposure within a few times a month or indirect exposure within a few times a week)
G4: Low exposure: 3 persons (indirect exposure a few times a month)
The distribution of the four exposure groups G1, G2, G3 and G4 by seniority in the workplace, we deduce the following results: (Table N° 7)

- Limited exposure to moderate with a seniority of less than 10 years: 6 workers
- Important or very important exposure with seniority less than 10 years: 18 workers.
- Limited exposure to moderate with seniority above 10 years: 5 workers.
- Important or very important exposure with seniority above 10 years: 27 workers.

These results can be summarized as three levels of risk taking into account both the three parameters discussed below:

- The mode of exposure (direct or indirect)
- The frequency of exposure
- The length of the post *(Table N° 8)*

Seniority	Low	Moderate	High	Very important
[0-5]	1	1	2	1
[6-10]	0	4	3	12
[11-15]	2	2	1	12
[16-35]	0	1	6	8

Table 7. Distribution of the four exposure groups based on post seniority

Level of risk	Number	%
High	27	48,2
Moderate	23	41,1
Low	6	10,7

Table 8. Distribution of the population exposed based on the level of risk

3.3.1.3 Clinical characteristics of the exposed population:

- **Disease history**: Three quarters of the exposed population have no previous medical history.
- **Consumer Care**: Individuals exposed to solvents consume mainly analgesics (78.6%)
- **Tobacco use**: The majority of people exposed to solvents in this business are not smokers. (62.5%)
- **Alcohol**: 7.1% of workers consume occasionally alcohol.
- **Various neuropsychiatric symptoms**: They were noted in almost the exposed patients, at least one functional neuro event. The main found neuropsychiatric complaints neuropsychiatric disorders, sleep, memory problems, headaches and physical fatigue. Headaches occur mainly at the end of workday. Are sleep disorders are dominated by insomnia that manifests itself by early morning awakening. *(Table N°9).*

We note also that most people with a neuropsychological complaint (89.3 %) belongs to level 1 or 2 *(Table N° 10)*
Furthermore, we were able to identify two types of syndromes: the acquired syndrome of intolerance to vapours of solvent organic psychosyndrome and psychosyndrome that appears most frequently *(Table N° 11, Table N° 12)*

	Number	%
Memory impairment	36	64,3
Sleep disorders	56	100
Headache	32	57,1
Weakness and fatigue	33	58,9
Tightness in chest	22	39,3
Significant weakness of the whole body	30	53,6
Feeling faint	22	39,3
Tremor exaggerated hands	17	30,4
Tingling or tingling in hands or legs	20	35,7
Sexual dysfunction: frigidity in women and decreased libido in men	24	42,9

Table 9. Neuropsychiatric manifestations in exposed workers

Level of risk	Nombre	%
Level 1	27	48,2
Level 2	23	41,1
Level 3	6	10,7
Total	56	100%

Table 10. Distribution of presentations on functional neuro symptoms by level of risk

N°	age	Sex	Post	Seniority	Level of risk
1	26	M	Recyclage des solvants	5	2
2	48	M	Printing	27	1
3	49	M	Préparation of ink	18	1
4	49	M	Cleaning of a accessoires	27	1
5	47	M	Printing	23	1
6	30	M	Cleaning of a accessoires	5	1
7	37	M	Printing	5	1
8	45	M	Cutting	27	1

Table 11. Distribution of organic psychosyndrome by level of risk and according to socio-professional caracteristics

The 8 people identified as having solvent organic psychosyndrome are male and have a minimum of 5 years of exposure. They hold positions considered most at risk. Half of these men have they consulted a doctor other than the physician work. Among the eight cases of organic psycho syndrome, four were able to benefit of an neuropsychological exploration performed by the same psychologist and it was based on battery of psychometric tests or tests of intellectual efficiency which attach to assess mental deterioration and to search the organicity by offering tests exploring: *(Table N°13)*

- memory working (span, semantic, episodic verbal, immediate visual ...)

Neuropsychological complaints	Exposed (56)	Non exposed (18)	P
Memory Disorders	36	7	S <<0,05
Headache	32	2	S <<0,05
Sleep Disorders	56	8	S <<0,05
Physical fatigue	33	2	S <<0,05
Chest tightness	22	3	S <<0,05
Sensation of weakness throughout the body	30	4	S <<0,05
Fainting	22	1	S <<0,05
Tingling tingling of the hands or legs	20	3	S <<0,05
Tremor exaggerated hands	17	2	S <<0,05
Needs to return to work after a few days off	18	1	S <<0,05
Decreased sexual desire	24	2	S <<0,05

Table 13. Distribution of exposed and unexposed workers according to the neuropsychological encountered complaints

Cases	Post	Age	Seniority	Neuropsychological Exploration
Case 1	Cleaning agent pieces	30 years	5 years	• Deficit in working memory, semantic, and episodic verbal • Deficit of executive functions • State anxiety depression
Case 2	- Score printing machine - Cleaner Parts	49 years	27 years	- Deficit in working memory, semantic, and immediate verbal episodic verbal memory, Visio spatial - Deficit of executive functions
Case 3	Preparation of inks	49 years	18 years	- Alteration of executive functions - Depression
Case 4	recycling solvents platemaking Post	26 years	5 years	- Problems with verbal episodic memory - Trouble in working memory - behavior disorder - impulsive behaviour

Table 14. Distribution of exposed workers with organic psychosyndrome according the neuropsychological tests

- attention (selective, divided)
- executive functions (conceptual development, categorization, cognitive inhibition ability ..)
- constructive and Visio spatial functions (or Rey figure memory test of 15 words of Rey)
- personality

3.3.2 Analytical study

- Matching by age, sex and age: The difference regarding age, gender and seniority to the position between the two populations (exposed and unexposed) is not significant.
- Medical history: The two populations (exposed and unexposed) did not differ regarding disease history
- Lifestyle: The difference is not significant in terms of consumption of tobacco and alcohol between the exposed and unexposed
- Use of health care: there is no significant difference regarding the consumption of care.
- Functional neuropsychological symptoms: A significant difference was found between the two groups in the neuropsychiatric manifestations (P <<0.05) (*Table N° 14*)

4. Discussion

The solvents are a family of chemicals widely used in the formulation of many products used as industrial inks. In Europe, the use of solvents in printing inks represents 6% of their total industrial use. This expansion of the printing industry was accompanied by a remarkable use of these products could have adverse consequences on the health of workers in particular neuropsychiatric effects [5]

Indeed toxic encephalopathy due to solvents is described in more than half a century it has been reported in several studies concerned with some occupationally exposed groups. However, studies are not all consistent, the differences found between exposed and unexposed are minimal and abnormalities objectified by complementary neurophysiologic investigations are inconsistent.

At the individual level, the diagnosis is difficult because there are no specific additional review.

The neurotoxicity with exceeding doses of solvents is well known for a long time, but is there a neurological syndrome solvents at low dose? Several questions: what is the reality of the risk when exposed to low dose? How to evaluate and measure exposure? There is usually poly-exposure to many products, often mixed. These mixtures are they most often involved .....

The cutaneous or respiratory penetration influences such as the evolution toward dementia. Early detection is it possible? [6]

The choice of studying of the neurological effects of solvents emanates from the importance of this effect on the frequency, severity and speed of installation [7]

Thus, we were interested in this work in search of neuropsychological effects of solvent with 75 employees of a Tunisian company of package printing.

Solvent exposure was demonstrated at first by a qualitative with a job exposure matrix taking into account three parameters: the mode, frequency and duration of exposure. This assessment was completed in a second time by a quantitative assessment.

Qualitative assessment:

The inventory of used products was based on analysis of safety data sheets which revealed several types of solvents handled by the employee as follows: alcohol (methoxy-propanol ...) ethyl acetate ... ..

The study of work stations and interrogation of employees has allowed us to classify the workers: directly exposed, indirectly exposed and not exposed to solvents. Concentrations recorded by atmospheric monitoring for the following solvents: ethyl acetate, ethanol and iso-propanol was lower than the TWA and the same for the concentration of the mixture index, despite the absence of individual and collective preventive measures. This reduces the interest of regularly controlling atmospheric monitoring for the exposed workers. [7]

Socio-professional characteristics:

Our study population consists of 75 employees working for at least a year. It is a predominantly male population (sex ratio M / F = 5.48) with secondary mean age (40.6 years) and secondary education for the majority (47.3%). On a professional level, the majority are specialized or multipurpose workers (66.2%) with average seniority at the workplace (13.62 years).

Exposure to solvents:

Solvent exposure was assessed in our population according to the workstation. Two groups were identified: an exposed group (75.7%) and an unexposed group (24.3%). The exposure was assessed by establishing a matrix of exposure taking into account three parameters: the mode, frequency and duration of exposure and defining three exposure levels: high, moderate and low. This method was inspired by the method developed by the National Research and Safety (INRS) for the assessment of chemical hazards in the workplace. The majority of submissions (89.3%) have an exposure level considered high or moderate.

Both groups exhibited comparable regarding age, gender and seniority to the position and disease history.

Health care consumption was assessed by the number of medical consultations over the last 12 months and by taking regularly neuropsychic medication.

Almost two thirds of the exposed have consulted at least one doctor or specialist, the number of consultations is higher than that seen in unexposed but the difference is not significant as well as consumption of drugs that are basically painkillers.

Teyssier-cottec et al. in their study looking for witnesses of early achievement of the central nervous system by solvents and conducted among a population of 75 employees divided between the exposed and unexposed to solvents, found no significant difference between these two groups with regard to medical history and use of care [8].

Moreover, both exposed and unexposed groups are comparable in terms of tobacco and alcohol. Indeed the majority of exposed and unexposed are non smokers and non drinkers.Other studies have shown a greater tendency to smoking and / or consumption of alcohol by the solvent-exposed. Hogsted C showed through a meta-analysis that there is a higher consumption of alcohol by solvent-exposed [9].

Similarly, evaluation of effects of exposure to organic solvents at low doses on color vision in the paint industry has shown a homogeneous group of exposed and controls for alcohol consumption by cons Tobacco consumption seems to be higher among exposed conversely to what showed by the study directed by Teyssier and collaborator. [6,8].

Neuropsychological symptoms:

Epidemiological studies on solvents neurotoxicity date from about twenty years. They define a particular area of occupational epidemiology: neurotoxicological epidemiology. These studies have highlighted particular problems reaching cognitive function [10 13].
According to Hogstedt, the first cross-sectional study in this field was published in 1955 by Gran jeans and showed a relatively high frequency of subjective complaints of impairment of the autonomic nervous system and neurological and psychiatric symptoms among workers chronically exposed to trichlorethylene in concentration between 20 and 80 ppm. [9].
Lauwerys et al report that Scandinavian authors are the first to report that chronic exposure to vapors of various solvents can cause brain syndrome characterized by mental retardation and emotional disturbances which can lead in severe cases fortunately exceptional to dementia by brain atrophy (psycho-organic syndrome). Since then, literature has been enriched by studies (case reports, control, case control), who investigated the neurological effects associated with chronic exposure to solvents[14].
The terminology varies: "solvent induced psycho-organic syndrome (POS)", "organic solvent syndrome (PSO), chronic toxic encéphalopaty (CTE)" organic solvent encephalopaty (OSE), psycho-organic syndrome caused by solvents (SPO) "psycho-organic solvent syndrome (POS).
The existence of this syndrome is still controversial, however it is accepted by most authors.
Solvents can act even in small doses on the central and peripheral nervous system, initially due to their high liposolubulity and then intervene by various mechanisms such as increases in membrane fluidity that change the flow not only intra-and extracellular nutrient but also other toxic. Some solvents may also be metabolized in the nerve cells and cause structural and biochemical changes such us cellular destruction. Others act after hepatic metabolism and the passage of metabolites beyond the blood-nerve. [14]
Finally some solvents will affect the neuro-chemical mediators. The central effect of dopamine has been demonstrated in the phenomena of addiction to tobacco and other drugs likely to solvents. ... Indeed, there are about fifteen years, researchers have realized that all drugs that cause addiction in humans have in common the release a neurotransmitter called dopamine, which affects the "reward circuit". This circuit serves as a "barometer" for human to know at any moment his mental and physical condition.
The reward circuit is composed of different brain structures in relation (prefrontal cortex, septum, amygdala and hippocampus). These four structures are linked with each other. They receive dopaminergic terminals from the ventral tegmental area and they filter the information before sending them to the area that is the ultimate regulating hypothalamus. The hypothalamus receives only information previously processed by the structures of the reward circuit knows nothing of what was happening outside. The satisfaction felt by the individual is related to dopaminergic activation, regardless of the reality of its environment.
Dopamine release activates this reward circuit, generating a state of pleasure and satisfaction. It follows a change in the central nervous system that affects the overall balance and the relationship of the individual to his environment.
Dopamine release may occur several times per second. It is important that each train has a sense of liberation from the binding of the neurotransmitter on the postsynaptic receptor. To make this system work it requires that the body removes the neurotransmitter that has been issued. To do this, a process of reuptake into the neuron transmitter draws 80 to 85% of this neurotransmitter. [14]

It is certain that the adaptability of the nervous system is so great that the first changes are insensitive to the investigations of doctors.

It remains to define clearly the psycho-organic syndrome caused by solvents and its potential stages and identify harmonized and accurate diagnostic criteria.

In 1985, classification in 3 stages and diagnostic criteria were introduced by a joint working group of WHO and the Nordic Council of Ministers. The same year a slightly different classification was proposed by a working group in Raleigh, USA

According to WHO the organic psycho syndrome is the mildest form of chronic toxic encephalopathy to organic solvents, it evolves over several days or several weeks, but does not leave scars if exposure is stopped.

According to most experts, the diagnosis of an SPO is based on:

- Professional et extra-professional anamnestic data,
- Neuropsychological and neurobehavioral testing: different batteries are used to evaluate the functions and psychomotor cognitives including battery NCTB (Neurobehavioral core test battery) proposed by WHO to monitor neurotoxic professionals effects. It has been developed since 1983, includes a coherent set of readily available tests. It is validated in many countries. It is independent of the cultural heritage, gender, age and allows comparison between studies.
- Additional tests

Currently, several authors emphasize the harmonization of diagnostic procedures using standardized methods. Similarly, a precise definition of this syndrome is necessary. Meanwhile questionnaires can help doctors in their approach and include:

- The Q16 (Swedish) (16 questions) developed by Hogstedt and al in 1984 to detect groups of subjects exposed to neuro-toxic agents and individuals at risk for further consideration.
- Q18: Q16 amended
- EuroQuest (european questionnaire) developed in 1992 by the EURONEST network to enable comparison of neuropsychiatric symptoms between epidemiological studies. It is the fruit of reflection carried out by researchers in toxicology from the European community developed initially in English and was then translated in ten languages. It includes 83 items exploring three areas: neurological and psychosomatic symptoms (impaired memory, concentration, sleep and alertness, symptoms of acute poisoning, symptoms related to personality disorders and mood) [14].

Some authors emphasized the usefulness of Q16 used as a tool for diagnosis and control of the central nervous system dysfunction [5,11]. In our study, the questionnaire was inspired from that of Q16.

In our work all employees exposed to solvents reported at least one functional neuro event and a significant difference was found with the unexposed group ($P << 0.05$). This is consistent with the results published in the literature [16 - 19]

Our study showed that the number of people with at least a neurologic complaint increases with the level of risk.

The relationship between the degree of exposure and the importance of neuropsychiatric events was suggested in several studies [17-18,21-22] which joins the results of the study CNARAT and another conducted by Langau on 349 women working in the footwear industry and exposed mainly to n-hexane and toluene [23].

Our study has identified two neuropsychiatric syndromes:

- the acquired intolerance to solvent vapors
- The psycho-organic syndrome.

Acquired intolerance to solvents, which is a part of the syndrome of intolerance to chemical odor has been described by Gyntelberg in 1986. [7]

The syndrome has been selected to the following criteria:

- Having frequent headaches at the beginning of a working day or one to two hours after the onset of labor.
- Feeling of weakness of all the body with difficulty performing simple tasks. That it occurs even without physical effort or only during labor.
- Feeling faint sometimes. This sensation occurs even without physical effort and / or only during labor.

In our study, we observed two cases of intolerance syndrome to solvent vapors among expose group with a non significant difference between exposed and unexposed. Prost G, reported five cases of Acquired intolerance to solvents syndrome identified in various industries [24] Grimmer studied 30 cases of Acquired intolerance to solvents with a mean age of 38 years and with a slight male predominance [25]. The symptoms found in these studies are similar to those described in our patients.

A feature of the acquired intolerance to solvent syndrome is that the complaint is primarily against certain smells that are no longer supported and that trigger disorders and occur in the workplace and which disappear or improve over the weekends or work stoppages.

In our study, two patients with this syndrome has been observed had a high level of risk, but according to the literature this syndrome was observed in case of rather low exposure [25]

The diagnosis of organic solvent syndrome was selected to the following criteria:

- By returning often to see if the door is closed
- Having frequent headaches
- Having trouble sleeping
- Have the feeling of being constantly tired
- Have the feeling of weakness of all the body with difficulty performing simple tasks of decreased libido

In the literature, several studies have focused on the organic solvent psycho syndrome. [27-31]. Baker identifies 13 new cross-sectional studies since 1985 involving [20] different countries and the results converge to the fact that this syndrome affects the short-term memory and psychomotor functions.

Eight employees in our study had a solvent organic psychosyndrome (14.3% of the exposed population).

The confirmation of organic disease of the nervous system is a psychometric assessment including a battery of psychometric and clinical tests to assess the intellectual efficiency, as the battery NCTB (Neurobehavioral Core Test Battery) proposed by WHO for monitoring occupational neurotoxic effects, developed since 1983, whose contents were described by Regane in its study "interest of psychometric tests for the detection, of the neurobehavioral toxic effects of organic solvents. [21-30]

- Coherent set of easily usable tests
- Battery test paper and pencil
- Validated in many countries
- Independent of the cultural heritage, gender, age

- Comparison studies
- Easy Implementation
- Time for total shift is about 1 hour

In the literature, the evolution of organic solvent psychosyndrome is marked by a reversible lesion if the withdrawal is sufficiently early. By cons, from stage 2B, the prognosis is poor [5,27]. Moreover, a highly significant relationship between the organic solvent psycho syndrome and exposure to solvents has been revealed in the study by Soborg in 1993 [17]

5. Conclusion

The solvents occupy an important place in the sector of printing for flexible packaging but also in most other industrial sectors. Our study consisted of two parts: An environmental assessment and a medical section which consisted of carrying out a transversal epidemiological study based on a predetermined questionnaire, inspired of Q 16 usually used for the detection of neuro-psychological effects in workers exposed to solvents.

It appears from our study that the preventative measures are absent, analysis of safety data sheets revealed the presence of several types of solvents in the composition of inks and glues. Although we note that the average concentrations measured are below the requirement for the solvents tested in all units studied. Our study population is relatively young (40.6 years) male (sex ratio M / F = 5.48) with average tenure of 13.62 years. Neuropsychiatric complaints are frequent (100%) and significantly higher in exposed group with a very significant difference (<<0.05). The medical study identified two cases with the syndrome of acquired intolerance to solvent and eight cases of solvent organic psychosyndrome. Four cases with solvent organic psychosyndrome have received a neuropsychological exploration.

The gap between the assay results of atmospheric standards and the frequency of neuropsychiatric manifestations in exposed group incite us to reflect on the role of low doses in causing toxic effects.

Finally, in Tunisia as the medical reparation of neuropsychological damage remains difficult due to the lack of tables of occupational diseases, strengthening the technical and medical prevention is the only way that can reduce chemical risks and the socio-occupational-related neurological solvent exposure.

6. References

[1] A. Joffrey, F. van der Hoek, M. Maarten, Verberk and Gerard Hageman. Criteria for solvent-induced chronic toxic encephalopathy: a systematic review. Int Archiv occup Environ health 2000; 73:362-8

[2] Yu, Ignatius Tak-Sun MBBS, MPH; Lee, Nga Lan MPhil; Zhang, Xin Hua MD, PhD; Chen, Wei Qing MB, PhD; Lam, Yik Tsz MB, MMedSc; Wong, Tze Wai MBBS, MSc. Occupational Exposure to Mixtures of Organic Solvents Increases the Risk of Neurological Symptoms Among Printing Workers in Hong Kong. Journal of Occupational & Environmental Medicine 2004; 46: 23-330

[3] Larsen, Frode, Leira, Håkon Lasse. Organic Brain Syndrome and Long-term Exposure to Toluene: A Clinical, Psychiatric Study of Vocationally Active Printing Workers. Journal of Occupational Medicine 1988; 30: 875-8

[4] Roberta F., White PhD., Susan P., Proctor DSc., Diana Echeverria PhD., Janelle Schweikert PhD., Robert G., Feldman MD. Neurobehavioral effects of acute and chronic mixed-solvent exposure in the screen printing industry. American Journal of Industrial Medicine1995; 28: 221-31.

[5] Boillat MA. Troubles mentaux organiques d'origine toxique. Encycl Med chir, toxicologie, pathologie professionnelle 2004 ; 16-536-A70 : 5p 1

[6] Conso F, Labbe N, Yahoum A, Le syndrome psycho organique aux solvants : expérience d'une consultation de pathologie professionnelle. Arch Mal Prof 1995 ; 56 :142-44.

[7] Gyntelberg F ; verterlang, Fog P, Isâges H et Coll. Aquired intolerance to organic solvants and results of vestitular testing. Am J industr Med 1986.

[8] Teyssier-Cotte C, Mandereau L, Heman D, Moulin T, Niezborala M, Angomin M, Lamant J.M, Levyl P, Le pors. Recherche de témoins d'une atteinte précoce du sytème nerveux central par les solvants. XIIIème journée de médecine de travail, Besançon 1994 ; 138-40.

[9] Hogstedt C, Lundberg 1, Epidémiologie des risques professionnelle neurocomportementaux : expériences méthodologiques tirées de la recherche sur les solvants organiques. Arch Mal prof 1992 ; 53 : 499-506.

[10] Abecassis J.C, Bertrand A, Thomas Alliel S, Lulinshi F. Recherche de situation de l'exposition à des doses faibles de solvants dans le BTP, Arch des maladies professionnelles 1995 ; 56 : 127-28.

[11] Dally S, Garnier R Neurotoxicité des solvants à faibles doses d'exposition. Solvants et neuropathies : données cliniques. Arch des maladies professionnelles 1995 ; 56 : 101-04.

[12] Nayrat C, Paris C, Proust B, Creau Y, Nisse C, Ragazzini L, Cailard J.F. Exposition professionnelle à de faibles doses de solvants et effets neuropsychiques : une enquête transversale.

[13] Boillat MA. Troubles mentaux organiques d'origine toxique. EMC toxicologie-pathologie octobre 2004 ; 1 : 178-85. Document consulté sur internet le 16/02/2007

[14] Lawrys R, Hanfroid V, HOET P, LISON D. Toxicologie industrielle et intoxication professionnelle; 5ème édition 2007 ELSEVIER Masson.

[15] Ihring A, Triebig G, Dietz MC. Evaluation of a german version of the Q16 questionnaire for neurotoxic symptoms in workers exposed to solvents.

[16] Abecassis JC. Neurotoxicité des solvants à faibles doses d'exposition. Problèmes pratiques non rencontrés par me médecin du travail. Arch mal Prof 1995 ; 56 : 88-89.

[17] Chouanière D, Fontana JM, Wild P. Troubles neuro comportementaux liés à l'exposition prolongée au toluène. Arch Mal Prof 1998 ; 59 : 330-32.

[18] Chouanière D, Nenan D. méthodes épidémiologiques pour l'étude de la neurotoxicité des solvants ; Arch Mal Prof 1995 ; 56 : 144-22.

[19] De Haro L, Jouglard J, Jahjah F, Sainty JM, Chave B, Cura B. Psycho syndrome aux solvants et recherche du syndrome d'apnée du sommeil. Arch Mal Prof 1995 ; 56 : 145-147.

[20] Baker EL. A review of recent research on health effects of human occupational exposure to organic solvent. J Occup Med 1994 ; 36 : 1079-92.

[21] Soborg P.A, Bruhn P, Gyldensted C, Melgaard B. Chronic painters syndrome. Chronic toxic encephalopathy in house painters. Arch Neurol Scand 1979 ; 60 : 149-56.

[22] Eller N, Netterstorm B, Laursen P. Risk of chronic effects on the central nervous system at low toluene exposure. Occup Med 1999 ; 1 : 389-95.
[23] Langauer-Lewowicka H, Wocka- nareck T, Braszczynoker Z et coll. Correlation of psychological and neurological changes with indicators of exposure of workers in a shoe factory to glue solvents. Med Pr 1983 ; 34 : 397-04.
[24] Prost G, Tell JP, Beget A, Davezies P, Normand JC, Intolérance acquise aux solvants, Discussion pathologisue à propos de cinq cas. Arch Mal Prof 1992 ; 53 : 369-70.
[25] Grimmer A, Gerant C, Dupas D, Conso F. Syndrome d'intolérance aux douleurs chimiques : A propos de 30 cas. Arch Mal Prof 1995 ;56 :149-50.
[26] Reygane A, Fayada C, garnier R. Intérêt des tests psychométriques d'une batterie de psychologie du travail pour le dépistâge des effets toxiques neuro comportementaux des solvants organiques. Arch Mal Prof 1995 ; 56 :150-52.
[27] Dryson Ew, Ogden JA. Organic solvent induced chronic toxic encephalopathy : extent of recovery and associated factors, following cessation of exposure. Neuro Toxicology 2000 ; 21 :659-66.
[28] Juntunen J, Anti-Poika M, Tola S, Partanen T. Clinical prognosis of patients with diagnosed chronic solvent intoxication. Acta Neurol Scand 1982 ; 65 :488-03.
[29] Dick FD. Solvent neurotoxicity. Occup. Environ. Med. 2006 ; 63 : 221-26.
[30] Nijem K, Kristensen P, Thorud S, Al-Khalib A et Coll. Solvent exposures at shoe factories and workshops in hebron city, West Bank. Int J Occup Environ Health 2001 ; 7 : 182-88.
[31] Chenk R, Dick F, Seaton A . Health effects of solvent exposure among dochyard painters : mortality and neuropsychological symptoms. Occup Eviron. Med. 1999 ; 56 : 383-87.
[32] Normand JC, Gabet V, Davezies P, Chevallier V. Encéphalopathie toxiques chroniques aux solvants (1er cas français). Arch Mal Prof 1990 ; 51 :135-36.
[33] Lafores LJC, Falcy M. Neurotoxicité des solvants à faibles doses d'exposition. Classification. Arch Mal Prof 1995 ; 56 : 89-91.
[34] Moulin T, Cohe L. Neurotoxicité des solvants à faibles doses d'exposition. Diagnostic différentiel. Arch Mal Prof 1995 ; 56 : 104-09.
[35] Rouch I, Wil P, Fontana JM, Chouanière D. Evaluation of the french version of euroquest. Neurotoxicology, 2003 ; 24 :4-5 : 541-546.

17

Indoor Air Pollutants and the Impact on Human Health

Marios. P. Tsakas, Apostolos. P. Siskos and Panayotis. A. Siskos
Laboratory of Environmental Chemistry, National and Kapodestrian University of Athens
Greece

1. Introduction

The major area of public concern and government policy, in terms of the impact of air pollution on human health, continues to be outdoor air. However, over the last two decades, indoor air quality (IAQ) has caused increasing concern due to the adverse effects that it may have on human health. The term "indoors" is used in relative literature to refer to a variety of environments, including homes, workplaces, and buildings used as offices or for recreational purposes. In addition, a number of studies have been carried out to measure various compounds inside vehicles during commuting activities. Most people in the developed world spend up to 90% of their time in an indoor environment and up to 60% of the workforce work in an office. (Tsakas & Siskos, 2010; McCurdy et al., 2000; Ashford & Caldart, 2008; Andersson & Klevard Setterwall, 1996) Decreased ventilation rates for energy conservation, along with increased use of synthetic materials in buildings, have resulted in increased health complaints from building occupants (Siskos, 2003). Many indoor pollutants are either known, or suspected to be, allergens, carcinogens, neurotoxins, immunotoxins or irritants, while all may contribute to sick building syndrome (SBS). The set of health symptoms associated with SBS includes nasal, ocular and generalised diseases. According to various studies performed in public buildings by the National Institute of Occupational Safety and Health (Soldatos et al., 2003), the three most significant symptoms that were experienced in more than 70% of the buildings are dry eyes, dry throat and headaches.
The IAQ and the presence of air pollutants in indoor environment is a worldwide issue, since many governments and environmental institutes have faced this serious phenomenon. Starting in the 1990s in Japan, tightly sealed buildings with low ventilation rates have been constructed. This, combined with the use of some new types of building materials has often resulted in IAQ problems. Many inhabitants suffering with SBS and multiple chemical sensitivity (MCS), have been reported (McCurdy et al., 2000; Zhang & Niu, 2004). As a result the Japanese Ministry of Health, Labour and Welfare have introduced indoor air guidelines for a range of VOCs including HCHO based on hazard assessments (Shinohara et al., 2009).
The importance of IAQ has also been recognised in Europe and has been identified as an important element within the European Collaborative Action (ECA) (ECA, 1998) and the European Environment and Health Action Plan (Dimitroulopoulou et al., 2006). In America, the State of California has adopted an active programme for the last two decades aiming to the reduction of indoor air pollution, which has led to a range of policy instruments (Waldman & Jenkins, 2004). Over recent years, important steps have been made towards

setting IAQ standards and guidelines in the UK (Dimitroulopoulou et al., 2006) and recently, the UK Department of Health Committee on the Medical Effects of Air Pollutants (Short, 2001), launched a guidance document on the effects of indoor air pollutants. In many developing countries, exposure to indoor air pollution causes a major health burden (Committee on the Medical Effects of Air Pollutants, 2004). Increased concern regarding indoor air quality especially in the last two decades has led to a number of studies and meetings on the subject. For example, in Greece many researchers have conducted significant studies about IAQ issue (Siskos & co-workers, 2001, 2003, 2005, 2010; Helmis & co-workers, 2007, 2009; Santamouris et al., 2001). With increasing concern in relation to health effects, in recent years the problem has come into sharper focus. Additional new sources of contaminants are being introduced, which haven't been measured before. In several countries, studies have been undertaken, in some cases involving comprehensive investigations of the factors governing air quality, so that effective control measures ranging from the setting of minimum ventilation standards, to controlling, or even banning, certain products such as urea-formaldehyde foam insulation or unvented paraffin or gas heaters. It is nevertheless recognized that some of the responsibility for maintaining acceptable and healthy indoor air quality will continue to rest with building owners and occupants of buildings.

Some studies have revealed a variety of contaminants of indoor air including odorous, non-odorous gases and vapours, and particles, and although there were suggestions that some of these contaminants could be responsible for health effects, proving causal relationships is exceedingly difficult even where elevated levels of potentially toxic substances exist (World Health Organization [WHO], 1989; Perry & Kirk, 1986; WHO, 1986; Priorities for Indoor Air Research and Action, 1991).

2. Indoor air pollutants and their sources

There are many indoor air contaminants, which can be separated based on their effects on human health, the frequency of their appearance, their usual concentration levels, their sources etc. This chapter is focused, primarily, on those species common to indoor and outdoor air environments and those who are measured more often in indoor environments.

2.1 Radon

The main source of indoor radon is its immediate parent radium-226 in the ground of the site and in the building materials (Nero, 1988, 1989). Outdoor air also contributes to the radon concentration indoors, via the ventilation air. Tap-water and the domestic gas supply are usually radon sources of minor importance, with a few exceptions. In most situations it appears that elevated indoor radon levels originate from radon in the underlying rocks and soils (Castren et al., 1985). This radon may enter living spaces in dwellings by diffusion or pressure driven flow if suitable pathways between the soil and living spaces are present. It should be noted, however, that in a minority of cases elevated indoor radon levels may arise due to the use of building materials containing high levels of radium-226. Examples of such materials, used in some buildings, are by-product gypsum, alum shale and volcanic tuffs.

The United Nation Scientific Committee on the Effects of Atomic Radiations (UNSCEAR) has made a very simple model to try to estimate the relative contribution of these sources: for a "typical" house, with a radon concentration of 50 Bq/m^3 at ground floor, the contributions of soil, building materials and outdoor air are, respectively, 60%, 20% and

20%, while for the upper floors in high rise buildings, where the radon concentration-is estimated to be "typically" 20 Bq/m^3, these values become 0%, 50% and 50% (UNSCEAR, 1993).

2.1.1 Soil

For those who live close to the ground, e.g. in detached houses or on the ground floor of apartment buildings without cellars, the most important radon source is radium in the ground.

The radium concentration in soil usually lies in the range 10 Bq/kg to 50 Bq/kg, but it can reach values of hundreds Bq/kg, with an estimated average of 40 Bq/kg (UNSCEAR, 1993). Typical radon concentrations in soil gas range from 10000 Bq/m^3 into 50000 Bq/m^3. The potential for radon entry from the ground depends mainly on the activity level of radium-226 in the subsoil and its permeability with regard to air flow. Example of terrains with a high radon potential are alum shales, some granites and volcanic rocks, due to high concentrations of radium-226 and the presence of eskers (gravel, sand and rounded stone deposited from subglacial streams during the ice ages), all these being characterised by high permeability. The ground could also be contaminated with waste tailings from uranium or phosphate mining operations with enhanced activity levels (Tyson et al., 1993).

The ingress of radon from the soil is predominantly one of pressure-driven flow, with diffusion playing a minor role (de Meijer et al., 1992). The magnitude of the inflow varies with several parameters, the most important being the air pressure difference between soil air and indoor air, the tightness of the surfaces in contact with the soil on the site, and the radon exhalation rate of the underlying soil. If there is no airtight layer between the basement and the ground, the underpressure indoors causes radon to be drawn in from the ground under the building. Underpressure occurs in most houses if either the adjustment of inlet and outlet of air in forced ventilation systems or the outdoor air supply for vented combustion appliances is inappropriate. The underpressure may be considerable for all types of ventilation systems when the inlet air is restricted too much. The tightness of the structures has to do with e building regulations and techniques and is very dependent on cracks, openings and joints. Structures are hardly ever so airtight that radon inflow is completely prevented. For example, to get a radon daughter concentration of less than 100 Bq/m^3 EER in a house with a volume of 500 m^3 and a ventilation rate of 0.5 air changes per hour, not more than 1 m^3 per hour must be allowed to leak into the house if the radon gas concentration in soil air is about 50000 Bq/m^3. Such values are quite typical.

2.1.2 Building materials

Building materials are generally the second main source of radon indoors, while in the Seventies they were considered the principal one (UNSCEAR, 1977; Meyer et al., 1986). Radon exhalation from building materials depends not only on the radium concentration, but also on factors such as the fraction of radon produced through material release, the porosity of the material and the surface preparation and finish of the walls. In general, no action needs to be taken concerning traditional building materials. Typical values for radium and thorium content in building materials are 50 Bq/kg or less (Nuclear Energy Agency Organisation for Economic Co-operation and Development - NENOECD, 1979). Building materials containing by-product gypsum (UNSCEAR, 1982) and concrete containing alum shale (Swedjemark & Mjones, 1984) may have much higher radium concentrations. The activity concentrations in brick and concrete may also be high if the raw materials have been taken from locations with high levels of natural radioactivity. Examples of such natural materials, used in some buildings, are volcanic tuffs and pozzolana (Sciocchetti et al., 1983; Campos Venuti et al., 1984; Battaglia et al., 1990), where radium and thorium content can reach some hundreds of Bqlkg. Other measurements of radioactivity content and exhalation of building materials are reported in NENOECD (1979).

Building materials are the main sources of radon-220 (also called "thoron") in indoor air. Due to its short half life (55 s), thoron originating in soil in effect is usually prevented from entering buildings and therefore makes negligible contribution to indoor thoron levels. For this reason and due to the greater difficulties of measurement, thoron concentration measurements are very much fewer than those for radon. Although the indoor thoron concentrations are usually low (Cliff, 1992; UNSCEAR, 1993), in some cases the doses due to this isotope and its daughters are significant and comparable to those due to radon- 222 (Sciocchetti et al., 1983, 1992; Guo et al., 1992; Bochicchio et al., 1993; Doi & Kobayashi, 1994).

2.1.3 Outdoor air

Outdoor air usually acts as a diluting factor, due to its normally low radon concentration, but in some cases, as in high rise apartments built with materials having very low radium content, it can act as a real source. The radon concentration in outdoor air is mainly related to atmospheric pressure, and (in case of non-perturbative weather) it shows a typical oscillating time pattern, with higher values during the night.

Until a few years ago the average level of radon gas concentrations in the atmosphere at ground level was, in most cases, assumed to be of the order of few Bq/m^3 -e.g. in the range of 4 to 15 Bq/m^3 in USA (Gesell, 1983), but more recent measurements seem to indicate higher values, reaching some tens of Bq/m^3 (Hopper et al., 1991; Robé et al., 1992; Bochicchio et al., 1993; Deyuan, 1993; Grasty 1994; Price et al., 1994). Quite high radon concentrations in the outdoor air have been reported near substantial radon sources, such as mine tailings (Tyson et al., 1993), or in the case of particular weather conditions, such as thermal inversion or very low precipitation (Grasty, 1994).

Ambient air over oceans has very low values (~ 0.1 Bq/m^3) of radon concentrations, due to the minimum presence of radium in the sea water and the high solubility of radon in water at low temperatures. Therefore radon concentration in outdoor air of islands and coastal regions is generally lower than in continental countries, e.g. United Kingdom and Japan have an average outdoor air value of ~4 Bq/m^3.

Taking into account recent measurements, the mean value of outdoor radon concentrations adopted by UNSCEAR in its last report has been changed from 5 to 10 Bq/m^3 for continental areas and somewhat less in coastal regions (UNSCEAR, 1993).

2.1.4 Tap water

In wells drilled in rock the radon concentrations of water may be high. When such water is used in the household, radon can be partially released into the indoor air, causing an increase in the average radon concentrations. In a few regions, such as Finland and Maine (USA), the tap water from wells drilled in rock has been shown to contribute significantly to radon concentrations indoors. Radon concentrations in tap-water from deep wells can range from 100 kBq/m^3 to 100 MBq/m^3 (UNSCEAR, 1988). The indoor radon concentrations in these regions may already be high due to high rates of radon entry from the ground. The world average radon concentration in all types of water supplies is assumed to be 10 kBq/m^3 (UNSCEAR, 1993).

2.1.5 Domestic gas

In some regions, natural gas used for cooking and heating contains elevated concentrations of radon, which is released on combustion. Normally this source is insignificant, and can be monitored at transmission and distribution points. Typically the radon level in natural gas is about 1000 Bq/m^3. Natural gas, as it is usually supplied, contains gas from a number of wells and fields and thus can vary over time, depending on the proportions supplied by different sources (UNSCEAR, 1993).

2.2 Oxides of nitrogen

2.2.1 NOx

A large number of studies of NO and NO_2 have been carried out in many different indoor air environments (Finlayson-Pitts, 1999; Pitts et al., 1985). Because of air exchange, indoor levels are generally higher when outdoor levels increase (Hoek et al., 1989; Rowe et al., 1991; Hisham & Grosjean, 1991; Spengler et al., 1994; Weschler et al., 1994; Baek et al., 1997). However, enhanced indoor levels can be found when combustion sources are present. These include gas stoves, paraffin heaters, water heaters, and cigarette smoke (Wade et al., 1975; Marbury et al., 1988; Ryan et al., 1988; Petreas et al., 1988; Hoek et al., 1989; Pitts et al., 1989; Spengler et al., 1994; Levy et al., 1998). While combustion generates primarily NO, the focus indoors has been on NO_2 because of its health impact. Again, the use of gas stoves was highly correlated with indoor NO_2, with an indoor/outdoor concentration ratio of 1.19 for homes with a gas range compared to 0.69 for those without a gas stove. The ratio was even higher for homes with a paraffin space heater, 2.3 compared to 0.85 without such a heater (Levy et al., 1998). Both the indoor and outdoor concentrations of NO_2 were higher in cities where at least 75% of the homes had gas stoves; for example, the mean outdoor NO_2 concentration in such gas-intensive cities was 38 ± 20 ppb, compared to 14 ± 6 ppb in cities where fewer than 25% of the households had gas stoves installed. High concentrations of NO_2 have also been measured in indoor skating rinks where the use of ice resurfacing machines powered by propane, gasoline, or diesel fuel results in significant emissions (e.g., Brauer & Spengler, 1994; Brauer et al., 1997; Pennanen et al., 1997). Mean concentrations of NO_2 of ~200 ppb have been reported, with some rinks having concentrations up to 3 ppm! The indoor-to-outdoor ratios of the arithmetic mean concentrations varied from about 1 to 41, with an overall mean of 20. In the absence of such sources of NOx, indoor and outdoor concentrations are quite similar (Weschler et al., 1994), since removal of NO and NO_2 indoors, e.g., on surfaces, is relatively slow. However, as it has been discussed shortly, although the surface reaction of NO_2 is relatively slow, it is still of interest since it generates nitrous acid (HONO). Different surfaces found inside homes have been found to have different removal rates for NO_2. In short, there is a variety of evidence that there are higher levels of NO_2 indoors when combustion sources are present and that the concentrations generated indoors can be quite substantial in some circumstances. One word of caution is in order, however, particularly in regards to earlier measurements of NO_2.

2.2.2 HONO and HNO_3

HONO is formed by the reaction of NO_2 with water on surfaces. The reaction is usually represented as

$$2NO_2 + H_2O \rightarrow HONO + HNO_3 \quad (1)$$

Although the detailed mechanism is not known; gaseous HNO_3 is not generated in equivalent amounts, something which has been attributed to its remaining being adsorbed on the surface. This overall reaction occurs on a variety of surfaces in the laboratory and hence might be expected to also occur on surfaces in other environments, such as homes. This, indeed, is the case. (Pitts et al., 1985) first used differential optical absorption spectrometry (DOAS) to establish unequivocally that NO_2 injected into a mobile home forms HONO. Interestingly, the dependence of the rate of HONO generation on the NO_2 concentration was similar to that measured in laboratory systems, consistent with production in, or on, a thin film of water adsorbed on surfaces. A number of studies have confirmed that the behaviour is similar to that in laboratory systems; i.e., the rate of production of HONO increases with NO_2 and with relative humidity. Indoor levels of HONO as high as 8 ppb as a 24-h average and 40 ppb as a 6-h average have been reported in normal, in-use buildings and homes (Febo & Perrino, 1991; Spengler et al., 1993; Weschler et al., 1994). The ratio of HONO to NO_2 indoors can be quite large, up to ~0.15 (e.g., Febo & Perrino, 1991; Brauer et al., 1990, 1993; Spengler et al., 1993). This can be compared to typical values of a few percent outdoors. High levels of HONO (up to ~ 30 ppb) have also been measured in automobiles in use in polluted urban areas, and again, the ratio of HONO to NO2 was quite large, ~0.4, compared to 0.02-0.03 measured outdoors in the same study (Febo & Perrino, 1995). The generation of NO was attributed by Spicer and co-workers to a reaction of gaseous NO_2 with adsorbed HONO:

$$NO_2(g) + HONO(ad) \rightarrow H^+ + NO^-_3 + NO_3(g) \quad (2)$$

The same process was hypothesised to explain some time periods in a commercial office building when indoor NO actually exceeded outdoor NO (Weschler et al., 1994). As is the case in laboratory systems, equivalent amounts of HNO_3 are not observed as might be expected from the stoichiometry of reaction (1), likely due to HNO_3 remaining on the surface after formation and/or being taken up by surfaces. The accumulation of nitrate on indoor surfaces in a commercial building has been reported by Weschler and Shields (1996) and attributed to the formation and uptake of HNO_3 via reactions of NO_3 and/or oxidation of nitrite (i.e., adsorbed HONO) in an aqueous surface film. Subsequently, it was shown that HONO is also directly emitted by gas stoves (Pitts et al., 1989). In a house used for investigating indoor air pollution that had natural gas fueled appliances (a convective heater, a radiant heater, and a range with four burners), both the surface reaction of NO_2 and the direct combustion emissions contributed significantly to the measured indoor HONO. When an appliance was operational, the contribution of direct emissions was the more important source (Spicer et al., 1993). In short, the "dark reaction" of NO_2 with water on surfaces is ubiquitous and occurs not only in laboratory systems but also indoors. The combination of this heterogeneous reaction with combustion sources of HONO can produce significant concentrations of HONO indoors. As a result, there is a concern regarding the health impacts of nitrous acid, not only because it is an inhalable nitrite but also because it is likely the airborne acid present in the highest concentrations indoors.

2.2.3 CO and SO_2

As for NOx, combustion sources such as gas stoves and paraffin heaters can be significant sources of indoor CO. The ratio of indoor to outdoor concentrations of CO in homes using gas stoves has been measured to be 1.2-3.8 (Wade *et al.*, 1975), with the highest ratios found close to the source. Similarly, higher CO levels indoors compared to outdoors have been reported for restaurants in Korea, with those using charcoal burners as well as gas giving much higher concentrations (Baek *et al.*, 1997). In buildings where motor vehicle exhaust can be entrained from outdoors or attached parking garages, elevated indoor CO levels may also result (Hodgson *et al.*, 1991). On the other hand, in homes and offices where there was no direct indoor source of CO, the indoor to- outdoor ratio was about one, and sometimes less. For example, in Riyadh, Saudi Arabia, CO concentrations were measured indoors and outdoors; the indoor to- outdoor ratio varied from 0 to 2, but was typically below one (Rowe *et al.*, 1989). There have been a number of measurements of CO in the "indoor environment" of automobiles. Given that cars are major CO sources in urban areas, one might expect higher concentrations of CO during commutes and this is indeed the case. Typical CO concentrations of ~9-56 ppm have been measured inside automobiles during commutes in major urban areas (Flachsbart *et al.*, 1987; Koushki *et al.*, 1992; Ott *et al.*, 1994, 1995; Dor *et al.*, 1995; Fernandez-Bremauntz & Ashmore, 1995). This can be compared to peak outdoor levels of ~ 10 ppm in highly polluted urban areas. Thus, a significant enhancement of CO inside automobiles during commutes is common. For example, Chan *et al.* (1991) report a ratio of the in-vehicle CO concentration to that outdoors of ~ 4.5 in Raleigh, North Carolina. As is the case for CO, SO_2 levels indoors and outdoors tend to be similar if there are no combustion sources indoors.

2.2.4 Volatile Organic Compounds (VOCs)

Volatile organic compounds (VOC) are ubiquitous components not only of ambient air but also of indoor air environments, including offices, commercial and retail buildings, and homes (Shah & Singh, 1988; Finlayson-Pitts, 1999). There are three sources/categories for VOC: (1) entrainment of air from outside the building, (2) emissions from building materials, and (3) human activities inside buildings. As might be expected, given the nature of the sources, a very large variety of organic compounds have been identified and measured indoors (e.g., Brown *et al.*, 1994; Crump, 1995; Kostiainen, 1995). These numbers in the hundreds of different compounds, with the particular species and their concentrations depending on the particular sources present as well as the air exchange rates. Some of the compounds associated with the three sources: entrainment from outdoors, emissions from building materials, and anthropogenic activities - are now briefly reviewed.

Entrainment of air from outdoor sources: Entrainment of outdoor air through ventilation systems brings with it the species found in ambient air. Some of them, such as HNO_3, can be removed on surfaces such as those in air conditioning systems, and hence the indoor concentrations tend to be lower than those outdoors. Others such as NO tend to have similar concentrations indoors and outdoors if there are no significant combustion sources indoors (e.g., Weschler *et al.*, 1994). In the case of hydrocarbons, the concentrations of compounds that do not have significant indoor sources tend to be about the same as the outdoor concentrations. For example, Lewis and Zweidinger (1992) measured VOC in 10 homes in winter and showed that the concentrations of ethene, benzene, 2-methylpentane, methylcyclopentane, 2,2,4-trimethylpentane, and 2,3-dimethylbutane indoors were within experimental error of those outdoors. There are, however, some specific outdoor sources that can lead to higher concentrations of certain VOCs indoors than in the general outdoor air environment. For example, gases generated in landfills or from petroleum contamination can migrate through the soil and groundwater to adjacent buildings and homes to give larger indoor concentrations, particularly in basements and crawl spaces, than otherwise expected (Moseley & Meyer, 1992; Hodgson *et al.*, 1992; Fischer *et al.*, 1996). In one such case, the total hydrocarbon concentration was measured to be 120 ppm in a crawl space beneath the floor of a school where petroleum contamination was present from adjacent sources, compared to < 80 ppb outdoors (Moseley & Meyer, 1992). Although concentrations in various rooms were lower, they were still elevated compared to outdoors, ranging from 0.13 to 3.4 ppm. The use of pesticides *outside* buildings can also lead to enhanced concentrations of these compounds indoors. For example, Anderson and Hites (1988) measured the concentrations of chlorinated pesticides indoors and found elevated levels inside, e.g., a factor of 7 times higher for y-chlordane compared to outdoor levels. One home that had the highest indoor concentrations had been treated with chlordane about a decade earlier, presumably by subsurface injection from which the pesticide migrated into the house through cracks in the basement walls. Enhanced levels of chlorpyrifos were observed indoors in homes where soil surrounding the home had been treated on a regular basis. Another source of VOC is motor vehicle emissions, which can be drawn into buildings from outdoors or parking garages (e.g., Perry & Gee, 1994; Daisey et al., 1994). For example, motor vehicles were major sources (responsible for > 75%) of 12 of 39 individual compounds measured in a dozen buildings by Daisey *et al.* (1994). Of the 12 compounds, 5 were alkanes and 7 were aromatics. Similarly, Baek *et al.* (1997) report that vehicle emissions are important VOC sources indoors in Korea during the summer in homes and offices, as has been reported in the United States (e.g., Hodgson *et al.*, 1991; Daisey *et al.*, 1994).

Building materials: Emissions associated with building materials are major contributors to indoor levels of VOC. New buildings often have higher concentrations of certain compounds compared to older buildings. For example, enhanced levels of n-dodecane, n-decane, and n-undecane, the xylenes, and 2-propanol have been measured in new buildings, and the total VOC concentration is generally larger (by factors of 4-23) compared to established buildings (Brown *et al.*, 1994). Kostiainen (1995) identified more than 200 individual VOCs indoors in 26 houses. In addition, they compared the VOC concentrations in normal houses to those where complaints of odours or illness had been registered. A number of different VOCs were present at increased concentrations in the houses with complaints compared to the normal houses; these included a variety of aromatic hydrocarbons, methylcyclohexane, n-propylcyclohexane, terpenes, and chlorinated compounds such as 1,1,1-trichloroethane and tetrachlorethene. Carpets are a major source of VOCs in homes. For example, Sollinger *et al.* (1993, 1994) have identified 99 different VOCs emitted from a group of 10 carpet samples, and Schaeffer *et al.* (1996) identified more than 100 different VOCs emitted from the carpet cushion alone. Emissions come not only from the carpet fibres but also from the backing materials and the adhesives used to bind the carpet to the backing. As a result, the individual compounds emitted by carpets can vary substantially, depending on the carpet construction. Many of the compounds emitted are known to be used in the manufacturing processes (e.g., e-caprolactam is used in Nylon-6 production) and /or are common solvents Emissions of VOC from carpets tend to decrease with time and increase with temperature.

The dependence of VOC emissions from building materials on relative humidity is more complex, with some emissions increasing with relative humidity, but others not. For example, Sollinger *et al.* (1994) report that the VOC emissions from carpets did not change with relative humidity over the range from 0 to 45% RH. On the other hand, the emissions of formic and acetic acids from latex paints have been reported to increase dramatically with relative humidity; for example, for one paint sample the emission rate for acetic acid almost tripled when the relative humidity was changed from 4-5% to 5-23% (Reiss *et al.*, 1995b). A number of different aldehydes have been measured indoors (Crump & Gardiner, 1989; Lewis & Zweidinger, 1992; Zhang *et al.*, 1994; Daisey *et al.*, 1994; and Reiss *et al.*, 1995a), some of which are directly emitted and some of which are formed by chemical reactions indoors of VOCs such as styrene. Of these, there is an enormous amount of evidence for direct emissions of HCHO from building materials. Interest in formaldehyde emissions and levels in homes and other buildings stems from its well-known health effects, which include possible human carcinogenicity and eye, skin, and respiratory tract irritation (Feinman, 1988). Formaldehyde is emitted from urea-formaldehyde foam insulation as well as from resins used in reconstituted wood products such as particleboard and plywood (Meyer and Reinhardt, 1986); urea-formaldehyde resins comprise about 6-8% of the weight of particleboard and 8-10% of mediumdensity fiberboard (Meyer and Hermanns, 1986). Other sources include permanent press fabrics (such as draperies and clothing), floor finishing materials, furniture, wallpaper, latex paint, varnishes, some cosmetics such as fingernail hardener and nail polish, and paper products (Kelly, 1996; Howard *et al.*, 1998a, 1998b).

Many measurements of HCHO have been made in indoor air environments. In conventional homes, average concentrations are typically about 10-50 ppb (Stock, 1987; Zhang *et al.*, 1994; Reiss *et al.*, 1995a). Sexton *et al.* (1989) measured concentrations of HCHO in 470 mobile homes in California and found geometric mean concentrations of 60-90 ppb, although maximum values of over 300 ppb were recorded in some cases. In a similar study in Wisconsin, levels up to 2.8 ppm were measured (Hanrahan *et al.*, 1985). Higher levels are typically found in mobile homes because of the reconstituted wood products (e.g., particleboard and plywood) used in their construction. Interestingly, HCHO does not appear to be a significant product of natural gas combustion, as levels in dwellings with and without gas stoves turned on are not significantly different (e.g., Pitts *et al.*, 1989; Zhang *et al.*, 1994). Temperature is again an important determinant of HCHO levels.

Human activities: There are many sources of VOCs associated with human activities in buildings. For example, mixtures of C_{10} and C_{1}~ isoparaffinic hydrocarbons, which are characteristic of liquid process copiers and plotters, have been identified in office buildings in which these instruments were in use (Hodgson *et al.*, 1991). Emissions of a number of hydrocarbons and aldehydes and ketones have been observed during operation of dry-process copiers; these include significant emissions of ethylbenzene, o-, m-, and p-xylenes, styrene, 2-ethyl-l-hexanol, acetone, nnonanal, and benzaldehyde (Leovic *et al.*, 1996). Enhanced levels of acetaldehyde in an office building in Brazil were attributed to the oxidation of ethanol used as a cleaning agent (Brickus *et al.*, 1998), although levels outdoors were also enhanced due to the use of ethanol as a fuel. Pyrocatechol has been measured in an occupational environment where meteorological charts are mapped on paper impregnated with this compound (Ekinja *et al.*, 1995), and p-dichlorobenzene is observed when mothballs containing this compound are in use (e.g., Tichenor *et al.*, 1990; Chang and Krebs, 1992). Elevated concentrations of the n-C_{13} to n-C_{18} alkanes and branched-chain and cyclic analogs were measured in a building having a history of air quality complaints; the source was found to be volatilisation from hydraulic fluids used in the building elevators (Weschler *et al.*, 1990). Enhanced levels of chlorinated compounds have been observed indoors due to human activity as well. For example, increased levels of perchloroethylene have been observed from unvented dry-cleaning units (Moschandreas & O'Dea, 1995) and volatilisation of chlorinated organics such as chloroform from treated tap water can occur (McKone, 1987). Other sources include the use of household products. For example, chloroform emissions have been observed from washing machines when bleach containing hypochlorite was used (Shepherd *et al.*, 1996). It is interesting that emissions of organics associated with the use of washing machines are decreased when the machine is operated with clothes inside (Howard and Corsi, 1998). Of course, activities such as smoking result in enhanced levels not only of nicotine (e.g., Thompson *et al.*, 1989) but also of a variety of other gases associated with cigarette smoke (e.g., California Environmental Protection Agency, 1997; Nelson *et al.*, 1998). For example, using 3-ethenylpyridine as a marker for cigarette smoke, Heavner *et al.* (1992) estimated that 0.2-39% of the benzene and 2-49% of the styrene measured in the homes of smokers was from cigarette smoke. Humans emit a variety of VOCs such as pentane and isoprene (e.g., Gelmont *et al.*, 1981; Mendis *et al.*, 1994; Phillips *et al.*, 1994; Jones *et al.*, 1995; Foster *et al.*, 1996). In addition, emissions from personal care products have been observed. Decamethylcyclopentasiloxane (D5), a cyclic dimethylsiloxane with five Si-O units in the ring, and the smaller D4 analog, octamethylcyclotetrasiloxane, are used in such products as underarm deodorant and antiperspirants at concentrations up to 40-60% by weight (Shields and Weschler, 1992; Shields *et al.*, 1996). Increased concentrations of D5 have been measured in offices and are correlated to human activity, as expected if personal care products were the major source (Shields and Weschler, 1994). In some cases, increased concentrations attributable to emissions from silicone-based caulking materials were also observed (Shields *et al.*, 1996). The use of pesticides indoors can lead to very large concentrations not only of the pesticide but of the additional VOCs used as a matrix for the pesticide, which represent most (>95%) of the mass of the material as purchased. For example, Bukowski and Meyer (1995) predict that VOC concentrations immediately after the application of a fogger could reach levels of more than 300 mg m^{-3}!

2.2.5 Ozone

Because O_3 decomposes on surfaces, indoor levels are usually lower than those outdoors due to the decomposition that occurs as the air passes through air conditioning systems and impacts building surfaces (Reiss *et al.*, 1994; Finlayson-Pitts, 1999). The measured ratio of indoor-to-outdoor concentrations of ozone vary from 0.1 to 1, but are typically around 0.3-0.5 (e.g., Druzik *et al.*, 1990; Hisham & Grosjean, 1991; Liu *et al.*, 1993; Weschler *et al.*, 1989, 1994; Gold *et al.*, 1996; Jakobi & Fabian, 1997; Avol *et al.*, 1998; Drakou *et al.*, 1998; Romieu *et al.*, 1998). Buildings with low air exchange with outside air tend to have lower ratios, ~0.1-0.3 (Druzik *et al.*, 1990; Weschler *et al.*, 1994; Romieu *et al.*, 1998). For example, Gold *et al.* (1996) estimate that at outdoor ozone concentrations of 170 ppb in Mexico City, the indoor-to-outdoor ratio of O_3 at a school was 0.71 ± 0.03 with the windows and doors open, which maximised the exchange with outside air, 0.18 ± 0.02 with the windows and doors closed and the air cleaner off, and 0.15 ± 0.02 with the windows and doors closed and the air cleaner on. There are some additional sources of O_3 indoors. These include dry-process photocopying machines, laser printers, and electrostatic precipitators (e.g., Leovic *et al.*, 1996; Wolkoff, 1999). Indeed, it is not unusual to detect O_3 by its odour during operation of some copy machines and laser printers. In the "indoor environment" in cars, ozone levels tend to be significantly less than in the surrounding area. For example, Chan *et al.* (1991) report that in-vehicle O_3 concentrations during commutes in Raleigh, North Carolina, were only about 20% of those measured in the local area at a fixed station. There are several contributing factors to these low concentrations. One is that NO concentrations are higher near roadways, so that O_3 is titrated to NO_2 by its rapid reaction with NO. A second is that O_3 can decompose on the surfaces of the automobile air conditioning system. A similar titration effect has been observed inside homes where there are combustion sources of NO.

2.2.6 Particles

With the epidemiological studies suggesting increased mortality associated with particles, there has been increasing interest in indoor particle concentrations compared to outdoor levels (Finlayson-Pitts, 1999). A number of studies have examined this over the years and are summarised in a review by Wallace (1996). In general, if there are no indoor sources of particles, the levels indoors tend to reflect those outdoors. For example, application of a mass balance model to measurements of indoor and outdoor particle concentrations in Riverside, California, indicated that 75% of $PM_{2.5}$ and 65% of PM_{10} in a typical home were from outdoors (Wallace, 1996). Similar conclusions were reached by Koutrakis *et al.* (1991, 1992) for homes in two counties in New York. For example, they report that 60% of the mass of particles in homes is due to outdoor sources. However, the contribution to various individual elements in the particles varies from 22% for copper to 100% for cadmium. There are some differences in indoor levels of particulate matter in areas with low outdoor compared to high outdoor levels. In the case of high outdoor levels, the indoor concentrations tend to be somewhat lower than those outdoors; for example, Colome *et al.* (1992) report that the ratio of indoor-to-outdoor median concentrations of PM_{10} is 0.7 in residences in southern California. On the other hand, when outdoor levels are low, indoor levels tend to be higher Night time mass concentrations indoors tend to be smaller than those during the day, probably because of the decreased activity. Interestingly, when individuals wear personal exposure monitors to measure their actual exposure to particles, the measured mass concentrations tend to be higher than those measured with fixed monitors located indoors. A major source of increased particles indoors is cigarette smoking. (e.g. Spengler *et al.*, 1981; Quackenboss *et al.*, 1989; Neas *et al.*, 1994). In addition to the contribution to the *mass* concentrations of indoor particles, cigarette smoke is of concern because of the mutagens, carcinogens, and toxic air contaminants that are emitted (Löfroth *et al.*, 1991; Chuang *et al.*, 1991; California Environmental Protection Agency, 1997; Nelson *et al.*, 1998). Thus, a variety of both gaseous and particulate polycyclic aromatic hydrocarbons (PAH) and compounds (PAC) have been identified in buildings with cigarette smoke (Offermann *et al.*, 1991; Mitra & Ray, 1995). Indeed, in the homes of smokers, almost 90% of the total PAH was from tobacco smoke (Mitra and Ray, 1995). Higher levels of mutagenic particles have also been shown to be associated with indoor air containing cigarette smoke (e.g., Lewtas *et al.*, 1987; Löfroth *et al.*, 1988, 1991; Georgiou *et al.*, 1991). Other significant sources identified in a number of studies are cooking, the use of paraffin heaters, wood burning, and humidifiers. For example, a study carried out under the auspices of the U.S. Environmental Protection Agency, the TEAM study (Total Exposure Assessment Methodology), indicated that an increase in PM m of ~10-20 / $\mu g\ m^{-3}$ could be attributed to cooking (Wallace, 1996). This source will obviously depend on the amount of cooking, the types of cooking, and the ventilation. For example, Löfroth *et al.* (1991) measured emissions of particles ranging from 0.07 to 3.5 mg per gram of food cooked, depending on the particular food. Baek *et al.* (1997) measured indoor and outdoor concentrations of particles in homes, offices, and restaurants in Korea and report ratios of 1.3, 1.3, and 2.4, respectively. The higher value in restaurants, even those using only gas and not charcoal, suggests a significant contribution from cooking. Paraffin heaters can be significant sources of particles under some circumstances. For example, paraffin heaters were reported to contribute to indoor $PM_{2.5}$ in homes in Suffolk County, New York, but not Onondaga County; wood stoves and fireplaces and gas stoves did not contribute in either case (Koutrakis *et al.*, 1992; Wallace, 1996). A similar conclusion was reached in a study of eight mobile homes in North Carolina (Mumford *et al.*, 1990). However, it should be noted that even where paraffin heaters do not contribute significantly to particle *mass* concentrations, they may still be important in terms of health effects. This is because of the composition of the particles emitted, which include polycyclic aromatic compounds and other mutagenic species, as well as sulfate (Traynor *et al.*, 1990). For example, Traynor *et al.* (1990) studied the emissions from unvented paraffin space heaters and identified a number of PAHs (naphthalene, phenanthrene, fluoranthene, anthracene, chrysene, and indeno[c,d]pyrene) and nitro-PAHs (1-nitronaphthalene, 9-nitroanthracene, 3-nitrofluoranthene, and 1-nitropyrene), in addition to a host of other gaseous species.

Baek *et al.* (1997) also reported increased levels of a number of gases indoors in homes and offices in Korea due to the use of paraffin heaters. In studies of indoor air in eight mobile homes, Mumford *et al.* (1991) identified the PAHs and nitro- PAHs measured in emissions from paraffin heaters by Traynor *et al.* (1990), as well as a number of compounds that may be animal carcinogens, such as cyclopenta[c,d]pyrene, benz[a]anthracene, benzofluoranthenes, benzo-[a]pyrene, and *benzo[ghi]perylene.* While the mass concentrations of PM10 did not increase with the paraffin heater on in six of the eight homes studied, the particles in five of the homes had increased mutagenicity using TA98 with or without $9 added. In short, not only the mass emissions but also the nature of the compounds emitted must be taken into account in assessing the health effects of indoor particles.

Where indoor heating and cooking involves the use of coal or biomass, indoor particle concentrations can be extremely large. For example, Florig (1997) and Ando *et al.* (1996) report that in China typical indoor total suspended particle (TSP)concentrations can be in the range from 250 to 900/μg m^{-3} in homes using coal and 950-3500 / μg m^{-3} in those using biomass fuels. These levels can be compared to annual average outdoor concentrations of 250-410 / μg m^{-3}. The high concentrations associated with coal burning combined with the mutagenic nature of the emissions have been suggested to be responsible for enhanced lung cancer in China (Mumford *et al.*, 1987). Similarly, Davidson *et al.* (1986) measured TSP concentrations of 2900-42,000/ μg m^{-3} in homes in Nepal that used biomass fuels, compared to outdoor levels of 280 / μg m^{-3}. For particles with diameters less than 4 /μm, the levels ranged from 870 to 14,000/ μg m^{-3}. Similar conclusions regarding the relative indoor and outdoor concentrations have been reached in studies of office and commercial buildings. For example, Ligocki *et al.* (1993) measured indoor and outdoor concentrations of particles and their components at five museums in southern California. The indoor-to-outdoor ratios of particle mass varied over a wide range, depending to a large extent on the ventilation and filtration systems in use. Ratios varied from 0.16 to 0.96 for particles with diameters less than 2.1/μm and from 0.06 to 0.3 for coarse particles with diameters greater than this.

2.2.7 Microbial pollutants

Microbial pollution is a risk to health and is associated with allergic illnesses. Published results indicate that 20% of the population can be sensitised by airborne fungal spores in the

UK, while 40% of the inspected houses in Germany suffer from mould-related problems (Waubke & Kusterle, 1990). The medical consequences of immune response, allergic reactions, endotoxins, mycotoxins, and epidemiology have been extensively studied by Miller (1990), Morey (1990), Gravensen et al. (1990) and Burge et al. (1990). Similarly, Legionnaires ' disease and Pontiac fever are associated with wet cooling towers and domestic hot-water systems in complex buildings.
Accordingly to the official published figures, some 560,000 people need treatment because of indoor pollution due to mites and mould in damp houses (House of Commons Environment Committee, 1991). Indoor airborne allergic components come from two sources: outdoor airborne spores moving inside and allergic components originating inside the dwelling. The source of biological growth within buildings is associated with moisture and the formation of microclimates; it also depends upon the type of the buildings and their ventilation. Mould fungi thrive on surfaces on which there is nourishment and suitable humidity, for example on damp water pipes, windows and walls in kitchens and bathrooms, in central air-conditioning systems, circulation pumps, blowers, ventilation ductwork and air filters, central dehumidifiers, and inside damp structures. Allergenic substances can be airborne and inhaled, such as pollen, fungus and dust, digested, such as mouldy food or drink. Investigations suggest that airborne allergies cause more problems throughout the world than all other allergies combined. Additionally, cross-infection from patient to patient is of great concern in hospitals. The medical field that treats allergies recognises the following allergenic diseases: asthma, allergic rhinitis, serous otitis media, bronchopulmonary aspergillosis, and hypersensitivity pneumonitis.
Allergic load and cocktail effect: For some people, an allergic reaction in the indoor environment may be triggered by non-biological factors, such as chemicals or other indoor air pollutants, emotional stress, fatigue or changes in the weather. These factors burden allergic people further if they are suffering from allergic reactions to biological contaminants. This combination is known as 'allergic load'. Microbial contaminants propagated within the health care establishment are particularly aggressive to patients due to reduced immune system resistance.
Recently, attention has been focused on the cocktail effect of chemicals present in indoor air. Volatile organic compounds may be produced from the use of wood preservatives and remedial timber treatment chemicals, moth-proof carpets, fungicides, mouldicide-treated paints, furnishing materials such as particle board and foamed insulation which may emit formaldehyde. Biological pollutants alone or in synergetic effect with any of the above-mentioned volatile organic compounds may produce symptoms such as stuffy nose, dry throat, chest tightness, lethargy, loss of concentration, blocked, runny or itchy nose, dry skin, watering or itchy eyes or headache in sensitive people. The 'sick building syndrome' (SBS) or tight building syndromes may arise from a variety of causes. Because of the uncertainties about the causes of SBS and the rising levels of health related problems in buildings there is an increasing use of the term building-related illness (BRI) to cover a range of ailments which commonly affect building occupants.

2.2.8 Asbestos and manmade mineral fibres

Asbestos is known to cause a number of diseases after occupational exposure (Brown & Hoskins, 1993). Before the hazards associated with the inhalation of these mineral fibres were understood these exposures were often very large with frequent reports of dust clouds so great that visibility in the workplaces was considerably reduced. This type of exposure is quantitatively quite different from those in the general environment that have provoked a response which in some quarters approaches hysteria. In the USA at least there is massive expenditure on asbestos removal, management and litigation.

Asbestos is a collective, trivial, name given to a group of highly fibrous minerals that are readily separated into long, thin, strong fibres occurring on sufficient large bulk deposits for their industrial exploitation. Asbestos minerals were usually used for their insulating properties, or in a composite, where they added strength, as in cement, or increased friction, as in brake shoes. Chrysotile, or white asbestos has counted for over 90% of the world trade in asbestos minerals. It is a serpentine mineral while the others (amosite (brown asbestos); crocidolite (blue asbestos); anthophyllite; tremolite; and actinolite) are all amphibole minerals. Amphibole asbestos has grater acid and water resistance than chryusotile and was used where these properties made it more suitable. Sometimes users would be unaware of the differences between the types of asbestos and so different minerals could have been used for a single application.

Recently the concern over the health effects of asbestos has been extended to another group of fibrous materials- the man-made mineral fibres (MMMF). While this term is self-explanatory a variety of types are produced with diverse chemical compositions, properties and uses. While sometimes referred to as 'asbestos substitutes' the majority of uses for the manmade fibres are relatively novel and ones for which the natural fibres are unsuitable. For example refractory ceramic fibres are resistant to considerably higher temperatures than are any of the natural fibres. The development of synthetic fibrous insulation materials has been given a great impetus in recent years by the need for more thermally efficient buildings and industrial processes.

MMMF can be made from most types of glass, from rock such as basalt, diabase and olivine and from various types of slag. Ceramic fibres can be made from kaolin or from pure silica and other oxide starting materials. The MMMF have been classified into four broad groups based on the manufacture and use: continuous filament glass fibre made by extrusion and winding processes, insulation wool (including ceramic fibre), and special purpose fibres. The non-continuous fibres are made by dropping molten material onto spinning disks or by air or steam jet impingement on a stream of the molten material. They contain a wide range of fibre sizes and are contaminated by small glassy balls called shot which often account for 50% of the product by weight.

3. Factors that influence exposure to indoor air pollutants

3.1 General

The types and quantities of pollutants found indoors vary temporally and spatially. Depending on the type of pollutant and its sources, sinks and mixing conditions, its concentration can vary by a factor of 10 or more, even within a small area.

Human mobility constitutes an important kind of complexity in the determination of exposure to air pollutants. Human activity patterns differ between midweek and weekend, between one season and another, and between one part of one's life and another. Activity patterns determine when and how long one is exposed to both indoor and outdoor pollutants. Therefore, in reviewing the factors that influence air-pollution exposures, we have specifically separated them into two major components: time (activity) and concentration (location).

Information on the time spent in various activities is summarised first, and then the variations in concentration often encountered in different locations. Unfortunately, most of the studies discussed were not longitudinal and thus do not offer information on seasonal differences in time spent indoors and outdoors or on regional differences in activity patterns.
Outdoor concentrations of pollutants and rates of infiltration affect the concentrations to which people are exposed indoors. Building construction techniques, as they vary geographically, and their effect on pollution infiltration are particularly important. But the measurement techniques available are limited; the need for additional studies is discussed. The rates of infiltration on a neighbourhood scale have been studied by only a few researchers. Although their work has focused on energy conservation, their findings can easily be applied to the study of impact on indoor pollution.
Patterns of human behaviour and activity determine the time spent in any specific location, and thus knowledge of them is essential in estimating exposures of populations to pollutants. As indicated by Ott (Ott, 1995), a large number of variety of studies in which data on human activities were collected from population samples have been completed over the past 50 ye.
When one examines the literature on human activities, the term "time budget" ("zeitbudget", "budget de temps") is encountered often. A time budget produces a systematic record of how time is spent by a person in some specified period, usually 24 h. It contains considerable detail on a person's activities; including the locations in which the activities take place (Michelson, 1973).
One way of obtaining time budget information from the populations surveyed is to ask each respondent to maintain a diary of his or her activities over a 24-h period or longer. In another approach, the so-called "yesterday" survey approach, the interviewer asks each responder about his or her activities on the preceding day.
Several summaries of the historical development of time-budget research have been published (Chapin, 1974; Converse, 1968; Ottensman, 1972). Ott (Ott, 1995) discussed the literature on activity patterns in the context of estimation of exposure to air pollution.
Owing to the small number of field monitoring studies, the geographic distribution of indoor air pollutants has not been determined. However, it is instructive to review the geographic distribution of the major factors that affect variations in the concentrations of pollutants and their impact on the quality of the indoor environment. Outdoor air quality, air-infiltration rates, and sources of emission of indoor air pollutants are the major factors.
Outdoor air quality has been studied with respect to some pollutants, and the geographic distribution of these few pollutants is well understood. Descriptive statistics published annually by EPA and state and local air-quality agencies furnish much scientific information useful in discerning regional and local differences in concentrations of carbon monoxide, total suspended particles, ozone, NOx, sulfur dioxide, sulfates, and others. It should be noted that the geographic distribution of some criteria pollutants has been studied and is easily accessible from the literature; information on non-criteria pollutants is sparse and often collected and analyzed by questionable methods.
Concentrations of chemically non-reactive pollutants in residences generally correlate with those outdoors. Distribution of indoor air quality is extremely difficult to describe on a geographic scale, because indoor air quality is determined by complex dynamic relationships that depend heavily on occupant activity and highly variable structural characteristics. Weather, which has a regional character, influences indoor air concentrations

of some chemicals, such as formaldehyde, and biologic contaminants, such as bacteria and molds. Therefore, the influence of relative humidity and other weather-related conditions affecting indoor environmental quality needs to be studied geographically. Research specifically addressed to geographic distribution of indoor air quality is needed.

Typically, the air-infiltration rate for American residences is assumed to be 0.5- 1.5 ach. This assumption is supported by the results of several energy and air-quality studies that experimentally determined the range of ventilation rates for typical residences to be between 0.7 and 1.1 ach (Moschandreas & Morse, 1979). However, the sample that yielded the data is small, and statistical documentation for such statements is not strong.

The quality of indoor air is a function of outdoor air quality, emission from indoor sources, air-infiltration rates, and occupant activity is likely to vary within each metropolitan and suburban area, is indeed within each neighbourhood. Within a metropolitan area, it has been shown that an urban complex leads to the so-called urban heat reservoir (American Society of Heating, Refrigerating and Air-Conditioning engineers. ASHRAE, 1972). Urban characteristics-- such as city size, density of buildings, and population-- correlate with such meteorological factors as temperature, pressure and wind velocity (Gibson, & Cawley, 1977; Kostiainen, 1995). The urban heat island affects both urban pollution patterns and meteorological characteristics that affect the infiltration rates of buildings. Thus, although the exact nature of the impact on indoor air quality is not known, it is fair to expect that the heat island to have an impact on the indoor environment that is likely to be adverse. Also, the variations due to mechanical ventilation, structural differences, and air infiltration may vary within a neighbourhood as a function of such factors as house orientation, tree barriers, and terrain roughness.

Occupant activity, air-infiltration rates, the indoor sources of pollutants and their chemical natures are some of the factors that cause variations within a city. A study (Moschandreas et al., 1980) in the Boston metropolitan area obtained indoor air samples from 14 residences under occupied "real-life" conditions for 2 week each. The indoor air character not only was driven by outdoor concentrations, but was greatly affected by other factors, such as indoor activities.

Wind speed, temperature difference, pressure differential, terrain characteristics (roughness and barriers, such as trees and fences), building orientation, and structure characteristics may be affected by the location of one residence relative to another within a neighbourhood. The indoor air quality of an individual building is often characterised by the 24-h average for the concentration of one pollutant measured at one sampling location. Because the activity patterns of persons are such that more time is spent in some indoor areas than in others, the question arises (Moschandreas et al., 1978): "Do indoor zones (independent areas) with distinct pollutant patterns exist?" At issue here is whether sampling from one monitoring zone is sufficient to characterise the air quality of an entire building.

In an extensive analytic study of indoor air quality, Shair and Heitner (1974) assumed that there are no pollutant gradients in the indoor environment. The experimental database of Moschandreas and co-workers (1980) verified that the gradients in concentrations of several gaseous pollutants in the residential environment are negligible. J.D. Spengler, R.E. Letz, J.B. Ferris, Jr., T. Tibbets, and C. Duffy reported (at the annual meeting of the Air Pollution Control Association, 1981) on weekly nitrogen dioxide measurements in 135 homes in Portage, Wisconsin. On the average, kitchen concentrations were twice those in bedrooms in homes that had gas stoves. A study of the air quality in a scientific laboratory by West (1977) showed an almost uniform distribution of an intern tracer continuously released in the room. Similar experiments performed by Moshandreas et al. in residential environments showed that equilibrium is reached throughout a house within an hour. Episodic release of sulphur hexafluoride tracer gas also illustrates this point. The source location was the living room; adjacent locations were the kitchen and the hall. Episodic release of this inert gas in 24 residences was followed by uniform indoor distributions within 30 min (Moschandreas et al., 1978; Peterka & Cermak, 1977). The one-zone concept does not require instantaneous mixing, because it is based on the behaviour of hourly average pollutant concentrations.

Moschandreas and associates (1980) used a different database derived from the monitoring of 14 indoor environments in the Boston metropolitan area. Analysis of variance was used to reach the following conclusions:

- Pollutants (ozone and sulphur dioxide) generated principally outdoors have little or no interzonal statistical difference indoors.
- Pollutants with strong indoor generation have interzonal statistical differences in residences with gas facilities and offices, but not in electric-cooking residences. In general, the observed differences are not large, and the health differences are not expected to be serious.
- Depending on indoor activity and outdoor episodic pollutant activity, the indoor arithmetic 24-h average may or may not adequately represent the variation of hourly indoor concentrations.
- Although more than one zone would be preferable, hourly pollutant concentrations obtained from one indoor zone adequately characterise the indoor environment.

The most important factors that influence exposure to indoor air pollutants are the one described under. It should be noticed that these conclusions are not applicable to short-lived pollutants. Contaminants associated with tobacco smoke, bathroom odours, allergens, and other pollutants related to dust are expected to vary considerably in a given residence. Additional documentation is needed to determine the extent of this variation.

3.2 Site characteristics

The characteristics of a building site that influence indoor air quality are addressed as three related subjects: air flow around buildings, proximity to major sources of outdoor pollution, and type of utility service available.

The air flow around a building has been shown to be determined by the local characteristics of the geometry of surrounding buildings (Peterka & Cermak, 1977), the location and type of surrounding vegetation (White, 1995), the terrain (Geiger, 1965), and the size and shape of the building itself. Pollutants can be transferred by the air flow from the street level, over the façade of the building and onto the roof (Cermak, 1976). Field tests of isolated buildings have been used to develop scaling coefficients for both isothermal and and stratified cases of surface wind pressures, turbulence, and dispersion (Davenport, 1960.). Air flow around the building creates low pressure on the leeward side and/or the sides adjacent to the windward face, as well as the roof. Air pollutants released from stacks, flues, vents, and cooling towers in the region can re-enter the building through make-up air intakes for ventilation (Cermak, 1976).

Trees and forests have been generally studied as shelter belts in an agricultural context. Shelter belts affect air flow around buildings. When an air current reaches a shelter belt, part of it is deflected upward with only a slight change in velocity, part passes through the crowns of the trees with very low velocity, and part is deflected beneath the canopy with rapidly decreasing velocity (Federer, 1971). The changes in velocity of air flow outside may change the infiltration rate and thus affect indoor air quality.

The location of a building relative to a major outdoor pollution source can affect indoor air quality. For example, buildings near major streets or highways often have high carbon monoxide and lead concentrations, owing to the infiltration of these pollutants.

The type of utility service available is also related to the site of the building and may affect the character of its indoor environment. The availability of particular fuels (e.g., natural gas and oil) influences the types and concentrations of pollutants (e.g., combustion products) emitted by space-and water- heating. Service moratoria, development timing, and development scale are institutional elements that contribute to the variability of utility services and thus can affect indoor air quality.

3.3 Occupancy

Occupancy factor that affect indoor air quality include the type and intensity of human activity, spatial characteristics of a given activity, and the operation schedule of a building.

Several human activities-such as smoking, cleaning and cooking- generate gaseous and particulate contaminants indoors. The number of occupants of a space and the degree of their physical activity (i.e., metabolic rate at rest or under intense activity) are related to the production of various pollutants, such as carbon dioxide, water vapour, and biologic agents. If the only source of indoor carbon dioxide is that caused by occupants, ventilation rates may be proportional to the number of people and their metabolic rates (Mclntyre, 1980). Although studies have shown no constant relationship between carbon dioxide concentrations and the concentrations of other pollutants, carbon dioxide concentration is often used as a general indicator of the adequacy of ventilation in an occupied space.

Building occupancy is often expressed as occupant density and the ratio of building volume to floor area. The importance of occupancy in indoor air quality is illustrated by the fact that the choice of natural or mechanical ventilation is based on occupant density and the spatial characteristics of the building under consideration.

Occupancy schedule and associated building use may affect the type, concentration, and time and space distribution of indoor pollutants. Because most buildings are unoccupied for substantial portions of each day, the manipulation of "operating schedule" is a means of controlling energy use (American Institute of Architects Research Corporation. Phase Two Report for the Development of Energy Performance Standards for New Buildings, 1979). Efforts to conserve energy through the design of ventilation systems can result to the degradation of indoor air quality. However, detailed studies relating ventilation capacity, occupancy schedules, energy requirements, and indoor air quality have only recently been implemented.

3.4 Design

Elements of building design that affect the indoor environment include interior-space design (space planning), envelope design, and selection of materials.

The evolution of space planning in many building types has resulted in flexibility in assigning functions to specific locations. However, this flexibility is accompanied by a decrease in the ability to predict exposure to air pollutants. In particular, "open-plan" offices and schools have serious technical problems of redundant service distribution, limited acoustic control, incomplete air diffusion, and incomplete pollutant dispersion indoors, compared with "fixed-plan" floor layouts.

Evaluation of the success of a floor plan in achieving space efficiency, structural economy, and energy efficiency is usually in terms of net area per occupant and ratio of net usable area to total area. Explicit planning for environmental quality must be included to ensure that spatial arrangements are acceptable to the occupants.

A building's structural envelope consists of both primary elements -foundations, floors, walls, and roofs- and secondary "skin" elements -facings, claddings, and sheathing. To various degrees, the function of these is to maintain the integrity of the structure under the stresses caused by structural load, wind pressure, thermal expansion, precipitation, earth movement, and fire. The integrity of the building envelope is a major consideration in uncontrolled air movement into and out of the building -usually referred to as "infiltration". This is a major factor in indoor air quality. There has been no systematic survey of infiltration rates of buildings in the United States. The dominant factor in determining a building's infiltration rate is the total area of effective leakage, as measured with fan pressurisation. Following the leakage area in importance are the terrain and shielding near the building, the mean climatic conditions during heating (or cooling) periods, and the building height (Sherman, 1981). There is much evidence (Dickerhoff et al., 1980), both in the United States and in Europe, that houses in mild climates are "very leaky", whereas houses in severe climates are "tight".

Greater height of a building increases the "stack effect", or updraught, and exposes the building to higher wind speeds. Thus, higher wind pressures drive air through existing openings, referred to as "leakage", increasing the infiltration rate.

The dominant building factors that determine infiltration have not been identified, but a catalogue of leakage openings found in typical structures is as follows:

- Walls: Leakage around sill plates (the openings at the bottom of wallboard), electric outlets, plumbing penetrations, and headers in attics for both interior and exterior walls.
- Windows and doors: Window type is more important than manufacturer in determining window leakage. This source of leakage tends to be overrated; it contributes only about 20% of the total leakage of a house
- Fireplaces: This includes dampers, glass screens, and fireplace caps.
- Heating and cooling systems: The variables include combustion air for furnaces, dampers for stack air draft, air-conditioning units, and location of ductwork.
- Vapour barrier and insulation penetrations.
- Utility accesses: This includes recessed lighting and plumbing and electric penetrations leading to attic or outside.
- Terminal devices in conditioned space: This includes leakage of dampers, especially those for large air-handling systems.
- Structural types: Examples are drop ceilings above cupboards or bathtubs, prism-shaped enclosures over staircases in two-story houses, and elevator and utility shafts that lead from basement to attic.

Wall and ceiling materials and floor finishes are the constituents of the building interior. Modular components, weight, strength, thermal insulation, thermal stability, sound insulation, fire resistance, ease and speed of installation and ease of maintenance are among the criteria considered in the selection of materials for walls, ceiling and floors. But emphasis on first cost, ease of installation, maintenance and long service life has also led to the use of materials that may be sources of indoor contaminants.

3.5 Operations

Depending on the type of ownership (owner-occupied or developer-owned), building operation may vary considerably, and this variation may have an impact on indoor air quality. "Building operation" pertains to the following elements of a building: the building envelope, service and plant, building facilities, equipment and landscaping. Cleaning, preventive maintenance, and replacement and repair of defects are also included in building operation. The staff responsible for building operation includes management, engineering, and custodial personnel. The care responsibilities are operation of the heating, ventilation, and air-conditioning systems and building services, such as hot water, lighting and power distribution. Building operation has an impact on indoor air quality in numerous ways, but the magnitude of this impact is not known.

4. Health effects of indoor air pollution

Indoor air pollution, apart from the health impact, has socio-economic costs. The potential economic impact of poor indoor air quality is quite high, and has been estimated to be in the order of tens of billions of ECU per year in Western Europe. This includes costs of medical care, loss of income during illness, days lost due to illness, poor working performance and lower productivity. Labour costs are significantly greater per square metre of office space than energy and other environmental control costs (ECA, 1989). In the US, the loss in productivity for each employee which is attributable to IAQ problems is currently estimated to be 3% (14 minutes/day) and 0.6 added sick days annually. Other estimates have been made by calculating the impact of IAQ on productivity. For instance, in Norway, the authorities estimate that the costs to society related to poor IAQ are in the order of 1 to 1.5 billion ECU per year or about 250 - 350 ECU per inhabitant. This estimation only includes costs related to adverse health effects requiring medical attention and does not include reduced working efficiency or job-related productivity losses. Thus, from an economic consideration, remedial action to improve indoor air quality is likely to be cost effective even if an expensive retrofit is required.

As far as it concerns the health effects on IAP, it is very interesting to present the methods of studying health effects, the criteria for the assessment of the impact of IAP on the community and the diverse effects of IAP on human health (ECA, 1991).

4.1 Methods of studying health effects

Methods of studying health effects of indoor pollutants can be grouped into three broad categories:

a. Human studies, subdivided into observational and experimental studies.

Epidemiological studies of pollutants are mostly observational, i.e. the investigator has no means of experimentally exposing humans to pollutants, or of allocating subjects to exposed and unexposed groups. Critical issues are therefore the validity and precision of exposure assessment, and the control for confounding factors in these studies. Recent developments have stressed the importance of reducing exposure misclassification, and of studying restricted, well defined, homogenous populations to address these issues. The main advantage is that humans are studied under realistic conditions of exposure. By themselves, observational epidemiological studies are not usually sufficient to support causality of an observed association, so that additional information is needed from other types of studies. Experimental studies are among these; however, these are only suitable for studying moderate, reversible, short term effects in persons who are healthy or only moderately ill. Their main advantage is that exposure conditions and subjects election are under the control of the investigator.

b. Animal studies, which can be subdivided into a number of categories depending on their length (acute, subchronic, chronic) or end-point (morbidity, mortality, carcinogenicity, irritation, etc.). Here, the investigator has full control over exposure conditions and health effects studied. However, the principle limitations lie in the fact that extrapolation from the studied animal species to man is always necessary. Also, while in human populations health effects with low incidences are often of interest (e.g., specific cancers), it is not feasible to study very large groups of animals to detect these low incidences. In practice, therefore, animal experiments are often carried out using very high experimental doses to compensate for the relatively small number of animals used and as a consequence, an additional extrapolation from high to low doses is also often necessary.

c. In vitro studies, in which effects of pollutants on cell or organ cultures are studied. These studies have the advantage that they are less costly than animal studies, and that results can generally be obtained in a shorter period of time. They are useful for studying mechanisms of action, but it is not usually possible to predict effects on whole organisms from their results in a quantitative way.

4.2 Criteria for the assessment of the impact of IAP on the community

The process of risk characterisation for indoor pollutants occurs through several phases: hazard identification, exposure assessment, dose-effect evaluation, and finally qualitative and quantitative risk assessment. The final product of this process may be an individual risk estimate per exposure unit or the evaluation of the incidence of the concerned effects in a given population. The risk characterisation through a multi-stage process as described above is particularly informative because, by dividing the analysis of the scenario of each pollutant into steps, it allows the separate recognition of the importance of each variable in the scenario and the prediction of the changes of frequency or severity of effects obtainable by modifying (increasing or decreasing) exposure.

For some types of IAP, our understanding of human health risk is well defined. For most indoor air pollutants, however, the risk assessment process has its limitations.

First, it has been applied successfully only to individual pollutants for which information is available for exposure and dose-response relationships and for which the effect is clear, certain, and measurable, such as mortality and cancer. Little progress has been made in applying the risk assessment process to environmental issues involving pollutant mixtures or effects for which the causes are difficult to ascertain precisely, such as in heart disease, allergic reactions, headache, and malaise. A different approach is needed for the assessment and characterisation of the risks associated with most indoor air pollutants.

A basic and simple criterion for assessing the importance of the health risk related to indoor pollution makes reference to the severity of the effect concerned and to the size of the population affected. Important issues for the community may come from severe health impacts, particularly when affecting a large segment of the population. Minor impacts, such as those related to discomfort or annoyance may, however, become important when a large number of individuals in the community are concerned.

4.3 The impact of IAP on humans' health

4.3.1 Respiratory health effects associated with exposure to IAP

Several effects on the respiratory system have been associated with exposure to IAP. These include acute and chronic changes in pulmonary function, increased incidence and prevalence of respiratory symptoms, augmentation of pre-existing respiratory symptoms, and sensitisation of the airways to allergens present in the indoor environment. Also, respiratory infections may spread in indoor environments when specific sources of infectious agents are present, or simply because the smaller indoor mixing volumes allow infectious diseases to spread more easily from one person to the next. The latter mechanism is particularly operative in schools, nursery schools, etc.

Observed changes in pulmonary function due to exposure to, e.g., tobacco smoke in the home, have mostly been due to acute or chronic airway narrowing leading to obstruction of air flow. This is measured as a reduction in the quantity of air that can be exhaled in one second after deep inspiration (FEVI), and a limitation in the various measures of air flow such as Peak Expiratory Flow (PEF), Maximum Mid Expiratory Flow (MMEF), and Maximum Expiratory Flow at x% of Forced Vital Capacity (MEFx). In growing children, it has also been suggested that lung development could be impaired by exposure to IAP.

Asthma, manifested by attacks of excessive airway narrowing leading to shortness of breath and wheezing, can be caused or aggravated by exposure to allergens at home, but it has also been associated with exposure to substances such as nitrogen dioxide and environmental tobacco smoke (ETS). Bronchitis, manifested in inflammatory changes in the airways and mucus hypersecretion has been linked to high levels of ambient air pollution in the past, and to exposure to ETS in the home in recent studies. Respiratory symptoms which have been associated with exposure to indoor air pollutants are symptoms mostly related to the lower airways such as cough, wheeze, shortness of breath and phlegm.

In contrast to the occurrence of chemical pollutants in indoor air, attention to which has grown considerably over the past two decades, the role of infectious agents in indoor air has been known for a long time. Infectious agents can be involved in the inflammatory conditions rhinitis, sinusitis, conjunctivitis and sinusitis, in pneumonia, in asthma and in alveolitis.

4.3.2 Allergic diseases associated with exposure to IAP

Allergic asthma and extrinsic allergic alveolitis (hypersensitivity pneumonitis) are the two most serious allergic diseases caused by allergens in indoor air. Allergic rhinoconjunctivitis and humidifier fever are other important diseases; it is not clear if or how the immunological system is involved in humidifier fever.

Allergic asthma is characterised by reversible narrowing of the lower airways. Pulmonary function during an attack shows an obstructive pattern in serious cases together with reduced ventilation capacity. Allergic asthma may be caused by exposure to indoor air pollutants, either acting as allergens or as irritants. Immunological specific IgE sensitisation to an airborne allergen is a major component of this disease, but non-specific hypersensitivity is also important for the asthmatic attacks occurring on exposure to irritants in the indoor air.

The prevalence of asthma varies considerably from country to country. Although asthmatic attacks seldom lead to death, the costs of medical care are considerable in terms of hospital admissions, medication, and lost work days.

Allergic rhinoconjunctivitis is also an IgE-mediated disease, but while asthma occurs in all age groups, allergic rhinoconjunctivitis is especially prevalent among children and young adults. The main symptoms are itching of the eye andlor the nose, sneezing, watery nasal secretion and some stuffiness of the nose. The severity of the symptoms varies with the exposure to the allergen. Individuals often suffer from both allergic asthma and allergic rhinoconjunctivitis and are seldom sensitive to only one allergen. Aeroallergens from house dust mites, pets, insects, moulds, and fungi in the indoor air have been shown to be associated with allergic asthma and/or rhinoconjunctivitis. Extrinsic allergic alveolitis, also called hypersensitivity pneumonitis, is characterised by recurrent bouts of pneumonitis or milder attacks of breathlessness and flu-like symptoms. Studies of the pulmonary function during an acute episode will usually show a restrictive pattern with a decreased diffusion capacity. The disease is believed to be an inflammatory reaction in the alveoli and bronchioles involving circulating antibodies and a cell-mediated immunological response to an allergen. For example it occurs in farmers as a result of handling mouldy hay ("farmer's lung") and in pigeon breeders due to bird droppings. However, the disease has also in a few cases been associated with exposure to IAP, most frequently related to humidifiers in homes and offices contaminated with bacteria, fungi, or protozoans.

Allergic asthma and extrinsic allergic alveolitis resolve with cessation of exposure to the allergen, but continued exposure in sensitised patients may result in permanent lung damage and death from pulmonary insufficiency.

Humidifier fever is a flu-like illness involving the immune system, in which X-ray abnormalities are usually absent. The exact cause is not clear. The disease may occur among persons exposed to humidification systems contaminated with microbial growth. The symptoms typically occur 4-8 h after the exposure on the first day back at work after a weekend, but resolve within 24 h. Despite continuous exposure the disease does not recur until after the next weekend. Even though pulmonary changes are seen during attacks of humidifier fever, the disease does not lead to permanent lung damage.

4.3.3 Cancer and effects on reproduction associated with exposure to IAP

Lung cancer is the major cancer which has been associated with exposure to IAP (radon or ETS). Asbestos exposure has been linked to cancer in workers and also in workers' family members, presumably due to asbestos fibres brought into the home on workers' clothing. However, there are no studies associating asbestos exposure in homes or public buildings from asbestos used as a construction material to the development of cancer. Effects on human reproduction have been associated with exposure to chemicals in the environment, but it is as yet unclear to what extent (if any) exposure to IAP is involved

4.3.4 Sensory effects and other effects on the nervous system associated with IAP

Sensory effects are defined as the perceptual response to environmental exposures. Sensory perceptions are mediated through the sensory systems and result in a conscious experience of smell, touch, itching, etc. Sensory effects are typically observed in buildings with indoor climate, problems because many chemical compounds found in the indoor air have odorous or mucosal irritation properties. Most indoor air chemicals with a measurable vapour pressure will be odorous when the concentration is high enough.

Sensory effects are important parameters in indoor air quality control for several reasons. They may appear as: (1) adverse health effects on sensory systems (e.g., environmentally-induced sensory dysfunctions); (2) adverse environmental perceptions which may be adverse per se or constitute precursors of disease to come on a long term basis (e.g., annoyance reactions, triggering of hypersensitivity reactions); (3) sensory warnings of exposure to harmful environmental factors (e-g., odour of toxic sulfides, mucosal irritation due to formaldehyde); (4) important tools in sensory bioassays for environmental characterisation (e.g.. using the odour criterion for general ventilation requirements or for screening of building materials to find those with low emissions of volatile organic compounds).

The senses responding to environmental exposure are not only hearing, vision, olfaction and taste, but also the skin and mucous membranes. As pointed out by WHO (1989), many different sensory systems that respond to irritants are situated on or near the body surface. Some of these systems tend to respond to an accumulated dose and their reactions are delayed. On the other hand, in the case of odor perception the reaction is immediate but also very much influenced by olfactory fatigue on prolonged exposures.

Responders are often unable to identify a single sensory system as the primary route of sensory irritation by airborne chemical compounds. The sensation of irritation is influenced by a number of factors such as previous exposures, skin temperature, competing sensory stimulation, etc. Since interaction and adaptation processes are characteristic of the sensory systems involved in the perception of odour and mucosal irritation, the duration of exposure influences the perception. Humans integrate different environmental signals to evaluate the total perceived air quality and assess comfort or discomfort. Comfort and discomfort by definition are psychological and for this reason the related symptoms, even when severe cannot be documented without using subjective reports. Sensory effects reported 10 be associated with IAP are in most cases multisensory and the same perceptions or sensations may originate from different sources. It is not known how different sensory perceptions are combined into perceived comfort and into the sensation of air quality. Perceived air quality is for example mainly related to stimulation of both the trigeminus and olfactorius nerves.

Several odorous compounds are also significant mucosal irritants, especially at high concentrations. The olfactory system signals the presence of odorous compounds in the air and has an important role as a warning system. In the absence of instrumentation for chemical detection of small amounts of some odorous vapours, the sense of smell remains the only sensitive indicator system. It is well known that environmental pollution can affect the nervous system. The effects of occupational exposure to organic solvents can be mentioned as an example. A wide spectrum of effects may be of importance, ranging from those at molecular level to behavioural abnormalities. Since the nerve cells of the CNS typically do not regenerate, toxic damage to them is usually irreversible. The nerve cells are highly vulnerable to any depletion in oxygen supply.

4.3.5 Cardiovascular effects associated with IAP

Increased mortality due to Cardiovascular Diseases (CVD) has been associated with exposure to ETS in some groups of non-smoking women married to smokers. Some investigators have also addressed the question whether total mortality is influenced by exposure to ETS, but results have been contradictory. As any effect on mortality would not be expected to occur until after many years of exposure, a problem in these types of study is the accuracy and reliability of the exposure classification. Attempts have also been made to

relate ETS to electrocardiographic abnormalities and cardiovascular symptoms, but results have been inconclusive.
Carbon monoxide (CO) exerts its influence primarily through binding to the haemoglobin (Hb) in blood. The affinity of CO to Hb is about 200 times higher than the affinity of oxygen to Hb, so that at relatively low levels of CO in the air. Oxygen is replaced by CO. The percentage of Hb bound to CO (O/O carboxyhaemoglobin) is a measure of recent exposure to CO. Organs with a high oxygen demand, such as the heart and the brain, are particularly susceptible to a reduced oxygenation caused by CO exposure. Early effects include reduction of time to onset of chest pain in exposed, exercising heart disease patients. At higher levels of exposure, myocardial infarctions may be triggered by CO.

5. Basic control strategies

There are some basic control methods for lowering concentrations of indoor air pollutants (Ashford & Caldart, 2008), which are described bellow:
Source Management includes source removal, source substitution, and source encapsulation. Source management is the most effective control method when it can be practically applied. Source removal is very effective. However, policies and actions that keep potential pollutants from entering indoor are even better than preventing IAQ problems. Source substitution includes actions such as selecting a less toxic art material or interior paint than the products which are currently in use. Source encapsulation involves placing a barrier around the source so that it releases fewer pollutants into the indoor air (e.g., asbestos abatement, pressed wood cabinetry with sealed or laminated surfaces). Local Exhaust is very effecting on removing point sources of pollutants before they can disperse into the indoor air by exhausting the contaminated air outside. Well known examples include restrooms and kitchens where local exhaust is used. Other examples of pollutants that originate at specific points and that can be easily exhausted include science lab and housekeeping storage rooms, printing and duplicating rooms, and vocational/ industrial areas such as welding booths. Ventilation through use of cleaner (outdoor) air to dilute the polluted (indoor) air that people are breathing. Generally, local building codes specify the quantity (and sometimes quality) of outdoor air that must be continuously supplied to an occupied area. For situations such as painting, pesticide application, or chemical spills, temporarily increasing the ventilation can be useful in diluting the concentration of noxious fumes in the air. Exposure Control includes adjusting the time of use and location of use. An example of time of use for school students would be to strip and wax floors on Friday after school is dismissed, so that the floor products have a chance to off-gas over the location of use deals with moving the contaminating source as far as possible from occupants, or relocating susceptible occupants. Air Cleaning primarily involves the filtration of particles from the air as the air passes through the ventilation equipment. Gaseous contaminants can also be removed, but in most cases this type of system should be engineered on a case-by-case basis.

6. Conclusions

As it has been clearly proven above, indoor air pollution is a major public concern issue, which can be characterised as "global environmental phenomenon". Also, it is obvious, that the causes which created this domestic environmental problem, such as modern way of living, decreased ventilation rates for energy conservation or increased use of synthetic materials in buildings, are not expected to be reduced (it is more probable that they are going to be increased). Nevertheless, the task of reducing levels of exposure to air pollutants is rather complex. It begins with an analysis to determine which chemicals are present in the air, at what levels, and whether likely levels of exposure are hazardous to human health and the environment. It must then be decided whether an unacceptable risk is present. When a problem is identified, mitigation strategies have to be developed and implemented so as to prevent excessive risk to public health in the most efficient and cost-effective way. In addition, analyses of air pollution problems are exceedingly complicated. Some are national in scope (such as the definition of actual levels of exposure of the population, the determination of acceptable risk, and the identification of the most efficient control strategies), while others are of a more basic character and are applicable in all countries (such as analysis of the relationships between chemical exposure levels, and doses and their effects). So, it is very essential for governments of all countries- especially the governments of the more developed ones- to adopt and implement these policies, in order to effectively face this worldwide issue in combination with the need of energy saving, the use of new building materials and the modern trend of living.

7. References

American Institute of Architects Research Corporation. (1979). Phase Two Report for the Development of Energy Performance Standards for New Buildings. *Report to U.S. Department of Housing and urban Development and U.S. Department of Energy*, Vol.197

American Society of Heating, Refrigerating and Air-Conditioning engineers. (1972). ASHRAE Handbook of fundamentals. *New York: American Society of Heating Refrigerating and Air-Conditioning Engineers, Inc.*, Vol.688

Anderson, D. J., & Hites, R. A. (1988). Chlorinated Pesticides in Indoor Air. *Environ. Sci. Technol.*, Vol.22, No 6 (June 1988) pp. 717-720, ISSN 0013-936X

Andersson, H.E.B. & Klevard Setterwall, A. (Eds). (1996). *The energy book. A resume of present knowledge and research,* The Swedish council for building research, ISBN 91-540-5744-2, Stockholm, Sweden

Ando, M., Katagiri, K., Tamura, K., Yamamoto, S., Matsumoto, M., Li, Y. F., Cao, S. R., Ji, R. D. & Liang C. K. (1996). Indoor and Outdoor Air Pollution in Tokyo and Beijing Supercities. *Atmos. Environ.*, Vol. 30, No 5 (1996), pp. 695-702, ISSN 1352-2310

Ashford, N. A. & Caldart, C.C. (2008). *Environmental Law, Policy, and Economics.Reclaiming the Environmental Agenda*, ISBN-10:0-262-01238-3, ISBN-13:978-0-262-01238-6, MIT.

Avol, E. L., Navidi, W. C. & Colome, S. D. (1998). Modeling Ozone Levels in and around Southern California Homes. *Environ. Sci. Technol.*, Vol. 32, No 4 (February 1998),pp. 463-468, ISSN 0013-936X

Baek, S.-O., Kim, Y.-S. & Perry, R. (1997). Indoor Air Quality in Homes, Offices, and Restaurants in Korean Urban Areas--Indoor/Outdoor Relationships. *Atmos. Environ.*, Vol. 31, pp. 529-544, ISSN 1352-2310

Battaglia, A., Capra, D., Queirazza, G. and Sampaolo, A. (1990). Radon exhalation rate in building materials and fly ashes in Italy, INDOOR AIR '90, *Proceedings of the 5th International Conference on Indoor Air Quality and Climate,* Vol.3, pp. 47-52. Toronto, 29 July - 3 August

Bochicchio, F., Campos Venuti, G., Felici, F., Grisanti, A., Grisanti, G., Moroni, F., Nuccetelli, C., Risica, S. and Tancredi, F. (1993). Characterisation of some parameters affecting the indoor radon exposure of the population, *Proceedings of the First International Workshop on Indoor Radon Remedial Action*, Rimini, Italy, 27th June - 2th July 1993

Brauer, M., Ryan, P. B., Suh, H. H., Koutrakis, P., Spengler, J. D., Leslie, N. P.& Billick, I. H. (1990). Measurements of Nitrous Acid inside Two Research Houses. *Environ. Sci. Technol.*, Vol. 24, No 10 (October 1990), pp. 1521-1527, ISSN 0013-936X

Brauer, M. & Spengler, J. D. (1994). Nitrogen Dioxide Exposures inside Ice Skating Rinks. *Am. J. Public Health*, Vol. 84, No 3 (March 1994), pp. 429-433, ISSN 0090-0036

Brauer, M., Lee, K., Spengler, J. D., Salonen, R. O., Pennanen, A., Braathen, O. A., Mihalikova, E., Miskovic, P., Nozaki, A., Tsuzuki, T., Rui-Jin, S., Xu, Y., Quing-Xiang, Z., Drahonovska, H., & Kjaergaard, S. (1997). Nitrogen Dioxide in Indoor Ice Skating Facilities: An International Survey. *J. Air Waste Manage. Assoc.*, Vol. 47, (October 1997), pp. 1095-1102, ISSN1047-3289

Brickus, L. S. R., Cardoso, J. N. & De Quino Neto, F. R. (1998). Distributions of Indoor and Outdoor Air Pollutants in Rio de Janeiro, Brazil: Implications to Indoor Air Quality in Bayside Offices. *Environ. Sci. Technol*, Vol. 32, No 22 (November 1998) pp. 3485-3490, ISSN 0013-936X

Brown, R. C. & Hoskins, J. A. (1993). Asbestos and Man-made Mineral Fibres. MRC toxicology Unit, Woodmansterne Road, Surrey, ISBN 0-948411-07-4, London

Brown, S. K., Sim, M. R., Abramson, M. J. & Gray, C. N. (1994). Concentrations of Volatile Organic Compounds in Indoor Air—A Review. *Indoor Air*, Vol. 4, No 2 (June 1994), pp. 123-134, ISSN 0905-6947

Bukowski, J. A. and Meyer, L. W. (1995). Simulated Air Levels of Volatile Organic Compounds Following Different Methods of Indoor Insecticide Application. *Environ. Sci. Technol*, Vol. 29, No 3 (March 1995), pp. 673-676, ISSN 0013-936X

Burge, P.S., Jones, P. & Robertson A.S. (1990). Sick Building Syndrome: environmental comparisons of sick and healthy buildings. *Indoor Air*, Vol. 90, 1, pp. 479-84

California Environmental Protection Agency, Office of Environmental. (1997). Health Hazard Assessment. Health Effects of Exposure to Environmental Tobacco Smoke. *Final Report*, September 1997.

Campos Venuti G., Colilli, S., Grisanti A., Grisanti G., Monteleone, G., Risica, S., Gobbi, G., Leogrande, M.P., Antonini, A. and Borio R. (1984). Indoor exposure in a Region of Central Italy, *Radiat. Prot. Dosim*, Vol. 7 (1-41,271-274)

Castrén, O., Voutilainen, A., Winqist, K. and Mäikeläinen, I. (1985). Studies of high indoor radon areas in Finland. *Science of the Total Environment*. Vol.45, (October 1985), pp. 311-318, ISSN 0048-9697

Cermak, J. E. (1976). Nature of air flow near buildings. *ASHRAE Trans*, Vol. 82 (Pt.1), pp. 1044-1054

Chan, C.-C., Ozkaynak, H., Spengler, J. D. & Sheldon, L. (1991). Driver Exposure to Volatile Organic Compounds, CO, Ozone, and NO2 under Different Driving Conditions. *Environ. Sci. Technol.*, Vol. 25, No 6 (June 1991), pp. 964-972, ISSN 0013-936X

Chang, J. C. S., & Krebs, K. A. (1992). Evaluation of Para-Dichlorobenzene Emissions from Solid Moth Repellant as a Source of Indoor Air Pollution. *J. Air Waste Manage. Assoc.*, Vol. 42, No 9 (September 1992), pp. 1214-1217, ISSN1047-3289

Chapin, F.S., jr. (1974). Human activity patterns in the City: Things People Do in time and in space. New York: John Wiley & sons, Inc., p. 272, ISBN 10: 0-471-14563-7, ISBN 13: 9780471145639

Chuang, J. C., Mack, G. A., Kuhlman, M. R. & Wilson, N. K. (1991). Polycyclic Aromatic Hydrocarbons and Their Derivatives in Indoor and Outdoor Air in an Eight-Home Study. *Atrnos. Environ.*, Vol. 25B, pp. 369-380, ISSN 1352-2310

Cliff, K.D. (1992). Thoron daughter concentrations in UK-homes, *Radiat. Prot. Dosim.*, Vol.45(1-4) Supplement, pp. 361-366

Colome, S. D., Kado, N. Y., Jaques, P. & Kleinman, M. (1992). Indoor-Outdoor Air Pollution Relations: Particulate Matter Less Than 10 / zm in Aerodynamic Diameter (PM 10) in Homes of Asthmatics. *Atmos. Environ.*, Vol. 26/1, pp. 2173-2178, ISSN 1352-2310

Committee on the Medical Effects of Air Pollutants: Guidance on the Effects on Health of Indoor Air Pollutants 2004; Department of Health, London

Converse, P.E. (1968). Time budgets. In D. L. Sills, Ed. *International Encyclopedia of the Social sciences*. Vol. 16. pp. 42-47 New York: The Macmillan Company and the Free Press

Crump, D. R. & Gardiner, D. (1989). Sources and Concentrations of Aldehydes and Ketones in Indoor Environments in the UK. *Environ. Int.*, Vol. 15, pp. 455-462, ISSN0160-4120

Crump, D. R., (1995). Volatile Organic Compounds in Indoor Air," in Issues in *Environmental Science and Technology* (R. E. Hester and R. M. Harrison, Eds.), Chap. 4, pp. 109-124, Royal Chem. Soc., Letchworth, UK

Daisey, J. M., Hodgson, A. T., Fisk, W. J., Mendell, M. J. & Ten Brinke J. (1994). Volatile Organic Compounds in Twelve California Office Buildings: Classes, Concentrations, and Sources. *Atmos. Environ.*, Vol. 28, No. 22 (December 1994), pp. 3557-3562, ISSN 1352-2310

Davenport, A. G. (1960). A rationale for determination of design wirx: velocities. *Proceedings of the ASCE Journal of the Structures Division*, Vol. 86, pp. 39-66

Davidson, C. I., Lin, S.-F., Osborn, J. F., Pandey, M. R., Rasmussen, R. A. & Khalil, M. A. K. (1986). Indoor and Outdoor Air Pollution in the Himalayas. *Environ. Sci. Technol.*, Vol. 20, No. 6 (June 1986), pp. 561-567, ISSN 0013-936X

De Meijer, R.J., Stoop, P. and Put, L. (1992). Contribution of radon flows and radon sources to the radon concentration in a dwelling, *Radiat. Prot. Dosim.* Vol.45 (1-4) Supplement, pp. 439-442

Deyuan, T. (1993). Indoor and outdoor air radon concentration level in China", INDOOR AIR '93, *Proceedings of the 6th International Conference on Indoor Air Quality and Climate*, Vol.4, pp. 459-463, Helsinki, Finland, July 4-81993

Dickerhoff, D., Grinsrud, D. T. & Shohl, B. (1980). Infiltration and Air-Conditioning: A Case Study. Lawrence Berkeley Laboratory *Report LBL-11674*. Berkeley, Cal.: Lawrence Berkeley Laboratory

Dimitroulopoulou, C., Ashmore, S.R., Hill, M.T.R, Byrne, M.A. & Kinnersley, R.(2006). INDAIR: A probabilistic model of indoor air pollution in UK homes. *Atmos Environ*, Vol. 40, pp. 6362-6379, ISSN 1352-2310

Doi, M. and Kobayashi, S. (1994). Characterization of Japanese wooden houses with enhanced radon and thoron concentrations, *Health Phys.*, Vol.66, No3, pp. 274-282; ISSN 0017-9078

Dor, F., Le Moullec, Y. & B. Festy. (1995). Exposure of City Residents to Carbon Monoxide and Monocyclic Aromatic Hydrocarbons during Commuting Trips in the Paris Metropolitan Area. *J. Air Waste Manage. Assoc.*, Vol. 45 (February 1995), pp. 103-110, ISSN 1047-3289

Drakou, G., Zerefos, C., Ziomas, I. & Voyatzaki, M. (1998). Measurements and Numerical Simulations of Indoor 0 3 and NO X in Two Different Cases. *Atmos. Environ.*, Vol. 32, pp. 595-610

Dremetsika, A.V., Siskos, P.A. & Bakeas, E.B. (2005). Determination of formic and acetic acid in the interior atmosphere of display cases and cabinets in Athens Museums by reverse phase high performance liquid chromatography. *Indoor Built. Environ.*, Vol.14, No:1, pp. 51 -58

Druzik, J. R., Adams, M. S., Tiller, C. & Cass, G. R. (1990). The Measurement and Model Predictions of Indoor Ozone Concentrations in Museums. *Atmos. Environ.*, Vol. 24A, pp. 1813-1823, ISSN 1352-2310

Ekinja, I., Grabaric, Z. & Grabaric, B. S. (1995). Monitoring of Pyrocatechol Indoor Air Pollution. *Atmos. Environ.*, Vol. 29, pp. 1165-1170, ISSN 1352-2310

European Collaborative Action(ECA). (1998). Indoor air quality and its impact on man. *Report No 15*, ISBN 92-827-01 19-0

European Collaborative Action (ECA). (1991). indoor air quality & its impact on man (formerly COST Project 61 3). *Environment and Quality of Life Report No. 10*, Effects of Indoor Air Pollution on Human Health

Febo, A. & Perrino, C. (1991). Prediction and Experimental Evidence for High Air Concentration of Nitrous Acid in Indoor Environments. *Atmos. Environ.*, Vol. 25A, pp. 1055-1061, ISSN 1352-2310

Febo, A. & Perrino, C. (1995). Measurement of High Concentration of Nitrous Acid inside Automobiles. *Atmos. Environ.*, Vol. 29, pp. 345-351, ISSN 1352-2310

Federer, C. A. (1971). Effect of trees in modifying urban microclimate, pc, 23-28. In S. Little, and J. H. Noyes, Eds. Trees and Forests Urbanizing Environment. Amherst, Mass.: University of Massachusetts Cooperative Extension Service

Feinman, S. E. (1988). Formaldehyde Sensitivity and Toxicity, *CRC Press*, Boca Raton, FL

Fernandez-Bremauntz, A. A., & Ashmore, M. R. (1995). Exposure of Commuters to Carbon Monoxide in Mexico City. Measurement of In-Vehicle Concentrations. *Atmos. Environ.*, Vol. 29, pp. 525-532, ISSN 1352-2310

Finlayson-Pitts, B.J. & Pitts, J.N. (1999). Chemistry of the upper and lower atmosphere Academic Press, 0-12-257060-x

Fischer, M. L., Bentley, A. J., Dunkin, K. A., Hodgson, A. T., Nazaroff, W. W., Sextro, R. G. & Daisey, J. M. (1996). Factors Affecting Indoor Air Concentrations of Volatile Organic Compounds at a Site of Subsurface Gasoline Contamination. *Environ. Sci. Technol.*, Vol. 30, No. 10 (September 1996), pp. 2948-2957, ISSN 0013-936X

Flachsbart, P. G., Mack, G. A., Howes, J. E. & Rodes, C. E. (1987). Carbon Monoxide Exposures of Washington Commuters. *JAPCA*, Vol. 37, No. 2, pp. 135-142, ISSN 0894-0630

Florig, H. K. (1997). China's Air Pollution Risks. *Environ. Sci. Technol.*, Vol. 31, pp. 274A-279A

Foster, W. M., Jiang, L., Stetkiewicz, P. T. & Risby, T. H. (1996). Breath Isoprene--Temporal Changes in Respiratory Output after Exposure to Ozone. *J. Appl. Physiol.*, Vol. 80, pp. 706-710

Geiger, R. (1965). The climate near the ground. Cambridge, Mass.: *Harvard University Press*, 611pp

Gelmont, D., Stein, R. A. & Mead, J. F. (1981). Isoprene--The Main Hydrocarbon in Human Breath. *Biochem. Biophys. Res. Commun.*, Vol. 99, No. 4 (April 1981), pp. 1456-1460, ISSN 0006-291X

Georghiou, P. E., Blagden, P., Snow, D. A., Winsor, L. & Williams, D. T. (1991). Mutagenicity of Indoor Air Containing Environmental Tobacco Smoke: Evaluation of a Portable PM-10 Impactor Sampler. *Environ. Sci Technol.*, Vol. 25, No. 8 (August 1991), pp. 1496-1500, ISSN0013-936X

Gesell, T.F. (1983). Background atmospheric Rn-222 concentrations outdoors and indoors: a review, *Health Phys.*, Vol.45, pp.289-302, ISSN 0017-9078

Gibson, U. E. & Cawley, R. E. (1977). The heat pump solar collector interface—A practical experiment. *Appliance Eng.* Vol. 11(4), pp. 68-71

Gold, D. R., Allen, G., Damokosh, A., Serrano, P., Hayes, C. & Castillejos, M. (1996). Comparison of Outdoor and Classroom Ozone Exposures for School Children in Mexico City. *J. Air Waste Manage. Assoc.*, P. B. Vol. 46 (April 1996), pp. 335-342, ISSN 1047-3289

Grasty, R.L. (1994). Summer outdoor radon variations in Canada and their relation to soil moisture, *Health Phys.* Vol.66, No.2, pp.185-193, ISSN 0017-9078

Gravesen, S., Larson, L., Gyntelberg, F. & Skov, P. (1990). The role of potential immunogenic components of dust (MOD) in the sick building syndrome. *Indoor Air*, Vol. 90, 1, pp. 9-14, ISSN 0905-6947

Guo, Q., Shimo, M., Ikebe, Y. and Minato, S. (1992). The study of thoron and radon progeny concentrations in dwellings in Japan, *Radiat. Prot. Dosim.*, Vol.45(1-4) Supplement, pp. 357-359

Halios, C.H. & Helmis, C. G. (2007). On the estimation of characteristic indoor air quality parameters using analytical and numerical methods. *Science of the Total Environment*, Vol.381, pp. 222-232

Hanrahan, L. P., Anderson, H. A., Daily, K. A., Eckmann, A. D. & Kanarek, M. S. (1985). Formaldehyde Concentrations in Wisconsin Mobile Homes. *JAPCA*, Vol. 35, pp. 1164-1167, ISSN 0894-0630

Heavner, D. L., Ogden, M. W. & Nelson, P. R. (1992). Multisorbent Thermal Desorption/Gas Chromatography/Mass Selective Detection Method for the Determination of Target Volatile Organic Compounds in Indoor Air. *Environ. Sci. Technol.*, Vol. 26, No. 9 (September 1992), pp. 1737-1746, ISSN 0013-936X

Helmis, C.G., Tzoutzas, J., Flocas, H.A., Panis, V., Stathopoulou, O.I., Assimakopoulos V.D., Halios, C.H., Apostolatou, M., Sgouros, G. & Adam, E. (2007). Indoor Air Quality in a Dentistry Clinic. *Science of the Total Environment*, Vol.377, pp. 349-365

Helmis, C.G., Assimakopoulos V.D., Flocas, H.A., Stathopoulou, O.I., Sgouros G. & Hatzaki M. (2009). Indoor air quality assessment in the Air Traffic Control Tower of the Athens Airport, Greece. *Environmental Monitoring and Assessment*, Vol.148, pp. 47-60

Hisham, M. W. M. & Grosjean, D. (1991). Sulfur Dioxide, Hydrogen Sulfide, Total Reduced Sulfur, Chlorinated Hydrocarbons, and Photochemical Oxidants in Southern California Museums. *Atmos. Environ.*, Vol. 25A, pp. 1497-1505, ISSN 1352-2310

Hodgson, A. T., & Daisey, J. M. & Grot, R. A. (1991). Sources and Source Strengths of Volatile Organic Compounds in a New Office Building. *J. Air Waste Manage. Assoc.*, Vol. 41, No. 11 (November 1991), pp. 1461-1468, ISSN 1047-3289

Hodgson, A. T., Garbesi, K., Sextro, R. G. & Daisey, J. M. (1992). Soil-Gas Contamination and Entry of Volatile Organic Compounds into a House near a Landfill. *J. Air Waste Manage. Assoc.*, Vol. 42, No. 3 (March 1992), pp. 277-283, ISSN 1047-3289

Hoek, G., Brunekreef, B. & Hofschreuder, P. (1989). Indoor Exposure to Airborne Particles and Nitrogen Dioxide during an Air Pollution Episode. *JAPCA*, Vol. 39, pp. 1348-1349, ISSN 0894-0630

Hopper, R.D., Levy, R.A., Rankin, R.C. and Boyd, M.A. (1991). National ambient radon study, *Proceedings of the International Symposium on Radon and Radon Reduction Technology*, Philadelphia, PA (USA)

House of Commons Environment Committee. (1991). Indoor Pollution, *6th report, HMSO*, London.

Howard, C., & Corsi, R. L. (1998a). Volatilization of Chemicals from Drinking Water to Indoor Air: The Role of Residential Washing Machines. *J. Air Waste Manage. Assoc.*, Vol. 48, (October 1998), pp. 907-914, ISSN 1047- 3289

Howard, E. M., McCrillis, R. C., Krebs, K. A., Fortman, R., Lao, H. C. and Guo, Z. (1998b). Indoor Emissions from Conversion Varnishes. *J. Air Waste Manage. Assoc.*, Vol. 48, (October 1998), pp. 924-930, ISSN 1047- 3289

Jakobi, G. and Fabian, P. (1997). Indoor/Outdoor Concentrations of Ozone and Peroxyacetyl Nitrate (PAN). *Int. J. Biometeorol.*, Vol. 40, No. 3 (May 1997), pp. 162-165, ISSN 0020-7128

Jones, A. W., Lagesson, V. & Tagesson, C. (1995). Origins of Breath Isoprene. *J. Clin. Pathol.*, Vol. 48, No. 10 (October 1995), pp. 979-980, ISSN 0021-9746

Kelly, T. J. (1996). Determination of Formaldehyde and Toluene Diisocyanate Emissions from Indoor Residential Sources. *California Air Resources Board, Research Division*, 2020 L Street, Sacramento, CA 95814, Final Report, Contract Number 93-315, November, 1996.

Kostiainen, R., (1995). Volatile Organic Compounds in the Indoor Air of Normal and Sick Houses. *Atmos. Environ.*, Vol. 29, pp. 693-702, ISSN 1352-2310

Koushki, P. A., Al-Dhowalia, K. H. & Niaizi, S. A. (1992). Vehicle Occupant Exposure to Carbon Monoxide. *J. Air Waste Manage. Assoc.*, Vol. 42, No. 12 (December 1992), pp. 1603-1608, ISSN 1047-3289

Koutrakis, P., Brauer, M., Briggs, S. L. K. & Leaderer, B. P. (1991). Indoor Exposures to Fine Aerosols and Acid Gases. *Environ. Health Perspect.*, Vol. 95, pp. 23-28

Koutrakis, P., Briggs, S. L. K. & Leaderer, B. P. (1992). Source Apportionment of Indoor Aerosols in Suffolk and Onondaga Counties, New York. *Environ. Sci. Technol.*, Vol. 26, No. 3 (March 1992), pp. 521-527, ISSN 0013-936X

Leovic, K. W., Sheldon, L. S., Whitaker, D. A., Hetes, R. G., Calcagni, J. A. & Baskir, J. N. (1996). Measurement of Indoor Air Emissions from Dry-Process Photocopy Machines. *J. Air Waste Manage. Assoc.*, Vol. 46, pp. 821-829, ISSN 1047-3289

Levy, J. I., Lee, K., Spengler, J. D. & Yanagisawa, Y. (1998). Impact of Residential Nitrogen Dioxide Exposure on Personal Exposure: An International Study. *J. Air Waste Manage. Assoc.*, Vol. 48, pp. 553-560, ISSN 1047-3289

Lewis, C. W., & Zweidinger, R. B. (1992). Apportionment of Residential Indoor Aerosol VOC and Aldehyde Species to Indoor and Outdoor Sources, and Their Source Strengths. *Atmos. Environ.*, Vol. 26A, pp. 2179-2184, ISSN 1352-2310

Lewtas, J., Goto, S., Williams, K., Chuang, J. C., Petersen, B. A. & Wilson, N. K. (1987). The Mutagenicity of Indoor Air Particles in a Residential Pilot Field Study: Application and Evaluation of New Methodologies. *Atmos. Environ.*, Vol. 21, pp. 443-449, ISSN 1352-2310

Löfroth, G., Ling, P. I. & Agurell, E. (1988). Public Exposure to Environmental Tobacco Smoke. *Mutat. Res.*, Vol. 202, pp. 103-110, ISSN 0027-5107

Ligocki, M. P., Salmon, L. G., Fall, T., Jones, M. C., Nazaroff, W. W. & Cass, G. R. (1993). Characteristics of Airborne Particles inside Southern California Museums. *Atmos. Environ.*, Vol. 27A, pp. 697-711, ISSN 1352-2310

Liu, L.-J. S., Koutrakis, P., Suh, H. H., Mulik, J. D. & Burton, R. M. (1993). Use of Personal Measurements for Ozone Exposure Assessment: A Pilot Study. *Environ. Health Perspect.*, Vol. 101, pp. 318-324

Löfroth, G., Stensman, C. & Brandhorst-Satzkorn, M. (1991). Indoor Sources of Mutagenic Aerosol Particulate Matter: Smoking, Cooking, and Incense Burning. *Mutat. Res.*, Vol. 261, pp. 21-28, ISSN 0027- 5107

Marbury, M. C., Harlos, D. P., Samet, J. M. & Spengler, J. D. (1988). Indoor Residential NO 2 Concentrations in Albuquerque, New Mexico. *JAPCA*, Vol. 38, pp. 392-398, ISSN 0894-0630

McCurdy, T., Glen, G., & Smith, L. (2000). The National Exposure Research Laboratory's Consolidated Human Activity Database. *J Exposure Analysis Environ Epidemiol*, Vol. 10, No. 6 (November 2000), pp. 566-578, ISSN 1559-0631

McKone, T. E. (1987). Human Exposure to Volatile Organic Compounds in Household Tap Water: The Indoor Inhalation Pathway. *Environ. Sci. Technol.*, Vol. 21, No. 12 (December 1987), pp. 1194-1201, ISSN 0013-936X

Mclntyre, D. A. (1980) Indoor Climate. London: Applied Science Publishers Ltd. 443 pp, ISBN 0853348685

Mendis, S., Sobotka, P. A. & Evler, D. E. (1994). Pentane and Isoprene in Expired Air from Humans--Gas Chromatographic Analysis of a Single Breath. *Clin. Chem.*, Vol. 40, No. 8 (August 1994), pp. 1485-1488, ISSN 0009-9147

Meyer, B. & Hermanns, K. (1986). Formaldehyde Release from Wood Products. An Overview. in Formaldehyde Release from Wood Products, ACS Symposium Series 316, Chapter 1, pp. 1-16, *Am. Chem. Soc.*, Washington, DC, ISBN 13: 9780841209824

Michelson, W. (1973). Time-budgets in environmental research: Some introductory considerations, pp. 262-268. In W. F. E. Preiser, Ed. *Environmental Design Research. Vol. 2. symposia and Workshops*. Fourth International EDRA Conference. Stroudsburg, Pa.: Dowden, Hutchinson, and Ross, Inc.

Miller, J. (1990). Fungi as contaminants in indoor air. *Indoor Environment,* Vol. 90, 5, pp. 51-64

Mitra, S., & Ray, B. (1995). Patterns and Sources of Polycyclic Aromatic Hydrocarbons and Their Derivatives in Indoor Air. *Atmos. Environ.*, Vol. 29, pp. 3345-3356, ISSN 1352-2310

Morey, A. (1990). Biological contaminants in the Indoor Environment ASTM. ISBN-EB: 978-0-8031-5131-4

Moschandreas, D. J. & Morse S. S. (1979). The relationship between energy conservation measures and exposure to indoor toxic pollutants. *Paper No. 27 b.*, presented at the American Institute of Chemical Engineers 87th National Meeting, Boston, Massachusetts, August 19-22, 1979

Moschandreas, D. J., Ed. (1978). Indoor Air pollution in the Residential Environment. Vol. II. Field Monitoring Protocol, Indoor Episodic Pollutant Release Experiments and Numerical Analyses. *U.S. Environmental Protection Agency Report No. EPA-600-7-78-229b.* Research triangle Park: U.S. Environmental Protection Agency, Environmental Monitoring and Support Laboratory, 1978, 240 pp.

Moschandreas, D. J., Zabransky, J. & Pelton, D.J. (1980). Comparison of Indoor-Outdoor concentrations of Atmospheric Pollutants. *GEOMET Report No. ES-823.* Palo Alto, Cal.: Electric Power Research Institute

Moschandreas, D. J., & O'Dea, D. S. (1995). Measurement of Perchloroethylene Indoor Air Levels Caused by Fugitive Emissions from Unvented Dry-to-Dry Dry Cleaning Units. *J. Air Waste Manage. Assoc.*, Vol. 45 (February 1995), pp. 111-115, ISSN 1047-3289

Moseley, C. L., & Meyer, M. R. (1992). Petroleum Contamination of an Elementary School: A Case History Involving Air, Soil-Gas, and Groundwater Monitoring. *Environ. Sci. Technol.*, Vol. 26, No. 1 (January 1992), pp. 185-192, ISSN0013-936X

Mumford, J. L., He, X. Z., Chapman, R. S., Cao, S. R., Harris, D. B., Li, X. M., Xian, Y. L., Jiang, W. Z., Xu, C. W., Chuang, J. C., Wilson, W. E. & Cooke, M. (1987). Lung Cancer and Indoor Air Pollution in Xuan Wei, China. *Science,* Vol. 235, No. 4785 (January 1987), pp. 217-220, ISSN 0036-8075

Mumford, J. L., Williams, R. W., Walsh, D. B., Burton, R. M., Svendsgaard, D. J., Chuang, J. C., Houk, V. S. & Lewtas, J. (1990). Indoor Air Pollutants from Unvented Paraffin Heater Emissions in Leaderer, B. P., P. M. Boone, and S. K. Hammond, "Total Particle, Sulfate, and Acidic Aerosol Emissions from Paraffin Space Heaters. *Environ. Sci. Technol.*, Vol. 24, No. 6 (June 1990), pp. 908-912, ISSN 0013-936X

Neas, L. M., Dockery, D. W., Ware, J. H., Spengler, J. D., Ferris, B. G. Jr. & Speizer, F. E. (1994). Concentration of Indoor Particulate Matter as a Determinant of Respiratory Health in Children. *Am. J. Epidemiol.*, Vol. 139, No. 11 (June 1994), pp. 1088-1099, ISSN 0002-9262

Nelson, P. R., Kelly, S. P. & Conrad, F. W. (1998). Studies of Environmental Tobacco Smoke Generated by Different Cigarettes. *J. Air Waste Manage. Assoc.*, Vol. 48 (April 1998), pp. 336-344, ISSN 1047-3289

NENOECD (Nuclear Energy Agency Organisation for Economic Co-operation and Development) (1979). Exposure to Radiation from the Natural Radioactivity in Building Materials, Expert Group Report, OECD, Paris

Nero, A.V. Jr. (1988).In:Radon and its decay products in indoor air: an overview, In *Radon and its decay products in Indoor Air,* W.W. Nazaroff and A.V. Nero Jr. edit., pp.1-56, Wiley Interscience, ISBN 0-471-62810-7

Nero, A.V. Jr. (1989). Earth, Air, Radon and Home. *Physics Today.* Vol. 42, No 4,(April 1989), pp. 32-39, ISSN 0031-9228

Offermann, F. J., Loiselle, S. A., Hodgson, A. T., Gundel, L. A. & Daisey, J. M. (1991). A Pilot Study to Measure Indoor Concentrations and Emission Rates of Polycyclic Aromatic Hydrocarbons. *IndoorAir*, Vol. 4, pp. 497-512, ISSN 0905-6947

Ott, W., Switzer, P. & Willits, N. Carbon Monoxide Exposures inside an Automobile Traveling on an Urban Arterial Highway. (1994). *J. Air Waste Manage. Assoc.*, Vol. 44 (August 1994), pp. 1010-1018, ISSN 1047-3289

Ott, W. R. (1995). Human activity patterns: A review of the literature for air pollution exposure estimation. SIMS Technical report. Stanford Cal.: University, Department of statistics

Ottensman, J. R. (1972). Systems of Urban Activities and Time: An interpretive Review of the Literature. *An Urban Studies Research Paper*. Chapel Hill, N.C.: University of North Carolina, Center for Urban and Regional Studies, 1972. 45pp.

Pennanen, A. S., Salonen, R. O., Aim, S., Jantunen, M. J. & Pasanen, P. (1997). Characterization of Air Quality Problems in Five Finnish Indoor Ice Arenas. *J. Air Waste Manage. Assoc.*, Vol. 47 (October 1997), pp. 1079-1086, ISSN 1047-3289

Perry, R. & Kirk, P.W. (1986). Indoor and Ambient Air Quality. *Selper, London*

Perry, R., & Gee, I. L. (1994). Vehicle Emissions and Effects on Air Quality: Indoors and Outdoors. *Indoor Environ.*, Vol. 3, pp. 224-236

Peterka, J. A. & Cermak, J. E. (1977). Turbulence in Building wakes, pp. 447-463. In K.J. Eaton, Ed. *Proceedings of the fourth international conference on wind effects on buildings and structures*, Heathrow, 1975. New York: Cambridge University Press, ISBN 0-521-20801-7

Petreas, M., Liu, K.-S., Chang, B.-H., Hayward, S. B. & Sexton, K. (1988). A Survey of Nitrogen Dioxide Levels Measured inside Mobile Homes. *JAPCA*, Vol. 38, pp. 647-651, ISSN 0894-0630

Phillips, M., Greenberg, J. & Awad, J. (1994). Metabolic and Environmental Origins of Volatile Organic Compounds in Breath. *J. Clin. Pathol.*, Vol. 47, No. 11 (November 1994), pp. 1052-1053, ISSN 0021-9746

Pitts, J. N., Wallington, T. J., Biermann, H. W. & Winer, A. M. (1985). Identification and Measurement of Nitrous Acid in an Indoor Environment. *Atmos. Environ.*, Vol. I9, pp. 763-767, ISSN 1352-2310

Pitts, J. N., Biermann, H. W., Tuazon, E. C., Green, M., Long, W. D. & Winer, A. M. (1989). Time-Resolved Identification and Measurement of Indoor Air Pollutants by Spectroscopic Techniques: Gaseous Nitrous Acid, Methanol, Formaldehyde, and Formic Acid. *JAPCA*, Vol. 39, pp. 1344-1347, ISSN 0894-0630

Price, J.G., Rigby, J.G., Christensen, L., Hess, R., LaPointe, D.D., Ramelli, A.R., Desilets, M., Hopper, R.D., Kluesner, T. and Marshall, S. (1994). Radon in outdoor air in Nevada, *Health Phys*. Vol.66, No.4, pp. 433-438, ISSN 0017-9078

Priorities for Indoor Air Research and Action. (1991). International Conference, Montreux.

Quackenboss, J. J., Lebowitz, M. D. & Crutchfield, C. D. (1989). Indoor-Outdoor Relationships for Particulate Matter: Exposure Classifications and Health Effects. *Environ. Int.*, Vol. I5, pp. 353-360, ISSN 0160-4120

Reiss, R., Ryan, P. B. & Koutrakis, P. (1994). Modeling Ozone Deposition onto Indoor Residential Surfaces. *Environ. Sci. Technol.*, Vol. 28, No. 3 (March 1994), pp. 504-513, ISSN 0013-936X

Reiss, R., Ryan, P. B., Tibbetts, S. J. & Koutrakis, P. (1995a). Measurement of Organic Acids, Aldehydes, and Ketones in Residential Environments and Their Relation to Ozone. *J. Air Waste Manage. Assoc.*, Vol. 45 (October 1995), pp. 811-822, ISSN 1047-3289

Reiss, R., Ryan, P. B., Koutrakis, P. & Tibbets, S. J. (1995b) Ozone Reactive Chemistry on Interior Latex Paint. *Environ. Sci. Technol.*, Vol. 29, No. 8 (August 1995), pp. 1906-1912, ISSN 0013-936X

Rob, E, M.C., Rannou, A. and Le Bronec, J. (1992). Radon measurement in the environment in France, *Radiat. Prot. Dosim.* Vol.45(1-4) Supplement, pp. 455-457

Romieu, I., Lugo, M. C., Colome, S., Garcia, A. M., Avila, M. H., Geyh, A., Velasco, S. R. & Rendon, E. P. (1998). Evaluation of Indoor Ozone Concentration and Predictors of Indoor-Outdoor Ratio in Mexico City. *J. Air Waste Manage. Assoc.*, Vol. 48 (April 1998), pp. 327-335, ISSN 1047-3289

Rowe, D. R., A1-Dhowalia, K. H., & Mansour, M. E. (1989). Indoor- Outdoor Carbon Monoxide Concentrations at Four Sites in Riyadh, Saudi Arabia. *J. Air Waste Manage. Assoc.*, Vol. 39, No. 8 (August 1989), pp. 1100-1102, ISSN 1047-3289

Rowe, D. R., Al-Dhowalia, K. H. & Mansour, M. E. (1991). Indoor- Outdoor Nitric Oxide and Nitrogen Dioxide Concentrations at Three Sites in Riyadh, Saudi Arabia. *J. Air Waste Manage. Assoc.*, Vol. 41, No. 7 (July 1991), pp. 973-976, ISSN 1047-3289

Ryan, P. B., Soczek, M. L., Spengler, J. D. & Billick, I. H. (1988). The Boston Residential NO 2 Characterization Study: I. Preliminary Evaluation of the Survey Methodology. *JAPCA*, Vol. 38, pp. 22-27, ISSN 0894-0630

Salmon, L. G., Nazaroff, W. W., Ligocki, M. P., Jones, M. C. & Cass, G. R. (1990). Nitric Acid Concentrations in Southern California Museums. *Environ. Sci. Technol.*, Vol. 24, No. 7 (July 1990), pp. 1004-1012, ISSN 0013-936X

Santamouris, M., Papanikolaou, N., Livada, I., Koronakis, I., Georgakis, C., Argiriou, A. & Assimakopoulos D. (2001). On the impact of Urban climate on the energy consumption of buildings. *Solar Energy*, Vol.70, pp. 201-216

Schaeffer, V. H., Bhooshan, B., Chen, S.-B., Sonenthal, J. S. & Hodgson, A. T. (1996). Characterization of Volatile Organic Chemical Emissions from Carpet Cushions. *J. Air Waste Manage. Assoc.*, Vol. 46 (September 1996), pp. 813-820, ISSN 1047-3289

Sciocchetti, G., Clemente, G.F., Ingrao, G. and Scacco, F. (1983). Results of a survey on radioactivity of building materials in Italy. *Health Phys.* Vol.45, No.2 (August 1983), pp.385-388 ISSN 0017-9078

Sciocchetti, G., Bovi, M., Cotellessa, G., Baldassini, P.G., Battella, C. and Porcu, I. (1992). Indoor radon and thoron surveys in high radioactivity areas of Italy, *Radiat. Prot. Dosim.*, Vol.45(1-4) Supplement, pp. 509-513

Sexton, K., Petreas, M. X. & Liu, K.S. Formaldehyde Exposures inside Mobile Homes (1989). *Environ. Sci. Technol.*, Vol. 23, No. 8 (August 1989), pp. 985-988, ISSN 0013-936X

Shah, J. J. & Singh, H. B. (1988). Distribution of Volatile Organic Chemicals in Outdoor and Indoor Air. *Environ. Sci. Technol.*, Vol. 22, No. 11 (November 1988), pp. 1381-1388, ISSN 0013-936X

Shair, F. H., & Heitner, K. L. (1974). Theoretical Model for relating indoor pollution concentrations to those outside. *Environ. Sci. Technol*, Vol. 8, No. 5 (May 1974), pp. 444-451, ISSN 0013-936X

Shepherd, J. L., Corsi, R. L. & Kemp, J. (1996). Chloroform in Indoor Air and Wastewater: The Role of Residential Washing Machines. *J. Air Waste Manage. Assoc.*, Vol. 46 (July 1996), pp. 631-642, ISSN 1047-3289

Sherman, M. H. (1981). Air Infiltration in Buildings. Berkeley, Cal.: University of California, Ph.D. Dissertation,. (*published as Lawrence Berkeley Laboratory Report LBL-1U712*)

Shields, H. C., & Weschler, C. J. (1992).Volatile Organic Compounds Measured at a Telephone Switching Center from 5/30/85-12/6/88: A Detailed Case Study. *J. Air Waste Manage. Assoc.*, Vol. 42, No. 6 (June 1992), pp. 792-804, ISSN 1047-3289

Shields, H. C., Fleischer, D. M. & Weschler, C. J. (1996). Comparisons among VOCs Measured in Three Types of U.S. Commercial Buildings with Different Occupant Densities," *Indoor Air,* Vol. 6, pp. 2-17, ISSN 0905-6947

Shinohara, N., Kai, Y., Mizukoshi, A., Fujii, M., Kumagai, K., Okuizumi, Y., Jona, M. & Yanagisawa, Y. (2009). On-site passive flux sampler measurement of emission rates of carbonyls and VOCs from multiple indoor sources. *Building Environ,* Vol. 44, pp. 859 - 863, ISSN 0360-1323

Short, S. (2001). The issues and implications of setting and applying Indoor Air Quality Guidelines. *IEH Seminar, University of Leicester,* http://www.le.ac.uk/ieh/pdf/IAQmtgrep.

Siskos, P.A., Bouba, K.E. & Stroubou, A.P. (2001). Determination of selected pollutsnts and measurements of physical parameters of the evaluation of indoor air quality in school buildings in Athens, Greece. *Indoor Built. Environ.*, Vol.10, pp. 185 -192

Siskos, P.A. (2003). An experimental Investigation of the Indoor Air Quality in Fifteen School Buildings in Athens, Greece. *International Journal of Ventilation,* Vol. 2, No. 3, pp. 185-201, ISSN 1473-3315

Soldatos, A.P., Bakeas, E.B. & Siskos P.A. (2003). Occupational exposure to btex of workers in car parkings and gasoline service stations in Athens, Greece. *Fresenius Environmental Bulletin,* Vol.12, No.9, pp. 1064 -1070

Sollinger, S., Levsen, K. & Wfinsch, G. (1993). Indoor Air Pollution by Organic Emissions from Textile Floor Coverings. Climate Chamber Studies under Dynamic Conditions. *Atrnos. Environ.*, Vol. 27B, pp. 183-192, ISSN 1352-2310

Sollinger, S., Levsen, K. & Wiinsch, G. (1994). Indoor Pollution by Organic Emissions from Textile Floor Coverings: Climate Test Chamber Studies under Static Conditions. *Atrnos. Environ.*, Vol. 28, pp. 2369-2378, ISSN 1352-2310

Spengler, J. D., Dockery, D. W., Turner, W. A., Wolfson, J. M. & Ferris, B. G. (1981). Long-Term Measurements of Respirable Sulfates and Particles inside and outside Homes. *Atrnos. Environ.*, Vol. 15, pp. 23-30, ISSN 1352-2310

Spengler, J. D., Brauer, M., Samet, J. M. & Lambert, W. E. (1993). Nitrous Acid in Albuquerque, New Mexico, Homes. *Environ. Sci. Technol.*, Vol. 27, No. 5 (May 1993), pp. 841-845, ISSN 0013-936X

Spengler, J., Schwab, M., Ryan, P. B., Colome, S., Wilson, A. L., Billick, I. & Becker, E. (1994). Personal Exposure to Nitrogen Dioxide in the Los Angeles Basin. *J. Air Waste Manage. Assoc.*, Vol. 44 (January 1994), pp. 39-47, ISSN 1047-3289

Spicer, C. W., Kenny, D. V., Ward, G. F. & Billick, I. H. (1993). Transformations, Lifetimes, and Sources of NO2, HONO, and HNO 3 in Indoor Environments. *J. Air Waste Manage. Assoc.*, Vol. 43 (November 1993), pp. 1479-1485, ISSN 1047-3289

Stock, T. H. (1987). Formaldehyde Concentrations inside Conventional Housing. *JAPCA*, Vol. 37, pp. 913-918, ISSN 0894-0630

Swedjemark, G.A. and Mjönes, L. (1984). Radon and radon daughter concentration in Swedish Homes, *Radiat. Prot. Dosim.* Vol. 7 (1-4), pp 341-345

Tamura, G. T. (1975). Measurement of air leakage characteristics of house enclosures. *ASHRAE Trans*, Vol. 81(Pt. 1), pp. 202-211

Thompson, C. V., Jenkins, R. A. & Higgins, C. E. (1989). A Thermal Desorption Method for the Determination of Nicotine in Indoor Environments. *Environ. Sci. Technol.*, Vol. 23, No. 4 (April 1989), pp. 429-435, ISSN 0013-936X

Tichenor, B. A., Sparks, L. A. White, J. B. & Jackson, M. D. (1990). Evaluating Sources of Indoor Air Pollution. *J. Air Waste Manage. Assoc.*, Vol. 40, No. 4 (April 1990), pp. 487-492, ISSN 1047-3289

Traynor, G. W., Apte, M. G., Sokil, H. A., Chuang, J. C., Tucker, W. G. & Mumford J. L. (1990). Selected Organic Pollutant Emissions from Unvented Paraffin Space Heaters. *Environ. Sci. Technol.*, Vol. 24, No. 9 (September 1990), pp. 1265-1270, ISSN 0013-936X

Tsakas, M. & Siskos, P. (2010). Indoor Air Quality in the Control Tower of Athens International Airport, Greece. *Indoor and Built Environment*, Vol. 0, pp. 01-06, ISSN 1420-326X

Tyson, J.L., Fairey, P.W. and Withers, C.R. (1993). Elevated radon levels in ambient air, INDOOR AIR '93, Proceedings of the 6th International Conference on Indoor Air Quality and Climate, , Vo1.4, pp.443-448. Helsinki, Finland, July 4-8,1993

UNSCEAR (United Nations Scientific Committee on the Effects of Atomic Radiation) (1977). Sources and effects of ionizing radiation, United Nations ed., New York, E.77.IX.1.

UNSCEAR (United Nations Scientific Committee on the Effects of Atomic Radiation) (1993). Sources and effects of ionizing radiation, United Nations ed., New York, E.94.IX.2.

Wade, W. A., Cote, W. A. & Yocom, J. E. (1975). A Study of Indoor Air Quality. *JAPCA*, Vol. 25, pp. 933-939, ISSN 0894-0630

Waldman, J. & Jenkins, P. (2004). Indoor air exposure research and reducing indoor air exposures in California. In: Krzyzanowski, M., Jantunen, M., Bartonova, A., Oglesby, L., Kepahalopoulos, S., Kotzias D, (Eds.). Role of Human assessment in air quality management. *Report EUR 21052, European Communities, Luxembourg*, pp. 66–71

Wallace, J. C., Brzuzy, L. P., Simonich, S. L., Visscher, S. M. & Hites, R. A. (1996). Case Study of Organochlorine Pesticides in the Indoor Air of a Home. *Environ. Sci. Technol.*, Vol. 30, No. 9 (August 1996), pp. 2715-2718, ISSN 0013-936X

Wallace, L. (1996). Indoor Particles: A Review. *J. Air Waste Manage. Assoc.*, Vol. 46 (February 1996), pp. 98-126, ISSN 1047-3289

Waubke, N.V. & Kusterle, W. (1990). Mould infestations in residential buildings. Paper presented at the *1st symposium on mould infestations, Innsbruck*, January 1990

Weschler, C. J., Shields, H. C. & Naik, D. V. (1989). Indoor Ozone Exposures. *JAPCA*, Vol. 39, pp. 1562-1568, ISSN 0894-0630

Weschler, C. J., Shields, H. C. & Rainer, D. (1990). Concentrations of Volatile Organic Compounds at a Building with Health and Comfort Complaints. *Am. Ind. Hyg. Assoc. J.*, Vol. 51, pp. 261-268, ISSN 0002-8894

Weschler, C. J., Shields, H. C., & Naik, D. V. (1994). Indoor Chemistry Involving O3, NO, and NO 2 as Evidence by 14 Months of Measurements at a Site in Southern California. *Environ. Sci. Technol.*, Vol. 28, No. 12, pp. 2120-2132, ISSN 0013-936X

Weschler, C. J. & Shields, H. C. (1996). Chemical Transformations of Indoor Air Pollutants, in Indoor Air '96 (S. Yoshizawa, K. Kimura, K. Ikeda, S. Tanabe, and T. Iwata, Eds.). *Organizing Committee of the 7th International Conference of Indoor Air Quality and Climate*, Vol. 1, pp. 919-924, Nagoya, Japan

West, D.L. (1977). Contaminant dispersion and dilution in a ventilated space. *ASHRAE Trans*, Vol.83 (Pt. 1), pp. 125-140

White, R.F. (1995). Landscape development and natural ventilation. Effect of moving air on buildings and adjacent areas. *Landscape Archit*, Vol. 45, pp. 72-81, ISSN 0023-8031

Wolkoff, P. (1999). Photocopiers and Indoor Air Pollution. *Atmos. Environ.*, Vol.33, pp. 2129-2130, ISSN 1352-2310

World Health Organization. (1986). Regional Seminar on Human Exposure to Outdoor and Indoor Air Pollution and its Effects on Health. *PEPAS, Kuala Lumpur*

World Health Organization (1989). Indoor air quality: Organic Pollutants. *Copenhagen, EURO 111*

Zhang, J., He, Q. & Lioy, P. J. (1994). Characteristics of Aldehydes: Concentrations, Sources, and Exposures for Indoor and Outdoor Residential Microenvironments. *Environ. Sci. Technol.*, Vol.28, No. 1 (January 1994), pp. 146-152, ISSN 0013-936X

Zhang, L.Z., & Niu, J.L. (2004). Modeling VOCs emissions in a room with a single-zone multi-component multi-layer technique. *Building Environ*, Vol.39, pp. 523 - 531, ISSN 0360-1323

18

Moisture and Estimation of Moisture Generation Rate

Tao Lu, Xiaoshu Lu and Martti Viljanen
Department of Civil and Structural Engineering,
Aalto University School of Science and Technology
Finland

1. Introduction

Buildings are exposed to outside weather, which cannot remain free of water in liquid and vapour phases. "Except for structural errors, about 90% of all building construction problems are associated with water in some way." by Lieff and Trechsel (Lieff & Trechsel, 1982). Moisture problem has become one of the most important factors decreasing indoor air quality (IAQ) and limiting the building service life. Over the last decade, moisture failures in building systems have reached billions of Euros in damages in Europe, many of which involved the deterioration of sheathing panels. Research has verified the existence of health problems in relation with moisture problems. In Finland, for example, it has been reported that 38% of detached houses and 25% of apartments had notable or significant moisture problems (Haverinen, 2002). The cost of repairing moisture damage that resulted in negative health effects was estimated as € 10 000 - 40 000 per case. Therefore, the prediction and control of moisture transport in buildings are import from both technical and health points of view.

Moisture can migrate into a building in several ways, depending on vapour, liquid or ice/snow. In general, three major moisture sources can be identified: outdoors (e.g. air humidity, precipitation, moisture in the ground), the indoors (e.g. humans, water use), and wet construction materials (Rousseau, 1984). Moisture accumulation into building structures may lead to physical, biological or chemical deterioration of building materials as well as the growth of some microbial or chemical contaminations (Salthammer, 1999; Andersson, 1997). One major problem caused by moisture is mould.

Moulds are the most typical form of fungus, which can be found anywhere. They can grow on virtually any organic substances as long as moisture and oxygen are present. Like other fungi, moulds reproduce through the production of spores. Mould spores continually waft through the air, both indoors and out-of-doors. Many types of moulds exist indoors, and almost all moulds could cause heath effects (through air), depending on the type and amount of mould present as well as the sensitivity of individual experiencing mould exposure. Certain health effects, such as irritation of eyes, nose and throat, dermatitis, exacerbation of asthma, and respiratory distress, have been reported to be associated with mould exposure (Verhoeff et al., 1995; Pope et al., 1993). Other effects such as fever, flu-like symptoms, fatigue, respiratory dysfunction, excessive and regular nose bleeds, dizziness, headaches, diarrhea, vomiting, liver damage, and impaired or altered inhalation have been identified in persons who have been exposed to mould (Croft et al., 1986; Jarvis, 1995; Smith & Moss, 1985). Some moulds even produce toxins (mycotoxins) (Jarvis, 1990; Burge, 1986; Yang, 1995).

Indoor mould was reported worldwide. A series of studies conducted in European countries, Canada and the United States showed that at least 20% of buildings had one or more signs of visible mould growth indoors on walls, floors or ceilings (Institute of Medicine, 2004). Similar reports also come from other areas of the world. A survey in rural Taiwan and China showed that 30.1% of visited people reported the presence of visible mould inside their houses (Yang et al., 1997). In a study of 98 houses in Japan, 15.6% was reported to have visible mould (Saijo et al., 2004). In another study in three China cities (Beijing, Guangzhou and HongKong), about 11% of parents of 10 902 schoolchildren reported visible mould on the ceilings and walls (Wong et al., 2004). Indoor mould was also reported to occur in school buildings, day-care centres, offices and other buildings (Mudarri & Fisk, 2007).

Excessive moisture is generally the cause of mould growth. Relevant experiments were committed to study the lowest moisture level for mould growth. Adan (Adan, 1994) found that moulds don't grow below a relative humidity of 80%. In Viitanen and Ritschkoff's work (Viitanen & Ritschkoff, 1991), they suggested relative humidity should be kept below 75% within a temperature range of 5-40 °C to avoid mould growth. This result was confirmed by Rowan et al. (Rowan et al., 1999). In a more detailed review work, Johansson et al. (Johansson et al., 2005) described that critical moisture conditions for mould growth were 75-90% for clean materials and 75-80% for contaminated or solid materials (Table 1).

Building material group	**Relative humidity (%)**
Wood and wood-based materials	75-80
Paper on plasterboard	80-85
Mineral insulation materials	90-95
Extruded and expanded polystyrene	90-95
Concrete	90-95

Source: Johansson et al. (2005)

Table 1. Critical relative humidity for various groups of materials.

Therefore, indoor moisture becomes the major factor on the presence of mould. Indoor moisture was also linked to the growth of Bacteria (Heinz & Mark, 2009). Hence, control of indoor moisture level becomes essential concerning IAQ. It is not uncommon in some climates to have moisture loads in excess of over 20 kg per day (Kerestecioglu & Gu, 1990). The load mainly comes from indoor activities like occupants, washing and cooking etc. In residential buildings, the largest single moisture load is from people's respiration and perspiration.

However, methods of evaluating indoor moisture generation rates in buildings are generally lacking. The difficulty lies on the fact that the moisture generation rate has no clear relationship with the number of occupants. Factors like cleaning and plants have big effect (Hite & Bray, 1948). Large background contributions from the foundation and soil have been reported (TenWolde, 1994). Typically, a trial-and-error method is adopted in evaluating the indoor moisture generation rate in calculating building heat and moisture transfer. Values of indoor moisture generation rate are adjusted in comparing the measured and the calculated moisture contents. The final selected value is based on the minimum difference of the moisture contents between the measured and the calculated. TenWolde (TenWolde, 1994) used this method and the mathematical model developed by the same author to analyse the indoor moisture generation rates for six manufactured houses during winter. Houses were occupied by one to two persons in each house and dimensions were varying from 4.3 × 20.1 to 4.9 × 23.2 m^2. The estimated average values of moisture generation rates were 32×10^{-6} to 75×10^{-6} kg s^{-1} per person. Very often, an experimental value is used.

In this chapter, a mathematical method in predicting indoor moisture generation rate is developed. The method is based on a series of measured indoor moisture contents. Using a Stirling interpolation fitting of the measured indoor moisture contents, the time-dependent indoor moisture generation rate can be determined. The moisture generation pattern can be obtained thereafter. The developed method provides an easy and concise way of determining indoor moisture generation levels, hence can be served as a useful tool in predicting building heat and moisture transfer.

This chapter is trying to accomplish two objectives. The first is to apply the mathematical method to predict the indoor moisture generation rate for a real test house. Results are analysed by the heat and moisture transfer model developed by Lu (Lu, 2002; Viljanen et al., 1999). With this model, the validation of the proposed mathematical method in deriving indoor moisture generation rate is studied. The second objective is to study the effect of indoor moisture generation patterns on the indoor moisture level using the developed heat and moisture transfer model.

The chapter is organised as follows: Following the introduction is a brief description of overall heat and moisture transfer model equations for the building system. The mathematical method used in predicting indoor moisture generation rate is given in details after the heat and moisture transfer model description. The validation of the developed model is conducted for a real test house. Finally, simulation results are presented, which mainly involve the study of the effect on indoor moisture behaviour of different moisture generation patterns.

2. Model equations for a building system

A building system is divided into building envelopes such as walls, roof, floor, foundation etc and building indoor air. The modelling work can be generalised as following main steps: 1) developing heat and moisture transfer model equations for building envelopes; 2) developing heat and moisture transfer model equations for indoor air and 3) numerically solving the system equations for building envelopes and indoor air at any time with outdoor conditions expressed as boundary conditions. Hence the transient temperature and moisture variations can be obtained for a building system under outdoor conditions. A complete heat and moisture transfer model for a building system is described by Lu in (Lu & Viljanen, 2000; Viljanen et al., 1999).

2.1 Model equations for building envelopes

The fundamentals of heat and moisture transfer in porous media (building envelopes) can be found in pioneering papers by Philip and DeVries (Philip & DeVries, 1957) and Whitaker (Whitaker, 1977). In modelling heat and moisture transfer in building porous materials,

Whitaker's REV method (Whitaker, 1977) is adopted. The general approach is to start with conservation equations, constitutive equations and state equations by means of the thermodynamics of the Gibbs relation. The conservation equations include the mass, the momentum and the energy conservation equations. In writing the heat and moisture transfer model equations for building envelopes, simplifications have been made. These include

i. In considering moisture transfer for building envelopes, convection effect is not included. Moisture transfer is described as Fick's generalised diffusion equation with water vapour concentration as a driving potential. This is justified by Liesen and Pedersen (Liesen & Pedersen, 1999) for example. However, in the case of existence of liquid phase, the transport of liquid by Darcy's law can be taken into account.
ii. Local thermodynamic equilibrium is assumed at every point of the material.
iii. For all moisture transport coefficients, any hysteresis effect is not included.
iv. The flux of heat or moisture at the interface of contacted materials has a continuous property.
v. Heat and moisture transfer equations are limited to one-dimensional for building walls and roof and two-dimensional for building floor, foundation and soil.
vi. Soil is a homogeneous material.
vii. At a depth of 5 m of the soil ground, the temperature is constant equal to 5°C. At certain depths, the relative humidity of soil is 100%.

Having the above assumptions the heat and moisture transfer equations for different components of building envelopes can be written down. For example, for building walls, we have

$$\rho_w C_p^w \frac{\partial T_w}{\partial t} = \frac{\partial}{\partial x}(\lambda_w \frac{\partial T_w}{\partial x}), \tag{1}$$

$$\frac{\partial c_w}{\partial t} = \frac{\partial}{\partial x}(D_w \frac{\partial c_w}{\partial x}). \tag{2}$$

Boundary condition: The interfaces between layers are treated separately and the conditions of continuity assumption (iv) are applied. Thus we have for example

$$\lambda_w^1 \frac{\partial T_w{}^1}{\partial x} |_{\text{interface}} = \lambda_w^2 \frac{\partial T_w{}^2}{\partial x} |_{\text{interface}}, \tag{3}$$

$$T_w^1 |_{\text{interface}} = T_w^2 |_{\text{interface}}, \tag{4}$$

$$D_w^1 \frac{\partial c_w{}^1}{\partial x} |_{\text{interface}} = D_w^2 \frac{\partial c_w{}^2}{\partial x} |_{\text{interface}}, \tag{5}$$

$$c_w^1 |_{\text{interface}} = c_w^2 |_{\text{interface}}. \tag{6}$$

For surfaces exposed to the outdoor and indoor air, we have for example

$$\lambda_w \frac{\partial T_w}{\partial x} |_{\text{surface,out}} = h_{T,w}^{\text{out}} \ (T_{out} - T_w |_{\text{surface,out}}) + q_w, \tag{7}$$

$$D_w \frac{\partial c_w}{\partial x} \Big|_{surface,out} = h_{m,w}^{out} (c_{out} - c_w \big|_{surface,out}) \tag{8}$$

and

$$\lambda_w \frac{\partial T_w}{\partial x} \Big|_{surface,in} = h_{T,w}^{in} (T_w \big|_{surface,in} - T_{in}), \tag{9}$$

$$D_w \frac{\partial c_w}{\partial x} \Big|_{surface,in} = h_{m,w}^{in} (c_w \big|_{surface,in} - c_{in}), \tag{10}$$

where q_w is the solar energy received by the exterior wall surface. Empirical values of the surface heat and moisture transfer coefficients, $h_{T,w}^{out}$ and $h_{T,w}^{in}$, are taken from RIL 117 (RIL 117, 1979). Lewis relation can be used to determine $h_{m,w}^{out}$ and $h_{m,w}^{in}$. Currently, raining effect is not included.

Similar equations of Eqs. (1)–(10) for building floor, foundation and soil are given with two-dimensional. In writing the boundary conditions, the heat and moisture fluxes normal to the boundary surfaces which are selected as far as possible from the interest building are assumed to be 0 in soil ground. Assumption (vii) is also employed. We are not going to address the equations here for space limitation. More details can be referred to Viljanen et al. (Viljanen et al., 1999).

For initial conditions for building envelopes, the temperature or the relative humidity is assumed at the steady state with indoor and outdoor temperatures or relative humidity. For soil, the initial temperature distribution is treated differently accounting for its big thermal capacity (Viljanen et al., 1999).

2.2 Model equations for building indoor air

The following physical condition is assumed for building indoor air:

viii. Indoor air is well mixed. Indoor temperature and moisture content are uniformly distributed.

The well-mixed indoor air model is generally not valid, however, the resultant error may be negligible depending on the model's application. For a normal residential house for example, the effect of the indoor air flow on the indoor temperature and moisture level is normally small if the ventilation rate is not very big. Hence a well-mix indoor air model is a good approximation of the practical situation.

The heat and moisture balance equations therefore can be written in the following forms:

$$V \rho_{in} C_p^{in} \frac{dT_{in}}{dt} = nV \rho_{in} C_p^{in} (T_{out} - T_{in}) + \sum Q_i, \tag{11}$$

$$V \frac{dc_{in}}{dt} = nV(c_{out} - c_{in}) + G + \sum M_i, \tag{12}$$

where ΣQ_i is the sum of heat transfer rates between building envelopes and indoor air (W), ΣM_i is the sum of moisture transfer rates between building envelopes and indoor air (kg s^{-1}) and G is the indoor moisture generation rate (kg s^{-1}).

To have a closed system of heat and moisture equations for building system, all material property data have to be taken into account. These data are taken from RIL 117 (RIL 117, 1979), some of which are based on the experimental fittings (Viljanen et al., 1999). Moreover, the time-dependent indoor moisture generation rate G has to be given in order to solve the equations. A mathematical method of calculating G is given in the following section.
The model partial differential equations, Eqs. (1)-(12), are discretized in space by the finite difference method. The resultant ordinary differential equations are discretized in time by the Crank-Nicolson scheme. The final algebraic equations are solved by the Newton iteration method. Transient temperatures and moisture contents for building envelopes and indoor air are then solved at any time under outdoor climatic conditions.

3. Calculation of indoor moisture generation rate

Eq. (12) gives the moisture generation rate as:

$$G = V\frac{dc_{in}}{dt} - nV(c_{out} - c_{in}) - \sum M_i. \tag{13}$$

To deduce G, the derivative of indoor moisture content c_{in} versus time, $\frac{dc_{in}}{dt}$, is needed.
However, direct differentiation of the indoor moisture contents can be unstable even though the data show smooth and continuous properties. The moisture generation rate increases exponentially when the moisture source releases the moisture and decreases exponentially after the source stops the release. Bennett et al. (Bennett et al., 1996) developed an approach in estimating emission rate from concentration data for formaldehyde measurement. A Stirling interpolation polynomial fitting was used in constructing the derivatives of contaminant concentrations. Applying this idea, indoor moisture contents c_{in} are interpolated by a Stirling polynomial, the derivatives can therefore be obtained by simply differentiating the smooth continuous polynomial.
Stirling interpolation polynomial is based on the average of the backward and forward Gauss interpolation polynomials. It uses a diagonal difference table with three successive points. It is a symmetrical approximation for the given values. Relevant formulas are shown in the appendix. Similarly, a Bessel interpolation polynomial can be used also. It is based on a diagonal difference table and uses four points. It is able to interpolate with data that are not evenly spaced apart. More details can be seen in Kunz (Kunz, 1957).
Another unknown value presented in Eq. (13) is the time-dependent moisture transfer rate ΣM_i between building envelopes and indoor air. In deriving the indoor moisture generation rate G, it is customarily assumed that ΣM_i is 0, for example in Bennett et al. (Bennett et al., 1996). However, it may affect the calculation accuracy. We propose two different methods in calculating ΣM_i here, namely the explicit and the numerical methods.
For the explicit method, we assume a steady-state condition in Eq. (13) and ignore the thermal lag between the envelopes and indoor air, so the moisture transfer rate M for any building envelope can be approximated as

$$M = A\delta(c_{out}-c_{in}), \tag{14}$$

where A and δ are the area and the overall moisture transfer coefficient of the building envelope respectively.

The implication behind the approximation Eq. (14) is that the moisture profile of the building envelope is given by a linear combination of c_{out} and c_{in}. For a thin, uniform construction material, the method can give a good estimation. However, for a general situation, Eq. (14) gives a poor estimation. One way of modifying it is to introduce more interpolation nodes for example c_{out}, c_1,...c_n, c_{in}. The Laplace transform can be used and Eq. (14) is reduced to a first order time lag corresponding to c_{out} and c_{in}. As it involves tedious basic mathematical manipulations, details are skipped here. More information can be found in Underwood (Underwood, 1999), where an approximated heat conduction equation was studied. It can be applied to moisture transfer equation.

For the numerical method, the developed program is adopted to predict ΣM_i at any time. One difficulty is that in calculating ΣM_i, the time-dependent moisture generation rate G is needed. Values of G can be approximated first by assuming $\Sigma M_i = 0$.

3.1 Model validation criteria

The Normalized Mean Squared Error (NMSE) and the Correlation Coefficient (r) are accepted as two key criteria to evaluate the fit to the identification data.

$$NMSE = \frac{\sum_{i=1}^{N}[\hat{x}_i - x_i]^2}{\sum_{i=1}^{N}[\bar{x}_i - x_i]^2} \quad (15)$$

$$r = \frac{N\sum_{i=1}^{N} z_i y_i - \sum_{i=1}^{N} z_i \sum_{i=1}^{N} y_i}{\sqrt{(N\sum_{i=1}^{N} z_i^2 - (\sum_{i=1}^{N} z_i)^2)(N\sum_{i=1}^{N} y_i^2 - (\sum_{i=1}^{N} y_i)^2)}} \quad (16)$$

where $\hat{x}_i$ and $\bar{x}_i$ are predicted value and the mean of measured ones (x_i) respectively, N is the number of samples, z_i and y_i are samples. By definition, the minimum of NMSE is zero, indicating the exact match between actual and estimated values. The higher NMSE, the worse is our estimation. The value of r is such that $-1 < r < +1$, where + and - signs are used for positive linear correlations and negative linear correlations, respectively. An r value of exactly +1 indicates a perfect positive fit while -1 means a perfect negative fit.

4. Simulation results and discussion

4.1 Validation of the heat and moisture transfer model

In order to apply the proposed mathematical method in predicting indoor moisture generation rate, a comparison of calculated indoor moisture content using the heat and moisture transfer model described in the last section with available measurement data for a test house without occupants was done. The result is shown in Fig. 1. In the simulation process, as the time-dependent ventilation rate was not measured, a fitted value of $n = 0.3\ h^{-1} = 8.3\times10^{-5}\ s^{-1}$ was used. It can be seen that good agreement is obtained with NMSE =0.44 and r=0.81. The relative difference between the calculated and the measured data is less than 10%.

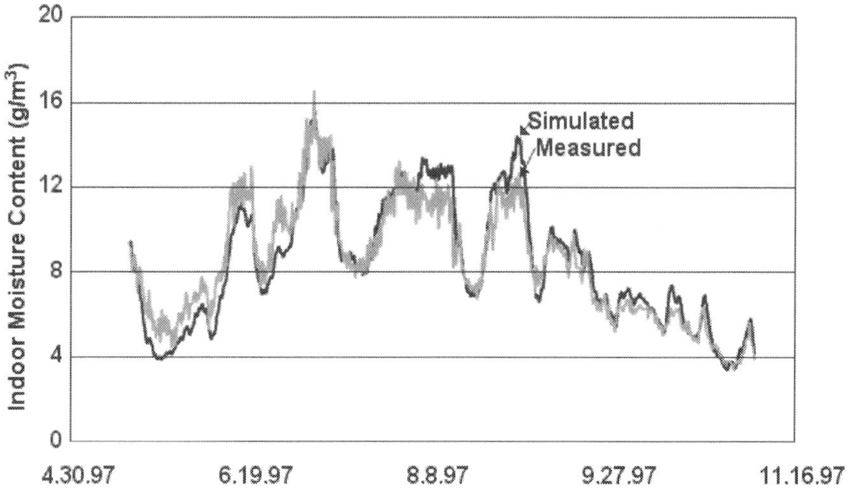

Fig. 1. Comparison between calculated and measured indoor moisture contents. Validation of the heat and moisture transfer model.

4.2 Validation of the indoor moisture generation rate model

To perform a validation of the proposed mathematical method in predicting indoor moisture generation rate, we measured a series of indoor temperatures and moisture contents for a real test house. The measurement period was from 24th of November to 2nd of December 1998 and the average ventilation rate was measured as $0.4\ h^{-1} = 1.1\times10^{-4}\ s^{-1}$. The test house is a one-storey detached house, a massive log house, with a dimension $7.6\times14.3\times2.5\ m^3$ located near Helsinki of Finland (Lehtinen et al., 1998). The floor area is about 110 m^2. The inner and outer walls, roof and floor are mainly composed of wood, mineral wool and concrete with thickness 200, 300 and 120 mm, separately. A schematic picture of the house dimensions and constructions is given in Figs. 2 and 3. The house was occupied by a family with two adults and a baby.

As reference figures, Figs. 4 and 5 display a comparison between the calculated and the measured indoor moisture contents and relative humidity. Indoor moisture generation rate was assumed to be 0. The measured indoor temperatures were used in the calculation, see Fig. 6. It can be seen that the calculated levels of indoor moisture contents and relative humidity are lower than those of the measured data. Indoor moisture generation rate has to be taken into account.

Table 2 lists some of the measurement data including outdoor and indoor moisture contents. The last column presents the calculated indoor moisture generation rate per unit volume using the method described in the previous section, see the appendix. Here we assume that the moisture transfer rate between building envelopes and indoor air is 0, i.e. $\Sigma M_i = 0$.

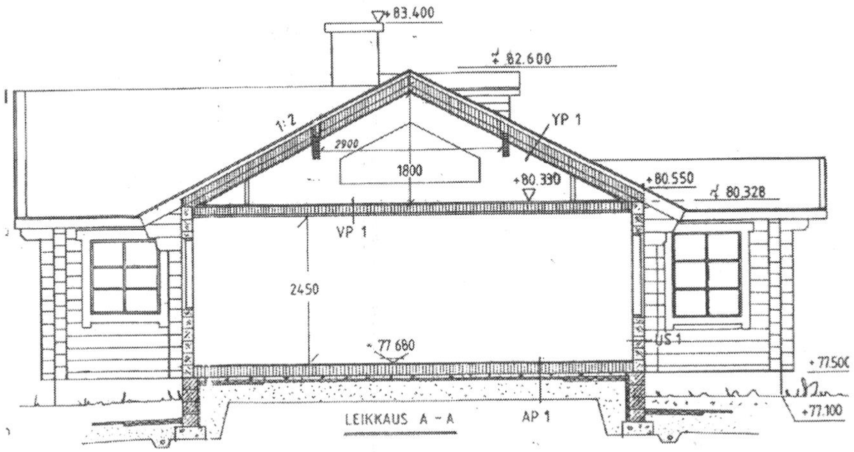

Fig. 2. Layout of cross-section view of the test house.

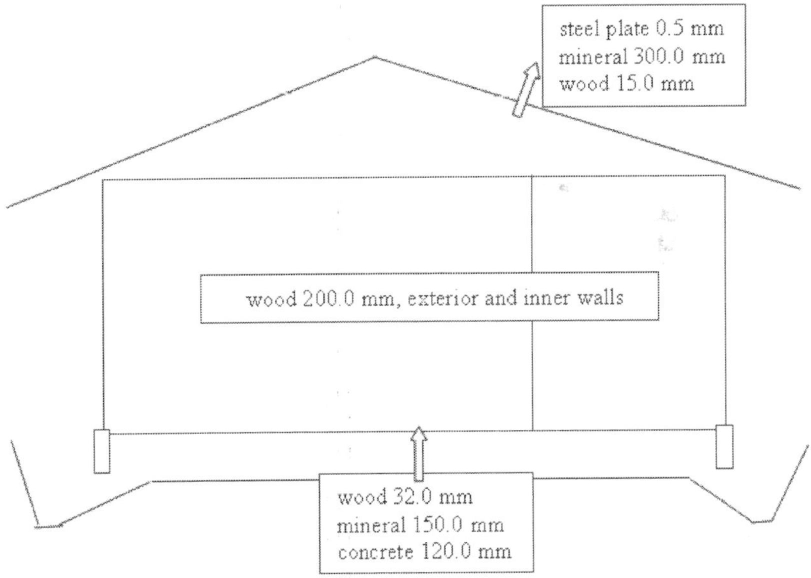

Fig. 3. Schematic picture of the constructions of the test house.

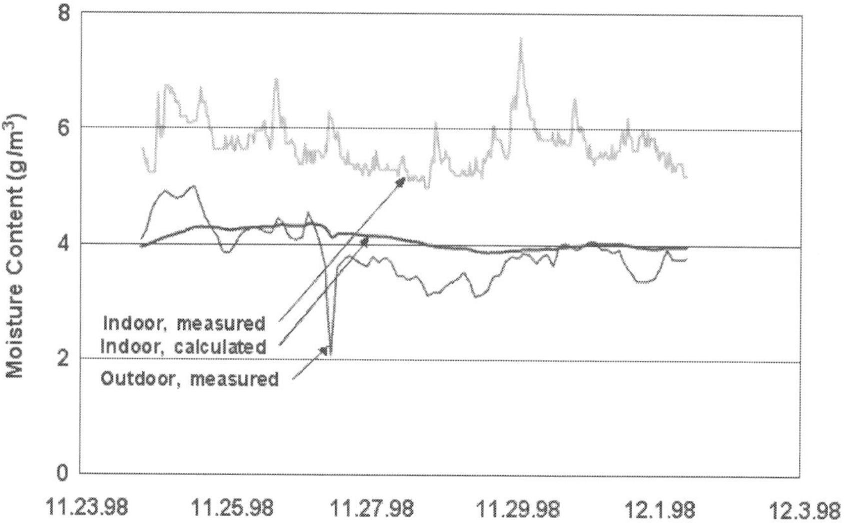

Fig. 4. Comparison between calculated and measured indoor moisture contents for the test house. Indoor moisture generation rate is assumed to be 0.

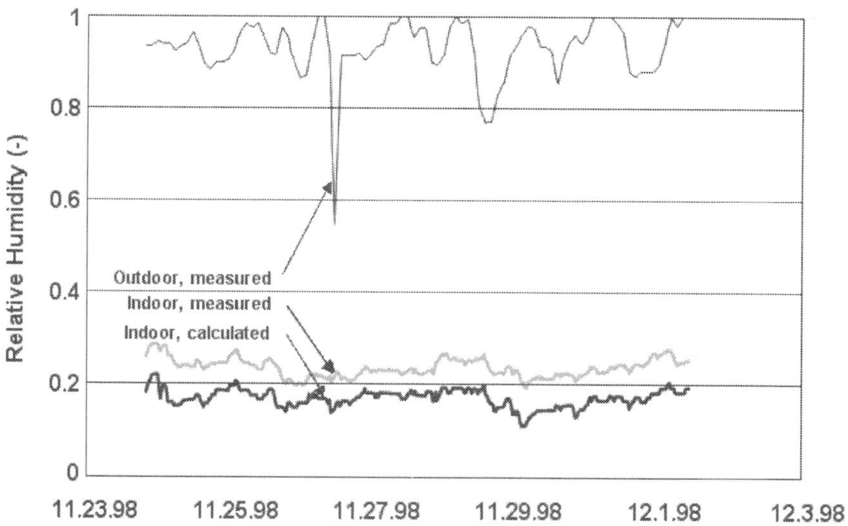

Fig. 5. Comparison between calculated and measured indoor relative humidity for the test house. Indoor moisture generation rate is assumed to be 0.

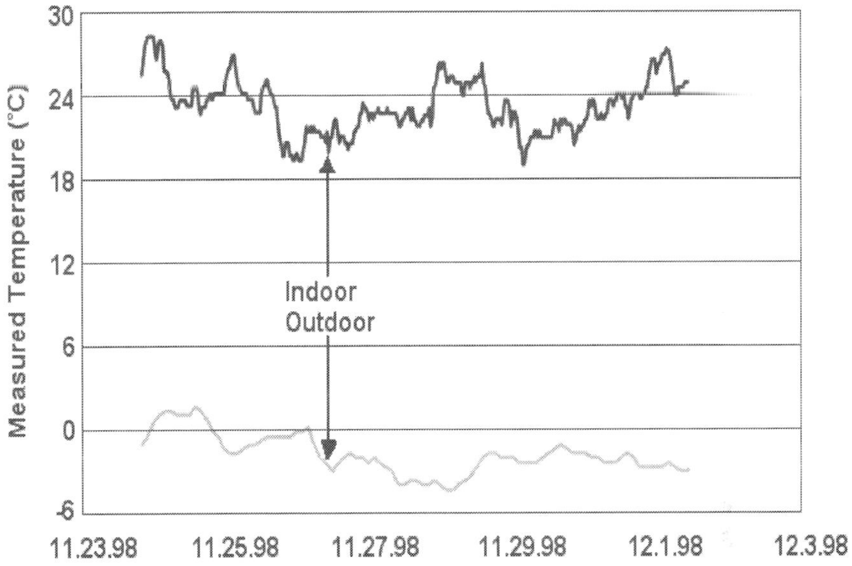

Fig. 6. Measured indoor and outdoor temperatures for the test house.

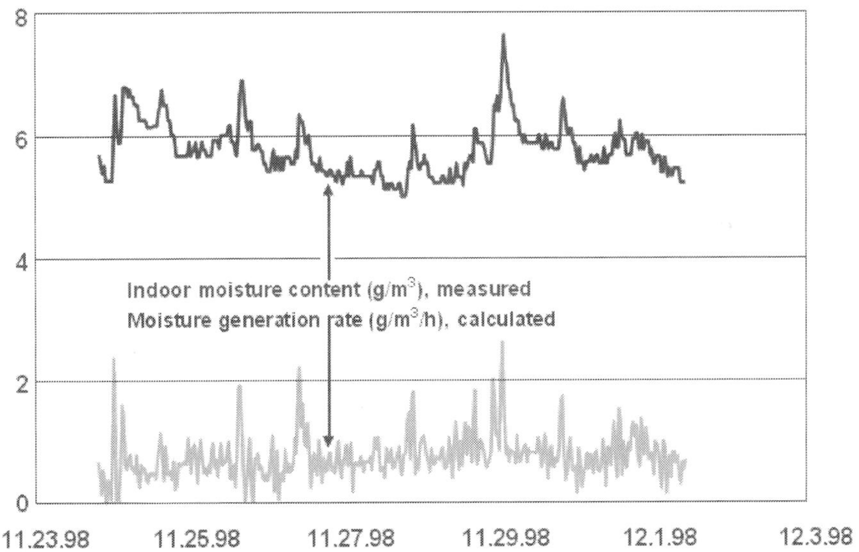

Fig. 7. Measured indoor moisture contents and calculated indoor moisture generation rate for the test house.

Fig. 7 shows the measured indoor moisture content and the calculated time-dependent indoor moisture generation rate for the test house. It can be seen that the trend of indoor moisture generation rate very much follows that of the indoor moisture contents. The rate values vary from 0 to 2.6 $gm^{-3}\,h^{-1} = 196 \times 10^{-6}$ kg s^{-1}. And the average rate value is 0.76 g m^{-3} $h^{-1} = 57\times10^{-6}$ kg s^{-1}. These values fall within the range of moisture generation rate reported by other researchers. For example, TenWold (TenWold, 1994) estimated the moisture generation rate in one- to two-person house between 63×10^{-6} kgs^{-1} and 75×10^{-6} kgs^{-1}in winter period.

Time (24th, Nov.)	c_{in} (gm^{-3})	$\frac{dc_{in}}{dt}$ ($gm^{-3}h^{-1}$)	c_{out} (gm^{-3})	$n(c_{out}-c_{in})$ ($gm^{-3}h^{-1}$)	$\frac{G}{V}$ ($gm^{-3}h^{-1}$)
12:00 am	5.67	0.2040	3.7521	-0.7672	0.9712
12:30 am	5.43	-0.5027	3.7260	-0.6816	0.1788
01:00 am	5.34	-0.0360	3.7096	-0.65214	0.6161
01:30 am	5.34	0.0160	3.7336	-0.6426	0.6586
02:00 am	5.34	0	3.7579	-0.6329	0.6329
02:30 am	5.34	-0.0270	3.774	-0.6264	0.5994
03:00 am	5.34	0.1350	3.7839	-0.6224	0.7575
03:30 am	5.43	0	3.7695	-0.6642	0.6642
04:00 am	5.34	-0.1350	3.7474	-0.6370	0.5020
04:30 am	5.34	0.0270	3.7226	-0.6470	0.6740

Table 2. Measured indoor moisture contents and calculated indoor moisture generation rate.

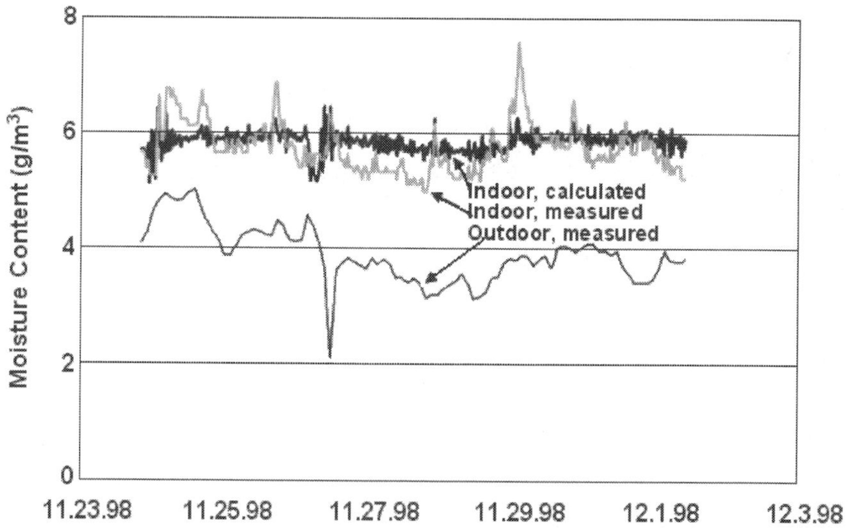

Fig. 8. Comparison between calculated and measured indoor moisture contents for the test house. Time-dependent moisture generation rate in Fig. 7 is applied.

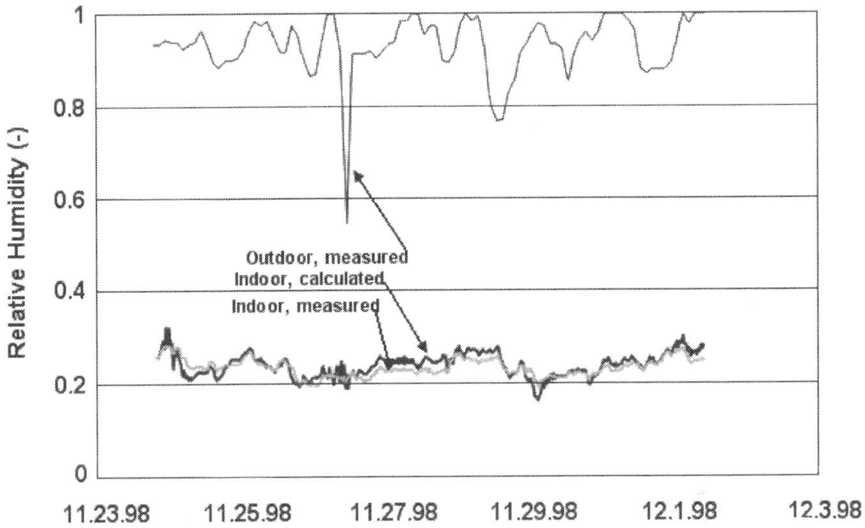

Fig. 9. Comparison between calculated and measured indoor relative humidity for the test house. Time-dependent moisture generation rate in Fig. 7 is applied.

From the obtained time-dependent indoor moisture generation rate in Fig. 7 we can see that the moisture generation pattern consists of different peaks, which are coincident with occupants' activities. Applying the result in Fig. 7, Figs. 8 and 9 present the results of the comparison between the measured and the calculated indoor moisture contents and relative humidity. Table 3 lists the model performances for results (Figs. 8 and 9). Good agreement is obtained.

Model	Normalized Mean Squared Error (NMSE)	Correlation Coefficient (*r*)	$\frac{\lvert calculated - measured \rvert^{a}}{measured}$
Calculated indoor moisture content *vs.* measured indoor moisture content (Fig. 8)	0.71	0.55	Maximum = 15%; Average= 5%;
Calculated indoor relative humidity *vs.* measured indoor relative humidity (Fig. 9)	0.38	0.83	Maximum = 15%; Average = 6%;

a The range of errors between calculated and measured.

Table 3. Model performances between measured indoor moisture content and calculated indoor moisture generation rate.

It is worth noting that the moisture generation rate in Fig. 7 was obtained by assuming the moisture transfer rate $\Sigma M_i = 0$. It seems this assumption influences the model accuracy in calculating indoor moisture content, which gives relatively low Correlation Coefficient (r=0.55). But, the error range is small and satisfactory (last column in Table 3). Using the program and the information of Fig. 7 we have calculated the moisture transfer rate per unit area $\frac{\sum M_i}{\sum A_i}$. Values vary from 0.02×10^{-9} kg m^{-2} s^{-1} to 0.13×10^{-6} kg m^{-2} s^{-1}. The average value is 0.03×10^{-6} kg m^{-2} s^{-1}. Considering the magnitude of the moisture transfer rate, it is neglected in deducing the moisture generation rate. Calculations would not make any principle difference if the effect of moisture transfer rate were taken into account. TenWolde (TenWolde, 1994) used a moisture transfer rate of 0.068×10^{-6} kg m^{-2} s^{-1} to match the calculated and the measured indoor relative humidity.

One interesting phenomenon is that the model gives very good agreement for indoor relative humidity (Table 3). As mentioned earlier, the model has assumed that the heat transfer rates between building envelope and indoor air were assumed as zero (ΣQ_i,, Eq. (11)). This assumption may have effects on model accuracy of both indoor temperature and moisture content estimates which is beyond statistical explanation. This guess is further proved by the following simulations (Section 4.3).

In the following, a number of simulations are conducted for the test house shown in Figs. 2 and 3. These calculations are concerned with the study of the effect on indoor moisture contents of different moisture generation rate patterns.

4.3 Effect of moisture generation rate patterns

Consider a varied moisture generation pattern shown in Fig. 7 and a constant moisture generation pattern. Figs. 10 and 11 present the calculated indoor moisture contents and relative humidity. For the constant moisture generation pattern, the rate value of 57×10^{-6} kg s^{-1} is assumed, which is predicted for the test house as an average rate value. It is observed that the indoor moisture content with the constant moisture generation rate is an average curve of that with the varied moisture generation rate in Fig. 10. Moreover, the difference is so small that it can be neglected. This conclusion is applicable to the indoor relative humidity in Fig. 11.

Next, assume that the test house is a commercial house with occupant activities existing only for certain working hours, for example from 9:00 a.m. to 5:00 p.m. for 8 hours. The moisture generation rate is assumed to be 171×10^{-6} kg s^{-1} during the working hours and 0 otherwise, which has a periodic pulse pattern. Note that the daily average moisture generation rate is still 57×10^{-6} kg s^{-1} as in Figs. 10 and 11. Figs. 12 and 13 show the comparison of the indoor moisture contents with these two different moisture generation rate patterns. Table 4 illustrates model performances. The difference is within a very small range, especially for indoor relative humidity.

Model performances are excellent in indoor relative humidity estimation. Results also show consistency with ones in Section 4.2, which evidences our guess. Neglecting moisture transfer between building envelope and indoor air can affect the model accuracy in calculating indoor moisture content, but the impact is limited and very small normally (Section 4.2). As such, the model may give relatively poor agreement for indoor moisture content which may go beyond statistics, such as Correlation Coefficient, but in general errors are small and acceptable (last columns in Tables 3 and 4). Furthermore, note that a relationship between two variables must be approximately linear when applying Correlation Coefficient. When the assumption of linearity is violated, the correlation will underestimate the strength of the relationship. In addition, the developed model in this

Model	Normalized Mean Squared Error (NMSE)	Correlation Coefficient (r)	$\frac{\lvert simulated - target \rvert^{a}}{target}$
Constant moisture generation rate *vs.* time-dependent moisture generation rate (indoor moisture content, Fig. 10)	0.74	0.54	Maximum = 15%; Average= 3%;
Constant moisture generation rate *vs.* time-dependent moisture generation rate (indoor relative humidity, Fig. 11)	0.12	0.94	Maximum = 15%; Average = 3%;
Moisture generation rate with constant pattern *vs.* Moisture generation rate with periodic pulse pattern (indoor moisture content, Fig. 12)	0.69	0.56	Maximum = 9%; Average = 4%;
Moisture generation rate with constant pattern *vs.* Moisture generation rate with periodic pulse pattern (indoor relative humidity, Fig. 13)	0.14	0.93	Maximum = 10%; Average = 4%;

a The range of errors between simulated and target.

Table 4. Model performances for the effect of moisture generation rate patterns

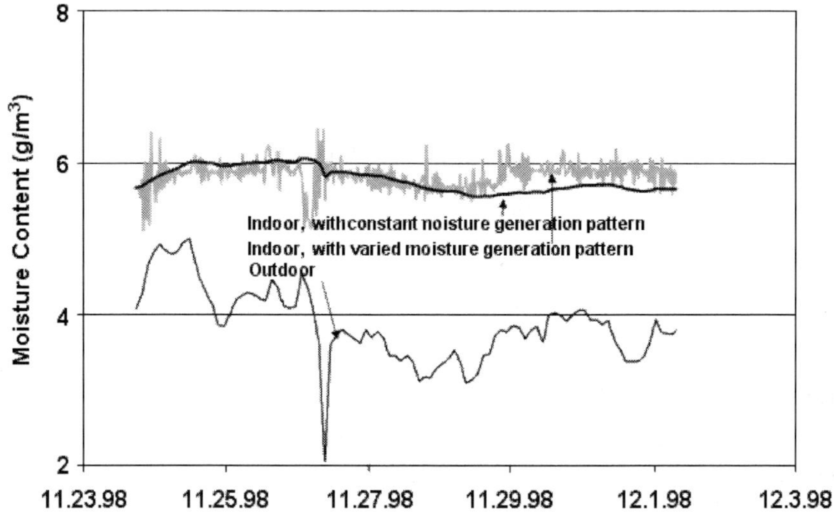

Fig. 10. Comparison of the effect on indoor moisture contents of different moisture generation patterns for the test house. Constant moisture generation rate and time-dependent moisture generation rate in Fig. 7 are applied.

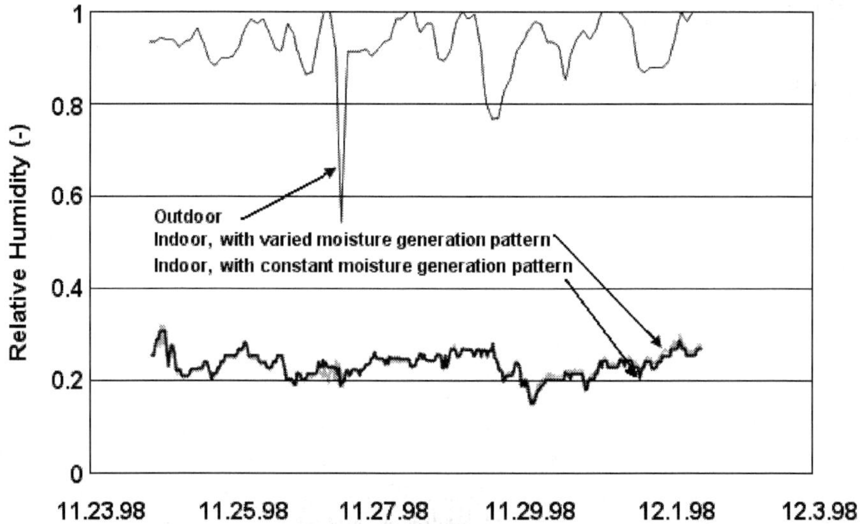

Fig. 11. Comparison of the effect on indoor relative humidity of different moisture generation patterns for the test house. Constant moisture generation rate and time-dependent moisture generation rate in Fig. 7 are applied.

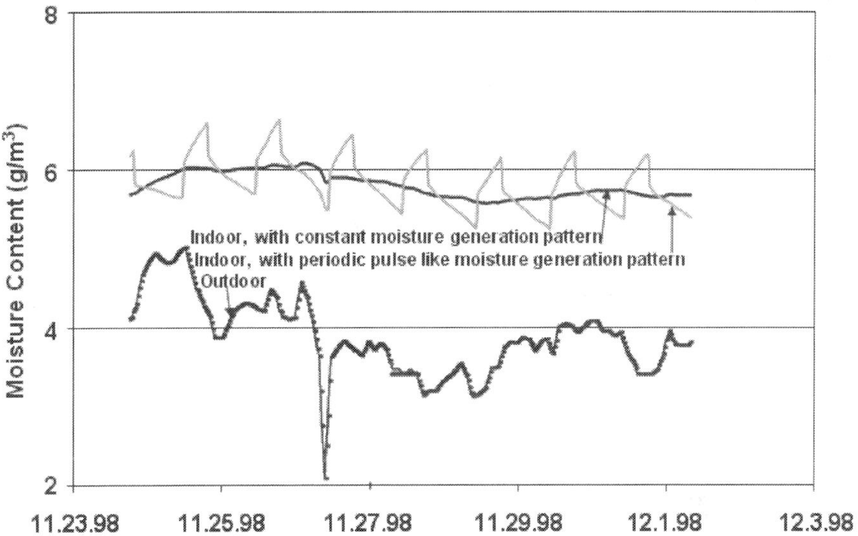

Fig. 12. Comparison of the effect on indoor moisture contents of different moisture generation patterns for the test house. Moisture generation rates with constant and periodic pulse patterns are applied.

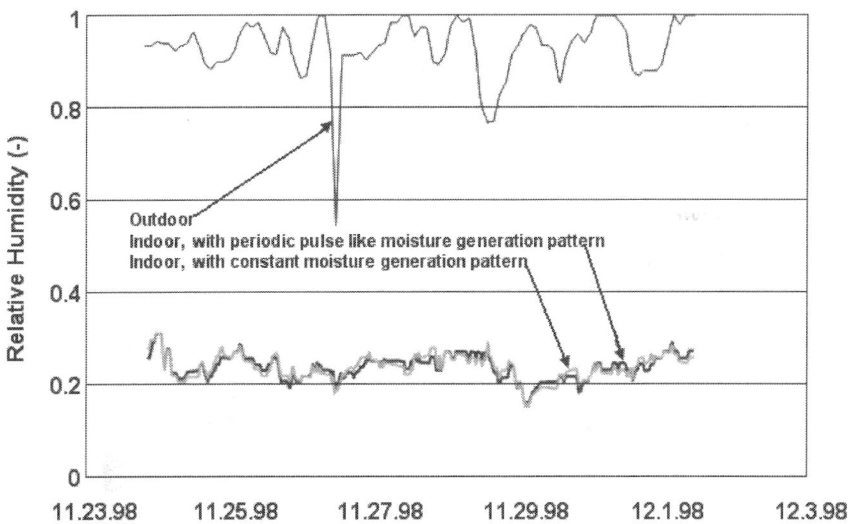

Fig. 13. Comparison of the effect on indoor relative humidity of different moisture generation patterns for the test house. Moisture generation rates with constant and periodic pulse patterns are applied.

paper (e.g. Eq. (13)) is physically verified in many related works (Lu, 2002). In this study, the differences between simulated and measurements are all in small margins (last columns in Tables 3 and 4), which further verify the model's usefulness in practice.
This is an interesting result as it can greatly simplify the calculation procedure. Note that any moisture generation pattern function can be approximated by a combination of periodic pulse functions, so the result actually implies that a constant moisture generation rate pattern is always 'enough' in calculating the indoor moisture content, especially in deciding indoor moisture level.
It is worth mentioning that Isetti et al. (Isetti et al., 1988) studied the effect on indoor relative humidity of three periodic pulse types of moisture generation patterns with different numbers of occupants in an office room. A similar result was obtained. A constant moisture generation pattern was not included in their paper.

5. Conclusion

The importance of the effect of the indoor moisture generation rate on accurately predicting moisture content for a building system has been demonstrated. Hence in studying the heat and moisture transfer for a building system, a method of precisely calculating indoor moisture generation rate is needed. To address the problem, this chapter proposes a mathematical method to derive time-dependent moisture generation rate function from a series of measured indoor moisture contents. The method incorporates the developed heat and moisture transfer model to calculate transient temperatures and moisture contents for building envelopes and indoor air exposed to any outdoor climatic conditions. The model is validated by a real test house. The effect on indoor moisture level of different moisture generation patterns is also investigated.
Through simulation results, we can draw the following main conclusions:

- The performance of indoor moisture generation rate has a big effect on indoor moisture level. In conducting the simulation work, this aspect has to be considered.
- Given a series of measured indoor moisture contents, it is possible to calculate time-dependent moisture generation rate using the method proposed in this chapter.
- A constant moisture generation rate model is generally not valid, but the resultant error may be negligible. This has important implication for simplifying a simulation procedure.

Finally, it is stressed that the present model is based on the condition of well-mixed indoor air moisture from the moisture sources. For some special cases especially right after the moisture release from a big moisture release source, special attention must be paid. Moreover, for a heavy moisture content room like sauna or bathroom, the effect of different moisture generation rate patterns may be too big to be neglected.

6. Acknowledgement

This research was supported by the Finnish Academy.

7. Appendix

Table 5 presents a difference table used in interpolating a series of measured data by a Stirling polynomial.

With equal space h, at point x = x_0, the approximation of the derivative of Stirling polynomial is give as (p127, Kunz, 1957):

$$\left[\frac{dy}{dx}\right]_{x=x0} = \frac{1}{h}\left(\frac{\Delta y_{-1}+\Delta y_0}{2} - \frac{1}{6}\frac{\Delta^3 y_{-2}+\Delta^3 y_{-1}}{2} + \frac{1}{30}\frac{\Delta^5 y_{-3}+\Delta^5 y_{-2}}{2} + \ldots\right). \quad (A.1)$$

Applying the above formula Eq. (A.1) to the indoor moisture content we get

$$\left[\frac{dc_{in}}{dt}\right]_{t=t0} = \frac{1}{\Delta t}\left(\frac{\Delta c_{in,-1}+\Delta c_{in,0}}{2} - \frac{1}{6}\frac{\Delta^3 c_{in,-2}+\Delta^3 c_{in,-1}}{2} + \frac{1}{30}\frac{\Delta^5 c_{in,-3}+\Delta^5 c_{in,-2}}{2} + \ldots\right). \quad (A.2)$$

Take Table 5 as an example, Eq. (A.2) gives the following result at time 2:30 am:

$$\left[\frac{dc_{in}}{dt}\right]_{2:30am} = \frac{1}{0.5}\left(\frac{0+0}{2} - \frac{1}{6}\frac{0+0.09}{2} + \frac{1}{30}\frac{0.09-0.45}{2}\right) = -0.0270 \text{ g m}^{-3}\text{h}^{-1}. \quad (A.3)$$

In Table 2, it shows that $\left[\frac{dc_{in}}{dt}\right]_{2:30am}$= -0.0270 g $m^{-3}h^{-1}$.

index no.	Time	c_{in}	Δc_{in}	$\Delta^2 c_{in}$	$\Delta^3 c_{in}$	$\Delta^4 c_{in}$	$\Delta^5 c_{in}$
-5	12:00 am	5.67					
			-0.24				
-4	12:30 am	5.43		0.15			
			-0.09		-0.06		
-3	01:00 am	5.34		0.09		-0.03	
			0		-0.09		0.12
-2	01:30 am	5.34		0		0.09	
			0		0		-0.09
-1	02:00 am	5.34		0		0	
			0		0		0.09
0	02:30 am	5.34		0		0.09	
			0		0.09		-0.45
1	03:00 am	5.34		0.09		-0.36	
			0.09		-0.27		0.9
2	03:30 am	5.43		-0.18		0.54	
			-0.09		0.27		
3	04:00 am	5.34		0.09			
			0				
4	04:30 am	5.34					

Table 5. Difference table.

Nomenclature

A area, m^2

c vapour content, kg m^{-3}

C_p specific heat, J $kg^{-1} K^{-1}$
D moisture diffusion coefficient, $m^2 s^{-1}$
G moisture generation rate, kg s^{-1}
h_m surface moisture transfer coefficient, m s^{-1}
h_T surface heat transfer coefficient, W $m^{-2} K^{-1}$
n ventilation rate, s^{-1}
q solar heat flux, W m^{-2}
Q heat transfer rate between envelope and indoor air, W
M moisture transfer rate between envelope and indoor air, kg s^{-1}
t time, s
T temperature, K
V volume, m^3
x space co-ordinate, m

Greek symbols

δ overall moisture transfer coefficient, m s^{-1}
λ thermal conductivity, W $m^{-1} K^{-1}$
ρ density, kg m^{-3}

Superscripts

in indoor air
out outdoor air

Subscripts

in indoor air
interface interface
out outdoor air
surface surface
w wall

8. References

Adan, OCG (1994). *On the fungal defacement of interior finishes [thesis]*, Eindhoven, Eindhoven University of Technology, Netherlands

Bennett, JS., Feigley, CE., Underhill, DW., Drane, W., Payne, TA., Stewart, PA., Herrick, RF., Utterback, DF. & Hayes, RB. (1996). Estimating the contribution of individual work tasks to room concentration: method applied to embalming, *American Industrial Hygiene Association Journal*, Vol. 57, pp. 599-609, ISSN 1542-8117

Burge, HA. (1986). Toxigenic potential of indoor microbial aerosols, *Fifth Symposium on the Application of Short-Term Bioassays in the Analysis of Complex Environmental Mixtures*, pp. 391-397, ISBN 0306427729, Sheraton University Center, Durham, NC., USA

Croft, WA, Jarvis, BB. & Yatawara, CS. (1986). Airborne outbreak of trichothecene toxicosis, *Atmos. Environ.*, Vol. 20, No. 3, pp. 549-552, ISSN 1352-2310

Haverinen, U. (2002). *Modeling moisture damage observations and their association with symptoms [thesis]*, National Public Health Institute, Department of Environmental Health, University of Kuopio

Heinz, RT. & Mark, TB. (2009). *Moisture Control in Buildings: The Key Factor in Mold Prevention,* ASTM International, ISBN 978-0-8031-7004-9, West Conshohocken, Pennsylvania, USA

Hite, SC. & Bray, JL. (1948). Research in humidity control *Research Series No. 106,* Purdue University, Engineering Experiment Station, Lafayette, USA

Institute of Medicine (2004). *Damp indoor space and health,* National Academies Press, ISBN 0-309-09246-9, Washington, DC, USA

Isetti, C., Laurenti, L. & Ponticiello, A. (1988). Predicting vapour content of the indoor air and latent loads for air-conditioned environments: Effect of moisture storage capacity of the walls, *Energy and Buildings,* Vol. 12, pp.141-148, ISSN 0378-7788

Jarvis, BB. (1990). Mycotoxins and indoor air quality, *ASTM in Biological Contaminants in Indoor Environments,* pp. 201-204, ISBN 0-8031-1290-4, ASTM Symposium, Boulder, CO, USA, July 16-19, 1989

Jarvis, BB. (1995). Mycotoxins in the air: keep your buildings dry or the bogeyman will get you, *Proceedings from the International Conference: Fungi and Bacteria in Indoor Environments: Health Effects, Detection and Remediation,* pp. 35-44, ISBN 978-0970991508, Saratoga Springs, N.Y., USA, October 6-7, 1994

Johansson,p., Samuelson, I., Ekstrand-Tobin,A., Mjörnell,k., Sandberg,PI. & Sikander, E. (2005). *Microbiological growth on*

building materials – critical moisture levels. State of the art, Report 2005:11, SP Swedish National Testing and Research

Institute, Borås, Sweden

Kerestecioglu, A. & Gu, L. (1990). Theoretical and computational investigation of simultaneous heat and moisture transfer in buildings: "Evaporation and condensation" theory, *ASHRAE Transactions,* Vol. 96, No. 1, pp. 455-464, ISSN 0001-2505

Kunz, K.S. (1957). *Numerical analysis,* McGraw-Hill, ISBN 0-07-035630-0, New York, USA

Lehtinen, T., Viljanen, M., & Hänninen, J. (1998). *Massiivihirsistä tehtyjen ulkoseinien käyttö asuinrakentamisessa,* TKK-TRT-95, ISBN 951-22-4419-5, Helsinki University of Technology, Espoo, Finland

Lieff, M., & Trechsel HR. (1982). *Moisture migration in buildings : a symposium, Philadelphia, Pa., 6 Oct. 1980,* ASTM special technical publication / American Society for Testing and Materials 779, ISBN 0-8031-0605-X, West Conshohocken, Pa., USA

Liesen, R. & Pedersen, C. (1999). Modelling the energy effects of combined heat and mass transfer in building elements: Part 1-Theory, *ASHRAE Transactions,* Vol. 105, No. 2, pp. 941-953, ISSN 0001-2505

Lu, X. & Viljanen, M. (2000). On controlling indoor thermal and moisture content for an occupied building, *Proceedings of the 7th international conference on air distribution in rooms,* pp. 951-956, ISBN 0080430171, Reading, UK, July 9-12, 2000

Lu, X. (2002). Modelling heat and moisture transfer in buildings – (I) model program, *Energy and Buildings,* Vol. 34, pp. 1033-1043, ISSN 0378-7788

Mudarri, D. & Fisk, WJ. (2007). Public health and economic impact of dampness and mold, *Indoor air,* Vol. 17, pp. 226-235, ISSN 0905-6947

Philip, JR. & DeVries, DA. (1957). Moisture movement in porous materials under temperature gradients, *Trans. Am. Geophys. Union,* Vol. 38, pp. 222-232, ISSN 0002-8606

Pope, AM., Patterson, R. & Burge, H. (1993). *Indoor Allergens. Assessing and Controlling Adverse Health Effects*, National Academy Press, ISBN 978-0-309-04831-6, Washington, D.C., USA

RIL 117. (1979). Lämmön- ja kosteudeneristys, *Suomen Rakennusinsinöörien Liitto,* Finland

Rousseau, MZ. (1984). Sources of moisture and its migration through the building enclosure, *ASTM Standardization News 1984*, Vol. 12, No. 11, pp. 35-37, ISSN 1094-4656

Rowan, NJ., Johnstone, CM., McLean, RC., Anderson, JG. & Clarke, JA. (1999). Prediction of toxigenic fungal growth in buildings by using a novel modelling system, *Applied and Environmental Microbiology*, Vol. 65, pp. 4814–4821, ISSN 0099-2240

Saijo, Y., Kishi, R., Sata, F., Katakura, Y., Urashima, Y., Hatakeyama, A., Kobayashi, S., Jin, K., Kurahashi, N., Kondo, T., Gong, YY. & Umemura, T. (2004). Symptoms in relation to chemicals and dampness in newly built dwellings, *International Archives of Occupational and Environmental Health,* Vol. 77, pp. 461-470, ISSN 0340-0131

Smith, JE. & Moss, MO. (1985). *Mycotoxins Formation, Analysis, and Significance,* John Wiley and Sons, ISBN 978-0471906711, NY., USA

Salthammer, T. (1999). *Organic Indoor Air Pollutans,* WILEY-VCH, ISBN 3-527-29622-0, NY., USA

Andersson MA., Nikulin M., Koljalg U., Andersson MC., Rainey F., Reijula K., Hintikka EL. and Sakinjoja-Salonen M. (1997). Bacteria, molds and toxins in water-damaged building materials, *Applied and Environmental Microbiology,* Vol. 63, pp. 387-393, ISSN 0099-2240

TenWolde, A. (1994). Ventilation, humidity, and condensation in manufactured houses during winter, *ASHRAE Transactions,* Vol. 100, No. 1, pp. 103-115, ISSN 0001-2505

Underwood, CP. (1999). *HVAC Control Systems: Modelling, analysis and design,* ISBN 978-0419209805, E & FN SPON, London, UK

Verhoeff, AP., Van Strien, RT., Van Wijjnen, JH. & Brunekreef B. (1995). Damp housing and childhood respiratory symptoms. The role of sensitization to dust mites and mold, *Am.J. of Epidemiology,* Vol. 141, No. 20, pp. 103-110, ISSN 0002-9262

Viitanen, H. & Ritschkoff, A.C. (1991). Mould growth in pine and spruce sapwood in relation to air humidity and temperature, *Report No. 221*, Swedish University of Agricultural Sciences, Department of Forest Products Uppsala, Sweden

Viljanen, M., Bergman, J., Grabko, S., Lu, X., & Yrjölä, R. (1999). *Ensuring the long service life of unheated buildings,* TKK-TRT-100, ISBN 951-22-4715-1, Helsinki University of Technology, Espoo, Finland

Whitaker, S. (1977). Simultaneous heat, mass and momentum transfer- A theory of drying, *Adv. Heat Transfer*, Vol. 13, pp. 119-203, ISSN 978-0-12-020033-7

Wong, GW, Ko, FW., Hui, DS., Fok, TF., Carr, D., von Mutius, E., Zhong, NS., Chen, YZ. & Lai, CK. (2004). Factors associated with difference in prevalence of asthma in children from three cities in China: multicentre epidemiological survey, *British Medical Journal*, Vol. 329, pp. 486-488, ISSN 0959-8138

Yang, CS. (1995). Understanding the biology of fungi indoors, *Proceedings from the International Conference: Fungi and Bacteria in Indoor Environments: Health Effects, Detection and Remediation,* pp. 131-137, ISBN 978-0970991508, Saratoga Springs, N.Y., USA, October 6-7, 1994

Yang, CY., Chiu, JF., Chiu, HF. & Kao, WY. (1997). Damp housing conditions and respiratory symptoms in primary school children, *Pediatric Pulmonology,* Vol. 24, pp. 73-77, ISSN 8755-6863

19

Volumetric Monitoring and Modeling of Indoor Air and Pollutant Dispersion by the Use of 3D Particle Tracking Velocimetry

Pascal Biwole[1], Wei Yan[2], Eric Favier[3],
Yuanhui Zhang[2] and Jean-Jacques Roux[3]
[1]*University of Nice Sophia Antipolis,*
[2]*University of Illinois at Urbana-Champaign,*
[3]*University of Lyon,*
[1,3]*France*
[2]*USA*

1. Introduction

Human beings spend most of their lifetime indoors. Monitoring indoor airflow in buildings is a matter of health safety, and energy savings. Regarding health safety it is proven that exposure to aerosol pollution like walls and painting material residues, gasses emitted through cooking and heating with biomass fuel, or agricultural residues can result in aggravated health damages for the occupants. In the same time, too high air exchange rates increase the heating demand in temperate and cold climates. On both cases it is important to control the mixing of the air and the pollutant dispersion to ensure the occupants' comfort and productivity.
Before monitoring air mixing and pollutants dispersion in rooms, one has to decide whether qualitative or quantitative data is needed. Qualitative measurements only allow recovering the trajectory and patterns of the airflow in rooms. This is usually done by the use of tracer gasses like fumes, incense, or neutrally buoyant particles with suitable light sources. For more precise purposes such as comfort parameters measurements (Fanger, 1970) or validation of CFD modeling, one need to get access to quantitative data.
Quantitative air measurement techniques may be divided into Eulerian and Lagrangian techniques. Eulerian techniques measure flow velocity at one or several fixed locations at a time, as a man on a bridge measures the velocity of a water stream going by below him. Eulerian methods used for indoor applications include hot wire and hot film anemometry, pulsed wire anemometry, ion anemometry, laser Doppler velocimetry, laser 2-focus velocimetry, particle image velocimetry (PIV), stereoscopic PIV, tomographic PIV and holographic PIV. Depending on the technique, the result can be a one, two, or three-dimension velocity vector or velocity vector field.
But Eulerian techniques suffer from two major drawbacks: First, many are intrusive since they involve probes which are inserted in the measured flow. The probes not only change the flow by their mere presence, but they also disturb the measure. For example, though hot

wire anemometers are widely used in indoor air research, they can produce a 50% error on low ascendant flows because the hot probe creates its own convection, which becomes predominant. Moreover, most hot wire anemometers cannot achieve quantitative measurements of indoor air speeds lower than 10cm/s. Hot-films can reach 2cm/s, but the velocity orientation then becomes unavailable. Second, most Eulerian techniques only yield point-wise velocity measurements and are therefore ill-adapted to indoor air flows because they are unsteady and highly three dimensional. For example, Laser Doppler velocimetry can only yield the 3D velocity field inside a measuring volume of a few millimetres large. This method is generally restricted to near-the-wall boundary layers measurements. Stereo-particle image velocimetry can recover the instantaneous 3D velocity of large fields, but only for particles situated inside 3 to 10mm-large laser sheets.
To cope with those impediments, scientists have tried to build Lagrangian measurement techniques for nearly three decades: instead of measuring fluid properties from a fixed measurement point, the goal is to actually ride the flow as on a boat, thanks to neutrally buoyant particles, and monitor the flow's fluctuations. The more particles we have in the fluid, the finer our understanding of its topology. Each individual seeded particle is followed through time, in order to get its trajectory. Consequently, Lagrangian techniques provide a better spatial resolution than Eulerian techniques. There are two main Lagrangian methods under development for indoor air applications:

1.1 Particle streak velocimetry

Particle streak velocimetry (PSV) uses neutrally buoyant particles whose displacement is seen as streaks by setting a long camera exposure time (Dimotakis et al., 1981). Trajectories are yielded when the dead time between two long exposures is very short, depending on the flow velocity. Each streak's pixel length and orientation can be calculated as the length and orientation of the major axis of the ellipse that has the same normalized second central moments as the streak region. Dividing the streak length by the exposure time gives the velocity. The third velocity component can be acquired by geometric reconstruction in a stereoscopic system of at least two cameras (see Figure 1). The particle streak direction can be obtained by setting one of the three cameras with a shorter exposure time (Scholzen & Moser, 1996).

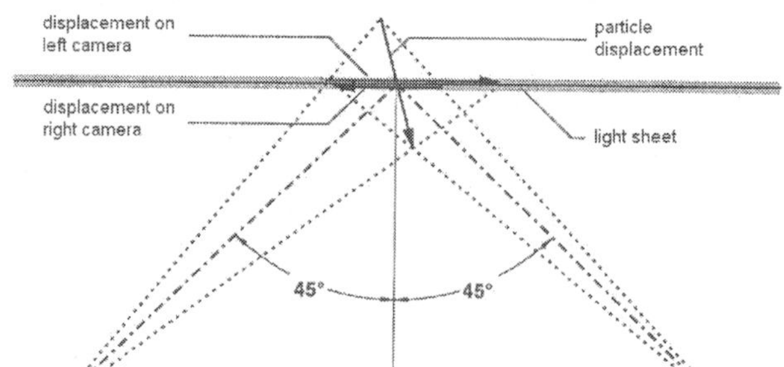

Fig. 1. Geometric reconstruction of the third displacement component in a stereoscopic system

Even though PSV has yielded promising results in indoor applications (Machacek, 2002; Sun & Zhang, 2003) the technique suffers from three structural limitations: Firstly, the flow velocity has to be high enough for particles to create streaks on camera image planes. This is not always possible with indoor air flow as a target. Secondly, many particle streaks may be bowed, especially where turbulence takes place. Additional approximation to reconstruct the streak length is then unavoidable. Thirdly, the velocity is always calculated as a mean over the length of the streak, thus somewhat limiting the spatial resolution of the method.

1.2 Particle tracking velocimetry

In particle tracking velocimetry (PTV), particles are detected as single points on each image by setting a very short camera exposure time. Velocity is calculated by dividing the displacement on object planes by the time between two exposures. PTV yields 3D trajectories by using at least two cameras. Depending on the algorithms used, either particles are first identified (spatial matching) then individually tracked (temporal tracking), or inversely 2D trajectories are first searched separately on each camera before being matched. A few schemes to achieve temporal tracking and spatial matching will be detailed later in this chapter.

The main drawback of 3D PTV is the difficulty of finding and tracking particles which overlap when the seeding density is strong. Therefore, densities need to be maintained low, typically about 0.005 particles per pixel for a system with three cameras (Maas et al., 1993). Other drawbacks of PTV are the limited number of suitable tracers and the fact that precisely measured 3D positions cannot be prescribed in advance. In spite of those drawbacks, PTV features a better spatial and temporal resolution than PSV. Calculated 3D velocities range from 0m/s to a maximum speed depending on the speed of the recording camera and on the intensity of the light source. Modern cameras go over tens of KHz but a very powerful light source is then needed to capture particle images.

The literature shows that over the past 15 years, most research on 3D PTV has been dedicated to volumes from Kolmogorov scales (Virant and Dracos 1997, Lee and Kim 2005) to centimetric scales (Suzuki et al., 2000). Small scale 3D PTV can track more than 1000 to 1500 particles. 3D PTV in air volumes over 1 m^3 has seldom been done. It raises new challenges in terms of illumination and camera positioning, but also in terms of particle size and localization. Pulsed lasers used in small-scale PTV (Adrian, 1991; Ouellette et al., 2006; Willneff and Gruen, 2002) cannot be used on larger volumes because the energy density of the light decreases rapidly when the beam is expanded. Nanometric and micronic particles used in small-scale PTV are extremely difficult to track in big volumes with a reasonable density. The use of at least three cameras positioned at large angles with respect to the other cameras is crucial to reducing measurement errors. In contrast to small-scale PTV, particle size and brilliance vary a lot since they are free of movement. In fact, particles close to the cameras create large blobs on the images.

The purpose of this chapter is to present the 3D particle tracking velocimetry (3D PTV) method applied to the monitoring of air displacements and pollutant dispersion in rooms. 3D PTV is searched in order to yield the three dimensional velocity and the trajectory of the air in a single zone versus time. Air exchange measurement and air leakage measurement are not within the scope of this chapter. Similarly, the measurement of airflow through ducts, fans and heat exchangers will not be covered herein, even though the 3D PTV may be extended to that use.

The layout of the chapter will be as follows: Section two will present the experimental set up required to perform 3D PTV in rooms. Special attention will be given to the choice of the tracer particles. Section three will briefly explain the layout of the most common 3D PTV algorithms. In particular, an image processing procedure to remove speckles from images of particles getting close to the cameras will be detailed. Section 4 will show a few examples of the results one might get when using the method. Some helpful guidelines will be provided in terms of camera and light positioning, depending on the room layout and wall color. Last, Section 5 will clearly state the limitations of the method, and present a few trends about the ongoing research on the subject.

2. 3D PTV experimental set-up

2.1 Choice of the tracer

The three major parameters for choosing a fluid tracer in any particle image velocimetry method are a neutral density with respect to the fluid, a detectability of the particles by the cameras, and a size and lifetime that suits the scale and duration of the flow characteristics to be measured. Other minor requirements are a low environmental impact (health hazards, corrosion on equipment, waste disposal) and an easy storage and manipulation. An extensive list of possible tracers for PIV tests was given by (Melling, 1997). For gaseous flows he proposes particles from olive oil, wheat oil, oil fumes, glass, polycrystalline, AL_2O_3, TiO_2, and ZrO_2. (Adamcyzyk & Rimai, 1988) also used nylon micro-balloons for 3D PTV in the air in a 5 x 5 x 5 cm section. All those tracers range from less than 1μm to 30 μm. For detecability purpose, they are always used with pulsed laser light. In spite of their good size for turbulence patterns visualization, the use of such minute particles is impossible in volumes as large as ours because they cannot be singled out and tracked.

In large scale air volumes with feeble pressure gradient, most researchers use helium filled soap bubble (Biwole et al. 2009; Kessler & Leith, 1991; Machacek, 2002; Müller & Renz, 1996; Okuno et al., 1993; Sholzen & Moser, 1996; Sun and Zhang, 2003; Suzuki & Kazagi, 1999; Zhao et al., 1999). The underlying idea is that a liquid film inflated with a lighter-than-air gas can produce a neutrally buoyant particle. Those particles fulfill most requirements mentioned above, except when studying small scale turbulence patterns because of their size (from 1.3mm to 3.8mm, Anonymous, 1988).

Fig. 2. Bright spot on a single helium filled soap bubble (from Machacek 2002)

The motion of a bubble can be derived from the equation of the movement of a small rigid sphere in a non-uniform and incompressible flow (Maxey & Riley, 1983):

$$m_p \frac{dv_p}{dt} = (m_p - m_f) g_v + m_f \left.\frac{Dv_f}{Dt}\right|_{Y(t)} - \frac{1}{2} m_f \frac{d}{dt}\left\{ v_p(t) - v_f\left[Y(t),t\right] - \frac{1}{10} a^2 \nabla^2 v_f\big|_{Y(t)} \right\} - \tag{1}$$

$$6\pi a \mu \left\{ v_p(t) - v_f\left[Y(t),t\right] - \frac{1}{6} a^2 \nabla^2 v_f\big|_{Y(t)} \right\} - 6\pi a^2 \mu \int_t^{\tau} \left(\frac{d/dt' \left\{ v_p(t') - v_f\left[Y(t'),t'\right] - \frac{1}{6} a^2 \nabla^2 v_f\big|_{Y(t')} \right\}}{\left[\pi \nu (t-t')\right]^{1/2}} \right) dt'$$

with m_p being the bubble mass, m_f the mass of fluid displaced by the sphere, v_p the speed of the buble, g_v the acceleration due to gravity, v_f the speed of the fluid, $Y\ (T)$ the center of the sphere at time t, a the radius of the sphere, μ and ν respectively the dynamic and kinematic viscosities of the fluid. If we make the assumptions that the flow is irrotational, that the bubbles remain spherical throughout their lifetime and that interactions among bubbles are negligible, Maxey and Riley equation simplifies into (Kerho & Bragg, 1994):

$$\begin{aligned} m_p \frac{dv_p}{dt} &= (m_p - m_f) g + m_f \frac{Dv_f}{Dt} - \frac{1}{2} m_f \left(\frac{dv_p}{dt} - \frac{Dv_f}{Dt} \right) \\ &+ \frac{1}{2} \rho C_D S \left| v_f - v_p \right| \left(v_f - v_p \right) + 9 m_p \sqrt{\frac{\mu \rho}{\pi D^2 \sigma^2}} \int_0^t \frac{d/dt' \left(v_f - v_p \right)}{\sqrt{t-t'}} dt' \end{aligned} \tag{2}$$

where D represents the diameter of a bubble, ρ the density of the air, σ the density of the bubble, C_D the coefficient of drag of the bubble, and S the apparent surface (half sphere surface). The left hand term of equation 1 represents the inertia force while the right hand terms are respectively the buoyancy force, the pressure force, the added mass, the drag term and the Basset force. If we assume that the slip velocity and the slip acceleration both present in the right-hand term of Eq. 2 are negligible, the equation becomes:

$$m_p \frac{dv_p}{dt} = (m_p - m_f) g + m_f \frac{Dv_f}{Dt} \tag{3}$$

The meaning of Eq. 3 is that the movement of a small particle in a flow depends primarily on the forces of pressure, inertia and gravitation. For a particle with neutral density with respect to the fluid, $m_p = m_f$. In that case, pressure forces are balanced by inertia forces. For a particle lighter than the fluid, we have $m_p < m_F$. The bubble will tend to deviate and to be elevated from the real streamlines. The constant diameter assumption is plausible given the weak air temperature and pressure gradients usually observed indoors. But when using helium filled bubbles as tracers for 3D PTV indoor, the experimenter have to keep in mind that the slip velocity and the slip acceleration between the bubble and the surrounding air may not be zero. This is a major assumption and still an open question.

2.2 Choice of the cameras

Three main features have to be taken into account when choosing a camera for flow visualization:

- The maximum frame rate: The frame rate is primarily chosen relatively to the maximum flow velocity. The quicker the flow, the higher the frame rate must be. In traditional PIV, the maximum time between two frames is calculated so that the fluid displacement is less than a quarter of the size of research windows. 3D PTV globally follows the same rule. The frame rate has also to be adjusted to the particle seeding density in order to help minimizing tracking ambiguities. The denser the seeding, the quicker the camera must be. For indoor air speeds, 100fps cameras generally allow a satisfying tracking of the flow.
- The camera resolution: It is chosen depending on the size of the particles employed, the size of the field to be visualized - the larger the field, the higher the resolution must be -, the illumination employed, and the background. In a case of poor contrast with the background or when the light sheet is very large, sensors featuring a high number of gray levels should be preferred, typically with at least 8 bits i.e. 2^8 of gray levels. Large pixel size cameras provide better sensitivity but poorer resolution than cameras with smaller pixel size. Color images are not preferred in PTV because the complexity of the algorithms is increased for a limited gain. Whatever the number of cameras, they must all be time-synchronous.
- The lenses: Fish-eyed lenses may be chosen (Biwole et al. 2008; Biwole et al. 2009) for indoor applications, provided the calibration procedure is powerful enough to calculate the distortion coefficients and reconstruct the scene.

2.3 Choice of the light source

The light source has to be strong and homogeneous enough for the cameras to see the light reflected on the tracer shells in every part of the measurement field. Especially in 3D PTV in large volumes, this must be true even for particles situated outside of the cameras object plane. Besides, the wavelength reemitted has to fit the spectral sensitivity of the recording sensor. Finally, light devices must produce low convective heat in order to keep the flow undisturbed. The light sources can be either pulsed or continuous. Pulsed lasers used in small scale PTV cannot be used on larger volumes because the energy density of the light decreases rapidly when the beam is expanded. For indoors applications, continuous type illumination solutions are usually preferred because it avoids synchronizing the light with the cameras. Arc lamps equipped with a cylindrical lens (Sholzen & Moser, 1996) or especially powerful halogen spot lamps (Machacek, 2002; Sun & Zhang, 2003; Biwole et al., 2009) are generally used. To reduce heat generation, the light sources can either be placed outside of the test room behind a glass panel, or switched on for only the few seconds of the recording.
To conclude on this section, tests *in situ* in presence of the particles, the cameras and the light sources are always necessary before settling for any specific equipment.

3. 3D PTV algorithms

3.1 Calibration

Before recording the seeded air, all the cameras must be calibrated. Calibration is the process of calculating the parameters taking part in the mathematical relationship between the 2D image coordinate system of each camera and a 3D real world coordinate system common to all cameras. Those parameters are actually the output of the calibration process.
Firstly, the coefficients yielded by camera calibration are the intrinsic parameters of each camera, i.e. the focal length, the radial and tangential distortion coefficients, the principal point

pixel coordinates and the skew coefficient defined as the angle between x and y pixel axes on the CCD (Charge-Coupled Devices) or CMOS (Complimentary Metal-Oxide Semiconductor) chip. Secondly calibration yields the extrinsic parameters of each camera, namely the rotation and translation matrices mapping each camera 3D coordinate system based at the center of the lens, to the common 3D coordinate system defined by the calibration target (see Figure 3). Thus, if Ti and Ri are respectively the 3×1 translation matrix and the 3×3 rotation matrix which transform camera i 3D reference frame XXci into the calibration target 3D reference frame XX, the relationship between the two coordinate systems reads:

$$XXci = Ri \cdot XX + Ti \quad (4)$$

Intrinsic and extrinsic parameters are calculated by minimizing the distance between actual specific points on a calibration target or a scene, and their simulated location based on the fitting of a mathematical camera model.

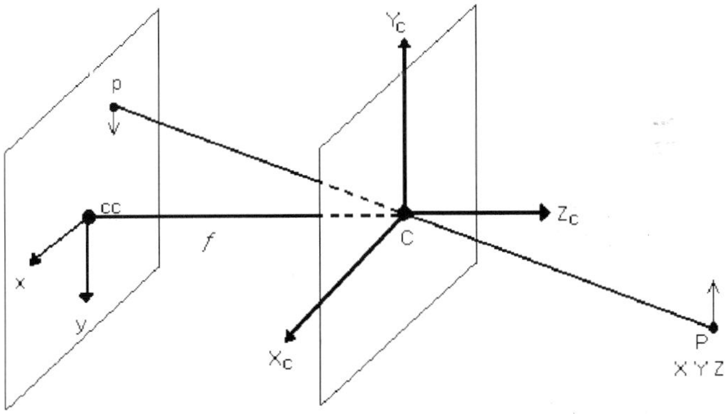

Fig. 3. Image plane coordinate system (cc,x,y) and camera coordinate system (C,Xc,Yc,Zc) in the pinhole camera model. Coordinates of point P(X,Y,Z) are given in a real world 3D coordinate system which origin is the origin of the calibration target.

There are roughly two different camera calibration methods (Zhang, 1999): Photogrammetric calibration and self-calibration. In photogrammetric calibration, a 3D object with precisely known features is observed. The calibration object generally consists of two or three perpendicular planes with square or round black and white patterns. Photogrammetric calibration can be done very accurately (Faugeras, 1993) but requires a very precise and expensive calibration set-up. Self-calibration doesn't require any calibration object. The calibration parameters are derived (Hartley 1994, Luong and Faugeras, 1997; Maybank & Faugeras, 1992) by observing a static scene from a moving camera. Yet flexible, this method is not yet mature (Bougnoux, 1998) and results obtained are not always reliable. Other calibration methods exist, including the vanishing points for orthogonal directions method (Caprile & Torre, 1990; Liebowitz & Zisserman, 1998) and the pure rotation calibration (Hartley 1994, Stein 1995).

The multi-camera calibration is done by recomputing the extrinsic parameters of all cameras from a single position of the calibration target simultaneously viewed by all the cameras (see Figure 4). After calibration, the cameras must not be moved during the whole recording session. This would cause a change in the value of the extrinsic parameters. Though camera calibration is usually performed prior to image acquisition, it can also be done after recording the particle images, provided the cameras have not been moved.

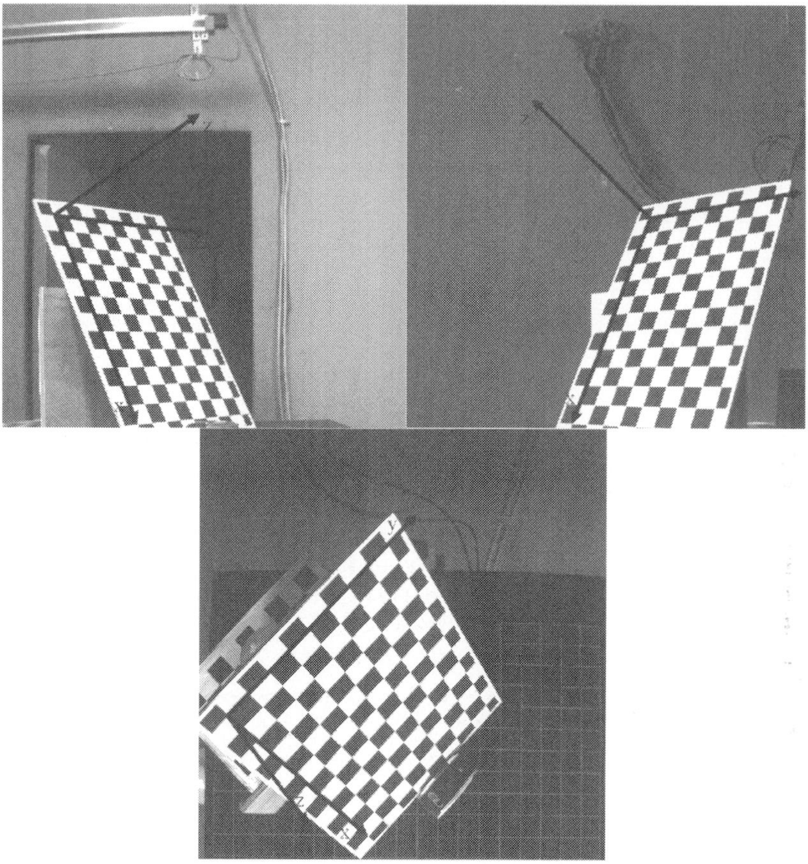

Fig. 4. The calibration target must be viewed simultaneously by all cameras when computing the common real world reference frame

3.2 Particle detection

The aim of the particle detection procedure is to compute the pixel coordinates of each particle center. Unlike traditional 3D PTV, PTV algorithms for large indoor air volumes must include a step where oversized particles images are removed from the images. Those oversized particle images are here referred to as "blobs". They are created by images of helium filled bubbles getting close to the cameras, since they are not constrained to remain inside a small delimited volume. Bubbles creating those blobs are generally out of the common field of vision. The particle detection processes usually includes the following steps:

1. Creation and subtraction of background from images.
2. Blob removal.
3. Calculation of pixel coordinates.

3.2.1 Creation and subtraction of background from images

The purpose of this step is to feed the tracking algorithm with grey level images of particles over totally black background images. The creation of the background is generally made by averaging a certain amount of images. Averaging the background is used to cope with the variations of continuous light intensity. It can be made in the presence of particles but conducting this process before introducing particles (or after all particles are gone) gives better subtraction results. When averaging the background in the presence of particles, it is worth averaging the whole set of particle images to be treated by the tracking algorithm. When averaging without particles, one should not use more than about 10 images otherwise the resulting image gradually becomes saturated.

After being created, the background is subtracted from each image. However, the result of the process generally does not permit a clear separation of particle images from residual noise. The main image processing functions usually used to remove noise are:

- Thresholding: This operation consists in retaining only the pixels which luminance is above a certain value determined empirically. Since noise has generally a luminance lower than particles, thresholding permits getting good images of particles. Finding the appropriate threshold can be made experimentally (the operation is rather quick) but it can also be implemented automatically, for instance by minimizing the interclass-variance of the segmented black and white pixels (Otsu, 1979).
- Removal of isolated pixels: Helium filled soap bubbles generally cover more than one pixel. Typically they create speckles of diameter three to six pixels. Therefore, isolated pixels can more than often be assimilated to remaining noise. They can be removed by eroding the image with structuring elements of 2-pixel lengths.

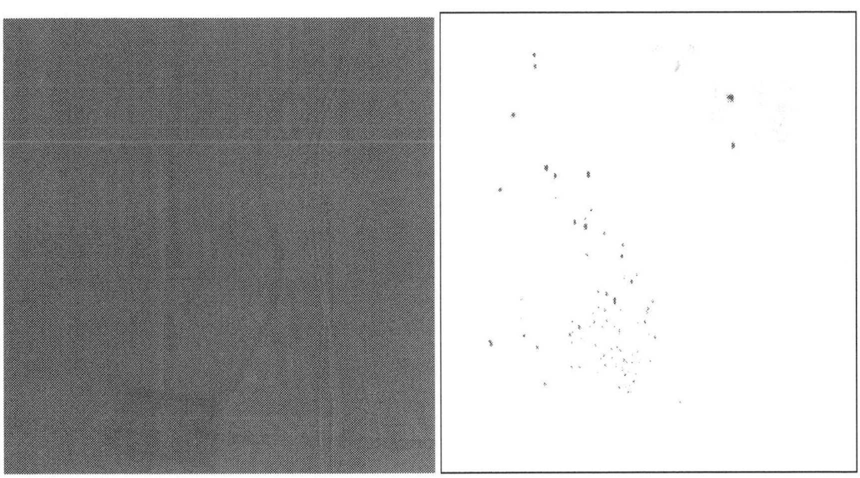

Fig. 5. Particle image before and after background and noise removal (the processed image is inverted for better clarity)

3.2.2 Blob removal

Blobs must be removed from the images because every single blob displays many isolated pixels, thus leading to false particle detection (see Figure 6). Removing round objects of a certain diameter from an image is a standard procedure in image processing called opening. An opening is an erosion followed by a dilation using the same structuring element for both operations. However, this operation generally gives poor results with images of helium filled bubbles. Since the particle images are neither perfectly round nor perfectly filled, the bubble shells can generally be seen after the processing, as shown in Figure 7. After removing the averaged background, blobs can be efficiently removed by the following procedure:

1. Binarize the image. The binary threshold level can be assigned empirically, or automatically calculated from iterative algorithms (Crouser et al. 1997; Otsu, 1979).
2. Fill-up and bridge all particle images in order to get homogeneous blobs (see Figure 8a). This filling-up and bridging may be made by iteratively dilating the image with structuring elements [1,1] and [1,1]T.
3. Erode the output image with a square structuring element of size the maximum diameter of a particle in the common camera's field of vision (see Figure 8b). Here particles of diameter equal or less than the maximum allowed are removed from the image. This maximum diameter value is determined empirically.
4. Dilate the resulting image with a square structuring element larger than the one previously used. Good results were achieved with a structuring element twice the size of the first one used (see Figure 8c).
5. Subtract the output image from the original unbinarized particle image (image before applying the first step of the procedure) to keep the Gaussian profile of the bubble images. The resulting image only contains particles of diameter equal or less than the size of the structuring element in step 3 (see Figure 9).

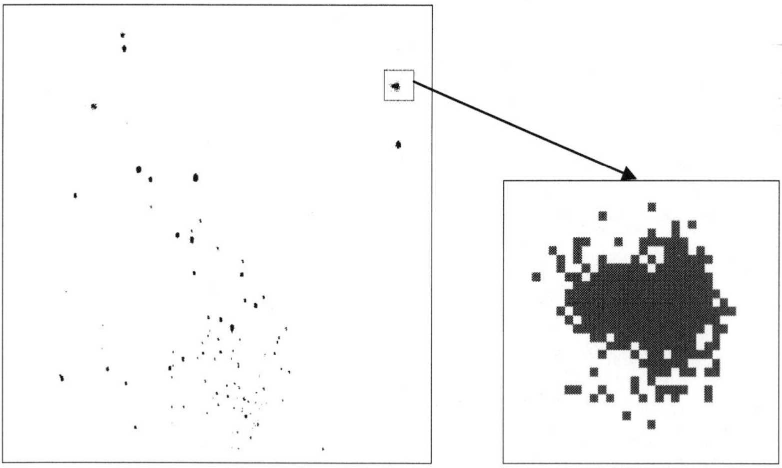

(a) (b)

Fig. 6. (a) Binarized particle image before blob removal (the image is inverted for clarity). (b) Standard over-large particle (blob). Blobs create many centroids leading to false detections

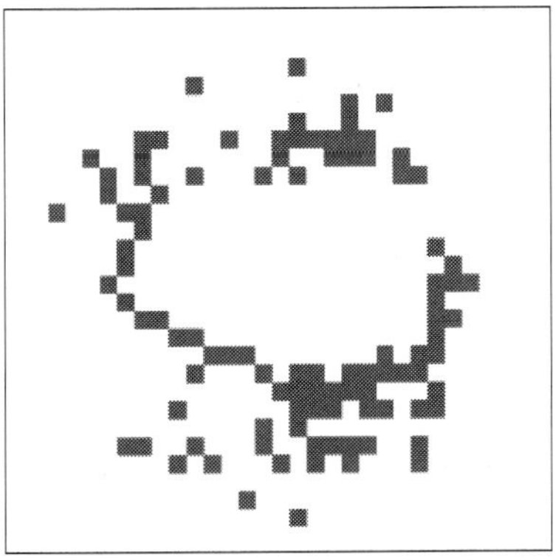

Fig. 7. Over-large particle image after standard image opening with an 8-pixel large disk-shaped structuring element. Bubble's shell is still visible

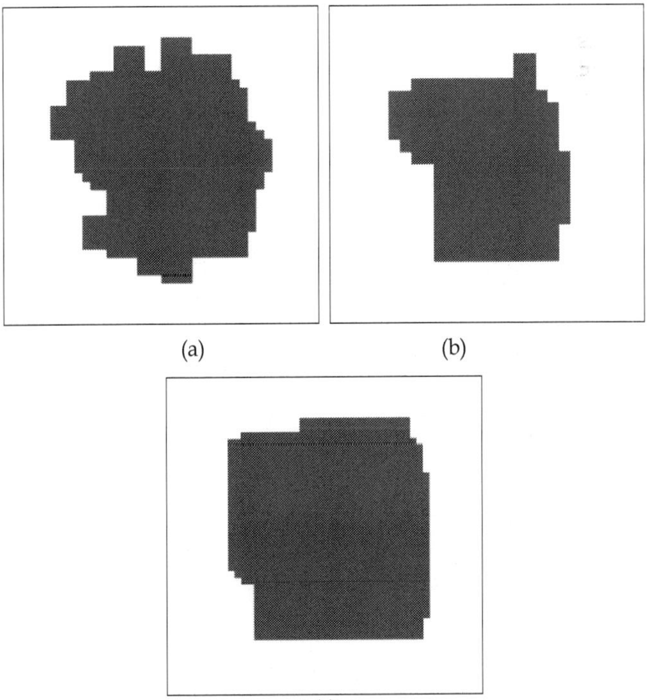

(a) (b)

(c)

Fig. 8. Proposed procedure for blob removal: (a) Blob filled and bridged. The displayed blob covers a 28x27 pixel region. (b) Same blob after an erosion with an 8-pixel square-shaped structuring element (16x15 pixel region). (c) Same blob after dilatation with a 16-pixel square-shaped structuring element (31x30 pixel region)

Fig. 9. Example of output image after the blob removal procedure. The image only contains particles of diameter less than 9 pixels

3.2.3 Calculation of particle centers pixel coordinates

Particle centers pixel coordinates are here referred to as "centroids". When calculating centroids from helium filled bubble images, template matching (Gruen & Baltsavias, 1988; Guezennec et al., 1994) gives poor results with soap bubbles because of the wide range of particle shapes and sizes after image processing. Hough transform (Hough, 1962) is ill-adapted to the smallness of the pixel area covered by average particles which varies from 2x2 to 8x8 pixels, depending on the distance from the cameras. Invariant second order grey moments method (Teh & Chin, 1988) works well when only two particle images are overlapping but fail when three particles create a larger speckle. Higher order moments are very noise sensitive. 2D Gaussian fitting (Mann et al., 1999; Nobach & Honkanen 2005) works well when particles intensity profile can be approximated by a Gaussian. In our case, a single particle often features two intensity peaks. In addition, Gaussian fitting is computationally costly and gives better results on large particle images. The same drawbacks work for neural network methods (Carosone et al., 1995) though those methods are robust in case of noisy images.

As shown in Figure 2, particles are seen as two (or only one) bright spots on the particle shell symmetric relative to the center of the sphere. This fact makes relevant the use of weight averaged methods to calculate the center of mass of each particle. For each particle, the coordinates (x_c, y_c) of the center or mass are given by:

$$x_c = \frac{\sum xI(x,y)}{\sum I(x,y)} \; ; \; y_c = \frac{\sum yI(x,y)}{\sum I(x,y)} \qquad (5)$$

where (x,y) are the pixel coordinates of each pixel belonging to the particle and $I(x,y)$ the pixel luminance. This method allows recognizing two connected bright spots as a single particle. Its main shortcoming is the creation of two centroids whenever the two bright spots are not connected (see Figure 10).

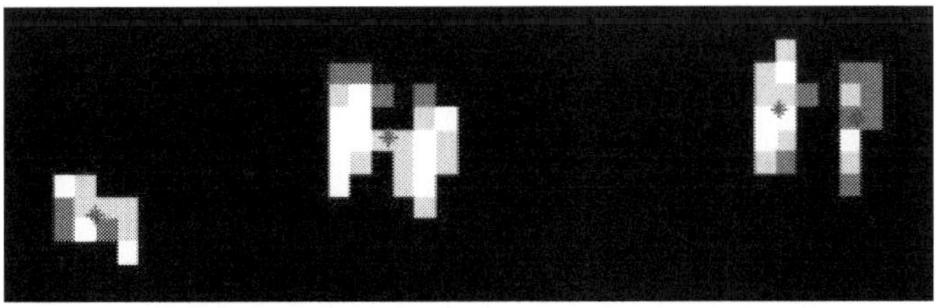

Fig. 10. Output of the center of mass calculation by weight averaging on three particles of the same image. Far left and center particles yield one centroid whereas far right particle displays two disconnected local maximum and therefore yields two centroids

As shown by (Ouellette et al., 2006), weight averaging methods may be less accurate than Gaussian fitting and neural network methods whenever particles are far enough from the camera to feature a single local maximum. Nevertheless, this method is efficient, readily implemented and rapid. Particle centers are given with sub-pixel accuracy (down to 1e-01) with derisive computation time using standard personal computers. Furthermore, in large volumes (over $1m^3$), overlapping cases are estimated to less than 5% of particle images for camera resolutions over 1024x1024 pixels.
Once we can accurately detect particle centers on each image, the next step is to establish their 2D or 3D trajectory from successive images. That is particle tracking.

3.3 Particle tracking

Particle tracking schemes used for 3D PTV in large scale indoor air applications do not differ from particle tracking schemes used in other applications. Particle tracking schemes can be divided into three main categories:

- Image plane based tracking schemes: Particles are tracked on each camera 2D image plane separately through time (temporal tracking). Afterwards, the resulting 2D trajectories are matched in 3D object space and the 3D coordinates are calculated (Biwole et al., 2009; Engelmann, 1998, 2000; Ge & Cha, 2000; Guenzennec et al., 1994; Jähne, 1997; Li et al. 2008; Wierzimok & Hering, 1993)
- Object plan based tracking schemes: Particle 3D coordinates are first calculated at each time step separately. Afterwards, the resulting set of time-ordered 3D coordinates is the only input for temporal tracking directly in object space. The correspondence problem is addressed at the first or only a few time steps of the image sequence (Kasagi & Sata, 1992; Maas, 1992; Maas et al., 1993; Malik et al., 1993; Nishino et al., 1989; Nishino & Kasagi, 1989; Papantonious & Dracos, 1989; Sata et al., 1994; Suzuki et al., 2000; Virant, 1996; Willneff & Gruen, 2002)

In spite of their high potential the object plan based tracking algorithms hold two main drawbacks: first, the repetitive error in the calculation of particle 3D coordinates aggregates as the trajectory gets longer. Second, the unsolved particle 3D correspondences create many discarded particles and broken trajectories. On both schemes, Particle tracking from one time frame to another is usually based on a set of criteria such as the minimum change in particle acceleration, the particle shape and the particle luminance. An additional cost

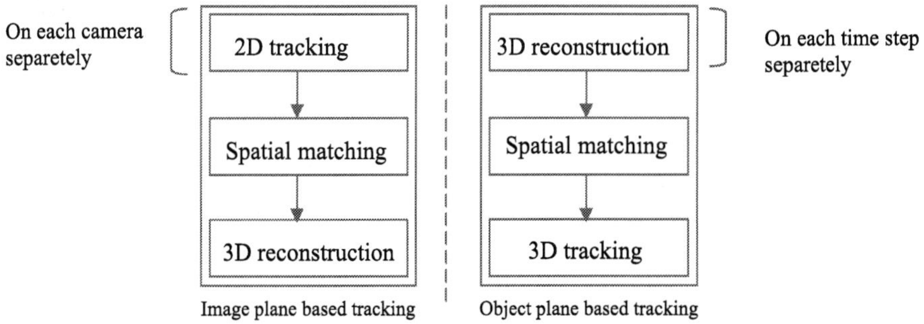

Fig. 11. Comparison of image and object plane based 3D tracking schemes.

function is applied in case any ambiguity arises. A detailed review of the particle tracking algorithms has been proposed by (Ouellette et al., 2006) and will not be repeated herein. The one proposed by (Biwole et al., 2009) is detailed here as an example:

3.3.1 2D tracking

Based on (Li et al., 2008) 2D tracking scheme, (Biwole et al., 2009) algorithm first utilizes a second order polynomial regression method to predict the center of the search region. The purpose of the regression is to minimize the changes in particle acceleration:

$$x_{pixeli} = \mathbf{a}t_i^2 + \mathrm{b}t_i + \mathrm{c} \tag{6}$$

where x_{pixel} stands for the pixel coordinates vector at instant time t and i stands for as much as 5 previous but not necessarily consecutive frames. Two by one constant vectors ***a***, ***b*** and c are acquired by least square fitting. Then, a cost function ϕ is used to resolve the conflicts within the search area:

$$\phi = \frac{\sqrt{\sum_{k=0}^{3} \left| D_k - G\tau_k - H \right|^2}}{\sqrt{\sum_{k=0}^{3} \left| D_k \right|^2}} \tag{7}$$

where D_k is the previously linked trajectory displacement between frames k and k+1, τ_k the middle time between frames k and $k+1$, and ***G*** and ***H*** the two by one constant vectors resulting from the linear regression of order 1 fitting D_k over frames 0 to 3:

$$D_k \approx G\tau_k + H \tag{8}$$

The cost function thus appears as a regression residual normalized by a geometrical mean displacement. Its physical meaning is also to minimize the changes in particle acceleration. (Biwole et al., 2009) 2D tracking scheme includes four additional features. First, a "cross-gap" strategy accounts for particles undetected in a single frame. Whenever a particle is absent from the search region at frame n+1, the regression and the search is extended to frame n+2 based on the estimate and the trajectory goes on if a suitable particle is found. If a suitable particle is not found, the trajectory is ended. Second, particles centroids are recomputed at each frame (instead of doing it once and tracking them in 2D space). It allows starting new trajectories as new particles enter the field of vision. Third, an overlapping trajectory detection step discards the shortest one of two trajectories with similar coordinates. This is done to compensate for the extra trajectories created by the previous step. Last, a bridging strategy connects parts of a same trajectory as follows: Let Te be a trajectory ending at frame t and Ts a trajectory starting at frame t or t+1. The cost function of a new trajectory composed of the last three positions of Te and the first three positions of Ts is calculated according to Equation 7. If the cost function is less than a preset parameter α (α is around 0.2), the bridge is validated.

This 2D tracking strategy is intended to yield longer trajectories than traditional ones. However, its computational cost in terms of space and time is higher than the cost generated by traditional algorithms. Going through the process with spatial matching at each time step would have been even more costly and error prone. This is why temporal tracking is done before spatial matching. As will be shown in section 5, 3D matching is later checked several times throughout the trajectory. A more detailed description of this 2D tracking scheme is given in (Li et al., 2008). The scheme was validated using simulated data of a jet flow impinging on a wall.

3.3.2 Stereo pair matching

Stereo pair matching also called spatial matching is first done using a three-camera arrangement. Firstly, the fundamental matrix of each pair of cameras is calculated. For two cameras 1 and 2, the fundamental matrix reads:

$$F_{12} = \left[T_1 - R_1 \cdot R_2^T \cdot T_2\right]_\times R_1 \cdot R_2^T \tag{9}$$

Matrixes R_1, T_1, R_2 and T_2 are extrinsic parameters given by calibration. The cross-product [ ]x is defined as:

$$[u]_\times = \begin{pmatrix} 0 & -u_3 & u_2 \\ u_3 & 0 & -u_1 \\ -u_2 & u_1 & 0 \end{pmatrix} \tag{10}$$

Two 2D trajectories are considered matched if 6 pairs of time-synchronous points, one in each trajectory, can be found verifying the following equation:

$$\begin{bmatrix} x_1^t & y_1^t & 1 \end{bmatrix} F_{12} \begin{bmatrix} x_2^t \\ y_2^t \\ 1 \end{bmatrix} \prec s \tag{11}$$

where (x_1^t, y_1^t) and (x_2^t, y_2^t) are normalized pixel coordinates from each trajectory, t an instant time (t = 1..6), and s is a threshold value. Ideally, the left term of Equation (8) should equal zero but it never does, due to experimental and computational errors. This is why s is generally given the value 1. Tests showed that if Equation (11) is validated for at least 6 time-synchronous pairs of points, it is also validated for the others points of the two trajectories. Because of the possible high length of the trajectories, this strategy brings additional reliability in contrast to traditional 3D PTV algorithms where spatial matching is done only once before temporal tracking.

At first, each trajectory is matched using all three fundamental matrixes. The remaining unmatched trajectories are then matched using only one fundamental matrix. Those trajectories come from particles whose displacement is seen by only two cameras. After these two processes, the remaining unmatched trajectories are discarded.

3.3.3 3D reconstruction

This section deals with the calculation of final 3D coordinates from multiple views of a trajectory point. Let f = (f1,f2)T be the focal length in pixels, cc =(cc1,cc2)T the vector containing the principal point pixel coordinates, β the skew coefficient (scalar), k = (k1,k2,k3,k4,k5)T the vector containing the image radial (k1,k2,k3) and tangential (k4,k5) distortion coefficients given by calibration. Let P be a real world point of which the coordinates in the reference frame of one camera are (Xc,Yc,Zc). Let be xn = (x,y) the normalized projection of P on the camera image plane. Normalized coordinates are defined as the pinhole projection coordinates obtained when using a unit focal length (see Figure 3). We can write:

$$\mathrm{x_n} = \begin{bmatrix} x \\ y \end{bmatrix} = \begin{bmatrix} X_c / Z_c \\ Y_c / Z_c \end{bmatrix} \tag{12}$$

Taking into account the relation between each camera 3D reference frame XXc (Xc,Yc,Zc) and the calibration target 3D reference frame XX (X,Y,Z) given in Equation (4), Equation (12) can be rewritten for each camera i as:

$$\begin{cases} x_i = \dfrac{R^i_{11}X + R^i_{12}Y + R^i_{13}Z + T^i_1}{R^i_{31}X + R^i_{32}Y + R^i_{33}Z + T^i_3} \\ y_i = \dfrac{R^i_{21}X + R^i_{22}Y + R^i_{23}Z + T^i_2}{R^i_{31}X + R^i_{32}Y + R^i_{33}Z + T^i_3} \end{cases} \tag{13}$$

where by abuse of notation, T_i and R_i are the same as in Equation (4), and (x_i, y_i) are the normalized pixel coordinates of the particle on camera i. With i = [1...n] (n cameras), equation (10) gives rise to an overdeterminated system of 2n equations for only 3 unknowns which is solved by a least squares method. When the calibration target 3D reference frame is different from the final reference frame XXo = (Xo,Yo,Zo), equation (10) reads for each camera:

$$\begin{cases} \left[x(R_{31}U_{11} + R_{32}U_{21} + R_{33}U_{31}) - (R_{11}U_{11} + R_{12}U_{21} + R_{13}U_{31})\right]X_0 + \\ \left[x(R_{31}U_{12} + R_{32}U_{22} + R_{33}U_{32}) - (R_{11}U_{12} + R_{12}U_{22} + R_{13}U_{32})\right]Y_0 + \\ \left[x(R_{31}U_{13} + R_{32}U_{23} + R_{33}U_{33}) - (R_{11}U_{13} + R_{12}U_{23} + R_{13}U_{33})\right]Z_0 \\ = -x(R_{31}V_1 + R_{32}V_2 + R_{33}V_3 + T_3) + (R_{11}V_1 + R_{12}V_2 + R_{13}V_3 + T_1) \\ \\ \left[y(R_{31}U_{11} + R_{32}U_{21} + R_{33}U_{31}) - (R_{21}U_{11} + R_{22}U_{21} + R_{23}U_{31})\right]X_0 + \\ \left[y(R_{31}U_{12} + R_{32}U_{22} + R_{33}U_{32}) - (R_{21}U_{12} + R_{22}U_{22} + R_{23}U_{32})\right]Y_0 + \\ \left[y(R_{31}U_{13} + R_{32}U_{23} + R_{33}U_{33}) - (R_{21}U_{13} + R_{22}U_{23} + R_{23}U_{33})\right]Z_0 \\ = -y(R_{31}V_1 + R_{32}V_2 + R_{33}V_3 + T_3) + (R_{21}V_1 + R_{22}V_2 + R_{23}V_3 + T_2) \end{cases} \tag{14}$$

where U and V are respectively the matrices of rotation and translation to map from XX to XXo similarly to Equation (4). U and V are identical for all cameras and are calculated from the experimental set-up. Normalized pixel coordinates are related to actual pixel coordinates by the following rationale:
Let $r^2 = x^2+y^2$. After including lens distortion, the new normalized point coordinate xd is defined as follows:

$$\mathrm{x_d} = \begin{bmatrix} x_{d1} \\ x_{d2} \end{bmatrix} = (1 + k_1.r^2 + k_2.r^4 + k_5.r^6)\mathrm{x_n} + \mathrm{d_x} \tag{15}$$

where d_x is the tangential distortion vector (Heikkilä & Silvén, 1997 camera model):

$$\mathrm{d_x} = \begin{bmatrix} 2k_3.x.y + k_4.(r^2 + 2x^2) \\ \\ 2k_4.x.y + k_3.(r^2 + 2y^2) \end{bmatrix} \tag{16}$$

The final pixel coordinates $\mathrm{x_{pixel}}$ of the projection of P on the image plane reads:

$$\mathrm{x_{pixel}} = \begin{bmatrix} x_p \\ y_p \end{bmatrix} = \begin{bmatrix} f_1.(x_{d1} + \beta.x_{d2}) + cc_1 \\ f_2.x_{d2} + cc_2 \end{bmatrix} \tag{17}$$

Thus, from the knowledge of final pixel coordinates on at least 2 image planes, real 3D coordinates can be calculated.

4. Example of results

This section shows some examples of application of 3D PTV for indoor air flow measurement. Some helpful guidelines are given in terms of camera and light positioning, depending on the room layout and wall color. The following performance indicators will be used:

- "Tracking density" ratio ξ: it is the ratio of mean particle spacing (in a nearest neighbor sense) to mean particle displacement between two consecutive frames. It is an indicator of the tracking difficulty (Malik et al., 1993).
- "Correct tracking" ratio γ2D (respectively γ3D): Proposed by (Li 2008), it is the number of 2D (respectively 3D) tracked positions which are identical to the actual 2D/3D particle positions divided by the total number of tracked positions. This ratio only deals with tracked trajectories and is an indicator of the tracking accuracy. The actual 2D particle trajectories are estimated for each camera by adding up all the camera frames after background subtraction. The deviation from actual 3D positions is estimated by backprojection of calculated 3D trajectories on each camera image plane thanks to calibration data. A particle tracked position is considered "correct" when it is either identical to the actual particle position or its deviation from the latter position is less than the radius of the actual particle.
- "Total tracking" ratio $E_{3Dtrack}$ (respectively $E_{2Dtrack}$): proposed by (Malik et al., 1993) as the number of correctly tracked 3D (/2D) trajectories divided by the estimated total number of input 3D (/2D) trajectories.

4.1 3D PTV in a light-gray walled room, low density seeding

The test-room *MINIBAT* of the National Institute for Applied Sciences of Lyon, France, has two experimental cells of dimensions 3.1mx3.1mx2.5m high each (see Figure 12). The 3D PTV set-up included three cameras Dalsa 4M60 set at 1024x1024 pixels and 100 fps each. Each camera was mounted with a 15mm Canon lens with a 4.8 aperture. All cameras were placed in experimental room noted (6) on Figure 12. The recording computer was located in the other experimental room (7) and the door between the two rooms was closed to prevent heating up of the flow. Camera 3 was fixed onto the ceiling while cameras 1 and 2 were fixed on the walls as shown in Figure 13.

Light was provided by four 1000W compact fluorescent lamps situated in the climatic chamber (3) and separated from the airflow by the simple glass partition (4) to prevent heating up the flow. All lights were set at full power, with not shading grid. Walls of the test room are not black but light gray; consequently, it was found that better contrast between particles and background was achieved by directing the lights towards the walls and not directly towards the particles; therefore, helium filled soap bubbles were indirectly illuminated by reflection of the light from walls and ceiling. The calibration target consisted of a planar checkerboard composed of black and white 30mm-large squares. The checkerboard had 12 horizontal squares and 8 vertical squares as shown in Figure 14. The calibration algorithm by (Bouguet, 2002) was used. Bubbles were released upward. Index ξ equaled 3.9, 8.1 and 6.3 respectively for cameras 1, 2 and 3, which corresponds to a low density seeding.

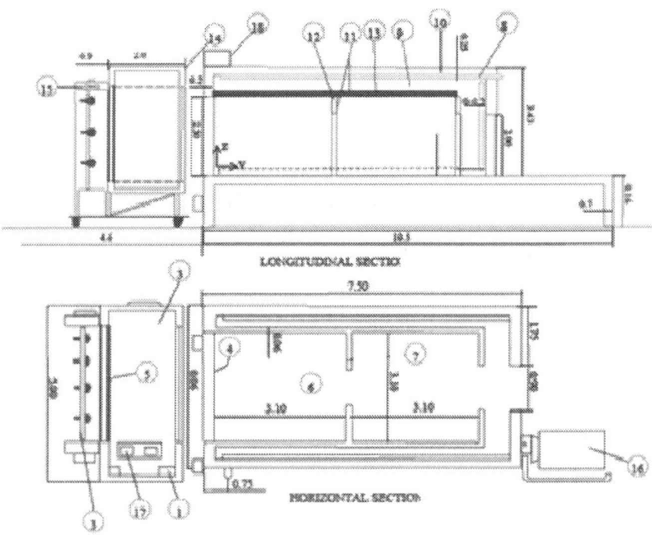

(1) cold conditioner; (2) spotlights; (3) climatic chamber;
(4) simple glass; (5) double glass; (6) experimental room; (7) experimental room;
(8) air pipe; (9) thermal guard; (10) insulated concrete;
(11) plaster plate ; (12) wood plate; (13) wood plate;

Fig. 12. Mock-up of the test-room MINIBAT. Lights were situated behind the glass (4) in the climatic chamber (3), cameras were situated in the experimental room (6) and the recording computer in experimental room (7)

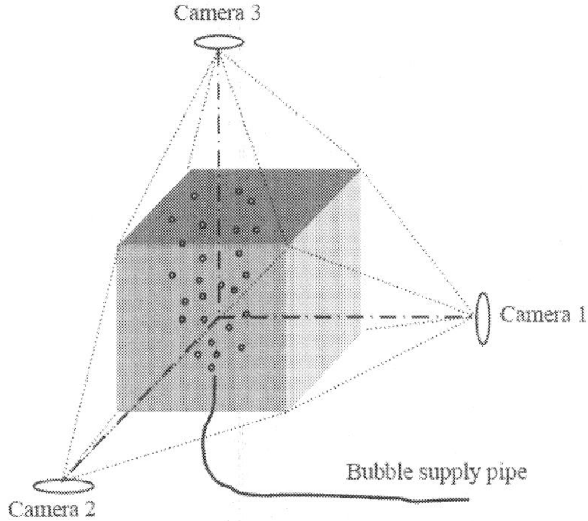

Fig. 13. Camera positioning for 3D PTV on an ascendant free flow.

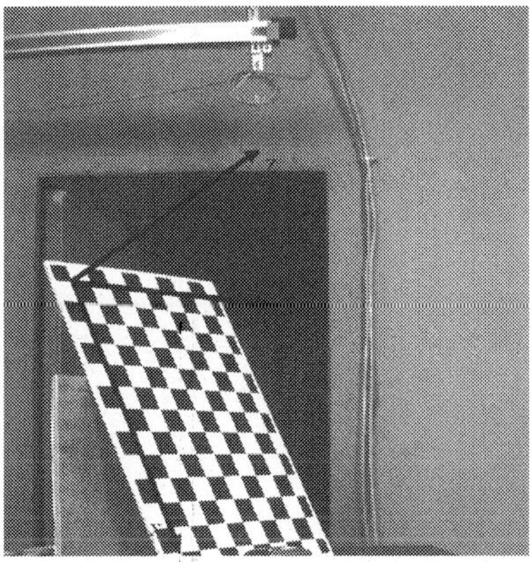

Fig. 14. Calibration target as observed from camera 2 viewpoint.

Full 3D displacement of particles over 10 frames is shown in Figure 15. The average $E_{2Dtrack}$ was 82% over the three cameras with $\gamma_{2D} = 1$. We found $E_{3Dtrack}$= 88% with $\gamma_{3D} = 0.9$. Having $E_{3Dtrack} > E_{2Dtrack}$ is readily explainable by the fact that some extra 3D trajectories were produced from particles seen by only two cameras at a time. It is also normal to have $\gamma_{3D} < \gamma_{2D}$ because of additional errors due to computational approximations when calculating the 3D coordinates.

For validation purpose, all 3D trajectories were projected back onto each camera image plane and compared with the real 2D bubbles trajectories obtained by adding up the original images. The resulting images for cameras 1 and 2 are shown on Figure 16. On the Figure, untracked white streaks are trajectories from particles seen by only one camera.

From 3D data, the calculated bubbles mean velocity was 0.375m/s with a minimum at 0.206m/s for bubbles far from the pipe nozzle, and a maximum at 0.651m/s.

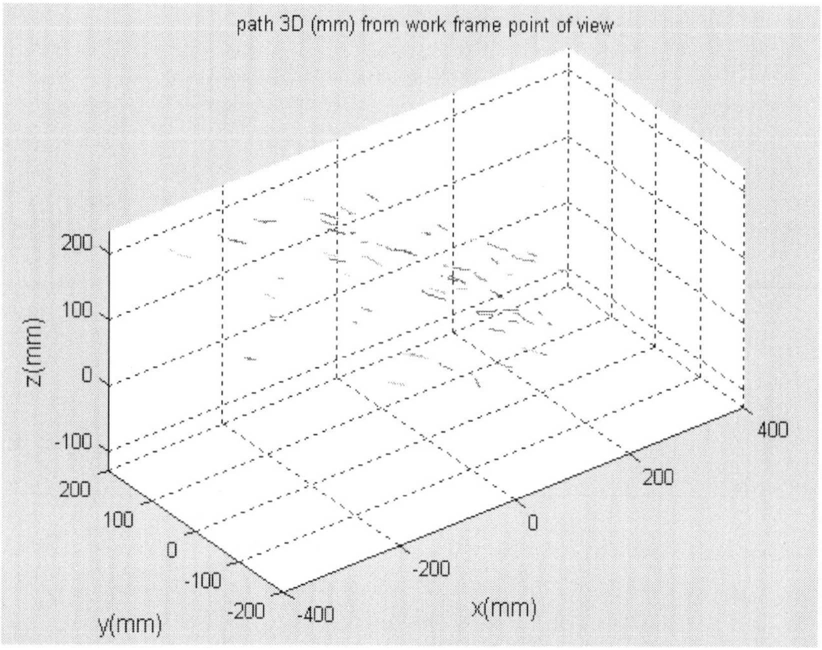

Fig. 15. 3D path (mm) of tracked particles. The orientation of axes is given by the calibration target.

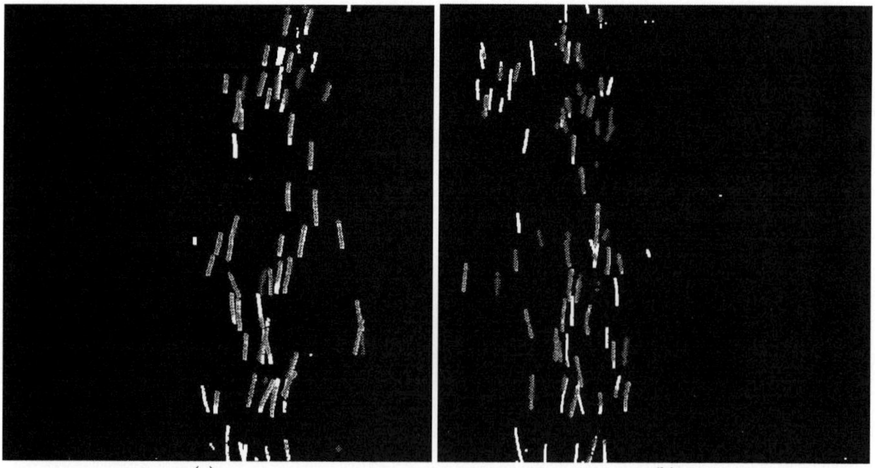

(a) (b)

Fig. 16. Comparison of real trajectories (white) versus backprojected 3D trajectories (blue) on cameras 1 (Figure 16a) and 2 (Figure 16b). Completely white tracks are particle trajectories that are seen by only one camera and therefore are not traceable in 3D space.

4.2 3D PTV in a black-walled room, high density seeding

Helium filled soap bubbles were released in the 5.5mx3.7x2.4m high *Room Ventilation Simulator* at the Bio Environmental Engineering laboratory, University of Illinois at Urbana-Champaign, USA. The bubbles were released from two converging pipes and the production was stopped 10s before the recording. To increase the measurement area, camera 3 was not situated directly above the measurement volume but at an angle of 45° as shown in Figure 17. The planar angle between each camera was approximately 120° with six 500W spotlights situated onto the horizontal bisectors. To prevent heating, the spotlights were turned on only for the two seconds of recording. Cameras were set at 30 fps. Calibration was done using the same calibration target as described above.

Index ξ equalled 3.3, 2.2, and 2.5 respectively for cameras 1, 2 and 3. Such values for ξ correspond to a difficult particle tracking situation (see Figure 18). The temporal tracking process yielded $E_{2Dtrack}$ = 670/1566 for camera 1, 314/889 for camera 2 and 581/1635 for camera 3 with an average $\gamma_{2D} = 0.9$. We found $E_{3Dtrack} \approx 714/1800$ with $\gamma_{3D} = 0.75$. Whereas the common view area was only 1.5mx1.5mx1m, the actual measured area was approximately 3mx3mx1.2m as shown on Figure 19. This is due to the fact that 3D coordinates are still calculated if the particles are seen by only two cameras. Figure 19 also shows the individual path of a bubble and validation by back-projection is shown in Figure 10-11. From the 3D data, the calculated bubbles mean velocity was found equal to 0.107m/s with a minimum at 0.015m/s and a maximum at 0.521m/s.

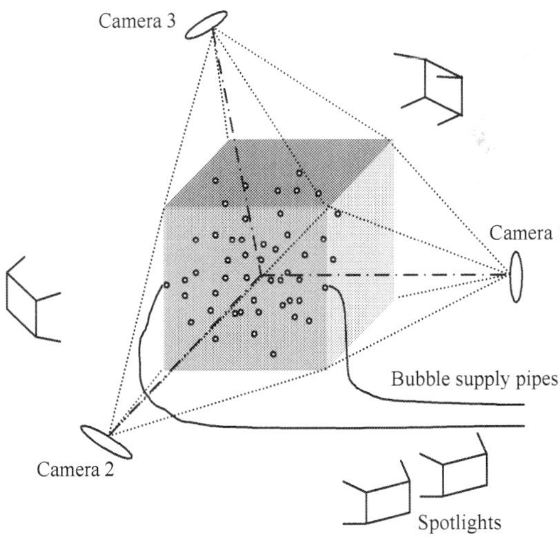

Fig. 17. Camera and light positioning for 3D PTV in a black-walled room.

Fig. 18. Particle images from a region of camera 1 after image processing (inverted and 100% zoomed in image).

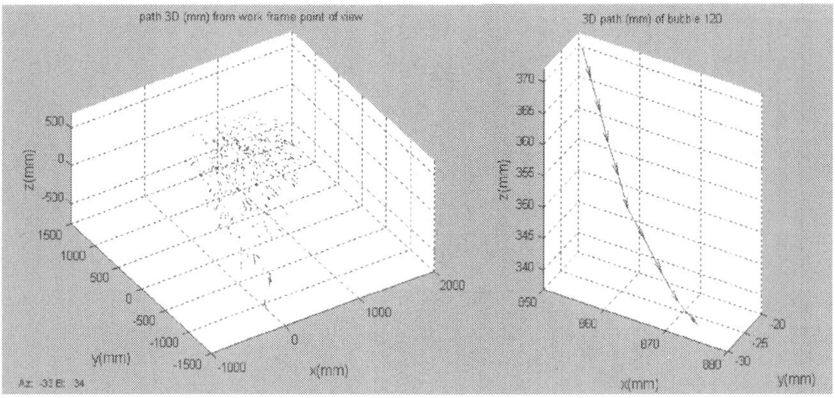

Fig. 19. 3D path of bubbles over 40 frames in high density case and 3D path of a single bubble over 10 frames.

4.3 Velocity distribution over a heat source

An electric heater was placed against one wall of the *Room Ventilation Simulator*. Helium filled bubbles were emitted from two pipes situated on both sides of the radiator. The bubble generation was not stopped during the recording to prevent their rarefaction due to the upward convection flow. In order to avoid any impact of the bubble initial velocity (5.33m/s) on the measured flow, the pipes nozzles were facing downwards. Therefore, the tracers were recorded from the radiator height of 90cm after rebounding on the floor.

One camera was placed parallel to the wall and looking downward at the field. The other two were facing the wall at a symmetrical angle of 45° as shown on Figure 20. To avoid illuminating the background (wall), the spotlights were also placed parallel to the wall on both sides of the heater. The targeted measurement field was the area above the heater. Therefore, the calibration target was placed over the heater as shown in Figure 21. The heating power was set at 600W. Eight T type thermocouples were set to record the temperature of the wall and of the air above the heater at 10cm from the wall. The temperature distribution was 28.5°C on the wall, 29.8°C above the heater and 29.2°C at the center of the room six feet from the wall and three feet from the floor.

Due to the chimney effect, many particles were driven out of the measurement field. Therefore, the tracking was performed in low density. Index ξ was 3.3, 3.9, and 4.8 respectively for cameras 1, 2 and 3. Results of 3D PTV over the heater are shown in figures 22 and 23. The measured area was approximately 1.5m x 0.9m x 1.5m high. Temporal tracking yielded E2Dtrack = 95/138 for camera 1, 73/90 for camera 2 and 61/65 for camera 3 with γ2D = 1. We found E3Dtrack ≈ 106/140 with γ3D = 0.95. The maximum velocity was 0.851m/s for a bubble situated above the heater.

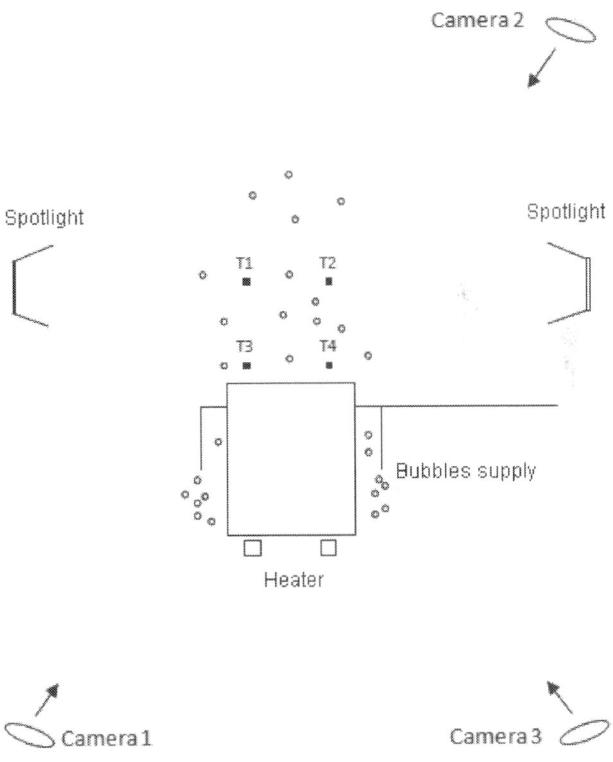

Fig. 20. Cameras and light sources positioned for 3D PTV over a heater

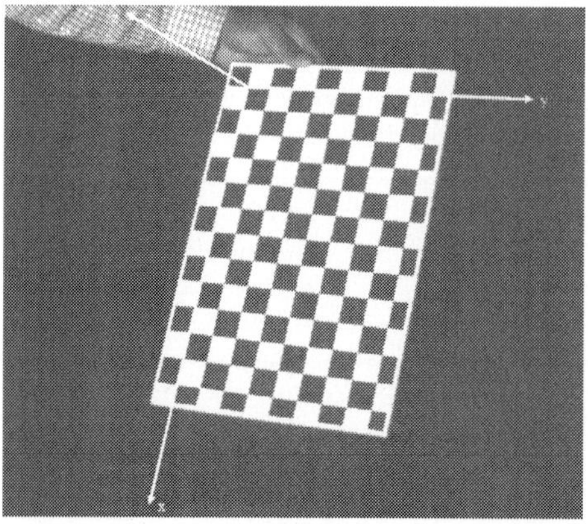

Fig. 21. Calibration target reference frame from camera 3 viewpoint

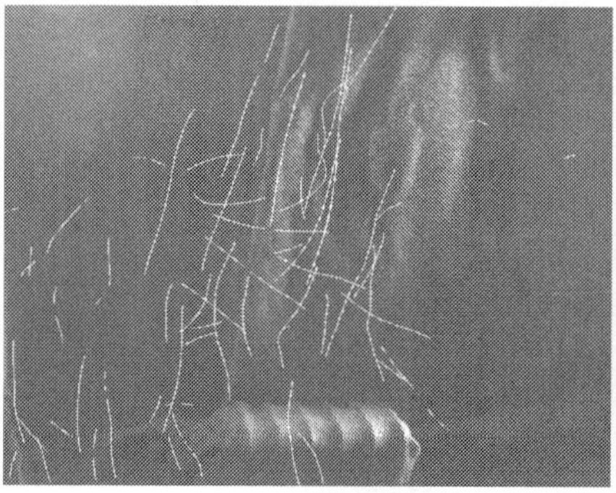

(a)

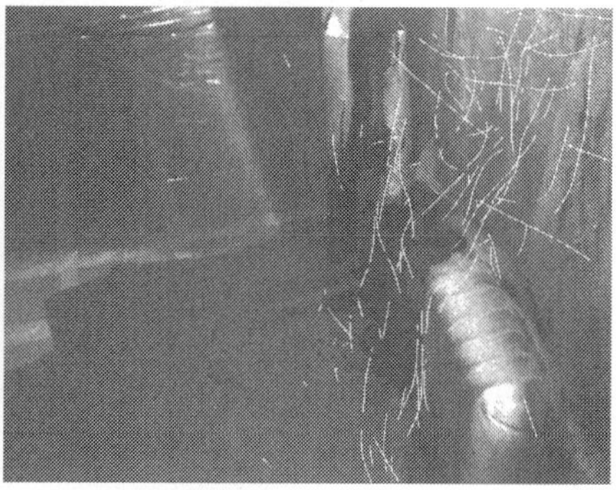

(b)

(c)

Fig. 22. White arrows show the bubbles 3D trajectories projected on image plane of cameras 1, 2 and 3 (respectively figures 17a, 17b and 17c). The measured area is approximately 1.5m x 0.9m x 1.5m high.

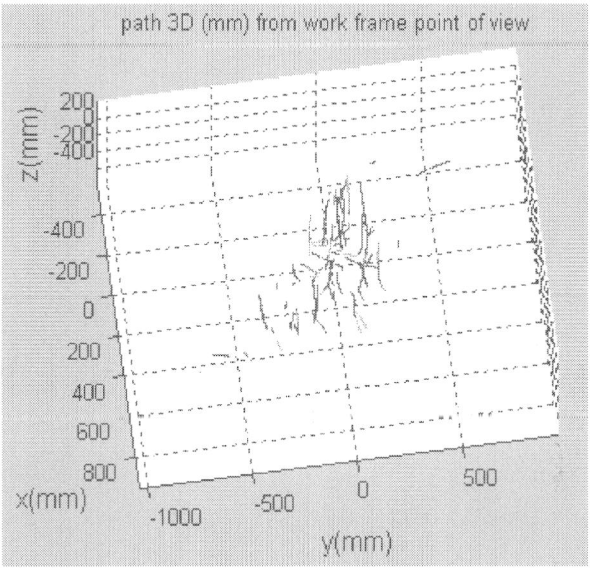

Fig. 23. 3D path of bubbles above the heater throughout 30 frames

4.4 Velocity distribution in an experimental aircraft cabin

The experimental aircraft cabin used is part of the Bio-Environmental Engineering laboratory test facilities. It is a full-scale, fully equipped five row section of a Boeing 767-300 with dummies to simulate passengers as shown in Figure 24. The cabin is 4m x 3m x 2m high. Helium filled bubbles were introduced from two pipes situated at the sides of the ceiling middle section. The aircraft ventilation system was on. Three cameras were placed outside of the cabin in a triangular pattern. They were directed convergently, with a large triangle base. The calibration target was a planar checkerboard composed of a 12 x 12 array of 6cm black and white squares as shown in Figure 25. Light was provided by eight 120W light bulbs. Two extra 500W spots were facing the cabin from the external side of the glass wall.

After bubble seeding, index ξ was 2.1, 2.3, and 2.3 respectively for cameras 1, 2 and 3. After temporal tracking, $E_{2Dtrack}$ averaged 0.4 over the three cameras with $\gamma_{2D} = 1$. Full 3D tracking yielded $E_{3Dtrack} = 0.63$ with $\gamma_{3D} = 0.7$. A total of 1083 particles were 3D tracked. Velocity profiles obtained (see Figure 26) were found similar to those obtained by (Wang et al., 2005) on the same experimental set-up. Especially, the same vortices due to recirculation of the air over the two aisles could be observed and 3D tracked. The mean velocity equaled 0.48m/s over the cabin, with a minimum at 0.018m/s which means quasi static air far from the ventilation nozzles. The 3D data enabled the tracking of the air trajectory and the precise identification of areas of minimum air displacement.

Fig. 24. Experimental aircraft cabin. Cameras positions are marked by white circles. The two ventilation inlets are marked by white arrows.

Fig. 25. Calibration target reference frame from camera 1 viewpoint

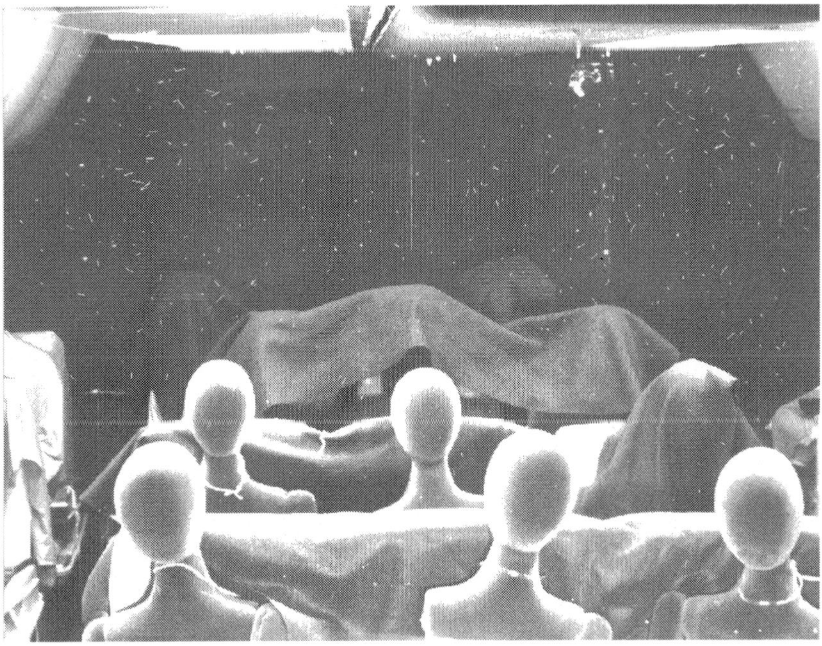

Fig. 26. Instantaneous 2D velocity in the aircraft cabin using the fast normalized cross-correlation temporal tracking scheme. Vortices due to recirculation of the air in the cabin can be seen over the aisles. University of Illinois at Urbana-Champaign, Department of Agricultural and Biological Engineering, Bioengineering Research Laboratory, USA

5. Conclusion and prospective research

Recent developments in large scale three dimensional particle tracking velocimetry have proven that this technique is suitable for the quantitative measurement en indoor airflow and pollutant dispersion. The technological breakthrough comes first from two aspects: first the measurement of the air velocity is no longer point-wise like in hot wire anemometry, or planar like in stereoscopic particle image velocimetry. On the contrary, the measurement is volumetric, with the 3D velocity of several points acquired simultaneously in large volumes. Second precise air and pollutants trajectories are acquired versus time. This chapter has presented typical experimental set-ups and algorithms that could be used to that purpose. The algorithm by (Biwole et al., 2009) has been detailed as an example. Around 1400 neutrally buoyant particles made of helium filled bubbles were tracked in volumes up to $3m^3$. The illumination was provided by halogen lamps. It was also shown that lighting up the helium filled bubbles by indirect reflection of the light on walls and ceiling allows a better particle detection on non black-walled rooms.

Current research on the subject aims at several objectives. Firstly, scientists try to increase the seeding density which is still limited to less than 2000 particle because of occlusion problems. A densely seeded volume would allow a finer understanding of the flow topology and a better modeling by CFD simulation. The measured indoor volumes can also be extended by juxtaposing several multi-camera 3D PTV arrangements. Secondly, researchers try to increase the range of measurable speeds by the design of algorithms coupling large scale 3D PTV with large scale three dimensional particle streak velocimetry. The latter technique comes in to capture the velocity of particles which cannot be detected as dots by the cameras because of their higher speed. In rooms, such speeds can be found at air inlets and outlets. Thirdly, researchers try to combining particles Lagrangian velocity with particles Lagrangian temperature. Trying to do so by monitoring the expansion of bubbles volume due to temperature is nearly impossible because of the low coefficient of volumetric thermal expansion of helium. The most realistic approach is to look for fluorescent bubble shells that visually and reversibly change colors at predetermined temperatures. In this regard, the use of thermochromic liquid crystals or of polythiophene films could be of some help.

6. References

Adamczyk, A.A. & Rimai, L. (1988). Reconstruction of a 3-dimensional flow field from orthogonal views of seed track video images. *Experiments in Fluids*, Vol.6, No.6, pp. 380-386

Adrian, R.J. (1991). Particle-imaging techniques for experimental fluid mechanics. Annual Review of Fluid Mechanics, Vol.23, pp. 261–304

Biwole, P.H.; Krauss G.; Favier, E; Rusaouen, G. & Roux J.-J. (2008). Non-stereoscopic 3D particle tracking velocimetry for full scale rooms, *Proceedings of Advanced building ventilation and environmental technology for addressing climate change issues*, Kyoto, Japan, October 2008

Biwole, P.H.; Yan, W.; Zhang, Y. & Roux, J-J. (2009). A complete 3D particle tracking algorithm and its applications to indoor airflow study. *Measurement Science and Technology*, Vol.20, 115403 (13 pp)

Bougnoux, S. (1998). From projective to Euclidean space under practical situation, a criticism of self calibration, *Proceedings of the 6th European conference on computer vision*, pp. 790-796, January 1998

Bouguet, J-Y. (2002). *Camera Calibration Toolbox for Matlab*, Available from http://www.vision.caltech.edu/bouguetj/calib_doc/

Caprile, B. & Torre, V. (1990). Using vanishing points for camera calibration. *The International Journal of Computer Vision* Vol.4, No.2, pp. 127-140

Carosone, F.; Cenedese, A. & Querzoli, G. (1995). Recognition of partially overlapped particle images using the Kohonen neural network. *Experiments in Fluids*, Vol.19, pp. 225–232

Crouser, P.D.; Bethea, M.D. & Merat, F.L. (1997). Unattenuated tracer particle extraction through time-averaged, background image subtraction with outlier rejection. *Experiments in Fluids*, Vol.22, No.3, pp. 220-228

Dimotakis, P.E.; Debussy, F.D. & Koochesfahani, M.M. (1981). Particle streak velocity field in two-dimensional mixing layer. *Physics of Fluids*, Vol.24, p. 995

Engelmann, D.; Garbe, C.; Stšhr, M.; Geissler, P.; Hering, F. & Jähne, B. (1998). Stereo Particle Tracking, *Proceedings of the 8th International Symposium on Flow Visualisation*, pages 240.1249.9, Sorrento, Italy, September 1-4, 1998

Engelmann, D. (2000). *3D-Flow Measurement by Stereo Imaging*, Dissertation, Rupertus Carola University of Heidelberg, Germany.

Faugeras, O. (1993). *Three-dimensionl computer vision: a geometric viewpoint*, MIT press, London, 1993, 663 p.

Fanger, P.O. (1970). *Thermal comfort: analysis and applications in environmental engineering*, Mc Graw Hill ed., New-York, 1970, 244p.

Ge, Y. & Cha, S.S. (2000). Application of Neural Networks to Spectoscopic Imaging Velocimetry. *AIAA Journal*, Vol.38, pp 487-492

Gruen, A. & Baltsavias, E. (1988). Geometrically constrained multi-photo matching. *Photogrammetric Engineering*, Vol.54, pp. 633-641

Guezennec, Y.G.; Brodkey, R.S.; Trigui, N. & Kent J.C. (1994). Algorithms for fully automated three-dimensional particle tracking velocimetry. *Experiments in Fluids*, Vol.17, pp. 209-219

Hartley, R. (1994). Self calibration from multiple views with a rotating camera, *Proceedings of 3rd European Conference on Computer Vision*, pp. 471-478, Stockholm, Sweden, May, 1994

Heikkilä J. & Silvén O. (1997). A Four step camera calibration procedure with implicit image correction, *Conference on Computer Vision and Pattern Recognition (CVPR'97)*, San Juan, Puerto Rico, 1997

Hough P. (1962). *Method and means for recognizing complex patterns*, U.S. Patent 3069654

Jähne, B. (1997). *Digitale Bildverarbeitung. 4. Außage*, Springer Verlag

Kasagi, N. & Sata, Y. (1992). Recent developments in three-dimensional particle tracking velocimetry. *Proceedings of the 6th International Symposium on Flow Visualization*, pp. 832-837, Yokohama, Japan, October, 1992

Kessler, M. & Leith, D. (1991). Flow measurement and efficiency modeling of cyclones for particle collection, *Aerosol Science and Technology* 15, pp. 8-18

Kerho, M.F. & Bragg M.B. (1994). Neutrally buoyant bubbles used as flow tracers in air, *Experiments in Fluids* 6, pp. 393-400

Lee, S.J. & Kim, S. (2005). Simultaneous measurement of size and velocity of microbubbles in opaque tube using
X-ray particle tracking velocimetry technique. *Experiments in Fluids*, Vol.39, No.3, pp. 492-497

Li, D.; Zhang, Y.; Sun, Y. & Yan W. (2008). A multi-frame particle tracking algorithm robust against input noise, *Measurement Science and Technology* 19, 105401 (11pp)

Liebowitz, D. & Zisserman, A. (1998). Metric rectification for perspective images of planes. *Proceedings of IEEE conference on computer vision and pattern recognition, Santa Barbara, California, June 1998*, pp. 482-488

Luong, Q.T. & Faugeras, O. (1997). Self-calibration of a moving camera from point correspondences and fundamental matrices, *The International Journal of Computer Vision* 22 (3), pp. 261-289

Maas, H.G. (1992) Complexity analysis for the establishment of image correspondences of dense spatial target fields, *International Advances of Photogrammetry and Remote Sensing*, XXIX (B5), pp. 102-107

Maas, H.G.; Gruen, A. & Papantonious, D. (1993). Particle tracking velocimetry in three-dimensional flows, Experiments *in Fluids* 15, pp. 133-146

Machacek, M. (2002). A Quantitative Visualization Tool for Large Wind Tunnel Experiments. PhD Thesis, Zurich: Swiss Federal Institute of Technology

Malik, N.A.; Dracos, Th. & Papantoniou, D.A. (1993). Particle tracking velocimetry in three-dimensional flows—part 2: Particle tracking. *Experiments in Fluids* 15, pp. 279–294

Mann J., Ott S. & Andersen J.S. (1999). *Experimental study of relative, turbulent diffusion*, Risø National Laboratory Report Risø-R-1036 (EN)

Maxey, M.R. & Riley, J.J. (1983). Equation of motion of a small rigid sphere in a nonuniform flow. *Physics of Fluids* Vol.26, No.4, pp. 883-889

Maybank, S.J. & Faugeras, O. (1992). A theory of self-calibration of a moving camera. *The International Journal of Computer Vision,* Vol.8, No.2, pp. 123-152

Melling, A. (1997). Tracer particles and seeding for particle image velocimetry. *Measurement Science and Technology,* Vol.8, pp. 1406-1416

Muller, D. & Renz, U. (1996). Determination of all airflow velocity components by a particle-image-velocimetry system, *Proceedings of 5th International Conference on Air Distribution in Rooms: ROOMVENT'96*, pp. 413-419, Yokohama, Japan, July 17-19, 1996

Nishino, K.; Kasagi, N. & Hirata, M. (1989). Three-dimensional Particle Tracking Velocimetry based on automated digital image processing. *Journal of Fluid Engineering,* Vol.111, pp. 384-391

Nishino, K. & Kasagi, N. (1989). Turbulence statistics in a two-dimensional channel ßow using a three-dimensional Particle Tracking Velocimeter, Proceedings on the Seventh Symposium on Turbulent Sheer Flows, Stanford University, August 21-23

Nobach, H. & Honkanen, M. (2005). Two-dimensional Gaussian regression for sub-pixel displacement estimation in particle image velocimetry or particle position estimation in particle tracking velocimetry. *Experiments in Fluids,* Vol.38, No.4, pp. 511-515

Okuno, Y.; Fukuda, T.; Miwata, Y. & Kobayashi, T. (1993). Development of three-dimensional air flow measuring method using soap bubbles. *Japanese Society of Automotive Engineers (JSAE) Review,* Vol.14, No.4, pp. 50-55

Otsu, N. (1979). A threshold selection method from gray-level histogram. *IEEE Transactions on Systems, Man, and Cybernetics,* Vol.9, No.1, pp. 62-66

Ouellette, N.T.; Xu, H. & Bodenschatz, E. (2006). A quantitative study of three-dimensional Lagrangian particle tracking algorithms. *Experiments in Fluids,* Vol.40, pp. 301–313

Papantoniou, D. & Dracos, T. (1989). Analyzing 3-Dimensional Turbulent Motions in Open Channel Flow by Use of Stereoscopy and Particle Tracking, In: *Advances in Turbulence* 2, Hernholz & Fiedler, Springer Verlag, Heidelberg

Sata, Y.; Sato, K.; Kasagi, N. & Takamura, N. (1994). Application of the three-dimensional particle tracking velocimeter to a turbulent air flow, In: *Proceedings of the 3rd Asian Symposium on Visualization*, Y., Nakayama, Springer-Verlag, pp. 705-709

Scholzen, F. & Moser, A. (1996). Three-dimensional particle streak velocimetry for room air flows with automatic stereo-photogrammetric image processing, *Proceedings of 5th International Conference on Air Distribution in Rooms: ROOMVENT'96*, pp. 555-562, Yokohama, Japan, July 17-19, 1996

Stein, G. (1995). Accurate internal camera calibration using rotation, with analysis of sources of errors. *Proceedings of the 5th International Conference on Computer Vision*, pp. 230-236, Cambridge, Massachusetts, June 1995

Sun, Y. & Zhang, Y. (2003). Development of a stereoscopic particle image velocimetry system for full-scale room airflow studies, Part II: experimental setup. *Transaction of the American society of Heating, Refrigerating and Air-Conditioning Engineers (ASHRAE)*, Vol.9, No.2, pp. 540-548

Suzuki, Y. & Kasagi, N. (1999). Turbulent air-flow measurement with the aid of 3-D particle tracking velocimetry in a curved square bend. *Flow, Turbulence and Combustion*, Vol. 3, pp. 415-442

Suzuki, Y.; Ikenoya, M. & Kasagi, N. (2000) Simultaneous measurement of fluid and dispersed phases in a particle-laden turbulent channel flow with the aid of 3-D PTV. *Experiments in Fluids*, Vol.29, pp. 185-193

Teh, Ch. & Chin, R. (1988). On image analysis by the method of moments. *IEEE Pattern Analysis and Machine Intelligence*, Vol.10, No.6, pp. 496-513

Wang, A.; Zhang, Y.; Sun, Y. & Wang, X. (2005). Experimental study of ventilation effectiveness and air velocity distribution in an aircraft cabin mock up. *Building and Environment*, Vol. 43, n° 3, pp. 337-343

Wierzimok, D. & Hering, F. (1993). Quantitative Imaging of Transport in Fluids with Digital Image Processing, In *Imaging in Transport Processes*, Begell House, pp 297-308

Willneff, J. & Gruen, A. (2002). A new spatio-temporal matching algorithm for 3D-ParticleTracking Velocimetry, *Proceeding of The 9th International Symposium on Transport Phenomena and Dynamics of Rotating Machinery*, Honolulu, Hawaii, USA, Feb 10-14, 2002

Virant, M. (1996). *Anwendung der dreidimensionalen "Particle Tracking Velocimetry" auf die Untersuchung von Dispersionsvorgšngen in Kanalstršmungen*, ETH Zürich - Dissertation Nr. 11678

Virant, M & Dracos, T. (1997). 3D PTV and its application on Lagrangian motion. *Measurement Science and Technology*, Vol. 8, 1539-52

Zhao, L.; Zhang, Y.; Wang, X.; Riskowski, G.L. & Christianson, L.L. (1999). Measurement of airflow pattern in ventilated spaces using particle image velocimetry, *Proceedings of ASAE/CSAE Annual International Meeting*, paper #994156, Toronto, Ontario, Canada, 1999

Zhang, Zhengyou. (1999). Flexible camera calibration by viewing a plane from unknown orientations, *Proceedings of the International Conference on Computer Vision (ICCV'99)*, pp. 666-673, IEEE 0-7695-0164-8/99, Kerkyra, Greece, September 20-27, 1999

20

Wind Driven Ventilation for Enhanced Indoor Air Quality

Jason Lien and N.A. Ahmed
*School of Mechanical and Manufacturing Engineering
University of New South Wales, Kensington, NSW 2052,
Australia*

1. Introduction

The last century has seen tremendous progress in technological development that is underpinning the progress of the modern day human civilization. Unfortunately, however, such progress is giving rise to unwanted problems of grave consequences that have the potential to destroy the very environment which sustains life. There is now a real eagerness and greater awareness in the general public to look for alternative natural energy systems and products to help alter the present energy use patterns that have lead to this dilemma.
United Nations' conferences in Rio in 1992 and Johannesburg in 2002 have brought the issue of environment sustainability to the forefront in international circles. Subsequently the organization has defined goals that buildings should seek to achieve in order to obtain green building recognition that include goals such as increase in reliability, increase in indoor air quality, decrease in natural resource use, considerable decrease of energy costs over the lifetime of the building, improving comfort due to improved energy efficiency in building and raise of employment as a result of increased activity in energy improvements in buildings. These benefits will theoretically take care of any type of increase (typically 3-5%) in construction costs and making improvements will have a direct positive impact upon life-cycle costs.
The present chapter is an attempt to advance the cause of sustainable living and improve the quality and comfort of human existence through the greater use of environment friendly wind driven ventilation. Most of the materials presented are based on the author's experimental and numerical computational works carried out at the School of Mechanical Engineering at the University of New South Wales of Australia with a particular focus on the development and production of highly efficient wind driven rotating ventilators for domestic and industrial use.

1.1 Significance of ventilation

In the past, the building codes and policies of most of the developed nations were implemented to ensure the safety and adequate sanitary conditions for occupants and protect the public from hazards such as fire, electrical problems, sewage disposal, etc. It is more recently that emphasis has been placed on energy efficiency and sustainable buildings. Issues have varied over time, but recently the catalyst for such policy changes has come about as a result of rising greenhouse gas emissions from energy consumption, as well as land use practices such as deforestation. Thus governments around the world are grappling with ideas regarding building policies and the potential benefits to not only the people that use them but the environment in which the buildings exist.

Generally speaking, a movement or circulation of the air within an enclosed space in any ventilation system is essential to ensure that the temperature and humidity be maintained within a range that allows adequate evaporation of perspiration from the skin. There is evidence to show, that the deleterious effects result largely from interference with the heat-regulating mechanism of the body. Lack of air currents and the increase in relative humidity and temperature, especially noticeable in crowded or poorly ventilated places, prevent normal evaporation of perspiration and loss of heat from the surface of the skin. Under requirements to maintain a safe working environment, many dwellings and factories now need adequate fresh air exchange to remove gaseous, process emissions and/or heat buildup. The high priority placed on indoor air quality from health considerations has prompted New York in the USA to pass legislation effective from December, 2008 to require landlords to notify tenants and building occupants of indoor air test results.

Good ventilation is a key element in maintaining a healthy environment for people within the confines of a building. Failure to provide adequate ventilation in a building can result in problems with moisture, unpleasant smell, lack of oxygen, and unacceptable content of poisons gases such as CO which cause medical conditions for tenants. Contaminants such as formaldehyde or radon can also accumulate in poorly ventilated homes, causing health problems. Due to fact that every person's resistance to pollutant varies, it is hard to obtain an accurate quantitative measurement of the impact of ventilation on human beings (Hanssen 1997). However, some method was proposed based on the consideration of the factors such as direct medical costs and lost earnings due to major illness as well as increased employee sickness days and lost productivities while on the job (European Collabrative Action report 2000). For example, Brooks and Davis (1992) calculated that productivity loss in the United States of America attributed directly to indoor air quality (IAQ) is around 14 minutes per day; Pillgram Larsen (1991) estimated loss in Norway is of 1 to 1.5 billion of euro per year which is equivalent of 250 to 350 euro per inhabitant. For Finland, a more detailed estimation was given by Seppanen and Palonen (1998) where the potential economic impact of indoor air pollution was estimated to be tens of billions of dollars per calendar year (see Table 1.1).

Building Type	Maintenance Costs (Billion Finnish Marks)	Energy Costs (Billion Finnish Marks)	Costs of poor indoor climate (Billion Finnish Marks)
Houses	12.7	7.2	7.0 (Mould allergies)
High-rise residential building/office buildings/schools/hospitals and other buildings	37.3	7.0	10.6 (Radon cancers, decreases in efficiency, poor learning, hospital infections)

Table 1.1. Cost due to poor indoor air quality

Ventilation is also an effective means of ameliorating the internal environment (Billington 1982). This is usually achieved in three prime motive forces: mechanical means(fans, pumps, HAVCS), heat (stack effect) and wind pressure. The use of mechanical ventilation system consumes a significant amount of energy in a building. For example, electricity bills for residential buildings in the United States of America is approximately US$140.8 billion in 2006 (U.S. department of energy 2006), and the energy consumed by heating, ventilating and air conditioning system (HVAC) accounts for 40 to 60 percent of the total energy. For example, the energy cost has risen 30% in the state of New South Wales since 2009 (IPART website 2009). Due to the rising fossil and energy costs, people's awareness have increased and attention is focusing towards the use of an alternative solution. Under such concern, natural ventilation becomes a viable and sustainable method in providing suitable environment and is increasingly considered in many building designs.

2. Brief overview on different modes of ventilation

In general, natural ventilation can be induced by cross ventilation or stack ventilation and they can be either passive or active in nature. Cross ventilation technique use the natural wind force to direct the air movement through a building. When wind impact on a building it produces positive pressure on the windward side and relative negative pressure on the leeward side and these difference in pressures drive the airflow through the building. In order to achieve good cross ventilation effect, openings such as windows, doors and other openings are required at the different facings of the building. The stack ventilation technique utilizes the temperature between inside and outside of a building. When the room temperature is greater than outside, the warm indoor air will rise and exit and cooler, denser air from below can enter.

The natural ventilation can be further enhanced by wind driven ventilation technique. These techniques can be either passive or active in nature however the primary driving force for providing ventilation are primarily caused by wind.

2.1 Passive wind driven ventilation techniques

2.1.1 Window openings/vents

This is probably the simplest form of its kind. Through strategically placement of openings, the potential natural ventilation in a building can be optimized to cut down the need for artificial cooling (Mochida et. al. 2005, Karave et. al. 2005). In conjunction with the window openings, static vents can also be placed at the planned location where window openings are not required. There are various of types of vents that are commercially available as shown in Fig. 2.1. Due to the relative cost effective and sustainable benefits, a large number of experimental and numerical investigations have been conducted to study the benefit of this technique (Greeno 1997, Heiselberg et al., 2001, Olsen & Chen 2003, Faye et. al. 2005, Asfour & Gadi 2007).

The low-tech nature and manual operability of window openings/static vents technique are acceptable in domestic dwellings. However, a major limitation of naturally ventilated technique in general is the unpredictability of the driving force. Wind pattern and availability for different region varies according to seasons and more over for large buildings some physical limitations exists. Hence its application on its own cannot be considered to be a primary ventilation method.

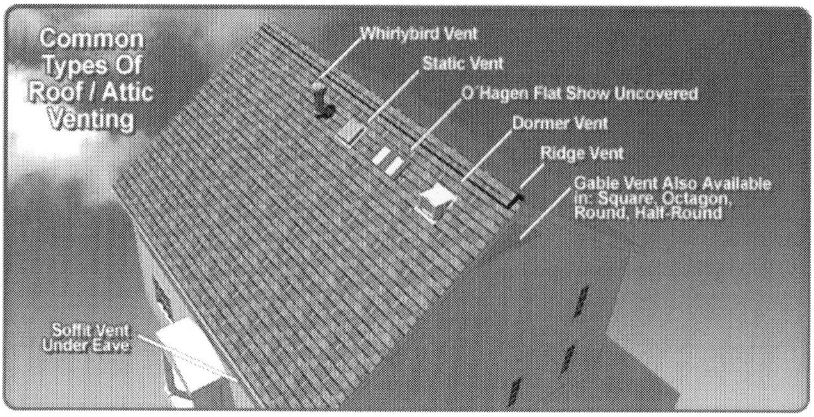

Fig. 2.1. Common types of roof/attic venting

2.1.2 Atria and courtyards

Courtyard is an effective way of providing natural ventilation and can be found in many architectural designs across the globe for thousands of years. A courtyard provides a relatively enclosed space to channel and direct airflow into some openings and results in convective natural ventilation within and around the building. Some studies have been conducted to study the effect of atria and courtyards on energy performance (Rajapaksha et. al. 2003, Aldawoud & Clark 2008). These studies generally confirmed that the wind driven ventilation can be enhanced by using atria and courtyards and are more effective for low rise building. However, the passive nature of this technique indicates that its effectiveness depend largely on the availability of its driving force.

2.1.3 Wind towers

Wind towers have been used in conjunction with courtyard and atria for centuries in many middle-eastern architectures. They work on the principle of both wind driven and stack effect ventilation. Wind enters via the windward side that has a positive pressure coefficient and was redirected through the tower into the building. These winds disperse through the building and exit through relative negative pressure areas. The exit paths can be multiple, either through the tower or openings that was carefully designed. Multiple wind towers have also been used to maximize the effect as can be seen from Fig. 2.2. Wind towers can also benefit for modern buildings and have been the subject of several studies (Bansal et. al. 1994, Bahadori 1994, Kolokotroni et. al. 2002, Badran 2003) through the following considerations: (1) Tower with evaporated cooling columns can increase the cooling potential of incoming air. (2) Natural ventilation principle can dictate the extract height and cross section of the towers to accommodate for the targeted number of occupants. (3) Solar collectors can be used to enhance the stack effect ventilations in times or areas of little wind.
A modern modification of wind towers is a device known as the wind catcher. These devices capture the wind at roof level and direct it down to the building. The working principle and their application of these devices are shown in Fig. 2.3.

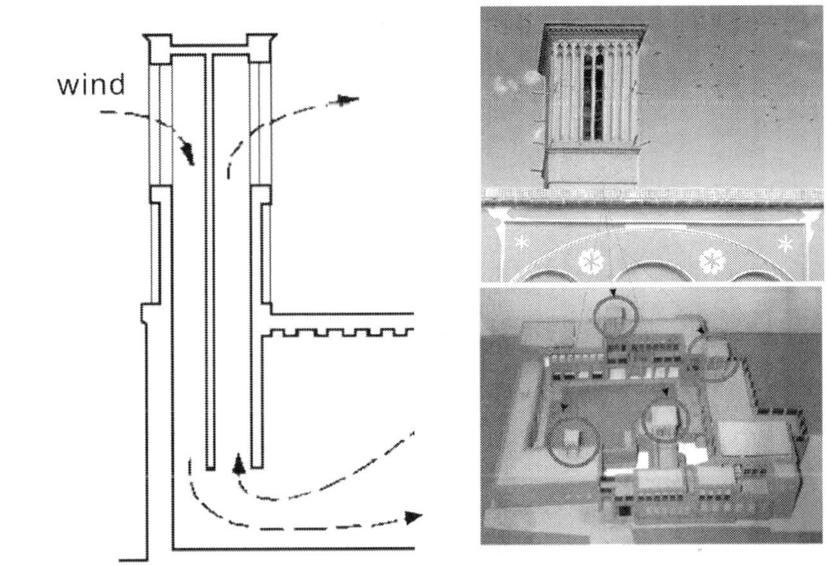

Fig. 2.2. Wind tower design in Middle Eastern building

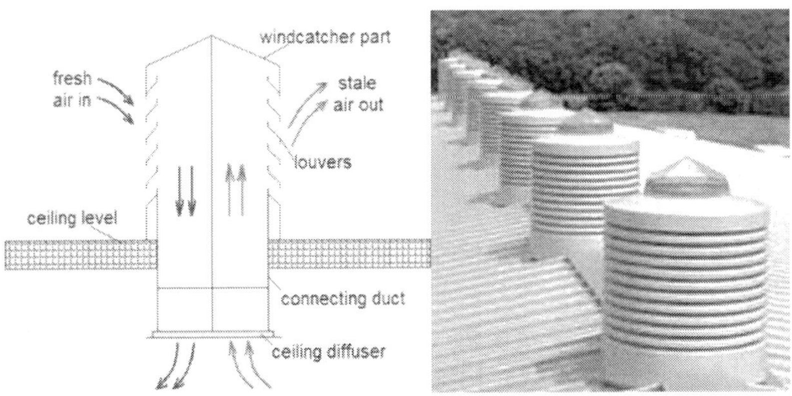

Fig. 2.3. Operation principle and application of wind catcher system

Since the introduction of this technique in 1990 there have been a number of studies associated with its performance both experimentally and numerically (Kolokotroni 2002, Li & Mak 2005, Elmualim 2006). These studies points put that the performance of a wind catcher system is influenced by a number of factors: the outdoor wind; temperature difference between indoor and outdoor environment and location of other openings and windows.

The studies surrounded the wind tower technique not only indicates that the principle works, more importantly they highlighted the needs of redundancy via another source when using wind driven ventilation technique such as active system.

One major concern of the passive natural ventilation technique is that the driving forces are variable and the minimum required ventilation rate is hard to be ensured in still air conditions (no wind or stack effects). Hence several alternative engineering solutions are formulated to maximize and control the effect of natural ventilation. For example, the roof mounted rotating turbine ventilators can be used or in conjunction with the natural ventilation techniques. The rooftop ventilators are commonly found in most countries of the world where they are mounted on the roof and rotate with the slightest breeze and generate a powerful updraft to effectively pump the air out of the room.

2.2 Active wind driven ventilation techniques

2.2.1 Turbine ventilators

Turbine ventilators are available commercially with various blade designs, sizes and materials. They usually rotate in its vertical axis to create updraft inside the turbine which extracts air and in the absence of wind, they can still facilitate ventilator using the stack effects. These types of installations are usually on the top of the roof where higher wind speeds are available for most buildings and these are shown in Fig. 2.4.

Fig. 2.4. Various blade design of turbine ventilators

The concept of a turbine ventilator was originally patented in 1929 by Meadows and it's commercial production was started by Edmonds of Australia since 1934. Although these types of devices have been on the market for many decades, there were only limited quality aerodynamic researches available in open literature.

The first series of investigation into the aerodynamics of a turbine ventilator was reported by Lai (2003). His flow visualization study on three different seized turbine ventilators of 6, 14 and 20 inches respectively showed the airflow pattern around a turbine ventilator. Another experimental study by Flynn and Ahmed (2005) also confirmed the flow pattern. These studies suggested that the larger diameter ventilators induce greater mass flow extraction rates.

Fig. 2.5. Installation of multiple stacked ventilators at UNSW

The performance studies of a turbine ventilator are usually associated with the air extraction rates (Revel 1998, Khan et. al. 2008). These studies suggested that the performances of such devices are significantly influenced by the blade design as well as the construction material and the sizes significantly affect producing higher exhaust mass flow rate. On the other hand, the aerodynamics forces acting on a turbine ventilator have also been studied by Rashid and Ahmed (2003). It was concluded that the operation of a ventilator without the effect of inclined rooftop is more efficient at lower wind speeds. At higher wind speeds larger flow separation was induced on the blades and this reinforced the need of greater attention to optimize blade designs so that they are capable of operating over a wide range of wind speeds.

2.2.2 Hybrid ventilators

As the performance of the traditional turbine ventilators largely depend on the local wind conditions and the ventilation rates are cannot be guaranteed, hence various of forms of hybrid system are purposed to boost its performance to comply with the increasing building

regulation. In this section only three examples of ventilation systems that rely purely on wind power or renewable power source are discussed.
The first such ventilator system considered has been proposed by Lai (2006). This ventilator combines the traditional turbine ventilator with a small DC fan in the base duct which is powered by solar cell as presented in Fig. 2.6. This system was found to be effective in enhancing the ventilation rate at low wind speeds.

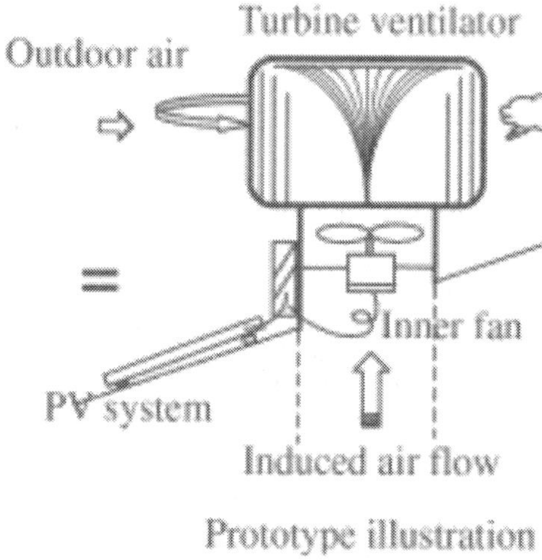

Fig. 2.6. Prototype turbine Ventilator (Lai, 2006)

The second ventilator system considered uses a different approach. It can be easily demonstrated that any obstruction in the throat of a wind ventilator can produce significant noise and reduce vent performance, and this performance reduction can be as high as 40% (West 2001, Khan et. al. 2008) indicates than under normal wind load. The UNSW research team under an Australian Research Council Grant in active collaboration with CSR Edmonds, Pty Ltd, therefore, proposed a model (Ahmed, 2010), the EcoPower, which has no motor and fan blade in the throat of the vent. The model utilises an electronic commutating (EC) motor installed in the head of the ventilator to enable motorised boost during periods of low wind speed or special ventilation needs. The motor can also be programmed and activated by digital measures, such as temperature, humidity gas concentration level etc or by manually switch. The research behind this device also indicates a significant improvement of ventilation rate at low wind speeds. The EcoPower is now commercially produced and marketed worldwide by CSR Edmonds, Pty Ltd.
The third approach mentioned here is the hybrid system proposed by Arup. This device incorporates a vertical axis wind turbine that produces renewable energy to drive the fan which is located directly underneath the turbine to produce the desired ventilation. A photograph of this device is presented in Fig. 2.8. This first full scale model was constructed and installed at the Harrare International School in Zimbabwe. It was found that the room

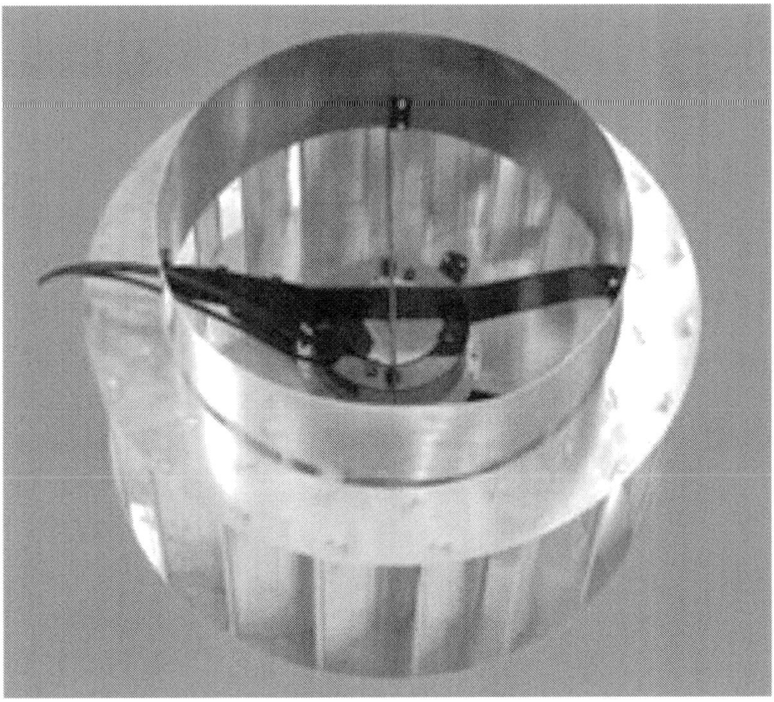

Fig. 2.7. EcoPower Ventilator (CSR Edmonds, Pty Ltd)

Fig. 2.8. Vertical axis Wind Extractor (Arup, Zimbabwe)

temperature in the classroom is significantly cooler (up to 80 C) during the day than without the system installed. However there is very little academic literature on the flow rates and performance associated with this device.

3. Towards enhanced wind driven ventilation technique

In this section some recent studies on means to enhance the performance of wind driven ventilator by the authors at University of New South Wales have been summarised.

3.1 Physical experimentation

The physical experimentation were conducted in a 76mm diameter open return, 0.2% turbulence intensity open test section wind tunnel of the Aerodynamics Laboratory of the University of New South Wales. The experimental setup is shown in Fig. 3.1.

A five-hole pressure probe was used in this investigation. Recent advances in using multi hole pressure probe techniques (Pisasale & Ahmed, 2002-2004), makes it possible to obtain the velocity vector and static pressure field in a highly three dimensional flow field. A step by step procedure of how to obtain the velocity vector and pressure terms can be obtained is detailed in reference 21. A Furness FCO510 micromanometer and a box consisting of seven channels of Honeywell transducers were used to measure the pressures on the five-hole pressure probe. The manometer and the pressure transducer provided measurement in the range that varied from -2000 Pascal (Pa) to 2000 Pa with an accuracy of ±0.1Pa.

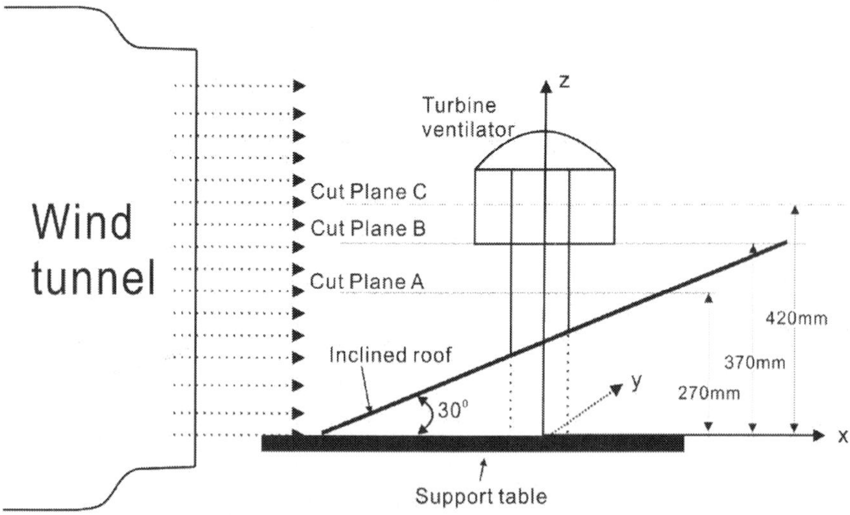

Fig. 3.1. Schematic of experimental setup

Apart from the forces obtained directly from the force balance, pressure and skin friction distribution were used to determine a flow features of the system. Since these parameters contribute largely to the aerodynamic forces acting on a body. In the present investigation, the effects of the inclination angles on the performance of a rotating ventilator were investigated in a wind tunnel. The Reynolds number, Re_x (= xU_∞/υ), used was based on, x, the distance measured from the leading edge of the plate, U_∞, the free stream velocity and υ, the kinematic viscosity of the fluid. Re_x values ranged between 7 x 10^4 and 1 x10^6.

3.1.1 Effect of inclination angle on forces on a ventilator

The force components in the x, y and z directions as well as the total force acting on the rotating ventilator at different inclination angles were examined against free stream velocities and presented in Figures 3.2a to 3.2d.

From these Figures, the force acting on the turbine ventilator is found to increase with free stream velocity. The increasing trend is more pronounced in the x direction signifying that the inclination angle has greater effect on F_x which largely contributes to the drag than on the side and normal components of the force that contribute to both lift and drag (Rashid & Ahmed, 2003).

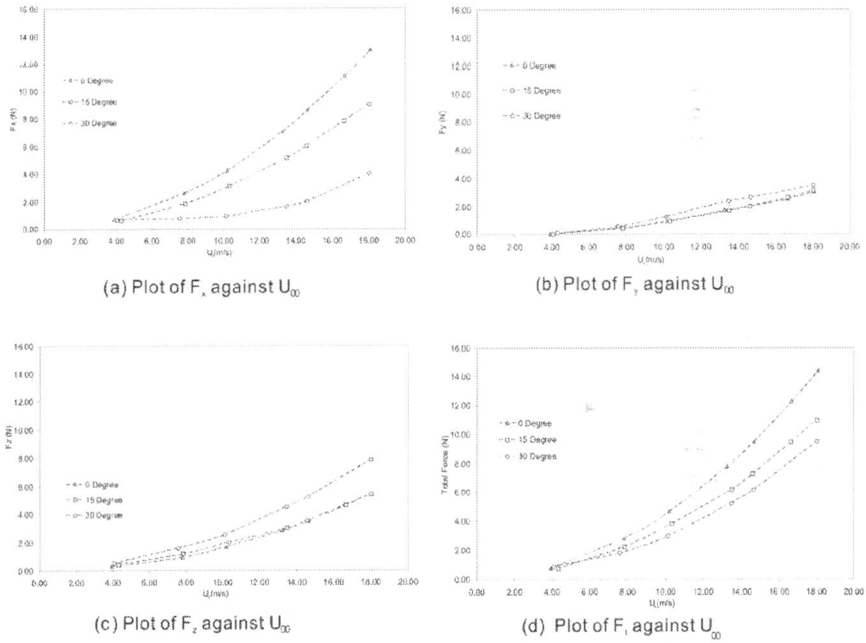

Fig. 3.2. Force acting on the turbine ventilator at various inclination angles.

Some additional features may be elucidated if the total and its component forces (F_t, F_x, F_y, F_z) are examined in non dimensional forms or force coefficients (C_t, C_x, C_y, C_z) against Reynolds number of the flow (Re_d) based on the rotor diameter. The total force and its components were normalised by $0.5\rho U_\infty^2 A$, where ρ is the density of air, U_∞ is the free stream velocity and A is the total wetted area of the ventilator. The wetted area was calculated based on the total area exposed to the flow (see Fig. 3.1). The results are shown in Figures 3.3a to 3.3d.

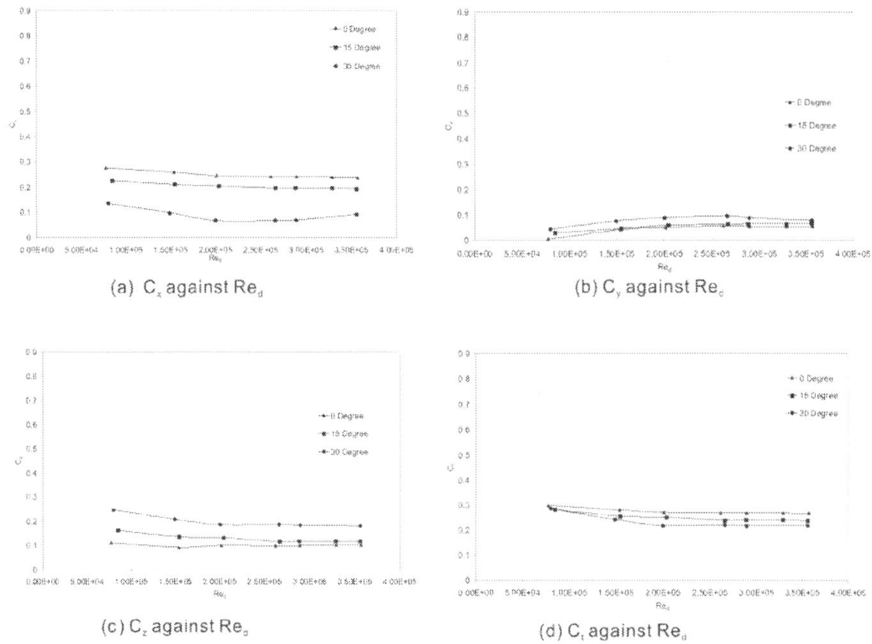

Fig. 3.3. Coefficient of force at various inclination angles.

It is evident that the C_x value decreased with the increase in inclination angle while the C_z value showed the opposite trend. The inclination angle, however, showed little or minimum effect on the C_y values. Consequently, the total force coefficient, C_t, acting on the ventilator reduced with increases in inclination angle and converged towards a constant value beyond Reynolds number of $2x10^5$. This is significant as it implies that the force acting on the ventilator is independent of the Reynolds number beyond a certain rotational speed and may be highly relevant during the operation of a turbine ventilator at lower wind speed. A hybrid system that incorporates other power source such as solar power may provide additional means to efficiently overcome the inertia and friction forces acting on the body to initiate adequate rotation at low wind speed.

3.1.2 Effect of inclination angle on rotational speed of a ventilator

The overall effect of the roof inclinational angle on the operation of a turbine ventilator was further examined against its rotational speed. The results are shown in Fig. 3.4. The results indicate that the rotation speed of the turbine ventilator showed a linear relationship with the free stream velocity (Flynn & Ahmed, 2005; Rashid & Ahmed,2003; and Khan et al., 2008). Additionally, it also indicates that the high inclination angle lowers the rotational speed of the ventilator that is more noticeable at higher free stream velocity suggesting that the inclination angle has minimal impact on the total mass flow extracted at lower wind speeds. This is also a significant observation since the operation of these ventilators are designed for low wind speeds and as such will not suffer greatly from adverse effect of the high inclination angle.

The Australian regulation stipulates that for the safety of operation, the turbine ventilator must be able to withstand wind loads from speeds of 200km/hr without suffering deformation. As such loads are directly proportional to ventilator rotation, at high wind speed safety in operation become more of a priority than mass extraction rate. Thus, the lowering of rotation due to increasing inclination angle of roof may thus be usefully exploited to reduce wind loads and hence extend the safety margin.

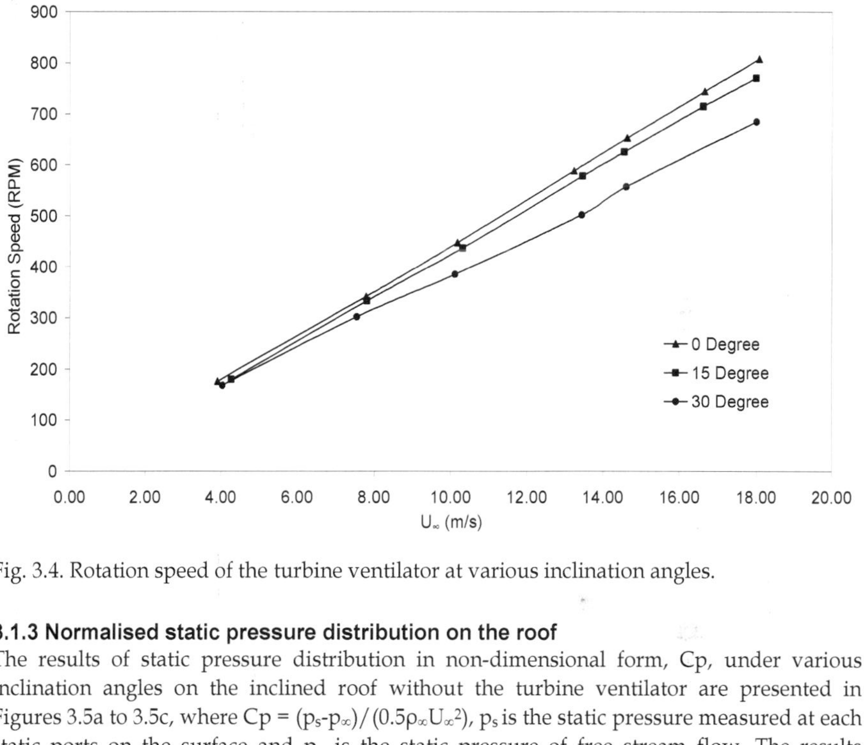

Fig. 3.4. Rotation speed of the turbine ventilator at various inclination angles.

3.1.3 Normalised static pressure distribution on the roof

The results of static pressure distribution in non-dimensional form, Cp, under various inclination angles on the inclined roof without the turbine ventilator are presented in Figures 3.5a to 3.5c, where $Cp = (p_S - p_\infty)/(0.5\rho_\infty U_\infty^2)$, p_s is the static pressure measured at each static ports on the surface and p_∞ is the static pressure of free stream flow. The results obtained when the turbine ventilator present are shown in Fig. 3.5d to 3.5f. An advantage of such representation is that the Cp values can range from +1 to any negative value. A positive Cp values generally suggests that the flow is slowing down until reaching the limiting value of +1 or the stagnation or zero velocity condition, while any negative Cp value means that the flow is experiencing acceleration, with higher negative Cp signifying higher acceleration.

The inclined roof at its leading edge imposed pressure gradient to the oncoming flow and forced it to slow down producing a Cp distribution that had values moving towards +1 with approximately a symmetrical distribution in the spanwise direction. The Cp values at the leading edge for 0, 15 and 30 degrees of inclination were found to be 0.05, 0.19 and 0.42 respectively that decreased towards zero value in the streamwise direction.

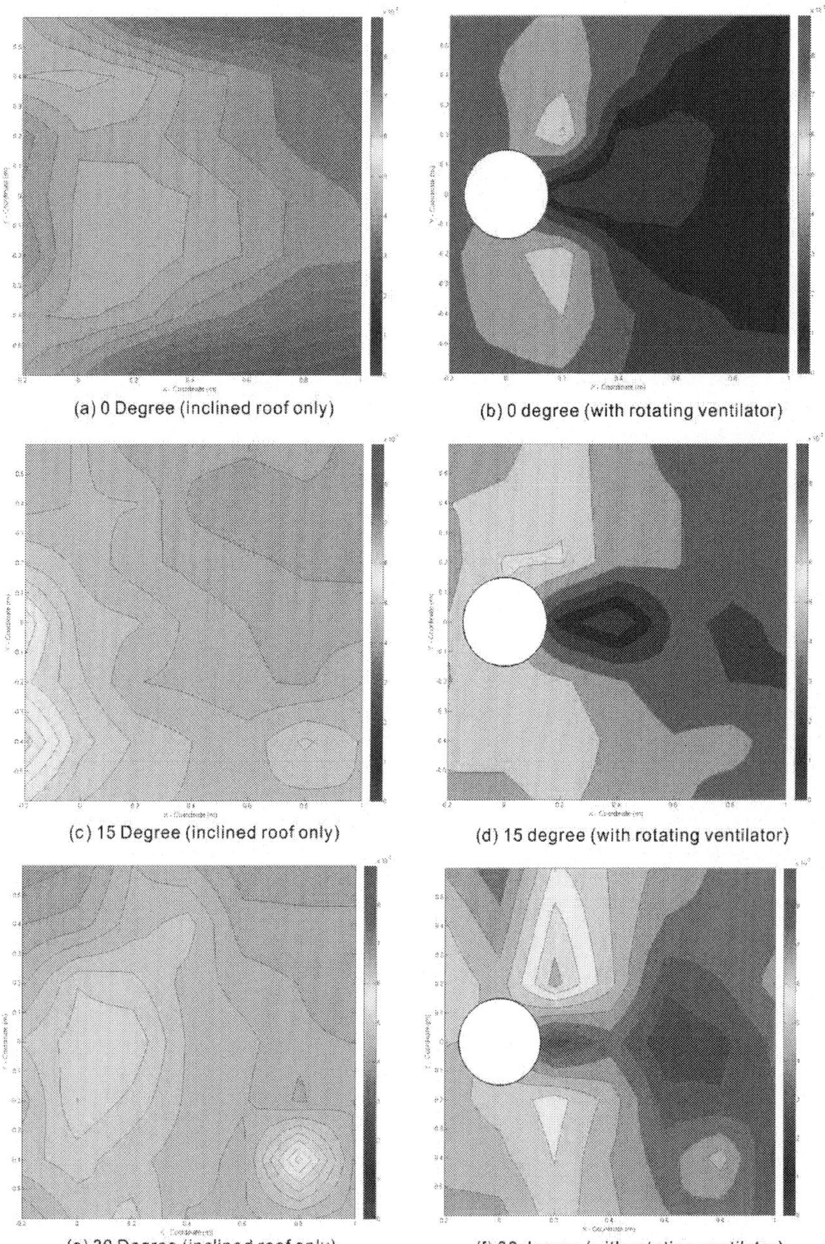

(a) 0 Degree (inclined roof only)

(b) 0 degree (with rotating ventilator)

(c) 15 Degree (inclined roof only)

(d) 15 degree (with rotating ventilator)

(e) 30 Degree (inclined roof only)

(f) 30 degree (with rotating ventilator)

Fig. 3.5. Static Pressure distribution on the roof, where $U_\infty = 10$ m/s

On the inclined plane, the addition of the ventilator imposed further pressure gradient that slowed the incoming flow even further. From Figures 3.5d to 3.5f, the Cp values at the leading edge increased to more positive Cp values due to the presence of the ventilator while maintaining symmetrical distribution for the upstream flow as expected. Similar trends were also observed for flows at 5m/s suggesting that the Reynolds number appeared to have negligible effect on the overall pressure distribution.

3.1.4 Skin friction distribution on the roof

Recent advances (Pisasale & Ahmed, 2002-2004, Lien & Ahmed, 2003) in using multi hole pressure probe have greatly aided in the quest of measuring the velocity and surface skin friction distribution in a complex flow field. Following the procedure outlined in Lien & Ahmed (2003), a five hole pressure probe was used to measure the skin friction distribution at each measuring station.

The skin friction distributions on the roof at different inclination angles 'without the ventilator' are shown in Figures 3.6a to 3.6c and 'with the ventilator' present are shown in Fig. 3.6d to 3.6f.

From tests conducted at the two wind tunnel speeds of 5m/s and 10m/s and at the three different inclination angles of 0, 15, and 30 degrees, the skin friction distributions were found to become progressively lower downstream of the ventilator. This reduction in skin friction values appears to be a consequence of reduced free stream velocity due to the momentum deficit occurring behind a bluff body. Since skin friction varies with square of the free stream velocity (Baskaran & Bradshaw, 1993), it is not surprising that the trend is more apparent at higher wind speeds.

On the other hand, with the ventilator placed on the roof, the skin friction distribution on the roof increased significantly with increases in the roof inclination angle. Since the skin friction signifies losses in the flow, it is expected that the increase of this quantity appears to emanates from the overall reduction of the forces acting on the ventilator that produces lower rotational speed.

3.2 CFD modelling

After gaining insights of flow physics around the ventilator and on the inclined roof, Computational Fluid Dynamic software is also utilised to extend the performance study of rooftop ventilator under the effect of inclined roof.

The turbine ventilator used in this investigation is a Hurricane H100 produced by CSR Edmonds Australia Ltd. It consists of a rotating portion (Rotor) with 8 curved blades and a stationary portion in the form of a cylindrical base. The dimensions of the various components of this ventilator are shown in Fig. 3.7.

This numerical simulation of flow characteristics of the turbine ventilator used in wind tunnel tests were then performed using FLUENT. The equations for the conservation of mass, momentum and turbulence scales are solved in FLUENT using the control volume method in a three-dimensional body-coordinate system.

The geometry is created in Computer Aided Three-dimensional Interactive Application (CATIA) and exported to GAMBIT (the pre-processor) for meshing. For modelling of the rotating motion of the turbine ventilator, the multiple reference frame (MRF) meshing in FLUENT was adopted as they have been successfully applied in flows with rotating objects (Luo et al., 1994). This meshing allows multiple moving reference frames to be solved in a

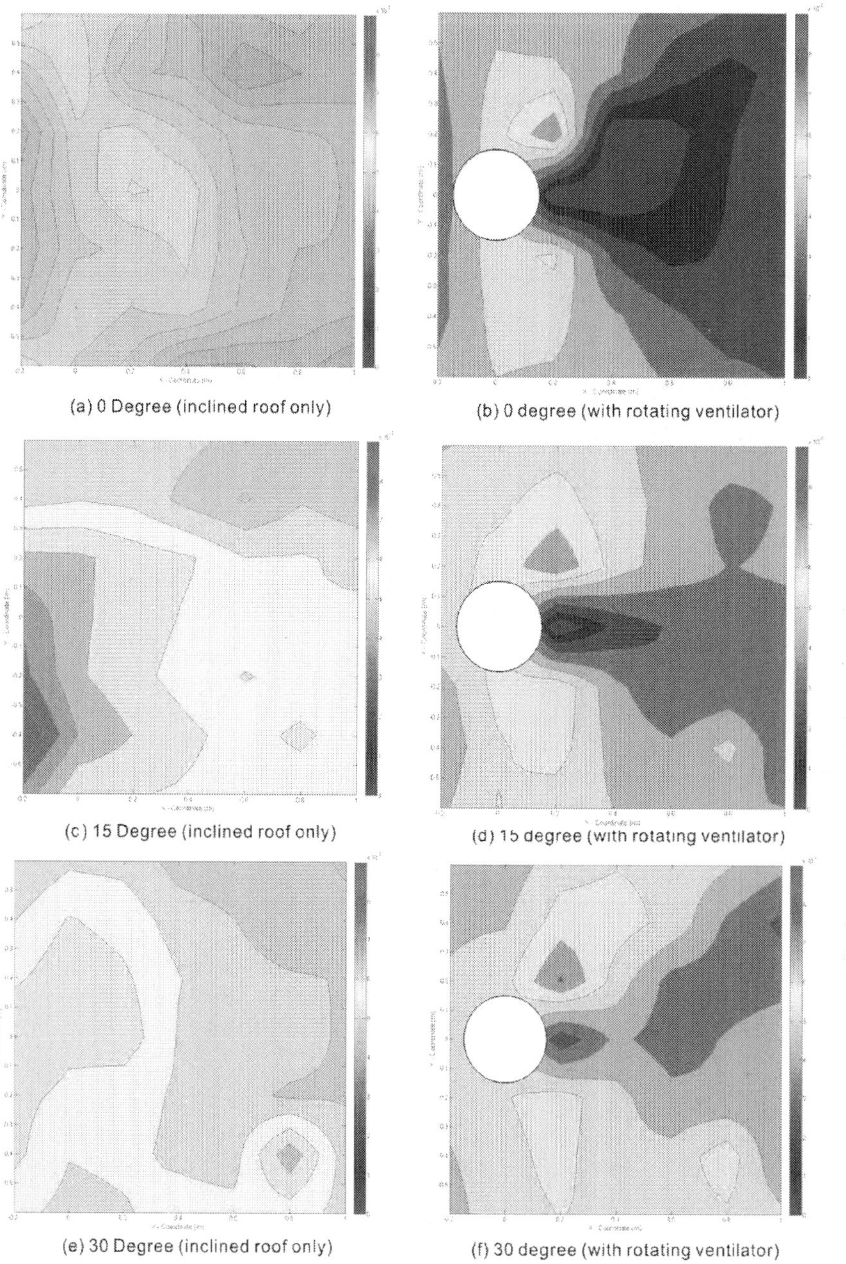

(a) 0 Degree (inclined roof only)
(b) 0 degree (with rotating ventilator)
(c) 15 Degree (inclined roof only)
(d) 15 degree (with rotating ventilator)
(e) 30 Degree (inclined roof only)
(f) 30 degree (with rotating ventilator)

Fig. 3.6. Skin friction distribution on the roof, where $U_\infty = 10$ m/s

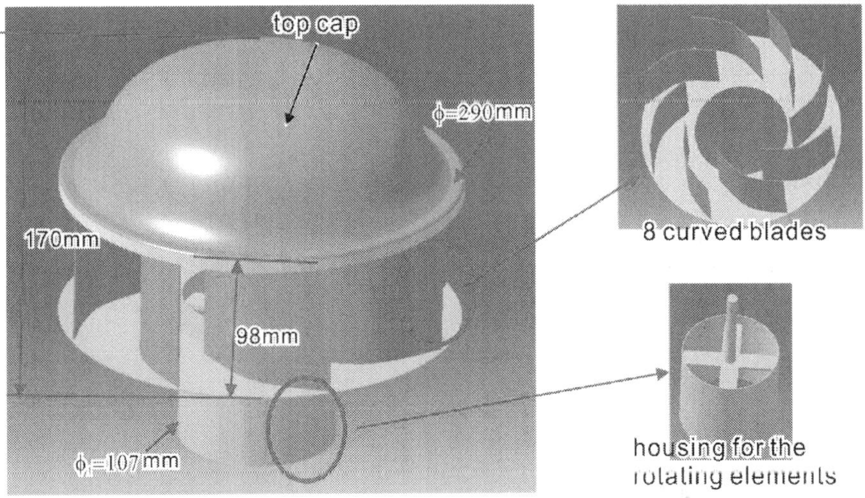

Fig. 3.7. Modelling of turbine ventilator

single domain. MRF is normally used for steady state condition and the individual cell zone can move at different rotational speeds and provide solution using the moving reference frame equations. Grid sensitivity test was performed by examining the effect of different number of mesh grids and the total number of grids was determined when grid independence was established. The final total number of grids used in the present simulation was determined to be approximately 1,200,000. The maximum and minimum grid volumes were approximately 1.8×10^{-4} m^3 and 2.4×10^{-9} m^3 respectively. Mesh quality was checked in GAMBIT to ensure the skewness of the three dimensional grids was less than 0.8.

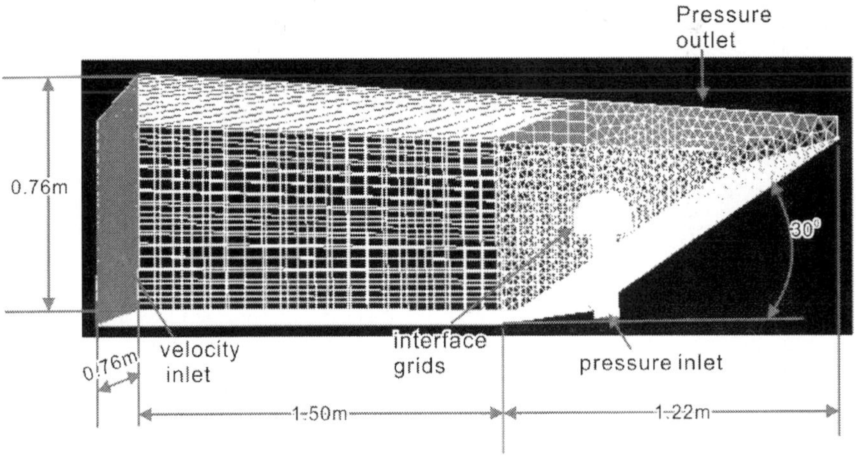

Fig. 3.8. Modelling of domain and dimension

3.2.1 Performance investigation of rooftop turbine ventilator

The major function of a rooftop turbine ventilator is to create suction and move air out from an enclosed building. The performance of a wind driven ventilator is generally gauged against its exhaust mass flow rate. However, before exhaust flow rate was determined, an attempt was made to explore the nature of the flow associated with the rotating ventilator particularly those inside the rotor.

CFD results can provide clear image of the external and internal flow on the entire ventilator in terms of the three dimensional path lines. This provided a qualitative description of the general flow pattern. This is shown in Fig. 3.9. Two distinct streams of the flow can be observed. The left hand-side flow follows the direction of the turbine rotation while the flow direction of the right-hand side flow is contrary to the turbine rotation. The right hand side flow becomes the defected flow field that acts against the rotation of the turbine. This results in the formation of secondary circulation between the separated flow and the blades on the right-hand side flow.

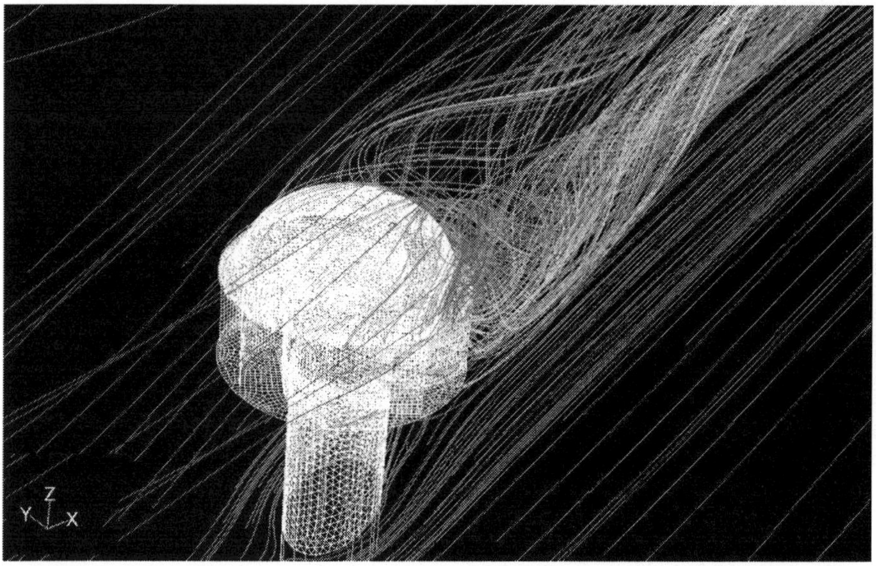

Fig. 3.9. Three dimensional path line of the flow associated with the rotating ventilator at 10m/s.

The CFD software can further observe the features of the behaviour of the flow as they are extracted and exhausted by the ventilator as shown in Fig. 3.10. The path lines of the flow at different wind speeds within the rotor and downstream in the wake are shown. The rotor cap is removed from this Fig. for greater clarity. It is clear that at various wind speeds tested, the rotation of the turbine creates updraft.

The flow interactions within the rotor can be shown using vector plots in three dimensions and the results are shown in Fig. 3.11a to 3.11c. The path line in the duct (base) and the rotor also indicates that the flow swirls upwards and mixes with the secondary circulation of the wake region downstream of the ventilator.

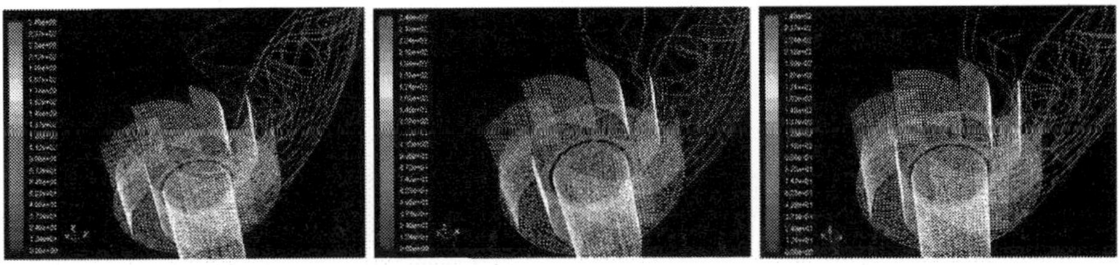

(a) Free Stream Velocity = 5m/s (b) Free Stream Velocity = 10m/s (c) Free Stream Velocity = 15m/s

Fig. 3.10. Three dimensional path lines of flow within and downstream of the rotor.

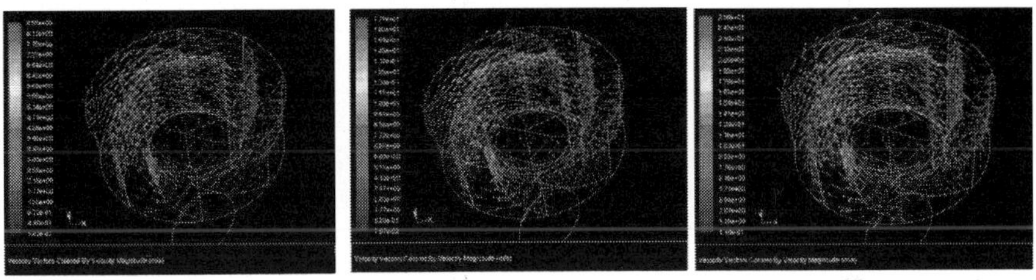

(a) Free Stream Velocity = 5m/s (b) Free Stream Velocity = 10m/s (c) Free Stream Velocity = 15m/s

Fig. 3.11. 3D velocity vector inside the ventilator

It is evident that as the ventilator rotates in the clockwise direction, the air flow extracted through the base swirls up and mixes with the oncoming wind entering the ventilator from the left. The resulting mixed air exhausts from the right half of the ventilator. These Figures also show that with increasing wind speed there is a corresponding increase in swirl flow within the rotor suggesting subsequent increase in exhaust mass flow rate.

To provide further insight, the flow features within and around the rotor at the two cut planes of B and C at three wind speeds are presented in Figures 3.12 and 3.13 respectively. The results are given as two dimensional vector plots.

Figures 3.12 and 3.13 indicate that with increasing wind speeds, the flows at the left side of the ventilator are higher than at the right side. The extracted air swirls in clockwise direction. Figures 3.13a to 3.13c further indicate that the wake behind the rotor increases in size with increases in wind speeds that produces the increases in suction and hence the increases of swirl component within the ventilator. These observations are in good agreement with the flow visualisation experiments conducted by Flynn &Ahmed (2005) and Lai (2003).

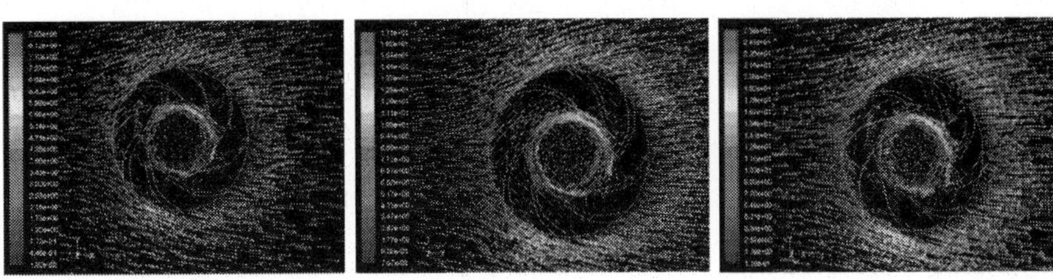

(a) Free Stream Velocity = 5m/s (b) Free Stream Velocity = 10m/s (c) Free Stream Velocity = 15m/s

Fig. 3.12. Velocity vector field at cut plane B at different wind speeds

(a) Free Stream Velocity = 5m/s (b) Free Stream Velocity = 10m/s (c) Free Stream Velocity = 15m/s

Fig. 3.13. Velocity vector field at cut plane C at different wind speeds

To explore the capability of the CFD modeling approach as an initial design tool, the exhaust flow rate was determined against a range of wind speeds between 3m/s to 15m/s. Numerical computation were performed by assigning boundary condition of 'pressure inlet' at the bottom of the pipe set as ambient pressure.

Once the exhaust mass flow rates for various cases were obtained, the effect of blade heights on performance of a turbine ventilator was explored. Three blade heights H, 1.5H and 2H were considered, with H = 98mm. The computed exhaust mass flow rates for varying wind speeds are shown in Fig. 3.14.

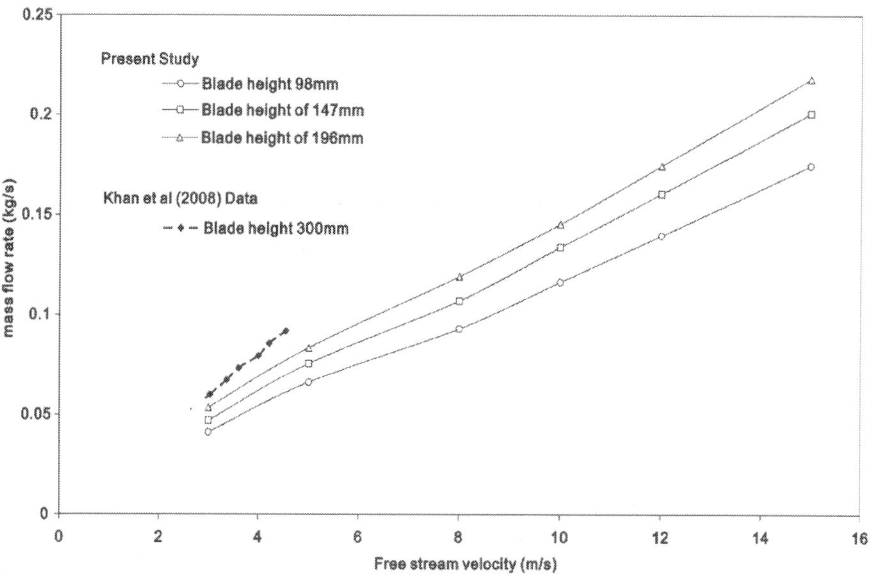

Fig. 3.14. Effect of free stream wind speed on exhaust mass flow rate

It is clear that the exhaust mass flow rate has a near linear relationship of increasing in value with increases in the incoming wind speed. The same trend has been observed by Khan et al., (2008) where four different geometry ventilators were tested in a wind tunnel.

Additionally, the wind tunnel results of Flynn & Ahmed (2005), Khan et al., (2008) indicate that the rotational speed of the ventilator also increases linearly with increases in wind

speed suggesting a linear increase in mass flow rate with increasing ventilator size (diameter or blade height). With 50% and 100% in the ventilator blade height, the improvement of exhaust mass flow rate was found to be 15% and 25% respectively.
Due to the absence of adequate performance data on turbine ventilator in open literature, the data from the experimental results of Khan et al., (2008) on Hurricane turbine ventilator of 300mm diameter and 300mm blade height were extrapolated to provide some comparison against the CFD results. The results are presented in Fig. 3.15.

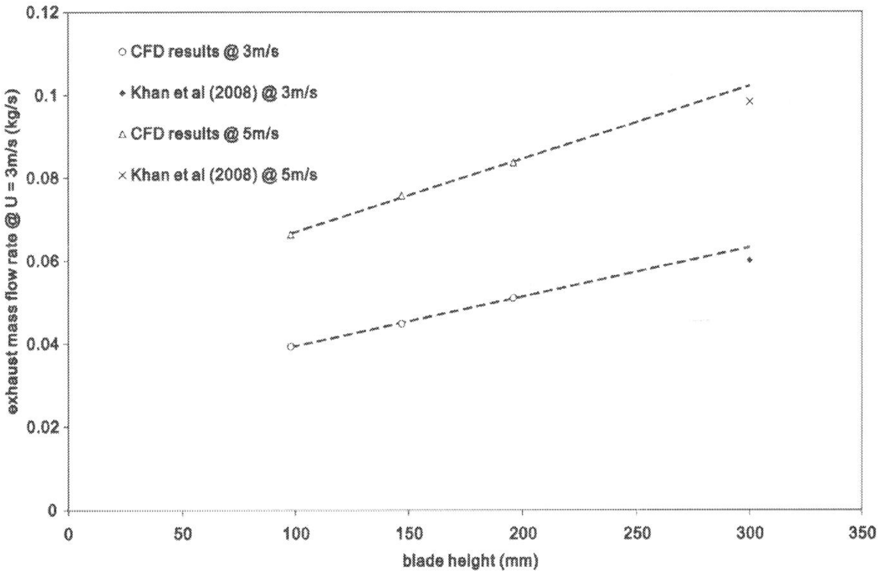

Fig. 3.15. Effect of blade height on exhaust mass flow rate (U_{∞}= 3m/s and 5m/s)

Again from Fig. 3.15, a linear relationship between exhaust mass flow rate with blade height was observed. A slight overprediction when compared against Khan et al.,(2008)'s data was observe. Given that the comparison was with extrapolated value and also for the fact that CFD results were from an inclined roof, the small discrepancies are not surprising. However, based on the observations that the similar trends were observed for the exhaust mass flow rate as well as the external and internal flows associated with a rotating ventilator on a 30^0 inclined roof, it appears that the roof inclined at such angle did not have a significant effect on the overall trend in ventilation performance of a rotating ventilator.

4. Conclusions

Wind driven ventilation systems that utilize wind as a natural energy to provide improved air quality within buildings have been considered in this chapter. Since the wind driven turbine ventilator has proved to be one of the most efficient such device, it has been the primary consideration in this chapter. However, the performance of wind driven ventilation depends largely on prevailing winds. This has necessitated efforts of the authors to

concentrate on exploring the basic flow physics around and within such wind driven ventilator that are detailed in this chapter. The results indicate that the efficiency and reliability of a ventilator operation may require boosting by a hybrid system by incorporating other power sources such as solar power to overcome the inertia and friction forces acting on the body to initiate ventilator rotation at lower wind speed.
Numerical analysis using computational fluid dynamics (CFD) was also employed to investigate both internal and external flows around and within a rotating ventilator. A promising conclusion that can be drawn from these studies is that CFD analysis could be used alongside physical experimentation, as a cost effective aid to enhance operational performance and safety features in the future design and development energy efficient natural wind or wind/hybrid powered ventilation systems to improve air quality and comfort within buildings.

5. Acknowledgements

The authors would like to acknowledge the Australian Research Council and CSR Edmonds Pty Ltd, Australia for providing the funding; Managing Director, Mr. Allan Ramsay and National Technical Director, Mr. Derek Munn, of CSR Edmonds and Technical Officer, Mr. Terry G Flynn, of Aerodynamics Laboratory of University of New South Wales for their assistance during the conduct of this study.

6. References

Ahmed, N.A. (2010). Wind-Solar Driven Natural Electric Hybrid Ventilators, *Wind Power*, InTech, ISBN: 978-953-7619-81-7.

Aldawoud, A. & Clark, R. (2008). Comparative analysis of energy performance between courtyard and atrium in buildings, *Energy and Buildings*, Vol. 40, pp. 209–214.

Asfour, O.S. & Gadi, M.B. (2007). A comparison between CFD and network models for predicting wind-driven ventilation in buildings, *Building and Environment*, Vol. 42, pp. 4079–4085

Badran, A.A. (2003). Performance of cool towers under various climates in Jordan, *Energy & Buildings*, Vol. 35, pp. 1031–1035.

Bahadori, M.N. (1994) Viability of wind towers in achieving summer comfort in the hot arid regions of the Middle East, *Renewable Energy*, Vol. 5, pp. 879–892.

Bansal, N.K.; Mathur, R. & Bhandari, M.S. (1994). A study of solar chimney assisted wind tower systems for natural ventilation in buildings, *Building & Environment*, Vol. 29, pp. 495–500.

Baskaran, V. & Bradshaw, P. (1993). Effect of bluff body wakes on skin friction and surface heat transfer in a turbulent boundary layer. *Journal of Wind Engineering and Industrial Aerodynamics*, Vol. 49, pp. 269-278

Billington, N.S. (1982) Energy efficient domestic ventilation systems for achieving acceptable indoor air quality, 3rd AIC Conference, September 20-23 1982, London, UK

Brooks, B.O. & Davis, W.F. (1992). Understanding indoor air quality, CRC Press Inc., Boca Raton, Fla.

Elmualim, A.A. (2006). Dynamic modeling of a wind catcher/tower turret for natural ventilation, *Building Serv. Eng. Res. Technol.*, Vol. 27, pp. 165-182.

Faye, C. et al (2005). Spitler, Heating, *Ventilating and Air Conditioning: analysis and design (6th edition)*, John Wiley and Sons, Hoboken, NJ.

Flynn, T.G. & Ahmed, N.A. (2005). Investigation of Rotating Ventilator using Smoke Flow Visualisation and Hot-wire anemometer. *Proc. of 5th Pacific Symposium on Flow Visualisation and Image Processing Whitsundays, Australia*, Paper No. PSFVIP-5-214.

Greeno, R. (1997). *Building Services, Technology and Design*, Longman Publishing, Edinburgh

Hanssen, S.O. (1997). Economical consequences of poor indoor air quality and its relation to the total building operational costs, *EuroFMlIFMA Conference & Exhibition*, Turin, Italy, 1. - 3.

Heiselberg, P.; Bjorn, E. & Nielsen, P.V. (2001). Impact of open windows on room airflow and thermal comfort, *International Journal of Ventilation*, Vol.1, pp. 91–100.

Karave, P.; Stathopoulos, T. & Athienitis, A.K. (2005). Wind induced natural ventilation analysis, *Solar Energy*, Vol. 81, pp. 20–30.

Khan, N.; Su, Y.; Riffat, S.B. & Biggs, C. (2008). Performance testing and comparison of turbine ventilators. *Renewable Energy*, Vol. 33, pp. 2441-2447.

Kolokotroni, M.; Ayimomitis, A. & Ge, Y.T., (2002). The suitability of wind driven natural ventilation towers for modern offices in the UK: a case study, *World Renewable Energy Congress*, Vol. 8

Lai, C. (2003) Experiments on the ventilation efficiency of turbine ventilator used for building and factory ventilation. *Energy Buildings*, Vol. 35, pp. 927-932.

Lai, C.M. (2006) Prototype development of the rooftop turbine ventilator powered by hybrid wind and photovoltaic. *Energy & Buildings*, Vol. 38, pp. 174-180.

Li, L. & Mak, C.M. (2005). The assessment of the performance of a wind catcher system using CFD, *Building & Environment*, vol. 42, pp. 1135–1141.

Lien, S.J. & Ahmed, N.A. (2003). Application of multi-hole pressure probes in skin friction determination. *AIAA conference paper; San Francisco*: AIAA-2006-3659

Luo, J.Y.; Issa, R.I. & Gosman, A.D. (1994). Prediction of impeller-induced flows in mixing vessels using multiple frames of reference. *I Chem E Symposium Series*, Vol. 136, pp. 549-556.

Mochida, A. et al. (2005). Methods for controlling airflow in and around a building under cross ventilation to improve indoor thermal comfort, *Journal of Wind Engineering and Industrial Aerodynamics*, Vol. 93, pp. 437–449

Olsen, E. & Chen Q. (2003). Energy consumption and comfort analysis for different low energy cooling systems in a mild climate, *Energy & Buildings*, Vol. 35, pp. 561–571.

Pillgram Larsen, G (1991). 300000 pa skolebenken for a bedre innemijoet, BE-nytt nr 2, Statens Bygningstekniske Etat, Oslo(In Norwegian)

Pisasale, A. & Ahmed, N.A. (2002a). Theoretical calibration for highly three-dimensional low-speed flows of a five-hole probe, *Meas. Sci. Technol*, Vol. 13, pp. 1100–1107.

Pisasale, A. & Ahmed, N.A. (2002b). A novel method of extending the calibration range of five hole probe for highly three dimensional flows, *J. of Flow Measurement and Instrumentation*, Vol. 13(1), pp. 23-30.

Pisasale, A. & Ahmed, N.A. (2003). Examining the effect of flow reversal on seven-hole probe measurements, *AIAA Journal*, Vol. 41(12), pp. 2460-2467.

Pisasale, A. & Ahmed, N.A. (2004). Development of a functional relationship between port pressures and flow properties for the calibration and application of multihole probes to highly three-dimensional flows, *Experiments of Fluids,* Vol. 36, pp. 422–436.

Rajapaksha, I.; Nagai, H. & Okumiya, M. (2003). A ventilated courtyard as a passive cooling strategy in the warm humid tropics, *Renewable Energy,* Vol. 28, pp. 1755–1778.

Rashid, D.M.H. & Ahmed, N.A. (2003). Study of Aerodynamic forces on a rotating wind driven ventilator. *Journal of wind engineering,* Vol. 27, pp. 63-72.

Revel, A. (1998). Testing of 2 wind driven roof ventilators, available from www.edmonds.com.au (accessed 13/03/2007).

Seppanen, O. & Palonen, J. (1998). The effects of indoor climate on national economy in Finland in billoon FIM, FISIAW 1998 SIY Rapportti 10, ISSN 1237-1866

West, S. (2001). Improving the sustainable development of building stock by the implementation of energy efficient, climate control technologies, *Building & Environment,* Vol. 36, pp. 281–289.

21

Improving the Quality of the Indoor Environment Utilizing Desiccant-Assisted Heating, Ventilating, and Air Conditioning Systems

M.D. Larrañaga[1], E. Karunasena[2], H.W. Holder[3], M.G. Beruvides[2] and D.C. Straus[4]
[1]*Oklahoma State University*
[2]*Texas Tech University*
[3]*SWK, LLC*
[4]*Texas Tech University Health Sciences Center*
USA

1. Introduction

Throughout history, diverse cultures and societies have appreciated the importance of a clean and healthy environment (Bardana, Montanaro et al. 1988). In western societies today, most people spend over 90% of their time indoors (Teichman 1995; EPA 2003). Reports involving buildings with indoor air-related problems have appeared increasingly in the medical and scientific literature, although this problem has been with humans for centuries. Sick building syndrome (SBS), a common term for symptoms that result from individuals' exposure to poor IAQ (IAQ), was first recognized as an important problem that affects occupants in certain buildings in 1982. The first official study of SBS to examine more than one structure was published in 1984 (Finningan, Pickering et al. 1984). SBS has been difficult to define, and no single cause of the malady has been identified (Hodgson 1992). Early studies showed that many of the reported causes of SBS included undesirably high levels of known respiratory irritants such as nitrogen and sulfur dioxides, hydrocarbons, and particulates (National Academy of Sciences 1981), known or suspected carcinogens such as asbestos, radon, formaldehyde, and tobacco smoke (Sterling 1984), or chemicals released by new building materials. Although fungal spores are ubiquitous in the indoor and outdoor environments, they are now generally accepted as important causes of respiratory allergies (Solomon 1975; Bernstein 1983). Although no single cause for the symptoms induced by poor IAQ likely exists, the presence of fungi, spores, and fungal growth in sick buildings has become consistently associated with this problem (Miller 1992; Mishra 1992; Cooley 1998).

Since Biblical times, indoor fungal contamination and IAQ have been a cause for concern in society. Leviticus 14:33-45, in the Old Testament, states that a house needs to be cleaned and ridded of mildew to be clean and free of mold (Heller, Heller et al. 2003). If the house cannot be cleaned, then it is to be torn down. This is similar to the manner in which IAQ issues associated with fungal contamination are addressed today. Air pollution became an important environmental issue during the industrial revolution, and several instances of deadly environmental air pollution are cited in the literature. In the early 1900s, building design and construction underwent radical change in the United States; the introduction of air conditioning with cooling coils and forced ventilation, new construction materials, new lighting and heating standards, insulation, and similar advances changed the way buildings were designed and constructed. The new construction designs and methods allowed architects and engineers to build quality structures at lower costs. Schools of engineering and architecture in the United States then began to teach the new building designs and construction methods to students. Thus, thousands of years of collective architectural and engineering design experience, specifically natural ventilation design, were erased from the educational system in the United States.

World War II was a pivotal point in IAQ. At approximately the same time as building design and construction was experiencing radical change, American society also experienced radical change. America began to transform from an agricultural economy wherein the majority of time was spent outdoors, to a service and manufacturing economy, where the majority of time was spent indoors. Since 1945, a phenomenal transformation has occurred: Americans and people in other industrialized nations now spend more than 90 percent of their time indoors (Teichman 1995; EPA 2003). The Arab Oil Embargo of 1973-74 and the resulting energy supply shortages dramatically changed the public's attitude toward environmental control, because building energy usage had a monetary impact on every American. The energy crisis resulted in a tighter building design that produced more energy-efficient homes by closing the windows, redesigning the buildings, circulating re-heated and/or re-cooled air, and reducing the amount of fresh outside air brought into buildings. This "tight" building design resulted in an improvement of buildings and HVAC system energy efficiencies. This improvement in efficiency, however, was paralleled by an increase in SBS complaints. SBS is a term used to describe situations in which building occupants experience health and comfort effects that appear to be linked to time spent in a building, but for which no specific illness or cause can be identified. Indicators of SBS include headache; eye, nose, or throat irritation; dry cough; dry or itchy skin; dizziness and nausea; difficulty in concentrating; fatigue; sensitivity to odors; and other symptoms. Another important indicator was the relief of symptoms when complainants leave the building for extended periods of time (EPA 1991).

The first study to compare indoor versus outdoor fungal morphology was conducted in 1904 by Saito in Japan (Saito 1904); this work was followed by Rostrup in Denmark (Rostrup 1908) and Peyronel in Italy (Peyronel 1919). Since the 1930s, medical specialists in the allergy and immunology community recognized molds as being allergenic and capable of both exacerbating asthma and sensitizing patients (Bernton 1930; Flood 1931; Credille 1933; Conant 1936). With increasing awareness that poor IAQ may generate a variety of deleterious effects on human health, IAQ has become a serious public health concern (Samet 1990; Mishra 1992; Passon 1996; Hodgson, Morey et al. 1998; Sudakin 1998; Hagmann 2000; King and Auger 2002; Karunasena 2005; Karunasena E, Larrañaga MD et al. 2010). It has been estimated that more than 30% of the buildings in developed countries suffer from poor IAQ (World Health Organization 1983; Smith 1990). The Occupational Safety and Health Administration (OSHA) estimated that 30 to 70 million U.S. workers are affected by SBS (Bureau of National Affairs 1992).

Moisture problems have been encountered with increasing frequency both in family housing and in the workplace in the U.S. and Europe (Reijula 1996). Persistent water leaks and moist building materials inevitably lead to the growth of fungi and bacteria in these buildings. Several epidemiological studies suggested that dampness and fungal problems are present in 20% to 50% of modern homes (Dales 1991; Brunekreef 1992; Jaakkola 1993; Spengler 1994, Verhoeff 1995). Not only are dampness and fungi risk factors in the association between indoor dampness and respiratory symptoms (Hodgson, Morey et al. 1985; Dales 1991; Dales 1991b; National Academy of Sciences 1993; Flannigan 1994; Spengler 1994; Hodgson, Morey et al. 1998; Husman, Meklin et al. 2002), but damp homes tend to have higher levels of fungi than non-damp homes (Platt 1989; Verhoeff 1992). In addition, poorly maintained heating, ventilation, and air conditioning (HVAC) systems have been recognized as sources of microorganisms, including fungi.

Fungi are well known allergens that cause allergic rhinitis, allergic asthma, and hypersensitivity pneumonitis when inhaled (Salvaggio 1981; Tarlo 1988; Burge 1989; Flannigan 1994). Fungi also produce volatile organic compounds (VOCs) including alcohols, aldehydes, and ketones, which often produce moldy odors and can cause symptoms such as headaches, eye, nose and throat irritation, and fatigue (Tobin 1987; Flannigan 1991). Fungi also produce toxic metabolites called mycotoxins. Mycotoxins have been identified indoors or on materials indoors by several authors (Croft 1986; Hodgson, Morey et al. 1998; Nielsen, Hansen et al. 1998; Richard, Platnerr et al. 1999; Croft, Jastromski et al. 2002). Mycotoxins produced by *Stachybotrys chartarum* have been implicated in producing non-allergic respiratory symptoms in humans (Croft 1986; Johanning 1996; Andersson 1997; Croft, Jastromski et al. 2002) and have been shown to cause significant damage to cells of the neurological system in concentrations found indoors (Karunasena E, Larrañaga MD et al. 2010).

At present, no single environmental factor or group of factors has been established as the cause of SBS. Although fungal contamination in indoor environments has been shown to produce allergies in building occupants (Lehrer 1983; Licorish 1985; Verhoeff 1995), the role of fungi in SBS has become increasingly controversial. Numerous theories have been put forward (Mendell 1993). Along with the VOC theory, a heightened neurogenic inflammatory response to low-level chemical exposures has been suggested (Meggs 1993; Karunasena E, Larrañaga MD et al. 2010), while other theories have focused on particulates (Salvaggio 1994a) and physical factors (Levin 1995). Inadequate ventilation is a factor in all of these theories. Investigators who are more skeptical have emphasized the roles of psychosocial factors, stress, and gender (Stenberg 1994; Salvaggio 1994b; Bachmann 1995).

Today, children spend most of the day indoors, and because dampness in buildings has increased over the last decade, relationships have been identified between an increase in children's health symptoms, dampness, and fungal spores (Dill 1996; Li 1997; Garrett, Rayment et al. 1998). Children who attend school in buildings with dampness and fungal contamination have been shown to suffer higher rates of respiratory infections (Koskinen 1995). The literature suggests a strong association between the presence of fungal growth indoors and SBS (Cooley 1998). The Centers for Disease Control states that the inhalation of fungal spores inside buildings can cause allergic rhinitis, hypersensitivity pneumonitis, and exacerbate asthma (Redd 2002). Although associations have been made, far more research related to SBS, building-related illness (BRI), and IAQ is needed. Hodgson, et al., maintain that undesirable moisture levels indoors represent a public health concern inadequately addressed by building, health, or housing codes (Hodgson, Morey et al. 1998).

In warm and humid climates, conventional HVAC systems are incapable of adequately controlling humidity and simultaneously meeting the minimum fresh air requirements specified by the American Society for Heating, Refrigerating, and Air Conditioning Engineers' (ASHRAE) Standard 62, Ventilation for Acceptable IAQ. The bulk of a building's moisture load is carried by the incoming ventilation air, and simply drying the ventilation air will provide excellent humidity control at minimal cost (Harriman and Kittler 2001; Larrañaga, Beruvides et al. 2008). Drying the ventilation air via desiccant based cooling is a cost effective method of humidity control. A separate dedicated outdoor air system with humidity control is the simplest method and may be the only reliable method of meeting Standard 62 (Mumma 2001; Larrañaga, Beruvides et al. 2008). Controlling humidity is crucial for human comfort, minimizing adverse health effects associated with high humidity, and maximizing the structural integrity of buildings. The use of a desiccant pre-conditioner apparatus can improve the humidity control capabilities of HVAC systems arising from inherent evaporator coil limitations and can accommodate the minimum outdoor air ventilation rates specified by Standard 62 (Meckler 1994; Larrañaga, Beruvides et al. 2008).

Since the mid to late 1980s, desiccant based cooling systems have found increased applications as humidity control devices for non-industrial structures like schools, homes, hospitals, and commercial buildings (Hines 1992c; Harriman III, Witte et al. 1999; Larrañaga, Beruvides et al. 2008). Several authors have stated that the use of active desiccants enhances the quality of the indoor air by helping to maintain comfort criteria (temperature, humidity and ventilation) (Meckler 1994; Kovak 1997; Fischer and Bayer 2003), removing particulates and bioaerosols from the air (Hines 1992a; Kovak 1997), and removing chemical pollutants from the air (Hines 1992c; Popescu and Ghosh 1999). Several investigators explored the ability of solid and liquid desiccant materials to remove environmental tobacco smoke, particulates, radon, organic vapors, carbon dioxide, and several microorganisms responsible for the majority of nosocomial infections, including several bacteria and *Aspergillus niger*, from the airstream. Popescu and Ghosh used a fixed bed adsorber to simulate the operation of a rotary desiccant wheel, and showed that carbon dioxide and organic vapors were successfully removed by the desiccant materials (Popescu and Ghosh 1999). Hines, et al., used both fixed bed adsorbers in a column configuration for solid desiccants (Hines 1992a) and a packed bed absorber-stripper system for liquid desiccants (Hines 1992b) in their studies. Hines, et al., studied the desiccant removal capabilities of these column-type adsorbers and absorbers on carbon dioxide, volatile organic compounds (Hines 1992b), airborne particulates, environmental tobacco smoke, several bacteria, and *Aspergillus niger* (Hines 1992a). Kovak, et al., conducted a laboratory and field study of the capabilities of a solid-desiccant dehumidifier in removing seven microorganisms responsible for nosocomial infections (Kovak 1997).

This study quantified the removal capabilities of a rotary wheel solid-desiccant dehumidifier at removing selected IAQ-related fungal organisms from the airstream. While the above-mentioned authors studied the removal capabilities of solid and liquid desiccants and one rotary wheel solid desiccant dehumidifier, none explored the ability of a rotary wheel solid desiccant dehumidifier to remove IAQ-related fungal species from the air. Rotary wheel solid desiccant dehumidifiers in the honeycomb configuration are the most appropriate dehumidifier configuration for air-conditioning applications (Pesaran 1994). The use of active desiccation in warm and humid climates would result in energy savings from a reduction in latent cooling and an increase in sensible cooling, offsetting initial purchase costs (Dolan 1989; Pesaran 1994; Larrañaga, Beruvides et al. 2008).

Indoor environmental quality has been an issue throughout history. The interactions of a number of technological discoveries and historical events have resulted in the construction of buildings that make people sick. Every industrialized nation in the world has experienced an increase in asthma in the last 30 years. Asthma is an affliction of the industrialized world, and is not a prominent illness in third-world countries (Vogel 1997). The term sick-building syndrome was first used in the 1980s, when illnesses associated with buildings began surfacing throughout the industrialized world. SBS is a societal and public health issue that has become prevalent in today's world with a negative impact on the world's economy.

2. The economic impact of poor IAQ

The incidences of illnesses related to building occupancy have brought about a number of financial penalties, ranging from lower worker productivity to expensive lawsuits (Addington 2000). Poor IAQ has a negative effect on productivity, worker health, and morale. Direct evidence established in the literature stated that characteristics of buildings and indoor environments can directly affect worker health and productivity. Estimates (in 1996 dollars) of annual savings and productivity gains included $6 to $14 billion from reduced respiratory disease; $2 to $4 billion from reduced allergies and asthma; $15 to $40 billion from reduced symptoms of sick building syndrome; and $20 to $200 billion from direct improvements in worker performance. Building characteristics and indoor environments have been linked to SBS experienced by building occupants. The most common reported sufferers of SBS are office workers and teachers, who make up approximately 50 percent of the total workforce (64 million workers). SBS symptoms include irritation of the eyes, nose, skin and upper respiratory tract, increased airway infections, dizziness, nausea, headache, fatigue, bleeding from the nose, and lethargy. Cognitive impairment, memory loss, permanent lung damage, and other physiological effects have been correlated to SBS. Psychosocial factors are also known to influence the symptoms of SBS. Building factors such as the amount of fresh air ventilation, indoor lighting levels including sunlight, levels of chemical and microbial contamination, and indoor temperature and humidity have been shown to influence SBS symptoms (Gordon, Johanning et al. 1999; Fisk 2000).

SBS symptoms hinder work. Symptoms also can cause not only absences from work, but also visits to doctors and costly emergency medical services. The investigations, maintenance, relocation, legal fees, and insurance costs associated with SBS quickly rise to astronomical levels, and impose a societal burden. The quantification of SBS costs is extremely difficult, due to economic and psychosocial factors, decreases in productivity, legal costs, and other influences. However, several attempts have been made at quantification of the costs of SBS based on the Gross Domestic Product (GDP) associated with office-type-work. The GDP of the United States in 1996 was $7.6 trillion. On the basis of an estimated two percent decrease in productivity due to SBS symptoms for office-type-work, which has a GDP of $3.8 trillion, the annual nationwide cost of SBS symptoms is $76 billion. The potential financial benefits of improving U.S. indoor environments exceed costs by a factor of 18 to 47 (Bayer 2000). While poor IAQ places a heavy economic burden on the workplace, poor IAQ can also have a negative impact on school-aged children, and teachers (Handal, Leiner et al. 2004).

2.1 Asthma as epidemic

Mold and moisture indoors is a significant risk factor for asthma and the US Environmental Protection Agency identifies mold and moisture indoors as asthma triggers (Environmental Protection Agency 2010). Since 1970, a three-fold increase occurred in the incidence of asthma in the United States: 7 million cases in 1970 vs. 20 million cases in 2000. Persons with asthma collectively have more than 100 million days of restricted activity and 470,000 hospitalizations annually (Weiss 1992). Asthma is the most common chronic childhood disease, affecting 6.3 million children. 1 in 10 school-aged children has asthma (Environmental Protection Agency 2010), and the rate is rising more rapidly in children of preschool age than in any other age group. In 2000, there were nearly 2 million emergency room visits and approximately 500,000 hospitalizations due to asthma. Asthma symptoms that are not severe enough to require an emergency room physician visit can still be severe enough to prevent a child with asthma from living a fully active life. Asthma is the leading cause of school absenteeism due to chronic illness. During the past 20 years, the number of school absence days due to asthma has more than doubled. The CDC estimates 14 million school days were missed due to asthma in 2000 (EPA 2003).

The economic impact of asthma is staggering; During 1994, total US costs of asthma were $10.7 billion (Weiss, Sullivan et al. 2000), yearly treatment costs alone approach $6 billion (Cookson 1999), and asthma costs 5,000 deaths yearly with no signs of leveling off (Vogel 1997). The roots of asthma may be traced to heating of the bedrooms in homes, tight building design, and air conditioning. Forced air ventilation allowed homes to be heated or cooled throughout at all times of the year. Prior to the industrial revolution, homes were designed with a central room with heating capabilities (i.e. kitchen and living room) and plumbing. The surrounding rooms, most commonly bedrooms, were not heated, and would freeze through the cold months of the year. This yearly cycle of freezes in bedrooms allowed beddings to freeze, killing dust mites. Few indoor environments routinely freeze today. Thus, dust mites survive throughout the year in carpets and beddings, and are able to affect those with asthma. Heating and air conditioning system design has re-circulated the indoor air to maintain efficiencies, decreasing the amount of fresh air available in buildings.

2.2 The economics of poor IAQ and schools

Teachers and school-age children also suffer from SBS. One in five U.S. schools exhibits IAQ problems, and studies linking specific environmental conditions to student performance indicate impaired performance of students. A study of 627 Swedish secondary school pupils reported "impaired performances were more common in schools with lower air exchange rates, higher relative humidities, and higher concentrations of respirable dust, formaldehyde, VOCs, and total bacteria or molds. A relationship was demonstrated between subjective reports of impaired mental performance, measured indoor air pollutants, and low air exchange rate" (Bayer 2000). A similar study of 12 schools within the Denver, CO metropolitan area indicated an increased prevalence of nasal congestion, sore throat, headache, dustiness complaints, and red and watery eyes in schools with certain ventilation system types (Kinshella, Van Dyke et al. 2001). In a similar study of 85 schools in Canada, it was shown that children with allergies who displayed allergic-type symptoms during school were disproportionately in the below-average category for academic achievement (Landrus 1990). Several other studies demonstrated a link between student absenteeism and IAQ factors. Additionally, there seemed to be a link between unsatisfactory IAQ and the proportion of a school's students from low-income households.

IAQ and SBS appeared especially important in schools because (1) children are developing physically and affected by pollutants to a greater degree than adults, (2) the number of children with asthma has risen approximately 49 percent since 1982, (3) children below the age of 10 have three times as many colds as adults, (4) poor IAQ can lead to drowsiness, headaches, lack of concentration, and other symptoms, and (5) children have a higher rate of metabolism than adults and may ingest or inhale more air and surface contaminants than adults (Bayer 2000). Schools face separate epidemics: an epidemic of deteriorating facilities and an epidemic of asthma among children (Bascom 1997). Asthma is the principal cause of school absences, accounting for 20 percent of lost school days in elementary and high school (Richards 1986). Richards also states that allergic disease (nasal allergy, asthma, and other allergies) is the number one chronic childhood illness. It has been clearly established that SBS and poor IAQ affect productivity in a negative way. The factors associated with IAQ interact in a very complex relationship that sometimes requires extensive and diverse knowledge, experience, and diversity to solve. Ventilation, however, plays a key role as the underlying factor for SBS in the modern, sealed building with a controlled indoor environment (Cooley 1998).

2.3 Economic incentives for improving IAQ

In some cases, improving IAQ to acceptable levels can be quite expensive and uneconomical. In rare instances, structures are demolished rather than repaired or remediated because of poor IAQ. Potential savings from changes in building factors, that produced a 10 to 30 percent reduction in symptoms and associated costs, projected an annual savings of $2 to $4 billion, in addition to reducing the psychosocial and societal costs. Three general approaches for reducing allergy and asthma by changes in buildings are currently recognized: (1) control the indoor sources of the allergens and chemical compounds that cause symptoms or initial sensitization, (2) use cleaning systems to decrease the indoor concentrations of the relevant pollutants, and (3) modify buildings and IAQ in a manner that reduces viral respiratory infections among occupants (Fisk 2000).

Potential savings from changes in building factors creating a 20 to 50 percent reduction in symptoms and associated costs in office buildings projected an annual productivity increase on the order of $15 to $38 billion (Fisk 2000). Strong evidence existed that good IAQ can effectively increase health and productivity and the cost benefits associated with improving IAQ exceeded the cost of improving IAQ by a factor of 18 to 47. "For the United States, the estimated potential annual savings plus productivity gains, in 1996 dollars, are approximately $40 billion to $250 billion" (Fisk 2000). This evidence justified changes in (1) the components of building codes affecting IAQ and (2) company and institutional policies related to IAQ, (3) building design, operation, and maintenance to incorporate maintaining and promoting a desirable IAQ.

Fischer concluded that the payback period associated with a desirable indoor environmental quality is probably very short (Fischer 1996). He indicated that the many benefits listed would be recognized year after year, whereas the costs associated with providing the desirable indoor environmental quality are a one-time expense with minimal maintenance costs. The expected benefits—which included reductions in absenteeism and health care costs, positive impacts on productivity, alertness, drowsiness, allergies, and illness, avoidance of property damage and remediation, and reduced maintenance costs—quickly exceeded the initial expense associated with an improved indoor environment (Bayer 2000).

3. Causes of poor IAQ and desiccation as an IAQ control strategy

Indoor dampness, water damage, and fungi have been associated with respiratory complaints (Martin 1987; Andrae 1988; Brunekreef 1989; Dales 1991; Dales 1991b; Brunekreef 1992; Summerbell 1992; Jaakkola 1993; Joki 1993; Husman, Meklin et al. 2002), and with both allergic and non-allergic respiratory disease (Salvaggio 1981; Lehrer 1983; Licorish 1985; Flannigan 1994; Spengler 1994; Verhoeff 1995; Jarvis and Morey 2001) in several industrialized nations. Asthma was associated with 'damp houses and fenny countries' three centuries ago by Sir John Floyer (Sakula 1984). In a study of 48 U.S. schools, Cooley, et al., associated the presence of propagules of *Penicillium* and *Stachybotrys* on building surfaces with SBS (Cooley 1998). In a survey of 59 homes selected on the basis of previously measured mold levels in 400 houses, White correlated measurements of mold growth and immunological reactions of occupants, noting that lymphocytes from children are chronically activated, and immuno-regulation may be altered in households with mold growth (White 1995). The presence of moisture damage in schools was identified as a significant risk factor for respiratory symptoms in children based on data from microbial IAQ studies in 24 schools (Meklin, Husman et al. 2002) and in the home (Hyvärinen, Pekkanen et al. 2002), and airborne *Penicillium* and *Aspergillus* species was identified as a risk factor for asthma, atopy, and respiratory symptoms in children (Garrett, Rayment et al. 1998). In a population-based incident case-control study in South Finland of 521 adults with newly diagnosed asthma and 932 controls, Jaakkola, et al., found that 35.1% (95% confidence interval, 1.0%-56.9%) of the adult-onset asthma cases were related to workplace mold exposure indoors and that indoor mold problems constitute an important occupational health hazard (Jaakkola, Nordman et al. 2002). It is unlikely that the number of associations in several different countries is a result of chance. It is more likely these associations represent a combination of factors leading to poor ventilation, moisture damage, fungal contamination, and poor IAQ because of systemic and synergistic effects of contaminants within modern built environments (Passon 1996). One important building parameter in controlling indoor moisture and mold growth is the HVAC system, which if not properly designed, maintained, or operated, can cause poor IAQ.

3.1 HVAC systems as a cause of poor IAQ and desiccation as an IAQ control strategy

HVAC systems are essential to modern life and can provide healthy and comfortable indoor environments when properly installed, operated, and maintained (Batterman 1995). Sietz categorized the primary factors leading to building-related illness (BRI), and found that 53% of 529 IAQ evaluations conducted by NIOSH from 1971 through 1988 were associated with inadequate ventilation (Seitz 1990). NIOSH attributed more than half (52%) of the SBS cases to unsuitable facility ventilation systems (Bayer 2000). Additionally, conventional HVAC systems cannot adequately dehumidify air in warm and humid climates (Bayer 1992a; Fischer 1996; Davanagere 1997; Larrañaga, Beruvides et al. 2008) and it is not economically feasible to use only materials that are not susceptible to moisture damage. A systemic relationship between the HVAC system, outdoor air, and indoor environment exists when indoor relative humidities are high. A properly designed, functioning, and operating HVAC system can have a significant positive impact on reducing the number of SBS symptoms experienced within buildings.

The application of a control strategy that aids in removing organic materials and microorganisms from the air, while introducing fresh air into a building, can improve the

IAQ of a building and eliminate many problems associated with ventilation and lack of fresh air in buildings. Bayer suggested that IAQ improves when using active humidity control and continuous ventilation in schools (Bayer 2000). In a study of 10 schools in Georgia, Bayer states that of the five schools having conventional HVAC systems, none supplied outside air at the ASHRAE recommended 15 cfm/person. The schools having desiccant systems were delivering as much as three times more outside air, while maintaining equal or better control of the indoor relative humidity than the conventional systems. The average total volatile organic compound (TVOC) concentrations tended to be lower in schools having desiccant-based systems. The school showing the highest air exchange rate utilized a rotary desiccant system, and had the lowest carbon dioxide, TVOC, and airborne microbial concentrations, and the lowest average indoor relative humidity (Bayer 2000). In Phase II of the same project, Fischer and Bayer stated that increasing the air ventilation rate from 5 to only 8 cfm/student challenged the ability of the conventional systems to maintain the space relative humidity below the ASHRAE and ACGIH recommended 60% level. Increasing the ventilation rate of the conventional systems to the required 15 cfm/student allowed the space relative humidities routinely to exceed 70%. These data explained why all of the conventional HVAC system schools were designed and/or operated with only 6 cfm/student of outdoor air or less. The decreased ventilation rates were in direct response to the performance limitations of the conventional cooling equipment and contributed to the poor IAQ within the schools. Furthermore, the schools served by the conventional HVAC systems experienced absenteeism at a 9% greater rate than those served by the desiccant systems (Fischer and Bayer 2003). Kumar and Fisk proposed that the energy cost of providing additional ventilation may be more than offset by the savings that result from reduced employee sick leave, and that increasing ventilation rates above the minimum rates specified in ANSI/ASHRAE Standard 62, Ventilation for Acceptable IAQ, can yield substantial benefits, including the reductions of the incidence of allergy and asthma in building occupants (Kumar and Fisk 2002).

The general approaches for reducing allergy and asthma by changes in buildings are: (1) control the indoor sources of the allergens and chemical compounds that cause symptoms or initial sensitization, (2) use cleaning systems to decrease the indoor concentrations of the relevant pollutants, and (3) modify buildings and IAQ in a manner that reduces viral respiratory infections among occupants (Fisk 2000). Utilizing desiccant treatment to pretreat fresh air and maintain the desired airflows created benefits for building occupants by providing a means to meet all three general approaches for reducing allergy and asthma in buildings. The use of desiccation as an IAQ control strategy provided other benefits including an increase in sensible cooling and a decrease in latent cooling at the cooling coils.

IAQ studies found that molds, or bioaerosols, were primary link to building-related illness, infections, toxic syndromes, and hypersensitivity diseases (Kovak 1997). Outdoor air parameters can be controlled actively by pre-treating outdoor air prior to its entering a building. Studies have shown that forced desiccant treatment of air has been effective in reducing airborne levels of tobacco smoke and volatile organics (Hines 1992c; Kovak 1997; Popescu and Ghosh 1999). The objective of this research was to determine the effectiveness of a titanium dioxide/silica gel catalytic dehumidification system in removing IAQ-related bioaerosols from the air. The research consisted of subjecting a laboratory setup of the desiccation system to airborne concentrations of IAQ-related bioaerosols. Its success proved that adapting dehumidification technology for use in HVAC systems will allow for moisture control and removal of IAQ-related organisms from the air stream, and offered a viable control strategy for preventing moisture damage and mold growth in buildings. Furthermore, utilization of active desiccation in humid climates results in energy savings from a reduction in latent cooling and an increase in sensible cooling, offsetting initial purchase costs while providing an economic benefit.

Hines, et al., (1992), Kovak, et al., (1997), and Popescu and Ghosh (1999) showed that forced desiccant treatment of air effectively reduced airborne levels of bacteria, fungi, particulates from environmental tobacco smoke, and VOCs. Hines showed that a packed bed adsorber can act as a filter for airborne particulates by removing between 22% and 73% of particulate matter associated with environmental tobacco smoke (Hines 1992a). Kovak, et al., showed median reductions after exposure to desiccant based air conditioning (DBAC) systems in three field studies of 39%-64% for bacteria and 32%-72% for fungi (Kovak 1997). In laboratory tests, Kovak et al. showed an average reduction of 38% of seven organisms associated with nosocomial infections after exposure to a DBAC system (Kovak 1997). Popescu and Ghosh showed removal efficiencies of a desiccant bed to be 35% for formaldehyde, 70% for toluene, 29% for carbon dioxide, and 54% for 1,1,1-tricchloroethane using a packed bed adsorber configuration.

The above authors investigated removal capabilities of desiccant treatments in other configurations than this research utilized. For example, Hines et al. (1992) studied the desiccant setups in a packed bed filter-like configuration. Popescu and Ghosh (1999) simulated a rotating desiccant wheel using an experimental system consisting of a fixed bed adsorber, which acted as a filter. Kovak, et al., (1997) exposed a desiccant wheel to biological organisms that were both associated with nosocomial (hospital-induced) infections and inherently heat sensitive.

Fig. 1 shows a desiccant unit installed in a school in south Texas for the purposes of (1) drying the school, (2) preventing the structure from becoming wet, (3) reducing the amount of latent cooling by the cooling coils, (4) creating a dew point of 45 °F to prevent condensation at the cooling coils, within the HVAC unit, and within the building itself, and (5) providing fresh pre-conditioned air for the interior space. Most school facilities utilize packaged HVAC equipment designed for inexpensive, efficient cooling. This type of equipment is not designed to handle the continuous supply of outdoor air necessary to comply with ASHRAE-62, Ventilation for Acceptable IAQ. As a result, these schools are likely to experience IAQ problems (Fischer 1996; Larrañaga, Beruvides et al. 2008).

Fig. 1. Desiccant unit installed in a south Texas school to provide pre-conditioned fresh air for ventilation.

The desiccant setup in Fig. 1 is depicted in the flow diagram in Fig. 2. This unit is designed to provide the school with 100% outdoor make-up air while maintaining the indoor dew point below 45°F (7°C.). The use of 100% fresh make-up air helps maintain the building at positive pressure to minimize unplanned moisture infiltrations into the building. The desiccant wheel was constantly rotating and adsorbing moisture from the air stream on the process side, while moisture was removed from the wheel on the regeneration side. This provided a constant adsorption medium with no phase change. Heat is necessary to release the moisture from the desiccant wheel, which results in heating of the airstream and an energy penalty. However, the use of active desiccation saves energy costs by: 1) providing an enhanced occupant comfort at a lower cost, 2) improved humidity control resulting in sensible versus latent cooling, 3) equipment expenditures by allowing the downsizing of the evaporator coil and condensing units for comparable design loads, 4) allowing independent temperature and humidity controls, and 5) allowing higher temperature set points (Meckler 1994).

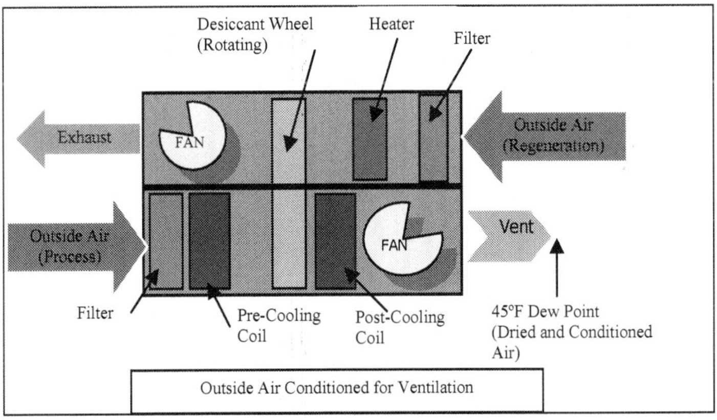

Fig. 2. Depiction of desiccant unit for treatment of 100% outside air with no treatment of return air from the building. This setup is typical of large commercial buildings or schools.

In units used for 100% outside air for ventilation (Fig. 2), or continuous ventilation, the outside air (process) is first filtered, cooled by the pre-cooling coil, dehumidified by the desiccant wheel, cooled again by the post-cooling coil, and delivered to the building's interior. This configuration does not allow for air within the building to be re-circulated, providing continuous fresh air to the system. Although the 100% make-up air configuration conditions more air, using more energy, than the typical desiccant setup depicted in Fig. 3 , it is preferred in very hot and humid climates for the protection of the building and building systems from moisture infiltration, condensation, and latent heat loads of occupants.

4. Materials and methods

There are five typical configurations for desiccant dehumidifiers: liquid spray tower, solid packed tower, rotating horizontal bed, multiple vertical beds, and rotating honeycomb (Harriman III 2002c). The most typical method of presenting solid desiccants to a high volume air stream is to impregnate the material into a lightweight honeycomb-shaped matrix

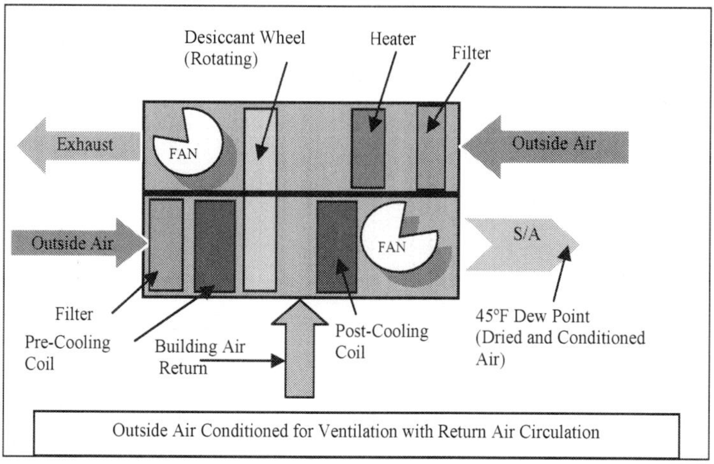

Fig. 3. Typical desiccant setup with return air from the building treated by the Post-Cooling coil. This setup is typical of small commercial buildings, schools, and residential buildings.

that is formed into a wheel (Harriman III, Witte et al. 1999). The rotating honeycomb wheel is a finely divided desiccant impregnated into a semi-ceramic structure, maximizing the surface area of the desiccant material. The appearance of the honeycomb wheel resembles corrugated cardboard that has been rolled up into the shape of a wheel. The air passes through the flutes formed by the corrugations, and the wheel rotates through the process and reactivation airstreams. The flutes served as individual desiccant-lined air ducts, which maximizes the surface area of the desiccant presented to the air stream. The rotating honeycomb wheel design has several advantages. The structure is lightweight and very porous. Different types of desiccants can be arranged into a honeycomb wheel configuration for different applications. The design allowed for laminar flow within the individual flutes, reducing air pressure resistance compared to packed beds. This allowed the honeycomb wheel to operate efficiently for low dew point and high capacity applications. The honeycomb wheels are very light, and their rotating mass is very low compared to their high moisture removal capacity. The result is an energy efficient unit (Harriman III 2002c). The design is simple, reliable, and easy to maintain. The design is the most widely installed of all desiccant dehumidifiers in ambient pressure applications (Harriman III 2002c). Additionally, the honeycomb design is the most appropriate dehumidifier configuration for air-conditioning applications (Pesaran 1994). A working desiccant honeycomb unit with environmental chambers at the outlets of both the regeneration side and process side of the unit was utilized to determine removal efficiencies. See Fig. 4.

The desiccant unit was modified with an additional heater (factory installed by Munters) and temperature controller (Omega CN132) to allow the operating temperature of the regeneration cycle (cycle that removes moisture from the wheel) to be varied between 100° (38°C) and 360° F (182°C). This allowed testing of the wheel at various increments of temperature. Spores were introduced using a 6-jet Collison nebulizer, a world standard for aerobiology research, that allowed for the aerosolization of fungal conidia for the purposes of testing the removal capabilities of the unit. The Collison nebulizer was operated at 40 psig, which generated a liquid generation rate of 16.5 ml/hr, allowing the experimenter to introduce a known concentration of fungi per volume of water as an aerosol into the test mechanism. For example, a fungal concentration of $1.25x10^5$ spores/ml provided an aerosol generation rate of $3.44x10^4$ spores per minute, allowing the experimenter to generate a known and constant airborne fungal concentration. Pressure was applied to the nebulizer with a Husky (model DK710700AV) Air Compressor.

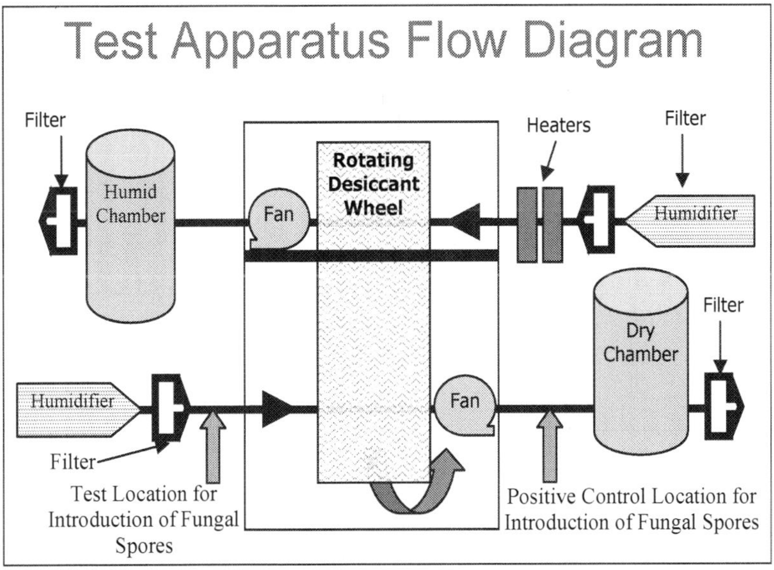

Fig. 4. Munters Dew-150 Desiccant Wheel Test Setup

The air was filtered with a HEPA Capsule prior to entering the Collison nebulizer. All microorganisms were concentrated, cultured, and speciated into a liquid solution. Lyophilization, a freeze-drying method under a vacuum, was used to concentrate spores (Virtis Freezemobile 6, Gardiner, NY). The organisms were then re-suspended in a known quantity of sterile water, and the concentration of spores determined through direct microscopic examination using a hemocytometer. The concentration was then manipulated by adding more water or spores to achieve the desired concentrations. The resulting fungal spore concentrates were frozen at -80°C. The test apparatus consisted of the Munters Dew 150 Desiccant Dehumidifier, four DeLonghi Dap-130 air purifiers with true high efficiency particulate air filters capable of capturing 99.97% of particles as small as 0.3 microns, and two environmental chambers (See Fig. 4). The environmental chambers were made of two 5-foot pieces of 24" agricultural high-pressure water pipe sealed at both ends with Plexiglas. Access doors were cut into each chamber, shelves installed, and a four plug-115V electrical outlet was installed to provide power for the sampling instruments. Each environmental chamber was sprayed twice with a Staticide ESD Clear Permanent Static Dissipative Coating to minimize the potential for electrostatic interference during the experiment. The environmental chambers were connected through metal duct to both the regeneration outlet (humid) and process outlet (dry) to permit the air exiting the desiccant unit to be sampled.

All air supplied to the desiccant unit was pre-filtered to remove microorganisms and other particulates from the air for quality control purposes. The air was again filtered upon leaving the environmental chamber so that microorganisms were not unnecessarily introduced into the laboratory. The air filters were modified to accept ducting so that the air could be supplied directly to the desiccant unit and filters and vice versa. Filters were connected to the apparatus via 5″ flex duct and are maintained under slightly positive pressure. Air supplied to the process air inlet was humidified to 60% relative humidity using a Holmes Cool Mist Humidifier (model HM3655) to simulate a humid environment.
Air sampling within the environmental chambers was conducted using Andersen Microbial Impactors and Allergenco MK-3 Microbial Air samplers. Samples were analyzed by microscopic analysis for total count or visual count readings for culturable samples. All culturable samples were cultured with potato dextrose agar. Data were analyzed using a two-factor analysis of variance with equal replication. The treatments were temperatures ranging from 60°-360° F (16°-182°C) at intervals of 100 °F (38°C). Temperature settings were determined based on the following information:

1. 60 °F (16°C) –Setting the desiccant unit at this temperature allows the unit to operate without heat and isolates the desiccant wheel from the effect of temperature. At this temperature, the desiccant wheel can be saturated so that no desiccation is occurring.
2. 160 °F (71°C) –Units utilizing low cost surplus heat generated from indirect sources such as steam boilers, engine-cooling jackets, refrigeration condensers, exhaust air, steam condensate, water heaters, etc., may operate at 160°F or less. The regeneration air must be either hotter or dryer than the process air (Harriman III 1999 p.28), allowing the desiccant unit to operate at lower temperatures. Using indirect energy sources can accomplish this task without the energy penalty of heating the regeneration air. These units are not suitable for low dew-point applications.
3. 260°F (127°C) –This temperature was chosen within the range necessary for low dew point applications, which may require reactivation temperatures as high as 250°F (121°C) to 275°F (135°C) (Harriman III, Humidity Control Design Guide, p. 211).
4. 360°F (182°C) –Setting the desiccant unit at this temperature allowed the unit to maximize the heat applied to the regeneration air stream and isolated effects due to temperature. At this temperature, the desiccant wheel could not be saturated by humid air and desiccation was maximized. Commercial desiccant units typically operate between 180°F (82°C) and 225°F (107°C) (Harriman III 1999 p. 29) and do not typically operate at this temperature.

To conduct an ANOVA for data generated by the two instruments (Allergenco™ and Andersen), a total of [(10 replications x 4 temperature settings x 3 tests (P, T1, T2) x 2 instruments) + (2 negative control samples x 2 chambers x 4 temperature settings x 2 instruments)] = a minimum of 272 samples must be taken per organism to complete each experiment. To test the three microorganisms *Aspergillus niger, Cladosporium cladosporioides,* and *Penicillium chrysogenum,* 816 samples must be taken, analyzed, and interpreted. Many simulation studies suggest that, generally, the central limit theorem holds for $n>30$ (Ott 1993), allowing the use of normal distributions in the analysis of data. If these data were shown to be non-normal and did not meet the assumption of normality via transformation, a non-parametric two-factor ANOVA was to be used. Non-normal data obtained from a two-factor experimental design should be analyzed non-parametrically by an extension of the Kruskal-Wallis test for single-factor analysis (Zar 1974), allowing the research hypotheses to remain unchanged.

4.1 Hypotheses

The general hypothesis stated that the active desiccant wheel removes statistically significant concentrations of airborne IAQ-related bioaerosols from the air supplied to a building. The sampling mechanism dictated whether results are expressed as viable or total number of spores per cubic meter of air. Viable particulate samplers are used to collect and assay airborne concentrations of aerobic species of culturable bacteria and fungi. All inertial impactors that use solid media produce data in colony forming units (CFU) (Ness 1991). A CFU is a viable fungal spore or bacterium capable of producing a mass of organisms, or colony that originates from a single cell or spore.

Data generated from the Andersen Impactor are expressed in CFU/m^3. Table 1 summarizes the testable hypothesis that this research will investigate from the data generated using the Andersen Impactor. Intertial slit impactors like the Allergenco MK-3 Microbial Air Sampler collect total concentrations of both viable and non-viable bioaerosols from the air. No distinction can be made between viable and non-viable airborne concentrations using slit impactors. The data generated by the Allergenco MK-3 Microbial Air Sampler are expressed as a total concentration of spores/m^3. Table 1 summarizes the testable hypothesis that this research will investigate from the data generated using the Allergenco MK-3 Microbial Air Sampler.

5. Results and analyses

A total of 816 samples were taken to complete the experiment. Tests for normality showed that several of the data sets did not meet the assumption of normality and several data sets contained outliers. When concerned over the normality effect and outliers, a separate ANOVA should be performed on both the original data and the ranks; should the two procedures differ, use of the rank transformation ANOVA is preferred because it is less

Testable Hypothesis for Data Generated by the Andersen Impactor		
Hypothesis	**Test Statistic**	**Variables** (airborne concentrations in colony forming units/m^3)
Hypothesis 1: No interaction between temperature and viable airborne fungal concentrations.	Interaction	Temperature is the temperature settings for each test. μ_P is the mean of the airborne concentrations of the positive control. μ_{T1} is the mean of the airborne concentrations of the dry-chamber test.
Hypothesis 2: No difference on the airborne concentration of viable spores due to temperature.	H1: $\mu_{T1} = \mu_P$ Ha: $\mu_{T1} \neq \mu_P$	μ_P is the mean of the airborne concentrations of the positive control. μ_{T1} is the mean of the airborne concentrations of the dry-chamber test.
Hypothesis 4: No difference on the airborne concentration of viable spores due to exposure to the desiccant wheel.	H1: $\mu_{T1} = \mu_P$ Ha: $\mu_{T1} \neq \mu_P$	μ_P is the mean of the airborne concentrations of the positive control. μ_{T1} is the mean of the airborne concentrations of the dry-chamber test.

Table 1. Summary of Testable Hypotheses for Data Generated by the Andersen Impactor.

Testable Hypothesis for Data Generated by the Allergenco MK-3 Microbial Air Sampler		
Hypothesis	**Test Statistic**	**Variables** (airborne concentrations in spores/m^3)
Hypothesis 1: No interaction between temperature and total airborne fungal concentrations.	Interaction	Temperature is the temperature settings for each test. μ_P is the mean of the airborne concentrations of the positive control. μ_{T1} is the mean of the airborne concentrations of the dry-chamber test.
Hypothesis 3: No difference on the airborne concentration of total spores due to temperature.	H1: $\mu_{T1} = \mu_P$ Ha: $\mu_{T1} \neq \mu_P$	μ_P is the mean of the airborne concentrations of the positive control. μ_{T1} is the mean of the airborne concentrations of the dry-chamber test.
Hypothesis 5: No difference on the airborne concentration of total spores due to exposure to the desiccant wheel.	H1: $\mu_{T1} = \mu_P$ Ha: $\mu_{T1} \neq \mu_P$	μ_P is the mean of the airborne concentrations of the positive control. μ_{T1} is the mean of the airborne concentrations of the dry-chamber test.

Table 2. Summary of Testable Hypotheses for Data Generated by the Allergenco MK-3 Microbial Air Sampler.

likely to be distorted by non-normality and unusual observations (Montgomery 2001). This experimental analysis was completed using a two-factor ANOVA for both the original and the ranked data. The results of the two analysis differed in several instances, specifically in differences in temperature and groups (control versus test).

5.2 Andersen impactor

Tests for normality were conducted to help determine protocols necessary for data analysis. The Kolmogorov-Smirnov (K-S) goodness of fit test was employed because it is particularly useful when sample sizes are small and when no parameters have been estimated. A large p-value indicates a good fit, while a small p-value indicates a poor fit (Banks 2001). Nine of the 12 *Aspergillus niger* data sets failed to meet the assumption of normality. Eight of 12 *Cladosporium cladosporioides*-Andersen Impactor and 4 of 12 *Penicillium chrysogenum*-Andersen Impactor data sets did not meet the assumption of normality. One two-factor ANOVA per organism tested was conducted for the data generated using the Andersen Impactor. The ANOVAs for each organism are described below.

5.2.1 *Aspergillus niger*-Andersen impactor data analysis

Post hoc analysis using the Student-Newman-Keuls (SNK) test for rank value showed that measured airborne concentrations were not significantly different at any of the four temperature settings. The SNK grouping for the positive control (μ_P) vs. dry chamber test (μ_{T1}) vs. humid chamber test (μ_{T2}) showed that the controls were significantly different from the dry chamber and humid chamber results at each temperature setting. The power of the *Aspergillus niger*-Andersen Impactor analysis for control vs. dry and humid chamber results is greater than 0.99.

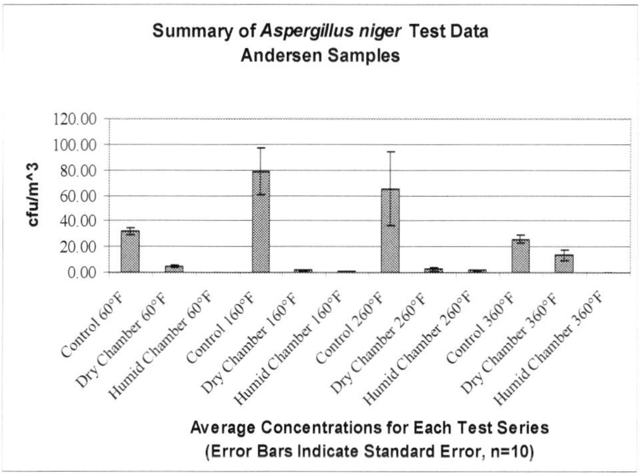

Fig. 5. Average airborne concentrations for *Aspergillus niger*-Andersen Impactor tests.

The results showed that there was interaction between that humid chamber and dry chamber data sets, there was no difference in the ranks due to temperature, and there was a significant difference between the mean ranks of the positive control (μ_P) and the mean ranks of the dry chamber tests (μ_{T1}) and humid chamber tests (μ_{T2}). Interaction occurred at the 160 °F (71°C) and 260 °F (127°C) between the dry chamber data ranks and the humid chamber data ranks. See Fig. 6 . The mean removal efficiencies are shown in Fig. 7 and can be described by the equation '$y = -0.0561x^3 + 0.2612\,x^2 - 0.2687x + 0.9153$' with a correlation coefficient (R^2) of one.

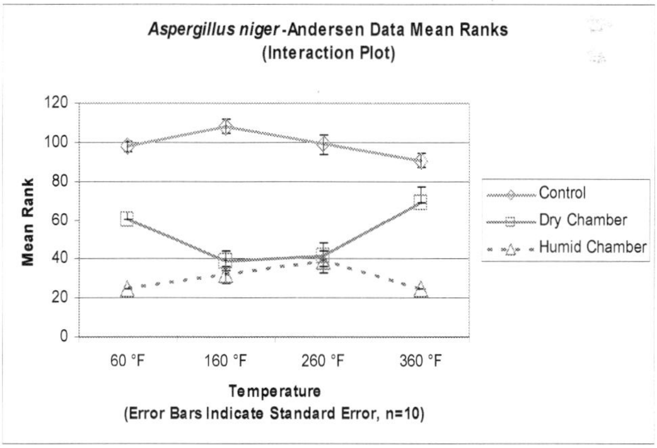

Fig. 6. Interaction Plot for the *Aspergillus niger*-Andersen Data Mean Ranks.

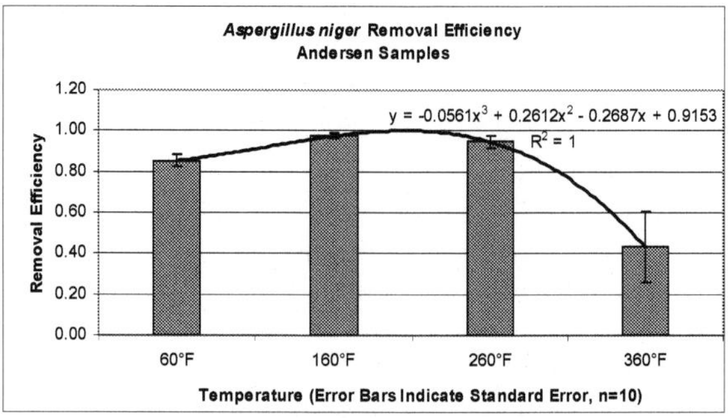

Fig. 7. *Aspergillus niger* Removal Efficiencies for Andersen Samples

5.2.2 Cladosporium cladosporioides-Andersen impactor data analysis

Post hoc analysis using the Student-Newman-Keuls (SNK) test for rank value showed that measured airborne concentrations were not significantly different at either of the four temperature settings. The SNK grouping for the positive control (μ_P) vs. dry chamber test (μ_{T1}) vs. humid chamber test (μ_{T2}) showed that the controls were significantly different from the dry chamber and humid chamber results at each temperature setting. The power of the *Cladosporium*-Andersen Impactor analysis for control vs. dry and humid chamber results is greater than 0.99. Fig. 8 summarizes the *Cladosporium cladosporioides*-Andersen Impactor data sets. These data show the mean airborne concentrations of the positive control versus the dry and humid chamber tests at each temperature setting. The results showed that there was no interaction, there was no difference in the mean ranks due to temperature, and there was a significant difference between the mean ranks of the positive control (μ_P) and the mean ranks of the dry chamber test (μ_{T1}) and humid chamber results (μ_{T2}). The mean removal efficiencies are shown in Fig. 9 and can be described by the equation '$y = -0.0053x^3 + 0.0475x^2 - 0.3278x + 1.18$' with a correlation coefficient (R^2) of one.

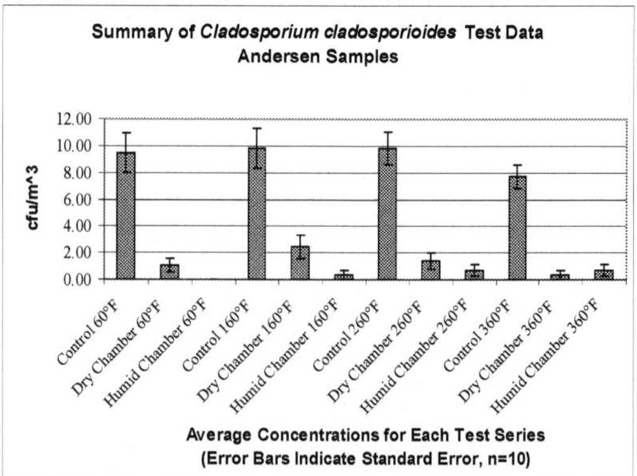

Fig. 8. Average airborne concentrations for *Cladosporium cladosporioides*-Andersen Impactor tests.

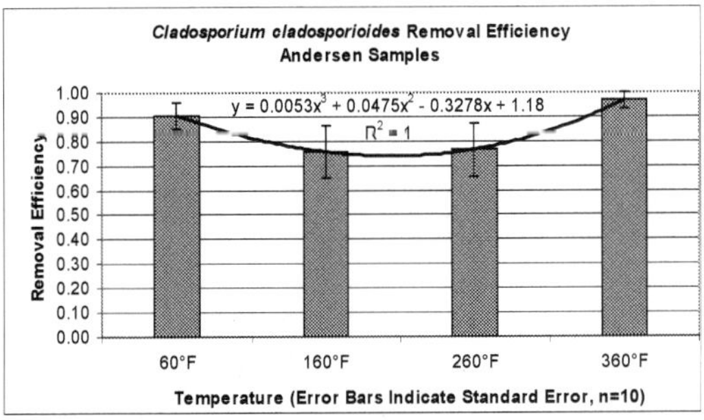

Fig. 9. *Cladosporium* removal efficiencies for Andersen samples

5.2.3 *Penicillium chrysogenum*-Andersen Impactor Data Analysis

Post hoc analysis using the Student-Newman-Keuls (SNK) test for rank value showed that measured airborne concentrations were not significantly different at any of the four temperature settings. The SNK grouping for the positive control (μ_P) vs. dry chamber test (μ_{T1}) vs. humid chamber test (μ_{T2}) showed that the controls were significantly different from the dry chamber and humid chamber results at each temperature setting. The power of the *Penicillium chrysogenum*-Andersen Impactor analysis for control vs. dry and humid chamber results is greater than 0.99. Fig. 10 summarizes the *Penicillium chrysogenum*-Andersen Impactor data sets. These data show the mean airborne concentrations of the positive control versus the dry and humid chamber tests at each temperature setting. The results showed that there was no interaction, there was no difference in the ranks due to temperature, and there was a significant difference between the mean ranks of the positive control (μ_P) and the mean ranks of the dry chamber test (μ_{T1}) and humid chamber tests (μ_{T2}). The mean removal efficiencies are shown in Fig. 11 and can be described by the equation '$y = -0.0728x^3 + 0.5584x^2 - 1.3065x + 1.5574$' with a correlation coefficient (R^2) of one.

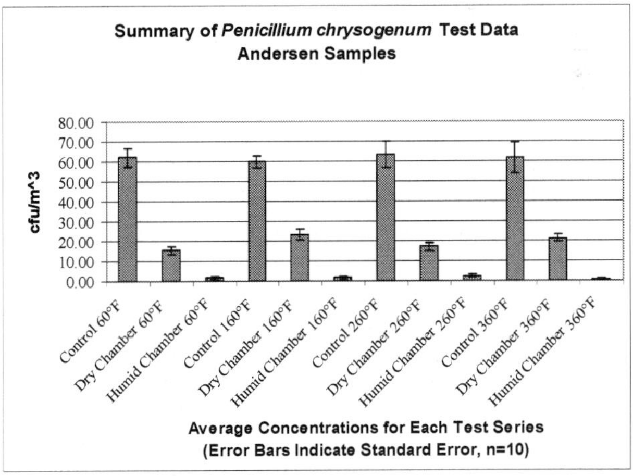

Fig. 10. Average airborne concentrations for *Penicillium chrysogenum* Andersen Impactor tests.

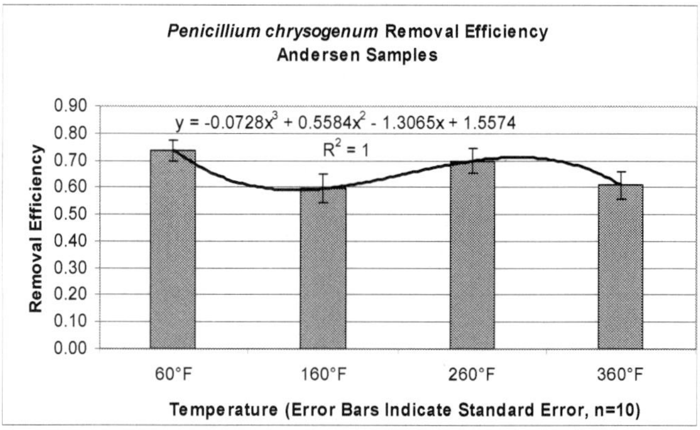

Fig. 11. *Penicillium chrysogenum* Removal Efficiencies for Andersen Samples

5.2.4 Combined removal efficiencies for andersen impactor data

The combined removal efficiencies of the *Aspergillus niger, Cladosporium cladosporioides, and Penicillium chrysogenum* data sets are summarized in Fig. 12 .

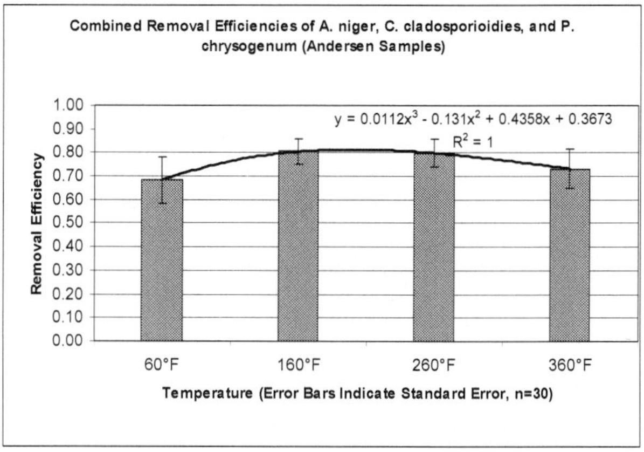

Fig. 12. Combined removal efficiencies of the *Aspergillus niger, Cladosporium cladosporioides, and Penicillium chrysogenum* –Andersen Impactor data sets.

5.3 Allergenco MK-III microbial air sampler

Tests for normality were conducted to help determine protocols necessary for data analysis. The K-S goodness of fit test was employed because it is particularly useful when sample sizes are small and when no parameters have been estimated. A large p-value indicates a good fit, while a small p-value indicates a poor fit (Banks 2001). Seven of the 12 *Aspergillus niger* data sets failed to meet the assumption of normality. Six of 12 *Cladosporium cladosporiodies*-Allergenco Microbial Air Sampler and 6 of 12 *Penicillium chrysogenum*-

Allergenco Microbial Air Sampler data sets did not meet the assumption of normality. One two-factor ANOVA per organism tested was conducted for the data generated using the Allergenco Microbial Air Sampler. The ANOVAs for each organism are described below.

5.3.1 *Aspergillus niger*-Allergenco microbial air sampler data analysis

Post hoc analysis using the SNK test for rank value showed that measured airborne concentrations were not significantly different at 160°F (71°C) and 260°F (127°C) or at 60°F (16°C) and 360°F (182°C). The two temperature groupings (160°F (71°C) and 260°F (127°C), 60°F (16°C) and 360°F (182°C)) were also statistically different. The SNK grouping for the positive control (μ_P) vs. dry chamber test (μ_{T1}) vs. humid chamber test (μ_{T2}) showed that the controls were significantly different from the dry chamber and humid chamber results at each temperature setting. The power of the *Aspergillus niger*-Andersen Impactor analysis for a temperature effect is 0.825, while the power of the analysis for control vs. dry and humid chamber results is greater than 0.99. Fig. 13 summarizes the *Aspergillus niger*-Allergenco Microbial Air Sampler data sets. These data show the mean airborne concentrations of the positive control versus the dry and humid chamber tests at each temperature setting. The results showed that there was interaction between that humid chamber and dry chamber data sets, there were differences in the ranks due to temperature, and there was a significant difference between the mean ranks of the positive control (μ_P) and the mean ranks of the dry chamber test (μ_{T1}). The interaction occurs at 160 °F (71°C) between the dry and humid chamber mean ranks. The mean removal efficiencies can be described by the equation '$y = 0.0662x^3 - 0.5048\ x^2 + 1.1476x + 0.1393$' with a correlation coefficient (R^2) of one. See Fig.s 4.10 AND 4.11.

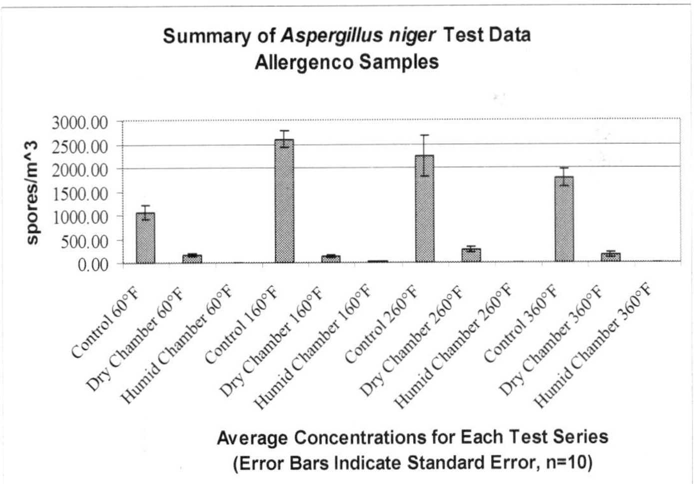

Fig. 13. Average airborne concentrations for *Aspergillus niger*-Allergenco Microbial Air Sampler tests

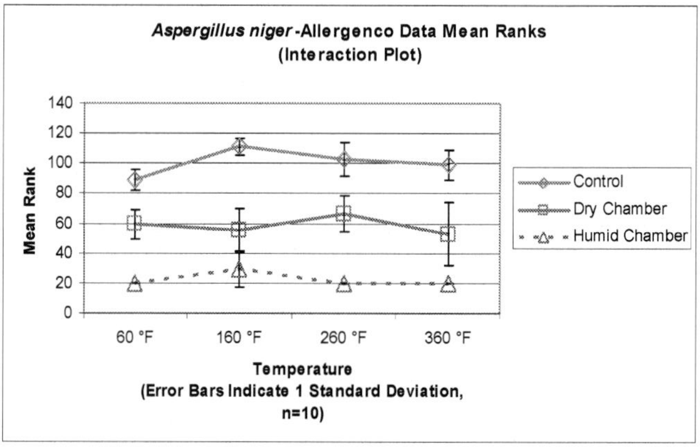

Fig. 14. Interaction Plot for the *Aspergillus niger*-Allergenco Microbial Air Sampler Data Mean Ranks.

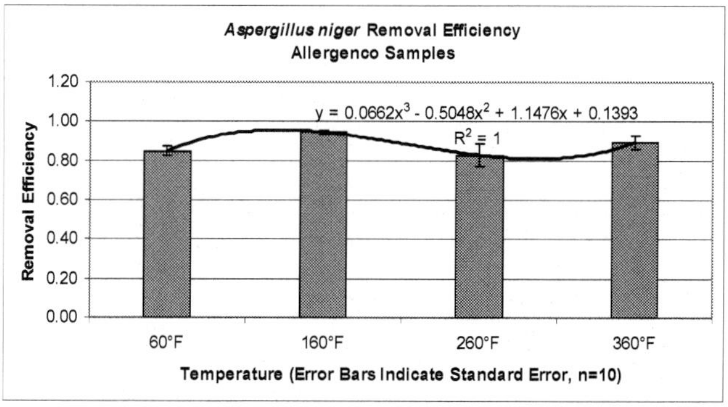

Fig. 15. *Aspergillus niger* Removal Efficiencies for Allergenco samples.

5.3.2 Cladosporium cladosporioides-Allergenco

Post hoc analysis using the SNK test for rank value showed that measured airborne concentrations were not significantly different. The SNK grouping for the positive control (μ_P) vs. dry chamber test (μ_{T1}) vs. humid chamber test (μ_{T2}) showed that the controls were significantly different from the dry chamber and humid chamber results at each temperature setting. The power of the *Cladosporium cladosporioides*-Allergenco Microbial Air Sampler analysis for control vs. dry and humid chamber results is greater than 0.99. Fig. 16 summarizes the *Cladosporium cladosporioides*-Allergenco Microbial Air Sampler data sets. Post hoc analysis using the SNK test for rank value showed that measured airborne concentrations were not significantly different at the four temperatures. The results showed that there was interaction between the dry chamber and humid chamber data sets, there was no difference in the ranks due to temperature, and there was a significant difference between the mean ranks of the positive control (μ_P) and the mean ranks of the dry (μ_{T1}) and humid chamber tests (μ_{T1}).

The mean removal efficiencies are shown in Fig. 17 and is described by the equation 'y = -0.033x³ + 0.22x² - 0.4149x + 0.9267' with a correlation coefficient (R^2) of one.

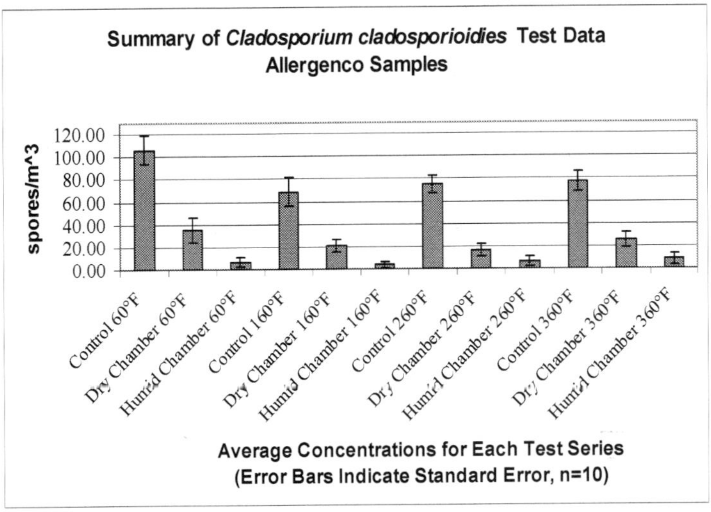

Fig. 16. Average airborne concentrations for *Cladosporium cladosporioides*-Allergenco Microbial Air Sampler tests.

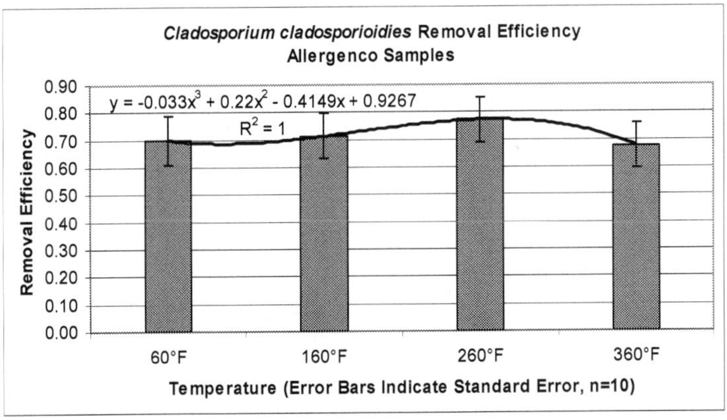

Fig. 17. *Cladosporium cladosporioides* Removal Efficiencies for Allergenco Samples.

5.3.3 *Penicillium chrysogenum*-Allergenco microbial air sampler data analysis

Post hoc analysis using the SNK test for rank value showed that measured airborne concentrations were significantly different at 360°F (182°C). The three remaining temperatures (60°F (16°C), 160°F (71°C), and 260°F (126°C)) were not statistically different. The SNK grouping for the positive control (μ_P) vs. dry chamber test (μ_{T1}) vs. humid chamber test (μ_{T2})

showed that the controls were significantly different from the dry chamber and humid chamber results at each temperature setting. The power of the *Penicillium chrysogenum*-Andersen Impactor analysis for a temperature effect was 0.55, while the power of the analysis for control vs. dry and humid chamber results was greater than 0.99. Fig. 18 summarizes the *Penicillium chrysogenum*-Allergenco Microbial Air Sampler data sets. The results showed that there was no interaction between that humid chamber and dry chamber data sets, there was a difference in the ranks due to temperature, and there was a significant difference between the mean ranks of the positive control (μ_P) and the mean ranks of the dry chamber test (μ_{T1}). The mean removal efficiencies are shown in Fig. 19 and can be described by the equation '$y = 0.0844x^3 - 0.7017x^2 + 1.9039x - 0.9933$' with a correlation coefficient (R^2) of one.

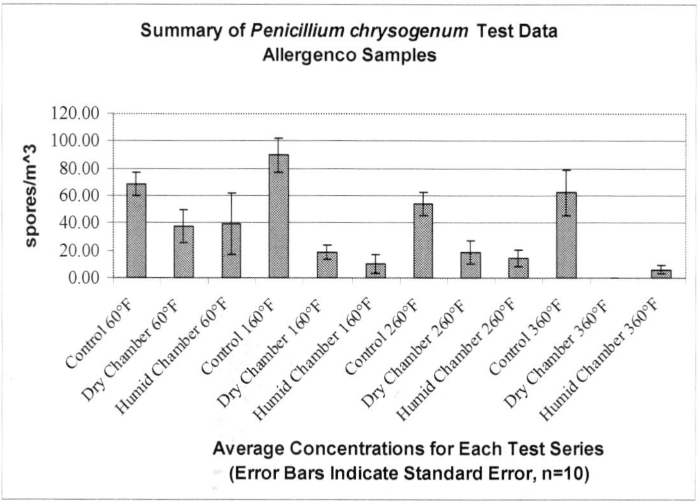

Fig. 18. Average airborne concentrations for *Penicillium chrysogenum*–Allergenco Microbial Air Sampler tests.

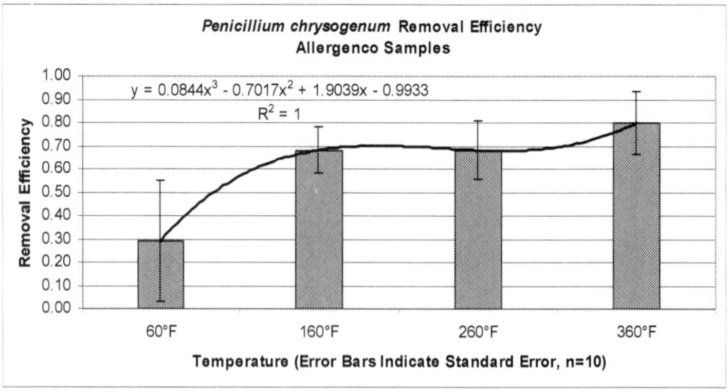

Fig. 19. *Penicillium chrysogenum* Removal Efficiencies for Allergenco Samples

5.4 Combined removal efficiencies for Andersen impactor data

The combined removal efficiencies of *Aspergillus niger, Cladosporium cladosporioides,* and *Penicillium chrysogenum* are summarized in Fig. 20 and can be explained by the polynomial regression equation '$y = -0.0132x^3 + 0.0912x^3 - 0.1913x + 0.8744$' with a correlation coefficient of one.

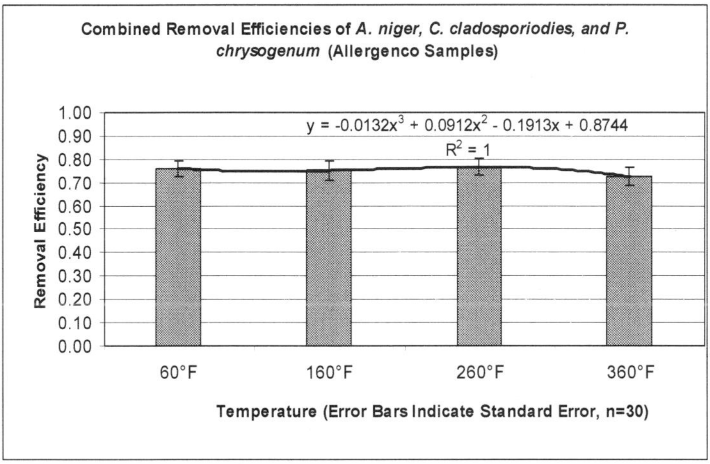

Fig. 20. Combined removal efficiencies of the *Aspergillus niger, Cladosporium cladosporioides, and Penicillium chrysogenum*-Allergenco Microbial Air Sampler data sets.

6. Summary of findings

The overall results of the hypothesis testing are summarized in Table 3. In all cases, the positive control (μ_P) was significantly greater than the dry chamber (μ_{T1}) results. Differences in temperature were found with the *Aspergillus niger* and *Penicillium chrysogenum* Allergenco data sets. Interaction was found within the *Aspergillus niger* data sets for both the Andersen Impactor and Allergenco Microbial Air Sampler.

Interaction of the *Aspergillus niger* data sets occurred in both cases between the mean ranks of the dry and humid chambers (μ_{T1} and μ_{T2}). Significant interaction did not occur between the mean ranks of the positive control and the mean ranks of the dry chamber test ($\mu_P > \mu_{T1}$), which were the parameters of interest when determining removal efficiencies and airborne concentrations delivered indoors. Differences due to temperature were identified within the *Aspergillus niger*-Allergenco Microbial Air Sampler data sets. See Table 3 above for the SNK grouping. A difference in temperature was also identified by the *Penicillium chrysogenum*-Allergenco Microbial Air Sampler data analysis with a statistical power (P) is 0.55, which is less than the commonly accepted value of 0.8. This means that there is 55% confidence that the temperature effect exists and a 45% confidence that the temperature effect does not exist. See the SNK grouping in Table 3 above. This indicates that the desiccant wheel may be more efficient at removing *Penicillium chrysogenum* spores from the airstream at 360°F (182°C). In both cases, the temperature difference occurred at one or both of the extremes of the temperature settings; the *Aspergillus niger*-Allergenco Microbial Air Sampler analysis showed that the total spore concentrations at 60°F (16°C) and 360°F (182°C) were different from those concentrations at 160°F (71°C) and 260°F (127°C), and the *Penicillium chrysogenum*-Allergenco Microbial Air Sampler analysis showed that the total spore concentrations at 360°F (182°C) were different from those concentrations found at 60°F (16°C), 160°F (71°C), and 260°F (127°C).

7. Discussion

The purpose of this research was to determine if the desiccant wheel was effective at removing statistically significant concentrations of IAQ-related microorganisms from the air it supplies to a building. In addition, the capabilities of the desiccant wheel at removing airborne concentrations of IAQ-related microorganisms at four separate temperature settings were explored. The purpose of exploring the removal capabilities at different temperature settings was to establish the mechanism of spore removal and generate prediction models of airborne removal efficiencies for the two sampling methods and three organisms tested. In two of the six two-factor analyses of variance, a temperature effects were significant.

In one instance the *Aspergillus niger*-Allergenco Microbial Air Sampler analyses showed that the mean of the ranks at 60°F (16°C) and 360°F (182°C) (60°F (16°C) and 360°F (182°C) were not statistically different) were statistically different from the mean of the ranks at 160°F (71°C) and 260°F (127°C) (160°F (71°C) and 260°F (127°C) were not statistically different). Both the 60°F (16°C) and 360°F (182°C) temperature settings were designed as controls to help determine the mechanism of spore removal by the desiccant wheel. This lack of a temperature effect between the two control temperatures indicates that the mechanism of removal was likely a mechanical filtration effect unaffected by the increasing magnitude of adsorption created by an increase in the reactivation temperature. This premise was supported by five of the six analyses. In the second instance of a significant temperature difference, the *Penicillium chrysogenum*-Allergenco Microbial Air Sampler analyses showed that the mean rank of the airborne concentrations at 360°F (182°C) was significantly different from the mean ranks at 60°F (16°C), 160°F (71°C), and 260°F (127°C), which were not statistically different ($P=0.55$). With a statistical power less than 0.8, however, the difference does not maintain the confidence necessary to validate the temperature effect. Therefore, the mechanism of removal appeared to be a mechanical filtration of spores resulting in a decrease in the airborne concentrations of viable and total concentrations of *Aspergillus niger, Cladosporium cladosporioides,* and *Penicillium chrysogenum* through the process air stream of the desiccant unit. The desiccant unit removed significant airborne concentrations of both viable and total spores for the three organisms in each of the six analyses. The results of this study showed that the desiccant unit significantly reduces the airborne concentrations of these organisms introduced into the indoor environment.

8. Conclusion

This study aimed to quantify the removal capabilities of a rotary wheel (honeycomb) solid-desiccant dehumidifier at removing selected IAQ-related fungal organisms from the airstream. For each organism, the reductions in airborne concentrations delivered to the dry chamber were significant. These results support the findings of several authors who have stated that the use of active desiccant technology enhances the quality of the indoor air by helping to maintain comfort criteria (temperature, humidity and ventilation) (Meckler 1994; Kovak 1997; Fischer and Bayer 2003), removing particulates and bioaerosols from the air (Hines 1992a; Kovak 1997), and removing chemical pollutants from the air (Hines 1992c; Popescu and Ghosh 1999). This study demonstrates the ability of the desiccant unit to remove IAQ-related microorganisms from the air. In addition, the study shows the removal capabilities are significant at the four temperatures tested. The ability of active desiccants to remove particulates, bioaerosols, chemical pollutants, and water vapor from the airstream delivered to a building provides a unique opportunity to view active desiccant technology as a viable control strategy for enhancing and maintaining a favorable IAQ in cooling climates.

Mold and other factors related to damp conditions indoors are linked to increased asthma symptoms in asthmatics and coughing, wheezing, and upper respiratory tract symptoms in otherwise healthy people; and damp indoor conditions may be associated with the onset of asthma, as well as shortness of breath and lower respiratory illness in otherwise healthy children. The Institute of Medicine calls for studies that compare various ways to limit moisture or eliminate mold growth indoors and to evaluate whether interventions improve the health of occupants. This study provides a foundation for exploring the feasibility of integrating desiccation technologies into existing HVAC system design for cooling climates for improving the IAQ within the built environment.

9. References

Addington, D. M. (2000). The History and Future of Ventilation. Indoor Air Quality Handbook. J. D. Spengler, Samet, J.M., McCarthy, J. New York, McGraw-Hill-Professional.

Andersson, M. A., Nikulin, M., Koljalg, U., Andersson, M.C., Rainey, F., Reijula, F., Hintikka, E.-L., Salkinoja-Salonen, M. (1997). "Bacteria, molds, and toxins in water-damaged building materials." Applied and Environmental Microbiology 63: 387-393.

Andrae, S., Axelson, O, Bjorksten, B., Fredriksson, M., Kjellman, N.M. (1988). "Symptoms of bronchial hyperrreactivity and asthma in relation to environmental factors." Archives of Disease in Childhood 63(5): 473-478.

Bachmann, M. O., Myers, J.E., (1995). "Influences on sick building syndrome symptoms in three buildings." Society, Science and Medicine. 40: 245-251.

Bardana, E., Jr., A. Montanaro, et al. (1988). "Building related illness." Clinical Reviews in Allergy 6(1): 61-89.

Bascom, R. (1997). Plenary paper: health and indoor air quality in schools - a spur to action or false alarm? Proc Healthy Buildings/IAQ' 97 American Society of Heating, Refrigerating and Air-conditioning Engineers, Bethesda, Md.

Batterman, S., Burge, H. (1995). "HVAC systems as emission sources affecting indoor air quality: A critical review." United States Environmental Protection Agency Research and Development(February): 1-39.

Bayer, C. W. (2000). "Humidity control and ventilation in schools." ASHRAE Journal(Summer).

Bayer, C. W., Crow, S.A.. (1992a). Odorous volatile emissions from fungal contamination. Atlanta, Proc IAQ'92, American Society of Heating, Refrigerating and Air Conditioning Engineers: 99-104.

Bernstein, R. S., Sorenson, W.G., Garabant, D., Reaux, C., Treitman, R.D. (1983). "Exposure to respirable, airborne *Penicillium* from a contaminated ventilation system: Clinical environmental and epidemiological aspects." American Industrial Hygiene Association Journal 44(3): 161-169.

Bernton, H. S. (1930). "Asthma due to a Mold - *Aspergillus fumigatus*." J.A.M.A 95(3): 189-192.

Brunekreef, B. (1992). "Damp housing and adult respiratory symptoms." Allergy: European Journal of Allergy and Clinical Immunology. 47(5): 498-502.

Brunekreef, B., Dockery, D.W., Speizer, F.E., Ware, J.H., Spengler, J.D., Ferris, B.G. (1989). "Home dampness and respiratory morbidity in children." American Review of Respiratory Disease 140: 1363-1367.

Bureau of National Affairs (1992). Indoor air - OSHA request for informatin for future regulation. Washington, D.C., Occupational Safety and Health Administration Report: 1097-1098.

Burge , H. A. (1989). "Indoor air and infectious disease." Occupational Medicine 4(4): 713-721.

Conant, N. F., Wagner, H.C., Rackemann, F.M. (1936). "Fungi Found in Pillows, Mattresses, and Furniture." The Journal of Allergy: 234-237.

Cookson, W. (1999). "The alliance of genes and environment in asthma and allergy." Nature 402: B5-B11.

Cooley, J. D., Wong, W.C., Jumper, C.A., Straus, D.C. (1998). "Correlation between the prevalence of certain fungi and sick building syndrome." Occupational and Environmental Medicine 55: 579-584.

Credille, B. A. (1933). "Report of a Case of Bronchial Asthma due to Molds." The Jouurnal of the Michigan State Medical Society 32(3): 167.

Croft, W. A., Jarvis B.B., Yatawara, C.S. (1986). "Airborne outbreak of trichothecene toxicosis." Atmospheric Environment 20(3): 549-552.

Croft, W. A., B. M. Jastromski, et al. (2002). "Clinical confirmation of trichothecene mycotoxicosis in patient urine." Journal of Environmental Biology 23(3): 301-320.

Dales, R., Zwanenburg, H., Burnett, R., Franklin, C. (1991). "Respiratory health effects of home dampness and molds among Canadian children." American Journal of Epidemiology 134(2): 196-203.

Dales, R. E., Burnett, R., Zwanenburg, H. (1991b). "Adverse health effects among adults exposed to home dampness and molds." American Rev Respiratory Disease: 505-509.

Davanagere, B. S., Shirey, D.B., Rengarajan, K., Colacino, F. (1997). "Mitigating the impacts of ASHRAE Standard 62-1989 on Florida Schools." ASHRAE Trans: 241-258.

Dill, I., Niggeman, B. (1996). "Domestic fungal viable propagules and sensitization in children with IgE mediated allergic diseases." Pedriatric Allergy Immunology 7: 151-155.

Dolan, W. H. (1989). "Desiccant Cooling Systems-A New HVAC Opportunity." Energy Engineering 86(4): 6-9.

Environmental Protection Agency (2010). "Managing asthma in the school environment."

EPA (2003). Managing asthma in the school environment. Washington, D.C., U.S. Environmental Protection Agency.

Finningan, M. S., C. A. C. Pickering, et al. (1984). "The sick building syndrome: prevalence studies." British Medical Journal 289: 1573-1575.

Fischer, J. C. (1996). Optimizing IAQ, humidity control, and energy efficiency in school environments through the application of desiccant-based total energy recovery systems. Proc IAQ'96, American Society of Heating, Refrigerating and Air Conditioning Engineers, Washington, D.C.

Fischer, J. C. and C. W. Bayer (2003). "Failing grade for most schools; report card on humidity control." ASHRAE Journal(May): 30-39.

Fisk, W. J. (2000). Estimates of potential nationwide productivity and health benefits from better indoor environments: an update. Indoor Air Quality Handbook. J. D. Spengler, Samet, J.M., McCarthy, J., Mcgraw-Hill.

Flannigan, B., McGabe, E.M., McGarry, F. (1991). "Allergenic and toxigenic microrganisms in houses." Journal of Applied Bacteriology. 70: 61S-73S.

Flannigan, B., Miller,J.D. (1994). Health implications of fungi in indoor environments-an overview. Health Implications of Fungi in Indoor Environments, Elsevier. Air Quality Monographs Vol. 2: 3-28.

Flood, C. A. (1931). "Observations on Sensitivity to Dust Fungi in Patients with Asthma." J.A.M.A 96(25): 2094-2095.

Garrett, M. H., P. R. Rayment, et al. (1998). "Indoor airborne fungal spores, house dampness and associations with environmental factors and respiratory health in children." Clinical and Environmental Allergy 28: 459-467.

Gordon, W. A., E. Johanning, et al. (1999). Cognitive impairment associated with exposure to toxigenic fungi. Bioaerosols, Fungi, and Mycotoxins: Health Effects, Assessment, Prevention and Control. E. Johanning. Albany, NY, Eastern New York Occupational Health Program: 94-98.

Hagmann, M. (2000). "A mold's toxic legacy revisited." Science 288: 243-244.

Handal, G., M. A. Leiner, et al. (2004). "Children symptoms before and after knowing about an indoor fungal contamination." Indoor Air 14: 87-91.

Harriman III, L. G., Ed. (2002c). The Dehumidification Handbook Second Edition. Amesbury, MA, Munters Corporation.

Harriman III, L. G., M. J. Witte, et al. (1999). "Evaluating active desiccant systems for ventilating commercial buildings." ASHRAE Journal(October 1999): 28-32.

Harriman, L. and R. Kittler (2001). Chapter 13 Dehumidifiers. Humidity Control Design Guide for Commercial and Institutional Buildings. L. G. Harriman III, Brundrett, G.W., Kittler, R. Atlanta, American Society of Heating, Refrigerating and Air Conditioning Engineers: 195-214.

Heller, R. M., T. W. Heller, et al. (2003). ""tsara'at," Leviticus, and the history of confusion." Perspectives in Biology and Medicine 46(4): 588-591.

Hines, A. L., and Ghosh, T.K. (1992b). Air Dehumidification and Removal of Indoor Air Pollutants by Liquid Desiccants: Investigation of Co-Sorption of Gases and Vapors as a Means to Enhance Indoor Air Quality - Phase II, Gas Research Institute GRI-92/0157.3.

Hines, A. L., and Ghosh, T.K., Loyalka, S.K., and Warder, R.C. Jr. (1992c). A Summary of Pollutant Removal Capabilities of Solid and Liquid Desiccants From Indoor Air: Investigation of Co-Sorption of Gases and Vapors as a Means to Enhance Indoor Air Quality, Gas Research Institute GRI-92/0157.1.

Hines, A. L., Ghosh, T.K., and Loyalka, S.K. (1992a). Removal of Particulates and Airborne Microorganisms by Solid Adsorbents and Liquid Desiccants: Investigation of Co-Sorption of Gases and Vapors as a Means to Enhance Indoor Air Quality - Phase II, Gas Research Institute GRI-92/0157.5.

Hodgson, M. (1992). "Field studies on the sick building syndrome." Ann NY Acad Sci 641: 21-36.

Hodgson, M., P. Morey, et al. (1985). "Pulmonary disease associated with cafeteria flooding." Archives of Environmental Health 40(2): 96-101.

Hodgson, M., P. Morey, et al. (1998). "Building-Associated Pulmonary Disease From Exposure to *Stachybotrys chartarum* and A*spergillus versicolo*r." Journal of Occupational and Environmental Medicine 40(3): 241-249.

Husman, T., T. Meklin, et al. (2002). Respiratory infection among children in moisture damaged schools. Indoor Air.

Hyvärinen, A., J. Pekkanen, et al. (2002). Moisture damage at home and childhood asthma - a case-control study. Indoor Air.

Jaakkola, J. J. K., Jaakkola, N., Ruotsalainen, R. (1993). "Home dampness and molds as determinants of respiratory symptoms and asthma in pre-school children." Journal of Experimental Analysis and Environmental Epidemiology. 3(S1): 129-142.

Jaakkola, M. S., H. Nordman, et al. (2002). "Indoor Dampness and Molds and Development of Adult-Onset Asthma: A population-Based Incident Case-Control Study." Environmental Health Perspectives 110(5): 543-547.

Jarvis, J. Q. and P. R. Morey (2001). "Allergic Respiratory Disease and Fungal Remediation in a Building in a Subtropical Climate." Occupational and Environmental Hygiene 16(3): 380-388.

Johanning, E., Biagini, R., Hull, D.L., Morey, P., Jarvis, B., Landsbergis, P. (1996). "Health and immunology study following exposure to toxigenic fungi (*Stachybotrys chartarum*)." International Archives of Occupational Environmental Health 68: 207-218.

Joki, A., Saano, V., Reponen, T., Nevalainen, A. (1993). Effect of indoor microbial metabolites on ciliary function in respiratory airways. Proc Indoor '93, Proceedings of the 6th International Conference on Air Quality and Climate, Helsinki, Findland.

Karunasena, E. (2005). "The mechanisms of neurotoxicity induced by a Stachybotrys chartarum tricothecene mycotoxin in an invitro model." A Dissertation in Microbiology and Immunology, Texas Tech University Health Science Center.

Karunasena E, Larrañaga MD, et al. (2010). "Building-associated neurological damaged modeled in human cells: a mechanism of neurotoxic effects by exposure to mycotoxins in the indoor environment." Mycopathalogia Dec(6): 377-390.

King, N. and P. Auger (2002). "Indoor air quality, fungi and health. How do we stand?" Canadian Family Physician 48: 298-302.

Kinshella, M. R., M. V. Van Dyke, et al. (2001). "Perceptions of indoor air quality associated with ventilation system types in elementary schools." Applied Occupational and Environmental Hygiene 16(10): 952-960.

Koskinen, O. (1995). "Reduced exposure to molds brings fewer respiratory symptoms." Indoor Air 5: 3-9.

Kovak, B., Heimann, P.R., Hammel J. (1997). "The Sanitizing Effects of Desiccant-Based Cooling." ASHRAE Journal: 60-64.

Kumar, S. and W. J. Fisk (2002). "IEQ and the impact on Employee Sick Leave." ASHRAE Journal: 97-98.

Landrus, G., Axcel, T. (1990). "Survey of asthma, allergy, and environmental sensitivity in an urban Canadian school system". Proceedings of the 5th International Conference on Air Quality and Climate, Toronto, Canada.

Larrañaga, M., M. Beruvides, et al. (2008). "DOAS & Humidity Control." ASHRAE Journal May 2008: 34-40.

Lehrer, S. B., Aukrust, L., Salvaggio, J.E. (1983). "Respiratory allergy induced by fungi." Clinical and Chest Medcine 4: 23-41.

Levin, H. (1995). "Physical factors in th indoor environment." Occupational Medicine 10: 59-95.

Li, C. S., Hsu, L.Y. (1997). "Airborne fungus allergen in association with residential characteristics in atopic and control children in a subtropical region." Archives of Environmental Health 52: 72-79.

Licorish, K., Novey, H.S., Kozak, P., Fairshter, R.D., Wilson, A.F. (1985). "Role of *Alternaria* and *Penicillium* spores in the pathogenesis of asthma." Journal Allergy Clinic and Immunology 76(6): 819-825.

Martin, C. J., Platt, S.D., Hunt, S.M. (1987). "Housing conditions and ill health." British Medical Journal 294: 1125-1127.

Meckler, M. (1994). "Desiccant-Assisted Air Conditioner Improves IAQ and Comfort." Heating/Piping/Air Conditioning(October 1994): 75-84.

Meggs, W. J. (1993). "Neurogenic inflammation and sensitivity to environmental chemicals." Environmental Health Perspectives 101: 234-238.

Meklin, T., T. Husman, et al. (2002). "Indoor air microbes and respiratory symptoms of children in moisture damaged and reference schools." Indoor Air 12(3): 175-183.

Mendell, M. J. (1993). "Non-specific symptoms in office workers: A review and summary of the epidemiologic literature." Indoor Air Quality 3: 227-236.

Miller, J. D. (1992). "Fungi as contaminants in indoor air." Atmospheric Environment 26A(12): 2163-2172.

Mishra, S. K., Ajello, L., Ahearn, D.G., et al (1992). ""Environmental mycology and its importance to public health." Journal of Medicine Veterinary Mycology 30(Supplement 1): 287-305.

Mumma, S. A. (2001). "Designing dedicated outdoor air systems." ASHRAE Journal(May): 28-31.

National Academy of Sciences (1981). Indoor Pollutants. Washington, D.C., National Academy Press.

National Academy of Sciences (1993). Indoor Allergens. Washington, D.C., National Academy Press.

Ness, S. A. (1991). Air Monitoring for Toxic Exposures-An Integrated Approach. New York, Van Nostrand Reinhold.

Nielsen, K. F., M. O. Hansen, et al. (1998). "Production of trichothecene mycotoxins on water damaged gypsum boards in Danish buildings." International Biodeterioration & Biodegradation: 1-7.

Ott, R. L. (1993). An Introduction to Statistical Methods and Data Analysis. Belmont, Ca, Duxbury Press.

Passon, T. J., Brown, J.W., Mante, S. (1996). "Sick-building syndrome and building - related illness." New and emerging pathogens(July): 84-95.

Pesaran, A. A. (1994). A Review of Desiccant Dehumidification Technology. Golden, Colorado, National Renewable Energy Laboratory NREL/TP-472-7010: 1-8.

Peyronel, B. (1919). "I germi atmosferici dei funghi con misselio." Diss. Padua 1914. f. Bakt. (2 Abt.) 49: 465.

Platt, S. D., Martin, C.J., Hunt, S.M., Lewis, C.W. (1989). "Damp housing, mould growth, and symptomatic health state." British Medical Journal 298: 1673-1678.

Popescu, M. and T. K. Ghosh (1999). "Dehumidification and Simultaneous Removal of Selected Pollutants from Indoor Air by a Desiccant Wheel Using a 1M Type Desiccant." Journal of Solar Energy Engineering 121(February 1999): 1-13.

Redd, S. C. (2002). "State of the Science on Molds and Human Health." Center for Disease Control and Prevention, U.S. Department of Health and Human Services.: 1-11.

Reijula, K. (1996). "Buildings with moisture problems - a new challenge to occupational health care." Scandinavian Journal of Work, Environment and Health 22: 1-3.

Richard, J. L., R. D. Platnerr, et al. (1999). "The occurence of ochratoxin A in dust collected from a problem household." Mycopathologia 146: 99-103.

Richards, W. (1986). "Allergy, asthma, ans school problems." Journal of School Health 56(4): 151-152.

Rostrup, O. (1908). "Nogle Undersogelser over Luftens af Svampekim." Bot. Tidsskr 29: 32.

Saito, K. (1904). "Untersuchungen über die atmospharischen Pilzkeime. I. Mittheilung." J. Coll. Sci. Tokyo. 18(Art. 5): 58.

Sakula, A. (1984). "Sir John Floyer's A Treatise of the Asthma (1698)." Thorax 39: 248-254.

Salvaggio, J. E. (1994a). "Inhaled particles and respiratory disease." Journal Allergy and Clinical Immunology 94: 304-309.

Salvaggio, J. E. (1994b). "Psychological aspects of environmental illness, multiple chemical sensitivity and building related illness." Journal Allergy and Clinical Immunology 94: 366-370.

Salvaggio, J. E., Aukrust, L. (1981). "Mold induced asthma." Journal Allergy and Clinical Immunology 68: 327-346.

Samet, J. (1990). "Environmental controls and lung disease." American Review of Respiratory Diseases 142: 915-938.

Seitz, T. A. (1990). NIOSH indoor air quality investigations: 1971 through 1988. The practitioner's approach to indoor air quality investigations. D. M. Weekes, Gammage, R.B. Cincinnati, OH, American Industrial Hygiene Association: 163-171.

Smith, R. C. (1990). "Controlling sick building syndrome." Journal of Environmental Health 53(3): 22-23.

Solomon, W. R. (1975). "Assessing fungus prevalence in domestic interiors." The journal of allergy and clinical immunology 56(3): 235-242.

Spengler, J., Neas, L., Nakai, S. (1994). "Respiratory symptoms and housing characteristics." Indoor Air 4: 72-82.

Stenberg, B., Eriksson, N., Hoog, J., Sundell, J., Wall, S. (1994). "The sick building syndrome (SBS) in office workers: a case-referent study of personal, psychosocial and building-related risk indicators." International Journal of Epidemiology. 23: 1190-1197.

Sterling, T. D., Arundel, A. (1984). "Possible carcinogenic components of indoor air quality, combustions by-products, formaldehyde, mineral fibers, radiation, and tobacco smoke." Journal of Environmental Science and Health(62): 185-230.

Sudakin, D. L. (1998). "Toxigenic fungi in a water-damaged building: An intervention study." American Journal of Industrial Medicine 34: 183-190.

Summerbell, R. C., Staib, F., Dales, R., Nolard, N., Kane, J., Zwanenburg, H., Burnett, R., Krajden, S., Fung, D., Leong, D. (1992). "Ecology of fungi in human dwellings." Journal of Medical Veterinary Mycology 30(1): 279-285.

Tarlo, S. M., Fradkin, A., Tobin, R.S. (1988). "Skin testing with extracts of fungal species derived from the homes of allergy clinic patients in Toronto, Canada." Clinical Allergy 18(1): 45-52.

Teichman, K. Y. (1995). "Indoor Air Quality: Research Needs." Occupational Medicine 10(1): 217-227.

Tobin, R. S., Baranowski, E., Gilman, A.P., Kuiper-Goodman, T., Miller, J.D., Giddings, M. (1987). "Significance of Fungi in Indoor Air: Report of a Working Group." Canadian Journal of Public Health. 78(S1): S1-S14.

Verhoeff, A. P., Van Strien, R.T., Van Wijnen, J.H., Brunekreef, B., (1995). "Damp housing and childhood respiratory symptoms: the role of sensitization to dust mites and molds." American Journal of Epidemiology 141: 103-110.

Verhoeff, A. P., Van Wijnen, J.H., Brunekreef, B., Fischer, P., Van Reenen-Hoekstra, E.S., Samson, R.A. (1992). "Presence of viable mold propagules in indoor air -relation to house dampness and outdoor air." Allergy: European Journal of Allergy and Clinical Immunology 47(2 Pt 1): 83-91.

Vogel, G. (1997). "New clues to asthma therapies." Science 276(13): 1643-1646.

Weiss, K. B., Gergen, P.J., Hodgson, T.A. (1992). "An economic evaluation of asthma in the United States." New England Journal of Medicine 326: 862.

Weiss, K. B., S. D. Sullivan, et al. (2000). "Trends in the cost of illness for asthma in the United States, 1985-1994." J Allergy Clin Immunol 106(3): 493-499.

White, J. H. (1995). Moldy Houses: Why they are and why we care. Ottawa, Ontario, Morrison Hershfield Limited: 1-65.
World Health Organization (1983). Indoor air pollutants: exposure and health effects. Copenhagen, World Health organization.
Zar, J. H. (1974). Biostatistical Analysis. Englewood Cliffs, NJ, Prentice-Hall, Inc.

22

Indoor Climate and Energy Performance in Typical Concrete Large-panel Apartment Buildings

Teet-Andrus Koiv and Targo Kalamees
Tallinn University of Technology
Estonia

1. Introduction

Energy consumption in buildings is between 20% - 40% of final energy (USA: 40%, EU: 37%, World: 24%), which is bigger than that in industry and transport. To improve energy efficiency, the EU Heads of State and Government has set a series of demanding climate and energy targets that must be met by the year 2020, known as the "20-20-20" targets: to reduce greenhouse gas emissions in the EU at least 20% below the 1990 levels, to produce 20% of the energy in the EU from renewable resources and to reduce primary energy use by 20% when compared with projected levels. The housing stock consists mainly of still existing old dwellings which strongly influence the energy performance of buildings. To improve energy performance the buildings need extensive renovation. Before the renovation process, it is necessary to obtain information about the current situation, i.e. about the indoor climate and the energy use. It should be emphasized that energy consumption in apartment buildings is closely connected with indoor climate.

The dwelling stock in Estonia is relatively adequate in size; however, the main problems concern quality and energy-efficiency in the dwellings. 71% of the population In Estonia occupy apartment buildings, the remaining population lives in detached or terraced houses (20%) and in farmhouses (9%). The majority (84%) of apartment buildings were erected after World War II, primarily between 1961-1990. One of the main types of building during that period was prefabricated concrete large-panel building, Fig. 1.

The thermal properties of the building envelope of these typical apartment buildings are modest

- Thermal transmittances
  - External wall 0.5…1.3 W/(m$^2\cdot$K)
  - Window3.0 W/(m$^2\cdot$K)
  - Roof 0.7…1.1 W/(m$^2\cdot$K)
- Air tightness of the building envelope: n_{50}: 6.1 h^{-1}, q_{50} 4.2 m^3/(h$\cdot$m^2)

The actual indoor temperature and humidity conditions are important data for assessing indoor climate and thermal comfort as well as energy consumption. To guarantee thermal comfort indoors, both the heating and the ventilation systems and the building envelope play a considerable role. In winter, continuous heating systems should keep the temperature according to the set point of thermostats. In old multi-storey prefabricated apartment

Fig. 1. Examples of typical apartment buildings composed of prefabricated concrete elements (left: type 1-464A-15, right: type 1-464 Д-84)

buildings, radiators are typically without thermostats and the system is typically equipped with standpipe valves. The one-pipe heat distribution system is typical of old multi-storey prefabricated apartment buildings. If the one-pipe system is properly installed and balanced, it can work satisfactorily. If improper modifications have been made and as the balancing of the heating system is complicated, the stability and the temperature level may cause problems during the heating season.

Air tightness of the building helps to avoid uncontrolled airflows through the building envelope. This can lead to problems related to the hygrothermal performance of the building envelope, energy consumption, health, performance of the ventilation systems, thermal comfort, noise, and fire resistance.

Almost all building envelopes have thermal bridges - locations where the thermal resistance of the assembly is locally lower. Due to lower temperatures on the thermal bridge, higher RH occurs. While surface condensation starts at the 100% of the RH, the limit value for RH in respect of mould growth is above 75% to 90% RH depending on the material. Thermal bridges lead to an increase in heat losses. If large poorly-insulated or non-insulated envelope areas exist, the surfaces will be cold in winter and may cause problems with thermal comfort due to cold draughts or radiation (in particular, asymmetric radiation).

The paper presents the results of the study of indoor climate and energy use and the factors that affect them in typical apartment buildings. As constructions and outdoor climate also influence the indoor climate and the energy use they should always be taken into account. The study is confined to the investigation of the problem in cold climate (Estonia) and to the apartment buildings erected from prefabricated concrete elements.

2. Methods

To give an overview of the indoor temperature and humidity conditions and energy use in old multi-storey apartment buildings composed of prefabricated concrete elements, long-term field measurements were carried out in 39 apartments in 20 blocks of flats in Estonia (Kalamees et.al. 2009).

The buildings were heated by district heating and a one-pipe radiator heating system. Radiators, as a rule, were not equipped with special thermostats, therefore individual control of the room temperature was impossible. Room temperature for the whole building was controlled in heat substations, depending on outdoor temperature. The data about energy use were collected from the building studied separately (room heating including heating of ventilation air), heating of domestic hot water (DHW), use of electric energy (for appliances, equipments, lighting, technical systems, etc.).

The temperature and the RH in the rooms were measured with data loggers at 1-h intervals over a 1-year period. The data loggers were located on the separating walls mainly in master bedrooms. The temperature range measured by the data loggers was between 20°C and +70°C with an accuracy of ±0.35°C and the RH range was between 5% and 95RH% with an accuracy of ±2.5RH%.

Indoor thermal conditions in the apartments studied were assessed on the basis of the target values from the standards (CR 1752, 1998, EN-15251, 2007). The II indoor climate category (normal level of expectation, for new buildings and renovations) and the III indoor climate category (acceptable, moderate level of expectation, for existing buildings) were selected for the comparison. The indoor temperatures during the winter season in the II indoor climate category must be in the range of +22±2°C and in the III indoor climate category +22±3°C. During the summer indoor temperature dependency on the outdoor temperature according to the standard (EN-15251, 2007) was used as old apartments are typically without mechanical cooling systems. The indoor RH values should be in the range of 25 % < RH < 45 % during the winter season and in the range of 30 % < RH < 70 % during the summer.

Based on the standard (EN-15251, 2007), various indicators of the indoor environment can be used for thermal evaluation purposes:

- Hourly criteria: the performance can be evaluated by calculating the number of actual hours or the % of the time when the criteria are met or are outside a specified range;
- Degree hours criteria: in respect of the thermal environment the degree hours outside the upper or lower boundary can be used as a performance indicator of a building for a warm or a cold season;
- Overall thermal comfort criteria (weighted PMV criteria): the time during which the actual PMV exceeds the comfort boundaries is weighted by a factor which is a function of the PPD. Starting from a PMV-distribution on a yearly basis and the relation between PMV and PPD, the following is calculated.

In this study hourly criteria and degree hour's criteria were used for the thermal evaluation.

The air tightness of each apartment was measured with the standardized (EVS–EN 13829, 2001) fan pressurization method, using "Minneapolis Blower Door Model 4" equipment (flow range at 50 Pa 25–7.800 m^3/h, accuracy ±3 %). To determine typical air leakage places and their distribution during the winter period, an infrared image camera was used FLIR Systems E320 (accuracy 2% or 2° C, measurement range; -20…+500° C). To compare air leakage of different apartments, the air flow rate at pressure difference 50 Pa was divided by the apartment's envelope area (including intermediate walls and floors) resulting air leakage rate at 50 Pa q_{50}, $m^3/(h \cdot m^2)$ or by the internal volume of the building resulting in the air change rate at 50 Pa, n_{50}, ach). As a rule, from each building the air tightness of one apartment was measured. If there were more measurements from one building, the average value was calculated to represent the air tightness of the building.

From the measurement results, the reference value of air tightness for different types of buildings was calculated. The reference value $q_{50,delc}$ (Eqn. 1) represents the median value

(50% fractile) with a confidence level of 90% for air tightness. The reference value of air tightness is applicable for energy calculations, when air tightness is not measured or the air tightness base value given in energy performance regulation is not suitable for use (too large or too small).

$$q_{50,decl} = q_{50} + k \cdot \sigma_{q_{50}} \text{, m}^3/(\text{h m}^2) \quad (1)$$

where: q_{50} is the mean value of air tightness of this building type, $m^3/(h\,m^2)$; k *is* a factor to take into account the median value with a confidence level of 90% (Eqn. 2), and $\sigma_{q_{50}}$ is the standard deviation of air tightness measurement results (Eqn.3).

$$k = \frac{1.645}{\sqrt{n}} \text{, -} \quad (2)$$

where: n is the number of measurements.

$$\sigma_{q_{50}} = \sqrt{\frac{\sum_{i=1}^{n}(q_{50,i} - q_{50})^2}{n-1}} \text{, m}^3/(\text{h m}^2) \quad (3)$$

where: $q_{50,i}$ is a single air tightness measurement result.
Additionally, an interview questionnaire was completed for each apartment, where the building characteristics, occupants' habits, typical perceptions, complaints and symptoms related to the indoor climate were included.
Indoor temperature and RH were continuously measured for each apartment building at 1-h intervals over a 1-year period. Outdoor climate was measured near the apartment buildings studied or climatic data were retrieved from the nearest weather station.

3. Energy use in typical apartment buildings

Most of the buildings were heated by district heating, a small part of them by local boiler houses and a one-pipe radiator heating system. Radiators as a rule were not equipped with special thermostats, therefore individual control of the room temperature was impossible. Room temperature for the whole building was controlled in heat substations depending on outdoor temperature. Data on energy use in the buildings studied were collected separately, room heating including the heating of the ventilation air, heating of domestic hot water (DHW), electricity use (for appliances, equipments, lighting technical systems etc.).
In spite of problems concerning the work of one-pipe heating systems the heat consumption in those buildings within the 35 years (1974-2009) has still decreased considerably.
The decrease has first of all taken place due to a decrease in the DHW consumption (Koiv, Voll & Hani, 2010, Koiv, Hani & Voll, 2009) and to a certain lowering of the internal temperature (the latter becoming possible owing to the use of automatic heat substations and the balancing of the heating systems) and partly by renovating the envelope elements of the buildings, all in all about 170 kWh/m^2 (per gross area of apartments), see Fig. 2. 115 kWh/m^2 of the decrease in the heat consumption was made up by a decrease in the DHW consumption, because by that time the DHW systems in bigger towns had mostly been

renovated. It can be seen that the 80s saw a very slight change in the heat consumption in apartment buildings, but after that it started decreasing very quickly. A decrease in the DHW consumption did not bring about any special problems.

A decrease in the heating costs became possible due to the renovation of the heat substations (boiler houses) and in part of the buildings also to that of the heating systems, as well as the renovation of the building envelope (first of all the windows).

By replacing the windows it was possible to economize on the heat consumption, but to a great extent it took place at the expense of decreasing the air change, which considerably deteriorated the indoor climate and often caused mould in apartments (Koiv, Voll, Mikola, Kuusk & Maivel, 2010).

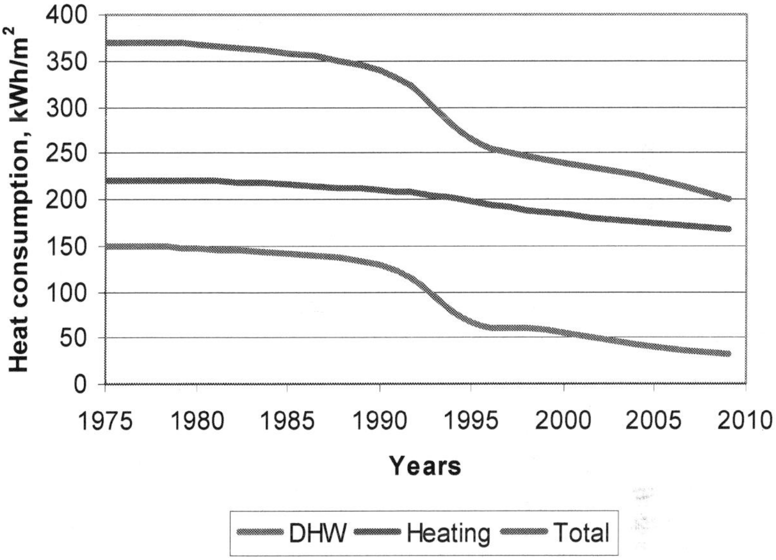

Fig. 2. Approximate heat consumption of DHW, heating (room heating and heating of ventilation air) and the total for the years 1975-2009 (per gross area of apartments)

4. Performance of ventilation

The ventilation system of the apartment buildings investigated was natural (passive stack) ventilation, Fig. 3. In some apartments kitchens were supplied with a hood. In all of the dwellings studied, windows could be opened for airing purposes. The air change in such buildings is unstable, being affected by external temperature, the height of the ventilation channel and also by the strength of the wind and its direction. Though mechanical ventilation has been the standard installation in new apartment buildings in Estonia during the last two decades, old apartment buildings have preserved natural ventilation due to the complexity of the ventilation renovation.

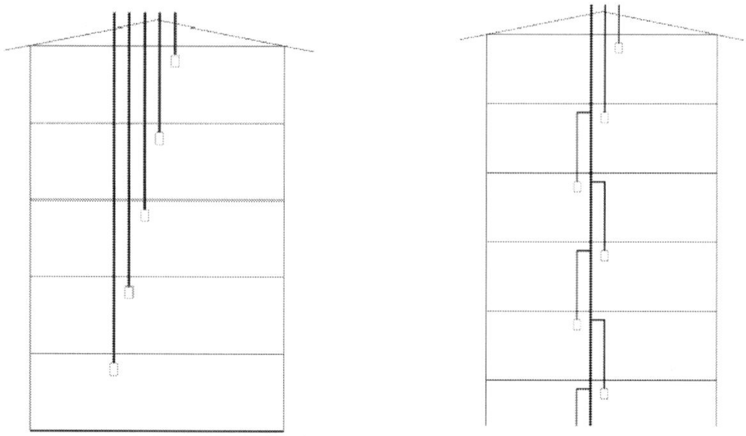

Fig. 3. The ventilation system of the apartment buildings investigated (left: up to a five-storey building, right: higher than a five-storey building)

It is practically impossible to achieve necessary airflows in apartments with natural ventilation. It is possible to use window airing, but this may worsen thermal comfort during the winter season. Because of worsened thermal comfort due to outdoor air inflow from stack (caused by wind) occupants seal the stack and ventilation outlets, Fig. 4.

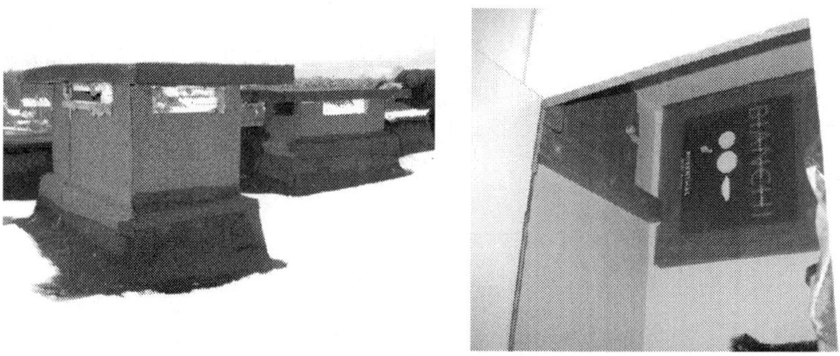

Fig. 4. Ventilation stack is closed by occupants due to cold air inflow caused by wind

Ventilation airflow in apartments is smaller than required (Fig. 5 left) and the CO_2 level is higher than the one permitted (Fig. 5 right). In such a situation the RH is also high and it is often accompanied by a rise of mould.

Replacing the old non-hermetic windows with more airtight ones in such apartment buildings results in dramatically worsened indoor air quality. The replacement of windows without renovating the ventilation results in lower air leakage and ventilation airflows, Fig. 6. In partly renovated apartment buildings problems with low air change often arise.

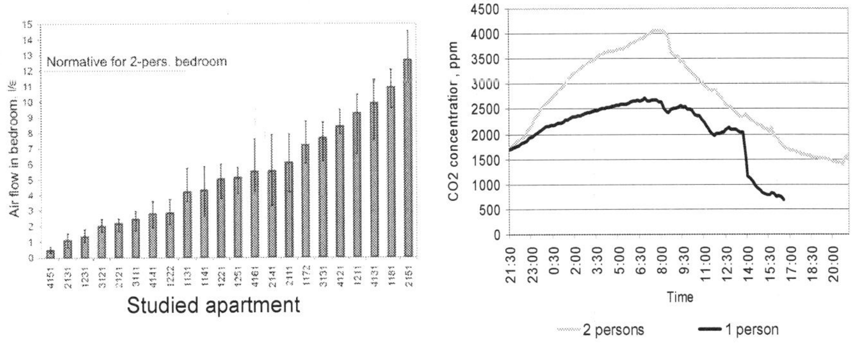

Fig. 5. Ventilation airflow (left) and CO_2 level (right) in bedrooms

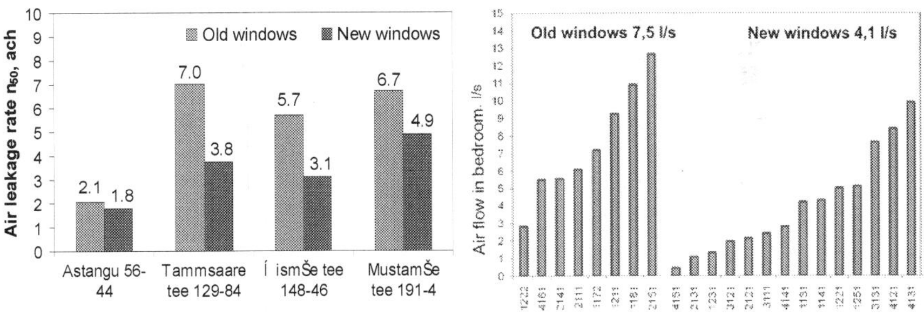

Fig. 6. The influence of the replacement of old windows on the air tightness of the apartment (before and after the replacement) (left) and on the air flow in bedrooms (right)

5. Indoor temperature and humidity conditions

5.1 Dependence of indoor temperature and humidity on outdoor temperature

To give an overall view of the indoor climate the dependence of the indoor temperature and the RH on the outdoor temperature was analysed. From each room at each average daily outdoor air temperature, all average daily indoor temperature values were selected, Fig. 7 left. From all the indoor temperatures at the corresponding outdoor temperature, the average value was calculated. Each individual thin solid lightly shaded curve in Fig. 7 (right) represents the average value from the average daily indoor temperature at the corresponding average daily outdoor temperature in one apartment. The dotted curve represents the average curve from all the apartments.

A one-pipe heat distribution system was installed in the apartments studied. As radiators were not equipped with thermostats, there was no possibility to control the room temperature. The room temperature was controlled in the boiler room based on the outdoor temperature. If the slope and the level of the control curve of the temperature of the supply water are correct, the room temperature does not depend strongly on the outdoor temperature. Fig. 8 (right) and

Fig. 8 (left) shows the measurement results from the buildings where the control curve is incorrect. Then during a cold period an apartment will be overheated. As occupants had no possibility to control the room temperature, the room temperature was typically lowered by opening the windows. Lowering temperature during winter-time by opening a window instead of using the thermostats of the radiators wastes energy.

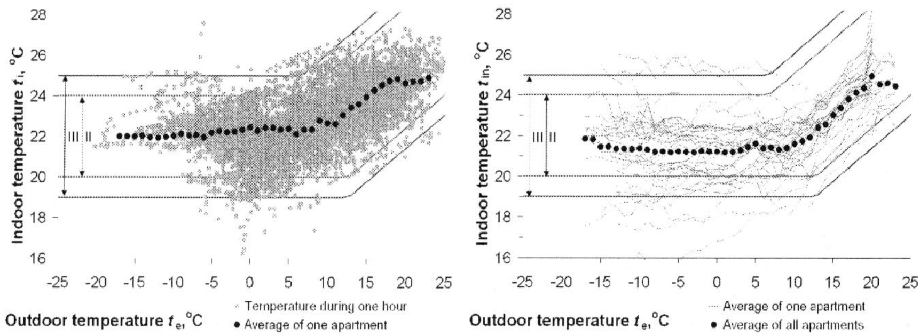

Fig. 7. Dependence of the indoor temperature on the outdoor temperature in one room (left) and the comparison of all rooms (right)

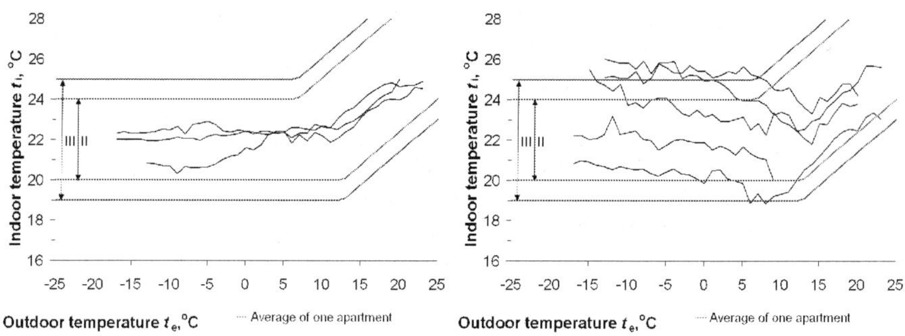

Fig. 8. Dependence of the indoor temperature on the outdoor temperature in buildings with correct (left) and incorrect (right) supply water temperature control curve

Similarly to the room temperature, the dependence of indoor RH on the outdoor temperature was also analyzed. In Fig. 9 (left) average daily indoor RH values from one room are divided according to the average daily outdoor air temperature. From all the indoor RH at the corresponding outdoor temperature, the average value was calculated (dotted line). Each individual thin solid lightly shaded curve in Fig. 9 (right) represents the average value of the average daily indoor RH at the corresponding average daily outdoor temperature in one apartment. The dotted curve represents the average curve of all the apartments. Even the average RH in the apartments studied stays within target values of RH in most of the apartments we can see large variations of indoor RH.

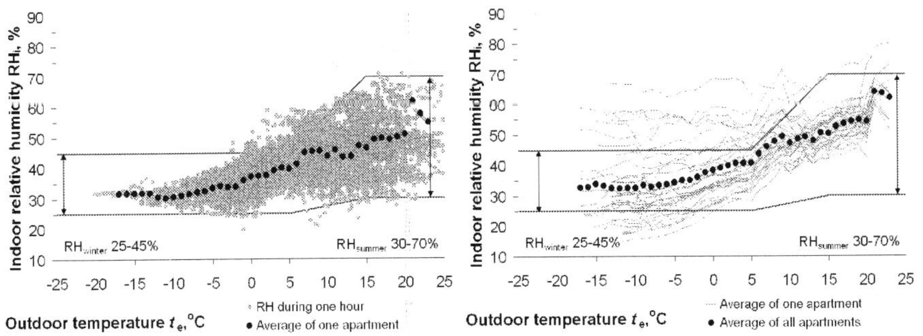

Fig. 9. Dependence of the indoor RH on the outdoor temperature in buildings with correct (left) and incorrect (right) supply water temperature control curve

5.2 Indoor temperature and humidity conditions during the winter season

Fig. 10 shows the indoor temperature (left) and the RH (right) in three sample apartments. All the temperature and the RH measurement results in the apartment buildings measured during the winter seasons are shown in Fig. 11. Each curve represents one apartment measured. Dotted curves show the average of all the rooms. Indoor climate during the winter season was analyzed based on the measurements during three winter months. The average indoor temperature during the winter season from all the apartments was +21.3 °C (min. average being +16.3 °C and max. average +25.8 °C) and the average indoor RH was 37 % (min. average being 23 % and max. average 65 %). Even the generally average temperature was between +19 and +22°C, larger variations in temperatures (Fig. 11 left) show problems related to the control of the heating system. The large deviation of the RH (Fig. 11 right), in addition to temperature variations, indicates that problems exist in the performance of the ventilation system.

The temperature in the apartments on the top floor was approximately one degree lower, (Fig. 12 left) and the RH was 10 % higher, (Fig. 12 right), than in the apartments on the lower floors. The significantly higher (P=0.03) RH on the upper floors is caused by higher humidity loads (lower air change rate due to smaller stack height). The type of the window had an insignificant influence on the room temperature and humidity.

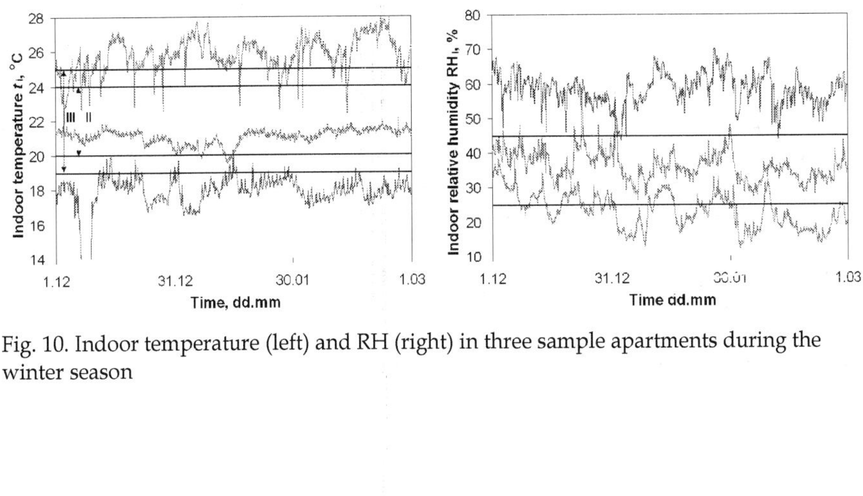

Fig. 10. Indoor temperature (left) and RH (right) in three sample apartments during the winter season

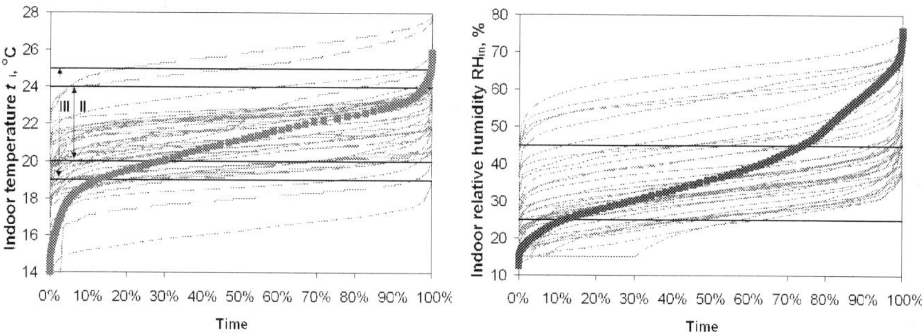

Fig. 11. The distribution of the indoor temperature (left) and the RH (right) in the winter season in Estonian old multi-storey apartment buildings composed of prefabricated concrete elements

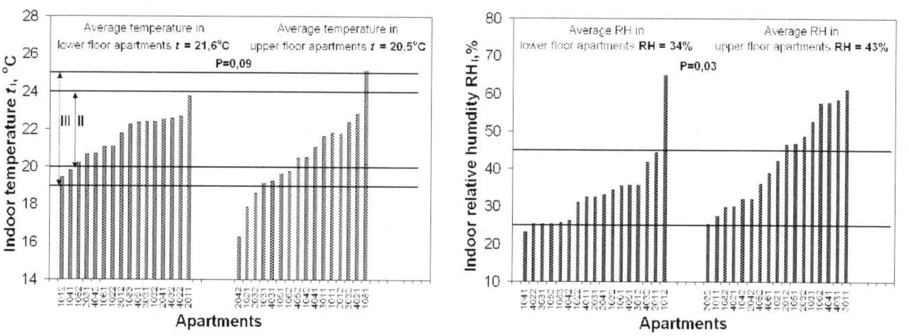

Fig. 12. The comparison of the average indoor temperature (left) and the RH (right) between the upper and lower floor apartments during the winter season

5.3 Indoor temperature and humidity conditions during the summer season

The average indoor temperature during the summer season of all the apartments was +23.4 °C (min. average being +20.2 °C and max. average +25.2 °C) and the average indoor RH was 52% (min. average being 43 % and max. average 70 %). All the temperature and RH measurement results in the dwellings measured during the summer seasons are shown in Fig. 13. Each curve represents one apartment measured. Dotted curves show the average from all the rooms.

Indoor temperature during summer depends mostly on the direction and properties of the window. Indoor temperature in rooms with windows facing the north was lower than those with windows facing the south, however, the difference was small (0.5 °C) and insignificant (P=0.4).

To assess and compare thermal comfort in apartments the number of hours during the summer season (1 June ... 31 August) in which the fixed air temperature of 27°C is exceeded was calculated for each apartment. In this study the amount of weighted excess hours 150 h during the summer season is used as the criterion of the acceptable thermal comfort. The limit of weighted excess hours 150 h during the summer season was exceeded in 10 % of the apartments.

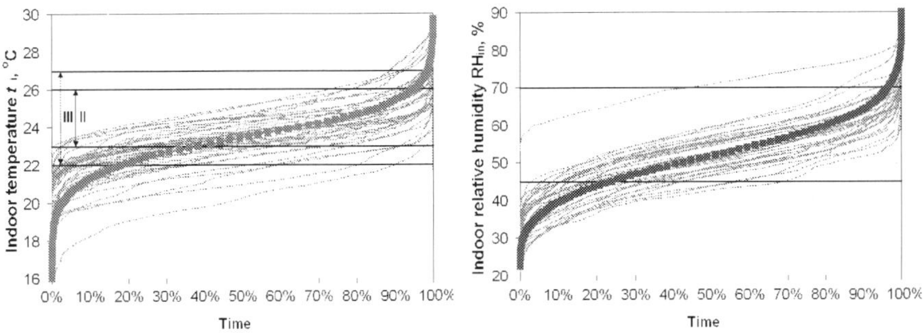

Fig. 13. The distribution of the indoor temperature (left) and the RH (right) in Estonian old multi-storey apartment buildings composed of prefabricated concrete elements during the summer season

5.4 Correspondence of the indoor thermal conditions to the target values

Target values for the room temperature and the RH in the average (II) and the lowest (III) indoor climate categories (CR1752, 1998, EN-15251, 2007) were selected to assess the indoor thermal conditions. Values outside the III category should only be accepted for a limited part of the year. Allowable rate for the target values of the indoor temperatures that may be exceeded was 5% of the measured time length.

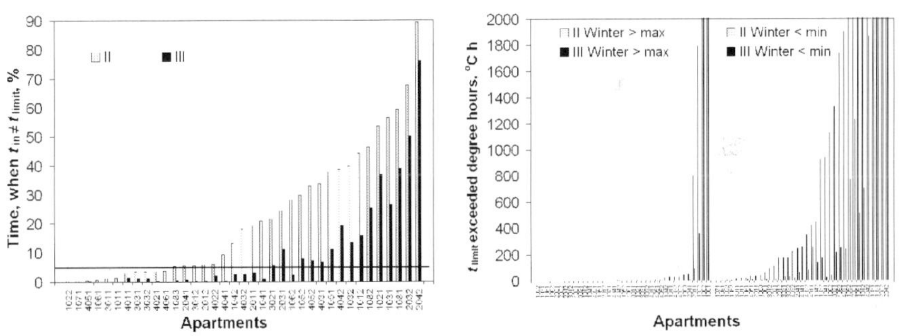

Fig. 14. Percentages of time when the indoor temperature exceeded the target values (left) and temperature target values exceeded degree hours during winter (right)

In 41% of the apartments the room temperature did not correspond to the indoor climate category III criteria allowing 5% excess, Fig. 14 (left). In 41% of the apartments the temperature was not within the target values during the winter season and in 14% of the apartments the temperature was not within the target values during the summer season. The target values of the temperature for the indoor climate category II were exceeded in 70% of the apartments (65% during the winter season and 42% during the summer season). The analysis of the degree hours outside the target values revealed that thermal problems in old multi-storey apartment buildings composed of prefabricated concrete elements relate to low room temperatures. Target values were exceeded at the highest rate there (Fig. 14 right).

6. Air tightness of the building envelope

In the apartments measured, the mean air leakage at the pressure difference of 50 Pa in the entire database was 4.2 $m^3/(h \cdot m^2)$, the minimum being 0.3 $m^3/(h \cdot m^2)$ and the maximum 15 $m^3/(h \cdot m^2)$. The mean air change rate at the pressure difference of 50 Pa from all the databases was 6.0 h^{-1}, the minimum being 0.4 ach and the maximum 24 h^{-1}. The entire database allowed analysing the influence of different external wall structures on the air tightness of the envelopes. The average values of air tightness and their reference values in different subdivisions are shown in Table 1. Older buildings were significantly more non-hermetic (Fig. 15). With these data it is possible to calculate the air infiltration rate and the results can be used for analysing energy performance.

Air leakage rate, q_{50}, $m^3/(h \cdot m^2)$		Air change @50Pa, n_{50}, h^{-1}	
Average	Reference value	Average	Reference value
4.2	4.7	6.0	6.8

Table 1. Results of air tightness measurements in apartment buildings composed of prefabricated concrete elements

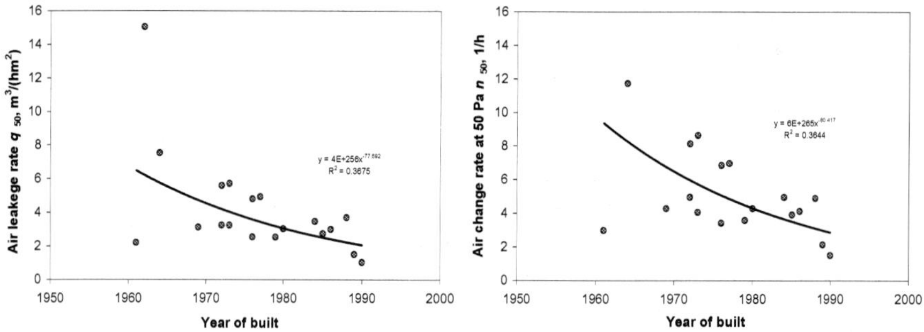

Fig. 15. Comparison of air tightness of building envelope depending on the age of the building

7. Principles of renovation for improving the indoor climate and energy efficiency

The overall living standards in the older part of the housing stock need to be improved to meet today's requirements in functional, urban design, architectural (visual, planning) and constructional-technical terms. Buildings are in an inadequate condition due to low requirements for energy performance in old building standards, a historical lack of attention to quality in construction materials and practices, poor record of operation, maintenance and regular renovation.

The needs for renovation can be viewed from the following aspects:

- Urgent repairs to guarantee safety of buildings (mechanical resistance and stability; safety in case of fire safety in use);
- Improvement of indoor climate (hygiene and health aspects);
- Improvement of energy performance of buildings and HVAC systems;
- Improvement of architectural planning, visual quality, overall living quality, and additional comfort.

In the process of renovation the indoor climate and energy performance of buildings in cold climate the following three components must be considered.

- Performance of ventilation;
- Hygrothermal performance of building envelope;
- Performance of heating systems.

7.1 Renovation of ventilation

In the current situation the renovation of the ventilation is inevitable. To improve the air change in old apartment buildings, it is necessary to use either central or more flexible individual mechanical ventilation. Typically mechanical exhaust ventilation is the easiest solution to implement. Challenges here are how to guarantee the thermal comfort during winter (one possibility is to combine fresh air inlets with radiators. It is possible to improve energy performance with heat pump (to heat up the domestic hot water). The mechanical exhaust ventilation solution is not suitable for all types of buildings (especially for buildings with a combined stack).

Balanced ventilation with room units is another possible solution to renovate the ventilation in old apartment buildings. Where to put room units (a little space); how to solve problems with sound pressure levels; where to put air channels (room height 2,5m); how to solve air flow in an apartment through existing doors are questions that need to be solved during the design process.

In addition to technical questions, also social/human questions need to be solved: occupants have different possibilities and motivations to pay the cost of renovation. Typically occupants do not accept any additional ventilation channels in their apartment. Understanding of the importance of ventilation is very poor.

In the following we will analyze possible solutions of the renovation of ventilation. They can be divided into:

- ventilation system without heat recovery,
- ventilation system with heat recovery.

Principal renovation solutions are proposed for a three-room apartment.

7.1.1 Ventilation system with heat recovery

Here are three main renovation solutions for the ventilation system:

1. Supply-exhaust air handling units (AHU) with heat recovery, Fig. 16 (left). Guarantees very good indoor climate if the maintenance of the AHU is good. Expenses on heating the air are small. Expenses on the maintenance and the consumption of electricity are bigger. The fitting costs are the biggest, about 4000 €. To avoid noise problems it needs careful installation and a good choice of the equipment. The height of standard

apartments being small (about 2.5 m), the owners of the apartments are against fitting the ventilation channels on the ceiling of rooms. Annual changing of filters is necessary.

2. Solution with room supply-exhaust AHU with heat recovery and with exhaust ventilators in WC, bathroom and kitchen, Fig. 16 (right). Good indoor climate is guaranteed. The efficiency depends on the control of exhaust ventilators. Compared with the previous solution the fitting costs are 2...3 times smaller. Needs annual changing of filters. In using effective equipment the consumption of electricity is small. Expenses on air heating are slightly bigger than in the previous solution.
3. Solution based on the exhaust air heat pump of the ventilation, Fig.17. Fresh air is sent to rooms through fresh air valves. A mechanical exhaust ventilation system together with cooling batteries to draw the heat of the exhaust air is built. The general working process of the system based on exhaust air heat pumps is the following:
   - exhaust air is removed from the kitchen and sanitary rooms of apartments,
   - a ventilator and a cooling battery are placed on the exhaust air channel; in the cooling battery the exhaust air cools down from 21°C to 7°C,
   - the water-glycol solution is directed from the cooling battery into the evaporator of the heat pump. With the help of the heat pump the temperature of the cooling agent is raised to 55°C. With the heat produced by the heat pump the DHW is heated and in low-temperature heating systems the temperature of the water is raised.

The efficiency of the system depends on the level of the initial air change in apartments and the temperature curve of the heating system as the temperature of the water leaving the heat pump is relatively low. Both the solution based on heat pump and the one with room supply-exhaust AHU may require additional heating coils in apartments in case this solution has not been taken into account in designing the system.

7.1.2 Solutions without heat recovery

Even nowadays the ventilation system without heat recovery is not preferable. In some cases the ventilation system is renovated without heat recovery (to reduce investment costs). The most important thing is to provide healthy living conditions for occupants. This requires minimum ventilation airflows in the apartment. Below we present main solutions for renovating the ventilation system without heat recovery.

1. Increasing the efficiency of natural ventilation by fitting fresh air valves into bedrooms and living-rooms (e.g. in the case of renovated windows). At least satisfactory air quality is achieved, but expenses on heating the ventilation air increase several times. It is important to find a suitable place, as, for example, behind the radiator or between the heating coil and the window sill. The cost of the renovation is small.
2. Mounting fresh air inlets in living- and bedrooms and exhaust air ventilators into WC, bathroom and kitchen. This solution guarantees acceptable indoor climate and control of the flow rates of the exhaust air. Owing to the control of the air change it is possible to decrease the costs on ventilation. The cost of the renovation is relatively small.
3. Fitting exhaust air ventilators above ventilation channels on the roof. This guarantees good indoor climate. The costs on heating the ventilation air are big (in new apartment buildings it makes up about 50% of the heating costs. In old apartment buildings in Estonia it has been used in single cases.

Fig. 16. Balanced ventilation solution with apartment AHU with heat recovery (left) and one room supply-exhaust AHU with heat recovery (right)

7.2 Renovation of the heating system

As radiators were not equipped with thermostats, it was not possible to control room temperature. Room temperature was controlled in the heat substation based on the outdoor temperature. If the slope and the level of the control curve of the temperature of the supply water are correct, room temperature does not depend strongly on the outdoor temperature, Fig. 8 left. Other problems concerning the heating system are:

- Incorrect water flow rate of the heating system or risers,
- Lack of direct room temperature control,
- Lack of maintenance and improper modifications of the heating systems.

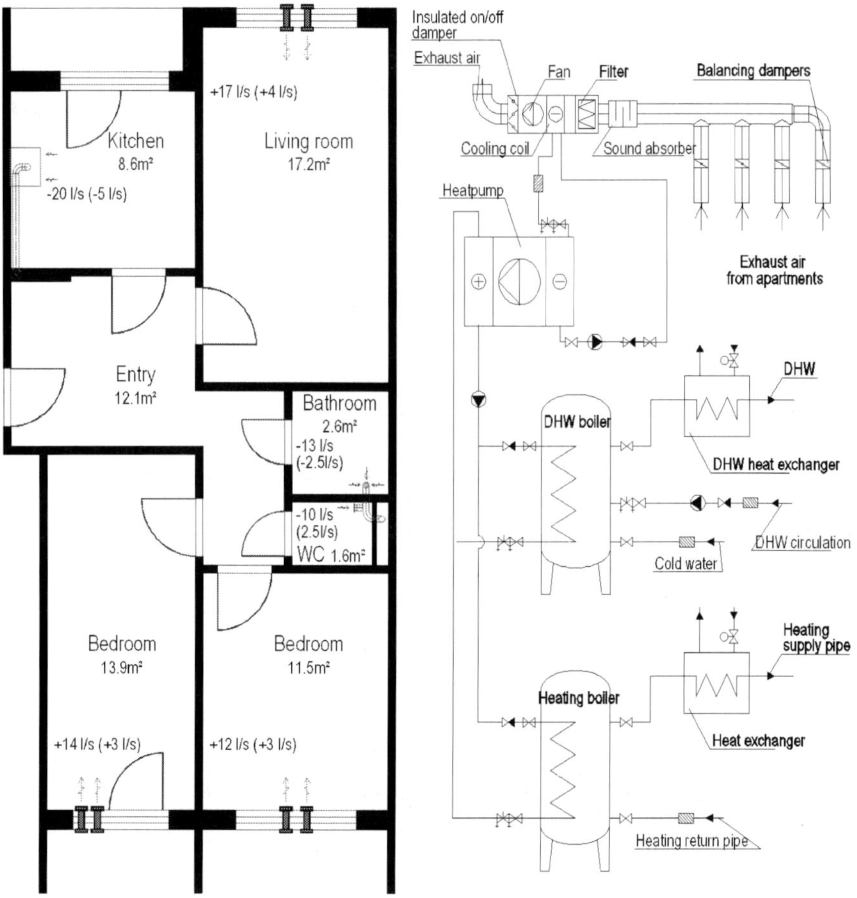

Fig. 17. Ventilation with exhaust air heat pump: solution in three room apartment (left) and the principal system (right)

To renovate the existing one-pipe heating systems it is necessary to find out the condition of the system, i.e. in which state are the heating coils and the pipes. This determines the solution of renovation:

- in case the heating coils and the pipes have exhausted their resources the heating system should be renovated completely, i.e. a new two-pipe heating system should be built,
- should the heating coils and the pipes be in a good condition (e.g. used are cast iron section radiators and the steel pipes still have a 15-20 year long lifetime) the connection unit of the heating coils of the one-pipe heating system should be renovated by making it controllable, Fig. 18.

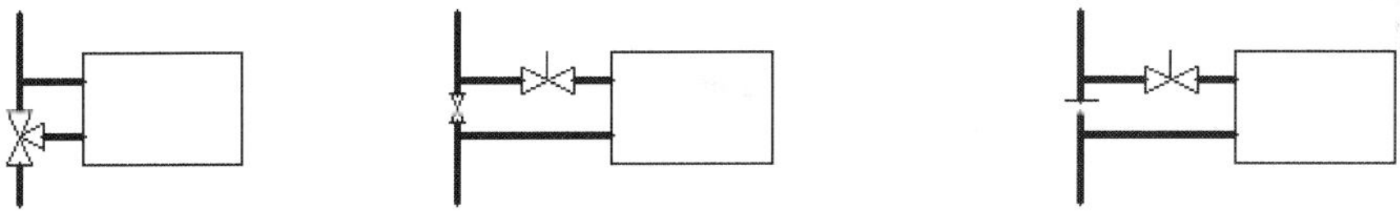

(a) (b) (c)

Fig. 18. Variants for making the connection unit of the heating coils of the 1-pipe heating system controllable:
a- with an adjustable 3-way thermostat valve
b- with a 2-way thermostat valve and an adjustable valve on bypass
c- with a 2-way thermostat valve and a drossel on bypass

7.3 Renovation of building envelope

Serious thermal bridges are a large problem in old apartment buildings composed of prefabricated concrete panels (Ilomets et. al., 2011). Mould growth and surface condensation on the internal surfaces of thermal bridges are unavoidable without additional external insulation and/or lowering internal humidity loads. Due to low frost resistance and carbonization of facades, it is necessary to protect them. It is economically viable to make the additional thermal insulation for walls and roofs.
Typical thickness of additional insulation is 10...15 cm for walls and 15...20 cm for roof, made from expanded polystyrene of mineral wool. During insulation works windows remain in their original place. Therefore the thermal bridge connects the window and the insulated external wall (Fig. 19, up) and windows look like embrasure if windows stay in the original place (Fig. 19, down).

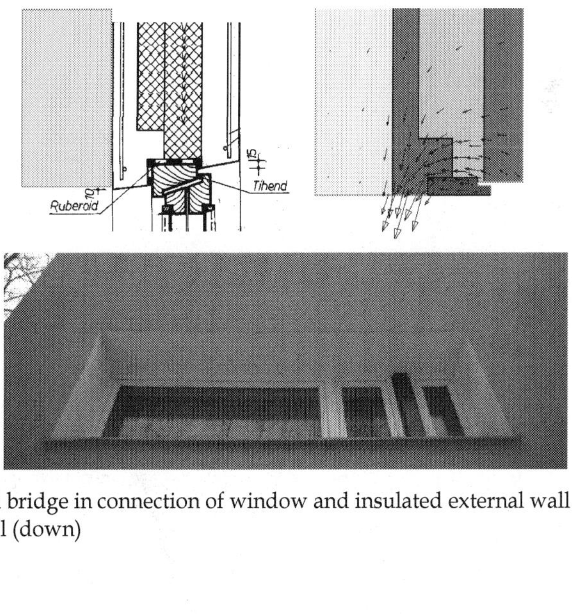

Fig. 19. Thermal bridge in connection of window and insulated external wall (up). Window in insulated wall (down)

Old walls are not flat. To get the final surface flat, air space may become between the original wall and additional insulation. If the air channel stays open from bottom and top, air flow through it can increase the thermal transmittance of the wall Fig. 20.

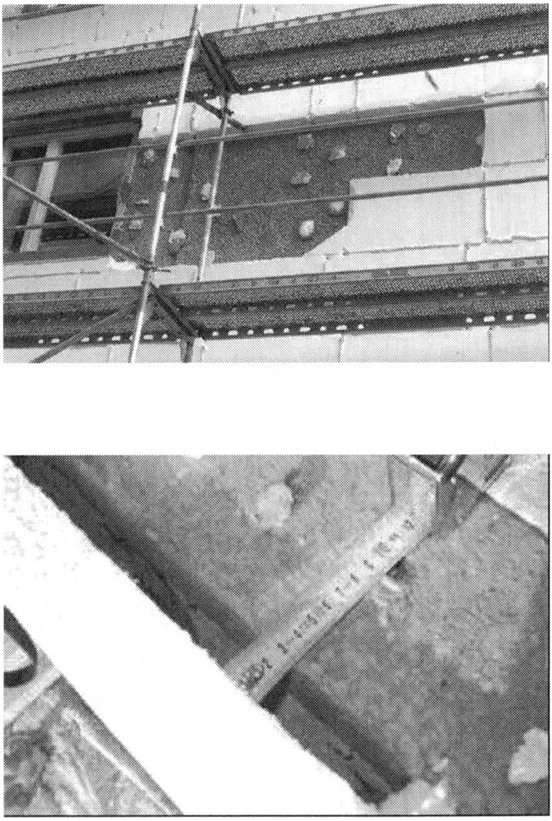

Fig. 20. Air space between the original wall and the additional insulation can cause air flow through it and the thermal transmittance of the wall may increase

Closed loggias generally stay without additional insulation. In the worst case this can cause extensive mould growth on the inner surface of the loggia's walls, Fig. 21.
According to the Estonian renovation practice, renovation works as a rule start in spring and finish in late autumn. The autumn is the typically rainy season in Estonia. The original wall may become wet before insulation works. Drying out moisture may cause much stronger hygrothermal loads than moisture diffusion (or convection). Drying out moisture can cause damage of the finishing layer of the external walls, Fig. 22.

Fig. 21. Closed loggia without additional insulation on walls (left). Mould on the inner surface of the loggia's walls before (left) and after (right) the renovation

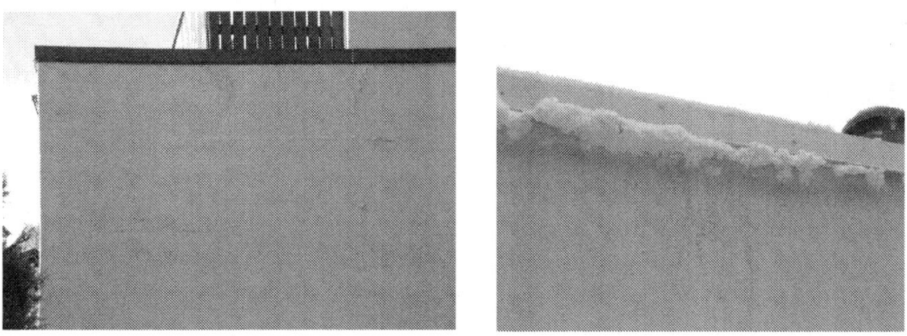

Fig. 22. Deterioration of finishing layer of exterior insulation finishing system due to drying out moisture

Complex renovation (ventilation + thermal envelope + heating systems) is not common in typical renovation practice. The main reasons for this are the lack of knowledge and higher momentary price. Non-complex renovation may result in the poor quality of indoor air and the expected energy savings remain smaller.

8. Discussion and conclusion

This study analyzed the indoor temperature and humidity conditions measured in field conditions in Estonian old multi-storey apartment buildings composed of prefabricated concrete elements, based on field measurements in 39 apartments in 20 buildings.

During the past 35 years immense changes have taken place in the heat energy consumption in typical apartment buildings. The average heat energy consumption in typical apartment buildings has decreased from about 370 kWh/m² to about 200 kWh/m² (per gross area of apartments). About 115 kWh/m² of this decrease was due to a decrease in DHW consumption and the rest due to a decrease in energy used for heating. The latter has decreased owing to the renovation of heat substations (boiler houses) and also to a partial renovation of the building envelope.

Despite nearly sufficient average temperature and humidity in the apartments studied, deviations in measurement results were high, indicating problems with heating system control and ventilation system performance. Correspondence of the indoor thermal conditions to the target values of the indoor standard was low. In 41% of the apartments the room temperature did not correspond to the indoor climate category III criteria. Target values were exceeded at the highest rate during the winter (heating) period. The main reasons for deviations from the target values are:

- incorrect control curve of the temperature of the supply water of the heating system,
- incorrect water flow rate of the heating system or risers,
- lack of direct room temperature control,
- difficulties in regulating heat output of heating coils in the one-pipe heat distribution system,
- poor maintenance and improper modifications of the heating and ventilation systems.

As the amount of weighted excess hours during the summer season was not high and the outdoor air temperature during the summer of 2008 corresponded well to the long-term temperature, we may conclude that overheating during the summer season is not a critical problem in old multi-storey apartment buildings composed of prefabricated concrete elements.

This study supports the results of an earlier study (Kalamees, 2006) that the heating season would change to the summer in Estonian dwellings at this +15 °C... +10 °C average daily outdoor temperature point. This boundary allows the target values of indoor climate conditions to be determined in buildings for winter (heating) and summer (cooling) seasons.

Comparison of the findings of this research into indoor climate in Estonian apartment buildings composed of prefabricated concrete elements with those of other studies in cold climate (Gustavsson, Bornehag & Samuelsson, 2004, Jenssen, Geving & Johnsen, 2002, Norlén & Andersson, 1993, Ruotsalainen & Säteri, 1992, Vinha, Korpi & Kalamees, 2009) may indicate that the indoor climate in Estonian apartment buildings composed of concrete elements is characterized by a larger deviation and lower temperatures and during the heating season by a larger deviation and a higher RH.

Indoor humidity load in the apartments studied was high: the moisture excess (the difference between indoor and outdoor air humidity by volume) at the 10% critical level was close to +7 g/m^3 during the cold period. This causes potentially serious moisture problems for the building envelope and health problems due to mould growth on the surfaces of the building envelope. The high indoor humidity loads were caused mainly due to the lack of ventilation and high occupancy. High indoor humidity loads are a major concern because of serious thermal bridges in the building envelope (Norlén & Andersson, 1993). Renovation of old apartment buildings composed of prefabricated concrete elements by:

- improving thermal insulation of the building envelope (roof and walls) (additional thermal insulation),
- renovation and rebalancing of heating systems with thermostats fitted to the radiators,
- improvement of ventilation (preferably supply-exhaust ventilation with heat recovery) is unavoidable to provide the healthy indoor environment for the occupants. As all these activities also improve the energy efficiency, the renovation is doubly advantageous.

Presented are recommendations for renovating ventilation in typical apartment buildings. They can be classified into systems without exhaust air heat recovery and with exhaust air heat recovery. Considering the prices of energy today the latter systems are more suitable: e.g. solutions based on exhaust air heat pump or apartment AHU.

9. Acknowledgment

The study has used the measuring data of the national research project "Technical condition and service life of Estonian apartment buildings composed of prefabricated concrete elements", which was carried out by the Department of Civil Engineering at Tallinn University of Technology. The financial support of the Credit and Export Guarantee Fund KredEx and Tallinn University of Technology are acknowledged.

10. References

CR 1752. (1998). Ventilation for buildings: Design criteria for the indoor environment. CEN.

EN-15251. (2007). Indoor environmental input parameters for design and assessment of energy performance of buildings- addressing indoor air quality, thermal environment, lighting and acoustics. CEN.

EVS–EN 13829. (2001). Thermal performance of buildings. Determination of air permeability of buildings. CEN.

Gustavsson,T., Bornehag, C-G. & Samuelsson I. (2004). Temperature, relative humidity and air exchange rate in 390 dwellings. *CIB W40 meeting, Glasgow, UK.*

Jenssen, J A., Geving, S. & Johnsen, R. (2002). Assessments on indoor air humidity in four different types of dwelling randomly selected in Trondhein, Norway. *Proceedings of the 6th Symposium on Building Physics in the Nordic Countries,* June 17–19, Trondheim, Norway.

Kalamees T., Õiger K. & Koiv T.-A.(2009). *Technical condition and service life of Estonian apartment buildings, built with prefabricated concrete elements.* Tallinn University of Technology (in Estonian), ISBN 978-9985-59-947-1.

Kalamees T. (2006). Indoor Climate Conditions and Ventilation Performance in Estonian Lightweight Detached Houses. *Indoor and Built Environment*, (15, 6), 555 - 569, ISSN 1423-0070.

Koiv, T.-A., Voll, H. & Hani, A. (2010). Domestic hot water consumption in educational premises, apartment and office buildings. *WSEAS TRANSACTIONS on Environment and Development*, Issue 1, vol 6, 2010, 54-63, ISSN 1790-5079.

Koiv, T.-A., Toode, A. & Hani, A. (2009). The influence of domestic hot water maximum consumption on the district heating network dimensioning. *WSEAS TRANSACTIONS on Environment and Development*, 1, vol 5, 2009, 104-108, ISSN 1790-5079.

Koiv, T.-A., Voll, H., Mikola, A., Kuusk, K. & Maivel, M. (2010). Indoor climate and energy consumption in residential buildings in Estonian climatic conditions. *WSEAS TRANSACTIONS on Environment and Development*, Issue 4, vol 6, 2010, 247-256, ISSN 1790-5079.

Norlén, U. & Andersson, K. (1993). The indoor climate in the Swedish housing stock. *Swedish Council for Building Research*, D10:1993. Stockholm, Sweden, ISBN 9 1-540-5569-5.

Ilomets, S., Paap, L. & Kalamees T. (2011). Evaluation of the thermal bridges of prefabricated concrete large-panel and brick apartment buildings in Estonia. *Proceeding of the 11th Symposium on Building Physics in the Nordic Countries*, 29 May - 2 June 2011, Tampere, Finland.

Ruotsalainen, R. & Säteri. (1992). Indoor Climate and the Performance of Ventilation in Finnish Residences. *Indoor Air*, No 2, 1992, 137-145, ISSN 0905-6947.

Vinha, J., Korpi, M. & Kalamees, T. (2009). *Air tightness, indoor climate and energy performance of dwellings*. Report 140. Tampere University of Technology (in Finnish), ISSN 978-952-15-2105-8, ISBN 1797-9161.

23

Air Quality in Rural Areas

J. P. Majra
Yenepoya University
India

1. Introduction

The immediate environment of man comprises of air on which depends all forms of life. Human beings need a continuous supply of air to exist. The requirement for air is relatively constant about 10-20m^3 per day (Park 2009). Air is a mechanical mixture of gases. The normal composition of external air by volume is approximately as follows Nitrogen 78.1%, Oxygen 20.93%, and Carbon dioxide 0.03%. The balance is made up of other gases which occur in traces e.g. argon, neon, krypton, xenon and helium. In addition to these gases, air also contains water vapor, traces of ammonia and suspended particulate matter such as dust bacteria, spores and vegetable debris. The external air has other compounds and elements along with the already mentioned. Air is rendered impure by a) respiration of man and animals b) decomposition of organic matter, c) combustion of coal, gas, oil etc. d) trade, traffic and manufacturing processes that give off dust, fumes vapors and gases.

Under ordinary conditions the composition of outdoor air is remarkably constant. This brought about by certain self cleansing mechanisms which operate in nature such as wind, sunlight, rain and plant life. Wind dilutes and sweeps away the impurities by its movements so that the impurities do not accumulate in any one place. Sunlight and heat oxidizes the impurities and kill the bacteria. Rain removes the suspended and gaseous impurities and the green plants utilize the carbon dioxide and generate oxygen. But when the rate of pollution becomes too high or the cleansing process becomes ineffective, the air gets polluted and it constitutes a health hazard.

Human occupancy and activity vitiate the air in the occupied rooms and gives a sense of discomfort to the occupants. Unless the vitiated in door air is replaced by the fresh air, it may adversely effect the comfort, health and the efficiency of the occupants. They may feel suffocation and complain of headache, drowsiness and inability to concentrate. There is also a risk of droplet infection and lowered resistance to diseases on prolonged exposure.

Thus the phenomenon called pollution is an inseparable consequence of the presence of man and his activities. The term air pollution signifies the presence in the ambient atmosphere of substances (e.g., gases, mixtures of gases and particulate matter) generated by the activities of man in concentrations that interfere with human health, safety or comfort, or injurious to vegetation and animals and other environmental media resulting in chemicals entering in the food chain or being present in drinking water and thereby constituting additional source of human exposure. The direct effect of air pollutants on plants, animals and soil can influence the structure and function of ecosystems, including self regulation ability, thereby affecting the quality of life (WHO 1987).

In the past, air pollution meant smoke pollution and it was limited to the urban areas. Today, air pollution has become more subtle and recognizes no geographic or political boundaries. The air pollution is one of the present day health problems throughout the world. The objective of this chapter is to highlight the public health importance of the so far neglected issue of quality of air in rural areas and to attract the attention of scientific community for further research on the subject.

2. Air quality in rural areas

Off late researchers, policy makers and governments has focused their attention on air quality in the urban areas only. Air quality in rural areas remains a neglected issue so far. The common belief is that rural areas are free from air pollution. On the contrary air quality in the rural areas all over the world and particularly in the developing countries may be more polluted than some of the urban areas. Rural areas suffer from out door air pollution as well as indoor air pollution. Major sources of out door air are indiscriminate use of insecticides/pesticides sprays and burning of wheat and paddy straw. This leads various health problems mainly affecting the respiratory and cardio-vascular system. In door air pollution exposes more people worldwide to health risks than out door air pollution. Air in big cities like Delhi in India and Xian in China, contains a daily average of $500/m^3$ of total suspended particulate, whereas, smoky houses in Nepal have peak level of $10000/m^3$ or more (World Bank 1993). Rural people in developing countries may receive as much as two-thirds of the global exposure to particulates. BaP the best known Poly-nuclear Aromatic Hydrocarbons (PAH) can increase to very high concentrations in door. BaP level of 6u/m3 were found in houses without chimneys in Southern China. In India BaP exposure averaged about 4u/m3 during cooking with biomass fuels (Smith et al. 1983). Owing to its carcinogenicity no safe level of PAH can be recommended. This contributes to the acute respiratory infection in the young children, chronic lung disease and cancer in adults and adverse pregnancy out comes in for pregnant women exposed during pregnancy. The third source of pollution affecting rural areas results from the transport of emissions far from their primary emission point—even continents away—and their conversion into health hazards. Some emissions such as hydrocarbons in particular, that are themselves relatively harmless, are converted to hazardous ones by sunlight and interactions with other pollutants.

3. Major air pollutants in rural areas

More than 100 substances which pollute air have been identified. They may be in the form of solids, liquids (vapors) or gases. They differ greatly from place to place depending upon the specific complex of contaminant source. They may be from natural sources or from anthropogenic sources (human activity). Whilst man-made pollution and poor air quality is major environmental concern, there are many natural sources of air pollution which are often much greater than their man-made counterparts'. Some of the important pollutants are given below.

3.1 Sulphur dioxide

It one of the several forms in which sulphur exists in the air. The main source of sulfur dioxide is the combustion of fuels containing sulfur. Fossil fuels, most notably coal and oil, contain varying amounts of sulfur according to their source but typically between 1% and 5%. On combustion, the sulfur in the fuel is converted almost quantitatively to sulfur dioxide. Sulfur is most abundant in the less volatile fractions of crude oil and hence shipping, which burns residual fuel oil, can be a very high emitter of sulfur dioxide. The sintering process used in metal smelting, which involves roasting metal sulfide ores in a stream of air, can also be a major mechanism of sulfur dioxide production. In less developed countries, however, unabated burning of coal and the use of fuel oils and automotive diesel with higher sulfur content are major sources of sulfur dioxide. Natural sources of sulphur dioxide include release from volcanoes, biological decay and forest fires. Actual amounts released from natural sources in the world are difficult to quantify. In 1983 the United Nations Environment Program estimated a figure of between 80 million and 288 million tonnes of sulphur oxides per year from natural sources compared to around 69 million tonnes from human sources world-wide. Further oxidation of SO_2, usually in the presence of a catalyst such as NO_2, forms H_2SO_4, and thus acid rain (WHO 2005).

3.2 Nitrogen oxides

In a process parallel to that of sulfur dioxide production during fuel combustion, nitrogen in fuels is converted to oxides of nitrogen in the combustion process (WHO 2010). Coal is the most important fuel in this context, as oil and gas contain much lower levels of nitrogen. However, there is a further process in which atmospheric nitrogen and oxygen are combined during high-temperature combustion to form oxides of nitrogen. This occurs in all high-temperature combustion processes and explains why road traffic and electricity generation tend to be among the predominant sources of these gases. Natural sources of nitrogen oxides include volcanoes, oceans, biological decay and lightning strikes. Estimates range between 20 million and 90 million tonnes per year nitrogen oxides released from natural sources compared to around 24 million tonnes from human sources worldwide.

3.3 Carbon monoxide

Carbon monoxide is one of the most common and widely distributed air pollutants. This is a gas formed during the incomplete combustion of carbon-containing fuels. While complete combustion leads to the formation of carbon dioxide, most combustion systems involve some fuel-rich regions in which a proportion of carbon is oxidized only to carbon monoxide. The most important example is the combustion of petrol in road vehicles. Estimates of man made carbon monoxide emission vary from 350 to 600 million tonnes per annum (WHO 1987). Some wide spread natural non-biological and biological sources have also been identified.

3.4 Ozone

Ozone is one of the strongest oxidizing agents. There are no significant anthropogenic emissions of ozone in the atmosphere. It is a secondary photochemical pollutant formed near ground level as a result of chemical reactions taking place in sunlight (Park 2009). About 10 to 15% of low level ozone, however, is transported from the upper atmosphere (called the stratosphere), where it is formed by the action of ultraviolet (UV) radiation on oxygen (the ozone layer). Atmospheric concentration of ozone has been observed higher in rural areas as compared to urban areas.

3.5 Carbonaceous particles

The particles emitted from burning fossil fuels and biomass, for example in diesel and petrol engines, are typically composed largely of carbon, both in the elemental form and as organic compounds of low volatility (WHO 2005). The elemental carbon is in the form of microcrystalline graphite formed from the build-up of carbon-containing free radicals into polycyclic aromatic structures within the flame. When these build only relatively small molecules they are emitted as polycyclic aromatic hydrocarbons, an important pollutant in their own right often associated with airborne particles (Smith et al. 1998). If larger graphite structures are created in the flame, these will be emitted as particles of elemental carbon. Such combustion systems also tend to emit hydrocarbons of low volatility, deriving for example from lubricating oils, and these will typically condense on to the carbon particles. Carbon present within organic compounds, rather than as elemental carbon, is referred to as organic carbon.

3.6 Non-carbonaceous primary particles

An important source of non-carbonaceous particles is fly ash, which comprises particles largely of mineral material freed from a fuel such as coal in a combustion source and carried into the atmosphere with the flue gases (WHO 2005). Purely mechanical processes such as quarrying can also create fragments of rock small enough to become suspended in the atmosphere and, as mentioned above, the action of the wind can suspend particles of soil and dust from land surfaces into the atmosphere. Construction and demolition activity can be an important source of coarser particles (Charron & Harrison, 2005), even in a street canyon where traffic emissions typically dominate. Natural sources of particulate matter are less important than man-made sources. These include volcanoes and dust storms. However, such sources do account for intense high particulate pollution episodes, occurring over relatively short times scales. It is not unknown for Saharan dust to be deposited in the UK after being blown thousands of miles.

3.7 Volatile organic compounds (VOCs)

VOCs comprise a very wide range of hydrocarbons, oxygenates, halogenates and other carbon compounds existing in the atmosphere in the vapour phase (WHO 2005). The predominant source is typically through leakage from pressurized systems (e.g. natural gas, methane) or evaporation of a liquid fuel such as benzene from the fuel tank of a vehicle. However, combustion of fossil fuels and incineration processes also give rise to combustion emissions containing some unburned or partially burned fuel fragments that are emitted in the form of VOC. Organic solvents, used for example in paints and adhesives, are designed to disperse in the atmosphere to allow the active ingredients to dry. VOCs are also naturally produced by plants and trees. Isoprene is a common VOC emitted by vegetation, and some believe it to be a more significant trigger for asthma, another allergic reactions than man-made irritants. Plant, grass and trees are also a source of pollen, which can act as triggers in some asthmatics. Pollen is in the air year round, but the concentration is highest during the growing season, from March to the first frosts in autumn.

4. Causes of air pollution in rural areas

Rural areas suffer from air pollution caused by both natural as well as because of human activities. Air pollution from Natural causes occurs when contaminants drawn from animals, plants and land resources get disseminated in the atmosphere in normal course. Natural pollution also results from forest fire, coal fires, volcanic eruptions, dust storms and sand storms. Natural contaminants include spores, pollens, moulds, fur, feather, hair, dander, dust, grit and other types of particulate matter. Human activities like burning solid fuel indoors and tobacco smoking apart from agriculture, mining, coal processing and cement making produce high levels of pollution.

Meteorological factors also play an important role in quality of air in rural areas. The level of atmospheric pollution at any one time depends upon meteorological factors, e.g., topography, air movements and climate. Winds help in the dispersal and dilution of pollutants. If the topography is dominated by mountains or tall buildings the winds become weak and calm and the pollutants tend to concentrate in the breathing zone. The vertical diffusion of the pollutants depends upon the temperature gradient. When there is rapid cooling of the lower layers of air (temperature inversion), there is little vertical movement of the air and the pollutants and water vapors tends remain trapped at the levels and the result is 'smog'. The temperature inversion which is more frequent during the winter months than in the spring or the summer is a threat to human health.

5. Sources of air pollution in rural areas

5.1 Natural sources

5.1.1 Air pollution from coal fires

Underground coal fires are consuming 20 to 30 million tons of coal a year, pumping tons of ash, carbon dioxide, methane, carbon monoxide and sulfur compounds into the atmosphere (Jeffrey Hays 2008). Coal produces thick, smoggy smoke. High sulphur coal is particularly nasty. It produces a rotten egg smell. Some of the fires have been burning for centuries. By one count there are 56 underground coals fires currently burning in China (Figure 1). The Coal fires produce as much carbon monoxide each year as all the cars in the United States. The underground coal fires are revealed by fumes and smoke that pour from cracks in the earth. The Wude coal field in Inner Mongolia, one of China's largest coal fields, is the home

Fig. 1. A Coal fire in China

of China's largest coal fire and some argue one of the world's worst environmental disasters. Sixteen of China's coal fires burn here, spewing out acrid clouds of sulfur dioxide. The fires at the Wude field are at a depth of between 110 to 220 feet. They advance about 100 feet a year. Workers try to extinguish the fires by starving them of oxygen by burying them under a 3-foot-layer of dirt. Pouring water on them produces dangerous methane gas, so workers pour a water-clay slurry into cracks instead of water if a dousing strategy is employed. Even when the fires are extinguished the ground can take years to cool down. Only 10 percent of China's coal underground fires are being fought. There have been some successes. In 2003, a centuries-old fire was extinguished near Urumqi after a four year battle. In 2009, a number of coal fires, one of which had been burning for 60 years, were put out in Xinjiang. The fires, which have been caused illegal mining and spontaneous combustion, had spread to more than 900,000 square meters and consumed 10 million tons of coal a year. The fires were put out through a coordinated plan of drilling, water injection and using earth to cut off oxygen.

5.1.2 Air pollution from volcanoes

One of the most important natural causes of air pollution is volcanic eruption. The most problematic gases emitted in a volcanic eruption include sulfur dioxide, carbon dioxide and hydrogen fluoride. In nearby places, sulfur dioxide gas can cause acid rain in nearby places and air pollution in downwind areas from the volcanic site. Besides the volcanic gases, there is also volcanic ash. Volcanic ash can move hundreds to thousands of miles downwind from a volcano. Fresh volcanic ash is gritty, rough, and at times corrosive. The ash can cause respiratory problems for young children, the elderly or those already with respiratory ailments. Volcanic eruptions can generate so much polluting gases and ash into the air that the sun's rays could be blocked, and land temperature in the affected area lowered, as with the Mount Pinatubo eruption in 1991. Kilauea Volcano on the Island of Hawai`i emits about 2,000 tons of sulfur dioxide (SO_2) gas each day during periods of sustained eruption (Figure 2).

Fig. 2. Kilauea Volcano on the Island of Hawai`i

Air pollution caused by sulfur dioxide and other volcanic gases became a frequent problem on the island in mid-1986, when the volcano's ongoing eruption, which began in 1983, changed from episodes of spectacular lava fountaining (shown here) to a nearly constant but quiet outflow of lava and gas (USGS Fact Sheet 2000).
People in areas downwind of the volcano began reporting a wide range of problems, including reduced visibility, health complaints, and damage to crops. The word "vog," an abbreviation for volcanic smog, was coined to identify this form of air pollution, which unfortunately has become a part of everyday life for people in Hawai`i. Vog is a visible haze consisting of gas plus a suspended mixture of tiny liquid and solid particles, called aerosol. The aerosol in vog is composed primarily of sulfuric acid and other sulfate compounds. Small amounts of several toxic metals, including selenium, mercury, arsenic, and iridium, have also been found in the volcanic air pollution coming from Kilauea.

Fig. 3. "Laze" another form of volcanic air pollution

Molten lava from Kilauea Volcano frequently flows through underground lava tubes to reach the Pacific Ocean, where it vigorously reacts with cold seawater to create large steam plumes laden with hydrochloric acid (USGS Fact Sheet 2000). These plumes, known as "laze", are another form of volcanic air pollution (Figure 3) and pose a local environmental hazard along the Island of Hawaii's southeast coast, especially to people who visit these ocean-entry sites.

5.1.3 Air pollution from forest fires

Forest fires are not uncommon, in remote regions, fires are often started by lightning, but in more populated areas people are the main cause. Following are some of the facts (Vegetation Fires Fact Sheet 2000)

- Fires in tropical monsoon forests recur every 1 to 3 years.
- In North America and Eurasia, between 5 and 20 million hectares (ha) of forest are consumed by uncontrolled fires every year.
- In tropical savannahs, it is estimated that over 300 million metric tons of vegetative matter burn annually.

Gas and particle emissions produced as a result of fires in forests and other vegetation impact the composition of the atmosphere. These gases and particles interact with those generated by fossil-fuel combustion or other technological processes, and are major causes of urban air pollution. They also create ambient pollution in rural areas. These fires are also sources of "greenhouse" and reactive gases. All forest fires emit carbon monoxide, sulfur dioxide, nitrogen dioxide, ozone, and particulate matter. Carbon monoxide, which is a poisonous gas, can be emitted in large amounts during forest fires. Particulates, which are mixtures of soot, tars, and volatile organic substances, either solid or liquid, are emitted in large quantities from forest fires. The particulates, which can be smaller than 2.5 micrometers in diameter, if deeply inhaled into the lungs, can damage lung tissues and cause respiratory and cardiovascular problems. Nitrogen oxides are released at temperatures greater than 1,500 degree centigrade. Therefore nitrogen oxides are released in significant quantities only during the most severe fires. Sulfur dioxide emissions are significantly less serious with forest fires, as concentration of the dioxide in most forest fuels is usually less than 0.2 percent. However, forest fire sites with "peat" and "muck soil" may be exceptions.

In 1997-98, forest fires in South-east Asia affected some 200 million people in Brunei Darussalam, Indonesia, Malaysia, the Philippines, Singapore and Thailand. Massive movements of population fleeing the fires and smoke added to the emergency, while the increase in the number of emergency visits to hospitals during the crisis demonstrated its severity (Vegetation Fires Fact Sheet 2000). The comparison of medical data reported during the 1997/1998 forest fire events in South-east Asia with corresponding data in 1995/1996 revealed the following impact of smoke on public health which is consistent with our knowledge of the effects of fine particles.

- The number of cases of pneumonia increased 5-25 times in South-east Kalimantan (Borneo) and 1.5-5 times in South Sumatra.
- The number of outpatient visits with respiratory diseases in Malaysia increased 2 to 3-fold.
- In September 1997 in Jambi (Sumatra), the number of reported cases of upper respiratory tract infections was 50% higher than in the previous month.

Health-related consequences of smoke from forest fires in the Americas have also been documented:

- During the fires in 1997 in Alta Foresta, Brazil, outpatient visits for respiratory disease increased 20-fold.
- During the 1993 California fires, a 40% increase in asthma and a 30% increase in emergency visits for chronic obstructive pulmonary diseases was recorded.

Early warning systems for fire and atmospheric pollution are essential components of fire and smoke management, and are based on space, ground and climate monitoring, as well as modeling. The use of "fire-weather" forecasts and assessment of vegetation dryness may also be included.

5.1.4 Dust/sand storms

Dust storms are a kind of severe natural disaster that frequently occurs in the arid and semiarid regions. Dust storms have local, national and international implications concerning global warming, and land degradation. They also impact human health. When the dust cloud moves downwind it inevitably passes through populated areas, contributing to the air pollution (Figure 4). As the dust settles over a populated area and people breathe in these tiny dust particles, those with asthma and other respiratory disorders will suffer. Dust particles have been shown to cause a wide range of respiratory disorders including chronic

bronchitis and lower respiratory illness. More sinister are the health related problems in areas where the dust is salt laden or is contaminated by toxins (Mattson & Nilsen 1996).
The 5.5 sand-dust storm (Yang Gengsheng 1996) that prevailed in the Hexi Corridor brought suspension dust to Lanzhou (hundreds of Kms further east) where atmospheric air quality was severely polluted and all factory workshops and office buildings had to turn on their lights during the day. Air inside rooms was full of mud smell and irritated the nose. Whitewash dusts floated everywhere. Respiratory diseases were spread. Particularly, the tailings dust exhausted from the metallurgy industry caused heavy metal pollution as these particles were entrained and transported.
Abbas et al. (2007) reported in a study on Impacts of Dust Storms on Air Pollution and Public Health in Sistan region in Iran that these caused air pollution and respiratory diseases have spread in the region. Due to all the villages are located in wind corridors' path and most often exposed to dust storms, therefore the number of patients in villages is more than the city. Also, drought and dust storms caused the agricultural land has transformed into wasteland. Hence, villagers live in complete poverty so they are not able to pay remedy costs. Most respiratory patients visit hospital during the summer season when the dust storms occur. Therefore dust storms have led to widespread damage and loss of life. It seems that the loss of life is more than property damage.

Fig. 4. Air pollution from a dust storm

There are many other air pollution causes like wind erosion, pollen dispersal, evaporation of organic compounds, and natural radioactivity. However, these causes are usually not significant. Although there is no doubt that the impact of natural causes for air pollution can sometimes be drastic, take for example volcanic eruptions, but they are usually not frequent.

5.2 Human sources of air pollution in rural areas

Agricultural activities are the major human source of air pollution in rural areas. Burning of stubble in the field after harvesting, threshing operation, grain dust and large scale use of tractors harvester, combines and diesels operated tube well are major factor contributing to air pollution. Agricultural use of organophosphorus and organochloride compounds creates a potentially serious air pollution problem in farming areas. Sprinkling of pesticides on farm from helicopters is another source of air pollution.

5.2.1 Air pollution due to burning of crop residue

It is estimated that 22289 Gg of rice straw surplus is produced in India each year out of which 13915 Gg is estimated to be burnt in the field. On an average, total amount of stubble generated for paddy and wheat per acre was around 23 and 19 quintals, respectively (Kumar & Kumar 2010).The two states namely Punjab and Haryana alone contribute 48 percent of the total and are subject to open field burning (Gadde et al. 2009).

Fig. 5. Burning of crop residue in northern India

The rice and wheat system (RWS) is one of the widely practiced cropping systems in northern India. About 90-95 percent of the rice area is used under intensive rice wheat system (RWS) in Punjab (Gadde et al. 2009). Widespread adoption of green revolution technologies and high yielding variety of seeds increased both, crop as well as crop residue. In the last few decades intensive mechanization of agriculture has been occurring and combine (The combine harvester, also called combine, is a machine that harvests grain crops. It combines into a single operation processes that previously required three separate operations, that is, reaping, binding and threshing) harvesting is one such input, particularly in the RWS. Note that in the RWS a short period of time is available between rice harvesting and wheat plantation and any delay in planting adversely affects the wheat crop. This coupled with combine harvesting compels the farmers to burn the residue to get rid of stubble left out after the harvest (Figure 5). Burning of straw emits emission of trace gases like CO_2, CH_4, CO, N_2O, NO_X, SO_2 and large amount of particulates which cause adverse impacts on human health. It is estimated that India annually emits 144719 Mg of total particulate matter from open field burning of rice straw (Gadde et al. 2009).

5.2.2 Air pollution due to pesticides/insecticide sprays

Agricultural use of organophosphorus and organochloride compounds creates a potentially serious air pollution problem in farming areas. Pesticides can enter the human body through inhalation of aerosols, dust and vapor that contain pesticides; through oral exposure by consuming food and water; and through dermal exposure by direct contact of pesticides with skin. Pesticides are sprayed onto food, especially fruits and vegetables, they secrete into soils and groundwater which can end up in drinking water and pesticide spray can also drift and pollute the air. Inappropriate pesticide application can lead to off-target contamination due to **spray drift** and "run-off" from plants, causing contamination of the bystanders, the soil, water courses and other environmental pollution.

Pesticide drift occurs when pesticides suspended in the air as particles are carried by wind to other areas, potentially contaminating them (Cornell University Pesticide fact sheet). Pesticides that are applied to crops can volatilize and may be blown by winds into nearby areas, potentially posing a threat to wildlife (National Park Service. US Department of the Interior, August 1, 2006). Also, droplets of sprayed pesticides or particles from pesticides applied as dusts may travel on the wind to other areas (US Environmental Protection Agency 2007), or pesticides may adhere to particles that blow in the wind, such as dust particles (Environment Canada 2001). Ground spraying produces less pesticide drift than aerial spraying does (Palmer et al. year of publication not available). Farmers can employ a buffer zone around their crop, consisting of empty land or non-crop plants such as evergreen trees to serve as windbreaks and absorb the pesticides, preventing drift into other areas. Such windbreaks are legally required in the Netherlands (Science Daily 1999). Pesticides that are sprayed on to fields and used to fumigate soil can give off chemicals called volatile organic compounds, which can react with other chemicals and form a pollutant called tropospheric ozone. Pesticide use accounts for about 6 percent of total tropospheric ozone levels (UC IPM Online 2006).

With placement (localized) spraying of broad spectrum pesticides, wind drift must be minimized, and considerable efforts have been made recently to quantify and control spray drift from hydraulic nozzles (Figure 6). On the other hand, wind drift is also an efficient mechanism for moving droplets of an appropriate size range to their targets over a wide area with ultra-low volume (ULV) spraying. Himel (1974) made a distinction between exo-drift (the transfer of spray out of the target area) and endo-drift, where the active ingredient (AI) in droplets falls into the target area, but does not reach the biological target. Endo-drift is volumetrically more significant and may therefore cause greater ecological contamination

(a) (b)

Fig. 6. Localized insecticide (a), pesticide (b) spraying

(e.g. where chemical pesticides pollute ground water). Although there has been much public concern and research into spray drift, several studies have concluded that point source pollution (allowing pesticides to enter water courses/groundwater following spillage of concentrate or after washing equipment) can cause the greatest harm to the environment. As with pesticide application in general, crop dusting is associated with a number of environmental concerns, including spray drift, soil contamination, water pollution, and occupational health concerns. In the U.S. in 1970, lawsuits and court cases involving spraying of pesticides, such as aerial application in commercial agriculture were a growing area in law, combining areas such as negligence, products liability, strict liability, statutory regulation and commercial law (William et al. 1970). Environmental and human rights issues associated with crop dusting is greatest in developing countries, where government oversight is weaker or absent, few safety practices are used, and chemicals are used that are banned in most developed countries.

5.2.3 Air pollution due to Grain dust

Grain dust is produced when cereal crops and maize are harvested, dried, moved, stored and processed (Figure 7). The dust includes bacteria, fungi, insects and possibly pesticide residues as well as dry plant particles. Inhaling grain dust can cause ill-health; for example asthma, bronchitis and grain fever. Some people can become sensitized to the dust; this means that any subsequent exposure, even at a low level, can result in nasal or eye irritation or trigger an attack of asthma. Grain dust may contain mould spores that, if inhaled, can cause the potentially fatal disease, Farmer's Lung. The maximum permissible exposure limit, measured over 8 hours, is 10 milligrams per cubic meter (mg/m3). The HSE

Process	Dust level measurement averaged over 8 hours	Comments
Combining (no cab)	18 to 41 mg/m3	2-4 times daily legal amount
Combining (with cab and air filtration)	0.2 to 2.5 mg/m3	1/4 of daily amount
Grain carting work	1 to 40 mg/m3	Up to 4 times legal daily amount
Grain drying	4 to 57 mg/m3	Up to almost 6 times legal amount
Milling and mixing	0.1 to 11 mg/m3	Can exceed daily legal amount

Table 1. Levels of dust in an operator's breathing zone.

(a) (b)

Fig. 7. Grain dust causing air pollution during wheat harvesting by a) threshing b) combine

information sheet no.3 (Controlling grain dust on farms 2002) below (table 1) gives the levels of dust in an operator's breathing zone.
Some people who work with high levels of grain dust over many years may develop chronic lung disease that is similar to that developed after years of smoking. This has been shown to persist even after removal from the exposure.

5.3 Indoor air pollution

Problems of indoor air quality are recognized as important risk factors for human health Indoor air is also important because people spend a substantial proportion of their time in buildings. In indoor environments, tobacco smoke and combustion of solid fuels for cooking and heating are the most significant sources. In addition, construction material, furniture, carpeting, air conditioning, and home cleaning agents and insecticides can also be significant sources of chemical and biological pollutants indoors.
It has been estimated that approximately half the world's population, and up to 90% of rural households in developing countries, still rely on biomass fuels (WRI, 1999). Although the portion of global energy derived from biofuel has fallen from 50% in 1900 to around 13% currently, this trend has leveled and there is evidence that biofuel use is increasing among the poor (WRI, 1999). Poverty is one of the main barriers to the adoption of cleaner fuels and slow pace of development in many countries implies that biofuels will continue to be used by the poor for many decades. Biomass fuel, or biofuel, refers to any plant or animal based material deliberately burned by humans. Wood is the most common biofuel, but use of animal dung and crop residues is also widespread. These fuels are typically burned indoors in a U-shaped construction made from mud (Figure 8), or simple household cook stoves, such as a pit, three pieces of brick (Figure 9), which burn these fuels inefficiently and are often not vented with flues or hoods to take the pollutants to the outside. Some countries, including China and South Africa also use coal extensively for domestic needs. Animal dung is on the lowest rung of the ladder progressing to crop residues, wood, charcoal, kerosene, gas, and finally electricity. People generally move up the ladder as socio-economic conditions improve. Other sources of indoor air pollution in developing countries include smoke entering the home from nearby houses, burning of forests, agricultural land and household waste, the use of kerosene lamps, and industrial and vehicle pollution.

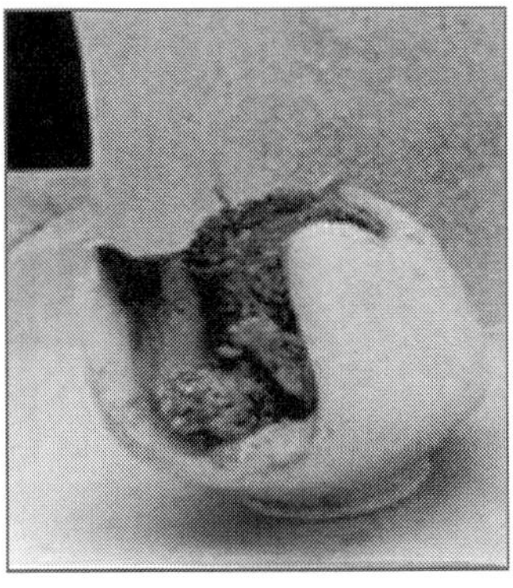

Fig. 8. U shaped household mud stove

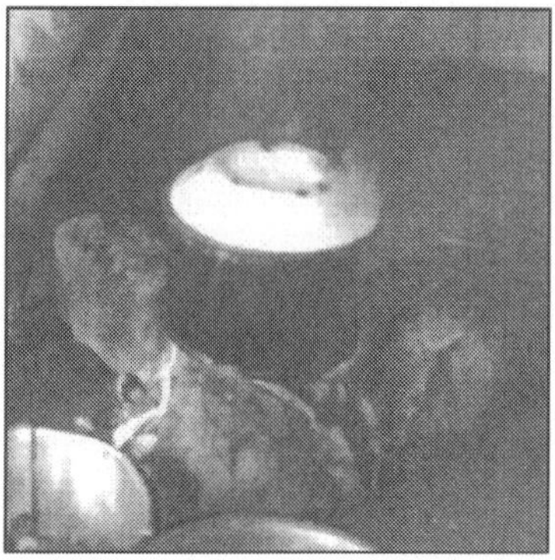

Fig. 9. Three brick cook stove

Environmental tobacco smoke (ETS), is another source of indoor air pollution exposures which can be expected to increase in importance in developing countries. It is important also to recognize that the open hearth (Figure 10) and resulting smoke often have considerable cultural and practical value in the home, including control of insects, lighting, drying food, fuel and housing materials and for flavouring foods (Smith, 1987).

Fig. 10. A women burning crop residue using house hold cook stove

Biomass smoke contains many thousands of substances, many of which damage human health. Most important are particulates, carbon monoxide, nitrous oxides, sulphur oxides (more with coal), formaldehyde, and polycyclic organic matter which include carcinogens such as benzo[a]pyrene (WHO 2010). Small particles of diameter less than 10 microns (termed PM10), and in particular those less than 2.5 microns (PM2.5), are able to penetrate deep into the lungs and appear to have the greatest health-damaging potential. In developing countries, individuals are typically exposed to these very high levels of pollution for between 3 and 7 hours each day over many years. During winter in the many cold and mountainous areas, exposure may occur over a substantial portion of each 24 hour period Cultural practices common in developing countries may promote exposure of infants, women, the elderly and the sick. Since it is the women who generally cook, therefore their exposure is much higher than that of men. Young children are often carried on their mother's back while she is cooking, so that from early infancy, children spend many hours breathing smoke.

5.4 Trans - boundary pollution

The rural areas have also to bear the effects of air pollution which is generated in the urban area and move towards the rural areas with the wind. It has been observed that the polluted air is not just the local phenomena rather the polluted air blows across the coutries as well continents across the globe. During the spring there are fierce dust and sand storms in the Gobi desert and northern and western China. As the dust and sand are blown eastward by westerly winds, they pick up air pollution particles, especially over heavily industrialized areas in northeast China like Shenyang, and carry them further east into South Korea and Japan and further east. South Koreans and Japanese also complain that the winds that carry the sulfur and nitrogen oxides also pick up heavy metals and carcinogens and dump them in their countries. Deaths attributed to yellow-sand-caused cardiovascular and respiratory diseases have been reported in South Korea. A group called Project Asia Brown Cloud—using aircraft and ground stations in China, South Korea and Japan—have observed that clouds of industrialized pollution that originate in Shenyang and merging with dust and sand clouds that originate in the Gobi Desert and Xinjiang (Jeffrey Hays 2008). Increasing amounts of pollution are being blown across the Pacific Ocean to North America. Pollutants carried by the jet stream can reach the United States in days. Pollutants from Asia reaching the United States are rising at a rate of 5 percent to 10 percent a year. By some estimates 25 percent of the air pollution in Los Angeles comes from China. American heath officials find this figure alarming when is tacked onto pollution that already exist in the United States (Jeffrey Hays 2008). Increases in amounts of ozone and fine particulate matter found in the air in the United States are blamed on "transpollution" from China. Scientists estimate that in 2010, on third of the ozone in Los Angeles will originate in Asia, much of it in China. Asian soot has been blamed for speeding up the thinning of Arctic ice and other polar environmental changes. Soot, dust and chemical pollutants from China have been captured in a weather observation stations on the summit of Mount Bachelor in Cascade Range in Oregon. Soot, dust, ozone and nitrous oxides can be detected by satellites moving across the Pacific. But just as the Americas worry about air pollution from Asia, Europeans worry about pollution carried by winds from the Americans and Asians worry about pollution carried by winds from Europe (Jeffrey Hays 2008).

6. Health impacts of air pollution in rural areas

Health impact of air pollution depends on the pollutant type (Table 2), its concentration in the air, length of exposure, other pollutants in the air, and individual susceptibility. Different people are affected by air pollution in different ways. Poor people,

Pollutant	Health Impact
Tobacco smoke	Tobacco smoke generates a wide range of harmful chemicals and is a major cause of ill health, as it is known to cause cancer, not only to the smoker but affecting passive smokers too, ranging from burning sensation in the eyes or nose, and throat irritation, to cancer, bronchitis, severe asthma, and a decrease in lung function.
Biological pollutants	These are mostly allergens that can cause asthma, hay fever, and other allergic diseases.
Volatile organic compounds	Volatile compounds can cause irritation of the eye, nose and throat. In severe cases there may be headaches, nausea, and loss of coordination. In the longer run, some of them are suspected to cause damage to the liver and other parts of the body.
Formaldehyde	Exposure causes irritation to the eyes, nose and may cause allergies in some people
Lead	Prolonged exposure can cause damage to the nervous system, digestive problems, and in some cases cause cancer. It is especially hazardous to small children.
Radon	A radioactive gas that can accumulate inside the house, it originates from the rocks and soil under the house and its level is dominated by the outdoor air and also to some extent the other gases being emitted indoors. Exposure to this gas increases the risk of lung cancer.
Ozone	Exposure to this gas makes our eyes itch, burn, and water and it has also been associated with increase in respiratory disorders such as asthma. It lowers our resistance to colds and pneumonia.
Oxides of nitrogen	This gas can make children susceptible to respiratory diseases in the winters.
Carbon monoxide (CO)	CO combines with haemoglobin to lessen the amount of oxygen that enters our blood through our lungs. The binding with other haeme proteins causes changes in the function of the affected organs such as the brain and the cardiovascular system, and also the developing foetus. It can impair our concentration, slow our reflexes, and make us confused and sleepy.
Sulphur dioxide. (SO2)	Sulphur Dioxide in the air is caused due to the rise in combustion of fossil fuels. It can oxidize and form sulphuric acid mist. SO2 in the air leads to diseases of the lung and other lung disorders such as wheezing and shortness of breath. Long-term effects are more difficult to ascertain as SO2 exposure is often combined with that of SPM.
Suspended Particulate Matter (SPM)	SPM consists of dust, fumes, mist and smoke. Lead is of major concern, others being nickel, arsenic, and those present in diesel exhaust. These particles when breathed in, lodge in our lung tissues and cause lung damage and respiratory problems. The importance of SPM as a major pollutant needs special emphasis as a) it affects more people globally than any other pollutant on a continuing basis; b) there is more monitoring data available on this than any other pollutant; and c) more epidemiological evidence has been collected on the exposure to this than to any other pollutant.

Table 2. Some of the indoor air pollutants and their health impacts

undernourished people, very young and very old, and people with preexisting respiratory disease and other ill health, are more at risk. Poor people are also more likely to suffer from ill health and disease, as they have less access to the health care. Air pollution has both acute and chronic effects on human health. Health effects range anywhere from minor irritation of eyes and the upper respiratory system to chronic respiratory disease, heart disease, lung cancer, and death. Air pollution has been shown to cause acute respiratory infections in children and chronic bronchitis in adults. It has also been shown to worsen the condition of people with preexisting heart or lung disease. In addition, evidence is now emerging of links with a number of other conditions, including asthma, tuberculosis, low birth weight, cataracts, and cancer of upper airways. Indoor smoke from solid fuels causes about 35.7% of lower respiratory infections, 22.0% of chronic obstructive pulmonary disease and 1.5% of trachea, bronchus and lung cancer (WHO 2002). Indoor air pollution may also be associated with tuberculosis, cataracts and asthma. In total, 2.7% of DALYs worldwide are attributable to indoor smoke, 2.5% in males and 2.8% in females.

Both short-term and long-term exposures have also been linked with premature mortality and reduced life expectancy. WHO estimates that exposure to indoor smoke from solid fuels may be annually responsible for about 1.6 million premature deaths in developing countries and 2.6 percent of the global burden of disease (WHO 2002). In India 3-5 percent of the national burden of disease is attributable to use of solid fuel (Smith 2000). Kumar & Kumar (2010) found that total annual welfare loss in terms of health damages due to air pollution caused by the burning of rice straw in rural Punjab (India) amounts to 76 millions. Measurement of health parameters also poses many challenges for the study of health impacts of air pollution. For example, in many developing countries, vital statistics and cause of death data are not routinely collected or are not reliable. Clinical data are also highly unreliable because there is often considerable selection due to differential access to health services and awareness of health problems. Multiple sources of health care, varying quality of services, poor diagnosis, and poor record keeping further complicate the study of health impacts.

7. Conclusion

While there has been considerable research on health impacts of tobacco smoking, and there is growing evidence on the effects of ambient air pollution, the research on health effects of indoor air pollution from household use of unprocessed solid fuels is particularly weak. There is need to strengthen both the quantity and quality of evidence linking air pollution and various health outcomes, especially for developing countries and for health conditions with weak or no evidence. This can be accomplished by measuring exposure levels more directly; by including clinical measures of disease outcomes; and by adequately accounting for social, behavioral, nutritional, and environmental confounding factors. In these efforts, there is need to use more powerful study designs, such as prospective cohort studies and randomized intervention trials, designed specifically for rural areas with weak data.

Many efforts to address air pollution have done little to alleviate its total impact. Installing chimneys to vent smoke from indoor stoves, for example, simply moves the pollution a few feet away, just adding to outdoor pollution. Moving industries from urban to rural areas similarly shifts the pollution from one environment to another. Future efforts to reduce the health effects of air pollution should consider a pollutant's total impact—in rural as well as urban areas.

8. Acknowledgement

Author is thankful to Mr. Yenepoya Abdulla Kunhi, Honorable Chancellor Yenepoya University, Dr. P. Chandramohan Honorable Vice Chancellor Yenepoya University and Dr. Janardhana Konaje, Registrar of Yenepoya University for sponsoring this chapter.

9. References

Abbas Miri, Hassan Ahmadi, Ahmad Ghanbari, Alireza Moghaddamnia. Dust Storms Impacts on Air Pollution and Public Health under Hot and Dry Climate. *International Journal Of Energy And Environment* Issue 2, Volume 1, 2007;101-105

Charron A, Harrison RM. Fine (PM2.5) and coarse (PM2.5–10) particulate matter on a heavily trafficked London highway: sources and processes. *Environmental Science & Technology*, 2005, 39:7768–7776.

Controlling grain dust on farms (2002). HSE *agriculture information sheet number 3* (revised) available at http://www.segurancaetrabalho.com.br/download/controlling-grain.pdf

Cornell University. Pesticides in the environment. Pesticide fact sheets and tutorial, module 6. *Pesticide Safety Education Program.* available at http://pmep.cce.cornell.edu/facts-slides-self/core-tutorial/module06/index.html

Environment Canada (September–October 2001), *Agricultural pesticides and the atmosphere.* available at http://www.ec.gc.ca/science/sandesept01/article3_e.html

Gadde, B.; Bonnet, S.; Menke, C.; Garivait, S. (2009) 'Air pollutant emissions from rice straw open field burning in India, Thailand and the Philippines', *Environmental Pollution* 157, 1554-58.

Himel, C.M. (1974) Analytical methodology in ULV. In: *Pesticide application by ULV methods British Crop Protection Council Monograph No. 11,* 112-119.

Jeffrey Hays (2008) *Air pollution in china.* Available at http://www.Factsanddetails.com/China/AIR POLLUTION IN CHINA - China Facts and Details.mht

Kumar P. and Kumar S. (2010) *Valuing the Health Effects of Air Pollution from Agricultural Residue Burning.* Available at http://www.cerdi.org/uploads/sfCmsContent/html/323/kumar.pdf

Mattson, J.O. and Nilsen, J. (1996)The transport of Saharan dust to southern Europe: a scenario. *Journal of Arid Environments*;1996; 32: 111- 119. National Park Service. US Department of the Interior. (August 1, 2006),

Sequoia & Kings Canyon National Park: Air quality -- Airborne synthetic chemicals. Nps.gov. available at http://www.nps.gov/seki/naturescience/air_synthchem.htm

Palmer, WE, Bromley, PT, and Brandenburg, RL. (year not available) *Wildlife & pesticides - Peanuts. North Carolina Cooperative Extension Service.* available at http://ipm.ncsu.edu/wildlife/peanuts_wildlife.html

Park K (2009)*Park's text book of preventive and social medicine.* M/s Banarasidas Bhanot Publishers, Jabalpur, India.20th ed.2009:639 Science Daily (November 19, 1999), *Evergreens help block spread of pesticide from crop fields.* Sciencedaily.com. available at http://www.sciencedaily.com/releases/1999/11/991119075237.htm

Smith KR, et al., (1983) Air pollution and rural biomass fuels in developing countries: a pilot village study in India and implications for research and policy, *Atmospheric environment,* 17: 2343-62

Smith KR (1987) *Biofuels, air pollution, and health: a global review.* Plenum, New York.

Smith KR (2000) National burden of disease in India from indoor air pollution. *Proceedings of the National Academy of Sciences* USA97(27:13286-93.)

Smith DJT, Harrison RM. (1998) Polycyclic aromatic hydrocarbons in atmospheric particles. In: Harrison RM, Van Grieken R, eds. Atmospheric particles. John Wiley & Sons, 1998:253–294 (IUPAC Series on *Analytical and Physical Chemistry of Environmental Systems,* Vol. 5).

UC IPM Online. (August 11, 2006), What's up, Doc? Maybe less air pollution. *Statewide IPM Program, Agriculture and Natural Resources, University of California.* Ipm.ucdavis.edu. available at http://www.ipm.ucdavis.edu/NEWS/carrot-news.html

US Environmental Protection Agency (September 11th, 2007), *Pesticide registration (PR) notice 2001-X Draft: Spray and dust drift label statements for pesticide products.* Epa.gov. available at http://www.epa.gov/PR_Notices/prdraft-spraydrift801.htm

U.S. Geological Survey (2000) *Fact Sheet 169-97 Online Version 1.1,* revised June 2000 Available at http://pubs.usgs.gov/fs/fs169-97/

Vegetation fires (2000). *Fact Sheet no.254 WHO,* Geneva 2000 Available at https://apps.who.int/inf-fs/en/fact254.html

William T. Birmingham; Jon L. Kyl, (1970)"Legal and Practical Aspects of Pesticide Spraying Cases", *Defense Counsel Journal,* volume 37, 1970.

WHO (1987) *Air Quality Guidelines for Europe.* WHO Regional publications, European Series No. 23, Copenhagen.

WHO (2002) *World Health Report 2002:reducing risks, promoting healthy life.* Geneva.

WHO (2005) *Air Quality Guidelines Global Update.* WHO Regional publications, Europe, Copenhagen.

WHO (2010) *WHO guidelines for indoor air quality: selected pollutants.* WHO Regional publications, Europe, Copenhagen.

World Bank (1993) *World Development Report,1993, Investing in health.* World Bank, Oxaford University Press.

WRI (1999) World Resource Institute, UNEP, UNDP, World Bank. 1998-99 *World Resources: a guide to the global environment.* Oxford University Press.

Yang Gengsheng (1996) Black Windstorm in Northwest China and Calamity Reduction Approaches, *Journal of Desert Research*, 16 (2).

24

CFD Analyses of Methods to Improve Air Quality and Efficiency of Air Cleaning in Pig Production

Bjarne Bjerg[1], Guo-Qiang Zhang[2] and Peter Kai[2]
[1]*University of Copenhagen*
[2]*University of Aarhus*
Denmark

1. Introduction

Release of contaminant gaseous as ammonia, carbon dioxide, sulphur hydrogen, methane and laughing gas is an inevitable consequence of livestock production, and degrades air quality in the production facilities and has a negative impact on both the short and long distance environment. Generally the knowledge on how the content of these gases influence animal performance, animal health and the health's of the farmers is deficient, but it is commonly accepted that the concentrations of especially ammonia, hydrogen sulphide and carbon dioxide should be kept below certain thresholds (CIGR 1984). Ammonia is of special interest because the concentrations inside the housing system in some cases reach critical levels and because it, in large parts of the world, is assumed to be the livestock production gas that has the largest negative impact on the surrounding environment.

Ammonia in livestock housing originates from animal excretion of faeces and urine. Especially urine has a large content of urea which readily is transformed into ammonia upon contact with the enzyme urease:

$$(NH_2)_2CO + H_2O \overset{\text{urease}}{\Rightarrow} 2NH_3 + CO_2 \tag{1}$$

The required enzyme urease is present in faeces and therefore the generation of ammonium takes place when urine brings together with faeces. Successively a part of the generated ammonia will be transformed into ammonium in the solution:

$$NH_3 + H_2O \Leftrightarrow NH_4^+ + OH^- \tag{2}$$

The lower pH-value in the solution the larger part of the ammonia will be transformed to ammonium and that part is bound in the solution. The remaining ammonia will evaporate to the air above with a velocity that depends on numerous properties in the solution and in the air above.

Due to the negative influence of ammonia on air quality in the housing system and on the surrounding environment it is of large interests to design housing systems that limit the evaporation of ammonia or enable efficient removal of the evaporated ammonia to maintain low concentration in the room and low emissions to the surrounding.

The purpose of the work presented in this chapter is to demonstrate the possibilities to design ventilation systems that reduce the ammonia concentration in pig housing and simultaneously reduce the required ventilation capacity. In addition it is demonstrated how improved ventilation system design including strategically located exhausts can be utilised to capture a large part of released ammonia in a relative small share of the entire air change, and how cleaning of that minor part of the total air change may contribute to a significant reduction of the total ammonia emission.

The analyses discussed in this work are based on Computational Fluid Dynamics (CFD) which is a branch of modelling tools that use numerical methods and algorithms to analyse problems that involve fluid flow. A general overview of the CFD methods used in relation to ventilated rooms or more specifically related to the agricultural industry lies beyond the scope of this chapter but can beneficially be found in Nielsen et al. (2007) and Norton et al. (2007). In this chapter it is prioritised to give an in depth description of the necessary or preferred methods to solve the challenges related to the special issues treated in this work.

The use of CFD methods involves a geometrical description and delimitation of the fluid volume - often called the domain - that has to be included in order to analyse the required flow phenomena. In pig production it is conmen to use mechanical ventilations system were the outdoor wind conditions have a relatively small influence on the indoor airflow, and therefore it is reasonable to delimitate the analysed spaces to indoor air volume.

In Denmark and a few other European countries it is common to equip pig housings with mechanical under pressure ventilations system with air intake through porous material in the ceiling. Typically these inlets consist of a layer of mineral wool between timber beams with a cement bonded wood wool plate beneath, see Fig. 1.

In such a system, the air is entering the room with a velocity as low as 0.005 to 0.05 m/s, and, consequently, draft will seldom be a problem if the system is designed in accordance

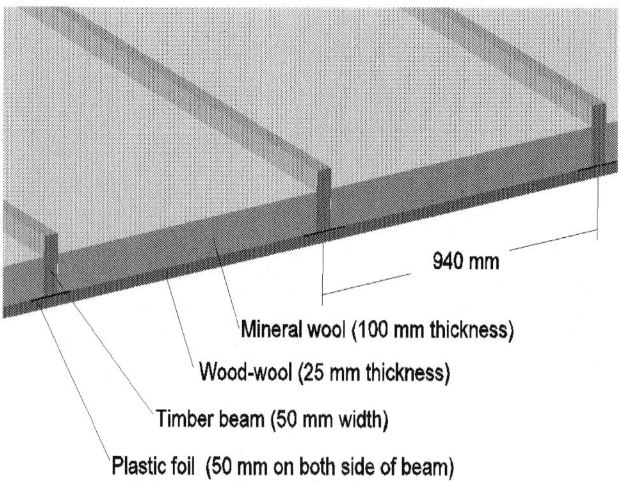

Fig. 1. Typical construction for diffuse air inlet through ceiling.

with the existence empirical knowledge. However, the empirical knowledge is not sufficient to optimize room layout in order to improve air distribution in the animal occupied zone or in order to reduce the release of ammonia and odorants from manure. CFD (Computational Fluid Dynamics) is an obvious tool for such tasks, but until now, the reported experience of using CFD in relation to air intake through porous material in the ceiling is limited.

CFD modeling requires detailed geometrical description of the flow domain and division of the domain into a comprehensive grid. The procedure is very time consuming if it is used in complex setups with pen partitions, equipment, slatted floor and animals. Wu & Gebremedhin (2001) performed CFD simulation of airflow around up to ten geometrically precise models of standing 600 kg Holstein cows. However, to simulate airflow in an animal room under operation, this method has the disadvantages that it is very time consuming to build such models, and that the animals are moving around and place themselves in different positions. A potential way to overcome this is to assume that the occupied zone consists of a porous media with a certain flow resistance. A similar method has been use to model airflow through slatted floor (Sun et al., 2004) and through partly open pen partitions (Wagenberg et al., 2004).

In this work the porous media assumption are utilized to avoid a detail geometrical modeling and complicated subdivision of the space around the animals the animals, and to model the condition in the porous inlet.

2. Analyses of a partly pit ventilation system as method to reduces ammonia emission from a pig production unit

To reduce ammonia and odour emission from livestock housing, the use of air cleaning systems is increasing. Available acid based exhaust air cleaning systems are capable of reducing ammonia emission through ventilation with approximately 95 percent (Melse & Ogink 2005), but due to the large amount of air that must be treated the investment and operating costs are high. In many cases a reduction of ammonia emission of 30 - 70 percent is sufficient to fulfil the requirement from legislation. Under northern Europe climate conditions the ventilation level is relatively low in a large part of the year resulting in high ammonia concentrations in exhaust air in those periods. This makes it possible to obtain a relatively large reduction of total ammonia emission with a cleaning system designed for a relatively small share of the total ventilation capacity.

The greater part of ammonia emission from pig production housing systems with slatted floor is released from the slurry surface. In this type of production units pit ventilation has been known as a method to improve indoor air quality, but it is seldom used due to increased energy cost and increased emission of ammonia and odour.

The scope of this work is to investigate the efficiency of a partly pit ventilation system with cleaning of the air exhausted from the pit. The investigation is based on theoretical calculations including CFD methods (Computational Fluid Dynamics) to predict airflow, and ammonia release and distribution in a growing pig unit.

2.1 Methods

The analyses were based on conditions in a typical Danish pig unit for growing pigs from 30 to 100 kg. The assumed unit included two rows of 2.4 m wide and 4.8 m long pens designed for 16 pigs each, see Fig. 2 and 3.

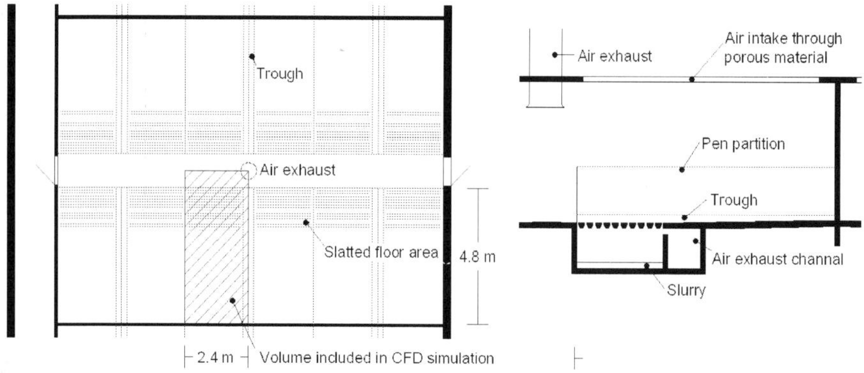

Fig. 2. Plan and cross section of assumed pig production unit.

One third of the pen area consists of slatted floor with a slurry pit beneath and the unit is equipped with negative pressure ventilation system with air intake through porous material in the ceiling. Along the slurry pit an assumed air exhaust channel makes it possible to evacuate a share of total exhaust air directly from pit, see Fig. 5. The maximum ventilation capacity was assumed to be 100 $m^3h^{-1}pig^{-1}$, and StaldVent 5.0© software (Strom & Morsing 2004) were used to calculated the distribution of expected ventilation levels during a year, see Fig. 4.

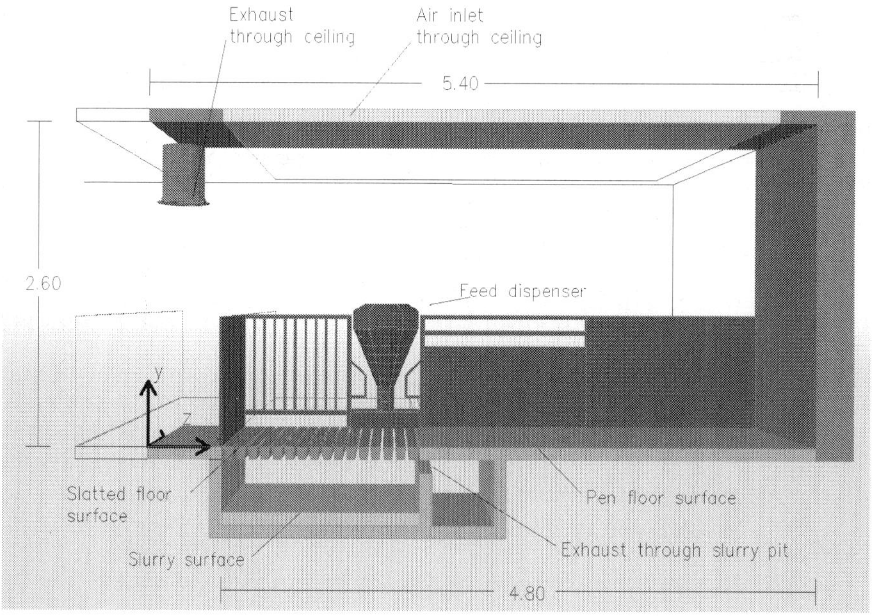

Fig. 3. Perspective and used system of coordinates in the assumed pig production unit.

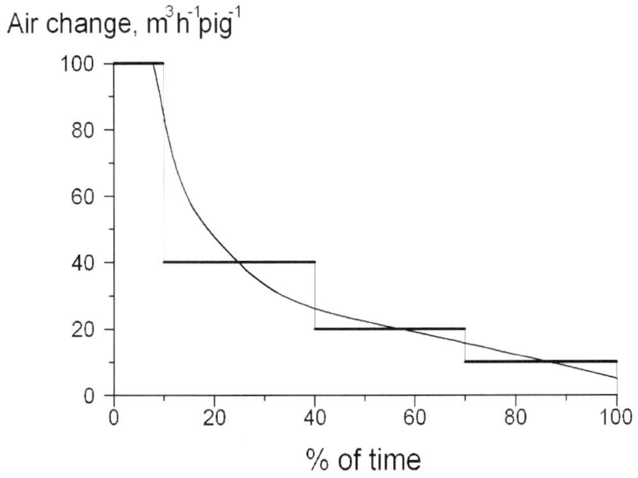

Fig. 4. Expected distribution of ventilation levels in a growing pig unit under Danish conditions (smooth curve), and four representative levels of ventilation (step curve) used in modelling of ammonia release and emission.

2.1.1 CFD modelling

The commercial CFD (Computational fluid Dynamic) code Fluent 6 (Anonymous 2006) were used to calculated air flow and ammonia distribution in one pen section of the room, see Fig. 2. The chosen section was divided into 37722 hexahedral cells and the grid on surfaces is shown in Fig. 5.

2.1.1.1 Porous media assumptions

Porous media boundary conditions were utilised to model airflow in three special regions of the used geometrical model:

- Porous material for air intake in the ceiling,
- Slatted floor partly covered with pigs, and
- Pigs in animal occupied zone pigs up to 0.6 m above floor level.

The flow resistance in each of the three directions (x, y and z) in the porous media regions were calculated using equation 1 (Bjerg et al., 2008; anonymous 2006).

$$\Delta p = 0.5 \cdot R_1 \cdot \rho \cdot v^2 + \mu \cdot R_2 v \tag{3}$$

where

Δp is pressure drop over the porous media, Pa,

R_1 internal risistance factor,

R_2 viscus risistance coefficient

ρ air density, kg/m^3 (1.2 kg m^{-3} at 20 °C)

v air velocity, through porous media, m/s

μ air viscosity, kg m^{-1}s^{-1} (1.8 · 10^{-5} kg m^{-1}s^{-1} at 20 °C)

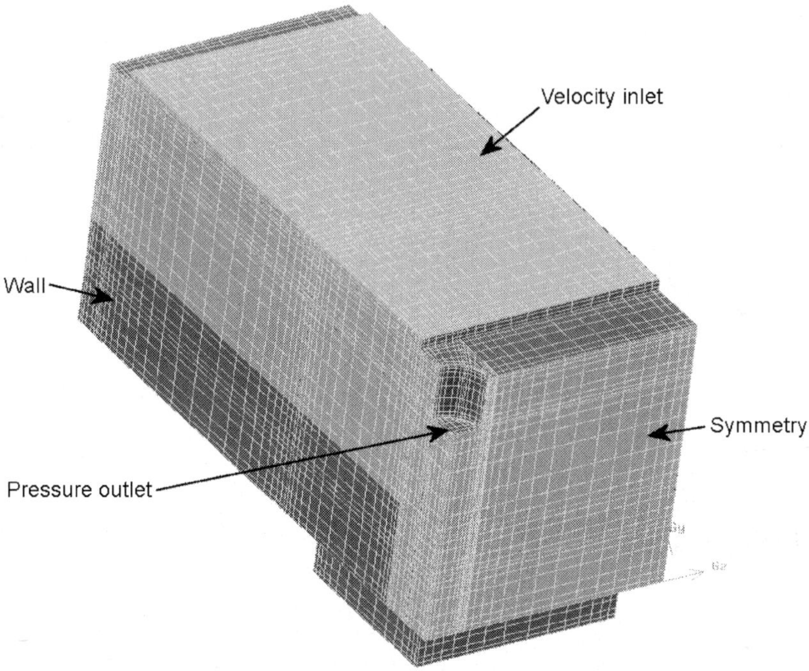

Fig. 5. Grid and boundary conditions on surfaces of the volume included in CFD modelling.

2.1.1.2 Flow resistance properties for air inlet

Internal resistance factor (R_1) and viscous resistance coefficient (R_2) through air inlet was determined from existing measurement of pressure drop through 100 mm mineral wool and 25 mm wood-wool (Anonymous, 1999). It appears from Fig. 1 that the plastic foil beneath the timber beams reduces the effective inlet area. But to avoid modelling each separate part of the inlet area in the subsequent CFD simulation the determined flow resistance parameters were adjusted to assure a realistic pressure drop if the inlet was modelled as a coherent area. Those adjusted flow resistance parameters appears from Table 1.

2.1.1.3 Flow resistance properties for slatted floor

Vertical flow resistance properties through slatted with and without assumed partly covering with pigs was determined in CFD simulation of airflow through detailed geometrical models of a segment of the floor, see Fig. 5.
The resulting relation between inlet velocity and the pressure drop over the floors were used for determination of the values of R_1 and R_2 using equation 3. The found values appear from Table 1.

	x-direction[1]		y-direction		z-direction	
	R_1	R_2	R_1	R_2	R_1	R_2
Porous ceiling[2]	4000	$5.4 \cdot 10^6$	4000	$5.4 \cdot 10^6$	4000	$5.4 \cdot 10^6$
Slatted floor[3]	-	-	40	11000	40	11000
Slatted floor with pigs[3]	-	-	80	15000	80	15000
Drained floor[3]	-	-	160	22000	160	22000
Drained floor with pigs[3]	-	-	4400	60000	4400	60000
AOZ, activity area[4]	0.4	400	0.4	400	0.4	400
AOZ, resting area[4]	1.3	1500	1.3	1500	1.3	1500

[1]System of coordinate orientation appears from Fig. 2.
[2]Resistance in horizontal directions assumed to be equal to resistance in vertical direction.
[3]Slats blocks for flow in x-direction and resistance in z-direction assumed to be equal to resistance in y-direction.
[4]Resistance in vertical direction assumed to be equal to resistance in horizontal directions.

Table 1. Overview over used flow resistance properties.

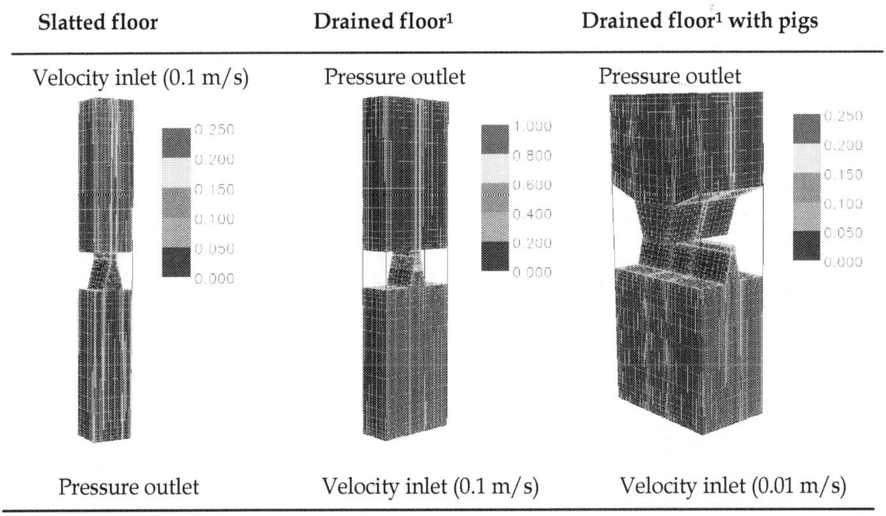

[1]Danish animal welfare legislation do not allow slatted floor with opening ratio of more than 10 % in the animal resting area, and the phrase "drained floor" is used for floor with opening ratio below 10 %.

Fig. 6. Grid and pressure distribution (Pa) on surfaces of simulation models for determination of flow resistance through slatted floor drained floor and drained floor with pigs.

2.1.1.4 Flow resistance properties for animal occupied zone.

The presence of animals will restrict the air flow in the Animal Occupied Zone (AOZ). The defined height of the AOZ is related to the expected height of a standing pig. Height of standing pigs can be calculated as 0.16(animal weight, kg)$^{0.33}$, m (Baxter 1984). For growing finishing pigs from 30 to 100 kg this results in heights from 0.49 to 0.73 m. In this work the determination of flow resistances is based on an assumed animal weight of 50 kg and a height of animal occupied zone of 0.6 m. The AOZ is assumed to be divided into resting including the third of the pen in largest distance from the slatted floor and an activity area including the remaining two thirds of the pen area. Based on 16 pigs in the pen following distribution of animal on staying and laying positions in resting area and activity area was assumed:

8 pigs laying in resting area corresponding to 2.1 pigs/m^2
1 pig standing in resting area corresponding to 0.26 pigs/m^2
5 pigs laying in activity area corresponding to 0.65 pigs/m^2
2 pigs standing in activity area corresponding to 0.26 pigs/m^2

The flow resistance parameter for flow through the AOZ was determined in CFD simulation on geometrically fairly detailed models as shown in Fig. 7. The model illustrates assumed conditions in the resting area with the density of 2.1 laying pigs and 0.26 standing pigs per m^2. The magnitude of the pigs corresponds to an average body weight of 50 kg. The pigs are constructed as the intersection of two dimensional silhouettes from three different directions. Unstructured grid was used in simulation of airflow around the pigs. Air velocity boundary conditions were assumed at the left end face and pressure outlet conditions were assumed in the right end face of model in Fig. 7.

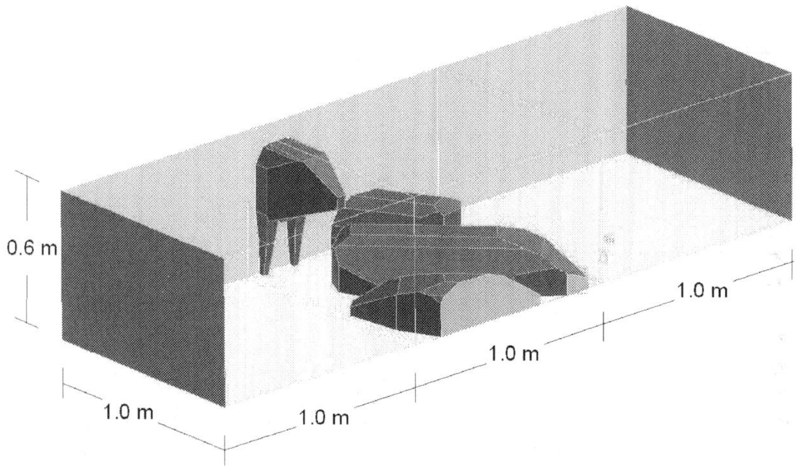

Fig. 7. Geometrical model used in simulation for determination of flow resistance through animal occupied zone with animal density of 2.1 laying and 0.26 standing 50 kg pig pr m^2. Velocity inlet and pressure outlet boundary condition were assumed at the left and right face, respectively.

The resulting relation between inlet velocity and the pressure drop around the animals were used for determination of the values of R_1 and R_2 using Equation 3. The found values appear from Table 1.

2.1.1.4 Ventilation configurations

Simulations were carried out with air intake of 10, 20, 40 and 100 $m^3h^{-1}Pig^{-1}$ corresponding to inlet velocities of 0.00411, 0.00822, 0.01644 and 0.0411 m/s, respectively. For each ventilation level simulations were carried out with 0, 5, 10 and 20 $m^3h^{-1}Pig^{-1}$ air exhaust from the pit.
For room exhaust pressure outlet conditions were assumed on the entry to the ventilation shaft through the ceiling, see Fig. 5. For pit exhaust velocity outlet conditions were assumed on the entry surface on the connections between the slurry pit and the under floor exhaust channel, see Fig. 2 and 3. Fig. 5 illustrates that symmetry boundary conditions were assumed at the surface where the modelled section was cut out of the entire room.

2.1.1.5 Ammonia release

As ammonia source were assumed a constant concentration on slurry surface, and a level of 340 ppm were chosen in order to obtain realistic ammonia concentrations in the room. Griskey (2002) state the diffusivity for ammonia in air at 295 °K and 101325 Pa to 1.8 x $10^{-5}m^2s^{-1}$, and this value was utilized in the CFD calculations.

2.1.1.6 Heat release

Releasing of convective heat due to animal heat production in animal occupied zone was included in simulations as a heat source of 200 Wm^{-3} in the volume from 0-0.6 m above the solid floor and of 100 Wm^{-3} above the slatted floor, corresponding to a total heat supply of 54 W per pig. Radiation heat release from pigs and transmission heat loss from the building was not taken into account.

2.1.1.7 Modeling of turbulence

Compared to the widely used Standard k-ε turbulence model (Launder & Spalding, 1974) the realizable k-ε turbulence model (Shih et al. 1995) has in earlier works (Bjerg et al., 2008) shown to be more suitable to handle turbulence in a room with air intake through porous material in the ceiling and, consequently, the realizable k-ε turbulence model was used in this work.

2.1.2 Calculation of ammonia emission

In estimation of yearly average ammonia concentrations and emissions it was assumed that the ventilation system runs with 100 percent of its capacity in 10 percent of time and with 40, 20 and 10 percent of capacity in 30 percent of time each, see Fig 4.
Calculation of air cleaning efficiency was based on the assumption that the ammonia will be removed to a level of 1 ppm in the treated air, which corresponds to the performance of commercial systems. A setup with treatment of all air from a unit without pit ventilation is used as reference case for estimating the efficiency of treating of pit exhaust or treating of a minor part of room (ceiling) exhaust.

2.2 Results and discussion

Fig. 8 shows simulated airflow and ammonia concentration distribution for the case with total ventilation of 40 $m^3h^{-1}Pig^{-1}$. The left picture show the case with a pit ventilation of 10 $m^3h^{-1}Pig^{-1}$, and the right picture show the case without pit ventilation.
It is clear that pit ventilation creates a significant reduction of ammonia concentration in the entire room. In the animal occupied zone (0 to 0.6 m above floor) the average concentration was reduced from 13.5 to 5.5 ppm corresponding to a reduction of nearly 60 percent.

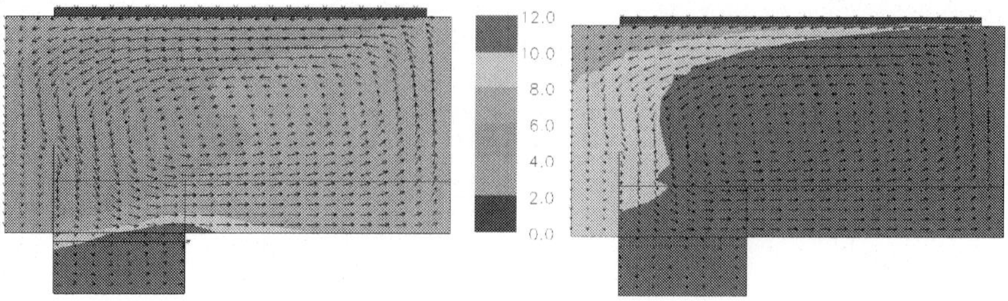

(a) (b)

Fig. 8. Simulated airflow and ammonia distributions (ppm) at total ventilation of 40 $m^3h^{-1}Pig^{-1}$, (a) with pit ventilation of 10 $m^3h^{-1}Pig^{-1}$, and (b) without pit ventilation.

Fig. 9 shows ammonia concentration in the exhaust air from the room at different levels of air change from the pit and totally.

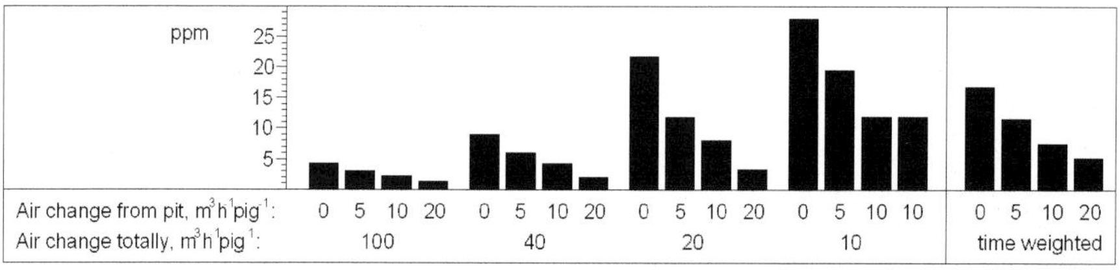

Fig. 9. Ammonia concentration in room exhaust at different pit ventilation and entire unit ventilation. Time weighted values, shown to the right in the graph, assumes total air change of 100 $m^3h^{-1}Pig^{-1}$ in 10 percent of time, and 40, 20 and 10 $m^3h^{-1}Pig^{-1}$ in 30 percent of time each.

As average during a year it is estimated that pit ventilation of 5, 10 and 20 $m^3h^{-1}Pig^{-1}$ will reduce ammonia concentration in the exhaust with 31, 55 and 68 percent, respectively, and inside the room the ammonia concentration will be reduced correspondently.
Fig. 10 shows total ammonia release and the emission from the room at different levels of air change from the pit and totally.

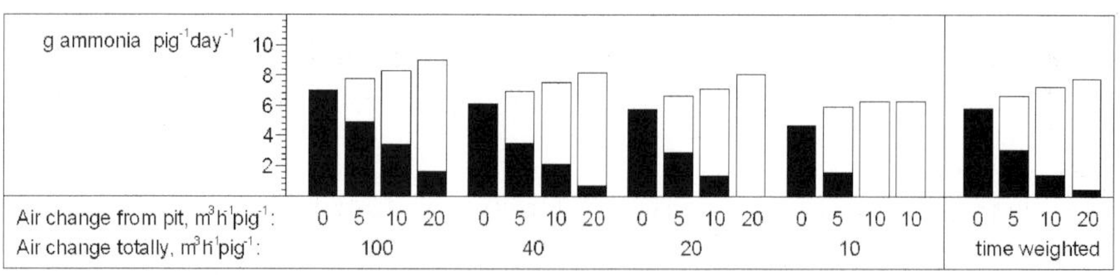

Fig. 10. Total ammonia release (aggregated bars) and emission from the room (black part of bars) at different levels of pit ventilation and entire unit ventilation. White part of bars show emission through pit exhaust assuming no air cleaning.

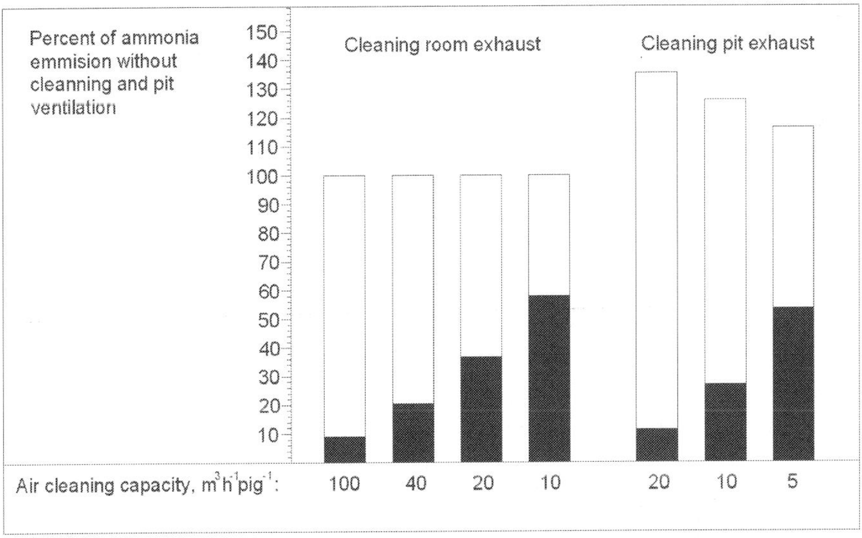

Fig. 11. Ammonia release (aggregated bars) and ammonia emission to the surroundings (black bars) at different air cleaning strategies. 100 percent corresponds to the ammonia emission from a unit with no air cleaning or pit ventilation. White part of bars shows removed ammonia.

It can be seen that an increased level of pit ventilation increased the ammonia release if cleaning system is not available. Using pit ventilations of 5, 10 and 20 $m^3h^{-1}Pig^{-1}$ during a year will result in increased ammonia emission of 16, 26 and 35 percent, respectively.

The simulation without pit ventilation showed that a decreased air change from 100 to 40 $m^3h^{-1}Pig^{-1}$ resulted in 13 percent decreased in ammonia emission. This can be compared with the result of a Danish investigation of the influence of air change on ammonia emission where the average air change were reduced from 80 to 40 $m^3h^{-1}Pig^{-1}$ (Lyngbye et al., 2006). In that investigation the room temperature was maintained by cooling the air inlet resulting in a reduction of the ammonia emission of 10 percent. The agreement with this investigation supports the assumption that a constant ammonia concentration on pit bottom surface is a reasonable simplification in CFD-modelling for analyzing the influence of ventilation strategy on ammonia emission.

Fig. 11 shows calculated performance of different cleaning strategies. For a unit with no pit ventilation it appears that cleaning of all exhaust air (100 $m^3h^{-1}Pig^{-1}$) reduce ammonia emission with about 90 percent. If the cleaning system is designed to handle 10 percent of the ventilation capacity the reduction of ammonia emission is 41 percent, and measured in relation to the capacity of cleanings system this strategy is 4.5 times as efficient as cleaning all the air. Cleaning of 10 percent of the air exhausted directly from the pit removes 73 percent of the ammonia emission and related to the cleaning capacity this is 8 times as efficient as cleaning all the air from the room. But due to increased ammonia release in the pit ventilated system the consumption of acid to reduce the ammonia emission with a certain amount will increase with about 35 percent.

3. Airflow pattern and ventilation efficiency

Ventilation systems in livestock buildings are designed to remove heat, moisture and containment gases in order to maintain temperatures and concentrations that are suitable for the animals. The airflow in pig units with diffuse air inlet through the ceiling is characterised by low air velocities primarily generated by the buoyancy force due to the convective heath release from the animals, and CFD-analyses have shown relative large spatial difference in temperature and concentrations in these systems, and consequently, there exists a potential to improve the ventilation effectiveness if the exhausts can be moved to locations with higher temperatures or concentrations.

Pigs are usually resting more than 80 percent of the time and the therefore the heat release will be most concentrated and generate a higher temperature and an upward air stream above the resting area. This has inspired to conduct an analysis of the influence of moving the exhaust to a location immediately above the resting area as it appears from Fig. 12.

The result presented in Fig. 13 shows that if the air change is maintained at 40 $m^3pig^{-1}h^{-1}$ then the general air temperature level will drop about three degrees. Ventilation systems in modern pig production facilities are usually designed to control the air change in a way that maintains a predefined temperature measured by a temperature sensor above the animals.

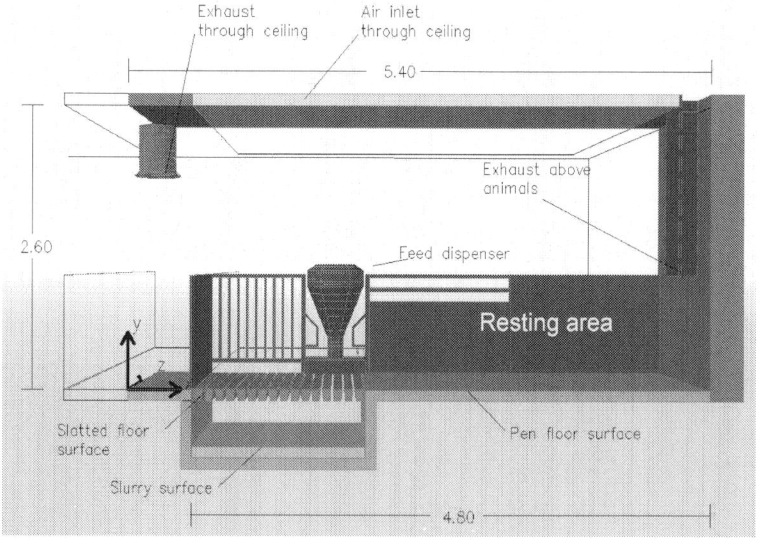

Fig. 12. Pig unit with two alternative ventilation exhausts (through ceiling or above animals).

The lower picture in Fig. 13 shows that the air temperature distribution becomes very similar to the ceiling exhaust case if the air change is reduced 40 % in the case with exhaust above the resting area. Additional analysis shows that a 40 % reduction of air change in combination with location of the exhaust above the resting area will not degrade the condition in relation to carbon dioxide or ammonia concentrations in the room. Potential benefits of such an arrangement will be reduced energy consumption for running the ventilation system, reduced emissions - where especially odour emission can be expected to decrease with decreased air change and finally if air cleaning of the entire air change are required it may result in significant savings in both investments and operation cost.

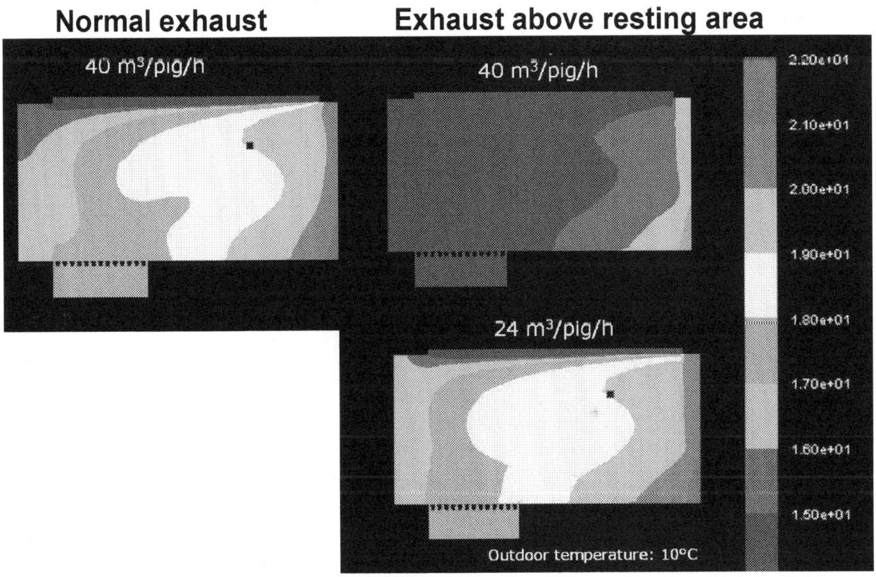

Fig. 13. CFD prediction of temperature distributions (°C) in a pig pen at different exhaust locations and air changes.

The CFD models discussed in this chapter have until this point been delimited to the condition in one pen only, and this simplification may naturally cause that important interactions between pens will be lost. As it is indicated in Fig. 2 a normal ceiling exhaust are usually design to remove air from a number of pens. To visualize the consequences of air streams between pens Fig. 14 show results from an analysis of 3 pens corresponding to one quarter of the pig unit shown in Fig. 2.

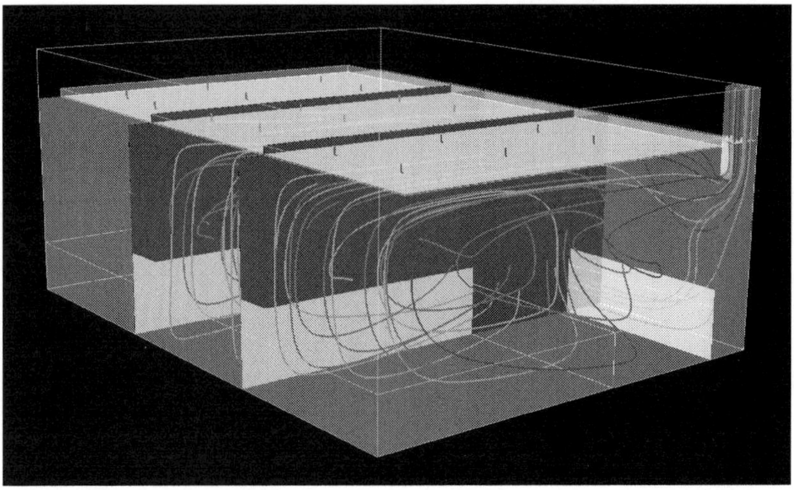

Fig. 14. Simulated air stream lines from inlet to exhaust for air entering a pig unit at 6 different locations above each of 3 pens.

The graph shows simulated stream lines for air entering the room at six locations above each pen. It appears that red stream lines entering the room above the pen in the largest distance from the exhaust pass through the animal occupied zone of al three pens, which has the negative consequence that pigs in one pen become exposed to airborne pathogens that might be released from pigs in another pen. The figure also shows that a large part of the air entering the room above the pen closet to the exhaust leaves without reaching the animal occupied zone, and as result significant difference in temperatures and concentrations between pens may occur. Subsequent analyses have shown that both problems can be avoided by installing an exhaust above the resting area in each pen.

4. Conclusions

The analyses, including CFD-methods (Computational Fluid Dynamics), showed that evacuating and cleaning of 10 percent of the total ventilation capacity from the pit may reduce the ammonia emission of the system with 73 percent, and the ammonia concentration in the room is significantly reduced. In a similar production system without pit ventilation cleaning of 10 percent of the ventilation capacity reduced the ammonia emission with 41 percent compared to no cleaning.

In addition analyses illustrated that CFD methods can be a very useful tool in the development of improved and more efficient ventilation systems.

5. References

Anonymous (2006) Fluent 6.3 User's Guide. Fluent Inc. Fluent Inc, 10 Cavendish Court, Lebanon, NH 03766, U.S.A.

Anonymous (1999) Rockwool A-diffusrulle og Heraklith staldplade-FI Luftintag. Testreport 922 in Danish. Research Bygholm, Denmark.

Baxter S. (1984) Intensive Pig Production. Granada, London Toronto Sydney New York, 588 pp.

Bjerg, B.; Zhang, G. & Kai, P. (2008). Porous media as boundary condition for air inlet, slatted floor and animal occupied zone in numerical simulation of air flow in a pig unit. Conference *Proceedings CD paper OP-1520. AgEng2008. International Conference on Agricultural Engineering, Hersonissos, Crete, Greece, June 23-25 2008.*

CIGR (1984) Report of Working Group on Climatization of Animal Houses. CIGR and the Scottish Farm Building Investigation Unit, Aberdeen, 72 pp.

Griskey, R. C. (2002). Transport Phenomena and Unit Operations. John Weley and Sons, Inc., New York. 448 pp.

Launder, B. E. & Spalding, D. B. (1974). The numerical computation of turbulent flows. *Computer Methods in Applied Mechanics and Engineering*, 3, 269-289

Lyngbye, M.; Riis, A. L. & Feilberg, A. (2006). Luftskiftets betydning for lugt- og ammoniak emission fra slagtesvinestalde. (The significance of air change on odor and ammonia emission from growing pig units) Meddelelse 756 in Danish. Danish Pig Production. 15 pp.

Melse, R. W. & Ogink, N. W. M. (2005): Air scrubbing techniques for ammonia and odour reduction at livestock operations: Review of on-farm research in the Netherlands. *Transactions of the ASAE*: 48, 2303-2313.

Nielsen, P. V.; Allard, F.; Awbi.; Dawidson, L. & Schälin, A. (2007) Computational Fluid Dynamics in Ventilation Design. Rehva Guidebook no 10. ISBN 2-9600468-9-7

Norton, T.; Sun, D.W., Grant, J.; Fallon, R. & Dodd, V. (2007). Applications of computational fluid dynamics (CFD) in the modelling and design of ventilation systems in the agricultural industry: a review, *Bioresour. Technol.* 98, pp. 2386–2414.

Shih, T.-H.; Liou, W. W.; Shabbir, A. & Zhu J. (1995) A new k-ε eddy viscosity model for high Reynolds number turbulent flows. *Computer Fluids*. Vol 24, No 3, pp 227-238.

Strom, J. S. & Morsing, S. (2004). Getting started with StaldVent 5.0. *Report from Danish Institute of Agricultural Sciences, Research Centre Bygholm, Horsens Denmark*, 24 pp.

Sun, H.; Keener, M. K.; Deng, W.& Michel, F. C. Jr. (2004). Development and validation of 3-D CFD models to simulate airflow and ammonia distribution in a High-Rise hog building during summer and winter conditions. *Agricultural Engineering International: the CIGR Journal of Scientific Research and Development*. Manuscript BC 04 004. VI December, 2004.

Wagenberg, A. V.; Bjerg, B.; Bot, G. P. A. (2004). Measurements and simulation of climatic conditions in the animal occupied one in a door ventilated room for piglets. *Agricultural Engineering International: The CIGR Journal of Scientific Research and Developmen*, Manuscript BC 03 020, Vol VI.

Wu B; Gebremedhin K G (2001). Numerical simulation on flow field around an animal using 3-D body-fitted coordinate system. *Journal of Thermal Biology*, 26, 563-573.

25

Air Quality in Horse Stables

Lena Elfman[1], Robert Wålinder[1], Miia Riihimäki[2] and John Pringle[2]
[1]*Department of Medical Sciences,*
Occupational and Environmental Medicine, Uppsala University
[2]*Department of Clinical Sciences,*
Equine Internal Medicine, Swedish University of Agricultural Sciences,
Sweden

1. Introduction

A large number of people are highly engaged in the equine industry around the world, especially in industrialised countries. For example, in the USA about 4.6 million Americans are directly involved in the equine industry, and in Sweden the industry provides full-time employment to over 30 000 people. Horse riding is the second biggest sport in Sweden after soccer, engaging approximately 6% of the population, and is by far the biggest sport for girls. Consequently, many people spend a considerable portion of their day in stable environments, either as workers in the care and training of horses, or in their leisure time.

The type of buildings used for stabling horses can vary widely. Some are newly built and thus designed solely for housing horses, whereas others have been converted or "retrofitted" from barns used for other farming activities, such as housing cattle or for feed storage. However, regardless of design these buildings have many common requirements for indoor environment, including selection of type of feed and bedding, and the need for continual removal of faeces and urine. Moreover, the building ventilation must provide adequate exchange of fresh air to the horses' living space to regulate the stable temperature and remove moisture, stable gases and organic debris produced by the horses and their activities.

There is no need to remove excess heat in stables in the winter in northern climates. However, this puts more focus on ventilation, which must rid the stables of moisture, odour, and ammonia that have built up in the more closed environment of the stable. Moisture originates from horse's respiration, faeces and urine, and other stable activities such as showering the animals, and general stable cleaning. Accompanying this moisture build-up is increased risk of condensation, intense odour, and increased ammonia release from bedding that ventilation must remove. How much ventilation should be provided? Ventilation of one air change per hour (ACH) means that the total volume of air in the stable is replaced in one hour. Guidelines for horse stables suggest providing between 4 to 8 air changes per hour to reduce mould spore contamination, minimize condensation, and reduce moisture, odour, and ammonia accumulation (Horse Stable Ventilation, 2003). Thus, for the mean of 6 ACH a stable's ventilation system must be capable of a complete air change every 10 minutes.

With the inevitable seasonal appearance of subzero winter weather, horse owners occasionally work against these principles of ventilation in an unheated building by trying to maintain warm stable conditions for their personal comfort during animal care activities. In addition, horses in active competition throughout winter may also be kept in warmer housing, which is why many stables prevent a seasonal change to a thicker, winter hair coat. Thus, during winter conditions, the inflow of cold outside air is limited by the closure of windows, blockage of ventilation channels, or even closure of entire sections of the stable to provide a more comfortable (personal) working environment. This, however, can adversely impair the removal of excess moisture, stable gases and organic air particles that should be replaced by fresh outside air via ventilation.

The outside air is presumed to be the optimal in air quality for respiratory health, and ventilation of stables aims to approach outside-air quality for both the horse, and the horse owner. Curiously, there is recent evidence that horses spending time outdoors in cold climates can acquire airway inflammation (Robinson et al., 2006). During winter with subzero temperatures, the air holds very little water (≤ 2 g/m^3). Thus the horse's respiratory system must both warm and humidify the inhaled air before it reaches the alveoli. In particular, exercise in cold air has been associated with up-regulation of inflammatory cytokines in the airways (Davis et al., 2005; Riihimäki et al., 2008a) and increased inflammatory cells in the lung in actively training race-horses during winter (Riihimäki et al., 2008a). However, despite the theoretical disadvantages of the horse breathing cold air, the bulk of evidence suggests that many components of stable air, and its gaseous and particulate matter, such as endotoxin, are perhaps greater risk factors for induction of airway inflammation in both the stabled horses (Pirie et al., 2003; Berndt et al., 2010) and for those people working in the stable (Tutluoglu et al., 2002).

The air of farm buildings housing large animals contains a wide variety of gases and organic dusts from microbial, plant, and animal sources. The organic dust may contain bacteria that are pathogenic or non-pathogenic, living or dead, and fungi, high molecular weight allergens, bacterial endotoxins, 1-3-β-glucans, pollen, and plant fibres (Douwes et al., 2003). Two major pro-inflammatory components of organic dusts include bacterial endotoxin, a component of the cell wall of gram-negative bacteria, and mould 1-3-β-glucans (Douwes et al., 2003). Endotoxin has been shown to be present in the air of animal housings in amounts presumed to be a respiratory hazard to both humans and livestock animals (Pomorska et al., 2007). However, apart from proposed guidelines (Donham et al., 1989) there are no officially accepted limits for allowable endotoxin content in the air for either farm animals or humans. Even less is known regarding the critical threshold for 1-3-β-glucans before impacting on airway health. However, it is well known that exposure to the variety of organic dust particles, microorganisms and endotoxins present in different farm animal stable buildings can cause chronic airway diseases in both humans and horses. In particular, recurrent airway obstruction (RAO, heaves), which is an animal model of asthma, and the syndrome of inflammatory airway disease (IAD) in horses are major, non-infectious, largely environmental causes of reduced performance and persistent coughing in stabled horses. For the first-named disease, endotoxin appears to act in synergy with other airway challenges to induce excessive airway inflammation even with low dust challenge (Pirie et al., 2003).

Stable environments can also be influenced through selection of type of feed, bedding and sanitary procedures, any or all of which can have a complex influence on stable-air quality (McGorum et al., 1998). For example, when assessing the influence of bedding on concentration of ammonia in the stable air, wheat straw was preferred over other materials such as wood shavings or straw pellets (Fleming et al., 2008a, 2008b). In contrast, bedding material low in inhalable dust, such as shredded paper, appeared to be preferable to straw bedding as the former was associated with reduction of markers of airway inflammation in horses with RAO (Wyse et al., 2005). However, in practical terms use of paper as bedding in horse stalls is associated with cumbersome handling and manure disposal. Peat moss is another popular stable bedding product, touted as being low in organic dust. However, depending on its source this product has been shown to vary widely in dustiness and hygienic quality (Airaksinen et al., 2005). Even local differences of bedding between boxes within a stable can influence the overall respirable particle load in the common stable airspace (Clements & Pirie, 2007). Clearly the various forms of bedding in a stable and the differences in bedding between boxes within a stable can have a substantial, yet elusively defined effect on the stable air dust load, and consequently on the airway health of the horses and people present.

The type of feed and method of feeding can also influence the amount of dusts being inhaled via stable air. For example, providing a low-dust feed in the form of ensilage has been shown to reduce the markers of airway inflammation in stabled horses (Wyse et al., 2005). In addition, even short-term immersion of dusty feed (hay) can reduce the air particle load in the stabled horses' breathing zones (Clements & Pirie, 2007). It is not known if these reductions in dust from horse feed are also important to the respiratory health of people.

Even the methods used to routinely clean the stable, irrespective of bedding or feed type, can have a substantial influence on the stable air quality. It is advisable to carry out work when there are no horses in the stables because activities in the stable (feeding, mucking out, and sweeping) cause increased levels of ammonia and airborne particle concentrations (Fleming et al., 2008a, 2008b; Sadegh et al., 2009). However, while common sense would suggest that daily cleaning and replacement of bedding, or at least daily removal of manure, would minimize accumulation of undesirable stable organic and gaseous materials, this may not be the case. Fleming and colleagues (Fleming et al. 2008a) have recently found evidence to the contrary - that, for a 2-week period, regardless of bedding type, the least build-up of particles less than 10μm in diameter, and of ammonium, was achieved when there was no mucking out or even manure removal. Presumably a build-up of urine during the course of the trial increased the moisture of the bedding material, which led to increased binding of organic particles that were subsequently not released into the atmosphere. If ones' goal is solely to minimize stable air respirable dust then, based on these results, it would be advisable to dispense with a daily mucking-out regimen for week-long periods and simply add additional bedding regularly to absorb urine. However, the choice of mucking-out regimen should not be based solely on single, selected measurements of factors such as ammonia and particle concentrations since the growth of pathogenic germs and fungi would certainly be favoured if the same bedding were maintained for some weeks. Moreover, specific composition of the stable air particles, in particular endotoxin and glucans, are likely to be far more important to respiratory health than increased stable air particles or manure gases. To help clarify which factors in the stable air are important to respiratory health, ventilation capacity, organic dusts, microorganisms and/or fungi were measured and correlated to the respiratory health of the horses and people breathing the stable air daily.

1.1 Aim

The aim of this chapter is to describe differences in indoor air in horse stables under winter and summer conditions. An intervention study is also described with measurements of stable air quality before and after installation of mechanical ventilation in a previously naturally ventilated riding school stable. In the same stable, detailed measurements of stable air quality were performed together with measurements of respiratory health of stable personnel and stabled horses. The hypothesis is that selected components of stable air will be correlated to indices of respiratory health in people and in stabled horses spending considerable time in the stable environment. Another aim is to determine if these indices are improved after changing the ventilation system of the stable.

2. Material and methods

2.1 Study designs

Seasonal changes in air quality were studied in a conventional racing stable with natural ventilation. The indoor stable environment, personnel and horses were investigated three times, first in the winter stabling period (February, year 1), second after the intervening summer (September, year 1), and the third time in the following winter (March, year 2). The second step was to perform an intervention study by introducing a mechanical ventilation system in a riding school stable.

2.2 Hygiene measurements in stable environment

Environmental monitoring was generally performed during the morning routines in the stables, when the personnel were cleaning the boxes or stalls by mucking-out the faeces and urine. Fresh straw was put in for bedding. Horses were typically fed three times a day with hay/haylage, pelleted fodder and oats: in the morning before going out to the fields, in the afternoon before training sessions, and in the evening. A second sampling period was performed in the afternoon, when horses were groomed before training sessions and made ready for riding, and when people and horses were going in and out sporadically.

Hygienic measurements included real-time monitoring (24h) of carbon dioxide (CO_2), temperature, relative humidity (RH%), ultra-fine particles and particulate matter (PM10, particle size 10 μm). Temperature, RH% and CO_2 were generally logged every minute with a Q-Trak™ (model 8550, TSI Inc., Minnesota, USA). Monitoring of ultrafine particles, which are defined as <0.1 μm particles, were performed with a P-Trak™ (model 8525, TSI Inc., Minnesota, USA). Airborne dust concentrations of PM10 were measured with a Dust-Trak™ aerosol monitor (model 8520, TSI Inc., Minnesota, USA) (Riihimäki et al., 2008a; Elfman et al., 2009).

Passive sampling is a convenient method for ammonia measurements. Passive samplers (IVL, The Swedish Environmental Research Institute, Gothenburg, Sweden) were put out at approximately 1.5m above the floor for 24 hours at three times during a week (Kirchner et al., 1999). Indoor samples were compared with outdoor sampling for 1 week at different distances from the stable (1-100m). Samples can be analysed by e.g. IVL, Sweden and values are expressed as $\mu g/m^3$.

Air sampling was performed with pumps for sampling of horse allergen, total and respirable dust, airborne microorganisms and products thereof such as endotoxin and 1-3-β-glucan. Generally, three samples for each factor were collected during a 4-7 hour sampling round. Pumps (SKC Inc., Eighty Four, PA, USA) were run with a flow of 2L/min, and placed at 1-1.5m above ground at representative places in the stable.

Airborne total and respirable dust were collected on a membrane filter in a cassette (∅ 25 mm, pore size 0.8 μm). For respirable dust, a metal cyclone (SKC Inc., USA) was used before the filter cassette. The pump was attached to the person's clothing and the filter head was placed at the breathing zone. For horses, the pump was put on a lunging-girth on the horse's back and the filter head was attached on the halter close to the horse's muzzle. Airborne dust samples were analysed by a gravimetric method and the organic portion were calculated after combustion of the filter and weighing of the remaining inorganic material. Results were expressed as mg/m^3.

Collection of microorganisms and endotoxins in the air were performed with cassettes with a sterile nucleopore filter (∅ 25 mm, pore size 0.4 µm). Surface sampling was performed with a Scotch-brite® or tape on the walls in stalls or boxes (Wålinder et al., 2011). The total concentration of airborne and surface bacteria and fungi was analysed by the CAMNEA method (Palmgren et al., 1986) based on acridine orange staining and epifluorescence microscopy. The detection limit for viable organisms was 30 colony forming units (CFU) per m^3 of air. Analysis of endotoxin was generally performed with the kinetic turbidimetric Limulus test. Results were expressed as ng/m^3 or EU/m^3. Analysis of 1-3-β-glucan was made with the Limulus test with glucan-specific lysate in the chromogenic kinetic version and results were expressed as ng/m^3 (Cape Cod Inc., MA, USA and Endosafe, Charles River Endosafe, Charleston, USA).

Horse allergen particles were collected with a pump fitted with an IOM-filter holder (SKC Inc., USA) with a fluoropore membrane filter (pore size 1.0 µm, Type FA, Millipore, USA). Airborne samples were analysed with an ELISA method using monoclonal antibodies for horse allergen (Mabtech, Nacka, Sweden) and results were expressed as Units/m^3 of air (Elfman et al., 2008).

2.3 Investigation of human respiratory health

2.3.1 Subjects

The first study was performed in a race-horse stable with 13 employees (Table 1). Initially, all personnel agreed to participate but, due to high turnover among the stable personnel, only one person fulfilled all three rounds of this study. The intervention study was performed in a riding school with 7 personnel and approximately 320 students aged 7 to 18 years attending the riding school (Table 1). All personnel agreed to participate, but one employee dropped out during the first sampling round and one left the workplace before the second sampling time point. Riding students older than 12 years, and who spent more than one day per week at the riding school, were asked to participate in the investigation. In the end, seven students gave their written consent to participate after permission was also obtained from their parents.

	Race-horse stable	Riding-school stable	
Number of	Personnel	Personnel	Students
Subjects	*n=13*	*n=7*	*n=7*
Males/Females	*6/7*	*1/6*	*1/6*
Mean age (range) years	*33 (22-66)*	*34 (21-51)*	*14 (13-17)*
Smokers	3	*0*	*1*
Atopics	*7*	*0*	2
Doctor-diagnosed asthma	*1*	*0*	*0*
Signs of bronchial obstruction*	3	*0*	*0*

*Signs of bronchial obstruction via increased PEF variability during daily measurements for two weeks

Table 1. Personal characteristics of stable personnel and riding school students.

2.3.2 Questionnaire studies

Personnel in the race-horse stable and the riding school stable as well as the riding-students were asked to answer a questionnaire regarding health issues. The questions were about current or earlier allergic diseases including asthma and medication, as well as current symptoms of eczema, eye irritation, nasal congestion and dyspnoea during the week prior to the investigation. Furthermore, they were asked to give their ratings of smell and dustiness in the stable.

2.3.3 Pulmonary function tests

Lung volumes and peak expiratory flow (Table 2) were measured by dynamic spirometry using a Spirobank G (MIR, Rome, Italy) (Figure 1) according to the American Thoracic Society standards (ATS, 1995).

Fig. 1. Portable spirometer (Spirobank G) for pulmonary function tests. Photo: Robert Wålinder

Parameter	What is measured?
FVC	Forced vital capacity (L)
FEV1	Forced vital capacity in one second (L)
PEF	Peak expiratory flow (L/min)
PEF-variability	Repeated PEF measurements, four times per day during two weeks (expressed as total variability between max. and min. in percent or coefficient of variation)
NO	Concentration of nitrogen oxide in exhaled air (ppb)

Table 2. Pulmonary function test parameters.

Subjects were also instructed to make four daily measurements of peak expiratory flow (PEF) for two weeks, using an electronic disposable flow meter (Piko-1, Medeca Pharma, Uppsala, Sweden) (Table 2). In the intervention study performed at the riding school, concentrations of nitrogen oxide (NO) in exhaled air were measured with a NIOX MINO (Aerocrine AB, Solna, Sweden) (Figure 2).

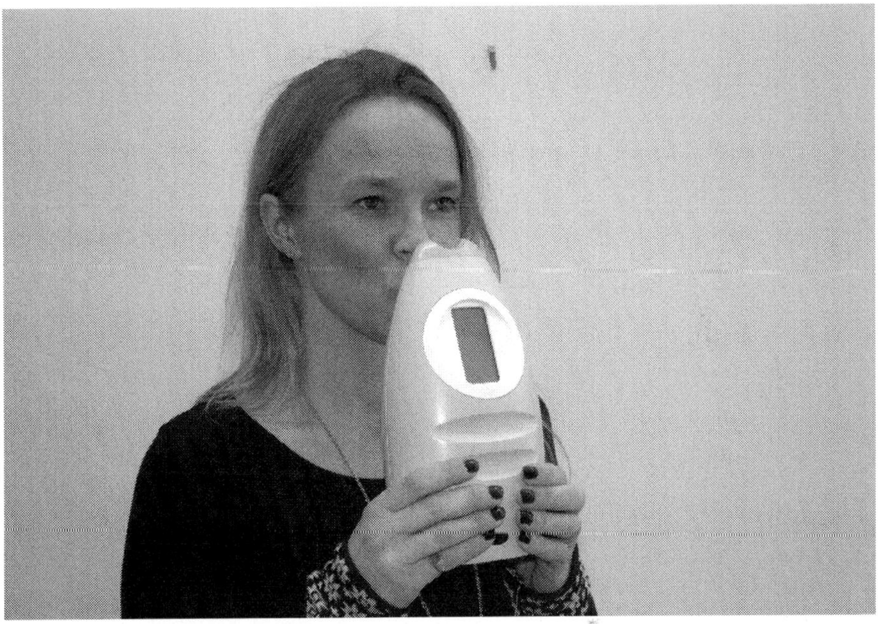

Fig. 2. Measurement of nitrogen oxide in exhaled air (NIOX MINO). Photo: Robert Wålinder

2.3.4 Nasal lavage and analysis of biomarkers

Different reactions in the nasal mucosa to environmental exposures can be explored by lavage of the nose and subsequent analysis of proteins in the lavage fluid (Table 3). All the personnel at the two horse stables therefore underwent lavage of the nasal cavity. Five millilitres of isotonic saline solution were flushed back and forth five times via a nasal olive attached to a 20 ml syringe. The lavage fluid was kept on ice until centrifuged twice at 800xg and 1200xg, respectively (Wålinder et al., 1998). The remaining supernatant was frozen at -20 °C until biochemical analysis. Biomarkers analyzed included eosinophil cationic protein (ECP), which is a marker of eosinophil activity (Venge et al., 1987), myeloperoxidase (MPO) from neutrophils in the mucosa (Venge, 1994), lysozyme a marker of neutrophil activity and secretion from parasympathetically innervated mucosal glands (Raphael et al., 1989), and albumin, which is a marker of capillary leakage of plasma proteins (Raphael et al., 1991). Analyses of ECP, MPO and lysozyme were made by radio-immunoassays (RIA) and albumin by nephelometry at the University Hospital of Uppsala.

ECP	Eosinophil cationic protein: the concentration reflects the presence and activity of eosinophil granulocytes
MPO	Myeloperoxidase: the concentration reflects the presence and activity of neutrophil granulocytes
Lysozyme	Is mainly released by parasympathetic stimulation of glandular cells, and the concentration is a measure of the secretory activity of the nasal mucosa. Smaller amounts are also released by neutrophils, macrophages, and monocytes.
Albumin	The concentration reflects the vascular leakage and also glandular secretion.

Table 3. Biomarkers in nasal lavage reflect different reactions in the nose

1.4 Investigation of horses

There are several techniques used in research and practice to detect, quantify, and characterize the inflammatory process in the respiratory tract of horses. These include physiological measurements. There are also various analyses aimed to examine cellular and non-cellular immunologic components. Diagnosis of airway inflammation in horses is determined based on a combination of history, clinical examination, endoscopy and, in many instances, bronchoalveolar lavage (BAL) cytology (Derksen et al., 1985; Tremblay et al., 1993). Results from pulmonary function tests are used in research and for clinical purposes for disease monitoring and in grading of disease severity (Derksen et al., 1985). Additionally, exhaled breath condensate has been introduced to evaluate inflammation markers in the lung (Wyse et al., 2005). In the search for potential allergens involved in RAO, an intra-dermal skin test and serum IgE analysis have been evaluated, but these have not proven to be clinically reliable in horses (Lorch et al., 2001; Lebis et al., 2002). Bronchial brushings, BAL, tracheal lavage, exhaled breath condensate, and even lung tissue samples have been shown to be of value in the investigation of the inflammatory process in the equine lung (Bureau et al., 2000; Gerber et al., 2003; Wyse et al., 2005). Imaging techniques, such as radiographic examination of equine lung, however, have shown to be of little value for the diagnosis of lower grade inflammation (Mazan et al., 2005).

Tissue samples from the airways provide an opportunity to further study pathological changes related to structural and cellular morphology. The trans-thoracic lung biopsy, the classical way to obtain tissue material from the distal airways from RAO horses, (Schatzmann et al., 1974; Pusterla et al., 2007) has recently been complemented by thoracoscopically guided pulmonary wedge resection (Lugo et al., 2002; Lugo et al., 2006). However, both these techniques are invasive, and the former may be associated with fatal complications (Savage et al., 1998). Endobronchial biopsies, which are a safer alternative, have been used increasingly in human medicine as a complementary sample to BAL and induced sputum. This sampling method is used for studying the underlying pathogenesis of, for example, asthma and chronic obstructive pulmonary disease, and evaluation of effects of different treatments (Jeffery, 1998; Jeffery et al., 2003; Behndig et al., 2006). The utility of bronchial biopsy for study of RAO in horses has been reported recently (Ainsworth et al., 2006; Riihimaki et al., 2008a).

In the first study, we investigated respiratory health in race horses, stabled in a conventional stable environment, during different seasons (Riihimaki et al., 2008a; Elfman et al., 2009). Twelve standard-bred trotters (8 mares, 3 geldings, 1 stallion) aged 3-7 years were included in the study. Detailed clinical and respiratory examinations were performed including the following methods: pulmonary auscultation with and without re-breathing bag, cough provocation, routine venous and arterial blood sample analysis, upper and lower airway endoscopy, and BAL fluid collection under endoscopic guidance (Figure 3 and 4). The recovered BAL fluid was used for cytological examination and quantitative real-time multiplex PCR-analysis for cytokine mRNA expression of interleukins (IL) IL-6, IL-8 and IL-10. In the second study, the effects of stable air quality on equine airways were evaluated before and after installation of mechanical ventilation. From the riding school stable, twelve mixed-breed horses were included in the investigation. The respiratory health of horses was based on detailed clinical and respiratory examinations, including chest auscultation following re-breathing, scored from normal to markedly abnormal values. Routine blood sample analysis for white and red blood cell parameters and plasma fibrinogen was performed in order to detect any infections and/or general inflammatory processes. All horses were examined by an upper and lower airway endoscopy and BAL at each sampling occasion. The BAL samples were analysed for cytology and cytokine mRNA expression as stated above.

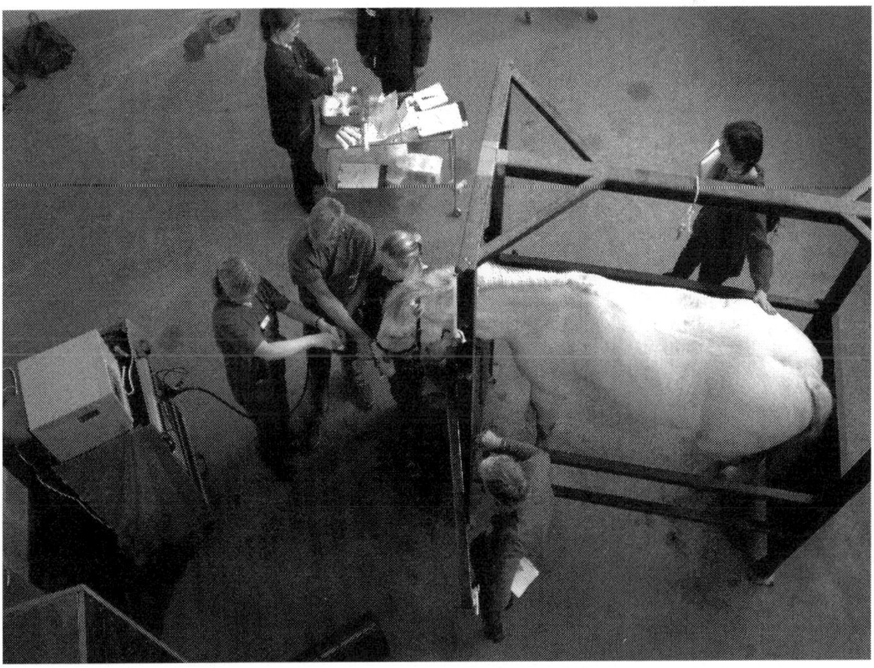

Fig. 3. Endotracheal bronchoscopy in a horse. Photo: Roland Thunholm

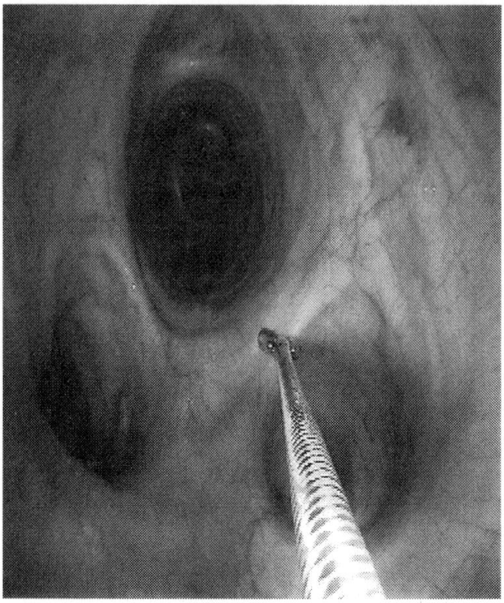

Fig. 4. Bronchoscopic view in a horse. Photo: John Pringle

3. Results

3.1 Hygienic measurements

3.1.1 Seasonal variations in air quality

Indoor climate in a stable can be monitored by using logging instruments. A typical example of logging temperature, relative humidity (RH%) and CO_2 over a week is shown in Figure 5. During winter, the outdoor temperature can be far below zero, but the aim is to keep indoor temperature above zero to prevent water freezing in the stable. The CO_2-levels in the stable generally fluctuate from about 250-300 ppm (the same as outdoor levels), when horses are not inside, to about 1500-3000 ppm during the night, depending on how many horses are kept in the stable and the type of ventilation. For horses, the upper acceptable limit value is 3000 ppm according to regulations on farm animal housing (Swedish Animal Welfare Agency, 2004), while for humans the recommended indoor value is 1000 ppm. In stables without insulation, as seen in Figure 5, the relative humidity should not exceed the outdoor RH% by more than 10% units. In stables with insulation, the relative humidity during winter should not exceed 80%, unless the temperature in the stable is less than 10 °C.

Results from logging particles in the air over a 24-hour period in a stable with three horses are shown in Figure 6. The activities that typically take place in the stable cause ultrafine particles to vary more and stay airborne longer than PM10 particles. Many factors affect the levels of particles, including horse and human activities, type of feed and bedding material. In this example the bedding material was a relatively new type of pelleted pinewood saw dust, which is watered to give a comfortable permanent bed. This type of bedding material is slightly less dusty than peat bedding, but there are no significant differences between the two types of bedding.

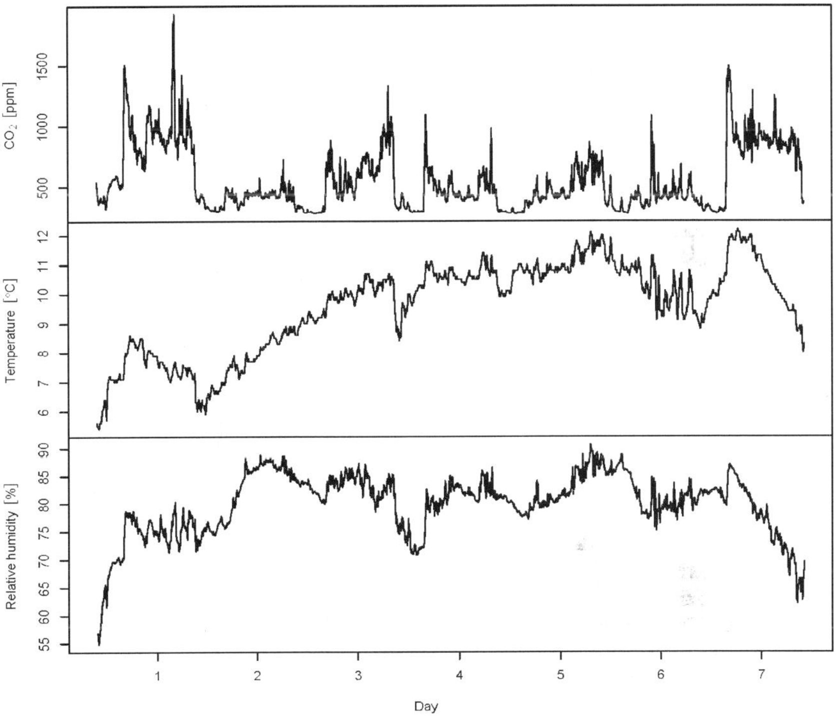

Fig. 5. Temperature, RH% and CO_2 in a stable logged over a week in winter with a Q-Trak™

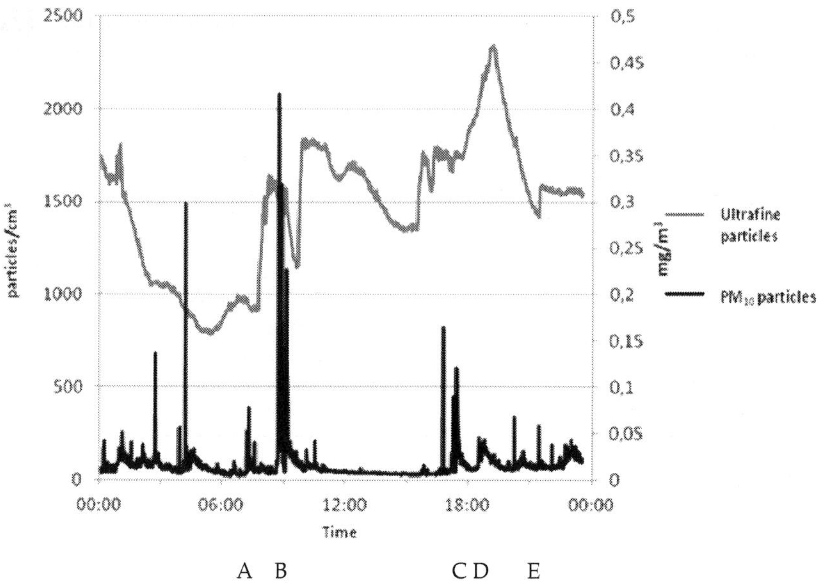

Fig. 6. Ultrafine particles (P-Trak) and PM10 particles (Dust-Trak) logged during 24 hours A=horses let out; B=measurements in stable; C=cleaning the stables from urine and faeces D= horses let into stable; E= evening feed

Results from total and respirable dust measurements are shown in Figure 7, and show variation between seasons: higher in winter when doors and windows are closed and lower in summer when they are kept open. These results are from a stable housing 18 horses and without mechanical ventilation. The bedding was straw. Horses were fed three times a day with haylage and pelleted fodder. The organic dust level was approximately 70% of total dust (range 0.4 - 0.8 mg/m³), which is well below the Swedish hygienic limit value for humans (5 mg/m³) (Swedish Work Environment Authority, 2005) or the Dutch proposed occupational exposure limit of nuisance dust which is the same as for horses (10 mg/m³). According to Sadegh and co-workers (Sadegh et al., 2009) dust, endotoxin and 1-3-β-glucan are considerable in horse stables, while bacterial and fungal exposures are moderate. They reported that sweeping the floor is the predominant task that explains high levels of dust, endotoxin, and 1-3-β-glucan, and feeding the horses is a specifically important contributor for high 1-3-β-glucan levels.

Levels of airborne bacteria were slightly increased in February and September of year 1, but were normal in March of year 2, while fungi were slightly increased at all three sampling times compared to reference environments without microbial damage (Figure 8a). Samples from the indoor walls showed slightly increased levels of bacteria in February, year 1, but were normal in September, year 1 and March, year 2, compared to unaffected building materials. Levels of fungi were slightly increased on the inner wall surfaces at all three sampling times (Figure 8b). Endotoxin levels were lower in winter (February year 1, median 5 ng/m³) and increased somewhat in the summer (September year 1, median 14 ng/m³). Levels of 1-3-β-glucan, on the other hand, were higher in the winter sampling time

(February year 1, 1.85 ng/m^3), and lower in summer (September year 1, 1.21 ng/m^3) (Riihimäki et al., 2008).

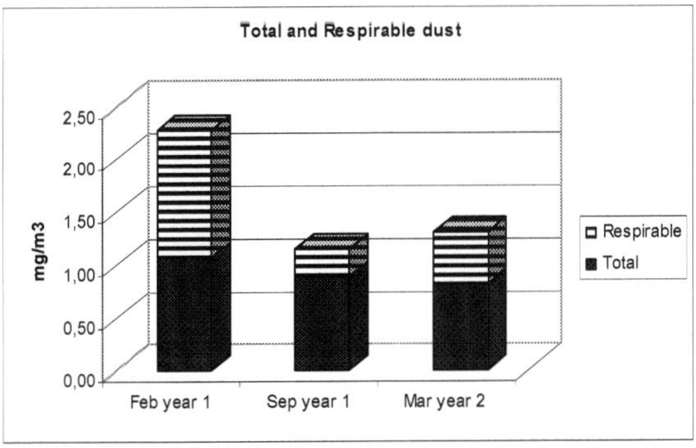

Fig. 7. Levels of total and respirable dust at three sampling time points.

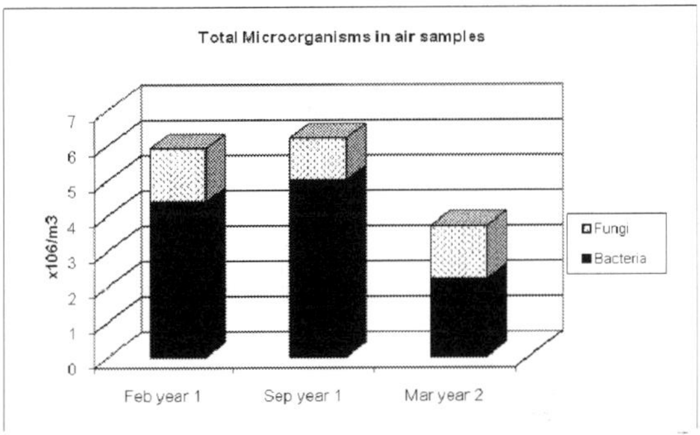

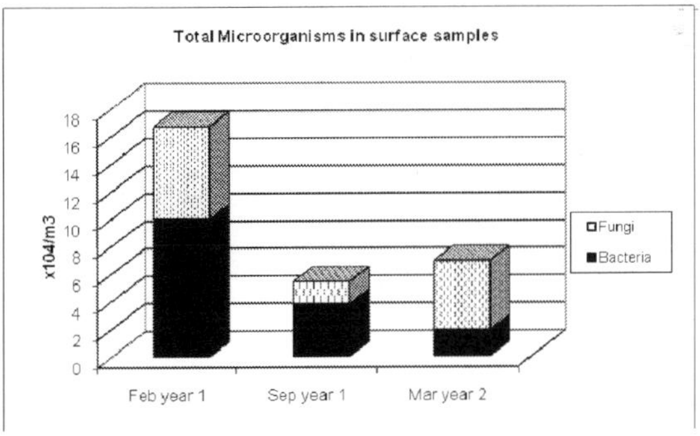

Fig. 8. a) Levels of total microorganisms in air samples; b) levels of total microorganisms in surface samples, at three sampling points: Feb, year 1; Sep, year 1; and Mar, year 2.

3.1.2 Bedding material and influence on air quality

The choice of bedding material affects the air quality in stables. Some materials are dustier than others and the capability to reduce ammonia varies a lot (Fleming et al., 2008a, 2008b; Airaksinen et al., 2005). We have gained some experience by evaluating two bedding materials: peat and pelleted pine sawdust. The latter material is also used for heating houses. The pellets are distributed over the floor in the stall and then watered until they fall apart and form a soft bed. This material can retain a lot of water or urine, without feeling wet, and is a very suitable material for permanent bedding. The following hygienic measurements were investigated: particles, temperature, air humidity, carbon dioxide, volatile organic compounds, horse allergen and ammonia. Samples from the two bedding materials were also analyzed for microorganisms.

The results showed that there were no differences in levels of PM10-particles (<10 μm) or ultrafine (<1μm) particles (Figure 6), when data had been adjusted for temperature and air humidity. The concentration of carbon dioxide did not differ between the bedding materials. The level of volatile organic compounds in the air was higher when pelleted wood was used than with peat, and δ-limonen was only detected with pelleted wood. After four weeks, high amounts of microorganisms were present in both bedding materials, with levels in peat 10-100 times higher than in pelleted wood. The conclusion is that there are no significant differences between these two bedding materials regarding amount of dust in the stable environment. Peat tends to give a better stable environment due to higher absorption of ammonia and lower levels of volatile organic compounds, while wood pellets have other advantages: they are easier to handle, use half the storage space, do not freeze in winter, and are the most cost-effective product.

3.1.3 Hygienic measurements before and after installation of mechanical ventilation

The most striking effect of installing mechanical ventilation in a stable was the reduction of carbon dioxide. The CO_2-level in a riding school stable with only natural ventilation was 950 ppm (median), with a range of 990 - 1280 ppm. At the summer sampling time in August, when indoor stable conditions were the best possible, with open windows and doors, the median CO_2-level was only slightly reduced (800 ppm, range 670 - 900 ppm). After installation of a balanced supply and exhaust mechanical ventilation system, the CO_2-levels had improved to 510 ppm (range 350 - 750 ppm).

Other parameters that improved with mechanical ventilation included ammonia, horse allergen and ultrafine particles. The median level of ammonia dropped from 3200 μg/ m^3 before intervention to 1330 μg/ m^3 after intervention. The median horse allergen level was 5 170 U/m^3 in the winter with natural ventilation, which dropped to 790 U/ m^3 with mechanical ventilation, which was almost as good as in the intervening summer sampling time point of 750 U/ m^3. The median level of ultrafine particles was 8000 particles (pt)/cm^3 in the winter before intervention and 5 400 pt/ cm^3 after intervention (Figure 9), while the median summer value was 1 500 pt/cm^3.

The factors that did not significantly change after intervention were total and respirable dust. The median total dust was 210 μg/m^3 in February before intervention, and 220 μg/m^3 in the winter after intervention, while the median respirable dust level was 100 μg/m^3 the winter before and 130 μg/m^3 the winter after intervention. Microorganisms in air samples showed almost the same level for bacteria (approx 20 x 10^5/m^3), while fungi increased from 3.5 x 10^5/m^3 before to 6.6 x 10^5/m^3 after intervention. Median endotoxin levels in air samples increased after intervention from 96 to 275 EU/m^3. We have no real explanation for this increase, but it could depend on increased levels of endotoxin due to handling of slightly mouldy hay or straw. Glucan levels were not investigated in this study (Wålinder et al., 2011).

We believe that it is best to compare before and after intervention (for example, installation of mechanical ventilation) during the winter stabling period, since this is the season with the most challenging indoor climate. Despite the need for air exchange, doors and windows have to be kept closed most of the time to avoid freezing of drinking water and drainage pipes. In the study presented here we chose to install an automatic system regulated by the indoor temperature in the stable. That means that the airflow runs at maximum (2 200 L/s) at high temperature and at a minimum of 400-500 L/s when the temperature is very low, otherwise the indoor temperature would go below zero. Ventilation systems have to be run this way in cold temperate climates since it is too expensive to warm up the incoming air. In many stables, ventilation systems are even shut down in winter to reduce the inlet of cold air and the result is often bad air quality with a lot of dampness kept indoors. This means that seasonal variations, such as differences in hygienic measurements between winter and summer, are often bigger than differences seen before and after installation of mechanical ventilation. In summer, stable doors and windows are kept open and horses spend most of their time outdoors in pastures, which is why the indoor stable environment is usually at its best. Levels of microorganisms can be elevated reflecting the generally higher outdoor levels in late summer, i.e. end of August to beginning of September, in temperate climates in the northern hemisphere.

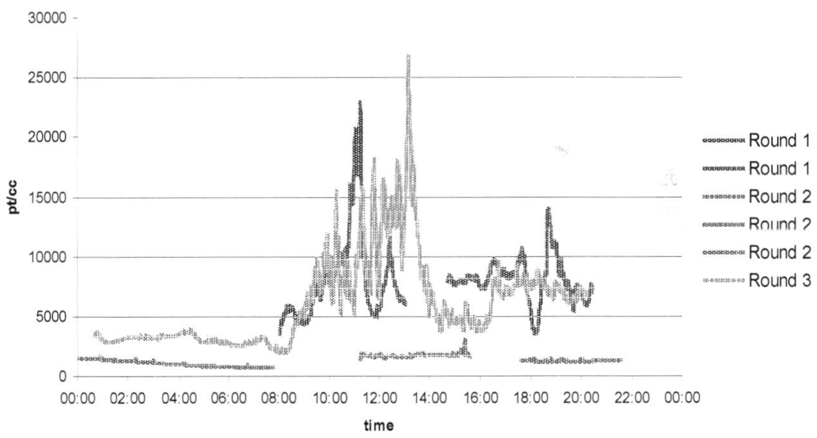

Fig. 9. Levels of ultrafine particles (P-Trak) logged over 24 hours. Round 1=in winter before intervention with only natural ventilation, Round 2= in summer before intervention, Round 3=in winter after installation of mechanical ventilation.

3.2 Air quality in stables and effects on human airways

3.2.1 Respiratory health in personnel at a conventionally managed race-horse stable

In this study one of the subjects reported a history of asthma. However, 3 out of 13 subjects had increased PEF-variability (CV>20%), indicating bronchial obstruction. These individuals also had increased levels of ECP in nasal lavage, indicating allergic inflammation equivalent to allergic asthma. Only two of the personnel reported work-related airway symptoms. Compared with teachers (Wålinder et al., 2000) the stable personnel in the present study (9 out of 13) also had increased levels of both lysozyme and myeloperoxidase (MPO) in the nasal mucosa (Table 4).

Biomarker	Race-horse stable personnel	Riding school students	School teachers	
No of subjects	n=13	n=7	Males n=59	Females n=352
ECP (μg/L)	<2	<2	<2	<2
MPO (μg/L)	77	26	20	10
Lysozyme (mg/L)	5.5	3.7	2.0	1.7
Albumin (mg/L)	6	<5	<5	<5

Table 4. Comparison of biomarker concentrations (median) in nasal lavage in personnel (adults) at a race-horse stable, students at a riding school stable, and among teachers in schools.

3.2.2 Respiratory health in personnel and riding-students before and after installation of mechanical ventilation in a stable

All personnel and riding-students, enrolled in the study at the riding school stable, had normal pulmonary function. None reported a history of asthma, but two of the students had hay-fever (Table 5). This means that all people included in this study had better respiratory health than the personnel at the race-horse stable. A growth-related increase of pulmonary function among the riding-students (forced vital capacity, FVC, increased as a mean 0.3 L in the students during the study period of 13 months) was observed. There was a slight, but not significant, decrease in PEF-variability in both personnel and riding-students (Table 5).

	Personnel n=7		Riding-students n=7	
	Before	After	Before	After
PEF (L/min) (SD)	426 (65)	407 (25)	387 (50)	416 (40)
PEF-variability (CV)	9.7%	3.5%	10.3%	7.2%
NO (ppb)	12	11	8	13

Table 5. Pulmonary function values before and after intervention.

Exhaled NO was low before, and was mainly unaltered after intervention. Levels of biomarkers of inflammation in nasal lavage from the personnel were somewhat lower in comparison with results from the personnel at the race-horse stable (Table 4) but did not change after intervention. Although air exchange increased after intervention, the mean ratings of smell and dustiness were slightly higher (not significantly) among personnel and students. The number of subjects having eye irritation, nasal obstruction, dyspnoea, or eczema was either slightly lower or unchanged after intervention (Table 6).

Symptoms	No. of subjects with symptoms before n=14	No. of subjects with symptoms after n=12
Eye irritation	5	4
Nasal obstruction	13	7
Dyspnoea	2	1
Eczema	3	3

Table 6. The prevalence of subjects having symptoms of eye irritation, nasal obstruction, dyspnoea, or eczema before and after installation of mechanical ventilation in a horse stable.

Apart from a growth-related increase of pulmonary function among the riding-students, intervention did not show any significant health effects on the airways. Although the hygienic measurements did show increased air exchange rate, reduced dust, ammonia and horse allergen levels, the absence of measurable health effects after intervention could be due to either a low statistical power, because of a small study-population, or a healthy-worker effect in this stable, or both. It could also depend on the shortness of the interval between installation of the mechanical ventilation and the follow-up.

3.3 Air quality in stables and effects on horse airways

The aim was to examine possible effects of seasonal changes, that is summer versus winter conditions, on horse airways. Despite the complexity of the study we were able to detect a significant up-regulation of IL-6 mRNA expression during both winter periods compared to summer, which coincided with a trend of increased neutrophils in BAL fluid. It is, however, unclear whether the increased IL-6 mRNA expression in BAL during winter was an effect of the stable environment and/or exercise in cold weather conditions. We also found significant alterations in IL-10 mRNA expression between summer and one winter. The change was, however, detected in samples from only one of the two winters. Furthermore, there was no correlation between cytokine mRNA expression and BAL cell cytology. There were no significant seasonal differences in any other clinical parameters, including blood parameters, PaO_2 and subjectively-graded appearance of mucous in trachea (Riihimaki el al., 2008a).

Unfortunately, it was discovered later that three out of twelve horses showed transient pulmonary eosinophilia when the horses were sampled in late summer (Riihimaki et al., 2008b). However, this was probably related to factors other than stable environment. Our initial research plan and hypotheses were based on the twelve horses, but since three horses had obvious pulmonary eosinophilia they were not included in the analyses. It was not possible to perform measurements in the following summer due to a change in ownership of the stable. It would also have been valuable to have data on race-horses not being stabled, but kept in pastures over the summer.

3.3.2 Respiratory health in horses before and after installation of mechanical ventilation in a stable

In horses, a significant decrease in accumulation of tracheal mucous was the strongest indication of reduced airway inflammation after intervention (Wålinder et al., 2011). Cytological results from BAL were normal except for in two horses that showed a slightly increased percentage of neutrophils after intervention, and one of the horses also had increased respiratory rate and respiratory sounds. Of the cytokines investigated, the expression of IL-6 mRNA in BAL cells was significantly lower in the winter after intervention compared to the winter before. The cytokine IL-6 is a pre-inflammatory marker and indicator for exposure to stable environment per se. It has also been shown to be associated with increased mucous secretion (Chen et al., 2003; Gerber et al., 2003). Therefore, a decreased amount of mucous in trachea together with a lower expression of IL-6 mRNA in BAL cells may indicate improvement of horse airway health after intervention. However, after intervention, one of nine horses had increased respiratory rate and respiratory sounds, and this horse, together with another animal, had a mild elevation in BAL neutrophils after intervention. Therefore, the effect of intervention in the stable on clinical outcome in horses is not clear cut.

4. General discussion

The battery of methods that we have presented here for monitoring indoor air quality in stables seems to be adequate for the task. Many of the methods have been used for extended times and have been well validated. However, one should be aware of the fact that there are many pitfalls in pursuing studies like this. It is well known that there is a large day-to-day variability and also a substantial measurement error in many hygienic measurements (O'Meara & Tovey, 2000). There is a high turnover of both personnel and horses, especially in race-horse stables, which makes follow-up studies very cumbersome and difficult to perform.

In the future, we will hopefully progress to better and quicker techniques to identify, for example, moulds using DNA-based technology. Dust from large animal farming is a complex mixture, and factors other than endotoxin and glucan, such as muramic acid and 3-hydroxy fatty acid, may contribute to respiratory inflammation and these components should therefore be added to the list of markers (Poole et al., 2010).

Regarding the horses being examined, for most management systems in our northern climates stabled horses spend far more time indoors than their keepers on a daily basis. Thus, they are likely to have a larger cumulative exposure to the variety of potentially irritating components of stable air and dusts. Much of the research on air quality in relation to respiratory problems in the horse has had its focus on the asthma like disease recurrent airway obstruction (RAO), mentioned earlier, in which as yet poorly defined components of stable air trigger airway inflammation solely in selected susceptible horses, while stablemates breathing the same stable air remain clinically normal. While desirable, defining specific allergens or irritants in this stable air that triggers airway inflammation in these RAO horses has yet to be identified. Moreover, initiation of airway inflammation (and obstruction) appears to occur in concert with the presence of endotoxin in the stable air, which interacts synergistically with other ill defined components of stable and hay dusts to greatly enhance the inflammatory effects of stable air on RAO susceptible horses (Pirie et al., 2003). For these highly reactive horses, there are as yet no clear specific markers of stable air quality that can be used to distinguish stables with "poor " air quality from those in which the horse with RAO will not overreact. Additionally, even clinically normal horses during stabling have more recently been shown to develop lower airway inflammation (Holcombe et al., 2001), in which endotoxin has been incriminated as one key inflammatory component (Berndt et al., 2010). Unfortunately, specific guidelines for air quality in stables for the resident horses, even if clinically normal are not available.

Turning to the actual measurement of airway problems in relation to air quality, the sensitivity and precision of methods used for the horse lag considerably behind those available to physicians for their human subjects. For example, the use of forced expiratory volume measurements to detect small changes in pulmonary function over time, as done on the human subjects in our studies, is not feasible in the horse, as it requires conscious cooperation and patient compliance in performing respiratory maneuvers on command. Alternative pulmonary function measurements in the horse that do not require patient compliance, such as simple measurement of changes in intrapleural pressure, or the noninvasive but more sophisticated measurements using forced oscillation techniques (Young & Tesarowski, 1994) that we have in our laboratory are in sufficiently sensitive to small but clinically relevant changes in pulmonary function in relation to changes in stable air quality.

On the other hand, measurement of airway inflammation in the horse is becoming all the more sensitive (Riihimäki, 2008; Berndt et al., 2010) and in the future will likely provide better objective data on the true state of airway health or inflammation in the stabled horse. Moreover, in horses we are able to obtain samples (bronchoalveolar lavage cytology, bronchial biopsy (Riihimäki, 2008)) that can directly identify alterations in level of pulmonary inflammation, but for ethical reasons are not feasible in human subjects.

Unfortunately, field studies aiming to determine the influence of stable air quality on the horse respiratory health are plagued with confounding external factors, such as changes in bedding or feeding regimens for horses in adjacent stalls in the same stable, the potential for occurrence of subclinical respiratory viral infections or transient parasitic infections in competition horses (Riihimäki et al., 2008b). Even the influence of cold air and exercise can influence gene regulation of inflammation in the horses' airways (Davis et al., 2005; Riihimäki et al., 2008a). Such uncontrolled external facets in these field studies of the horse are difficult to control, and cloud the scientific interpretation of results. As such, future studies on the influence of the quantitative and qualitative indices of stable air on the respiratory health of the horse need to ensure more rigorous control and standardization of the actual stable environment, both in terms of bedding and feed throughout the stable, as well as of the timing and intensity of exercise regimen.

For those working and living on the horse farm, whether the farm environment is a risk factor for developing allergic disease and asthma is under debate. There are conflicting epidemiological results that state, on one hand, that children growing up in farm environments have less allergic disease, especially if they are growing up on a farm with animals (Bråbäck et al., 2004; Ege et al., 2011). On the other hand, there are studies showing that farmers are among the occupations having the highest prevalence of asthma (Lembke et al., 2004). Furthermore, exposure to organic dust may cause rhinitis, asthma, bronchitis, or intrinsic lung disorders such as farmer's lung/allergic alveolitis. There are several studies now showing that working in a stable can have adverse health effects on the respiratory function similar to those experienced by farmers (Kristiansen & Lahoz, 1991; Mackiewicz et al., 1996; McGorum et al., 1998; Tutluoglu et al., 2002; Elfman et al., 2009).

A majority of the personnel in the race horse stable had higher levels of lysozyme and MPO in nasal lavage than white collar workers. Since lysozyme is a marker of nasal mucosal secretion this could be a reaction to the relatively high levels of airborne dust present in the stables. MPO, a marker of neutrophil activity, may indicate a response to high exposure to bacteria in straw bedding and horse dung. Today, transmission of parasites, worms and infections from horses to humans is rare. Instead the exposures to organic dust, gases, microbes, mould and storage mites in the stable environment might have a greater health impact. Since 3 out of 13 personnel had objective signs of asthma without actually having reported asthma, this may indicate neglect of respiratory symptoms and disease among this category of workers in the race-horse stable. Asthma is the most common chronic disease among children and young adults, affecting about 5-10% of school children and 6-7% of adults, of which 30-50% is attributed to allergy (The National Board of Health and Welfare, 2009). Therefore, screening of asthma and allergy among stable-workers may be a way of detecting sensitive individuals who can then be informed of preventive measures to reduce worsening of their disease.

Primary prevention in the stable should involve reduced exposure to organic dust, irritating gases, and microbial agents. Feeding and cleaning activities can be done in a way that creates less dust in stable air. Also ventilation is important, where the installation of mechanical ventilation could increase the air exchange rate and reduce particles and gaseous substances in stable air (Wålinder et al., 2011).

Secondary preventive measures of stable-related airway disease could involve pre-employment testing of stable workers followed by medical advice for those with a risk of developing airway-related diseases, such as hay fever and asthma. The examination could include a medical history, especially of allergy and airway disease, physical examination of the airways, and skin disorders such as eczema. This could be followed by a pulmonary function test and an allergen test. This regime may also be applicable for riding-school students who often spend a lot of time in the stables. The purpose of these investigations would be to protect persons with atopy and airway disease from impairment caused by the stable environment. This screening could be of importance since a possible disregard among stable personnel having respiratory problems has been described (Elfman et al., 2009).

In order to protect sensitive inhabitants living in the neighbourhood of horses from the spread of horse allergens and smell, there is a recommended distance between horse establishments and nearby dwellings. In Sweden, the official recommendation since 1989 is 500m (National board of housing, building and planning 1998), but not much was known about dispersion of horse allergen in ambient air at that time and no assays were available for detection of horse allergen. Later studies have shown that horse allergens do not disperse more than about 50-100m from source, which is from stables and fields where horses are kept (Emenius et al., 2001; Elfman et al., 2008). On days with increased wind speed and with prevailing winds, horse allergen levels can sometimes be detected up to 200-450m from the source. Other importance factors for spread of horse allergens are typography, vegetation and buildings (Elfman et al., 2008). In fact, dense vegetation can also be used to reduce dispersion of horse allergen to nearby houses. The best choice of vegetation for shielding is conifers, which can reduce particle levels by 15-25%, compared with 3-17% by deciduous trees (Freer-Smith et al., 2004).

5. Conclusions

Many people spend a considerable time in stable environments, either as workers in the care and training of horses, or in their leisure time. Therefore, the indoor stable environment must provide adequate exchange of fresh air to the horses' living space to regulate the temperature of the stable and remove moisture, stable gases and organic debris produced by the horses and stable activities.

The most striking effect of installing mechanical ventilation in a stable was a reduction of the carbon dioxide level by half. Other parameters that improved with mechanical ventilation included levels of ammonia, horse allergen and ultrafine particles. Many factors affect the levels of particles, including horse and human activities, type of feed and bedding material.

In a conventionally-managed race-horse stable, 3 out of 13 workers had increased PEF-variability (CV>20%), indicating bronchial obstruction. These individuals also had increased levels of ECP in nasal lavage, indicating allergic inflammation equivalent to allergic asthma. Only two of the personnel reported work-related airway symptoms. After, installation of mechanical ventilation in a stable, there was a slight, but not significant, decrease in PEF-variability in both personnel and riding-students. The absence of measurable health effects after intervention may be due to a low statistical power because of a small study-population, a healthy-worker effect in this stable, or both.

In horses, we could see a significant up-regulation of IL-6 mRNA expression during both winter periods compared to summer, which coincided with a trend of increased neutrophils in BAL fluid. It is, however, unclear whether the increased IL-6 mRNA expression in BAL during winter was an effect of the stable environment and/or exercise in cold weather conditions. After intervention, a significant decrease in accumulation of tracheal mucous was the strongest indication of reduced airway inflammation. Of the cytokines investigated, the expression of IL-6 mRNA in BAL cells was significantly lower in the winter after intervention compared to the winter before. Therefore, a decreased amount of mucous in trachea together with a lower expression of IL-6 mRNA in BAL cells may indicate improvement of horse airway health after intervention.

This paper contributes to the identification of suitable biomarkers to monitor the indoor horse stable environment and respiratory health in humans and horses. An improved management of horse stable climate will be beneficial for the airways of workers, riding-students and horses spending considerable time in confined stabling systems with variable air quality.

6. Acknowledgements

This work has received financial support from The Swedish Research Council Formas, The Swedish Farmers´ Foundation for Agricultural Research and The Foundation for Horse Research. Thanks are also expressed to Mabtech, Stockholm, Sweden, who generously supplied the antibody reagents for the development of the horse allergen ELISA.

7. References

Ainsworth, D. M., Wagner, B., Franchini, M., Grünig, G., Erb, H. N., & Tan, J.-Y. (2006). Time-dependent alterations in gene expression of interleukin-8 in the bronchial epithelium of horses with recurrent airway obstruction. *American Journal of Veterinary Research, 67*(4), 669-677. doi:10.2460/ajvr.67.4.669

Airaksinen, S., Heiskanen, M.L., Heinonen-Tanski, H ., Laitinen, J. & Linnainmaa, M. (2005). Variety in dustiness and hygiene quality of peat bedding. *Ann Agric Environ Med,* Vol.12, No.1, pp. 53-9

ATS. (1995). American Thoracic Society Medical Section of the American Lung Association, Standardization of spirometry 1994 update. *Am J Respir Crit Care Med,* Vol.152, pp. 1107-1136

Behndig, A. F., Mudway, I. S., Brown, J. L., Stenfors, N., Helleday, R., Duggan, S. T., Wilson, S. J., et al. (2006). Airway antioxidant and inflammatory responses to diesel exhaust exposure in healthy humans. *The European Respiratory Journal: Official Journal of the European Society for Clinical Respiratory Physiology, 27*(2), 359-365 doi:10.1183/09031936.06.00136904

Berndt, A., Derksen, F. J. & Robinson. N.E. (2010). Endotoxin concentrations within the breathing zone of horses are higher in stables than on pasture. *Veterinary Journal,* Vol. 183, No.1, (London 1997), pp. 54-57

Bråbäck, L., Hjern, A. & Rasmussen, F. (2004). Trends in asthma, allergic rhinitis and eczema among Swedish conscripts from farming and non-farming environments. A nationwide study over three decades. *Clinical and Experimental Allergy: Journal of the British Society for Allergy and Clinical Immunology,* Vol.34, No.1, pp. 38-43

Bureau, F., Bonizzi, G., Kirschvink, N., Delhalle, S., Desmecht, D., Merville, M. P., Bours, V., et al. (2000). Correlation between nuclear factor-kappaB activity in bronchial brushing samples and lung dysfunction in an animal model of asthma. *American Journal of Respiratory and Critical Care Medicine, 161*(4 Pt 1), 1314-1321

Chen, Y., Thai, P., Zhao, Y., Ho, Y., & DeSouza, M. (2003). Stimulation of airway mucin gene expression by interleukin (IL-17) through IL-6. *J Biol Chem, 2778,* 17036-17043

Clements, J. M. & Pirie, R.S. (2007). Respirable dust concentrations in equine stables. Part 2: the benefits of soaking hay and optimising the environment in a neighbouring stable. *Research in Veterinary Science,* Vol.83, No.2, pp. 263-268

Davis, M.S., Malayer, J.R., Vandeventer, L., Royer, C.M., McKenzie, E.C. & Williamson, K.K. (2005). Cold weather exercise and airway cytokine expression. *Journal of Applied Physiology,* Vol.98, No.6, (Bethesda 1985), pp. 2132-2136

Derksen, F J, Scott, J. S., Miller, D. C., Slocombe, R. F., & Robinson, N E. (1985). Bronchoalveolar lavage in ponies with recurrent airway obstruction (heaves). *The American Review of Respiratory Disease, 132*(5), 1066-1070

Donham, K., Haglind, P., Peterson, Y., Rylander, R. & Belin, L. (1989). Environmental and health studies of farm workers in Swedish swine confinement buildings. *British Journal of Industrial Medicine,* Vol.46, No.1, pp. 31-37

Douwes, J., Thorne, P., Pearce, N. & Heederik, D. (2003). Bioaerosol health effects and exposure assessment: progress and prospects. *The Annals of Occupational Hygiene,* Vol.47, No.3, pp. 187-200

Ege, M.J., Mayer, M., Normand, A-C., Genuneit, J., Cookson, W.O.C.M., Braun-Fahrländer, C., Heederik, D., Piarroux, R. & von Mutius, E. (2011). Exposure to environmental microorganisms and childhood asthma. *The New England Journal of Medicine,* Vol.364, No.8, pp. 701-709

Elfman, L., Brannstrom, J., & Smedje, G. (2008). Detection of horse allergen around a stable. *International Archives of Allergy and Immunology,* Vol.145, No.4, pp.269-276

Elfman, L., Riihimäki, M., Pringle, J., & Wålinder, R. (2009). Influence of horse stable environment on human airways. *Journal of Occupational Medicine and Toxicology*, Vol.25, No.4, (May 2009), 10.1186/1745-6673-4-10 [doi]

Emenius, G., Larsson, P.H., Wickman, M., & Härfast, B. (2001). Dispersion of horse allergen in the ambient air, detected with sandwich ELISA. *Allergy*, Vol.56, No.8, pp.771-774

Fleming, K., Hessel, E.F. & Van den Weghe, H.F.A. (2008a). Evaluation of Factors Influencing the Generation of Ammonia in Different Bedding Materials Used for Horse Keeping. *Journal of Equine Veterinary Science*, Vol.28, No.4, pp.223-231

Fleming, K., Hessel, E.F & Van den Weghe, H.F.A. (2008b). Generation of Airborne Particles from Different Bedding Materials Used for Horse Keeping. *Journal of Equine Veterinary Science*, Vol.28, No.7, pp. 408-418

Freer-Smith, P.H., El-Khatib, A.A., & Taylor, G. (2004). Capture of particulate pollution by trees. A comparison of species typical of semi-arid areas (Ficus nitida and Eucalyptus globulus) with European and Northern American species. *Water, Air and Soil Pollution*, Vol.155, pp.173-187

Gerber, V., Robinson, N E, Venta, R. J., Rawson, J., Jefcoat, A. M., & Hotchkiss, J. A. (2003). Mucin genes in horse airways: MUC5AC, but not MUC2, may play a role in recurrent airway obstruction. *Equine Veterinary Journal*, *35*(3), 252-257

Holcombe, S. J., Jackson, C., Gerber, V., Jefcoat, A., Berney, C., Eberhardt, S. & Robinson, N.E. (2001). Stabling is associated with airway inflammation in young Arabian horses. *Equine Veterinary Journal*, Vol.33, No.3, pp. 244-249

Horse Stable Ventilation. (2003). Pennsylvania State Agricultural and Biological Engineering. http://www.abe.psu.edu/extension/factsheets/g/

Jeffery, P. (1998). Structural alterations and inflammation of bronchi in asthma. *International Journal of Clinical Practice. Supplement*, *96*, 5-14

Jeffery, Peter, Holgate, S., & Wenzel, S. (2003). Methods for the assessment of endobronchial biopsies in clinical research: application to studies of pathogenesis and the effects of treatment. *American Journal of Respiratory and Critical Care Medicine*, *168*(6 Pt 2), S1-17. doi:10.1164/rccm.200202-150WS

Kirchner, M., Braeutigam, S., Ferm, M., Haas, M., Hangartner, M., Hofschreuder, P., Kasper-Giebl, A., Römmelt, H., Striedner, J., Terzer, W., Thöni, L., Werner, H. & Zimmerling, R. (1999). Field intercomparison of diffusive samplers for measuring ammonia. *Journal of Environmental Monitoring*, Vol.1, No.3, pp. 259-265

Kristiansen, J. D., & Lahoz, A.X. (1991). Riding-school lung? Allergic alveolitis in an 11-year-old girl." *Acta Paediatrica Scandinavica*, Vol.80, No.3, pp. 386-388

Lebis, C., Bourdeau, P., & Marzin-Keller, F. (2002). Intradermal skin tests in equine dermatology: a study of 83 horses. *Equine Veterinary Journal*, *34*(7), 666-671

Lembke, B., Janson, C., Norbäck, D. & Rask-Andersen, A. (2004). High risk of adult-onset asthma and work-related wheeze in farmers despite low prevalence of asthma in young farmers. *The International Journal of Tuberculosis and Lung Disease: The Official Journal of the International Union Against Tuberculosis and Lung Disease*, Vol.8, No.11, pp. 1285-1291

Lorch, G., Hillier, A., Kwochka, K. W., Saville, W. A., & LeRoy, B. E. (2001). Results of intradermal tests in horses without atopy and horses with atopic dermatitis or recurrent urticaria. *American Journal of Veterinary Research, 62*(7), 1051-1059

Lugo, J., Stick, J. A., Peroni, J., Harkema, J. R., Derksen, Frederik J, & Robinson, N Edward. (2002). Safety and efficacy of a technique for thoracoscopically guided pulmonary wedge resection in horses. *American Journal of Veterinary Research, 63*(9), 1232-1240

Lugo, J., Harkema, J. R., deFeijter-Rupp, H., Bartner, L., Boruta, D., & Robinson, N Edward. (2006). Airway inflammation is associated with mucous cell metaplasia and increased intraepithelial stored mucosubstances in horses. *Veterinary Journal (London, England: 1997), 172*(2), 293-301. doi:10.1016/j.tvjl.2005.04.018

Mackiewicz, B., Prazmo, Z., Milanowski,J., Dutkiewicz, J. & Fafrowicz, B. (1996). [Exposure to organic dust and microorganisms as a factor affecting respiratory function of workers of purebred horse farms] (in Polish). *Pneumonologia I Alergologia Polska: Organ Polskiego Towarzystwa Ftyzjopneumonologicznego, Polskiego Towarzystwa Alergologicznego, I Instytutu Gruźlicy I Chorób Płuc,* Vol.64, Suppl 1, pp. 19-24

Mazan, M. R., Vin, R., & Hoffman, A. M. (2005). Radiographic scoring lacks predictive value in inflammatory airway disease. *Equine Veterinary Journal, 37*(6), 541-545

McGorum, B.C., Ellison, J. & Cullen, R.T. (1998). Total and respirable airborne dust endotoxin concentrations in three equine management systems. *Equine Veterinary Journal,* Vol.30, No.5, pp. 430-434

National board of housing, building and planning. (1998). General advice for: Better space for work - planning of work areas in relation to environment, health and security. Report from The Swedish National board of housing, building and planning. ISBN 91-7147-399-8

O'Meara, T. & and Tovey, E. (2000). Monitoring personal allergen exposure. *Clin Rev in Allergy & Immunology,* Vol.18, No.18, pp. 341-395

Palmgren, U., Ström, G., Blomqvist, G. & Malmberg P. (1986). Collection of airborne micro-organisms on Nucleopore filters, estimation and analysis-CAMNEA method. *J Appl Bacteriology,* Vol.61, pp.401-406

Pirie, R.S., Collie, D.D.S., Dixon, P.M. & McGorum, B.C. (2003). Inhaled endotoxin and organic dust particulates have synergistic proinflammatory effects in equine heaves (organic dust-induced asthma). *Clinical and Experimental Allergy,* Vol.33, No.5, pp. 676-683

Pomorska, D., Larsson, L., Skórska, C., Sitkowska, J. & Dutkiewicz, J. (2007). Levels of bacterial endotoxin in air of animal houses determined with the use of gas chromatography-mass spectrometry and Limulus test. *Annals of Agricultural and Environmental Medicine,* Vol.14, No.2, pp. 291-298

Poole, J.A., Dooley, G.P., Saito, R., Burrell, A.M., Bailey, K.L., Romberger, D.J., Mehaffy, J. & Reynolds, S.J. (2010). Muramic acid, endotoxin, 3-hydroxy fatty acids, and ergosterol content explain monocyte and epithelial cell inflammatory responses to agricultural dusts. *Journal of Toxicology and Environmental Health. Part A,* Vol.73, No.10, pp. 684-700

Pusterla, N., Whitcomb, M. B., & Wilson, W. D. (2007). Internal abdominal abscesses caused by Streptococcus equi subspecies equi in 10 horses in California between 1989 and 2004. *The Veterinary Record, 160*(17), 589-592

Raphael, G.D., Igarashi, Y., White, M.V. & Kaliner, M.A. (1991). The pathophysiology of rhinitis. V. Sources of protein in allergen- induced nasal secretions. *J Allergy Clin Immunol,* Vol.88, pp. 33-42

Raphael, G.D. Jeney, E.V., Baraniuk, J.N., Kim, I., Meredith, S.D. & Kaliner M.A. (1989). Pathophysiology of rhinitis. Lactoferrin and lysozyme in nasal secretions. *J Clin Invest,* Vol.84, pp. 1528-35

Riihimäki, M., Raine, A., Elfman, L. & Pringle, J. (2008a). Markers of respiratory inflammation in horses in relation to seasonal changes in air quality in a conventional racing stable. *Canadian Journal of Veterinary Research = Revue Canadienne De Recherche Vétérinaire,* Vol.72, No.5, pp. 432-439

Riihimäki, M., Lilliehöök, I., Raine, A., Berg, M. & Pringle, J. (2008b). Clinical alterations and mRNA levels of IL-4 and IL-5 in bronchoalveolar cells of horses with transient pulmonary eosinophilia. *Research in Veterinary Science,* Vol.85, No.1, pp. 52-55

Riihimäki, M. (2008). Inflammatory response in equine airways. Acta Universitatis agriculturae Sueciae, Vol. 2008:12.

Robinson, N E, Karmaus, W., Holcombe, S. J., Carr, E. A., & Derksen, F J. (2006). Airway inflammation in Michigan pleasure horses: prevalence and risk factors. *Equine Veterinary Journal, 38*(4), 293-299

Sadegh, M., Wouters, I.M., Wouters, I., Houben, R. & Heederik, D.J.J. (2009). Exposure to inhalable dust, endotoxins, beta(1-3)-glucans, and airborne microorganisms in horse stables. *Ann Occup Hyg,* Vol.53, pp. 595-603

Savage, C. J., Traub-Dargatz, J. L., & Mumford, E. L. (1998). Survey of the large animal diplomates of the American College of Veterinary Internal Medicine regarding percutaneous lung biopsy in the horse. *Journal of Veterinary Internal Medicine / American College of Veterinary Internal Medicine, 12*(6), 456-464

Schatzmann, U., Straub, R., Gerber, H., & Pauli, B. (1974). Percutaneous lung biopsy in the horse. *The Veterinary Record, 94*(25), 588-590

Swedish Animal Welfare Agency. (2004). Regulations on farm animal housing. DFS 2004:17

Swedish Work Environment Authority. (2005). Occupational exposure limits against air contaminants. AFS 2005:17

The National Board of Health and Welfare. (2009). *Folkhälsorapport 2009 (in Swedish).* ISBN: 978-91-978065-8-9

Tremblay, G. M., Ferland, C., Lapointe, J. M., Vrins, A., Lavoie, J. P., & Cormier, Y. (1993). Effect of stabling on bronchoalveolar cells obtained from normal and COPD horses. *Equine Veterinary Journal, 25*(3), 194-197

Tutluoglu, B., Atiş, S., Anakkaya, A.N., Altug, E., Tosun, G.A., Yaman, M. (2002). Sensitization to horse hair, symptoms and lung function in grooms. *Clinical and Experimental Allergy,* Vol.32, No.8, pp. 1170-1173

Venge, P. (1994). Soluble markers of allergic inflammation. *Allergy,* Vol.49, pp. 1-8

Venge, P., Håkansson, L. & Peterson, C.G.B. (1987). Eosinophil activation in allergic disease. *Int Archs Allergy appl Immun,* Vol.82, pp. 333-337

Wyse, C.A., Skeldon, K., Hotchkiss, J.W., Gibson, G., Yam, P.S., Christley, R.M., Preston, T., Cumming, D.R.S., Padgett, M., Cooper, J.C. & Love, S. (2005). Effects of changes to the stable environment on the exhalation of ethane, carbon monoxide and hydrogen peroxide by horses with respiratory inflammation. *The Veterinary Record,* Vol.157, No.14, pp. 408-412

Wålinder, R., Norbäck, D., Wieslander, G., Smedje, G., Erwall, C. & Venge, P. (1998). Nasal patency and biomarkers in nasal lavage--the significance of air exchange rate and type of ventilation in schools. *International Archives of Occupational and Environmental Health,* Vol.71, No.7, pp. 479-486

Wålinder, R., Norbäck, D., Wieslander, G., Smedje, G., Erwall, C. & Venge, P. (2000). Acoustic rhinometry in epidemiological studies--nasal reactions in Swedish schools. *Rhinology. Supplement,* Vol.16, pp. 59-64

Wålinder, R., Riihimäki, M., Bohlin, S., Hogstedt, C., Nordquist, T., Raine, A., Pringle, J. & Elfman, L. (2011). Installation of mechanical ventilation in a horse stable - effects on air quality and human and equine airways." *Env Health Prev Med,* In press. Available from: http://www.ncbi.nlm.nih.gov/pubmed/21431789

Young, S.S. & Tesarowski, D. (1994). Respiratory mechanics of horses measured by conventional and forced oscillation techniques. *Journal of Applied Physiology* Vol76, No.6, (Bethesda1985), pp. 2467-2472